Periodic Table of the Elements

Key: group classification → 1 IA; atomic number → 1; oxidation state(s) → 1; symbol → H; atomic mass → 1.01

1 IA	2 IIA	3 IIIA	4 IVA	5 VA	6 VIA	7 VIIA	8 VIIIA	9 VIIIA	10 VIIIA	11 IB	12 IIB	13 IIIB	14 IVB	15 VB	16 VIB	17 VIIB	18 VIIIB
1 1 H 1.01																	2 He 4.00
3 1 Li 6.94	4 2 Be 9.01											5 3 B 10.8	6 ±4,2 C 12.0	7 ±3,5,4,2 N 14.0	8 -2 O 16.0	9 -1 F 19.0	10 Ne 20.2
11 1 Na 23.0	12 2 Mg 24.3											13 3 Al 27.0	14 4 Si 28.1	15 5,3 P 31.0	16 6,4,-2 S 32.1	17 ±1,3,5 Cl 35.5	18 Ar 39.9
19 1 K 39.1	20 2 Ca 40.1	21 3 Sc 45.0	22 4,3 Ti 47.9	23 5,4 V 50.9	24 3,6 Cr 52.0	25 2,3,4 Mn 54.9	26 3,2 Fe 55.8	27 2,3 Co 58.9	28 2,3 Ni 58.7	29 2,1 Cu 63.5	30 2 Zn 65.4	31 3 Ga 69.7	32 4 Ge 72.6	33 ±3,5 As 74.9	34 6,4,-2 Se 79.0	35 ±1,5 Br 79.9	36 Kr 83.8
37 1 Rb 85.5	38 2 Sr 87.6	39 3 Y 88.9	40 4 Zr 91.2	41 5,3 Nb 92.9	42 6,4 Mo 95.9	43 7 Tc (99)	44 4,3,2 Ru 101	45 3,4,2 Rh 103	46 2,4 Pd 106	47 1 Ag 108	48 2 Cd 112	49 3 In 115	50 4,2 Sn 119	51 ±3,5 Sb 122	52 6,4,-2 Te 128	53 ±1,5 I 127	54 Xe 131
55 1 Cs 133	56 2 Ba 137	57 3 La 139	72 4 Hf 178	73 5 Ta 181	74 6,4 W 184	75 7 Re 186	76 4,3,2 Os 190	77 4,3,6 Ir 192	78 4,2 Pt 195	79 3,1 Au 197	80 2,1 Hg 201	81 1,3 Tl 204	82 2,4 Pb 207	83 3,5 Bi 209	84 Po (209)	85 ±1 At (210)	86 Rn (222)
87 1 Fr (223)	88 2 Ra (226)	89 3 Ac (227)	104 4 Rf (261)	105 Db (262)	106 Sg (263)	107 Bh (264)	108 Hs (265)	109 Mt (268)	110 Uun (269)	111 Uuu (272)	112 Uub (277)		114 Uuq (289)		116 Uuh (289)		118 Uuo (293)

lanthanides	58 3,4 Ce 140	59 4,3 Pr 141	60 3 Nd 144	61 3 Pm (145)	62 3,2 Sm 150	63 3,2 Eu 152	64 3 Gd 157	65 3,4 Tb 159	66 3 Dy 163	67 3 Ho 165	68 3 Er 167	69 3,2 Tm 169	70 3,2 Yb 173	71 3 Lu 175
actinides	90 4 Th 232	91 5,4 Pa 231	92 6,5,4 U 238	93 5,4,6 Np (237)	94 4,5,6 Pu (244)	95 3,4,5,6 Am (243)	96 3 Cm (247)	97 3,4 Bk (247)	98 3 Cf (251)	99 3 Es (252)	100 3 Fm (257)	101 2,3 Md (258)	102 2,3 No (259)	103 3 Lr (262)

see Appendix A, Table A.1 for element names

Some Important Conversion Factors for Units of Measure (§1.C.1, §2.B)

Water

1 ppm = 1 mg/L = 1 g m^{-3} (1 ppm means 1 part per million [10^{-6}] *by mass*)
1 ppb = 1 μg/L = 1 mg m^{-3} (part per billion = 10^{-9})
1 ppt = 1 ng/L = 1 μg m^{-3} (part per trillion = 10^{-12})
1 M = 1 mole (mol) of species per liter (L) of solution (molarity)
1 eq/L = 1 mole of charge on ions (= 1 equivalent) per L of solution (normality)
Conversion from mass fraction (X_i) to mass concentration (C_i) for aqueous solutions:

$$C_i = \rho_s X_i$$

$$\rho_s = 0.998 \text{ g cm}^{-3} \text{ at } T = 293 \text{ K}$$

Mass fraction (X_i) equals ratio of species mass to total solution mass

Air

1 ppm means 1 part per million (10^{-6}) *by number* (mole fraction) or *by volume*
1 ppm ~ 1 mg m^{-3} (same magnitude, but not numerically equal!)
1 ppb ~ 1 μg m^{-3} (part per billion = 10^{-9})   1 ppt ~ 1 ng m^{-3} (part per trillion = 10^{-12})
Conversion from mole fraction (Y_i) to mass concentration (C_i):

$$C_i = \frac{P}{RT} \times MW_i \times Y_i$$

$$\frac{P}{RT} = 41.6 \text{ mol m}^{-3} \text{ at } P = 1 \text{ atm}, T = 293 \text{ K}$$

$$R = 82.05 \times 10^{-6} \text{ mol}^{-1} \text{ m}^3 \text{ atm K}^{-1} = 8.314 \text{ J mol}^{-1} \text{ K}^{-1}$$

Mole fraction (Y_i) equals ratio of species partial pressure to total gas pressure (P_i/P)

Some Properties of Liquid Water at 20 °C (§2.A, §3.B)

Property	Symbol	Value
Density	ρ	0.998 g cm^{-3} = 998 kg m^{-3}
(Dynamic) viscosity	μ	0.0100 g cm^{-1} s^{-1} = 1.0 × 10^{-3} kg m^{-1} s^{-1}
Kinematic viscosity	ν	0.0100 cm^2 s^{-1} = 1.0 × 10^{-6} m^2 s^{-1}
Specific heat (constant pressure)	C_p	4.182 J g^{-1} K^{-1}
Latent heat of melting (0 °C)	h_{melt}	334 J g^{-1}
Latent heat of evaporation	h_{evap}	2450 J g^{-1}
Vapor pressure	$P°_{H_2O}$	0.023 atm = 2337 Pa
Dissolved oxygen	DO_{sat}	9.0 mg/L (20 °C, 1 atm, 21% O_2)
Molar concentration	$[H_2O]$	55.6 M

Some Properties of Air at 20 °C, 1 atm (§2.B)

Property	Symbol	Value
Density	ρ	1.2 kg m^{-3}
(Dynamic) viscosity	μ	0.0181 g m^{-1} s^{-1}
Kinematic viscosity	ν	0.15 cm^2 s^{-1} = 1.5 × 10^{-5} m^2 s^{-1}
Specific heat (constant pressure)	C_p	1.005 J g^{-1} K^{-1}
Molecular weight (50% RH) [a]	MW_{air}	28.8 g mol^{-1}
Molar concentration	C_{air}	41.6 mol m^{-3}
Standard sea-level pressure	P_{atm}	101.3 kPa = 1 atm

[a]Composition: 77.2% N_2, 20.7% O_2, 1.2% H_2O, 0.9% Ar.

Environmental Engineering Science

William W Nazaroff and Lisa Alvarez-Cohen
Department of Civil and Environmental Engineering
University of California, Berkeley

WILEY

Environmental Engineering Science

Authorized reprint by Wiley India Pvt. Ltd., 4435-36/7, Ansari Road, Daryaganj, New Delhi – 110002.

Wiley also publishes its books in a variety of electronic formats. Some content that appears in print may not be available in electronic books. For more information about Wiley products, visit our website at www.wiley.com.

This edition is authorized for sale in the Indian Sub-continent only.

Reprint: 2014

Printed at : Beekam Print & Pack, Delhi

ISBN: 978-81-265-2450-1

Dedicated to

Ingrid, Rani, Alexis, and Daniela

Michael, Jason, and Ryan

and the extended Alvarez, Cohen, Hamann, and Nazaroff families

Preface

Over the past few decades, environmental engineering has emerged as a distinct academic discipline. At many universities, environmental engineering evolved from the study of water and wastewater treatment, a subdiscipline in civil engineering. Recently, the scope of environmental engineering has expanded to address a broader range of environmental issues. Most environmental engineering programs now include air quality and hazardous waste management in addition to water quality engineering.

Traditionally, the individual topics within environmental engineering have been taught as isolated subjects based on independently developed fundamental concepts. Even when multiple topics are addressed in a single course, they are commonly presented in a compartmentalized fashion. Although this situation is understandable given the historical development of the field, we believe that it is pedagogically undesirable.

There is a set of fundamental principles that serves as the foundation for the entire field of environmental engineering. These principles are based on scientific fundamentals: chemistry, biology, physics, and mathematics. However, courses in these subjects are not sufficient for students to clearly understand the scientific foundation of environmental engineering practice. We believe that students studying environmental engineering should be systematically reintroduced to the fundamentals in a manner that is tailored to the needs of environmental engineers and that is not too closely tied to any specific application. With this foundation, students can more rapidly learn and more thoroughly understand the main topics within environmental engineering. By following this approach, students will maximize their educational and professional adaptability.

We have written this book for three audiences. The primary targets are junior- or senior-level students in good engineering programs who are taking their first course in environmental engineering. Such students should have solid scientific and mathematical preparation typical of the lower-division requirements of engineering programs: calculus, differential equations, physics, and chemistry. Introductory courses in fluid mechanics and biology are also beneficial. For many students, this will be the first course geared to their professional objective of becoming a practicing environmental engineer. For others, this course will lay the groundwork for graduate study in the environmental engineering field. Another important audience for this book is the group of students who are entering an environmental engineering graduate program without strong, specific undergraduate preparation. Many students are admitted to such programs with degrees in mechanical or chemical engineering, chemistry, biology, mathematics, environmental sciences, and other fields. To rapidly prepare for graduate study, these students may formally take a course based on this text or may simply use the text as a guide for self-study. The third major audience for this text are students from other engineering disciplines who need a substantial overview of the principles and practice of environmental engineering so that they can practice professionally as, for example, chemical, civil, or mechanical engineers in a manner that is environmentally responsible.

At UC Berkeley, since 1989 we have been teaching a one-semester (45 lecture hours), upper-division course to undergraduates and beginning graduates in environmental engineering that embodies the approach presented in this text. We devote between one-half and two-thirds of the course to the principles of environmental engineering. The remainder of the course explores how these fundamental principles are applied. Depending on the instructor's interests, applications are selected from water quality engineering, air quality engineering, and hazardous waste management.

Because no available text supported our vision, we provided students with typed copies of our lecture notes as their primary reference. These notes served as the starting point for developing this text. The overall scope of this text represents 60–75 hours of lecture material. We have structured the chapters in the first half of the book so that the more advanced topics in each chapter occur toward the end and can be omitted without loss of continuity. Furthermore, the applications chapters are sufficiently modular to allow an instructor substantial latitude in selecting specific topics for a one-semester or one-quarter course. A typical one-semester, junior-level course might cover a majority of the material in Section I (Chapters 1–5) and a selection of the material in Section II (Chapters 6–8). Table P.1 illustrates the ways that we have used the text to teach semester-length courses, covering fundamentals first, followed by water quality and either air quality or hazardous waste applications. Alternatively, this text could be used to teach a two-quarter series covering fundamentals first and applications in the second quarter, as outlined in Table P.1. In addition, the layout of the book allows the flexibility to offer separate courses on any of the specific applications. Applied courses for which this text would be useful include upper-division offerings in water quality engineering, air quality engineering, and hazardous waste management.

Table P.1 Example Outlines for One-Semester, Two-Quarter, or One-Quarter Courses Using This Text

Length:	One semester	Quarter 1 of 2	Quarter 2 of 2	One quarter	One quarter	One quarter
Topic:	Environmental engineering	Environmental engineering fundamentals	Environmental engineering applications	Water quality engineering	Air quality engineering	Hazardous waste management
Week: 1	Chapter 1	Chapter 1	§§6.A–6.C	Chapter 1	Chapter 1	Chapter 1
2	Chapter 2	Chapter 2	§6.D	§§2.A, 2.C, 3.A	§§2.B, 2.C, 3.A	§§2.A, 2.C, 3.A
3	§§3.A, 3.B	§§3.A, 3.B	§6.E	§§3.B, 3.C	§§3.B, 3.C	§§3.B, 3.C
4	§§3.C, 3.D	§§3.C, 3.D	§7.A	§§3.D, 4.A	§§3.D, 4.A	§§3.D, 4.A
5	§§4.A, 4.B	Review and midterm	Review and midterm	Review and midterm	Review and midterm	Review and midterm
6	Review and midterm	§4.A	§7.B	§§4.B, 4.C	§§4.B, 4.C	§§4.B, 4.C
7	§§4.C, 4.D	§§4.B, 4.C	§§7.C, 7.D	§5.A	§5.A	§5.A
8	§5.A	§4.D	§§8.A, 8.B	§§6.A–6.C	§§7.A, 7.B	§§8.A, 8.B
9	§§6.A–6.C	§5.A	§8.C	§6.D	§7.C	§8.C
10	§6.D	§5.B	§§8.D, 8.E	§6.E	§7.D	§§8.D, 8.E
11	§6.E					
12	Review and midterm					
13	§7.A or §§8.A, 8.B					
14	§7.B or §8.C					
15	§7.C or §§8.D, 8.E					

To explore in depth the topics addressed in this text, we have deliberately omitted some subjects found in introductory environmental engineering texts. Our focus is on the quality of environmental fluids; we believe this theme is at the center of the environmental engineering profession. Topics such as solid waste disposal, noise pollution, and population growth, while important in their own regard, do not so completely share the common set of scientific principles that underlie the topics of water quality engineering, air quality engineering, and hazardous waste management.

The text contains many worked examples and end-of-chapter problems. These have been written to focus students' attention on how fundamental principles are applied to analyze environmental systems for the purpose of solving environmental problems. The problems stress conceptual mastery rather than rote application of the tools of engineering analysis. Many of the problems utilize real data. Some are even fun! A solutions manual is available for instructors.

William W Nazaroff
Lisa Alvarez-Cohen

Acknowledgments

It has been a long journey from first concept to completed book. Many people contributed to our efforts. With gratitude, we offer our thanks to them.

Our excellent teaching assistants provided useful feedback and wrote solutions to many problems: John Little, Dianne Gates, Shelly Miller, Mark Hernandez, Tracy Thatcher, Mike Van Loy, Tom Kirchstetter, Rula Deeb, Ruth Richardson, Nikki Lark, Glenn Morrison, Mark Sippola, De-Ling Liu, and Anna Steding. Our faculty colleagues provided useful information and suggestions, especially Bob Selleck, Jim Hunt, David Jenkins, and David Sedlak. Jil Geller, Tracy Thatcher, and Jim Hunt taught the course from various versions of the manuscript as it was developed. Sumeet Singh, Anna Liu, Kathleen Vork, Matt Dundas, and Ravi Arulanantham supported the book's development through a combination of library research, pointers to key references, and proofreading. Rose Nitzan and Karen Aczon provided secretarial support. We would also like to thank the anonymous reviewers who offered thought-provoking and useful feedback on early versions of the text. A large portion of the text was drafted during WWN's sabbatical leave in Haifa, Israel. He thanks the Technion for their support, and especially Prof. Yaacov Mamane for his graciousness as host of the visit. Additional text was drafted during LA-C's maternity and sabbatical leaves at home; she thanks her children for their long naps.

We have learned much from our students over the years. We are privileged to have worked with so many talented and enthusiastic people. Thanks!

CONTENTS

I FUNDAMENTALS

APPENDICES

1

Overview

1.A WHAT IS ENVIRONMENTAL ENGINEERING SCIENCE?

In modern, industrialized society, it is easy to lose sight of the dependence of human life on the natural environment. Yet even modest reflection reveals that the elemental and chemical composition of the earth, together with energy from the sun, constitutes all of the raw materials that support life.

Throughout the technological age, many engineering disciplines have focused on utilizing the earth's resources. Mining engineers have located, extracted, and refined materials from mineral deposits. Petroleum and chemical engineers have developed useful products from gas and oil fields. Mechanical and electrical engineers have developed technologies to convert natural resources into useful services such as electricity. Agricultural engineers have refined methods of producing food, using the sun's energy plus nutrients in fertilizers and soils as their raw materials. Civil engineers have devised systems for capturing, storing, treating, and delivering water for human uses. These and other engineering disciplines have also provided an infrastructure for moving goods and services developed from natural materials from producers to users.

These engineering efforts have emphasized the extraction, conversion, supply, and use of the earth's resources. Less attention has been given to the generation and release to the environment of contaminants from these and other activities. However, as the level of human industrial activity has grown, it has become clear that the technological foundation of modern society cannot be sustained unless we devote explicit, substantial attention to contaminant releases and their impact on the quality of the environment.

Environmental engineering has emerged as a distinct technological discipline that focuses on environmental contaminants and their impacts. Although use of the title "environmental engineering" began recently, this field draws on a long history of practice. Much of its early history targeted human wastes, following recognition of their disease-carrying properties. Documented concerns about water quality appear early in human history, as illustrated by recommendations, dated about 2000 B.C., to purify water by boiling and filtering it (Baker, 1981). The ancient Romans were aware of the benefits of water quality management, as evidenced by the extensive aqueduct system built to provide the Roman Empire with fresh water. Roman ruins also reveal evidence of sophisticated wastewater collection and transmission systems. In the middle of the nineteenth century, the field of sanitary engineering emerged with the specific mission to provide clean drinking water and to collect, treat, and appropriately dispose of wastewater.

Documented concerns over pollutant emissions to air from human technological activities date back at least to the seventeenth century, when Evelyn decried the foul air of urban England:

> *The weary Traveller, at many Miles distance, sooner smells, than sees the City to which he repairs. This is that pernicious Smoake which sullyes all her Glory, superinducing a sooty Crust or Furr upon all its lights, spoyling the moveables, tarnishing the Plate, Gildings and Furniture, and corroding the very Iron-bars and hardest Stones with those piercing acrimonious Spirits which accompany its Sulphure; and executing more in one year, than exposed to the pure Aer of the Country it could effect in some hundreds.*
>
> John Evelyn, *Fumifugium,* 1661

The source of this air pollution problem was the poorly controlled combustion of low-grade fuels, such as sulfur-containing coal, for light industry, domestic heating, and cooking. Despite early recognition of the problem, substantial, focused efforts to control urban air pollution began only in the 1950s. Much of the world's population still suffers severe air pollution problems caused by combustion effluents from poor-quality fuels. And this is only one of many air quality problems that receive attention from environmental engineers.

In the United States and other developed nations, the decade of the 1970s marked the beginning of a massive effort to improve environmental quality, building on the momentum of public awareness and concern that had grown over the previous few decades. Politically, this concern was expressed in the United States through federal legislation that established the Environmental Protection Agency (EPA), created mandates for the restoration and protection of air and water resources, and focused attention on the characterization and remediation of waste sites that had been contaminated by improper practices. The technological activities spurred by this legislation have generated much of the momentum for the emergence of the modern field of environmental engineering.

Overall, environmental engineering is concerned with the quality and availability of environmental resources and with the waste streams that impact them. Environmental resources include all natural materials in those parts of the earth-atmosphere system where life exists. However, environmental engineering focuses its greatest efforts on the major *environmental fluids*—water and air. Soils also attract the attention of environmental engineers, but to a lesser extent than water and air. In fact, much of the environmental engineering concern with soils occurs because of the roles that they play in water and air quality. The focus on environmental fluids can be understood by considering the routes of exposure to environmental contaminants. Most frequently, exposure involves contact with an environmental fluid. Consider how a person might come into contact with hazardous materials. There are three significant potential pathways: (1) breathing air; (2) ingesting water, other fluids, and foods; and (3) direct contact of materials with skin. On a daily basis, an average adult inhales 10–20 kg of air and ingests 1–3 kg of liquids plus ~1 kg of food, while coming into direct contact with only small quantities of soil (a few grams or less). Thus, based simply on the contact quantities, the exposure potential is greatest for air and water.

Environment, as used in this book, includes both natural and constructed settings. Of particular interest are environments in which life exists, known collectively as the *biosphere*. When we refer to water and air as environmental fluids, we include water in a municipal distribution system, wastewater in a treatment plant, and air in a building as well as water and air in the natural environment.

The compounds present in environmental fluids will be variously referred to as *constituents*, *impurities*, *species*, *contaminants*, and *pollutants*. *Constituent* is the broadest term and includes all materials present in the fluid, including the basic fluid molecules. We will use *impurities* or *species* to refer to anything other than the dominant background fluid. Impurities may or may not be harmful. For example, dissolved oxygen is generally present as a species in water, but not a harmful one. On the other hand, sulfur dioxide is an impurity in air that can cause adverse health effects. *Contaminants* and *pollutants* are used more or less interchangeably to denote that a species has some undesired consequences associated with its presence. Thus, all contaminants and pollutants are considered impurities, and all impurities are considered constituents of environmental fluids.

A *pristine environment* is a setting in which the composition of the environmental fluids is not significantly impacted by human activities. Pristine environments are not free of contaminants, and it is *not* the goal of environmental engineering to create a pristine world. For example, volcanic eruptions spew ash and acidic gases into the atmosphere with devastating effects on the local environment. Lightning produces ozone and triggers forest fires that generate other air pollutants, such as soot, carbon monoxide, and nitrogen oxides. Even if the establishment of pristine environments were a desirable goal, it is entirely impractical, given the current scale of human population.

Instead, the central mission of environmental engineering is to develop and apply scientific knowledge through technology to minimize adverse effects that are associated with contaminants in environmental media. To carry out this mission, environmental engineers conduct a spectrum of activities. They assess levels of environmental contamination. They design and operate treatment processes and emission control facilities to meet environmental quality standards. They devise control strategies, deciding what sources should be reduced by what amounts to meet environmental quality goals. Environmental engineers also help draft environmental standards. Although establishing environmental standards is a political activity, the input of environmental engineers is essential to the development and presentation of accurate information on the technical options for control and the associated costs.

Environmental engineers must understand the fundamentals that govern the concentrations of contaminants in water, air, and other media. Environmental engineers must also have knowledge of past and current engineering practice. The importance of understanding fundamentals is amplified by the fact that many control measures mimic processes that occur in nature. Engineering solutions to current and future environmental problems will also require innovation. Knowledge, understanding, and inspiration are the prerequisites for innovation.

Environmental engineering and environmental science are distinct disciplines that share many features. The primary objective in environmental science is to improve our understanding of natural processes. Environmental engineers strive to *use* that understanding to develop and apply technologies that will maintain or improve environmental quality. To properly serve their role, engineers need an understanding of environmental science that is substantial yet not necessarily as deep as that of the environmental scientists.

This text is entitled *Environmental Engineering Science* because it addresses both science and engineering in application to environmental quality. The topics included in this book serve the goal of technological management of environmental quality. This book differs from many first texts in environmental engineering in that it strives to place the knowledge of environmental engineering practice on a firm scientific foundation. The book focuses on the two most important environmental fluids—air and water—and on the waste streams that impact them. These topics share a common set of scientific principles and embody the dominant branches of environmental engineering.

Section I represents the core of the text. It presents descriptions of the fundamental processes that govern the concentrations and fates of contaminants in air and water, both in natural and engineered systems. Chapter 1 summarizes some of the accomplishments and future challenges in environmental engineering and introduces central concepts. Chapter 2 describes characteristics of water, air, and the species of concern in environmental engineering. Chapter 3 addresses contaminant transformation processes, presenting both a summary of the fundamental principles and a discussion of those processes of specific interest in environmental engineering. Chapter 4 discusses the transport of contaminants within fluids in the environment. The first section closes with Chapter 5, which introduces important modeling approaches for predicting the impact of transport and transformation processes on contaminant concentrations. Throughout Section I, examples and problems are presented that are drawn from water quality engineering, air quality engineering, and hazardous waste management.

Environmental quality problems and their engineering solutions are more systematically presented in Section II. This section draws heavily on the principles developed in Section I. The material is organized into three chapters that address, respectively, water quality engineering (Chapter 6), air quality engineering (Chapter 7), and hazardous waste management (Chapter 8).

At the end of the book, several appendices are provided. Appendix A contains tables of useful data for environmental engineering analysis. Appendices B and C provide introductory background on the topics of ionizing radiation and organic chemistry, respectively. Appendix D summarizes some important tools of mathematical analysis used in environmental engineering. Appendix E contains additional details about transformation processes to supplement Chapter 3. Appendix F summarizes important U.S. federal regulations concerning environmental quality.

1.B DOMAINS OF ENVIRONMENTAL ENGINEERING

This section provides a brief overview of the three branches of environmental engineering addressed in this book. The goals are to highlight some of the history, summarize what has been accomplished, and point out some of the challenges currently faced by the profession. These three branches of environmental engineering are described more substantially in Section II.

1.B.1 Water Quality Engineering

Water quality engineering is traditionally divided into two subspecialties: water treatment and wastewater treatment. In water treatment, the goal is to take water from a source and subject it to treatment processes necessary to make the water suitable for its intended use. The specific treatment processes vary according to both the water source and the application. Important water sources are surface waters, including rivers and lakes, and groundwater. The dominant uses of treated water are for municipal purposes (drinking water), for industrial applications, and for agricultural irrigation. Wastewater treatment begins once the water is used. Wastewater is collected and subjected to physical, chemical, and biological treatment processes before being returned to the environment. The specific processes needed for wastewater treatment vary according to both how the water was used and where the wastewater will be discharged. In practice, water treatment engineering is often considered an entirely separate discipline from wastewater treatment engineering. However, most of the underlying principles are the same, and the distinction has become blurred. The blurring is partially a result of the incidental and purposeful recycling associated with limited surface water availability and the expansion of the domain of water quality engineering to include broader concerns about natural waters. Water quality engineers now address issues of contaminant fate and transport, aquatic ecology, effects of urban runoff on estuarine environments, and other topics that extend beyond the traditional boundaries of water and wastewater treatment systems.

For much of its history, when water and wastewater treatment were the dominant emphases of environmental engineering, the field was known as "sanitary" or "public health" engineering. These titles reflected the close coupling among water quality, sanitation, and public health discovered in the middle of the nineteenth century. Before that time, poor sanitation led to the contamination of drinking water and food by human and animal wastes. No satisfactory means were in place for disposing of the human wastes generated in cities, and water supplies were widely polluted. Epidemic outbreaks of cholera, typhoid, and dysentery in European cities were common.

A few key developments during the middle of the nineteenth century precipitated the onset of sanitary engineering. Sir Edwin Chadwick was charged with investigating the general state of ill health in urban England. His report argued that the health problems were the result of poor sanitation and that proper urban sanitation required a clean water supply and a proper drainage system for removing human wastes. Of course, this is the approach used today. However, at the time, Chadwick's recommendations were largely unheeded. Then, in 1854, a new outbreak of cholera occurred in London, ultimately causing 10,000 deaths. John Snow, a medical doctor, prepared maps from the addresses of the cholera victims. One map showed a strong clustering of cases in the vicinity of a municipal water pump on Broad Street (see Figure 1.B.1). Snow eventually convinced city officials to remove the pump handle, forcing people to collect their water from other sources. Quickly the number of new cholera cases in

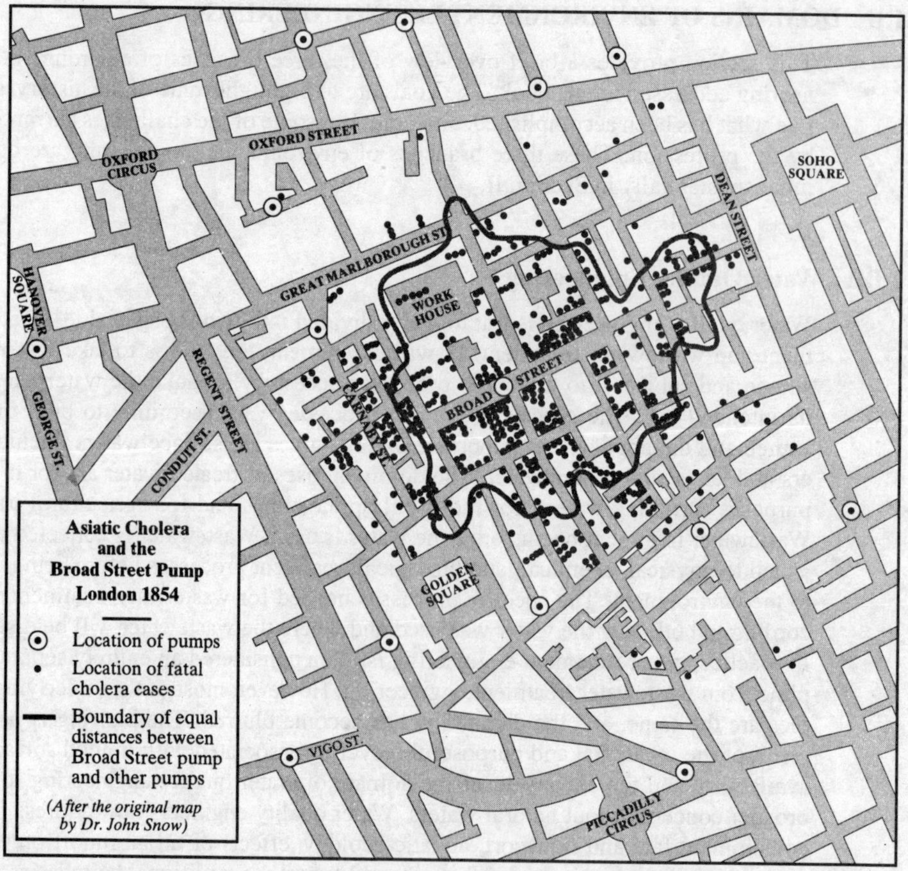

Figure 1.B.1 Map showing cholera victims clustered within the vicinity of the Broad Street pump in London, 1854 (Cosgrove, 1909).

that vicinity showed a marked decrease. This was a compelling demonstration of the relationship between water and disease. Later, in the 1870s, Robert Koch showed that infectious diseases were caused by microorganisms. By this time, Chadwick's recommendations had begun to be implemented in England and in some cities elsewhere in Europe and in North America.

Two key drinking water treatment technologies for preventing disease transmission are filtration and chemical disinfection. Filtration was originally introduced as a municipal treatment process in Paisley, Scotland, in 1804 (Baker, 1981). In that system, water was passed through beds of coarse gravel and then fine-grain sand in which suspended particles were trapped. At the time, the link between water and disease was not yet understood, so the primary purpose of filtration was to make the water more clear. Rapid sand filtration, the technology currently in use, was introduced into municipal water treatment in the United States in the 1880s (Baker, 1981). Although the details have changed, sand filtration remains an important treatment technology today.

Chemical disinfection involves adding a chemical oxidizer to water that kills microorganisms. Chlorine compounds are the most common drinking water disinfectants used in the United States. In Europe and Canada, ozone is a widely used chemical disinfectant. The first large-scale use of chlorine for drinking water disinfection occurred in 1908 at the Jersey City Water Works (Baker, 1981). The first application of ozone as a potable water disinfectant occurred in the Netherlands in 1893 (Haas, 1990).

Even ancient civilizations understood the importance of clean drinking water. The following quote, derived from Sanskrit sources ca. 2000 B.C., demonstrates this insight (Baker, 1981): "It is directed to heat foul water by boiling and exposing to sunlight and by dipping seven times into it a piece of hot copper, then to filter and cool in an earthen vessel." Disinfection can be achieved both by boiling and by ultraviolet radiation (sunlight is a relatively weak source of UV radiation). In addition, copper is a biocide still used in reservoirs, and filtration removes fine particles that can harbor disease-causing microorganisms.

Modern systems for the collection and treatment of human wastes were introduced in the middle of the nineteenth century. Before then, a common practice was to have city residents place human waste ("night soil") in buckets along the streets. Workers emptied the waste into horse-drawn tanks known as "honey wagons." The waste was then transported to rural areas and distributed onto agricultural land as fertilizer. Flush toilets and sanitary sewers began to be widely implemented in the middle of the nineteenth century (Hart-Davis, 1997). Initially, the raw wastes were discharged directly from sewers to receiving waters. Of course, this practice led to gross pollution and health problems. In the late nineteenth century, treatment of wastewater began. By the 1920s, processes were introduced that are in common use today.

Thus, water quality engineering has historically had two primary objectives: to provide drinking water that looks and tastes good, and to prevent the waterborne transmission of infectious disease. A third important objective has more recently emerged as a goal of water quality engineering: to protect the natural environment from the adverse effects of wastewater contamination.

In the industrialized parts of the world, these goals have largely been met. However, sporadic outbreaks of waterborne disease still occur, and wastewater discharges still have adverse effects on many receiving waters. Moreover, the success achieved in many industrialized countries has not been realized everywhere. In fact, much of the world's population still lacks basic sanitation facilities and access to clean drinking water. In the developing world, it is estimated that waterborne diseases cause millions of childhood deaths annually and the loss of billions of hours of worker productivity. Progress toward solving these problems, which are greatest in rural areas and among the urban poor, is slow.

In developed countries, the mission of water and wastewater treatment engineers has expanded to encompass a full range of public health and aesthetic issues associated with water. In addition to particle removal and chemical disinfection, the processes used by a modern municipal treatment plant that prepares drinking water from a surface-water source may include aeration to remove objectionable gases, chemical precipitation to reduce hardness, treatment with activated carbon to remove trace organic molecules, and fluoride addition to help prevent tooth decay. Wastewater may be subjected to sedimentation to remove coarse solids, biological treatment to oxidize the waste organics, chemical disinfection to kill pathogenic organisms, and chemical or biological treatment steps to remove nutrients such as nitrogen and phosphorus.

In the United States, public concern about water quality has been very strong for several decades. This concern has been expressed in part through federal legislation. The 1986 Safe Drinking Water Act requires the Environmental Protection Agency to set standards and monitoring requirements for a large number of contaminants. As of 1999, federal drinking water standards were in place for 79 contaminants or contaminant groups, including organic chemicals (54), inorganic chemicals (16), radionuclides

(3), and microorganisms (6) (Appendix F). Meeting these and future standards will require substantial effort from environmental engineers.

Attention is also focused on water quality in the natural environment. Wastewater discharges from all uses—municipal, industrial, and agricultural—are being subjected to progressively more stringent restrictions, expressed as contaminant discharge limits or treatment process requirements. As point-source wastewater discharges become better controlled, the entry of pollutants into aquatic environments from so-called nonpoint sources becomes more important. Current concerns include the contamination of aquatic environments by fertilizers and pesticides in runoff from agricultural lands, by oils and particulate matter from urban streets, and by atmospheric deposition of acids. The contamination of groundwater resources, especially by hazardous wastes, is also of great concern. In addition, as water resources are being used to their limit, interest is growing in treating wastewater for reuse. These and other problems will keep water quality engineers busy for many decades.

1.B.2 Air Quality Engineering

Air pollution problems caused by the combustion of fuels have been common throughout human history. In fact, it is reasonable to believe that the use of fires in caves by prehistoric humans created the first air pollution problems. In the absence of adequate ventilation, products of incomplete wood combustion, such as carbon monoxide, aldehydes, and polycyclic aromatic hydrocarbons, would have created irritating and even hazardous conditions. Of course, given the many other challenges of life in those times, air quality problems were probably not a major concern for prehistoric people.

The combustion of coal in England caused air pollution problems even before the industrial revolution. In the thirteenth century, to reduce smoky conditions over London, King Edward I issued a ban on the use of "sea coal." The industrial revolution was accompanied by severe air pollution problems of the same nature as those found in many parts of the world today. Extensive combustion of coal led to emissions of a spectrum of pollutants, especially particulate matter and sulfur oxides, that impaired visibility, caused extensive soiling, and affected respiratory health. Emissions from other industrial processes, such as those associated with metals industries, also contributed to air pollution problems.

Apart from occasional complaints and largely unenforced edicts, little was done about air pollution until the 1950s. Before serious action was taken, several infamous episodes occurred in which adverse meteorological conditions contributed to intense localized air pollution. The first widely publicized event occurred in December 1930 in the Meuse Valley, Belgium. Hundreds became ill and 60 people died during the three-day episode. The following month, a similar situation occurred in the vicinity of Manchester, England, killing hundreds. In Donora, Pennsylvania, heavily polluted conditions during a several-day period in 1948 caused 6000 of the 14,000 inhabitants to become ill and killed 20. However, it was not until the largest such episode that effective corrective efforts began. In December 1952 a heavily polluted fog blanketed London for four days. Visibility conditions were so bad that, according to a popular account, a doctor who received an emergency call one night had to rely on a blind man to guide him to the patient's home (Wise, 1968). This episode is estimated to have caused 4000 deaths. The British Parliament then passed clean air legislation that initiated a program to reduce the burning of highly polluting coal.

In cities dominated by a few large industries, an important air pollution tool was the exhaust stack. By venting the air pollutants far above the ground, atmospheric dispersion processes could be counted on to reduce pollutant concentrations before they reached the ground, where humans would be exposed. This approach worked adequately in areas with sparse industry but not, ultimately, in areas of high industrial activity.

During the 1940s, in the Los Angeles area, a different type of air pollution problem was emerging. The London episode and others like it occurred in the winter, when fuel use was highest and pollutant dispersion due to air movement was weakest. These cases featured high levels of sulfur oxides and particulate matter. In Los Angeles, air pollution problems peaked during the long, warm days of summer. Breathing and visibility were impaired by an eye-watering, lung-tightening haze that became known as photochemical "smog" (*sm*oke + f*og*). The problem began to gain attention during the 1940s, but its causes were not understood until Haagen-Smit and collaborators demonstrated that the main irritants, such as ozone and nitrogen dioxide, were not emitted directly from sources but rather were formed in the atmosphere by photochemical reactions (Haagen-Smit, 1952). Emissions of hydrocarbons and nitric oxide (NO) from motor vehicles, industrial activities, and other sources combined with the Southern California sunshine to create the problem.

In the United States, extensive efforts have been made to improve urban air quality. Sulfurous smog problems like the 1948 episode in Donora have not recurred. Fuel substitution, process changes, and waste-gas treatment have greatly reduced emissions from industrial activities and power generation. Likewise, progress has been made in California and elsewhere in reducing the frequency and intensity of photochemical smog episodes. However, healthful air has not been fully achieved in the United States. Approximately 25 percent of the U.S. population live in counties that do not meet the air quality standard for one or more of the six pollutants whose concentrations are regulated nationwide (see §7.A). In Los Angeles, which has had the most severe air pollution problems in the United States, levels of ozone exceeded the federal air quality standard more than 100 days per year through the early 1990s (Lents and Kelly, 1993).

Problems similar to these exist elsewhere in the world. Eastern Europe and China have sulfurous smog problems from the widespread uncontrolled use of coal as a fuel. Mexico City has experienced some of the highest ozone concentrations on record.

Several additional air pollution problems that have begun to receive attention in the last few decades are briefly summarized next.

Acid Deposition

Some contaminants emitted into the atmosphere tend to be converted to acidic species. The deposition of excessive amounts of these species can cause economic and ecosystem damage. Most of the effort in acid deposition research and legislation in the United States has focused on sulfur-containing species produced from the extensive combustion of coal for electricity generation in the Midwest, which leads to acidified precipitation in the eastern states.

Stratospheric Ozone Depletion

Chlorofluorocarbons (CFCs) are widely used as the working fluids in air conditioners and refrigerators. They were also used as aerosol spray propellants and foaming agents. Through these uses, enormous quantities have been released to the atmosphere. CFCs are nonreactive in the lower atmosphere, but upon reaching the stratosphere they decompose because of exposure to ultraviolet radiation. The chlorine atoms that are released act in a catalytic cycle to destroy the stratospheric ozone that protects the earth's surface from ultraviolet radiation.

Hazardous Air Pollutants

One portion of the federal Clean Air Act amendments of 1990 explicitly designates 189 species as hazardous air pollutants (HAPs) (Appendix F). The legislation requires that sources emitting more than 10 tons per year of any single HAP or more than 25 tons per year of any combination of HAPs must apply maximum achievable control technology (MACT) to reduce emissions.

Most efforts on air pollutant control have focused on the emission sources. This is appropriate because the only true opportunities for ambient air pollution control occur at the source. Once pollutants are emitted into the atmosphere, there are no practical engineering techniques for removing them. However, human exposure is influenced as much by proximity to a source as by the quantity of pollutant emitted. One of the emerging challenges in air quality engineering is to improve our knowledge of how various pollution sources contribute toward human exposure to air toxics.

Biomass Cookstoves

Worldwide, the largest air pollution problem, in terms of human health impact, may be the exposure of women and young children to cookstove emissions in rural areas in developing countries (Smith, 1986). These stoves may be fueled with wood, agricultural residues, dung, or coal. Often, there is no flue or vent, and the combustion byproducts are emitted directly into the living space. Indoor air concentrations of carbon monoxide, suspended particles, and benzo(*a*)pyrene have been measured during cooking at levels much higher than those found in polluted urban air.

1.B.3 Hazardous Waste Management

Among the three branches of environmental engineering considered here, hazardous waste management is the youngest. The creation and rapid growth of the chemical industry during the twentieth century established the need for hazardous waste management. Figure 1.B.2 illustrates this point by showing the evolution in worldwide production of synthetic organic chemicals. Although the production of synthetic organic chemicals is growing rapidly, it is still small in comparison with the production and use of natural organic chemicals such as petroleum hydrocarbons, which constitute a large portion of the hazardous waste problem.

Hazardous wastes are generated by the extraction and refinement of raw materials and by manufacturing processes. Losses during transport and storage create additional wastes. Often, after the useful life of a material is expended, the product itself becomes a waste. During the middle of the twentieth century, large quantities of chemicals were produced and used with little or no attention to the proper handling of wastes. Practices that would be entirely unacceptable by today's standards were commonplace as recently as the 1970s. Examples include disposal of hazardous materials in ordinary municipal landfills; discharge of liquid hazardous wastes into storm drains or municipal sewer lines; storage of hazardous wastes in unprotected metal drums,

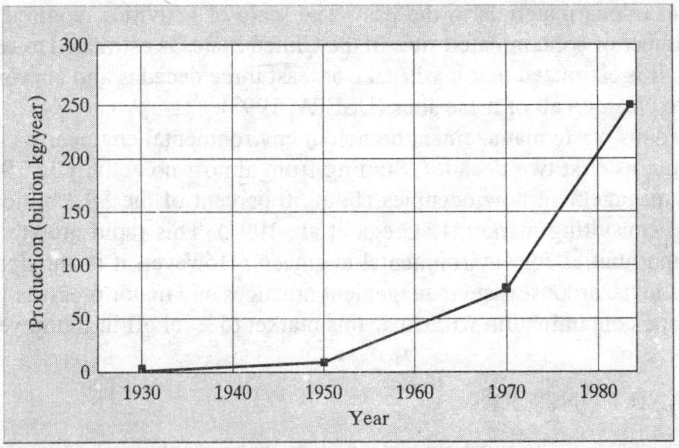

Figure 1.B.2 Global annual production of synthetic organic chemicals (Schwarzenbach et al., 1993).

which are subject to corrosion; and discharge of waste solvents or waste motor oil into soil or dry wells. These practices caused the contamination of a large number of sites with hazardous materials, and these impurities now threaten our water and air resources.

According to U.S. federal legislation, a waste is considered "hazardous" if it has any of these four characteristics:

- corrosivity (i.e., possesses very high or very low pH)
- ignitability
- reactivity (e.g., may cause an explosion)
- toxicity (i.e., causes harm to biological systems)

Wastes generated by specific operations are also classified as hazardous for regulatory purposes. Thus, waste liquids that contain concentrated inorganic acids or bases or significant concentrations of toxic metals are considered hazardous waste.

However, there is no precise boundary that defines hazardous waste. A working definition, modified from LaGrega et al. (1994), is as follows:

Hazardous wastes mean waste substances, typically in the form of solids, sludges, liquids, or containerized gases, which, by reason of their chemical activity or toxic, explosive, corrosive, or other characteristics, cause danger or likely could cause danger to health or the environment.

Radioactive and infectious wastes are excluded from these definitions of hazardous wastes primarily because of regulatory structures and the history of professional practice. However, the nuclear arms race has left the United States with an enormous radioactive waste problem. Infectious waste, generated primarily by the health care industry, is treated using some of the same technologies used for other hazardous wastes.

Hazardous waste management encompasses two broad goals. One goal is to develop and apply methods for the proper use, treatment, and disposal of materials to prevent future site contamination. Enormous progress has been made over the past two decades in the proper management of hazardous waste streams. The second goal is to identify and remediate waste sites contaminated by improper use, storage, and disposal of hazardous materials in the past. The scale of activities required is enormous. The number of contaminated sites in the United States is estimated to be greater than 200,000. It is estimated that it will take at least three decades and approximately \$200 billion to clean up all of these sites (USEPA, 1997).

The hazardous waste management branch of environmental engineering grew explosively during its first two decades. Starting from almost no activity in 1970, hazardous waste management now occupies about 50 percent of the \$8.2 billion yearly environmental consulting market (LaGrega et al., 1994). This rapid growth has created many opportunities for environmental engineers. However, it is anticipated that improvements in hazardous waste management practices and major progress in the remediation of site contamination will cause this market to level off in future years.

1.C CONTEXTS AND CONCEPTS

In this section, some of the most important ideas in environmental engineering science are introduced. These ideas are developed and applied throughout the remainder of the text.

1.C.1 Concentrations and Other Units of Measure

Environmental engineering is concerned with the quality of environmental fluids, as determined by the nature and amounts of contaminants within the fluids. This section defines the most important measures for quantifying the amounts of species in environmental fluids. The use of SI units is emphasized. Conversions from other engineering units are provided in Appendix A (Table A.2.a), which also provides a table that defines the abbreviations used for units (Table A.2.b) and a table listing the decimal prefixes applied to units (Table A.2.c).

Concentration

The concentration of a species in a fluid represents the amount of that species per volume of fluid. Amounts are commonly quantified either by mass or by moles, where one mole is equal to Avogadro's number of elements: $N_{av} = 6.02 \times 10^{23}$. Conveniently, the mass associated with one mole of a chemical element or compound is equal to its atomic or molecular mass, in grams. (See Appendix A, Table A.1, for a list of the atomic masses of the elements.) Thus, for example, 1 mole of carbon has a mass of 12.0 grams, and 1 mole of carbon dioxide (CO_2) has a mass of $12.0 + 2 \times 16.0 = 44.0$ grams. The abbreviation for *mole* is mol, and the abbreviation for *gram* is g. Fluid volumes are most commonly expressed in cubic centimeters (cm^3), liters (L), or cubic meters (m^3), where $1\ m^3 = 1000\ L = 10^6\ cm^3$. Thus, in water, common concentration units for contaminants are mg L^{-1} and mol L^{-1}. The latter concentration unit is given a special symbol, M, and is called the "molarity." In air, common units for specifying the concentration of species are μg m^{-3} and mol m^{-3}.

A special unit is defined for the concentration of charge associated with ions in water. We refer to the ionic charge concentration associated with some species as its normality (N), measured in units of equivalents per liter (eq L^{-1}), where one equivalent represents a net charge equal to one mole of electrons. So, for example, if 10^{-3} mol of table salt, NaCl, is completely dissolved in water to make 1 L of solution, the concentrations of Na^+ and Cl^- will both be 10^{-3} M and their normalities will both be 10^{-3} eq L^{-1}. If, on the other hand, 10^{-3} mol of $MgCl_2$ were dissolved in 1 L of water, the resulting concentration of Mg^{2+} would be 10^{-3} M but the normality would be 2×10^{-3} eq L^{-1}.

Mass Fraction and Mole Fraction

A common alternative to concentration for expressing the amount of a species in a fluid is the mass fraction or mole fraction. These measures represent the ratio between the amount of the species and the total amount of the solution (fluid plus species), both expressed with the same measure. Common units for these ratios are

%	percent	1 part species per 100 parts solution
‰	per mil	1 part species per 1000 parts solution
ppm	part per million	1 part species per 10^6 parts solution
ppb	part per billion	1 part species per 10^9 parts solution
ppt	part per trillion	1 part species per 10^{12} parts solution

For dilute solutions, in which species are present at ppm levels or less, the total amount of solution is well represented by the amount of background fluid.

These units are used both for air and for water (and for other materials, such as soil). By convention, however, the usage differs, and the difference is often not stated. In air, fractions are generally expressed on a *molar* or *volume* basis. (These are equivalent for an ideal gas; see Chapter 2.) In water and other condensed phases, the ratio or fraction is generally expressed on a *mass* basis. Thus, 5 ppb of benzene in air means that there are 5×10^{-9} moles (3×10^{15} molecules) of benzene in a mole of air; on the other hand, 5 ppb of benzene in water means that 1 g of water contains 5×10^{-9} g of benzene. Sometimes the units in air are followed by a subscript $_v$ to remind the reader that the measure denotes a volume ratio (e.g., ppm_v).

Partial Pressure

As an alternative to mass concentrations and mole fractions, species amounts in air may be expressed in terms of partial pressure. The partial pressure of a species in an ideal gas is equal to the mole fraction of the species multiplied by the total gas pressure; this will be discussed in detail in §2.B. So, for example, the partial pressure of 5 ppb of benzene in air at 1 atm is 5×10^{-9} atm.

Notation

Throughout the text, we will use the following notation. Consider a species represented by AB (often the chemical formula, but sometimes an abbreviated name). The mole fraction of that species in air or water will be designated by Y_{AB} and the mass fraction by X_{AB}. The molar concentration of the species in water will usually be denoted [AB]. The mass or molar concentration of the species in water or air may also be designated C_{AB}. The partial pressure of the constituent in air will be denoted P_{AB}.

Unit Conversions

It is important to become comfortable with converting species concentrations in fluids from one measure to another. Mass and molar concentrations are linked by molecular weight as follows:

$$C_i = MW_i[i] \tag{1.C.1}$$

where MW_i is the molecular weight of species i, and $[i]$ represents its molar concentration. For example, the mass concentration of 5 μg L^{-1} benzene (C_6H_6) in water can be represented as a molar concentration as follows:

$$C_B = 5\ \mu g/L$$

$$MW_B = 12 \times 6 + 1 \times 6 = 78\ g/mol$$

$$[B] = C_B \div MW_B = 0.064\ \mu M$$

Conversion from mass fraction to mass concentration is accomplished by multiplying by the solution density (ρ_s, with units of mass per volume), as follows:

$$C_i = \rho_s X_i \tag{1.C.2}$$

This conversion is especially simple in water because the density of fresh water is 1.0 g cm^{-3}. In environmental engineering, we generally deal with species that are present only at dilute levels, so that the density of aqueous solutions can usually be approximated by the density of fresh water. Consequently, the mass fraction of a species in water in units of ppm is the same as the mass concentration in mg L^{-1}, and the mass fraction in ppb is equivalent to the mass concentration in μg L^{-1}. So, for example, a mass fraction of 5 ppb of benzene in water is equivalent to a mass concentration of 5 μg L^{-1}. Example 1.C.1 illustrates unit conversion for species in water.

In air, the mole fraction of a species can be obtained by dividing the molar concentration of the species by the molar concentration of all of the molecules in air,

$$Y_i = \frac{C_i}{C_{air}} \tag{1.C.3}$$

with C_{air} given by the ideal gas law:

$$C_{air} = \frac{P}{RT} \tag{1.C.4}$$

where P is the air pressure, T is temperature, and R is the gas constant ($R = 82.05 \times 10^{-6}$ mol^{-1} m^3 atm K^{-1}). For typical conditions at the earth's surface, $T = 293$ K and $P = 1$ atm, so $C_{air} = 41.6$ mol m^{-3}. Thus, a species concentration of 41.6 μmol m^{-3} in air at $P = 1$ atm, $T = 293$ K corresponds to a mole fraction of 1 ppm. Because of the compressibility of air, the conversion from concentration units to mole fraction units varies with temperature and pressure.

EXAMPLE 1.C.1 ***Unit Conversion Exercise***

One gram of table salt (NaCl) is dissolved in pure water to make 1 L of solution. Determine the mass fraction, mass concentration, molarity, and normality of Na^+ in solution.

SOLUTION First, determine the number of moles of NaCl in solution. The molecular weight of NaCl is 23 + 35.5 = 58.5 g mol^{-1}; therefore, 1 g of NaCl contains 1/58.5 = 0.017 mol of Na^+ and 0.017 × 23 = 0.39 g of Na^+. Since the mass fraction of salt is small, it is reasonable to assume that the density of the solution is the same as the density of pure water, 1 g cm^{-3} or 1000 g L^{-1}. Therefore, the mass fraction of Na^+ in solution is (0.39 g)/(1000 g) = 390 ppm. The mass concentration is (0.39 g)/(1 L) = 390 mg/L. The molarity is (0.017 mol)/(1 L) = 0.017 M. The normality is 0.017 M × 1 (eq/mol = charge/ion) = 0.017 eq/L.

1.C.2 Material Balance

The conservation of matter is the most important principle in environmental engineering. The basic idea is simple: Matter is neither created nor destroyed by transformation and transport processes. Exceptions occur only in the case of nuclear reactions in which matter is converted to energy, or vice versa. For clarity, in the discussion that follows, nuclear reactions are excluded; however, the ideas can easily be extended to include that case.

When we apply the conservation-of-matter principle in environmental engineering, we often say we are using a "material balance." It is wise to check the result of an environmental engineering problem by considering whether it makes sense from a material-balance perspective.

We apply material balances to environmental systems containing fluids in open and closed containers. An open container permits the exchange of material with its surroundings; the contents of a closed container are isolated. A container may represent a physical vessel or tank, or it may simply be a control volume in space. Figure 1.C.1 illustrates typical containers and flows used to represent environmental systems.

Material balances are applied for both conserved and nonconserved properties of matter. An example of a conserved property is a chemical element. Conserved properties are not changed by transformation processes. Chemical reactions do not change the amount of a chemical element in a system. An example of a nonconserved property is a chemical compound, since reactions change the quantities of compounds.

Given these definitions of the components of environmental systems and the distinction between conserved and nonconserved properties, we may write the following statements of the material-balance principle:

MB1: The amount of a conserved property in a closed container does not change.

MB2: The rate of change in the amount of a nonconserved property within a closed container is equal to the net rate of production of that property within the container.

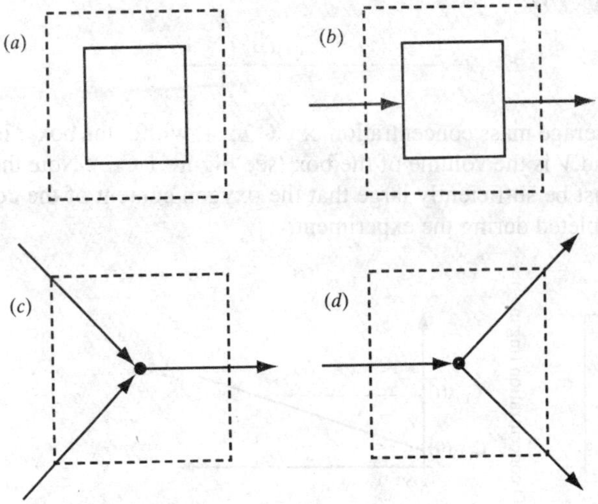

Figure 1.C.1 Model containers and flow paths used to represent portions of environmental systems: (*a*) closed container, (*b*) open container, (*c*) converging flow paths, and (*d*) diverging flow paths. The dashed lines represent the control volumes over which material balances are written.

Table 1.C.1 Properties for Which Material Balance Is Commonly Applied

- Moles or mass of water
- Moles or mass of air
- Moles or mass of a chemical element
- Moles or mass of a specific chemical compound
- Number or mass of suspended particles
- Number or mass of microorganisms
- Equivalents of electrical charge associated with ions
- Oxidation states associated with chemical elements

MB3: The rate of change in the amount of a conserved property in an open container is equal to the difference between the rate of flow of that property into the container and the rate of flow out of the container.

MB4: The rate of change in the amount of a nonconserved property in an open container is equal to the rate of flow in, minus the rate of flow out, plus the net rate of production within the container.

Some properties for which the material-balance principle is commonly applied are presented in Table 1.C.1. Examples 1.C.2 and 1.C.3 illustrate the application of the material-balance principle.

EXAMPLE 1.C.2 *Quantifying the Average CO Emission Rate from a Candle*

The material-balance principle can be applied to measure pollutant emission rates. For example, the emission rate of carbon monoxide from a candle can be measured by placing a lit candle in a closed container and measuring the CO concentration as a function of time. The rate at which the mass of CO in the air of the box increases is equal to the rate at which CO is emitted by the candle. Specifically, the CO emission rate can be determined as

$$E_{CO} = \frac{[C_{CO}(t) - C_{CO}(0)] \times V}{t}$$

where C_{CO} is the average mass concentration of CO in air within the box, t is the measurement period, and V is the volume of the box (see Figure 1.C.2). Note that the size of the container must be sufficiently large that the oxygen content of the container is not significantly depleted during the experiment.

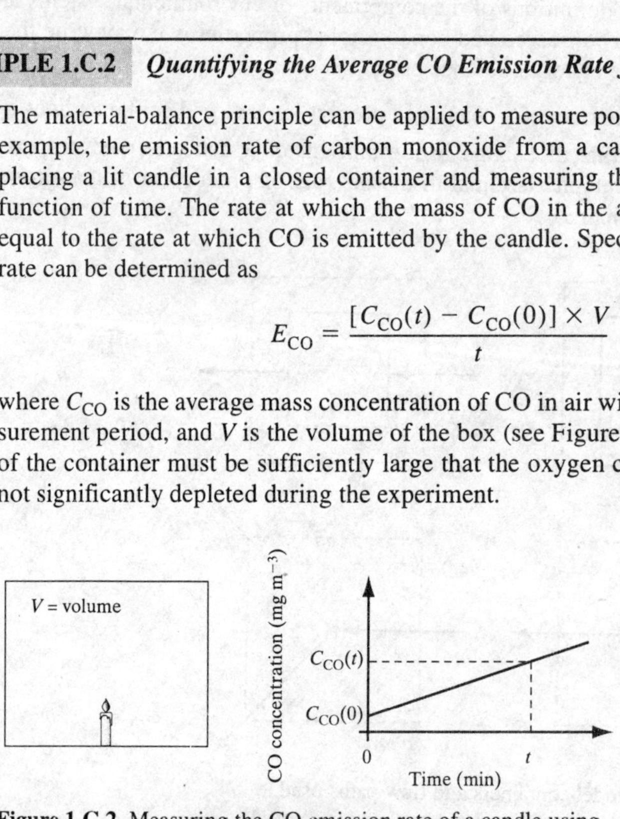

Figure 1.C.2 Measuring the CO emission rate of a candle using the principle of material balance.

EXAMPLE 1.C.3 ***Material Balances on Flows***

A wastewater stream containing a contaminant concentration C_w flows at a volumetric rate Q_w into a river. Upstream of the discharge, the contaminant concentration in the river is C_r and the volumetric flow rate of water is Q_r. Assuming complete mixing of the wastewater stream in the river at the point of discharge, what is the contaminant concentration immediately downstream?

SOLUTION The situation is represented as converging flows, as depicted in Figure 1.C.3. The material-balance principle can be applied twice—first on the mass flow rate of water and second on the mass flow rate of contaminant:

$$Q_{\text{mix}} = Q_r + Q_w$$

$$Q_{\text{mix}} C_{\text{mix}} = Q_r C_r + Q_w C_w$$

Therefore,

$$C_{\text{mix}} = \frac{Q_r C_r + Q_w C_w}{Q_r + Q_w}$$

The material balance for a diverging flow is somewhat different. The concentration of contaminants does not change in the fluid streams as they diverge, so the composition of each branch is the same as in the entering branch.

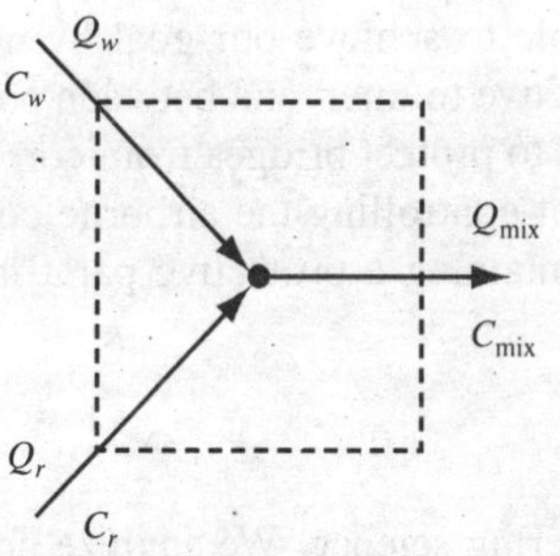

Figure 1.C.3 Application of material balance on contaminant in a converging flow.

1.C.3 Factors Governing Contaminant Concentrations

Although the concentrations of contaminants in environmental systems are influenced by many diverse factors, we may conveniently divide those factors into four groups, as depicted in Figure 1.C.4. Sources include direct emissions within the system and transport into the system from outside its boundaries. Transformation processes include all means by which species change in chemical or physical form within the system. Examples include phase-change processes, acid-base reactions, oxidation-reduction reactions, and dissolution. Transport processes include all mechanisms by which species move from one location to another within the system, such as by advection and diffusion. Removal mechanisms include transport out of the system. Some mechanisms that influence concentrations could be classified into more than one group. For example, a chemical reaction within a system might be considered a source of the product compounds within the system, a transformation process for the elements involved, and a removal mechanism for the reactant compounds.

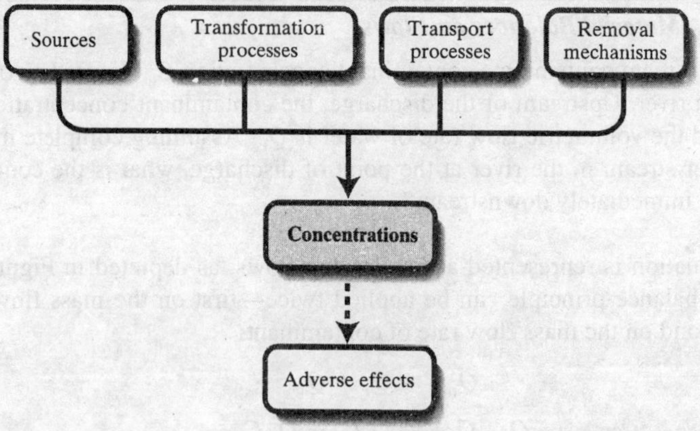

Figure 1.C.4 Four groups of factors control the concentrations of contaminants in environmental systems. In turn, the concentrations of contaminants have a major influence on adverse effects, although, as indicated by the dashed line, there may be other important factors.

Environmental engineers seek to understand the dependence of pollutant concentrations on controlling factors. Ideally, that understanding is quantitative and mechanistic since we must be able to predict how a system will respond to change in order to engineer solutions to environmental problems.

As stated in §1.A, the ultimate objective is not to control concentrations; instead, we seek to limit adverse effects. We may be able to achieve our goal by managing concentrations, but in some cases it is more effective to intervene between concentrations and effects. An example is the use of paints to protect bridges from corrosion. In theory, corrosion could be prevented by means of controlling the airborne concentrations of salt spray and corrosive gases, but maintaining a protective paint layer is a more cost-effective solution.

1.C.4 Engineering Analysis

Analysis is at the heart of environmental engineering science. We analyze both engineered and natural environmental systems, usually for one of two major purposes: to predict how they will behave or to explain why they behaved as they did.

Analyses may be conducted at several levels, from simple range-finding ("back of the envelope") calculations (Harte, 1988) to sophisticated numerical models. Often analyses are performed in a hierarchical, iterative fashion, progressing from simple estimation to more complex calculations. At each stage, the accuracy and precision of the results should improve. The analyses are terminated when the answer is known with sufficient confidence to justify whatever decision is required.

Regardless of the level of complexity of the calculation, engineering analyses are characterized by a set of common features. Typically, analysis involves three key steps, carried out in order:

1. Translate the physical system into a mathematical representation.
2. Solve the mathematical problem to obtain the result.
3. Interpret the significance of the result for the physical system.

The first of these steps is sometimes the most challenging. In environmental engineering, this step requires abstracting the most essential elements from a generally complex environmental system and representing them in a conceptual model. Often, the following substeps are taken to accomplish step 1:

a. Identify and specify symbols to represent the unknowns.

b. Identify and quantify the known parameters and variables in the system.

c. Identify and write relationships based on physical, chemical, or biological principles that link the unknowns to the knowns. For the problem to be fully specified, there must be one relationship for each unknown, and these relationships must be independent of one another.

Step 2 is purely mathematical. In this text, the mathematical problems typically involve solving sets of algebraic equations or relatively simple differential equations. In professional practice, it is sometimes necessary to solve more complex mathematical problems.

Step 3 is essential, although it is often overlooked. Think about this step at two levels. First, ask yourself, What is the significance of this result? (Or, What does this answer mean?) Often, an interesting qualitative conclusion emerges: A process is fast or slow, an impact is big or small, and so forth. Second, ask yourself this: Is this the result I expected? If the answer is yes, then you have a basis for increased confidence in your understanding. If the answer is no, then you should review your work and also think more carefully about the system. Either you made a mistake in the analysis, or your understanding needs adjustment. In either case, a valuable result has been achieved.

1.C.5 Control Opportunities

One of the major approaches used in the past to limit the environmental impacts of waste streams is described by the outdated expression "the solution to pollution is dilution." Air and water discharges were commonly sited downwind and downstream of urban centers. Tall stacks and long sewage outfalls were built to convey effluents far from city residents. This scheme works for a sparse population, since nature can tolerate small waste discharges. But as population density increases, the impact of emissions from one community on those downstream or downwind becomes unacceptably large. Human activities have reached a scale that is sufficient to cause regional and even global environmental impacts. It is certainly no longer the case that dilution suffices to solve environmental quality problems.

There is a chain of opportunities for applying environmental control. The activity that generates the waste is the first link in this chain. The next opportunity occurs when waste is present in a confined fluid stream, often at or near the waste-generating activity. The final opportunity occurs after the waste has been released to the environment. Although in theory environmental quality improvements can be achieved at any of these three points, in general it is more effective and efficient to apply control at earlier links in the chain. Applying pollution control measures at the generating process is known as *pollution prevention*. Control measures applied after the waste has been generated but before it is discharged are termed *end-of-the-pipe* treatment processes. Corrective measures applied after the waste has been released are known as *environmental restoration*. Typically, pollution prevention activities are specific to the particular industry or process involved and, although environmental engineers can contribute, they are seldom solely responsible. Consequently, this book focuses more on end-of-the-pipe and environmental restoration technologies. Nevertheless, it is very important for environmental engineers to vigilantly seek out pollution prevention opportunities.

In water quality engineering, environmental control is most frequently applied at two points: to water before it is delivered to the consumer and to wastewater before it is released to the environment. Air pollution control methods focus on the effluent streams that may lead to airborne release of pollutants. Because the air in an urban basin has a short residence time, the concept of environmental restoration has little meaning for local or regional air pollution. However, restoration is being explored for global air quality problems, where the natural time scale involved is decades rather than days or weeks. Two dominant types of effort are being pursued in hazardous waste management: site restoration, to correct for improper past practices, and management of current stocks and streams of hazardous materials, so that their uncontrolled release to the environment is minimized.

In addition to redirecting environmental control efforts higher up the chain of control opportunities, another important trend is emerging: the growing importance of distributed sources as contributors to environmental contamination. Early control efforts focused on the most obvious sources of environmental pollution: large industrial stacks, municipal and industrial wastewater streams, and improperly disposed of hazardous materials. As success has been achieved in controlling these major sources, their relative contributions to the total environmental release of contaminants have diminished. Conversely, the relative importance of uncontrolled sources of pollution has grown. Typically, these sources are associated with activities that are widely distributed in society. Each contribution may be small, but the cumulative impact can be large. Such sources are now recognized as important in urban air pollution. Therefore, in the Los Angeles area, for example, air quality regulations restrict the sales of oil-based paints, the use of lighter fluid for backyard barbecues, and the use of gasoline-powered lawnmowers. The goal of these regulations is to reduce the emissions of volatile organic compounds into the atmosphere because they contribute to photochemical smog. In water quality engineering, control measures are beginning to be developed to collect and treat urban runoff—the storm water that carries with it oil, grease, metals, pesticides, and other contaminants from urban land—to protect sensitive receiving waters.

1.C.6 Environmental Regulations

Most environmental engineering activities are directly or indirectly motivated by environmental regulations. Consequently, to gain a proper understanding of environmental engineering practice, we must have some understanding of environmental regulations, including why we need them and what forms they take. This section provides a brief overview of the philosophy behind environmental regulations. More information about environmental regulations can be found in Chapters 6–8 and Appendix F.

Releases of contaminants to the environment are inevitable, and the presence of some contaminants in the air we breathe and the water we drink cannot be avoided. Even pristine environments contain some species we would classify as pollutants, and many of these existed before humans appeared on the planet. Society cannot make a decision about whether or not there should be environmental contaminants; we can only decide what levels of pollution are acceptable.

Ideally, pollutant regulations are based on technical information concerning (1) health effects and other environmental costs of pollution and (2) technology and cost of control. But the costs and benefits of implementing pollution control measures cannot be quantified on a single scale, so human judgment must enter the decision-making process. The political process is made complex by many factors, two prominent ones being (1) that the technical information is incomplete, uncertain, and sometimes conflicting and (2) that interest groups (environmental activists, industry representatives, etc.) seek to influence the decision. In the United States, major federal legislation separately addresses water quality, air quality, and hazardous waste management. Many states also have extensive environmental regulations.

Why are environmental regulations needed at all? It is useful to think about this question in economic terms. Consider the case of air pollution caused by emissions from automobiles. Assume that there were no emission control requirements. The air in urban areas would be very heavily polluted and quite unhealthful. But even under these circumstances, no rational individual would act to reduce emissions from his or her car. To do so would cost a certain amount of money and yield, for that person, a negligible benefit, since he or she mostly breathes the emissions of other cars. This problem exists even if society's benefit is huge compared with an individual's cost.

Let's consider a hypothetical example for an urban area. Assume that there are 2 million cars in the area and 5 million people. For simplicity, assume that the air is well mixed throughout the air basin. Assume that it costs \$500 per automobile for a good emission control system. Assume that each emission control system would yield an average health benefit worth 1 cent (\$0.01) for each resident of the basin over the life of the car.

Would an individual voluntarily install such controls? Probably not. The cost of \$500 would greatly outweigh the individual's direct and near-direct benefit, which would be no more than several cents (for an entire family) or perhaps a dollar if the individual had many friends. Yet from society's perspective, the benefit/cost ratio is huge. For the cost of \$500/car, a benefit 100 times as large is realized (5×10^6 people $\times$ \$0.01 per person per car = \$50,000/car). Thus, it would be wise in this case for the society to act collectively to *require* each individual to pay for the cost of control.

This and the class of problems like it are sometimes called "the tragedy of the commons."* Environmental quality is a shared good. Although each of us, as individuals, can profit by avoiding the expense of proper waste management, that profit is gained at the expense of damaging the quality of the environment, however slightly, for everyone who shares it. When small damages are summed over the enormous number of individuals and activities in a region, the cumulative negative consequences can be much larger for all of us than the sum of our individual gains.

1.C.7 Precision and Accuracy

The concepts of precision and accuracy are important in all quantitative aspects of science and engineering. These are distinct ideas describing the degree to which our information about a system conforms to the truth. Figure 1.C.5 uses the metaphor of bullet holes in a target to illustrate the relationship between precision and accuracy. Information is precise if it can be repeatably determined with little variability. A higher level of precision implies a smaller level of variability. Accuracy describes the relationship between the average of several repeated determinations and the goal. The upper right target in the figure, denoted accurate but imprecise, reveals no shot that strikes within the center two rings of the target, yet the average is close to the goal. The large variability due to imprecision causes each shot to deviate from the center of the target in this case.

*The name refers to the "common" area in the center of a town. The expression "tragedy of the commons" was originally coined to describe the conflict between society's interests and individual interests with respect to the use of the commons for grazing livestock (Hardin, 1968).

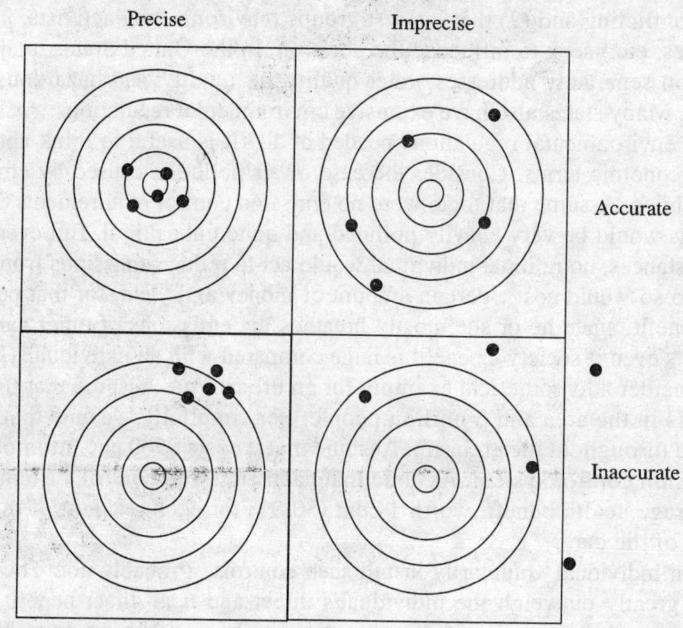

Figure 1.C.5 Sharpshooter's targets, illustrating the concepts of precision and accuracy.

Inaccuracy and imprecision are sometimes grouped under the heading of "errors," reflecting a limit on the quality of our knowledge. (In scientific usage, *errors* do not necessarily imply *mistakes*.) Improved precision and accuracy are desirable goals that usually require greater expenditure of effort or other resources. As engineers, we generally aim for answers that are precise and accurate enough for the application at hand. Because we seek to minimize costs, engineers do not generally seek accuracy or precision that is much greater than necessary.

By scientific convention, the precision of a numerical result is conveyed by the way it is written. Consider, for example, the numbers 1.0 and 1.000. Although mathematically equivalent, these numbers are very different in terms of implied precision. The number 1.0 indicates a result that is bounded by 0.95 and 1.05, whereas the number 1.000 implies a much more precise result that is bounded by 0.9995 and 1.0005. The uncertainty is 100 times larger in the former case than in the latter.

As a rule of thumb, quantitative results in environmental engineering rarely have less than 1 percent error and seldom have less than 10 percent error associated with them. Consequently, it is usually not justified for the result of an environmental analysis to be presented with more than three significant figures. Sometimes, as elaborated in the next section, we don't even know the first significant figure accurately; instead, we know only the appropriate power of 10.

1.C.8 Magnitudes: Length Scales and Characteristic Times

The objects of concern in environmental engineering span an enormous range of scales, as illustrated in Figure 1.C.6. The lengths of the objects depicted span almost 17 orders of magnitude, from roughly 10^{-10} m for the spacing between molecules in fluids to 10^{7} m for the diameter of the earth. On a mass scale, because mass varies in

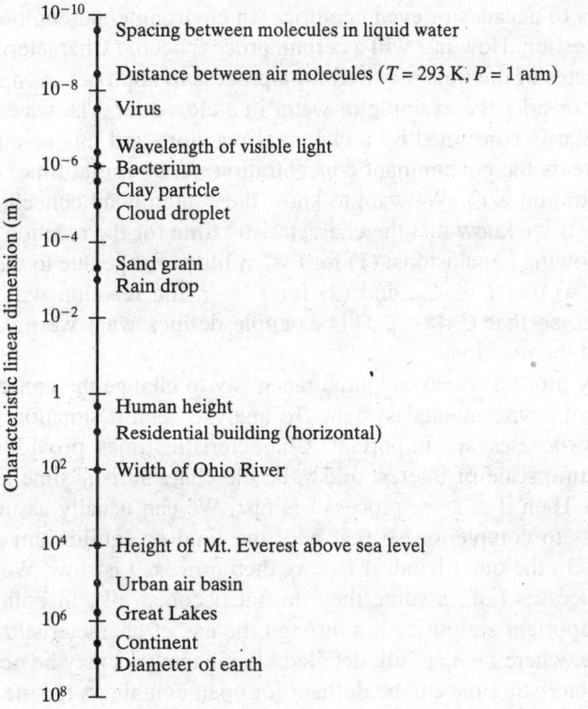

Figure 1.C.6 Length scales of some objects of interest in environmental engineering.

proportion to volume (i.e., as the cube of the linear dimension), the objects of interest vary over more than 50 orders of magnitude, from 3×10^{-24} g for a hydrogen molecule to 6×10^{27} g for the mass of the earth!

Often, in a preliminary analysis, we seek to determine only the magnitude of the answer. It may take very little effort to get the correct answer to within a factor of 10 compared with the work required to get the correct answer to within, say, 10–30 percent. To decide whether a process or parameter is important, we may need to determine only the magnitude of the result.

We will make extensive use of magnitude estimates in this book. In making these estimates, we use the symbol ~ to mean "has the same scale as" or "is of the same order of magnitude as." Although the idea of two parameters having the same order of magnitude defies precise definition, it is an enormously valuable concept. Here is an example that provides an approximate definition of the concept. Consider two parameters, a and b. If the ratio a/b lies within the range 0.33–3, we would conclude that a and b are of the same order of magnitude. If the ratio a/b is less than 0.03 or greater than 30, we would conclude that the parameters do not have the same magnitude. Between these bounds, that is, if a and b differ by a factor in the range 3–30, the decision of whether or not the parameters are of the same scale varies with circumstances. In this text, when a problem asks for a magnitude estimate, an answer that is accurate to within a factor of 3 is certainly correct.

We will frequently use magnitude estimates to assess the time-dependent behavior of environmental systems. We will refer to a magnitude estimate of the time scale associated with some process as a *characteristic time*, usually represented by the symbol τ.

The time scales over which dynamic processes operate in environmental fluids vary over an enormous range of scales, spanning at least 15 orders of magnitude, from microseconds or less to decades or even centuries. In environmental engineering, we regularly face the question, How fast will a certain process occur? Characteristic times are magnitude estimates of the time required for a process to approach completion.

To be specific, consider the example of water in a closed jar. The water contains some contaminant that is consumed by a chemical reaction, and this reaction is the only process that affects the contaminant concentration. At an initial time, $t = 0$, the contaminant concentration is C_0. We want to know the contaminant concentration, C, at some later time, t. If we know that the characteristic time for the reaction is τ, then we can draw the following conclusions: (1) for $t \ll \tau$, little change due to the reaction will have occurred, so that $C \sim C_0$; and (2) for $t \gg \tau$, the reaction will have approached completion, so that $C \ll C_0$. This example defines what we mean by the characteristic time of the reaction.

In general, many processes may act simultaneously to change the concentrations of contaminants in an environmental system. To analyze such a situation, we must first decide which processes are important. Characteristic times provide a useful guide. Let t be the time scale of interest and τ_i be the characteristic time associated with the ith process. Then if $t \gg \tau_i$, process i is fast. We can usually assume that a fast process proceeds to completion so that only the final or equilibrium condition need be considered. On the other hand, if $t \ll \tau_i$, then process i is slow. We can usually ignore slow processes (i.e., assume they do not occur at all). In both of these cases, we achieve important simplification through the use of characteristic times. In the intermediate case, where $t \sim \tau_i$, a full, detailed kinetic analysis may be necessary.

A specific characteristic time can be defined for open containers (frame *b* in Figure 1.C.1), the *characteristic residence time*. In water-filled systems, this measure is often referred to as the *hydraulic detention time*. It is a magnitude estimate of the average time that a fluid molecule will remain in the system before being removed by flow out of the system.

For such a system, we define a *stock* (denoted S) as the amount of the material in a system, and the *flows* (denoted F_i and F_o, for *flow in* and *flow out*, respectively) as the amounts per time entering or leaving the system, respectively (see Figure 1.C.7). The stocks and flows can apply to the fluids themselves or to constituents of the fluids. According to the material balance principle, if the property of interest is conserved, then the rate of change of the stock size is equal to $F_i - F_o$. At steady state, the flow in equals the flow out, and the size of the stock is constant.

The characteristic residence time, τ_r, for such a system is of the same magnitude as the ratio of the stock size to the rate of flow out of the system:

$$\tau_r \sim \frac{S}{F_o} \tag{1.C.5}$$

(Often $F_o \sim F_i$, so that either flow could be used in the denominator.) The stock size, S, usually has units of mass or number (e.g., moles); the corresponding units for the flow rates, F_i and F_o, are either mass/time or number/time. Thus, τ_r has units of time.

The meaning of τ_r is similar to the meaning of other characteristic times. Given τ_r for a specific system, we can say that (1) for some time period $t \ll \tau_r$, there will be

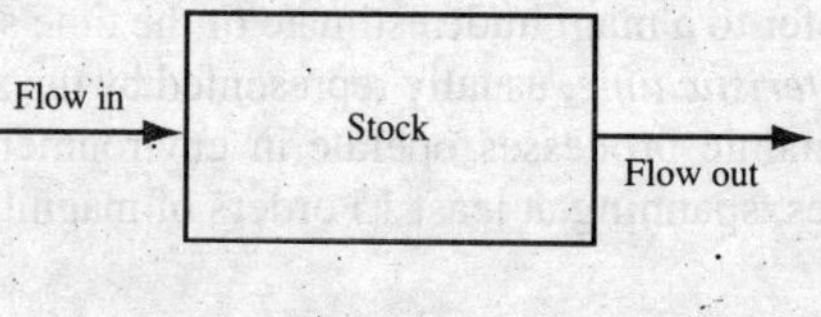

Figure 1.C.7 The relationship between stocks and flows.

very little removal of the stock from the system relative to the current state, and (2) for some time period $t \gg \tau_r$, the current stock will be almost completely removed from the system.

These ideas are illustrated by Examples 1.C.4 and 1.C.5. Example 1.C.4 shows that the concepts can be applied to systems that contain something other than fluids.

EXAMPLE 1.C.4 ***Characteristic Residence Times in Universities***

What is the characteristic residence time of undergraduate students at a university? Of faculty?

SOLUTION The "stock" of undergraduate students at a university is the current enrollment; the "flows" are the rates at which the students matriculate (flow in) or graduate (flow out). The characteristic time for undergraduate students is $\tau_{ug} \sim 4$ y, suggesting that for a university with a stable undergraduate population of ~4000, approximately 1000 students should matriculate (flow in) and graduate (flow out) each year. Note that this does not mean that all undergraduates complete their degrees in 4 years. Rather, it means that for $t \ll 4$ y (for example, 6 months), the makeup of the undergraduate population at a university will be largely unchanged. And, for $t \gg 4$ y (for example, 20 y), very few of the current undergraduates will still be undergraduates (or so one would hope!). At good universities, the characteristic residence time of faculty is $\tau_{fac} \sim 20$ y, a significant fraction of a career. The fact that $\tau_{ug} \ll \tau_{fac}$ suggests that students can be an important source of new ideas in a university, especially in rapidly evolving fields, such as engineering.

EXAMPLE 1.C.5 ***Characteristic Residence Time of Air in a Room***

A classroom contains 50 students and has a volume of 320 m^3. The recommended ventilation rate to avoid excessive CO_2 buildup is 8 L s^{-1} student^{-1}. What is the residence time of air in the room if the recommended ventilation rate is used?

SOLUTION In this case, we may express the stock S in units of volume, since (1) the space of this room is filled with air and (2) the air can be treated as roughly incompressible. Thus, $S = V = 320$ m^3. The flow rates are balanced at $F_i = F_o = (8 \text{ L s}^{-1} \text{ student}^{-1}) \times (50 \text{ students}) = 400 \text{ L s}^{-1} = 1440 \text{ m}^3 \text{ h}^{-1}$. Therefore,

$$\tau_r \sim \frac{S}{F} \sim \frac{320 \text{ m}^3}{1440 \text{ m}^3 \text{ h}^{-1}} = 0.22 \text{ h or } 13 \text{ min}$$

This value represents the characteristic residence time of an air molecule in the room.

REFERENCES

BAKER, M.N. 1981. *The quest for pure water*, vol. 1. 2nd ed. American Water Works Association, Denver.

COSGROVE, J.J. 1909. *History of sanitation*. Standard Sanitary Manufacturing Co., Pittsburgh.

HAAGEN-SMIT, A.J. 1952. Chemistry and physiology of Los Angeles smog. *Industrial Engineering Chemistry*, **44**, 1342–1346.

HAAS, C.N. 1990. Disinfection. In F.W. Pontius, ed., *Water quality and treatment: A handbook of community water supplies*. 4th ed. McGraw-Hill, New York, pp. 877–932.

HARDIN, G. 1968. The tragedy of the commons. *Science*, **162**, 1243–1248.

HART-DAVIS, A. 1997. *Thunder, flush, and Thomas Crapper.* Trafalgar Square Publishing, North Pomfret, VT.
HARTE, J. 1988. *Consider a spherical cow.* University Science Books, Mill Valley, CA, p. 259.
LAGREGA, M.D., BUCKINGHAM, P.L., & EVANS, J.C. 1994. *Hazardous waste management.* McGraw-Hill, New York.
LENTS, J.M., & KELLY, W.J. 1993. Clearing the air in Los Angeles. *Scientific American*, October, 32–39.
SCHWARZENBACH, R.P., GSCHWEND, P.M., & IMBODEN, D.M. 1993. *Environmental organic chemistry.* Wiley, New York.
SMITH, K.R. 1986. Biomass combustion and indoor air pollution: The bright and dark sides of small is beautiful. *Environmental Management*, **10**, 61–74.
USEPA. 1997. *Cleaning up the nation's waste sites: Markets and technology trends.* Report 542-96-005A. U.S. Environmental Protection Agency, April.
WISE, W. 1968. *Killer smog: The world's worst air pollution disaster.* Rand McNally, Chicago.

PROBLEMS

1.1 Units and unit conversion exercises

(a) What does the unit ppb mean when applied to a contaminant in water? How does that differ from the usage for air contaminants?

(b) The federal air quality standard for carbon monoxide (CO) is 35 parts per million based on a one-hour average. Convert this to an equivalent mass concentration in units of mg m^{-3}, assuming that $T = 293$ K and $P = 1$ atm.

(c) The maximum contaminant level for cadmium in drinking water is 0.005 mg/L. Convert this mass concentration to a mass fraction in units of parts per million (ppm).

1.2 Unit conversion exercises and exposure pathways

(a) In Contra Costa County, California, the median indoor airborne concentration of chloroform ($CHCl_3$) was found to be 0.4 μg m^{-3}. Convert this concentration to a mole fraction in parts per billion (ppb) assuming that $T = 293$ K and $P = 1$ atm.

(b) The same study estimated that the mean concentration of chloroform in drinking water is 42 μg/L. Convert this to a mass fraction in parts per billion (ppb).

(c) A typical adult inhales about 20 m^3 of air and ingests about 2 L of water per day. Assuming that the concentrations given in (*a*) and (*b*) are appropriate, compare the exposure to chloroform via inhalation (μg inhaled per day) with the exposure via ingestion (μg swallowed per day).

(d) For tetrachloroethylene (C_2Cl_4), the study found a mean concentration in water of 0.10 μg L^{-1} and a mean concentration in indoor air of 2.1 μg m^{-3}. Repeat part (*c*) for this species.

(e) Comment on the relative significance of breathing and drinking for exposure to these toxic chemicals.

1.3 Measures of vinyl chloride in water and air

Vinyl chloride (C_2H_3Cl) is a confirmed carcinogen that is sometimes found in drinking water and in the air that we breathe.

(a) If vinyl chloride is present in drinking water at a level of 80 ppb, what is the mass concentration of vinyl chloride in water (μg/L)?

(b) If vinyl chloride is present in air at a level of 80 ppb, what is the mass concentration of vinyl chloride in air (μg/L)? Assume $T = 293$ K, $P = 1$ atm.

1.4 Impurities in water and unit conversions

A brochure from a municipal water supply agency lists the following average concentrations of selected ions in drinking water:

Fe^{3+}	0.02 mg L^{-1}
Ca^{2+}	9.8 mg L^{-1}
Mg^{2+}	1.8 mg L^{-1}
K^+	0.6 mg L^{-1}
Na^+	4.6 mg L^{-1}

Compute the molarity and normality of each of these species.

1.5 Global CO_2 accumulation

An increase of CO_2 in the atmosphere may lead to global climate change. The mass of the atmosphere is 5.1×10^{18} kg. The mole fraction of CO_2 in the atmosphere is 350 ppm and is observed to be increasing at a rate of 0.5 percent per year. What is the rate of mass accumulation of CO_2 in the atmosphere?

1.6 Role of U.S. automobiles in global CO_2 accumulation

Human activities appear to be contributing to a significant increase in atmospheric CO_2 on a global scale. Of particular concern is the combustion of fossil fuels, which transforms fuel C into gaseous CO_2. Consequently, improving energy efficiency is considered to be important for reducing the global atmospheric impacts of human activities. Since a lot of fuel is burned in the U.S. auto fleet, let's consider whether improving automotive fuel efficiency might have a significant impact on global CO_2 accumulation.

(a) It is reasonable to assume that all of the C in gasoline is converted to CO_2 as it is burned. Use this assumption and the following data to estimate the CO_2 emission rate (metric tons per year) in the United States from private automobiles:

Number of vehicles (N)	120,000,000
Average gas mileage (M)	18 miles per gallon
Average miles traveled (D)	10,000 miles per year per car
Effective fuel composition	C_8H_{18}
Fuel density (ρ)	0.72 g cm^{-3}

(b) Compare the result of part (a) with that of Problem 1.5. Comment on the potential importance of regulations to improve the fuel efficiency of automobiles as a means of reducing CO_2 emissions.

1.7 Waste disposal from dry cleaning

The SpotBeGone Dry Cleaning Company uses tetrachloroethylene (C_2Cl_4, also known as perchloroethylene, or PCE) in its dry-cleaning operation. SpotBeGone buys one 250 L container of PCE per month and must dispose of PCE that does not evaporate during use. If SpotBeGone generally loses 30 percent of the PCE to evaporation, how often will the company have to pay to have its 0.7 m^3 disposal container emptied?

1.8 A water resources problem

Ecocity has two potential sources of water supply: groundwater and a small river fed by a mountain spring. The river supports a lot of fish and other wildlife, so the citizens do not want to use any more of it than necessary. Unfortunately, however, their groundwater has barium concentrations in excess of public health standards. Rather than invest money in a treatment plant to reduce the barium concentration of the groundwater to an acceptable level, the citizens decide to blend groundwater with the river water so that the standard is met. Determine the relative amounts of river water and groundwater that should be used, in m^3 per day.

Data

Standard for barium in drinking water	2 mg/L (maximum allowed)
Barium concentration in river water	0.5 mg/L
Barium concentration in groundwater	9 mg/L
Water supply needs for Ecocity	10 m^3 per day

1.9 Characteristic times and urban air quality
The Bay Area Air Quality Management District comprises nine counties. The total land area is 6094 square miles. Assume that a steady uniform wind blows from west to east at 3 m s^{-1} through the air basin.

(a) What is the characteristic residence time of air in the air basin?

(b) If pollutant emissions from human activities into the air basin were suddenly eliminated, how much time would be required for air in the basin to become pristine?

1.10 Cleansing time scale for a lake
A contaminant mass M is accidentally spilled into a lake of volume V. The contaminant undergoes no transformation processes (it is nonvolatile, does not biodegrade, etc.). A river flows into the lake at rate F_i. Water evaporates from the lake at rate F_e. Water flows out of the lake at rate F_o. All flows have dimensions of volume per time. What is the characteristic time needed for the contaminant to return to its background (prespill) level in the lake? Express your answer in terms of some combination of the variables M, V, F_i, F_e, and F_o. (*Hint*: It is not necessary to use all of these variables.)

1.11 Stocks, flows, and politics
The United States Senate is filled with politicians most of whom are either Democrats or Republicans. One day you are appointed chair of the Democratic Party and given responsibility for increasing the stock of Democrats in the Senate. When you take office, there are 50 Democrats and 50 Republicans in the Senate, each serving a 6-year term. Every 2 years an election is held in which one-third of the senators are either reelected or replaced. (*Note*: There is both an inflow and an outflow occurring.) Because you and the rest of the Democratic Party do a great job, Democrats win two-thirds of the contested seats for each of the next two elections. How much have you increased your stock after 4 years? (Although it would not actually be possible to elect a fraction of a senator, for simplification in this problem you may ignore that constraint—fractional senators are allowed.)

1.12 Residence time of a water molecule in a human
Using the concepts of stocks and flows, determine a characteristic residence time for a water molecule in an adult human male. Assume that the body's water content is constant with time.

Data

Body mass	70 kg
Fraction of water in body (by mass)	90%
Water intake	
Liquids	1500 g d^{-1}
Water in food	700 g d^{-1}
Metabolic production	300 g d^{-1}

1.13 Stocks, flows, and residence times in Lake Shasta
The Shasta Dam and reservoir in Northern California is a major component of the state's water resources management system. The capacity of the reservoir is given as 4,552,000 acre-feet. The average annual inflow (1922–1970) is 5,439,000 acre-feet. (*Note*: An acre-foot is a volume unit that represents an area of one acre [43,560 sq. ft.] times a depth of one foot.)

(a) Convert the capacity and flow rate information into units of m^3 and m^3 y^{-1}, respectively.

(b) Determine a characteristic residence time for a water molecule to remain in the reservoir, assuming that the reservoir level is unchanging.

(c) Imagine that a toxic chemical is spilled into Lake Shasta and mixes thoroughly with the water. The chemical is inert in the sense that it can be removed only with the water as it is extracted from the lake. What is the characteristic time needed for the water quality to approach its prespill condition?

(d) Based on a consumptive use rate of 1400 L $person^{-1}$ d^{-1}, how long could a fully charged Shasta reservoir supply the needs of all of the state, assuming that the inflow were suddenly stopped? (*Note*: Take the population of California to be 30 million.)

(e) A drought hits California, reducing the inflow by 50 percent. Compute the amount of time it would take to empty the reservoir if the outflow is maintained at the pre-drought flow rate.

(f) A major problem that limits the useful lifetime of reservoirs is the accumulation of silt, which is carried into the reservoir with the inflowing water. Assume that the average volume fraction of sedimenting silt in the inflowing water is 0.23 percent. Determine the period required for the capacity of Lake Shasta to be depleted by 50 percent owing to silt deposition.

1.14 Sulfur in the oceans

The mass fraction of sulfate (SO_4^{2-}) in seawater is 2.7 g kg^{-1}. The total mass of the oceans is 1.4×10^{21} kg.

(a) Compute the total mass of S in the oceans.

(b) Combine the result of (*a*) with the data below on sulfur flows to estimate the residence time of S in the oceans (Harte, 1988).

Transport of S from atmosphere to ocean surfaces	160×10^{12} g y^{-1}
River flow of S to the oceans	100×10^{12} g y^{-1}

1.15 Estimating magnitudes

For the following problems, accurate calculations are not required, but reasonable estimates are needed for unspecified parameters. Your answers can be brief but should clearly indicate the basis for your estimate as well as the result.

(a) How many moles of water are in an Olympic-sized swimming pool?

(b) How many moles of air are in a typical single-family home?

2

Water, Air, and Their Impurities

The first two sections of this chapter describe the physical and chemical properties of the most important environmental fluids, water and air. We also consider their abundance and, briefly, how they move through the biosphere. The third section of this chapter explores the chemical, physical, and microbiological impurities that are found in water and air. The emphasis in this section is on compounds that are of interest in environmental engineering.

2.A WATER AND THE HYDROSPHERE

Water has many remarkable properties. Its physical and chemical characteristics in combination with its great abundance have an enormous impact on the environmental conditions of the earth. Water plays an essential role in supporting life; consequently, the availability of water is often a critical sociopolitical issue. Managing the quality and quantity of water resources has been an important function for civil and environmental engineers since the early days of the profession.

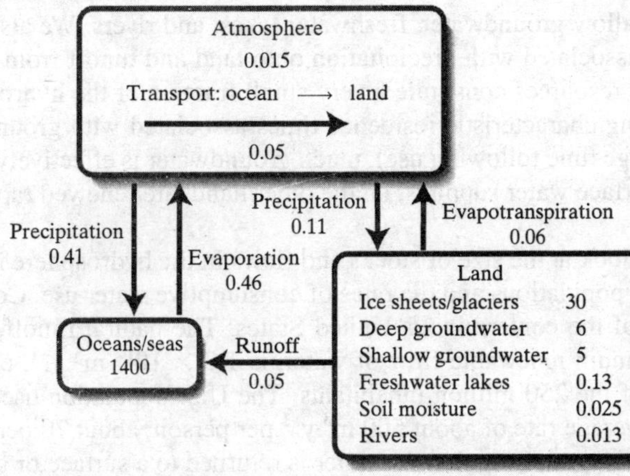

Figure 2.A.1 Global hydrologic cycle. The water stocks are expressed in units of 10^{15} m^3; the flows have units of 10^{15} m^3 y^{-1}. The dividing line between shallow and deep groundwater is drawn at about 800 m below the surface. (Data from P.R. Erlich et al., 1977.)

Hydrosphere and the Hydrologic Cycle

Water is the most abundant chemical in the biosphere and the most abundant liquid on the planet. The earth contains about 1.4×10^{18} m^3 of water in a liquid state, enough to uniformly cover the entire planet to a depth of 2.7 km (Harte, 1988). Oceans and seas cover about 70 percent of the earth's surface.

The stocks of water on earth are collectively known as the *hydrosphere;* the movement of water within and between compartments of the hydrosphere is called the *hydrologic cycle.* A global summary of the hydrologic cycle is presented in Figure 2.A.1.

The relative size of stocks and flows in the earth's water system can be combined to estimate the characteristic residence times of water in the different compartments of the hydrosphere, as is done in Table 2.A.1. The vast majority of the earth's water either is inaccessible to easy exploitation or, in the case of the oceans, has such high quantities of impurities that widespread use is impractical. The dominant water stocks

Table 2.A.1 Characteristic Residence Time of Water Molecules in Various Compartments of the Hydrosphere

Compartment	Characteristic residence time
Atmosphere	9 days
Rivers	2 weeks
Soil moisture	Months
Large lakes	10 years
Shallow groundwater	Tens to hundreds of years
Mixed layer of ocean	120 years
World ocean	3000 years
Deep groundwater	Up to 10,000 years
Antarctic ice cap	10,000 years

used by humans are shallow groundwater, freshwater lakes, and rivers. We also make great use of the flows associated with precipitation onto land and runoff from land to oceans. Together, these resources constitute a very small fraction of the hydrosphere. Because of the very long characteristic residence times associated with groundwater (implying a long recharge time following use), much groundwater is effectively a nonrenewable resource. Surface water supplies, on the other hand, are renewed rapidly by the hydrologic cycle.

It is informative to look at the size of stocks and flows in the hydrosphere in comparison with the size of populations and the rates of consumptive water use. Consider, for example, the case of the conterminous United States. The natural runoff, which approximates the maximum renewable flow of water, is 1.7×10^{12} m^3 y^{-1}, or about 6600 m^3 y^{-1} for each of the 250 million inhabitants. The U.S. population uses water for all purposes at an average rate of about 500 m^3 y^{-1} per person; about 70 percent of this use is nonconsumptive, meaning that the water is returned to a surface or groundwater source after use (Water Resources Council, 1978).* The ratio of consumptive water use to runoff for the conterminous United States is about 2.5 percent. This percentage is large enough to suggest that human activities at the present population scale can have continental-scale impacts on the hydrosphere.

However, looking at water from a continental perspective masks the important fact that the accessible freshwater stocks and flows are irregularly distributed in space and time. This issue is illustrated by considering the annual precipitation pattern for the United States. The eastern part of the country and the Pacific Northwest, with greater than 750 mm per year of precipitation, have, on average, adequate water resources. On the other hand, for most of the western United States, the availability of adequate water supplies is a major constraint for development. In California, extreme climate variability led to the development of an extensive system for managing water resources. California's population of 30 million uses a total of 57 billion m^3 of water per year, that is, about 1900 m^3 y^{-1} per person, about four times the national average (estimated for the year 2000 in Water Resources Council, 1978). The average natural stream flow from California to the oceans is of the same magnitude, approximately 90 billion m^3 per year. California's relatively high water use rate compared with the rest of the United States is explained by its very large agricultural industry, which relies heavily on irrigation. Most of the precipitation in California occurs in the northern half of the state during the winter. However, the demand is dominated by agricultural uses during the summer in the Central Valley and by the large population centers in Southern California. To balance the temporal and spatial mismatch between supply and demand, an extensive and complex system of reservoirs, aqueducts, and pumping stations has been developed throughout California. The reservoirs in the system have a storage capacity of approximately 70 billion m^3; the 22 major aqueducts have lengths totaling 3600 km and can move water at a maximum rate of 48 billion m^3 y^{-1} from available sites to areas of need (van der Leeden et al., 1990). In years with average precipitation, the system works fairly well. However, there are some problems. The creation of so many reservoirs in natural stream channels has had significant consequences for the natural environment. Also, about half of the total water use is drawn from groundwater supplies. Because the rate of withdrawal exceeds the rate of recharge (by about 10 percent), these supplies are being depleted. In addition, precipitation is highly irregular from year to year, and in periods of drought California's water management system is heavily strained.

*The total use includes cooling waters for electricity generation and industrial processes and other "nonconsumptive" uses. Total consumptive use of water in the United States averages about 180 m^3 person^{-1} y^{-1}.

Physicochemical Properties

Water is unique among abundant natural materials in that over the range of ordinary environmental temperatures and pressures it exists widely in all three phases: solid, liquid, and gas.

In combination with its great abundance, water's physical and chemical properties have enormous influence on the earth's climate. This influence depends on many characteristics that can mostly be grouped into thermal, mechanical, and optical properties.

Water's significant thermal and mechanical properties are summarized in Table 2.A.2. The latent heats and the specific heat describe the amounts of energy required to cause a phase change and to heat liquid water, respectively. For example, to melt 1 g of ice requires 334 J; to evaporate 1 g of liquid water requires a much larger quantity of energy, 2450 J, and to heat 1 g of water from the melting point (0 °C) to the boiling point (100 °C) requires about 420 J. The high latent heat of evaporation is very important in buffering the earth's environment against large temperature changes. Consider, for example, the different conditions in the humid tropics and the dry desert. In tropical regions, much of the sun's energy evaporates water (raising the humidity), whereas in arid regions, with little water to evaporate, the air temperature rises much higher.

The optical properties of water also play an important role in affecting climate. Bulk liquid water is transparent to visible light, so sunlight that strikes the surface of an ocean or lake penetrates and contributes its energy to the upper several meters of water. In contrast, when sunlight strikes an opaque solid surface, such as soil, the light energy is absorbed and transmitted by thermal conductivity to depths of only a few centimeters. Since the mass of water that is heated by sunlight is greater than the mass of soil (for equivalent surface areas), the temperature increase caused by sunlight is much lower for bodies of water than for soil surfaces. The differences in the interaction of sunlight with water and with land contribute significantly to our weather patterns.

The optical properties of atmospheric water also influence weather and climate. Clouds appear white because condensed water droplets are effective scatterers of visible light. (Clouds that appear gray or black have higher water content and so absorb a larger fraction of the incident visible radiation rather than scattering it.) Incoming solar radiation may be scattered by clouds back to space, reducing the heating of the earth's surface. In addition, both water vapor molecules and water droplets effectively absorb long-wavelength radiation that is emitted from the earth's surface, a process that tends to warm the atmosphere.

Table 2.A.2 Thermal and Mechanical Properties of Water[a]

Property	Symbol	Value
Density	ρ	$0.998\ g\ cm^{-3} = 998\ kg\ m^{-3}$
(Dynamic) viscosity	μ	$0.0100\ g\ cm^{-1}\ s^{-1} = 1.0 \times 10^{-3}\ kg\ m^{-1}s^{-1}$
Kinematic viscosity	ν	$0.0100\ cm^2\ s^{-1} = 1.0 \times 10^{-6}\ m^2\ s^{-1}$
Thermal conductivity	k_T	$0.59\ W\ m^{-1}\ K^{-1}$
Thermal diffusivity	α	$0.0014\ cm^2\ s^{-1} = 1.4 \times 10^{-7}\ m^2\ s^{-1}$
Specific heat (constant pressure)	C_p	$4.182\ J\ g^{-1}\ K^{-1}$
Latent heat of melting (1 atm, 0 °C)	h_{melt}	$334\ J\ g^{-1}$
Latent heat of evaporation	h_{evap}	$2450\ J\ g^{-1}$

[a]At 20 °C, except as noted.

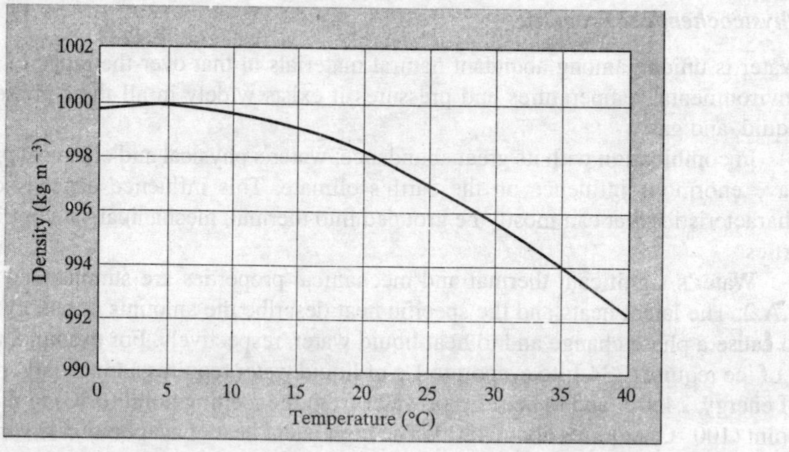

Figure 2.A.2 Density of freshwater, as a function of temperature.

For environmental purposes, water can be considered an incompressible fluid; that is, its density (ρ, units of mass per volume) is independent of pressure. Furthermore, for many applications, the density of liquid water can be treated as constant, at a value of $\rho = 1000$ kg m^{-3} = 1 kg L^{-1} = 1 g cm^{-3}. Given that the molecular weight of water is 18 g mol^{-1}, the corresponding molarity of water in the liquid phase is 1000 g L^{-1} ÷ 18 g mol^{-1} = 55.6 M. Impurities in freshwater are present at small levels, typically in the ppb to ppm range or less. Consequently, in most cases, the density of freshwater is assumed to be independent of the water's composition. The same is true even for wastewater streams.

However, in transport and mixing processes, even small density differences can be important. So, for these applications, the differences in water density as functions of composition and temperature may need to be considered. Figure 2.A.2 shows the density of freshwater as a function of temperature. The dependence of density on temperature has a major impact on mixing and on the temperature structure of lakes and reservoirs. Seawater is typically 2–3 percent more dense than freshwater. Figure 2.A.3 shows the density of seawater as a function of both salinity and temperature.

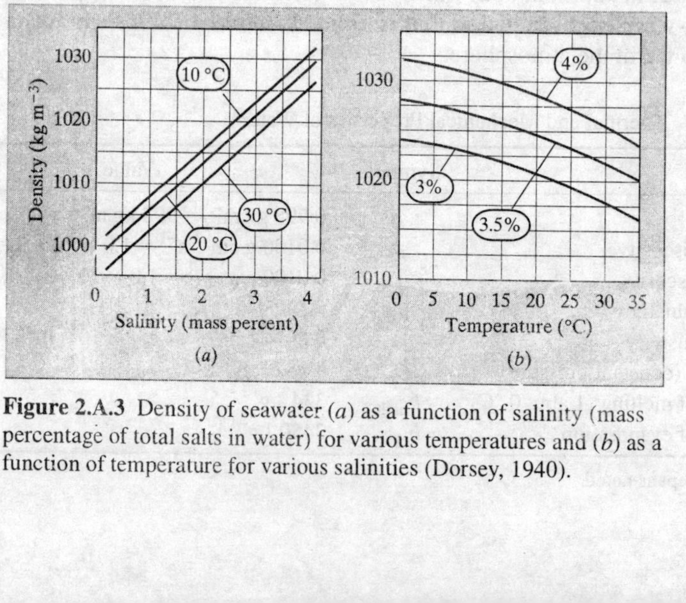

Figure 2.A.3 Density of seawater (*a*) as a function of salinity (mass percentage of total salts in water) for various temperatures and (*b*) as a function of temperature for various salinities (Dorsey, 1940).

The variation of density with salt content and temperature contributes to ocean currents. The stability of aquatic environments caused by density differences is discussed in §6.A.

Water possesses the unusual property (shared only with mercury) of being more dense as a liquid than as a solid. Compared with the range of liquid water densities (shown in Figure 2.A.2), the change in density from liquid to solid is quite large: The typical density of ice is only 920 kg m^{-3}. This property affects the environment in two important ways. First, consider what happens to lakes during winter in cold climates. The surface of the lake may freeze, but unless the lake is very shallow, the deep water will remain liquid. The layer of ice on the surface forms an insulating barrier that limits heat transfer from the lake water to the air and slows the rate of freezing. If, as with most solids, ice were denser than liquid water, then as it formed it would settle to the bottom. Liquid water would remain exposed at the surface. Heat transfer would continue rapidly with more ice forming and settling until finally the entire lake froze. If water and ice behaved in this manner, fish couldn't survive the winter and aquatic life in cold climates would have a very different character.

Shaping the natural and engineered landscape is the second important environmental effect of ice's low density. The freeze-thaw (expansion-contraction) cycle that is repeatedly experienced throughout the fall and spring in places with cold winters applies mechanical forces that can fracture rocks and damage concrete and asphalt roadways. Freezing water also bursts pipes.

More generally, the movement of water through the environment alters the physical landscape, sometimes in dramatic ways. The glacial sculpting of the Yosemite Valley represents an extreme example of the direct physical impact. Ocean surf and river scour cause physical erosion. Water is also a weak chemical solvent. Although the process is slow relative to human time scales, the dissolution of minerals into water as it passes through an environment contributes to chemical erosion.

A final important physical property of water is its viscosity, which affects the rate of movement of water through pipes and through soil, and also the rate of movement of suspended particles through water. Figure 2.A.4 shows that the viscosity of water varies by almost a factor of 3 over the typical range of liquid water temperatures. The

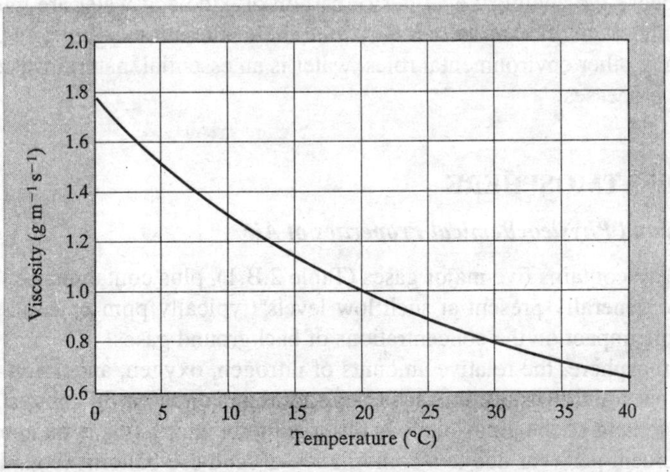

Figure 2.A.4 Dynamic viscosity, μ, of freshwater as a function of temperature.

dynamic (or absolute) viscosity (μ, with units of mass per length per time) is presented in this figure; the kinematic viscosity (ν, with units of length squared per time) is determined by dividing the dynamic viscosity by the fluid density.

Water Is Integral to Life

We may take it as a self-evident truth that we need water to live. Have you ever stopped to think why? Water is the dominant working fluid of our bodies. Our muscles and organs need oxygen to function, and they produce carbon dioxide as a waste product. These substances are transported through our bodies in blood, a water-based fluid filled with cells. Water is also used by the gastrointestinal and urinary systems for excretion of solid and liquid wastes including the parts of food that are indigestible and the decomposition products of proteins.

Water also plays an important role in the body's thermal regulation system, particularly in preventing overheating. We take advantage of the enormous cooling potential of evaporating water by perspiring. Consider a 70 kg adult whose core body temperature is 1 K too high. Approximating the heat capacity of the body to be the same as that of water, a total of about 300 kJ of energy must be liberated to reduce this temperature. This can be achieved by the evaporation of about 120 g (0.12 L) of water.

Finally, water functions as an important reactant in the food chain. Green plants produce carbohydrates (and liberate oxygen) by combining the carbon in CO_2 (from the air) with hydrogen in H_2O (delivered by the roots) plus energy in sunlight:

$$6\,CO_2 + 6\,H_2O + \text{sunlight energy} \rightarrow C_6H_{12}O_6 + 6\,O_2 \qquad (2.A.1)$$

Our bodies release the chemical energy stored in carbohydrates, such as glucose ($C_6H_{12}O_6$), in the food that we eat by combining them with the oxygen that we breathe:

$$C_6H_{12}O_6 + 6\,O_2 \rightarrow \text{metabolic energy} + 6\,CO_2 + 6\,H_2O \qquad (2.A.2)$$

Like the combustion of a hydrocarbon fuel, we produce CO_2 and H_2O as waste products that we exhale or otherwise release to the environment. Useful energy can be extracted because the chemical energies of carbon dioxide and water are much lower than the chemical energies of the carbohydrate molecule and oxygen. Thus, in addition to its many other environmental roles, water is an essential, natural waste product of metabolic processes.

2.B AIR AND THE ATMOSPHERE

Composition and Physicochemical Properties of Air

The atmosphere contains five major gases (Table 2.B.1), plus contaminants. The contaminants are generally present at such low levels (typically ppm or less), that they have negligible impact on the concentrations of background gases.

In the atmosphere, the relative amounts of nitrogen, oxygen, and argon in air are constant in space and over all time scales shorter than centuries. In contrast, the carbon dioxide content of the atmosphere is observed to be increasing at an approximate rate of 0.5 percent per year due to a combination of land use patterns (e.g., deforestation) and the large rate of fossil fuel combustion. The water vapor content of the atmosphere varies from a fraction of a percent to several percent.

Table 2.B.1 Composition of Background Air

Species	MW	Percent (dry air)
Nitrogen (N_2)	28	78.08
Oxygen (O_2)	32	20.95
Water (H_2O)	18	varies
Argon (Ar)	40	0.93
Carbon dioxide (CO_2)	44	0.035

In almost every circumstance of importance in environmental engineering, air can be treated as an ideal gas. This means that the molar concentration of gas molecules in air is a function of temperature and pressure, and independent of composition:

$$\frac{n}{V} = \frac{P}{RT} \tag{2.B.1}$$

where n is the number of moles of gas molecules contained in volume V, P is the pressure of the gas, T is the temperature of the gas (expressed on an absolute scale, such as degrees kelvin), and R is the ideal gas constant:

$$R = 8.314 \text{ J mol}^{-1} \text{ K}^{-1} = 82.05 \times 10^{-6} \text{ atm m}^3 \text{ mol}^{-1} \text{ K}^{-1} \tag{2.B.2}$$

So, for example, at a pressure of 1 atm and a temperature of 293 K, the concentration of gas molecules in air is 41.6 mol m^{-3}. This molar composition is the same for any ideal gas, independent of its composition. Note that 1.0 atm = 1.01325 × 10^5 Pa. The unit Pa is the abbreviation for pascal, the SI unit of pressure, and is equivalent to 1 N m^{-2} or 1 J m^{-3}. Recall that the temperature in K (degrees kelvin) is equal to the temperature in °C plus 273 K.

Gas concentrations in air are sometimes expressed in terms of their partial pressure, as an alternative to either the mass concentration or mole fraction. Dalton's law states that the total pressure exerted by a gas is equal to the sum of the pressures of each individual component. Combining Dalton's law with the ideal gas law, we obtain

$$P_i = \frac{n_i}{V} RT \tag{2.B.3}$$

where P_i is the partial pressure of gaseous species i and n_i is the number of moles of that species contained in volume V. Therefore, the ratio of the partial pressure of a gas to the total air pressure is equal to the mole fraction of that gas in air, Y_i:

$$\frac{P_i}{P} = \frac{n_i}{n} = Y_i \tag{2.B.4}$$

The effective molecular weight of air is the average molecular weight of the constituents weighted by their fractional abundance. The molecular weight of dry air is (see Table 2.B.1)

$$\begin{aligned} MW_{dry} &= 28 \times 0.7808 + 32 \times 0.2095 + 40 \times 0.0093 + 44 \times 0.00035 \\ &= 28.95 \text{ g mol}^{-1} \end{aligned} \tag{2.B.5}$$

Moist air has a lower effective molecular weight (and lower density) because water has a lower MW than does dry air. Let Y_{H_2O} be the mole fraction of water vapor in air. The molecular weight of moist air is then given by

$$MW_{moist} = 18 \times Y_{H_2O} + MW_{dry} \times (1 - Y_{H_2O}) \qquad (2.B.6)$$

Thus, if the mole fraction of water vapor in air is 0.03, then the molecular weight of the air would be 28.63 g mol^{-1}, that is, about 1.1 percent less than the molecular weight of dry air. Ironically, although we refer to hot, humid air as "heavy," it is in fact less dense than cool, dry air.

The mole fraction of any dry-air component in moist air is obtained as

$$Y_{i,moist} = Y_{i,dry} \times (1 - Y_{H_2O}) \qquad (2.B.7)$$

So, for example, the mole fraction of oxygen in air that contains 3 percent water vapor is $0.2095 \times (1 - 0.03) = 0.2032$.

Example 2.B.1 shows how to convert between the mass concentration of a constituent in air and its mole fraction.

EXAMPLE 2.B.1 *Relating Mole Fractions and Mass Concentrations of Gases*

Consider a gaseous species in air that has a molecular weight MW_i (g mol^{-1}). Let Y_i denote the mole fraction of this species.

(a) Derive an expression that, given the parameters Y_i, MW_i, temperature T (K), and total air pressure P (atm), yields the mass concentration of the species, C_i.

(b) Evaluate this expression for the case of oxygen in dry air ($Y_i = 0.2095$) at $T = 293$ K and $P = 1$ atm.

SOLUTION

(a) The mass concentration of any species can be written as the product of the molecular weight and the molar concentration:

$$C_i = MW_i \frac{n_i}{V}$$

where n_i represents the number of moles of the species in the volume. The mole fraction of the species can be written $Y_i = n_i/n$. Rearranging and substituting for n_i yields

$$C_i = Y_i \, MW_i \frac{n}{V}$$

Now we can substitute from the ideal gas law (equation 2.B.1) to obtain the desired result:

$$C_i = Y_i \, MW_i \frac{P}{RT}$$

(b) At $T = 293$ K, $P = 1$ atm, $P/RT = 41.6$ mol m^{-3}. For oxygen, $MW_i = 32$ g mol^{-1}, and in dry air $Y_i = 0.2095$, so $C_i = 279$ g m^{-3}.

The density of air is equal to the product of the molecular weight and the molar concentration:

$$\rho_{air} = MW \frac{n}{V} = MW \frac{P}{RT} \qquad (2.B.8)$$

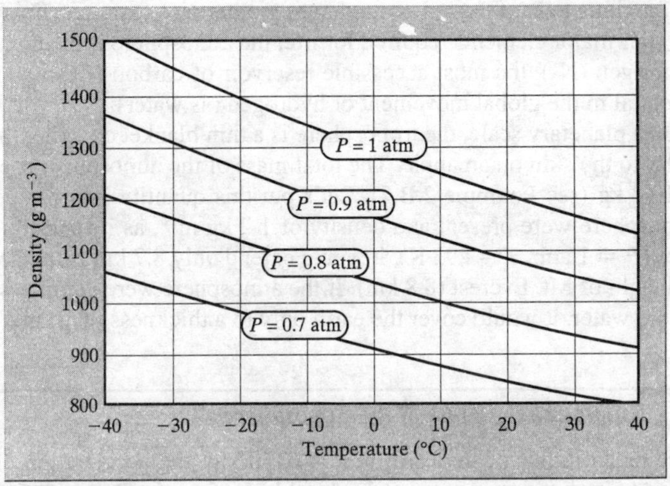

Figure 2.B.1 Density of dry air versus temperature at several pressures (P).

Air density is about three orders of magnitude smaller than liquid water density. Figure 2.B.1 shows that the density of dry air varies substantially over the range of temperature and pressure conditions that are encountered near the earth's surface.

Other important properties of dry, clean air are presented in Table 2.B.2. The presence of moisture and contaminants does not significantly affect these properties.

Structure and Function of the Atmosphere

The gaseous environment surrounding the earth is called the atmosphere. It is divided into layers; from the surface outward, they are the troposphere, the stratosphere, the mesosphere, the thermosphere, the exosphere, and the ionosphere. For environmental engineering, we are mainly concerned with the troposphere. The stratosphere is formally outside of the region where life exists and consequently might not be considered part of the domain of environmental engineering. However, the issue of stratospheric ozone depletion caused by human activities has emerged as a concern over the past few decades.

The atmosphere serves many important functions for the biosphere. It transmits much of the incoming sunlight but is relatively opaque to the outgoing infrared radiation, thereby increasing the average temperature of the earth's surface. Through winds and the evaporation and condensation of water, the atmosphere conveys energy from

Table 2.B.2 Thermal and Fluid-Mechanical Properties of Air[a]

Property	Symbol	Value
Density	ρ	$1.2\ \text{kg m}^{-3}$
(Dynamic) viscosity	μ	$0.0181\ \text{g m}^{-1}\ \text{s}^{-1}$
Kinematic viscosity	ν	$1.5 \times 10^{-5}\ \text{m}^2\ \text{s}^{-1}$
Thermal conductivity	k_T	$0.025\ \text{W m}^{-1}\ \text{K}^{-1}$
Thermal diffusivity	α	$2.08 \times 10^{-5}\ \text{m}^2\ \text{s}^{-1}$
Specific heat (constant pressure)	C_p	$1.005\ \text{J g}^{-1}\ \text{K}^{-1}$

[a]At $P = 1$ atm and $T = 293$ K.

the equatorial regions toward the poles, moderating otherwise extreme climate differences. Of the four major elements required for life, the atmosphere is the dominant repository of nitrogen (N_2), the most accessible reservoir of carbon (CO_2) and oxygen (O_2), and essential in the global movement of hydrogen (as water).

Viewed on a planetary scale, the troposphere is a thin blanket covering the earth's surface, similar to the skin on an apple. The total mass of the atmosphere is estimated to be 5.3×10^{18} kg (see Example 2.B.2). Let's put this quantity into perspective. If the entire atmosphere were present at a density of 1.2 kg m^{-3}, as it typically is at the earth's surface (P = 1 atm, T = 293 K), it would extend only 8.7 km from the surface, less than the height of Mt. Everest (8.8 km). If the atmosphere were compressed to the density of liquid water, it would cover the earth only to a thickness of 10 m!

EXAMPLE 2.B.2 ***Estimating the Mass of the Atmosphere***

Given that the mean radius of the earth is R = 6370 km, the acceleration of gravity is $g = 9.8$ m s^{-2}, and the pressure at the earth's surface is P = 1 atmosphere = 1.01325×10^5 N m^{-2}, determine the mass of the atmosphere.

SOLUTION We can use the principle of fluid statics. The air pressure at the earth's surface results from gravity acting on the mass of air in the atmosphere. Since the characteristic thickness of the atmosphere is very much less than the radius of the earth, we can treat gravity as a constant throughout the atmosphere.

The surface area of the earth is

$$A_{earth} = 4\pi R^2 = 5.1 \times 10^{14}\ \text{m}^2$$

The force of the atmosphere applied to the earth's surface is

$$F_{atm} = PA_{earth} = 1.01325 \times 10^5\ \text{N m}^{-2} \times 5.1 \times 10^{14}\ \text{m}^2 = 5.17 \times 10^{19}\ \text{N}$$

This force is equal to the mass of the atmosphere times the acceleration of gravity, so

$$M_{atm} = \frac{F_{atm}}{g} = \frac{5.17 \times 10^{19}\ \text{kg m s}^{-2}}{9.8\ \text{m s}^{-2}} = 5.3 \times 10^{18}\ \text{kg}$$

On the whole, air mixes fairly rapidly throughout the troposphere. The dominant influence is the heating of the earth's surface by incoming solar radiation. Air at the surface is heated, expands, and becomes buoyant. As the buoyant air rises, it expands in response to the lower atmospheric pressure at the elevated height. This expansion requires that work be done against the surrounding air pressure. The gases cool as their internal kinetic energy is transformed into the mechanical work of expansion. A typical rate of decrease of temperature with height throughout the troposphere is about 6.5 °C km^{-1}. Because the incoming solar intensity is greater near the equator than at the poles, the thickness of the troposphere also varies from a typical minimum of 5–6 km at the poles to a maximum of about 18 km at the equator.

Figure 2.B.2 shows in more detail the variation in temperature, pressure, and air density with height through the lower portion of the troposphere. These relationships are derived from the ideal gas law and the principle of fluid statics, assuming that temperature decreases linearly with height at a rate of 6.5 °C km^{-1}.

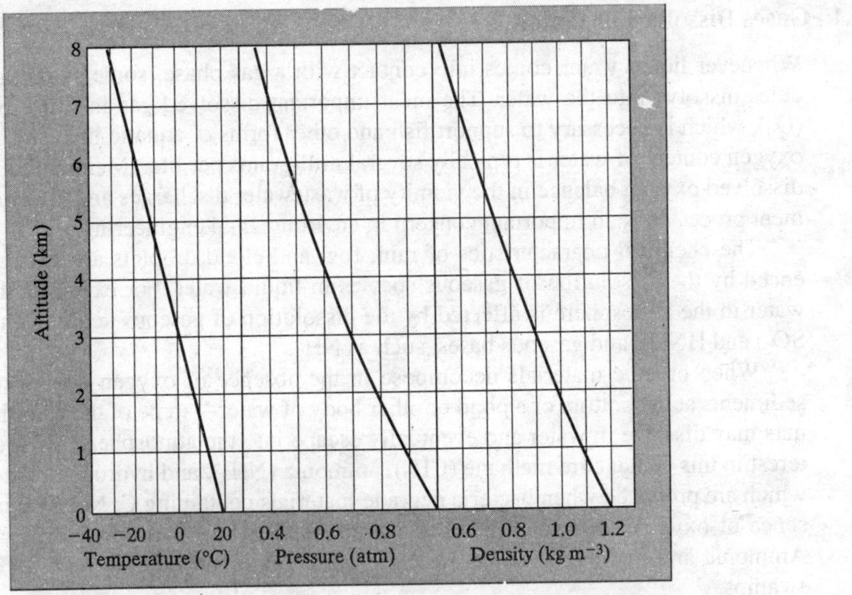

Figure 2.B.2 Typical dependence of temperature, pressure, and air density on height through the lower portion of the troposphere. The given altitudes range from sea level to near that of the tallest mountains on earth.

2.C IMPURITIES IN ENVIRONMENTAL MEDIA

Environmental engineers are concerned with the presence of impurities in water and air—their amounts, their effects, and how they can be controlled. The spectrum of impurities of interest is broad and not easily organized into a single logical structure. Some impurities are of interest only in water, others only in air, and some in both media as well as in soils. Some impurities are directly hazardous to human health. The health concerns vary markedly and include allergic reactions, short-term illnesses (e.g., respiratory infections and gastrointestinal disturbances), chronic illnesses (e.g., kidney or liver damage), acute toxicity, birth defects, cancer, and death. Other impacts are aesthetic, including the taste and clarity of water, the visibility of air, and odors. Some impurities, such as carbon dioxide in the atmosphere or selenium in sediments, are not of concern because of their direct health or welfare impact on humans, but rather because of their impacts on the environment. Finally, some species, such as dissolved oxygen in water, are beneficial rather than harmful.

The goal of this section is to provide an overview of the impurities in water and air that are of interest in environmental engineering. The coverage is intended to be broad, so that all major classes of impurities are considered, but not encyclopedic. Not every impurity of interest is discussed. This section focuses on defining the nature of the impurity, why it is of interest in environmental engineering, and how its levels are quantified. The factors that control impurity concentrations are a major theme for the entire first part of this text and are addressed in detail in subsequent chapters.

2.C.1 Gases Dissolved in Water

Whenever liquid water comes into contact with a gas phase, some of the gas molecules dissolve into the water. The most important dissolved gas in water is oxygen (O_2), which is necessary to support fish and other forms of aquatic life. The dissolved oxygen content of water is typically several milligrams per liter. Maintaining a proper dissolved oxygen balance in the vicinity of wastewater discharges and in certain treatment processes is an important concern in environmental engineering.

The chemical characteristics of rain, fog, and cloud droplets are strongly influenced by the dissolution of gaseous species in liquid water. For example, the pH of water in the atmosphere is affected by the dissolution of gaseous acids, such as CO_2, SO_2, and HNO_3, and gaseous bases, such as NH_3.

When organic materials decompose in the absence of oxygen—for instance, in sediments at the bottom of a pond or other body of water—some of the resulting products may dissolve in water and eventually escape into the atmosphere. Of specific interest in this context are methane (CH_4), ammonia (NH_3), and hydrogen sulfide (H_2S) which are produced when bacteria degrade materials containing C, N, and S in the absence of oxygen. Methane release to the atmosphere is a factor in global warming. Ammonia and hydrogen sulfide are toxic gases that contribute to the fetid smell of swamps.

The transfer of contaminants between air and water is widely used in pollution control equipment. For example, sulfur dioxide (SO_2) produced by burning coal can be removed from flue gas by dissolving it into an aqueous slurry (water plus lime or limestone) in a device known as a scrubber. Likewise, volatile organic compounds are removed from water by transferring them to the gas phase in a device known as an air stripper.

2.C.2 Water in Air

The water vapor content of air plays an important role in human comfort as well as in controlling the meteorological aspects of the atmosphere. If the water vapor content of air is too low, the moist tissues of our eyes, nose, and throat may feel dry and irritated. If the water vapor content is too high, our skin may feel excessively moist and clammy. Managing the moisture content of air in buildings is commonly practiced through humidification (adding water vapor to air) or dehumidification (removing water vapor from air).

The water vapor content of the ambient air also strongly influences human comfort in the outdoor air. Human activities have altered ambient humidities, especially in irrigated urban and suburban areas in arid climates, where the additional evapotranspiration adds to the atmospheric water vapor content. This modification has been an inadvertent consequence of urbanization, and there has been little discussion of reversing the phenomenon.

As outlined in §2.B, the presence and movement of water through the atmosphere has profound implications for the earth's energy balance and for climate.

In addition to the gaseous state, water can be present in air as a liquid or as a solid. Liquid water in the atmosphere plays an important role in the flow of energy through the atmosphere, in visibility, and in the transformation and removal of air pollutants. The liquid water content of air can be conveniently expressed on a volume fraction basis, that is, the volume of liquid water per volume of air. Also of interest is the size of the droplets in which the water is condensed. Table 2.C.1 summarizes the

Table 2.C.1 Droplet Radius and Volume Fraction of Liquid Water Suspended in Air under Different Atmospheric Conditions

System	Droplet radius (μm)	V_{liquid}/V_{air}
Haze	0.03–0.3	$(0.1–1) \times 10^{-11}$
Fog	10	$(0.2–2) \times 10^{-7}$
Clouds	10	$(0.1–3) \times 10^{-6}$
Rain	200–2000	$(0.1–1) \times 10^{-6}$

Source: Seinfeld and Pandis, 1998.

characteristic droplet size and liquid water content for various suspensions of water in the atmosphere.

2.C.3 Acids, Bases, and the Hydrogen Ion

Water is soluble in water. This surprising statement means that even in pure water, some of the water molecules dissociate to form the hydrogen ion (H^+) and the hydroxide ion (OH^-). The hydrogen ion concentration is such an important characteristic of water that a special measure, pH, is defined for it:

$$pH = -\log[H^+] \tag{2.C.1}$$

In this definition, $[H^+]$ has units of moles/liter and the logarithm is base 10. So, for example, water with a pH of 5 would have an H^+ concentration of 10^{-5} M. Water is considered to be *neutral* when its pH is 7. Water with lower pH values is considered *acidic*, and values of pH above 7 are considered *basic*.

The hydrogen ion concentration in environmental aqueous solutions can vary over many orders of magnitude. Acid fogs have been measured with pH values as low as 2.2 (Waldman et al., 1982), whereas in a water treatment process known as chemical precipitation, the pH is commonly raised to 11. In using pH, it is important to remember that a change by 1 unit equals an order-of-magnitude change in $[H^+]$.

The pH is very important in determining the concentration of many other species in water. It is sometimes called the "master variable" of water because of this role. In turn, pH is controlled by means of the relative amounts and strengths of acids and bases in solution. Table 2.C.2 lists important acid and base species for environmental engineering applications. The dependence of pH on the quantities and strengths of acids and bases in water is explored in §3.C.

Some acids (for example, CO_2, SO_2, H_2S, and HNO_3) and some bases (NH_3) are volatile and may be present in the gas phase as well as in the dissolved phase. However, the concept of hydrogen ion concentration and pH applies only to water, including condensed water dispersed in air, such as cloud droplets.

2.C.4 Inorganic Impurities

One useful classification of water and air impurities is inorganic versus organic. A functional definition of an organic molecule is any compound that contains carbon, with the following exceptions: CO, CO_2, HCO_3^-, and CO_3^{2-}. (We also exclude elemental carbon—diamonds, the dominant constituent of coal, graphite, and soot—from the organic class.) Within both inorganic and organic categories there exist broad ranges

Table 2.C.2 Acids and Bases Commonly Encountered in Environmental Engineering

Name	Chemical formula
Acids	
Perchloric acid	$HClO_4$
Hydrochloric acid	HCl
Sulfuric acid	H_2SO_4
Nitric acid	HNO_3
Phosphoric acid	H_3PO_4
Hypochlorous acid	$HOCl$
Sulfonic acid (sulfur dioxide [aq])	H_2SO_3
Carbonic acid (carbon dioxide [aq])	H_2CO_3
Hydrogen sulfide	H_2S
Bases	
Ammonia	NH_3
Calcium carbonate	$CaCO_3$
Lime	$Ca(OH)_2$
Sodium hydroxide	$NaOH$

of species. This section summarizes some of the more important aspects of inorganic impurities in air and water. An overview of organic impurities is presented in §2.C.5.

Ions in Water

In water, one major class of inorganic species is ions. These are atoms or molecules that possess an imbalance of electrons and protons in solution and therefore have a net electrical charge. Ions are classified according to whether their charge is positive or negative: *cations* have a net positive charge (an excess of protons relative to electrons), and *anions* have a net negative charge. (As a helpful mnemonic, you can think of an anion as "*a negative ion*"; it may also help to link the *t* in cation with the net "+" charge.)

The most common ions found in natural waters are listed in Table 2.C.3, along with representative concentrations found in natural seawater and freshwater. Note that

Table 2.C.3 Most Prevalent Ions in Natural Waters, Along with Typical Molar Concentrations

Ion	Symbol	Seawater (M)	River water (M)
Sodium	Na^+	0.47	0.23×10^{-3}
Magnesium	Mg^{2+}	0.053	0.15×10^{-3}
Calcium	Ca^{2+}	0.010	0.33×10^{-3}
Potassium	K^+	0.010	0.03×10^{-3}
Chloride	Cl^-	0.55	0.16×10^{-3}
Sulfate	SO_4^{2-}	0.028	0.07×10^{-3}
Bicarbonate	HCO_3^-	0.0024	0.86×10^{-3}

Source: Data from Stumm and Morgan, 1996.

ion concentrations are typically 100–1000 times higher in seawater than in freshwater. Bicarbonate is an exception, with freshwater and seawater levels differing only by a factor of 3.

Ions in water originate from the dissolution of electrically neutral molecules that either enter water from the gas phase or, more typically, come into contact with water as a solid. Removal of ions from freshwater can occur through transport to the oceans or by incorporation into solids and deposition as sediments, for example, onto the bottoms of lakes. Removal of ions from ocean water occurs primarily through uptake by marine organisms, followed by their settling, after death, to the bottom sediments.

The ionic composition of water is of interest for several reasons. The amounts of ionic species can directly determine the suitability of water for human consumption, as evidenced by the distinction between seawater and freshwater. Some toxic elements, such as certain metals, are commonly present in water as ions. Nutrients, a concern in wastewater and agricultural runoff, are also commonly present in ionic form. Several gross measures of water quality depend directly on the ionic composition of water. Three of these—ionic strength, hardness, and alkalinity—are discussed in subsections that follow. First, though, we summarize an important conservation principle that is helpful in solving problems that involve ions in water.

Electroneutrality The electroneutrality principle, derived from stoichiometry, states that an aqueous solution cannot possess a net electrical charge. The electroneutrality principle can be stated as follows:

> *The sum of the normalities of all cations in an aqueous solution must equal the sum of the normalities of all anions.*

Mathematically, the electroneutrality principle can be expressed in the following compact form:

$$\sum_{\substack{i,\text{ cations}\\ \text{and anions}}} z_i C_i = 0 \tag{2.C.2}$$

where z_i is charge per molecule on the ith ion, including a + sign for cations and a − sign for anions, and C_i is the molar concentration of the ith ionic species. Recall that the molar concentration of an ionic species times the number of excess charges per molecule equals the normality of the species (see §1.C.1).

Electroneutrality provides one algebraic equation that links the concentrations of all ionic species in a solution and is commonly applied in solving acid/base problems. It can also be used as a check on laboratory analyses of ionic species in water samples. First, the balance of positive and negative ions can be checked to see if any ions are missing. Second, a summation of the mass concentrations of all ions can be compared with the total dissolved solids (TDS) to determine whether any nonionic dissolved species, such as silicon dioxide, are present (see Example 2.C.1).

EXAMPLE 2.C.1 ***Electroneutrality in Seawater***

A sample of seawater is analyzed for ion content, and the following results are reported (Stumm and Morgan, 1996).

EXAMPLE 2.C.1 *Electroneutrality in Seawater (continued)*

Ion	Mass conc. (g L^{-1})	Molar conc. (M)
Na^+	10.8	0.468
Mg^{2+}	1.29	0.0532
Ca^{2+}	0.409	0.0102
K^+	0.399	0.0102
Cl^-	19.3	0.545
SO_4^{2-}	2.71	0.0282
HCO_3^-	0.146	0.0024

(a) Test whether electroneutrality is satisfied. Assume that the only other ions present are H^+ and OH^- and that their concentrations are negligible.

(b) Given that the measured TDS in this water is 35.2 g L^{-1} and that TDS measurements are good only to within 10–20 percent accuracy, are any other major dissolved species in this sample?

SOLUTION

(a) We sum the normalities of the cations and anions separately and compare the sums. The normality is obtained by multiplying the molar concentration by the magnitude of the ionic charge per molecule. The following table shows the results.

Ion	Molarity (M)	Normality (eq/L)
	Cations	
Na^+	0.468	0.468
Mg^{2+}	0.053	0.106
Ca^{2+}	0.010	0.020
K^+	0.010	0.010
Total cation normality		0.604
	Anions	
Cl^-	0.545	0.545
SO_4^{2-}	0.028	0.056
HCO_3^-	0.002	0.002
Total anion normality		0.603

The sums of the normalities of the cations and anions agree to within 1 part in 600, or to better than 1 percent. For routine environmental measurements, uncertainty is seldom less than 5 percent. Consequently, we conclude that the electroneutrality relationship is satisfied for this sample of seawater.

(b) To determine whether any significant nonionic species contribute to the dissolved solids, sum the mass concentrations of the ions and compare the result with the measured TDS. In this sample, the total ion concentration is 35.1 g L^{-1}, well within 10 percent of the reported TDS of 35.2 g L^{-1}, suggesting that no quantitatively significant species have been missed.

Ionic Strength The total amount of ions in water is commonly quantified as the *ionic strength*, defined by

$$I = \frac{1}{2}\sum_i (C_i z_i^2) \qquad (2.C.3)$$

where C_i is the molar concentration of the ith ionic species (M), and z_i is the number of net electrical charges associated with the ith species (–). The summation is carried out over all ions present in solution (see Example 2.C.2).

EXAMPLE 2.C.2 ***Ionic Strength of Seawater and River Water***

Compute the ionic strengths for the water samples presented in Table 2.C.3.

SOLUTION For seawater,

$$I = \tfrac{1}{2}(0.47 + 4 \times 0.053 + 4 \times 0.010 + 0.010 + 0.55 + 4 \times 0.028 + 0.0024)$$
$$= 0.70 \text{ M}$$

For river water,

$$I = \tfrac{1}{2}(0.23 + 4 \times 0.15 + 4 \times 0.33 + 0.03 + 0.16 + 4 \times 0.07 + 0.86) \times 10^{-3}$$
$$= 0.0017 \text{ M}$$

Ionic strength indicates the intensity of electrostatic interactions between ions in an aqueous solution. These interactions can affect the chemical transformation processes that occur in water. In some environmental engineering applications, the interactions are sufficiently weak to be ignored (see Appendix E, §E.1).

Hardness Water hardness is defined as the sum of the normalities of all multivalent cations (i.e., charge of +2 or greater). Hardness is a concern in municipal water supplies. Water that is too hard has a tendency to form solid precipitates known as scale that deposit on the inner surface of pipes and fittings. Excessive scale can seriously impede water flow through a distribution system. On the other hand, the deposition of some scale in pipes is desirable as a preventive measure against corrosion (see §3.D.2). Hardness of municipal water supplies is also an aesthetic concern: Solid precipitates from hard water can accumulate on bathroom surfaces and cookware. Soap does not form a good lather in hard water.

Usually, the main cations that contribute to drinking water hardness are Ca^{2+} and Mg^{2+}, but in some circumstances Fe^{3+} and Al^{3+} can also be important.

The primary measurement unit for expressing hardness is normality (or ionic charge per volume), that is, equivalents per liter. In engineering practice, hardness is often expressed in terms of the equivalent mass concentration of calcium carbonate ($CaCO_3$) that, dissolved in pure water, would produce the same hardness as found in the sample of interest. Given that $CaCO_3$ has a molecular weight of 100 g/mol and that each mole of $CaCO_3$ liberates two moles of cationic charge, the unit conversion is easy: A hardness of 1 meq/L is the same as a hardness of 50 mg/L as $CaCO_3$ (see Table 2.C.4).

Table 2.C.4 Hardness Classification of Water

Classification	Hardness (meq/L)	Hardness (mg/L as $CaCO_3$)
Soft	<1.5	<75
Moderately hard	1.5–3	75–150
Hard	3–6	150–300
Very hard	>6	>300

Source: Sawyer et al., 1994.

The hardness of water depends on contact with dissolvable minerals, among other factors. For example, water from freshly melted snow tends to be soft whereas groundwater tends to be hard or very hard.

The hardness of water is sometimes subdivided into carbonate and noncarbonate hardness. This division can be confusing because it does not depend on the hardness ions themselves, but rather on the anions present in solution. Conceptually, the difference between carbonate and noncarbonate hardness lies in the form of the solids that dissolved to generate the hardness. For example, carbonate hardness can be generated by the dissolution of calcium carbonate or magnesium carbonate, whereas noncarbonate hardness can be generated by the dissolution of calcium sulfate, calcium hydroxide, or calcium chloride. This distinction is important when chemical precipitation is used to reduce hardness (§6.D.4).

To determine the carbonate hardness (CH) and noncarbonate hardness (NCH) of a solution, first compute the total hardness (TH) and then compute the normality of the carbonate species: $N_c = [HCO_3^-] + 2[CO_3^{2-}]$. If $N_c < TH$, then $CH = N_c$ and $NCH = TH - CH$. Alternatively, if $N_c > TH$, then $CH = TH$ and $NCH = 0$. In either case, the sum of the carbonate and noncarbonate hardness must equal the total hardness. See Example 2.C.3.

EXAMPLE 2.C.3 ***Hardness of River Water***

Compute the total hardness, noncarbonate hardness, and carbonate hardness of the sample of river water presented in Table 2.C.3. Classify the water's hardness.

SOLUTION The total hardness is the sum of the normalities of the multivalent cations, in this case Ca^{2+} and Mg^{2+}:

$$TH = 2 \times 0.33 + 2 \times 0.15 = 0.96 \text{ meq/L}$$

This water would be classified as soft (see Table 2.C.4). The carbonate hardness and noncarbonate hardness depend on the concentration of carbonate, CO_3^{2-}, which is not given. Assume, as will be defended in Chapter 3, that the carbonate normality is much less than the bicarbonate normality for natural fresh waters. Then

$$N_c = [HCO_3^-] = 0.86 \text{ meq/L}$$

Since $N_c < TH$, we have

$$CH = N_c = 0.86 \text{ meq/L}$$

$$NCH = TH - CH = 0.10 \text{ meq/L}$$

Alkalinity Alkalinity refers to the capacity of water to neutralize acids. In most uncontaminated natural waters, the alkalinity is determined by the abundance of four ions: carbonate (CO_3^{2-}), bicarbonate (HCO_3^-), hydroxyl (OH^-), and hydrogen (H^+). For many practical engineering purposes, we can define the alkalinity by this expression:

$$A = [OH^-] + [HCO_3^-] + 2[CO_3^{2-}] - [H^+] \tag{2.C.4}$$

where the ionic species are measured in moles per liter and alkalinity is expressed in equivalents per liter.

In practice, alkalinity is measured by *titration*: A strong acid (such as hydrochloric) is slowly added to the water of interest until the pH diminishes to ~4.5. The amount of acid added (i.e., the moles of charge associated with the acid anion) per liter of water sample is the measured alkalinity.

To fully understand alkalinity, we need additional knowledge of acid-base reactions and the carbonate system. We will return to this topic in Chapter 3, once the necessary background information has been presented.

Inorganic Gaseous Pollutants

Many gaseous constituents are present in the atmosphere, especially in polluted urban air. Typically, gaseous contaminants are present at ppm levels or less, so that the major composition of the atmosphere (e.g., the oxygen concentration) is unaffected by the presence of impurities.

Four inorganic gases merit special attention as atmospheric impurities: carbon monoxide, nitrogen dioxide, sulfur dioxide, and ozone. In the United States, these are four of the six so-called criteria pollutants (the other two are condensed-phase materials—lead and particulate matter). The criteria pollutants have known health effects and are commonly observed in polluted urban air. Their regulatory status requires that airborne concentrations be routinely monitored. When concentrations exceed regulatory limits, control strategies must be developed to reduce levels to those deemed safe. The characteristics of these pollutants are discussed in §7.A.

Toxic Metals

The presence of metals in either air or water is a significant environmental concern. In water, metals may be in the form of ions, in inorganic molecular clusters known as metal complexes, or incorporated into organic molecules. In both air and water, metals may be found in small solid particles suspended in the fluid.

Metals that are subject to federal environmental regulation are listed in Table 2.C.5. All of these metals exhibit some adverse effects on human health at high doses. Ironically though, some regulated metals, such as chromium, copper, and selenium, are also trace elements that are an essential part of the human diet at low levels.

Brief sketches of four metals are provided in the following paragraphs.

Cadmium Cadmium is widely used in metal plating and is an active ingredient of rechargeable batteries. Smelting and refining operations are the dominant atmospheric sources; mine drainage and industrial wastewater are important sources of cadmium in water. Cadmium is present as an inadvertent trace constituent of galvanized pipe, so corrosion of water distribution pipes is another potential cause of human exposure. Cadmium's chemical similarity to zinc is thought to be at least partly the cause of its

Table 2.C.5 Toxic Metals Regulated in Water and Air by U.S. Federal Legislation[a]

Metal (symbol[b])	Regulated in water	Regulated in air
Antimony (Sb)	✓	✓
Arsenic (As)	✓	✓
Barium (Ba)	✓	
Beryllium (Be)	✓	✓
Cadmium (Cd)	✓	✓
Chromium (Cr)	✓	✓
Cobalt (Co)		✓
Copper (Cu)	✓	
Lead (Pb)[c]	✓	✓
Manganese (Mn)		✓
Mercury (Hg)	✓	✓
Nickel (Ni)	✓	✓
Selenium (Se)	✓	✓
Thallium (Tl)	✓	

[a] Under provisions of the Safe Water Drinking Act amendments of 1986 and Title II of the 1990 amendments to the Clean Air Act.
[b] Atomic symbol.
[c] Lead is regulated both as a criterion air pollutant and as a hazardous air pollutant.

adverse health effects: It substitutes for zinc in enzymes and interferes with their proper function. Cadmium causes high blood pressure and kidney damage. Inhalation exposure has induced lung tumors in laboratory rats. The EPA classifies cadmium as a probable human carcinogen.

Chromium Chromium is a trace constituent of ordinary soils. It is also a natural impurity in coal and may be released to the atmosphere when coal is burned. It is widely used in the manufacture of stainless steel. Chromium exists in two oxidation states in the environment, +III and +VI. (See §3.D for a discussion of oxidation states.) Cr(III) is an essential trace nutrient in human diets (deficiency results in metabolic disorders), whereas Cr(VI) causes a suite of adverse health effects, including liver and kidney damage, internal hemorrhage, respiratory disorders, and cancer. Environmental standards for chromium do not distinguish among the oxidation states in recognition of the possibility that Cr(III) may be converted to Cr(VI) in the environment.

Lead Lead was used extensively before its adverse health effects became well known. It was incorporated in pigments used in house paint and in glazes applied to dishware. Lead was also used in pipes and solder in water distribution systems. Probably the most important source of human exposure was from the use of tetraethyl lead, $(C_2H_5)_4Pb$, as a gasoline additive. Tetraethyl lead improved the performance of gasoline and helped lubricate the engine. When leaded gasoline was burned, most of the lead was emitted into the atmosphere as particulate matter, causing inhalation exposure of people nearby and contamination of soils and vegetation near roadways. In the United States, lead was gradually eliminated from gasoline over a period of two decades, beginning in 1973. Between 1984 and 1993, it is estimated that the lead emission rate into the atmosphere from human activities in the United States declined by almost 90 percent (U.S. Environmental Protection Agency, 1994). Airborne lead levels have diminished by about the same proportion. Ironically, the original reason for eliminating lead from gasoline was to reduce urban ozone concentrations, not airborne lead levels. Lead in automobile exhaust poisons catalytic converters and prevents them from reducing emissions of unburned gasoline vapors and nitrogen oxides. Although a substantial decrease in human exposure to lead was achieved by eliminating it from gasoline, we still have a legacy of environmental lead in the paint and pipes of old houses and in land near heavily used roadways.

A range of adverse health effects result from the accumulation of lead in the bloodstream, including anemia, kidney damage, elevated blood pressure, and central nervous system effects such as mental retardation. Infants and young children are especially susceptible to lead poisoning because they absorb ingested lead more readily than older humans. Lead is classified as a probable human carcinogen. Some historians believe that the use of lead for food and drink containers may have contributed to the gradual poisoning of the ruling class during the Roman Empire, hastening its downfall (Gilfillan, 1990).

Mercury An important source of environmental mercury is the combustion of coal. Mercury is also used in the manufacture of electrical components, in industrial production of chlorine and hydroxide, and as a fungicide in agricultural applications. Mercury has been added to paints to retard the growth of fungus, and some concern has arisen about environmental exposures from this application (Agocs et al., 1990). Mercury is also widely used in the amalgam of silver dental fillings, raising concern about chronic exposure due to the treatment of cavities. Even the potential for inhalation exposure from the cremation of corpses with dental fillings has been investigated (Mills, 1990).

Mercury poisoning is exhibited through central nervous system dysfunction. Mild cases may cause depression and irritability. Severe poisoning may cause paralysis, blindness, and birth defects. Historically, mercury was used as a stiffening agent in the manufacture of felt hats. The phrase "mad as a hatter" probably derives from poisoning symptoms exhibited as a result of this occupational exposure. The most notorious episode of mercury poisoning occurred in Minimata Bay in Japan (see §8.A.1).

2.C.5 Organic Impurities

Thousands of organic chemicals are found in air and water. Many are environmentally benign. Others cause one or more of a broad range of adverse effects. Some cause odor or irritation that is considered a nuisance. Others can cause, or at least increase the risk of, serious illness, including cancer and central nervous system damage. Some organic chemicals, such as polychlorinated biphenyls (PCBs) and DDT, have been found to be severely disruptive to ecosystems. Such chemicals may persist for long periods in the environment and may accumulate or become concentrated in certain environmental compartments, including food chains.

Organic compounds can cause taste, odor, and discoloration problems in drinking water, interfere with water treatment processes, cause oxygen depletion downstream of wastewater discharges, and be transformed from benign to hazardous compounds in treatment processes such as disinfection. In the atmosphere, organic gases play a pivotal role in the photochemistry that produces urban ozone. A subclass of organic compounds, chlorofluorocarbons (CFCs)—originally hailed as a significant technological achievement when introduced in the 1950s—were found after wide use to

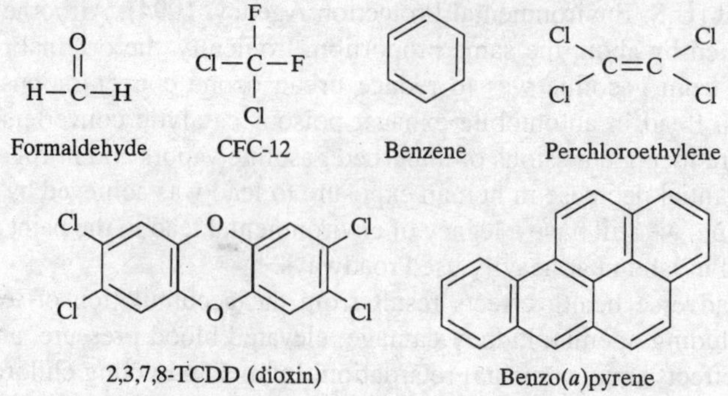

Figure 2.C.1 Structure of selected organic compounds of environmental concern.

damage the stratospheric ozone layer. Organic contamination plays a large role in our current problems with hazardous waste sites and industrial waste streams.

Specific Organic Impurities of Environmental Concern

In this section, brief sketches are provided of six specific organic impurities whose chemical structures are depicted in Figure 2.C.1. The goals are to introduce some of the more important problems associated with specific impurities and to illustrate the diverse range of uses, environmental actions, and adverse effects of organic compounds. For readers who lack background in organic chemistry, a brief introduction to environmental organic chemicals is provided in Appendix C.

Formaldehyde (HCHO) Formaldehyde is a widely used industrial chemical; annual production in the United States is about 10^9 kg. A major use of formaldehyde is as a component in resins used for bonding and laminating. Formaldehyde is also used as a preservative in cosmetics and toiletries and in the manufacture of lacquers, dyes, and plastics. In addition to emissions from industrial processes and products, formaldehyde is formed in photochemical smog by the degradation of other organics. It is also a by-product of incomplete combustion and so is found in wood and tobacco smoke and in auto exhaust.

Formaldehyde has a pungent odor and is irritating to mucous membranes at relatively low concentrations. It has been shown to cause nasal cancer in laboratory mice and rats. The evidence for formaldehyde as a cause of human cancer is inconclusive.

Chlorofluorocarbons Chlorofluorocarbons (CFCs) are a family of chemicals that typically contain one or two carbon atoms attached to a mix of chlorine, fluorine, and sometimes hydrogen atoms. The principal CFCs in the environment are CCl_3F, CCl_2F_2, and $CHClF_2$. CFCs were originally developed in the 1950s for use as the working fluid in refrigeration units (such as in refrigerators and air conditioners). For safety, this was an important development. Previously, toxic gases such as ammonia and sulfur dioxide were used as the working fluid in refrigerators and leaks caused hazardous conditions. CFCs exhibit very little toxicity, and they are very stable in the environment. In addition to their use as refrigerants, CFCs attained widespread use as aerosol propellants and foaming agents. Worldwide production reached 500 billion kg per year at the peak. Ironically, it is their stability that causes environmental problems. CFCs persist long enough in the atmosphere to be transported through the troposphere into the stratosphere. There, intense ultraviolet radiation causes them to decompose. The liberated chlorine atoms act in a catalytic cycle, destroying ozone. Stratospheric ozone serves the important function of absorbing ultraviolet radiation from the sun. Increasing its rate of destruction raises the risk of increasing the intensity of ultraviolet radiation that reaches the earth's surface. This radiation can cause damage to living cells; exposure of humans, for example, leads to an increased risk of skin cancer. By international agreement, CFCs are being phased out of use.

Benzene (C_6H_6) Benzene is the simplest member of a class of organic compounds known as aromatics. This class features a ring structure with six carbon atoms alternately joined by single and double bonds. Benzene is an important industrial chemical. It is used as an octane-boosting component in gasoline and as a solvent. Benzene is also a by-product of the incomplete combustion of hydrocarbon fuels. For example, it is an important constituent of tobacco smoke. Benzene is classified as a known human carcinogen, linked as a causative agent in leukemia, Hodgkin's disease, and lymphomas via inhalation exposure.

Because of benzene's toxicity and widespread use, its sources and human exposures have been extensively studied, revealing an important point about environmental exposures: The largest sources may not be the most important sources. Wallace (1989) has estimated that the "hundreds of thousands of tons of benzene emitted annually by all outdoor sources provide only about 20% of the total population exposure, whereas the mere 30 tons per year delivered in mainstream cigarette smoke accounts for 50% of the total population exposure." Although this statement may be surprising, the reason is clear enough: Essentially all of the benzene in mainstream cigarette smoke is inhaled by the smoker, whereas only a small fraction of the benzene emitted into the atmosphere is inhaled by anyone before being removed from the atmosphere. Wallace estimates that a 1.5-pack-per-day smoker inhales 1.8 mg of benzene daily, about 10 times the typical daily intake of nonsmokers.

Perchloroethylene (C_2Cl_4) Perchloroethylene (tetrachloroethylene), or PCE, belongs to a class of compounds known as halogenated organics or chlorinated solvents, organic molecules that contain one or more halogen atoms: chlorine, fluorine, iodine, and bromine. This class of compounds, which includes dichloromethane (CH_2Cl_2), trichloroethylene (C_2HCl_3), and trichloroethane ($C_2H_3Cl_3$), are widely used in industrial applications and as feedstocks for the production of other materials. In addition, a common use of PCE is for dry-cleaning clothes. Several chlorinated solvents are known or suspected human carcinogens and are federally regulated as both hazardous air pollutants and in drinking water. Much of the problem at hazardous waste sites in the United States has resulted from the contamination of soil or groundwater with chlorinated solvents. In fact, chlorinated solvents are the most commonly reported groundwater contaminants at hazardous waste sites.

Dioxin (2,3,7,8-TCDD) Dioxins are a group of compounds that have a double aromatic ring structure coupled by two oxygen atoms, with some chlorine atoms substituting for hydrogen on the aromatic rings (see Figure 2.C.1). The particular molecule of greatest environmental concern is known as 2,3,7,8-tetrachlorodibenzo-*p*-dioxin, where the numbers refer to the position of the Cl atoms on the ring structure. This name is commonly shortened to 2,3,7,8-TCDD, TCDD, or, more simply, dioxin. The last usage is not strictly correct, since dioxins are a class of compounds whose sources, environmental behavior, and effects vary.

Dioxins are not produced intentionally. They are formed as by-products in the manufacture of some pesticides and during the combustion of plastics and chlorinated solvents, creating serious concerns for the incineration of municipal, medical, and hazardous wastes. They may also be formed in trace amounts from the common practice of bleaching paper pulp with chlorine.

2,3,7,8-TCDD is considered to be one of the most highly toxic organic compounds produced by human activity, although the levels at which health effects occur are controversial. Dioxins don't decompose easily and are therefore relatively persistent in the environment. A panel of scientists convened by the USEPA in 1992 concluded that some wildlife is sensitive to dioxins and that significant ecosystem damage can result from even small releases.

Several well-publicized environmental incidents have involved dioxins. In 1976 an explosion occurred at a pesticide manufacturing plant in Seveso, Italy, exposing many people in the nearby area to TCDD. Unintentional exposure in Times Beach, Missouri, during the early 1970s caused the U.S. government to purchase all of the property in the community and evacuate the 2400 residents (see §8.A.1).

Benzo(a)pyrene ($C_{20}H_{12}$) and Polycyclic Aromatic Hydrocarbons Benzo(*a*)pyrene (BaP) is the best-known and most extensively studied member of the class of organic molecules known as polycyclic aromatic hydrocarbons (PAHs) or polynuclear aromatic compounds (PNAs). The distinctive structural characteristic of this class is the presence of multiple benzene rings fused together.

PAHs do not have significant industrial uses. They are formed by the combustion of hydrocarbon fuels under conditions of insufficient oxygen. Soot from diesel engines and smoke from wood fires, barbecues, or cigarettes tend to be rich in PAHs. Coal tar also contains high PAH levels. In terms of total human exposure, the most important source of BaP appears to be the use of wood or other biomass as a cooking fuel in the rural parts of economically less-developed countries. This type of exposure affects 500 million people, mostly women and young children. They are exposed to BaP concentrations that may be 1000 times as high as those encountered by nonsmokers in polluted urban environments.

BaP was one of the first chemicals shown to be carcinogenic in isolation. Studies demonstrating the carcinogenicity of BaP in laboratory animals were published in 1933 (Phillips, 1983). Polycyclic organic matter, of which PAHs are a subclass, is federally regulated as a hazardous air pollutant. The USEPA also regulates BaP in drinking water.

Aggregate Measures of Organic Compounds in Water and Air

Some environmental concerns associated with organic compounds do not result from direct toxic effects of specific species. For such concerns, measurement parameters have been defined that quantify the aggregate amount of organic material in an environmental fluid. Techniques to measure these aggregate parameters have been developed that are much more economical than the alternative of measuring the concentrations of all individual compounds, which may number in the hundreds. In water, the impact of organic compounds on dissolved oxygen content is a serious concern, while in air, the contribution of organic gases to photochemical smog and especially the generation of ozone in urban air is a major issue.

Oxygen Demand An early concern in environmental engineering resulted from the discharge of primary treated sewage effluents to lakes and rivers. (Primary treated wastewater contains organic compounds and nutrients but has undergone physical treatment to remove suspended particles.) Such discharges can disturb the dissolved oxygen content of the receiving water, seriously disrupting aquatic life. This fact led to the requirement that secondary (biological) treatment processes be applied to wastewater before discharge. The central issue can be summarized as follows:

- Wastewater contains organic compounds such as glucose ($C_6H_{12}O_6$).
- Microorganisms in the water use these organic compounds as food.
- While degrading these organic compounds, microorganisms consume oxygen (O_2) dissolved in the water.
- The O_2 used by the organisms is replenished by mass transfer from air (which has plenty of O_2).

If microorganisms use oxygen much faster than it can be replenished from the atmosphere, the oxygen content of the water can become significantly depleted. Even moderate depletion can impact the life of fish in the water.

When the concern with organic matter in wastewater focuses on the potential for oxygen depletion, the amount of organic matter is quantified in terms of the *oxygen demand* of the water, expressed in units of mass concentration (e.g., mg/L). The concept of specifying the amount of organic matter in terms of oxygen demand can be confusing. The oxygen demand of a sample of water specifies how much oxygen would be used, if it were available in unlimited quantity, to entirely degrade (or oxidize) all of the organic matter present in the water. Therefore, oxygen demand is unrelated to the amount of oxygen in the water. It depends only on the amounts and types of organic materials and the nature of the degradation products. As an example, consider water that contains only one organic material, glucose, at a molar concentration of 10^{-3} M = 180 mg/L. If the glucose were to be completely degraded to carbon dioxide and water, the total amount of oxygen required would be six moles for every mole of glucose, as can be seen from the following reaction:*

$$C_6H_{12}O_6 + 6\,O_2 \rightarrow 6\,CO_2 + 6\,H_2O \tag{2.C.5}$$

Since the molecular weight of oxygen is 32 g/mol, the total amount of oxygen required for the reaction in this sample of water is (10^{-3} mol glucose)/L × (6 mol O_2)/(mol glucose) × (32 g O_2)/(mol O_2) = 192 mg/L. We say that this sample of water, which contains 180 mg/L of glucose, has a *theoretical oxygen demand* of 192 mg/L.

For most water samples, it is impractical to compute the theoretical oxygen demand because it is expensive to measure each of the organic compounds. Furthermore, the true oxygen demand is less than the theoretical oxygen demand. Microorganisms are not capable of oxidizing all of the carbon in most environmental organic mixtures to CO_2; some of it would be used to increase their own mass (requiring less oxygen) and some would be resistant to degradation. Therefore, another measure of oxygen demand is defined: the *biochemical oxygen demand* (BOD). BOD is determined by measuring how much oxygen is actually used by microorganisms to oxidize the organic material in a water sample. The standard BOD test requires 5 days, an inconveniently long time. A more rapid (if somewhat less relevant) test has

*The amount of oxygen is determined by the requirement that the same number of molecules of each element appear on both sides of the reaction. Such use of a conservation-of-elements argument in balancing chemical reactions is referred to as "stoichiometry" and is discussed in Chapter 3.

been devised that measures the *chemical oxygen demand* (COD) of a water sample. Instead of using microorganisms to carry out the reactions, chemical reagents are added to the water to more rapidly oxidize or degrade the organic material. So although both BOD and COD measure how much oxygen is used to oxidize the organic matter in a water sample, we stress that both are indicators of how much *organic matter* is in the water, rather than how much *oxygen* is present. The significance of BOD in water quality engineering is explored more thoroughly in Chapter 3 (§3.D.5) and Chapter 6 (§6.A and §6.E).

Nonmethane Hydrocarbons (NMHC) The photochemical smog that impacts many urban areas is generated from three primary ingredients: nitrogen oxides, organic gases, and sunshine. The ensuing chemical reactions generate secondary pollutants such as nitric, sulfuric, and organic acids; ozone; aldehydes; and organic particulate matter. The production of these secondary pollutants depends on the total amount of organic gases in the air basin. Frequently, the level of reactive organic gases is expressed in terms of the mass or mole fraction of nonmethane hydrocarbons (NMHC), nonmethane organic gases (NMOG), or reactive organic gases (ROG). In each case, methane is specifically excluded. Methane is commonly a significant component of waste gas streams and of urban air; however, it is nonreactive on the residence time scale of air in an air basin, so it doesn't contribute significantly to the formation of photochemical smog. Reported in mole fraction units, the aggregate hydrocarbon measures are typically specified in ppmC or ppbC. These peculiar units specify the ratio of the number of carbon atoms in organic molecules in air (excluding those in methane) to the total number of air molecules. So, for example, a mole fraction of 10 ppb of butane (C_4H_{10}) would contribute 40 ppbC to the NMHC, since each butane molecule contains four carbon atoms. A key point to remember is that these measures are not designed to reflect the direct hazard of the organic materials, but rather the extent to which they can contribute to photochemical smog formation.

2.C.6 Radionuclides

Radionuclides are chemical elements with an unstable atomic nucleus. When radionuclides undergo radioactive decay, energy is released that can damage exposed tissue. At high doses, radiation exposure can cause acute illness and even death. At low dose rates more typical of environmental exposures, the primary concern is damage to the genetic material in cells, which may cause cancer or birth defects. General background information on ionizing radiation is presented in Appendix B.

Many radionuclides cause environmental concern. Chemically, radioisotopes behave almost identically to their stable counterparts. The transport and transformation processes that influence their environmental abundance are diverse and are largely governed by their position in the periodic table of elements.

Widespread concern with radiation in the environment began with the onset of atomic weapons testing toward the end of World War II. The early years of the atomic age featured many weapons tests conducted in the open atmosphere. These tests caused deposition (or "fallout") of radioactive debris in the nearby regions downwind. In addition, because of the penetration of the radioactive particles into the stratosphere, larger tests caused enhanced radioactive fallout on a global scale that persisted for years. The atmospheric abundance of gaseous ^{85}Kr also increased with weapons testing. Treaties that limited, and then banned, aboveground weapons testing were put in place during the 1960s and 1970s. Fallout from weapons testing is no longer a dominant global environmental issue. However, radioactive and hazardous wastes associated with nuclear weapons still raise many environmental issues, including decommissioning of undetonated weapons, long-term disposal of nuclear waste, and restoration of sites contaminated by weapons development.

Industrial uses of radioactivity, such as nuclear power production, also create major environmental concerns. Inadvertent radiation exposure or release of radionuclides to the environment can occur at every step in the process: mining, refining, transport, storage, use, and disposal. Exposures that result from routine operations are strictly regulated to be as small as possible. However, accidental releases have also occurred, such as the incident at the Three-Mile Island nuclear reactor in Pennsylvania in 1979 and the Chernobyl reactor accident in the former Soviet Union in 1986. In many countries, these incidents, perhaps coupled with the historical links between nuclear power and nuclear weapons, have caused enormous difficulties with public perception for the nuclear power industry. Radionuclides are also used for medical purposes, both as diagnostic tools and in the treatment of various disorders. The production, use, and disposal of medical radioisotopes also raise environmental concerns.

The greatest source of population exposure to ionizing radiation is not industrial, military, or medical uses, but rather the radionuclides that are naturally present in the environment. Some radionuclides that were formed when the earth was created are still present, widely distributed over the earth's crust. Other isotopes are naturally produced by the interaction of cosmic radiation with terrestrial materials. In terms of total human radiation dose, the most important source of radiation exposure arises from inhalation of the short-lived decay products of ^{222}Rn in indoor air (see Exhibit 2.C.1).

EXHIBIT 2.C.1 *The Problem of Indoor Radon*

Radon is created constantly in all soil and earth-based materials by the radioactive decay of radium, which is present as a trace element in the earth's crust. There are three naturally occurring isotopes of radon, of which ^{222}Rn has the greatest environmental significance.

Radon is a chemically inert gas. Being inert, some of the newly formed radon atoms escape from the solid mineral grains to which the radium is chemically bound (other atoms may be embedded too deeply in the solid to escape). During their several-day lifetime, the free radon atoms can migrate from soil to an environment where humans may be exposed.

Most exposure to radon occurs indoors, and the dominant source of indoor radon is the soil within a few meters of the building substructure. Minor pressure differences across the building shell can draw soil gas into the building through small openings in the foundation. These pressure differences are caused by factors such as wind blowing against a building and by indoor-outdoor temperature differences. The most successful radon control measures interrupt this entry pathway, for example, by depressurizing the soil pores beneath the building foundation so that air flows from the building into the soil, rather than from the soil into the building. Building materials and groundwater are also potentially significant sources of indoor radon, although they are less important than local soil.

Since radon is chemically inert, it does not pose a significant direct health threat. However, radon decay produces a series of three short-lived radioisotopes (the shaded isotopes in Figure 2.C.2) whose inhalation does pose a direct health hazard, increasing the risk of lung cancer. These elements are chemically reactive and readily adhere to surfaces. When inhaled, they may deposit onto respiratory tissues and undergo further radioactive decay to ^{210}Pb. The alpha particles emitted by the two isotopes of polonium cause the greatest damage.

EXHIBIT 2.C.1 ***The Problem of Indoor Radon (continued)***

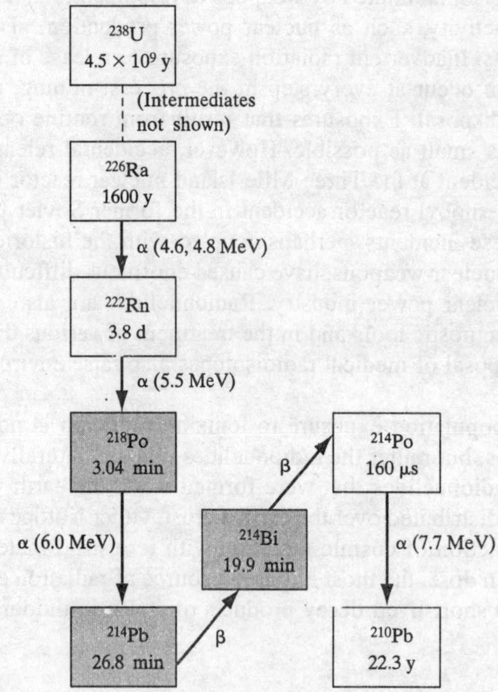

Figure 2.C.2 Radioactive decay chain from ^{226}Ra to ^{210}Pb. The half-life of each isotope is given in the box. Each arrow indicates the principal mode of decay and, for alpha particles, the associated energy released.

2.C.7 Compounds Causing Odor, Taste, or Color

In addition to health problems and ecological damage that may be caused by impurities, environmental engineering is concerned with the aesthetic quality of water and air. We want the water from our tap to be clear, colorless, and odorless. We want the air we breathe to be free of disagreeable odors and the sky to appear blue. In part, these desires are linked to health concerns: Some of the compounds in water and air that cause taste, odor, and color problems are harmful. On the other hand, some of the impurities that cause these problems exist at levels far below those that would prompt health concerns.

Table 2.C.6 summarizes some of the more common causes of aesthetic problems in water. As for air, the brown discoloration often seen in polluted urban atmospheres is caused by nitrogen dioxide, which absorbs blue light. Disagreeable odors in air may be caused by many different impurities. Common malodorous molecules include organic or inorganic sulfides, which have a skunky or rotten egg odor, and organic amines, which have the odor of stale fish. These compounds have very low odor thresholds.

2.C.8 Particulate Matter

The impurities in environmental fluids that we have considered so far have been either chemical compounds or elements. Suspended particles are another broad and impor-

Table 2.C.6 Common Impurities That Impair the Aesthetic Quality of Water

Impurity	Comment
	Taste
Metals	Examples are iron, manganese, copper, and zinc
Chlorine	From drinking water disinfection
Chlorinated organics	From disinfection or groundwater contamination
Total dissolved solids	≤1200 mg/L acceptable, ≤650 mg/L preferred
	Odor
Hydrogen sulfide	Rotten egg smell; anaerobic degradation of sulfur compounds
Geosmin	Earthy/musty smell from blue-green algae and actinomycetes
Phenolic odor	Chlorine from disinfection reacting with phenols in water
	Color
Iron and manganese	Red or brownish color; stains cloth
Humic and fulvic acids	From soil organic matter; reddish brown
Tannins	Dissolved plant material; brownish

tant class of environmental impurities. This group overlaps with the pollutant classes that we have already considered, since some chemical elements (e.g., toxic metals and radionuclides) and some organic compounds (e.g., polycyclic aromatic hydrocarbons) may occur in the particulate phase. Particulate matter also includes material that is chemically inert, such as soil dust, but may still have important effects if suspended in air or water. In air, suspended particles may be in either the liquid or the solid state. We refer to a suspension of particles in air as an *aerosol*. In water, the particles are typically solid (but may also be nonaqueous-phase liquids such as oil droplets), and their suspension in water is sometimes called a *hydrosol*. In each case, the individual particles may be pure substances or, as is more typical, complex mixtures of chemical elements and compounds. Particles may also include microorganisms, such as bacteria, algae, and viruses.

Figure 2.C.3 shows classification schemes for particle sizes in air and water. Also shown are size ranges for soil particles and major microorganism classes. Overall, the size range of interest for particles in environmental fluids varies approximately from small colloids or large molecular clusters (~10^{-3} μm) to sand grains and fog droplets (100 μm). Larger particles do not remain suspended in fluids for very long. Objects smaller than a nanometer (10^{-3} μm) are considered to be molecules or molecular clusters; however, the boundary between clusters and particles is imprecise.

In air, particles are typically divided into two broad classes, fine and coarse, with fine particles having a diameter less than 2 μm. Coarse particles tend to be generated by mechanical means or from combustion of fuels with a large ash (noncombustible) component. Dominant sources include tire and brake wear plus road and pavement dust from vehicles, wind-blown soil dust, sea spray, and agricultural and construction dust. Combustion sources include ash from coal and residential smoke from wood or solid fuels where this is a common source of home heating. Fine particles tend to be formed at the source or in the atmosphere by gas-to-particle conversion processes. A major transportation source is diesel soot. Atmospheric processes also contribute to suspended particles—the oxidation of hydrocarbons to low-vapor-pressure organics, and the formation of aerosol salts such as ammonium nitrate and ammonium sulfate.

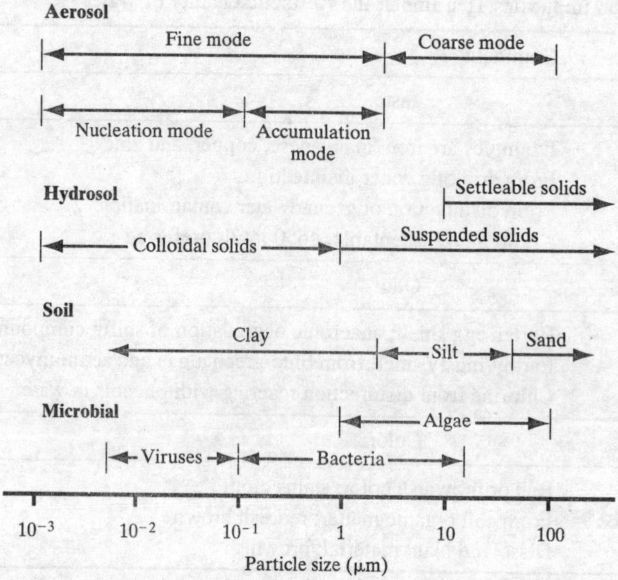

Figure 2.C.3 Approximate size of environmental particles in different media.

The fine particle mode is often subdivided into the nucleation (ultrafine) mode, with particle dimensions less than 0.1 μm, and the accumulation mode (0.1–2 μm). Fine particle emissions from sources into the atmosphere tend to be in the nucleation mode. In the atmosphere, these particles grow into the accumulation mode by coagulation and by condensation.

In water, particles are classified by size, first according to whether they readily settle under the influence of gravity, and second by whether they can be removed by passing the water through a common laboratory filter. Particles in water must be larger than about 10 μm in diameter to settle through distances of several centimeters in time scales of an hour or less. Particles in this class are called *settleable solids*. To be easily removed from water by filtration, particles must be larger than about 1 μm in diameter; this class of particles is called *suspended solids*. Particles that will pass through a filter with 1.2 μm pores are considered *colloidal solids*.

Most of the particle size range of environmental interest lies below the scale of ordinary human perception. For example, talcum powder is ground to a typical diameter of a few micrometers but is so fine that the grains feel smooth. The diameter of a human hair is typically 50–100 μm. Most environmental particles are extremely small from a human perspective. This, combined with the use of a logarithmic scale in Figure 2.C.3, makes it easy to overlook the enormous difference in size between a molecule and a small particle or between a small particle and a large particle. Table 2.C.7 stresses this point. The typical dimension of a soot or virus particle (0.1 μm) is about 250 times the diameter of a water molecule; the mass ratio between these two objects is about 16 million to 1. A sand grain, at 100 μm, is 1000 times the diameter of a soot particle but has a billion times as much mass. To put this comparison in more familiar terms, if a water molecule were of the same diameter as a baseball (~6 cm), the correspondingly scaled sand grain would have a diameter as large as the distance across a medium city (~15 km)! Because of this enormous range of scales, suspended particles

Table 2.C.7 Contrasting the Size and Mass of Molecules and Particles

Item	Characteristic dimension (μm)	Characteristic mass (pg)
H_2O molecule	0.0004	3×10^{-11}
Soot particle/virus	0.1	5×10^{-4}
Clay particle/bacterium	1	0.5
Silt particle/algae	10	500
Sand grain	100	5×10^{5}
Raindrop	1000	5×10^{8}

in a fluid do not behave like molecules; neither do large particles behave as small particles. Do not confuse particles with molecules!

Suspended particles can cause several environmental problems, among which are the following:

1. Decreased visibility (turbidity, haze, smog). By scattering light, suspended particles may impair the transmission of light. This is an aesthetic problem for both water and air.
2. Disease transmission. Suspended particles may be host to microorganisms such as bacteria, protozoa, and viruses. Through inhalation or ingestion, humans may come in contact with these disease-carrying agents.
3. Transport and delivery of toxic materials. Certain toxic elements, such as heavy metals (e.g., cadmium and lead), and toxic organics (e.g., PAHs and PCBs) are found in particulate form. Inhalation and ingestion are important routes of exposure for such compounds.
4. Acute respiratory health effects. A primary health concern with airborne particulate matter is impairment of the respiratory system. Exposure may have adverse effects on breathing, may cause or aggravate respiratory and cardiovascular disease, may impair the body's defense mechanisms against foreign materials, and may contribute to premature mortality. Recent epidemiological research indicates a surprisingly strong correlation between ambient air particulate matter levels and mortality, although the cause of this link is not understood.

Light Scattering and Turbidity

One of the distinctive features of particulate matter is its ability to efficiently scatter light. A given mass of material is much more effective at scattering light if that mass is distributed into small particles rather than existing either as a single large particle or as individual molecules in a fluid. Visible light has wavelengths in the range 0.4–0.7 μm. On a per-mass basis, particles of this approximate size are most effective in scattering light.

Turbidity refers to cloudiness, haziness, or murkiness of a fluid. As depicted in Figure 2.C.4, a turbid fluid does not permit light to be transmitted directly from an object to an observer. When objects are observed through a turbid fluid, they appear indistinct or fuzzy, if they can be seen at all.

The remarkable efficiency of light scattering by particles can be understood by considering a cloud. From common experience, we know that clouds scatter light very efficiently. Through the clear atmosphere we can easily see distances of 100 km, whereas visibility in a fog may be reduced to 10 m or less! What is remarkable is just

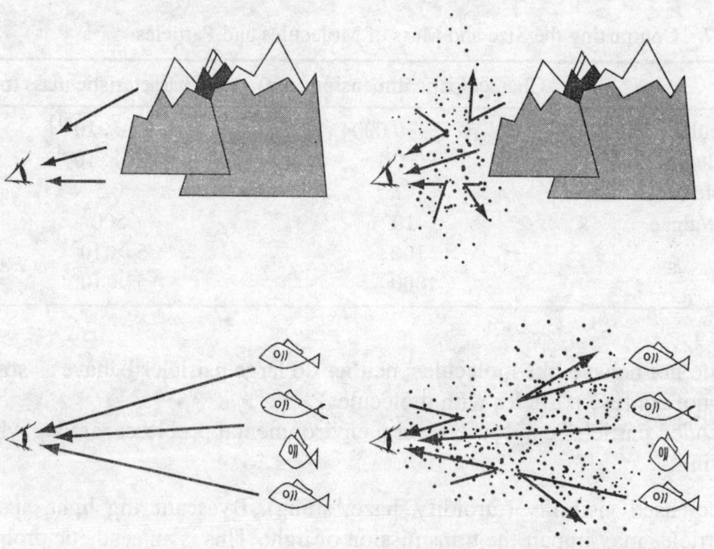

Clear fluid: Light reflected or emitted from object is transmitted directly to observer.

Turbid fluid: Light from object is scattered away from observer; other light is scattered to observer.

Figure 2.C.4 Particles suspended in a fluid scatter light, causing objects viewed through the fluid to be indistinct and also causing the fluid to appear murky. This property of a fluid is called turbidity.

how little liquid water a cloud contains.* Consider a cloud with a typical liquid water volume fraction of 10^{-6} (see Table 2.C.1) in which the water droplets have a radius of 10 μm. Then each cubic meter of air contains 1 cm^3 of condensed water, dispersed into 2.4×10^8 droplets. This may seem like a large number of droplets; however, the mass concentration of liquid water in the cloud is only 1 g m^{-3}, very much smaller than the mass concentration of gas molecules (1200 g m^{-3}) and also much smaller than the mass concentration of water *vapor* molecules (typically about 20 g m^{-3}). That the cloud appears white while the air is transparent indicates the enormously higher light-scattering efficiency of a condensed water droplet compared with the equivalent mass of water dispersed as gas molecules. Furthermore, cloud droplets are much larger than the wavelength of light and are not especially efficient light scatterers. Heavy haze from air pollution scatters light with comparable effectiveness as clouds, even though the mass concentration of condensed material is typically no more than the order of 100 μg m^{-3}, that is, about 10,000 times less than the mass concentration of liquid water in a typical cloud!

Environmental engineers are concerned with turbidity because it degrades the aesthetic quality of air and water. We are also concerned with atmospheric turbidity because it can pose a safety problem for ground and air traffic. Changes in turbidity levels can also affect ecosystems by altering the flow of electromagnetic (light) energy in air or water.

*Although this fact may be surprising, it is consistent with the observation that clouds tend to remain aloft. If the mass of condensed water were large compared with the mass of gas molecules, clouds would tend to settle under the influence of gravity.

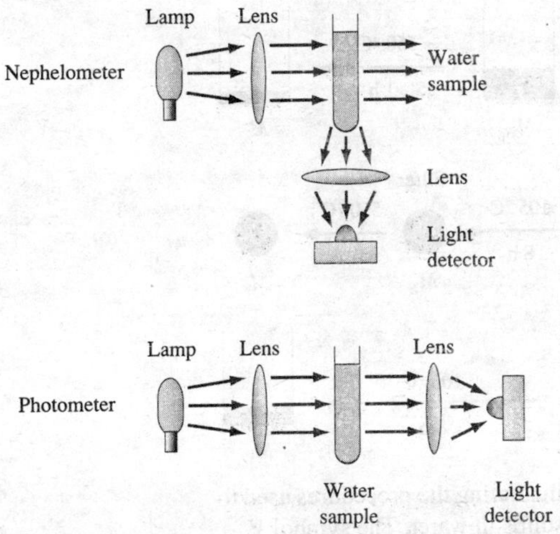

Figure 2.C.5 Turbidity of water and air can be measured with a nephelometer. The basic instrument shines a light on the fluid and measures the amount of light that is scattered. For very turbid aqueous solutions, more accurate measurements can be made by measuring the amount of light that is transmitted through a sample; this is the principle behind a photometer.

Light scattering is used to measure particulate impurities in air and water. For example, the performance of sand filters used to remove small particles from drinking water is monitored by measuring scattered light from samples of effluent water, using a nephelometer (see Figure 2.C.5). Light scattering is a sensitive means of detecting fine particles in air and is the basis for several commercial monitoring instruments.

Quantifying Particle Levels in Air and Water

Because particles are present as a distinct phase in a fluid, we must be careful in defining some basic terminology. The term *particle concentration* refers to the mass (or number) of particles per volume of fluid. The term *particle density* refers to the mass of the particle per particle volume.

In ordinary environments, the number of particles per volume of fluid is very large. In indoor air, for example, there are typically a thousand particles in every cubic centimeter, or about 10^{10} in a small room! Each particle may have a distinct size, shape, and chemical composition that reflects its history. There may be soot particles, which are agglomerates consisting of tiny spheres of elemental carbon. Particles of soil dust can also be found in the air. Small human skin flakes and clothing fibers are often present. Clearly, it is impractical to describe particles individually. Instead, we often use some aggregate measures to quantify the amount and nature of particles in a fluid. Common aggregate measures are described next.

Figure 2.C.6 illustrates the steps needed to quantify particulate matter in a water sample. The sample is divided into two portions, of volume V_A and V_D, respectively. The water from volume V_A is evaporated at a temperature just above water's boiling point. The mass of the residual, M_B, divided by the volume of the original sample, V_A, gives the concentration of *total solids (TS)* in the sample. The residual mass is then heated to 550 °C, causing most of the organic material to escape as a vapor. The difference between the sample masses, $M_B - M_C$, divided by V_A, gives the concentration of *total volatile solids (TVS)*. The second sample is passed through a filter with a 1.2 μm pore size. The material retained on the filter is dried at 105 °C and weighed. The particle mass on the filter, M_E, divided by the sample volume, V_D, gives the concentration of *suspended solids (SS)*. As with the first sample, the filter is heated at 550 °C, and the concentration of *volatile suspended solids (VSS)* is determined as the mass

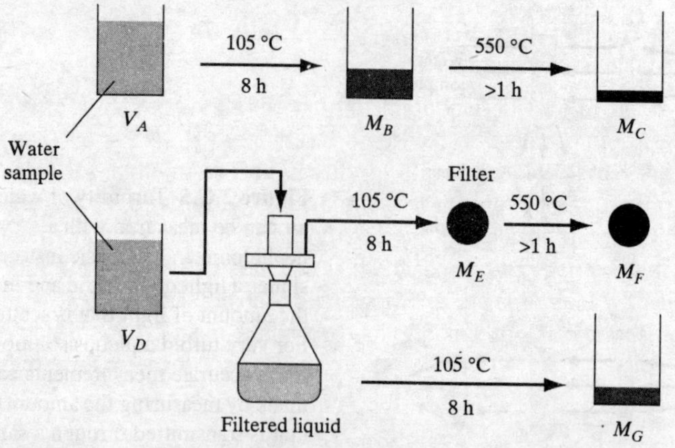

Figure 2.C.6 Schematic diagram illustrating the procedures used to measure aggregate components of solids in water. The symbol V represents measured water volume and the symbol M represents the measured mass of solid residual.

that escapes, $M_E - M_F$, divided by V_D. Finally, the water that passes through the filter is evaporated. The residual mass of ions and colloids, M_G, normalized by the sample volume, V_D, is the *total dissolved solids (TDS)*. Table 2.C.8 summarizes these definitions.

Common aggregate measures of airborne particles quantify the mass of particles per unit volume of air, often taking into account one or two particle size limits. Historically, the most common aggregate measure was the *total suspended particle (TSP)* concentration, which describes the total mass of particles of all sizes per volume of air. The first air pollution standard for particulate matter was based on TSP. Recognizing that TSP could be strongly influenced by large particles with relatively little environmental significance, especially for human health, current air quality standards for particles have an upper size limit. Such measures are referred to as PM_x (where x is

Table 2.C.8 Relationships for Computing Aggregate Particle Measures for Water

Parameter	Equation[a]
Total solids (TS)	$TS = \dfrac{M_B}{V_A}$
Total dissolved solids (TDS)	$TDS = \dfrac{M_G}{V_D}$
Suspended solids (SS)	$SS = \dfrac{M_E}{V_D}$
Total volatile solids (TVS)	$TVS = \dfrac{M_B - M_C}{V_A}$
Volatile suspended solids (VSS)	$VSS = \dfrac{M_E - M_F}{V_D}$

[a]See Figure 2.C.6 for an illustration of the mass and volume parameters.

some specific number), which stands for *particulate matter smaller than x micrometers in aerodynamic diameter.* Current air quality standards for particles are based on PM_{10}. The use of a second measure for regulatory purposes, which focuses on fine particles ($PM_{2.5}$), has been proposed (see §7.A.1).

2.C.9 Microorganisms

Microorganisms are too small to be clearly seen without the aid of a microscope. Although there is no precise definition, the size that separates microorganisms from other organisms is approximately 100 μm (0.1 mm, roughly the diameter of a human hair).

Microorganisms are common impurities in both water and air and are of great interest in environmental engineering. Certain organisms, called *pathogens,* can cause disease. Many diseases, some potentially fatal, are transmitted through water or air. Contamination by microorganisms or by their by-products can cause other adverse effects in air and water. For example, taste and odor problems in water supplies can be caused by microbial growth in the water distribution system. Exposure to dust mites (microscopic arachnids) and their by-products in indoor air is a common cause of respiratory allergy symptoms.

Microorganisms serve an important role in the natural environment in carrying out chemical processes known as oxidation and reduction reactions. This class of reactions tends to occur slowly under ordinary environmental conditions. Microorganisms cause these reactions to occur more rapidly by using proteins, called *enzymes*, as catalysts and, in so doing, are able to extract energy and materials needed for their growth and reproduction. The environmental impact of these processes from a human perspective can be beneficial or harmful. But, overall, these natural processes are essential in the cycling of key elements such as carbon, nitrogen, and sulfur through the environment. Examples of such processes include the decay of dead vegetation in forests and the conversion of atmospheric nitrogen (N_2) to a form that is usable by plants, such as nitrate (NO_3^-) or ammonia (NH_3).

Environmental engineers have harnessed the capabilities of microorganisms to rapidly carry out oxidation and reduction reactions for beneficial purposes. A major application is the biological treatment of wastewater. For example, in the activated sludge process, a dense broth of microorganisms is grown for the purpose of rapidly oxidizing organic material in wastewater, converting it to new microbial cells (which are further processed), carbon dioxide (which is released to the atmosphere), and water. The same processes would occur in nature, but when wastewater with a high concentration of organics is released to the environment, the microbial oxidation processes that occur in the receiving waters can cause severe oxygen depletion, disturbing the aquatic life. Biological treatment processes for hazardous wastes are also being developed. These applications are explored in greater detail in Chapters 6 and 8.

Microorganisms are a vast and diverse group of life forms. They can be classified according to several key characteristics, as summarized in Table 2.C.9. These classification categories are not restricted to microorganisms. For example, humans are heterotrophic (we obtain our carbon from food), chemotrophic (we degrade chemicals for energy), obligate aerobes (we need to breathe oxygen), and eukaryotic (our cells have a nucleus).

A general classification scheme that is based on differences in genetic material divides microorganisms into categories that arose from the evolutionary development of

Table 2.C.9 Classification Terms for Microorganisms

Characteristic	Label	Attribute
Principal carbon source	Autotrophic	Carbon dioxide
	Heterotrophic	Organic compounds
Energy source	Phototrophic	Sunlight through photosynthesis
	Chemotrophic	Conversion of chemicals
Relationship to oxygen	Obligate aerobe	Requires O_2 to grow
	Obligate anaerobe	Grows in the absence of O_2
	Facultative aerobe	Can grow with or without O_2
Cellular organization	Eukaryotic	Cell nucleus contains genetic material
	Prokaryotic	No cell nucleus; genetic material grouped in cell

cells (Figure 2.C.7). The main characteristics of these categories from an environmental engineering perspective are summarized below.

Prokaryotes

Prokaryote, a term that means "lacking a cell nucleus," describes two major types of single-celled microorganisms: the bacteria and the archaea. The bacteria are a metabolically diverse class of microorganisms, including autotrophs and heterotrophs, obligate aerobes, anaerobes, and facultative aerobes. While most bacteria are chemotrophs, one group of environmentally important bacteria, the cyanobacteria, can carry out photosynthesis. Some environmentally important chemotrophs include nitrifying bacteria and sulfur oxidizers, which obtain their energy by oxidizing ammonia and sulfide, respectively. Many archaea thrive in extreme environmental conditions such as hot springs, saline lakes, and highly acidic or alkaline waters. Methanogens are an important group of obligate anaerobes within the archaea. These cells live in anaerobic conditions by converting organic material to methane. Methanogens are used for the treatment of sludge in sewage treatment plants (§6.E.3).

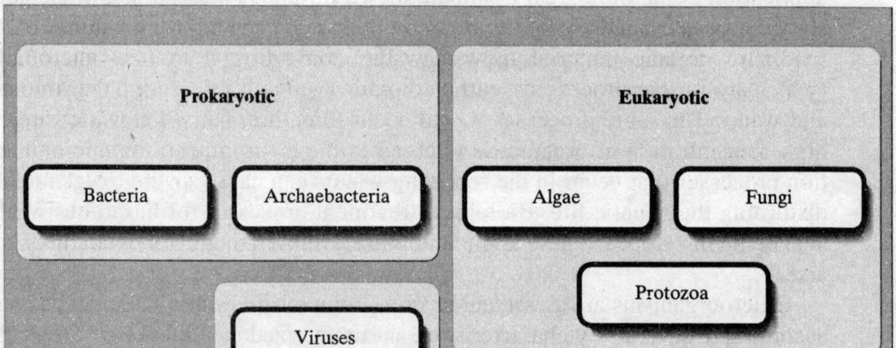

Figure 2.C.7 Microbial classification structure based on evolutionary differences in genetic material. Viruses are included without classification because they consist solely of genetic material and proteins, and are not considered living organisms.

Before the modern tools of molecular biology became available, one of the features used to classify microorganisms was their shape. The names of many cells are based on their shapes, as follows: *coccus* means "spherical," *bacillus* means "rod-shaped," *vibrio* means "curved," and *spirillum* means "spiral-shaped." Another broad classification that is often used is *gram-negative* or *gram-positive*, based on the outcome of a staining test developed in the late nineteenth century by Christian Gram.

In the course of roughly a decade beginning in 1985, the number of known prokaryotic microorganisms tripled, from 1500 to 4500. However, this is believed to be a very small fraction of all existing prokaryotes. Most human pathogens are classified as prokaryotic, and both bacteria and archaea are major players in biological environmental engineering applications.

Fungi

Fungi are eukaryotic, heterotrophic organisms that lack motility (the ability to move by self-generated propulsion). They are divided into three groups: molds, mushrooms, and yeasts. Fungi are typically aerobic and may be single-celled species or multicellular. Fungi are generally larger than prokaryotes, with typical cell dimensions in the range 1–30 μm. Along with prokaryotes, fungi are major actors throughout the environment in decomposing dead organic matter. One class, saprophytic fungi, is among the few groups of organisms capable of degrading lignin, a complex molecule that contributes to the structure of wood.

Yeasts are single-celled fungi. Molds grow as networks of filaments. Mushrooms grow as macroscopic organisms with fruiting bodies. Molds reproduce by means of spores, which are easily dispersed and often survive in conditions that would kill the active cells.

In persons with effective immune systems, very few fungi are pathogenic (common exceptions include the causative agents of athlete's foot, jock itch, and yeast infections). However, immunocompromised individuals are susceptible to respiratory tract invasions by *Aspergillus*, or common bread mold. Among otherwise healthy people, significant health problems have been associated with the inhalation of fungal bioaerosols in water-damaged buildings (Johanning et al., 1996).

Algae

Algae are eukaryotes and are almost all photoautotrophs, that is, cells that use chlorophyll to convert sunlight to chemical energy. Algae generally live in aquatic environments but can also be found in soils. Although most algae are microscopic, some are large, multicellular seaweeds.

Algae play an important role in the global cycles of carbon dioxide and oxygen and in supporting higher life forms in the oceans. Through photosynthesis, they remove carbon dioxide from the oceans and release oxygen. Consequently, additional carbon dioxide can dissolve into the ocean water, and oxygen can be released from oceans to the atmosphere. Marine algae are a major component of the phytoplankton that is at the base of the food chain in the oceans. Along with cyanobacteria (phototrophic prokaryotes), they are consumed by protozoa and microscopic animals, which are, in turn, eaten by fish.

Algae are also associated with a common water pollution problem, eutrophication, which refers to the addition of nutrients to an aquatic environment. One effect of eutrophication is severe depletion of dissolved oxygen in a water body. Algal growth is usually limited in water by the availability of phosphorus or fixed nitrogen. If these are in plentiful supply—for example, because of runoff of agricultural fertilizers or discharge of poorly treated sewage—then a massive "algal bloom" can result. When the bloom dies, the algae settle to the bottom of the body of water. Aerobic bacteria in the sediments convert organic carbon to CO_2, consuming dissolved oxygen in the process. Under extreme circumstances, the oxygen content of the lake water can become severely depleted, causing massive fish kills.

Protozoa

Protozoa are unicellular eukaryotic organisms found in marine and freshwater environments as well as in soils. Generally, they are chemotrophs and are motile, and they span an enormous range of sizes, from 2 μm to about 2 cm. Scientists have identified 40,000 species, which are subdivided into four groups largely distinguished by their means of locomotion. The Mastigophora move by waving a long whiplike structure known as a flagellum. Cells of the Rhizopoda or Sarcodina group are amoebic, and move by means of cytoplasmic streaming, allowing their fluid cellular material to flow over solid surfaces. The Ciliophora move by systematically waving cilia, fine hairlike structures. Finally, the Apicomlexa or Sporozoa group comprises a diverse mix of cells that are mostly immotile.

Protozoa constitute an important link in aquatic food chains. Benign protozoa inhabit the guts of cattle and certain termites, helping them digest cellulose in grass and wood, respectively. Protozoa are also responsible for some of the most important infectious diseases worldwide. Giardiasis, also known as "beaver fever," is a common malady of hikers who drink unfiltered stream water. Cryptosporidiosis is a disease that caused hundreds of thousands of illnesses and more than one hundred deaths due to a faulty water filtration system in Milwaukee in 1993. Protozoa also cause malaria, protozoan pneumonia, toxoplasmosis, African sleeping sickness, and amoebic dysentery.

Viruses

Viruses are not cellular organisms. Rather, they are fragments of genetic molecules (RNA or DNA) surrounded by a protein coat. It is debatable whether viruses are a life form, because they cannot reproduce, generate energy, or conduct other activities without a host cell. To reproduce, viruses infect and control systems within prokaryotic or eukaryotic organisms. Viruses are much smaller than other microorganisms (typical dimensions are 0.02–0.4 μm). Because of their size, they may escape removal by conventional water treatment techniques.

Presently, the main environmental significance of viruses is as agents of disease—human, animal, and plant. Pathogenic viruses can be transmitted via waterborne and airborne routes. Diseases caused by pathogenic viruses include the common cold and flu, mononucleosis, chicken pox, measles, mumps, herpes, hepatitis, viral pneumonia, polio, smallpox, and AIDS. In the longer term, viruses may play an important role in industrial processes and products and thereby have greater environmental significance. There is interest in the development of virus-based pesticides. For example, the baculovirus is being sprayed on cotton as an alternative to organic pesticides to control the bollworm.

Table 2.C.10 Some Infectious Diseases That Are Transmitted by Water or Air

Disease	Causative agent	Type of organism
	Waterborne	
Amebic dysentery	*Entamoeba histolytica*	Protozoa
Bacterial dysentery	*Shigella*[a]	Bacteria
Cholera	*Vibrio cholerae*	Bacteria
Cryptosporidiosis	*Cryptosporidium*	Protozoa
Gastroenteritis	*Escherichia coli, Campylobacter jejuni, Salmonella*	Bacteria
Giardiasis	*Giardia lamblia*	Protozoa
Hepatitis	Hepatitis A	Virus
Typhoid	*Salmonella typhi*	Bacteria
Viral gastroenteritis	Norwalk viruses, rotaviruses	Virus
	Airborne	
Chicken pox	Varicella-zoster virus	Virus
HARDS[b]	Hantavirus	Virus
Legionnaires' disease	*Legionella pneumophila*	Bacteria
Measles	Paramyxovirus	Virus
Pneumonic plague	*Yersinia pestis*	Bacteria
Tuberculosis	*Mycobacterium tuberculosis*	Bacteria

[a]Any of four species cause the disease: *S. dysenteriae*, *S. flexneri*, *S. boydii*, and *S. sonnei*.
[b]Hantavirus-associated respiratory distress syndrome.

Microbiological Pathogens in Water and Air

Some of the diseases that are transmitted by water or air pathways, along with their causative agents, are presented in Table 2.C.10. Many waterborne diseases are associated with water contamination by feces of either human or animal origin. Tests have been devised to rapidly determine whether a water supply has been contaminated with fecal matter. These tests are based on the use of coliform bacteria, including *Escherichia coli* (*E. coli*) as indicator organisms. Many strains of *E. coli* are normal and prevalent inhabitants of the intestinal tract of warm-blooded animals. If drinking water is contaminated by sewage or has had other recent contact with feces, then coliforms will be detected. Conversely, if no coliforms are detected, the water is presumed to be uncontaminated by sewage. Although many coliforms are harmless, the presence of these organisms indicates the potential for water contamination by more pathogenic organisms.

REFERENCES

AGOCS, M.M., ETZEL, R.A., PARRISH, R.G., PASCHAL, D.C., CAMPAGNA, P.R., COHEN, D.S., KILBOURNE, E.M., & HESSE, J.L. 1990. Mercury exposure from interior latex paint. *New England Journal of Medicine*, **323**, 1096–1101.

DORSEY, N.E. 1940. *Properties of ordinary water-substance*. Reinhold, New York, p. 248.

ERLICH, P.R., ERLICH, A.H., & HOLDREN, J.P. 1977. *Ecoscience: Population, resources, environment*. W.H. Freeman, San Francisco.

GILFILLAN, S.C. 1990. *Rome's ruin by lead poison*. Wenzel Press, Long Beach, CA.

HARTE, J. 1988. *Consider a spherical cow*. University Science Books, Mill Valley, CA.

JOHANNING, E., BIAGINI, R., HULL, D., MOREY, P., JARVIS, B., & LANDSBERGIS, P. 1996. Health and immunology study following exposure to toxigenic fungi (*Stachybotrys chartarum*) in a water-damaged office environment. *International Archives of Occupational and Environmental Health*, **68**, 207–218.

MILLS, A. 1990. Mercury and crematorium chimneys. *Nature*, **346**, 615.

PHILLIPS, D.H. 1983. Fifty years of benzo(*a*)pyrene. *Nature*, **303**, 468–472.

SAWYER, C.N., MCCARTY, P.L., & PARKIN, G.F. 1994. *Chemistry for environmental engineering*. 4th ed. McGraw-Hill, New York, Chapter 18.

SEINFELD, J.H., & PANDIS, S.N. 1998. *Atmospheric chemistry and physics: From air pollution to climate change*. Wiley, New York.

STUMM, W., & MORGAN, J.J. 1996. *Aquatic chemistry: Chemical equilibria and rates in natural waters*. 3rd ed. Wiley, New York.

U.S. ENVIRONMENTAL PROTECTION AGENCY. 1994. *National air quality and emissions trends report, 1993*. Report 454/R-94-026. Office of Air Quality Planning and Standards, Research Triangle Park, North Carolina, October.

VAN DER LEEDEN, F., TROISE, F.L., & TODD, D.K. 1990. *The water encyclopedia*. 2nd ed. Lewis, Chelsea, MI, pp. 646, 707.

WALDMAN, J.M., MUNGER, J.W., JACOB, D.J., FLAGAN, R.C., MORGAN, J.J., & HOFFMAN, M.R. 1982. Chemical composition of acid fog. *Science*, **218**, 677–680.

WALLACE, L.A. 1989. The exposure of the general population to benzene. *Cell Biology and Toxicology*, **5**, 297–314.

WATER RESOURCES COUNCIL, U.S. 1978. *The nation's water resources: 1975–2000. Vol. 1: Summary*. U.S. Government Printing Office, Washington, DC.

PROBLEMS

2.1 Density of water

Freshwater has a maximum density at $T = 4$ °C. What is the environmental significance of this fact?

2.2 Gaining a sense of scale

(a) How many moles are in 1 L of water? How many moles are in one L of air at $T = 293$ K, $P = 1$ atm?

(b) What is the average distance between gas molecules in air ($P = 1$ atm, $T = 293$ K)?

(c) What is the average distance between molecules in water (density = 1.0 g cm^{-3})?

(d) Considering only its water content, estimate the mass of O in the oceans. How does this compare with the mass of O in the atmosphere?

(e) Estimate the mass of the four major *elements* (N, C, O, and H) in the atmosphere. (*Hint:* Assume that these elements are primarily found in N_2, CO_2, O_2, and H_2O.)

2.3 Volume occupied by air

It seems that every student who took chemistry in high school remembers that one mole of air occupies 22.4 L. However, many students forget that this conversion is valid only for specific conditions.

(a) Under what conditions does one mole of air occupy 22.4 L?

(b) Typical environmental conditions are $T = 293$ K and $P = 1$ atm. For these conditions, what volume of space (L) is occupied by one mole of air?

2.4 Aspects of the hydrologic cycle

(a) Estimate the number of water molecules in the earth's atmosphere.

(b) Given that the mean residence time of a water molecule in the earth's atmosphere is 9 days, use the result of part (*a*) to estimate the flow rate of liquid water from the atmosphere to the earth's surface (use units of m^3 s^{-1}).

(c) If the flow that you determined in (*b*) occurred as rainfall uniformly over the earth's surface, what would be the mean annual rainfall (cm/y)?

(d) Look up the normal annual rainfall amount for your hometown. Compare your results with those in (c). Does your city receive more, less, or a comparable amount of rain compared with the earth's average?

2.5 Applying the ideal gas law
Anaerobic digesters are commonly used for processing sludge in wastewater treatment facilities. They are generally operated at 35 °C to optimize the rate of microbial conversion of sludge to methane and carbon dioxide. At a specific facility, the gases produced from this reaction are collected in a 1000 m^3 gas relief chamber, which requires the gas to reach a pressure of 3 atm before release.

(a) Given that the temperature within the gas collection chamber is 35 °C and that the composition of the gas is 65 percent methane (CH_4) and 35 percent carbon dioxide (CO_2), what mass (in kg) of each compound would be held within the container just as the release pressure (3 atm) is achieved?

(b) If the gas is allowed to cool to 25 °C before entering the gas relief chamber, compute the new mass of each compound within the chamber prior to release.

(c) If the rate of gas production is 400 kg d^{-1}, what is the characteristic residence time for a gas molecule within a 25 °C gas relief chamber operating with steady-state gas release?

2.6 Water, water, everywhere
Which contains more water, the air in a home (volume $V = 300$ m^3, relative humidity RH = 50 percent) or a 250 cm^3 drinking glass filled with water? Explain your reasoning.

2.7 Composition of humid air
Consider pristine air that contains 2 percent water vapor.

(a) What is the mole fraction of oxygen (O_2) in this air (%)?

(b) What is the effective molecular weight of this air (g/mol)? Report your results to a precision of three significant figures.

2.8 Cartesian diver
Consider the "Cartesian diver" depicted in the figure. It consists of an inverted glass test tube (the "diver") containing water and air. The test tube is contained within a flexible plastic 2 L soda bottle. The soda bottle is almost full of water but has a small air space at the top. The bottom of the test tube is open; the soda bottle is capped. As shown in (a), when the soda bottle is in its ordinary state, the diver floats

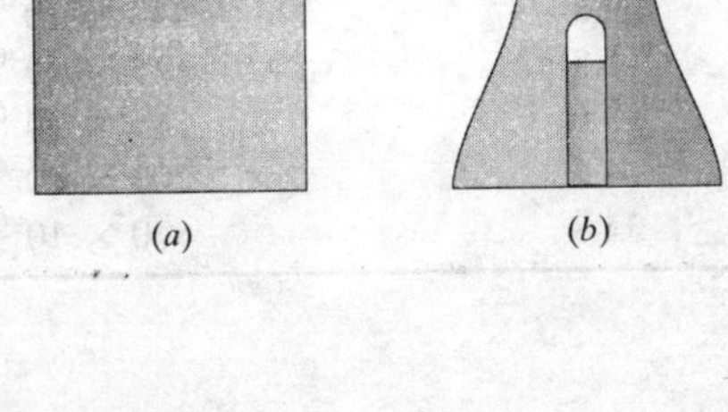

at the surface. When one squeezes the bottle as in (*b*), the diver descends to the bottom. Explain why the diver behaves this way.

2.9 Basic concepts about environmental contaminants

(a) What type of contaminant does BOD represent?

(b) In recent years the phosphate (PO_4^{3-}) content of detergents has been reduced. What environmental problem is associated with the presence of phosphates in detergents?

(c) Global atmospheric concentrations of CO_2 and CH_4 are observed to be increasing. Why is this a concern? What physical or chemical properties of CO_2 and CH_4 cause this potential problem?

2.10 An investigatory problem

Each of the following species is associated with an environmental problem. Describe briefly for each case (1) the nature of the environmental problem and (2) the major source(s) of the species. Use your skills in the library or on the Internet to find the answers.

(a) Asbestos

(b) Phenol (C_6H_5OH)

(c) Nickel (Ni)

2.11 1,3-Butadiene

The USEPA is currently revising its health risk assessment of 1,3-butadiene as an air pollutant. Use your investigation skills and do the research necessary to answer these questions.

(a) What is the chemical composition of 1,3-butadiene? Draw the chemical structure.

(b) What is (are) the main industrial use(s) of the chemical?

(c) What is the main source of 1,3-butadiene in urban air?

(d) 1,3-Butadiene is regulated as a "hazardous air pollutant." What is the major health hazard associated with 1,3-butadiene exposure?

2.12 Fundamental water quality parameters

A municipal water supplier reported concentrations of inorganic contaminants in treated drinking water as shown in the following table.

Ion	Mass concentration (mg/L)	Molarity (M)
Al^{3+}	0.10	3.7×10^{-6}
Ba^{2+}	0.012	0.09×10^{-6}
F^-	0.72	38×10^{-6}
NO_3^-	0.155	2.5×10^{-6}
Cl^-	4	113×10^{-6}
Cu^{2+}	0.005	0.08×10^{-6}
Fe^{2+}	0.02	0.36×10^{-6}
Mn^{2+}	0.002	0.04×10^{-6}
SO_4^{2-}	6.9	72×10^{-6}
Zn^{2+}	0.004	0.06×10^{-6}
Ca^{2+}	9.8	245×10^{-6}
Mg^{2+}	1.8	74×10^{-6}
K^+	0.6	15×10^{-6}
Na^+	4.6	200×10^{-6}
HCO_3^-	34.8	570×10^{-6}

(a) List in order the three species that contribute the most to hardness.
(b) Compute the total hardness associated with the three species listed in (*a*).
(c) List in order the five species that contribute the most to ionic strength.
(d) What is the total ionic strength caused by the five dominant species?

2.13 Impurities in water and unit conversions
A brochure from the Metropolitan Water District (Los Angeles) lists the following average concentrations of selected ions in drinking water just downstream of the Weymouth filtration plant. Additional information is provided on the permissible limits. For each of the three ionic species for which there are health-based standards, compute the molarity and normality that correspond to the measured concentration.

Name	Symbol	Concentration (mg/L)	Federal standard (mg/L)	Note
Aluminum	Al^{3+}	0.19	0.05–0.2	Aesthetic
Barium	Ba^{2+}	0.16	2	Health-based
Calcium	Ca^{2+}	69	—	
Chloride	Cl^-	91	250	Aesthetic
Fluoride	F^-	0.24	4	Health-based
Magnesium	Mg^{2+}	28.5	—	
Nitrate	NO_3^-	1.02	10	Health-based
Potassium	K^+	4.4	—	
Sodium	Na^+	96	—	
Sulfate	SO_4^{2-}	240	250	Aesthetic

2.14 Electroneutrality
The local drinking water has been found to contain the following mass concentrations (mg/L) of major ionic species. Other data reveal the total dissolved solids (TDS) content to be 43.7 mg/L. The pH is 8.0.

Anions		Cations	
Cl^-	8.1	Ca^{2+}	6.7
SO_4^{2-}	6.3	Na^+	2.4
F^-	0.4	Al^{3+}	0.1
NO_3^-	0.8	K^+	0.08
HCO_3^-	16.3	Fe^{2+}	0.06

(a) Use the principle of elcctroneutrality to determine whether a significant ion is missing from the list.
(b) Make an educated guess about what the missing ion would be and compute its mass concentration in this water. Can you confirm or reject your guess using the measured TDS? Explain.

2.15 River water quality
Given the characteristics listed below for a "typical" river, determine the following parameters related to drinking water quality. The water temperature is 25 °C.

(a) Total hardness (meq/L)
(b) Carbonate hardness (meq/L)
(c) Total dissolved solids (mg/L)
(d) Alkalinity (meq/L)
(e) Ionic strength (M)
(f) pH
(g) Check the electroneutrality balance.

Species	Concentration (mg/L)	Species	Concentration (mg/L)
Ca^{2+}	15	CO_3^{2-}	0.0085
Mg^{2+}	4.1	SO_4^{2-}	11.2
Na^+	6.3	Cl^-	7.8
K^+	2.3	NO_3^-	1
H^+	0.32 (μg/L)	SiO_2	13.1
HCO_3^-	58.4		

2.16 What are you drinking?
Listed below are four drinking water contaminants. The U.S. national standard for drinking water and the peak concentration in water supplied by a district utility are given for each contaminant. Use your skills as an engineering student to find out how these species come to be present in drinking water (i.e., what are the sources) and why they are regulated (i.e., what adverse effects they cause). Summarize your results in a brief narrative discussion (about two paragraphs on each compound). Cite your sources of information. (*Hint:* Start with Appendix F.)
(a) Aluminum: secondary standard, 0.05–0.2 mg/L; district water, 0.10 mg/L
(b) Barium: MCL, 2 mg/L; district water, 0.012 mg/L
(c) Nitrate: MCL, 10 mg/L; district water, 0.035 mg/L
(d) Arsenic: MCL, 0.05 mg/L; district water, 0.0006 mg/L

2.17 Water hardness
(a) Write a precise chemical definition of hardness.
(b) Which are the two most important species that contribute to hardness in treated public water?
(c) A sample of water contains the following cation levels. What is the total hardness?

Na^+, 0.2 mM Ca^{2+}, 0.3 mM Mg^{2+}, 0.1 mM K^+, 0.03 mM

(d) Calculate the total hardness, carbonate hardness, and noncarbonate hardness of water that contains the following ion concentrations (pH = 7).

Species	Concentration (mg/L)
Na^+	56
Ca^{2+}	40
Mg^{2+}	30
HCO_3^-	190
Cl^-	165
Al^{3+}	30

(e) Check the electroneutrality balance of the water sample in part (*d*).

2.18 Alkalinity

(a) What compound is the dominant contributor to alkalinity in natural waters?

(b) Alkalinity is a term commonly used to describe water quality. Define this term and describe the major implications (positive and negative) for its presence in water supplies.

(c) In a solution containing the following dissolved species, which would contribute positively to alkalinity? Explain.

Species: CO_3^{2-}, Na^+, H^+, PO_4^{3-}, Al^{3+}, H_2CO_3, Cl^-, Mg^{2+}, HCO_3^-, Fe^{2+}, OH^-, HS^-

2.19 Solids characterization

A sample of water is to be analyzed for solids content. Ten liters of water are passed through a filter initially weighing 15 g. After drying at 105 °C, the filter plus residual weighs 43 g. The filter is heated to 550 °C, and the filter plus residual then weighs 37 g. Another 10 L sample of the same water is evaporated, without filtering, yielding a residual mass of 40 g. This residual is heated to 550 °C and the remaining mass is reduced to 25 g. Determine the values of TS, TVS, SS, VSS, and TDS for this water.

2.20 Toward understanding turbidity

Consider two samples of water each containing 5 mg L^{-1} of suspended spherical particles. The particle density is 2.0 g cm^{-3}. The particles in sample A are 5 μm in diameter, and the particles in sample B are 15 μm in diameter.

(a) Prepare a table, as follows, that gives the particle number concentration and concentration of particle cross-sectional area for each sample (the latter parameter is the product of the cross-sectional area per particle times the particle number concentration).

	Sample A	Sample B
Concentration (# cm^{-3})	??	??
Area concentration (μm^2 cm^{-3})	??	??

(b) For particles whose size is significantly greater than the wavelength of light (~0.5 μm), the amount of light scattered is proportional to the suspended concentration of cross-sectional area associated with the particles. How much more turbid is sample A than sample B?

(c) What is a nephelometer and how is it different from a photometer?

2.21 The trouble with MTBE

The late 1990s saw a significant controversy emerge about the environmental consequences of the organic chemical methyl tertiary butyl ether (MTBE). To answer the following questions, you will need to do some investigation.

(a) In what common product is (was) MTBE used?

(b) What environmental benefit is supposed to result from the use of MTBE?

(c) What specific property of MTBE provides the benefit in (*b*)?

(d) What human exposure pathway(s) for MTBE is (are) attracting the most attention and concern? (An exposure pathway would describe how MTBE moves from the common product into air or water that humans would then inhale or ingest.)

(e) What are the suspected health consequences of low-level chronic exposure to MTBE?

(f) Are human health effects the only environmental concern regarding MTBE? Explain.

3

Transformation Processes

To engineer solutions to environmental problems, we must understand the factors that govern species concentrations in environmental fluids. Such an understanding is the foundation for predicting the concentrations and fates of pollutants in the environment. The same factors govern the effectiveness of treatment processes.

In this chapter, we consider transformation processes and their effects on species concentrations in environmental fluids. The term *transformation processes* is broad and includes all phenomena that alter the chemical or physical state of environmental impurities. The topic is too diverse to be covered comprehensively in a single chapter of a text. In fact, many excellent textbooks devote much of their attention to environmental transformation processes (Stumm and Morgan, 1996; Schwarzenbach et al., 1993; Snoeyink and Jenkins, 1980; Sawyer et al., 1994; Finlayson-Pitts and Pitts, 1988; Seinfeld and Pandis, 1998). Our goal here is to introduce transformation processes, exploring several major processes of environmental interest. We also introduce the methods and tools needed to analyze the effects of these processes on the concentrations of impurities in environmental fluids.

The chapter begins with the most important basic concepts of environmental chemistry: stoichiometry, equilibrium, and kinetics. Then three broad classes of transformation processes are described: phase-change phenomena, acid-base reactions, and oxidation-reduction reactions. Additional background information concerning equilibrium and kinetics is provided in Appendix E.

3.A GOVERNING CONCEPTS

Three central ideas serve as the foundation for environmental transformation processes. The first, *stoichiometry,* is the application of material balance to transformation processes. The second, *chemical equilibrium,* describes how species partition between phases and how elements partition among chemical species if some specific restrictions are met. One of these restrictions is that the system must be in steady state; that is, the species concentrations must be unchanging over time. The third idea, *kinetics,* deals with the rates of reactions and provides information on how species concentrations evolve. These three concepts are linked. Equilibrium and kinetic relationships must satisfy stoichiometry, and kinetic processes in a closed system ultimately lead to chemical equilibrium.

Before beginning, a comment should be made about the notation used for transformation processes. We can denote the process of water evaporation as

$$H_2O(l) \rightarrow H_2O(g) \tag{3.A.1}$$

We might term this the "forward reaction" for this system, where the reaction involves a phase change from liquid (l) to gas (g). The reverse reaction is then given by

$$H_2O(g) \rightarrow H_2O(l) \tag{3.A.2}$$

We use this chemical notation ($\rightarrow$) to emphasize that we are interested in the kinetics of the system. If our interest were principally in the equilibrium state, in which the rates of the forward and reverse reactions are equal, we could denote the transformation by

$$H_2O(l) \Leftrightarrow H_2O(g) \tag{3.A.3}$$

3.A.1 Stoichiometry

Stoichiometry is the application of the principle of material balance to a chemical transformation.

We may write a generic chemical reaction as follows:

$$b\,R_1 + c\,R_2 \rightarrow m\,P_1 + n\,P_2 \tag{3.A.4}$$

In this expression, R_1 and R_2 represent reactants, P_1 and P_2 represent products, and b, c, m, and n are the *stoichiometric coefficients*. This expression states that b molecules of R_1 combine with c molecules of R_2 to form m molecules of P_1 and n molecules of P_2.

Stoichiometry permits us to calculate the value of some stoichiometric coefficients based on the values of other coefficients. The principle of stoichiometry can be expressed in this manner:

A chemical reaction must conserve the number of atoms for each element involved in the reaction and the electrical charge associated with ions.

The mathematical result of this principle is a set of algebraic equations relating the stoichiometric coefficients in a reaction. Since stoichiometry applies to each element separately, there is one equation for each chemical element in the system, plus one for the charge associated with ions, if any. Stoichiometry is commonly used to determine the relative amounts of reactants that must be supplied for a given reaction, or to determine the relative amounts of products that are produced (see Example 3.A.1). Note, though, that stoichiometry does not supply any information about the forms of the molecules R_i or P_i, or about the rate of the chemical reaction.

EXAMPLE 3.A.1 ***Stoichiometry and Photosynthesis***

A general photosynthesis reaction may be written in the form shown below. Use the principles of stoichiometry to determine the coefficients *a–f*.

$$a\,CO_2 + b\,NO_3^- + c\,HPO_4^{2-} + d\,H^+ + e\,H_2O \rightarrow C_{106}H_{263}O_{110}N_{16}P_1 + f\,O_2$$

SOLUTION The problem has six unknown stoichiometric coefficients. Stoichiometry yields six algebraic equations that relate these coefficients—one each for the five elements involved in the overall reaction (C, O, H, P, and N) plus a sixth based on charge conservation. The algebraic equations are generated by equating the number of atoms of each element on the reactant and product sides of the reaction:

C: $a = 106$
O: $2a + 3b + 4c + e = 110 + 2f$
N: $b = 16$
H: $c + d + 2e = 263$
P: $c = 1$
+/−: $-b - 2c + d = 0$

The atom balances for carbon, nitrogen, and phosphorus, respectively, immediately determine *a*, *b*, and *c*. Given *b* and *c*, the charge balance specifies that $d = 18$. Given *c* and *d*, the balance on hydrogen yields $e = 122$. Finally, the oxygen balance is only satisfied if $f = 138$. So the balanced reaction is

$$106\,CO_2 + 16\,NO_3^- + HPO_4^{2-} + 18\,H^+ + 122\,H_2O \rightarrow C_{106}H_{263}O_{110}N_{16}P_1 + 138\,O_2$$

In mass terms, we can see from this balanced reaction that the production of 1 g of biomass by photosynthesis (effective molecular weight $= 106 \times 12 + 263 \times 1 + 110 \times 16 + 16 \times 14 + 1 \times 31 = 3550$ g/mol) produces $138 \times (1/3550) \times 32 = 1.24$ g of molecular oxygen.

All chemical reactions must satisfy the principle of stoichiometry. Specifically, they must conserve the number of atoms of each element and, consequently, the total mass of species in the system. However, chemical reactions do not necessarily conserve the number of molecules (see Example 3.A.2).

EXAMPLE 3.A.2 ***Stoichiometry in Chemical Kinetics***

Consider the following reaction in which sulfur trioxide combines with a water molecule to form sulfuric acid:

$$SO_3 + H_2O \xrightarrow{1} H_2SO_4$$

Does this reaction conserve elements? Yes, hydrogen, oxygen, and sulfur are present in the same amounts on the reactant and product sides of the reaction:

H: $2 \rightarrow 2$
S: $1 \rightarrow 1$
O: $3 + 1 \rightarrow 4$

Does this reaction conserve mass? Yes, the sum of the molecular weights of the species on the left, 98 g/mol, equals the molecular weight of the species on the right.

Does this reaction conserve molecules? No, two reactant molecules combine to form one product molecule.

3.A.2 Chemical Equilibrium

When a system is at chemical equilibrium, the partitioning of elements among chemical species and the partitioning of species among physical states satisfies thermodynamic relationships, regardless of the history of the system. This is a powerful notion but holds only under restricted conditions.

The underlying theory for chemical equilibrium is based on thermodynamics. The detailed development of equilibrium theory as given in thermodynamics texts (e.g., Sandler, 1989) is helpful but not essential for applying these concepts to environmental engineering problems. Some of the key ideas from chemical thermodynamics are summarized in Appendix E.

Strictly interpreted, the predictions derived using chemical equilibrium apply only when a system is at equilibrium. A system at equilibrium must satisfy these conditions:

1. It does not vary with time.
2. It is internally uniform (at least at the level of discrete subsystems).
3. There are no *net* flows of mass, heat, or species within the system or between the system and its surroundings.
4. The *net* rate of all chemical reactions is zero. The net rate of a reaction is the difference between the rates of the forward and reverse reactions.

We emphasize that equilibrium is a steady-state condition, not a static condition. That is, although the concentrations of impurities do not vary with time, transformation and transport processes may still be occurring. To the extent that they occur, they must be balanced so as to have no net effect.

The above conditions are restrictive and are seldom met in systems of practical interest. However, the predictions from chemical equilibrium apply as good approximations in many cases even when the strict requirements for equilibrium are not satisfied. The key issue is one of time scales. If the reactions are fast in comparison with other processes (such as the time period for measuring species concentrations), then the assumption of chemical equilibrium may give accurate predictions for nonequilibrium systems.

In most cases, the equilibrium state is unique, giving us enormous predictive power. When equilibrium conditions apply, we can say important quantitative things about a system with little input information. In Example 3.A.3, for instance, the equilibrium molar concentration of water vapor in the gas phase in a jar is predicted. This concentration depends only on temperature. It does not depend on other factors, such as the volume of liquid water or air in the jar, or on the surface area of the air-water interface. The prediction is accurate and reliable, independent of the system's history.

EXAMPLE 3.A.3 ***Water Vapor Concentration at Equilibrium***

Consider a sealed jar that is partially filled with pure liquid water and otherwise contains nitrogen gas. Assume that the temperature of the system is fixed at 293 K (20 °C). What is the steady-state molar concentration of $H_2O(g)$ in the gas phase above the liquid water?

SOLUTION At steady state, this isolated system will achieve chemical equilibrium, so the water vapor pressure in the jar will reach the saturation vapor pressure (that is, the relative humidity will be 100 percent). The saturation vapor pressure of water at $T = 293$ K is 2338 Pa (see §3.B.1). Note that 1 Pa = 1 N m^{-2} = 1 J m^{-3}. From the ideal gas law and the rule of partial pressures

$$\frac{n_{H_2O}}{V} = \frac{P_{H_2O}}{RT} = \frac{2338 \text{ Pa}}{8.314 \text{ J mol}^{-1} \text{ K}^{-1} \times 293 \text{ K}} = 0.96 \frac{\text{mol}}{\text{m}^3}$$

Note that to solve this problem we needed only a few pieces of information about the system: the general composition, the temperature, the appropriate thermodynamic data (saturation vapor pressure of water), and the fact that the system is in equilibrium.

Note, however, that if the jar is opened, this prediction is no longer valid. Also, equilibrium arguments cannot tell us how long we must wait after the jar is sealed for the equilibrium condition to be reached.

Equilibrium problems are systematically analyzed by first identifying all relevant species in the system. Whenever a compound is present in multiple phases, each phase should be considered as a separate species. Once the species are listed, an appropriate measure for each should be defined, such as concentration or partial pressure. Then one independent equation must be written for each species. These equations can be derived from chemical equilibrium relationships, material conservation, stoichiometry, electroneutrality, and fixed conditions specified in the problem statement. Recognizing how material conservation applies is often the most difficult aspect of these problems. Once the equations are defined, the goal is then to solve a system of algebraic equations. For small sets of equations, as encountered in this text, the best solution approach usually involves making substitutions to obtain one equation with a single unknown. The resulting equation may be a polynomial or may involve noninteger powers of the unknown species. The equation can usually be solved by straightforward methods (see Appendix D).

The mathematical form of chemical equilibrium relationships varies, depending on the nature of the reactions involved. Consider a simple case of compounds A, B, C, and D in water, which participate in the following reaction:

$$\mathrm{A} + \mathrm{B} \Leftrightarrow \mathrm{C} + \mathrm{D} \tag{3.A.5}$$

In kinetic terms, two reactions occur. The "forward" reaction involves a molecule of species A striking a molecule of species B, with a reaction occurring to produce molecules of species C and D. The "reverse" reaction involves a molecule of species C striking a molecule of species D and reacting to produce molecules of species A and B.

Consider the rate of the forward reaction. Molecules of A and B move about in the water, randomly striking each other. Each collision has some probability of producing a reaction to form C and D, and that probability depends on the energy of the collision and on the orientation of the molecules. Consider a system at a given, fixed temperature, T. The rate of the forward reaction is expected to be proportional to the rate of collisions between A and B. What happens to the reaction rate if the concentration of A is doubled? The rate should double. Likewise, doubling the concentration of B should double the rate at which A and B collide, thereby doubling the reaction rate. Thus, we expect the rate of reaction in the forward direction to be proportional to the concentrations of A and B:

$$R_f = k_f[\mathrm{A}][\mathrm{B}] \tag{3.A.6}$$

where the coefficient k_f depends, in general, on temperature. The rate of reaction, R_f, has units of moles per volume per time and represents the number of forward reactions that occur in the system per time and per water volume.

For the reverse reaction in our example, the same argument about the dependence of the reaction rate holds, and so we write for the rate of reaction in the reverse direction

$$R_r = k_r[\mathrm{C}][\mathrm{D}] \tag{3.A.7}$$

Now, here is a key point: At equilibrium, the rate of reaction in the forward direction equals the rate of reaction in the reverse direction (this is the meaning of equilibrium):

$$R_f = R_r \tag{3.A.8}$$

Thus, at equilibrium, we expect

$$k_f[\mathrm{A}][\mathrm{B}] = k_r[\mathrm{C}][\mathrm{D}] \tag{3.A.9}$$

or

$$\frac{[\mathrm{C}][\mathrm{D}]}{[\mathrm{A}][\mathrm{B}]} = \frac{k_f}{k_r} = K \tag{3.A.10}$$

The parameter K is known as the *equilibrium constant* of the reaction. Since k_f and k_r are constant for a fixed temperature, K is also constant at a fixed temperature. However, K generally varies with temperature, because k_f and k_r may vary independently with temperature. (Thus, referring to K as an "equilibrium *constant*" is potentially misleading.) Of course, a different value of the equilibrium constant applies for each distinct reaction.

When the stoichiometric coefficients differ from 1, the expected form of the equilibrium relationship is generalized as follows:

$$a\,\mathrm{A} + b\,\mathrm{B} \Leftrightarrow c\,\mathrm{C} + d\,\mathrm{D} \tag{3.A.11}$$

$$\frac{[\mathrm{C}]^c[\mathrm{D}]^d}{[\mathrm{A}]^a[\mathrm{B}]^b} = K \tag{3.A.12}$$

The forms of the equilibrium relationship given in equations 3.A.10 and 3.A.12 do not apply for all processes. In Example 3.A.3 we saw that for the case of vapor pressure in a sealed container, the equilibrium relationship was independent of the quantity of liquid water in the system. In subsequent sections, explicit forms of equilibrium relationships will be presented along with some explanations of cases that depart from the general forms.

In routine environmental engineering practice, the equilibrium constant is taken as a given parameter obtained from a technical reference. More fundamentally, the equilibrium constant depends on basic thermodynamic properties of the chemical species involved in the reaction. These properties are summarized in Appendix E.

All equilibrium reactions consist of forward and reverse reactions that proceed at identical rates. What happens to a system that is initially in equilibrium when some condition, such as temperature or the concentration of one reactant, is suddenly changed? The perturbed system is initially out of equilibrium. The rate of reaction in one direction exceeds that in the other direction, and this difference drives the system toward an equilibrium state. Such situations are frequently encountered in environmental engineering. To analyze them, we turn to the subject of kinetics.

3.A.3 Kinetics

We have seen that equilibrium relationships can be used to predict the partitioning of chemical elements among species and the distribution of chemical compounds among phases *if* a system is at equilibrium. But certain questions arise:

- How do we know when a system is at equilibrium?
- How much time is needed for a system to reach equilibrium?
- What can we say about systems that are not at equilibrium?

Kinetics can help us answer these questions. The topic of kinetics becomes important when a system is not at equilibrium because the time associated with one or more transformations or transport processes is comparable with time scales of interest, such as the residence time in a fluid environment.

In environmental engineering, some problems are encountered that are adequately described by equilibrium conditions. Others require consideration of rate-related processes governed by kinetics. An example of the former is acid-base chemistry in a well-stirred sample of water. An example of the latter is the daily cycle of ozone concentrations in an urban air basin. This section addresses the central aspects of transformation or reaction kinetics as applied in environmental engineering science. The rates of transport processes are discussed in Chapter 4.

Introduction to the Batch Reactor

Reactor models are used extensively in the analysis of environmental systems. The characteristics and use of these models are explored in detail in Chapter 5. Here we introduce one of the simplest models, the *well-mixed batch reactor*, as an aid to understanding the following discussion on chemical kinetics.

The batch reactor is a vessel that can be opened or sealed. The reactor is typically used by initially filling it with air or water plus specified quantities of the reactants of interest. Then the reactor is sealed for some period and reactions are allowed to proceed. Finally, the reactor is opened and the contents removed. In a well-mixed batch reactor, the species concentrations are assumed to be uniform throughout each fluid (hence the term *well-mixed*). It is called a *batch* reactor because the reactants are processed one batch at a time.

Reaction Rates, Rate Laws, and Reaction Order

We may denote a chemical kinetic reaction as follows:

$$a\,\mathrm{A} + b\,\mathrm{B} \xrightarrow{\gamma} c\,\mathrm{C} + d\,\mathrm{D} \tag{3.A.13}$$

Here the letters *a–d* represent stoichiometric coefficients, the letters A–D represent chemical species, and γ is an arbitrary symbol, typically an integer, that is used to identify the reaction.

Consider now a simple reaction in which one molecule of A reacts with one molecule of B to produce one molecule of C:

$$\mathrm{A} + \mathrm{B} \xrightarrow{1} \mathrm{C} \tag{3.A.14}$$

The *reaction rate* of reaction 1, denoted R_1, is defined as the number of occurrences of this reaction per unit time, usually per volume of fluid. So, the reaction rate may be expressed in units of concentration per time (e.g., moles L^{-1} d^{-1}). Each time reaction 1 occurs, it consumes one molecule of A and one molecule of B, producing one molecule of C in the process. Consequently, the changes in the concentrations of these species that result from reaction 1 are directly related to the rate of the reaction:

$$R_1 = -\left(\frac{d[\mathrm{A}]}{dt}\right)_1 = -\left(\frac{d[\mathrm{B}]}{dt}\right)_1 = \left(\frac{d[\mathrm{C}]}{dt}\right)_1 \tag{3.A.15}$$

In these equations brackets, [ ], are used to denote the molar concentrations of the species, and the subscript 1 denotes that these are the rates of change resulting only from reaction 1.

In the more general case the stoichiometric coefficients need not be equal to 1. So, for reaction γ (equation 3.A.13), the overall reaction rate is related to the rate of change of individual species concentrations by the following expression:

$$R_\gamma = -\frac{1}{a}\left(\frac{d[\mathrm{A}]}{dt}\right)_\gamma = -\frac{1}{b}\left(\frac{d[\mathrm{B}]}{dt}\right)_\gamma = \frac{1}{c}\left(\frac{d[\mathrm{C}]}{dt}\right)_\gamma = \frac{1}{d}\left(\frac{d[\mathrm{D}]}{dt}\right)_\gamma \tag{3.A.16}$$

The first cqual sign in this expression can be understood to define what we mean by the reaction rate. All of the other equal signs are a result of stoichiometry. Each time reaction γ occurs, *a* molecules of A and *b* molecules of B are consumed, producing *c* molecules of C, and *d* molecules of D.

This discussion provides the information needed to transform a chemical kinetic reaction from a chemical description to a mathematical description. We now know the relationship between the reaction rate and the rate of change of species concentrations. But, we do not yet know how to produce a mathematical expression for the reaction rate itself.

Here we introduce the idea of a *rate law.* This is an expression that links the reaction rate to the concentrations of the species involved. In almost all cases, reaction rates depend on the concentrations of the reactants, but not on the concentrations of the products.

For reaction γ, the expected form of the rate law is

$$R_\gamma = k_\gamma[\mathrm{A}]^\alpha[\mathrm{B}]^\beta \tag{3.A.17}$$

where k_γ is called the *rate constant,* and α and β are empirical coefficients. The rate constant, k_γ, is generally a function of temperature and, for gaseous reactions, may be a function of pressure. The parameters α and β are empirical coefficients, in general. However, if γ is an *elementary reaction,* meaning that it describes what happens mechanistically in a single reaction step, then usually $\alpha = a$ and $\beta = b$. On the other hand, if the reaction is not elementary, α and β must be determined experimentally.

Reactions may be described in terms of the *reaction order.* The overall reaction order is given by the sum of the exponents in the rate law. For example, in reaction γ the overall order is $\alpha + \beta$. We also say that reaction γ is order α in A and order β in B. To be more explicit, rate laws for γ that are zeroth, first, or second order are shown below:

Zeroth-order reaction:	$\alpha = \beta = 0$	$R_\gamma = k_\gamma$	(3.A.18a)
First-order reaction:	$\alpha = 1, \beta = 0$	$R_\gamma = k_\gamma[\mathrm{A}]$	(3.A.18b)
	or $\alpha = 0, \beta = 1$	$R_\gamma = k_\gamma[\mathrm{B}]$	(3.A.18c)
Second-order reaction:	$\alpha = \beta = 1$	$R_\gamma = k_\gamma[\mathrm{A}][\mathrm{B}]$	(3.A.18d)
	or $\alpha = 2, \beta = 0$	$R_\gamma = k_\gamma[\mathrm{A}]^2$	(3.A.18e)
	or $\alpha = 0, \beta = 2$	$R_\gamma = k_\gamma[\mathrm{B}]^2$	(3.A.18f)

Most but not all chemical reactions fit this conceptual model. Some processes, such as biological reactions, are the overall result of a complex combination of elementary reactions. For these cases, rate expressions (rate law, reaction order, and so on) are often determined *empirically* from reactor experiments and may take on different reaction forms from those described in this section. The case of microbial reaction kinetics, for example, is addressed in §3.D.

The units of the reaction rate, R, typically concentration per time, are independent of reaction order. From the forms of the rate laws shown above, we see that the units associated with the rate constant, k_γ, must vary according to the reaction order. So, typical units for a zeroth-order rate constant are concentration per time (e.g., mol L^{-1} d^{-1}), first-order rate constants have units of inverse time (e.g., d^{-1}), and second-order rate constants have units of the inverse product of concentration and time (e.g., mol^{-1} L d^{-1}).

Examples 3.A.4 and 3.A.5 illustrate how rate laws are used to formulate and solve kinetic problems.

EXAMPLE 3.A.4 *Solving a Simple Kinetic Problem*

Radon-222 is a naturally occurring radioactive gas formed by the decay of radium-226, a trace element in soil and rock. The radioactive decay of radon can be described by the elementary reaction

$$\text{radon-222} \rightarrow \text{polonium-218} + \text{alpha particle}$$

EXAMPLE 3.A.4 ***Solving a Simple Kinetic Problem (continued)***

The rate constant for this reaction is $k = 2.1 \times 10^{-6}\ s^{-1}$, independent of temperature. At time $t = 0$, a batch reactor is filled with air containing radon at concentration C_0. How does the radon concentration in the reactor change over time?

SOLUTION For $t \geq 0$, only one reaction influences the radon concentration, C_{Rn}. The change in radon concentration is related to the reaction rate by the expression

$$R = -\frac{dC_{Rn}}{dt}$$

Since radon decay is an elementary reaction, we expect the rate law to be first-order:

$$R = kC_{Rn}$$

The following expressions give the time rate of change in radon concentration and the initial condition:

$$\frac{dC_{Rn}}{dt} = -kC_{Rn}$$

$$C_{Rn}(0) = C_0$$

The differential equation can be solved by direct integration following rearrangement (see Appendix D, §D.1 for details) to obtain

$$C_{Rn}(t) = C_0 \exp[-kt]$$

The radon concentration decays exponentially toward zero with a characteristic time $\tau \sim k^{-1} = 5.5$ d (Figure 3.A.1).

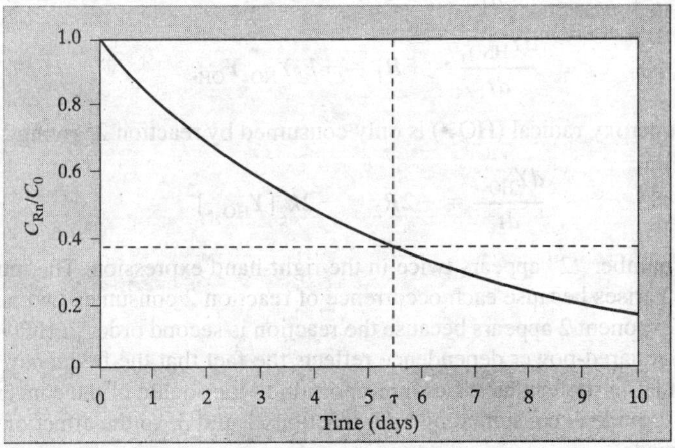

Figure 3.A.1 Radon concentration in a well-mixed batch reactor, changing in response to radioactive decay.

EXAMPLE 3.A.5 ***Effect of Reactions on Species Concentrations***

The following three elementary gas-phase reactions are among the many (hundreds) that occur in photochemical smog systems:

$$NO_2 + OH\bullet \xrightarrow{1} HNO_3$$

$$2\,HO_2\bullet \xrightarrow{2} H_2O_2 + O_2$$

$$NO_2 + O_3 \xrightarrow{3} NO_3 + O_2$$

(The symbol • denotes that the species is a radical, which, for present purposes, can be considered simply a highly reactive species. The hydroxyl radical, OH•, should not be confused with the hydroxide ion, OH^-, found in aqueous systems.)

Write differential equations to describe the net effect of these three reactions on the mole fractions of OH•, HNO_3, HO_2•, and NO_2, assuming that the reactions occur in a well-mixed batch reactor.

SOLUTION The rate laws for gas-phase chemical reactions can be written either in terms of the species concentrations or in terms of the mole fraction of the species. In either case, the form is similar.

Since these are elementary reactions, the rate laws can be expressed as

$$R_1 = k_1 Y_{NO_2} Y_{OH\bullet}$$

$$R_2 = k_2 (Y_{HO_2\bullet})^2$$

$$R_3 = k_3 Y_{NO_2} Y_{O_3}$$

The hydroxyl radical and nitric acid (HNO_3) are respectively consumed and produced only in the first reaction, so

$$\frac{dY_{OH\bullet}}{dt} = -R_1 = -k_1 Y_{NO_2} Y_{OH\bullet}$$

$$\frac{dY_{HNO_3}}{dt} = +R_1 = +k_1 Y_{NO_2} Y_{OH\bullet}$$

The hydroperoxy radical (HO_2•) is only consumed by reaction 2, giving

$$\frac{dY_{HO_2\bullet}}{dt} = -2R_2 = -2k_2 [Y_{HO_2\bullet}]^2$$

Note that the number "2" appears twice in the right-hand expression. The multiplicative factor of 2 arises because each occurrence of reaction 2 consumes two molecules of HO_2•. The exponent 2 appears because the reaction is second order in HO_2•. Fundamentally, this squared-power dependence reflects the fact that the frequency of collisions between HO_2• molecules varies in proportion to the square of the concentration.

Nitrogen dioxide is consumed by both reactions 1 and 3, so the effect of each reaction is summed:

$$\frac{dY_{NO_2}}{dt} = -R_1 - R_3 = -k_1 Y_{NO_2} Y_{OH\bullet} - k_3 Y_{NO_2} Y_{O_3}$$

Empirical Determination of Reaction Order and Rate Constant

In the development and design of treatment processes, or in research applications, an environmental engineer may be called upon to develop a rate law and to determine the rate constant of a kinetic process. In this subsection we explore how this may be accomplished for simple reaction systems using a well-mixed batch reactor.

Consider a system in which a single compound, A, is consumed by some transformation process. We wish to determine the reaction order and the rate constant for this process. The typical way to proceed is to establish an initial concentration of A in a batch reactor (A_0) and measure the change in concentration over time. This can be accomplished with a single experiment if we can monitor the concentration of A within the reactor without disturbing the process (e.g., by extracting small samples of fluid for analysis at regular intervals). Alternatively, a set of reactors can be operated in parallel and opened at different times for measurement.

Such experiments will generate a series of measurements of $A(t)$, all starting from the same initial condition A_0. By plotting the concentration against time, the reaction order and rate constant can be determined. We consider here the case of zeroth-, first-, and second-order reactions. The same general approach can be extended for other forms of the rate law.

The differential equations and solutions describing the change in concentration over time for zeroth-, first-, and second-order reactions are summarized below. The functions, $[A](t)$, were obtained by direct integration in each case.

Zeroth-order reaction:

$$\frac{d[A]}{dt} = -k_0 \qquad [A](0) = A_0 \quad \Rightarrow \quad [A](t) = A_0 - k_0 t \quad \text{for } 0 \le t \le \frac{A_0}{k_0} \tag{3.A.19}$$

First-order reaction:

$$\frac{d[A]}{dt} = -k_1[A] \qquad [A](0) = A_0 \quad \Rightarrow \quad [A](t) = A_0 \exp(-k_1 t) \tag{3.A.20}$$

Second-order reaction:

$$\frac{d[A]}{dt} = -2k_2[A]^2 \qquad [A](0) = A_0 \quad \Rightarrow \quad [A](t) = \frac{A_0}{1 + 2k_2 t A_0} \tag{3.A.21}$$

To test whether a zeroth-order reaction is the appropriate description of the reaction of interest, plot the time-dependent concentration, $[A](t)$, on a linear scale against time (Figure 3.A.2(*a*)). If the reaction is zeroth order, the data will conform to a straight line and the slope of this line will be $-k_0$. Typically, real data exhibit some scatter because of experimental imprecision, among other reasons. A technique such as linear regression can be used to obtain the best estimate of the slope (see Appendix D) and the associated experimental uncertainty.

To test for a first-order reaction, the logarithm of the concentration, $\log_{10}[A]$, is plotted against time. If we take the logarithm of both sides of the first-order kinetic solution, we obtain

$$\log_{10}([A](t)) = \log_{10}(A_0) - \log_{10}(\exp[-k_1 t]) = \log_{10}(A_0) - 0.43k_1 t \tag{3.A.22}$$

So, if the first-order description is correct, a plot of $\log_{10}([A](t))$ versus t will yield a straight line with a slope of $-0.43k_1$, as shown in Figure 3.A.2(*b*).

To test whether the data conform to a second-order reaction, plot the reciprocal of the concentration versus time. Rearranging the expression for $[A](t)$ in the second-

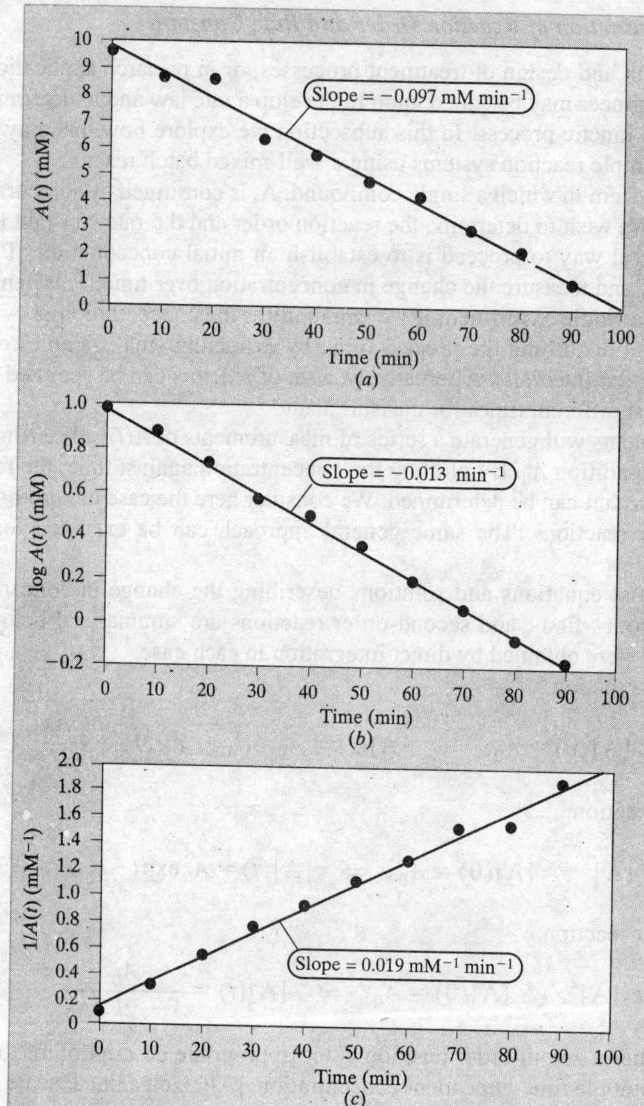

Figure 3.A.2 Hypothetical kinetic data. (*a*) The data conform to a zeroth-order reaction with a rate constant $k_0 = 0.097$ mM min^{-1}. (*b*) The compound decays by a first-order reaction with $k_1 = 0.03$ min^{-1}. (*c*) The compound decays by a second-order reaction with $k_2 = 0.0095$ mM^{-1} min^{-1}.

order case results in the following relationship:

$$\frac{1}{[\mathrm{A}](t)} = \frac{1}{A_0} + 2k_2 t \tag{3.A.23}$$

So if the second-order description holds, a plot of 1/[A] versus t will produce a straight line with a slope that is twice the desired rate constant, as shown in Figure 3.A.2(*c*).

Table 3.A.1 Characteristic Times of Transformation Processes[a]

Order	Kinetic description	Rate law	Characteristic time[b]
Zeroth	$A \rightarrow$ products	$\frac{d[A]}{dt} = -k_0$	$\tau_0 \sim \frac{A_0}{k_0}$
First	$A \rightarrow$ products	$\frac{d[A]}{dt} = -k_1[A]$	$\tau_1 \sim \frac{A_0}{k_1 A_0} = \frac{1}{k_1}$
Second	$A + A \rightarrow$ products	$\frac{d[A]}{dt} = -2k_{2a}[A]^2$	$\tau_{2a} \sim \frac{A_0}{2k_{2a}A_0^2} = \frac{1}{2k_{2a}A_0}$
Second	$A + B \rightarrow$ products	$\frac{d[A]}{dt} = \frac{d[B]}{dt} = -k_{2b}[A][B]$	$\tau_{2b} \sim \frac{\min(A_0, B_0)}{k_{2b}A_0B_0}$

[a]The initial concentrations of the reactants are A_0 and B_0 for [A] and [B], respectively.
[b]"$\min(A_0, B_0)$" indicates that the smaller (minimum) of A_0 and B_0 should be used.

Characteristic Time of Kinetic Processes and the Equilibrium Assumption

Each transformation process has associated with it a characteristic time. The characteristic time of a process can be thought of as a magnitude estimate of the time required for the process to approach completion, independent of the actions of other processes. By analogy with the stock-and-flow concept introduced in §1.C.8, the characteristic time for a transformation process can be estimated as the ratio of the initial quantity of a reactant divided by the initial rate of the reaction. The characteristic times for common reaction rate laws are presented in Table 3.A.1.

For first-order reactions, the characteristic time is a property of the rate constant only, and does not depend on species concentrations. For all other reaction orders, the initial quantity of the species influences the characteristic time of the process.

For a chemically reacting system, we can determine a characteristic time for each reaction to proceed to completion. To determine whether the kinetics of a particular reaction is important, we compare the reaction time scale, denoted τ_r, with the general time scale of interest, τ, for the overall problem. For example, in a treatment process, a time scale of interest might be the fluid residence time (stock divided by flow rate out) in the apparatus. The relative magnitude of the two time scales indicates the importance of including kinetics in any analysis, as follows:

Fast kinetics	$\tau_r \ll \tau$	Kinetics may be ignored; reaction proceeds rapidly to completion, or equilibrium conditions are effectively instantaneously achieved (e.g., acid-base reactions in a water treatment process)
Slow kinetics	$\tau_r \gg \tau$	Chemical transformations (and the associated kinetics) are slow enough to be neglected altogether (e.g., carbon monoxide oxidation in urban air)
Intermediate kinetics	$\tau_r \sim \tau$	We cannot make any useful simplifying approximations; we must consider in detail the effects of chemical kinetics on the system (e.g., ozone in urban air)

See Exhibit 3.A.1, which explores a simple case that serves to illustrate the key interactions between equilibrium and kinetics.

EXHIBIT 3.A.1 *Equilibrium and Kinetics of Phase Change*

A simple batch reactor system can be used to deepen our understanding of equilibrium and kinetics, and the relationship between them. The system we consider is a glass jar that can be sealed and maintained at a constant temperature (Figure 3.A.3). Half of the jar's volume is initially filled with pure liquid water; the other half is filled with dry nitrogen (N_2) gas at a pressure of 1 atm.

What happens in this system over time? According to what we can observe with our eyes, nothing happens. We wait and see that the vessel remains half filled with water.

However, molecular-scale processes are occurring. Water molecules on the surface of the liquid that possess sufficient kinetic energy escape from the liquid and enter the gas phase (i.e., evaporate). At the same time, some nitrogen gas molecules strike the surface of the water and dissolve into the liquid. Relative to the starting condition, these transformations will cause both the partial pressure of water vapor in the gas phase and the dissolved N_2 level in the aqueous phase to increase. The liquid water content and the gaseous nitrogen content likewise decrease, but by a fractional amount that is too small to be easily detected.

Let's focus our attention on the physical state of the water molecules (Figure 3.A.4). Provided the water temperature stays constant, water molecules will continue to evaporate (escape from the water surface into the gas phase) at a constant rate. As time proceeds, however, and the gas-phase water molecules become more abundant, some strike the surface of the liquid water and condense back into the liquid phase. The rate of condensation increases as the number of water molecules in the gas phase increases. Eventually, the rates of evaporation and condensation become balanced, and the state of chemical equilibrium is attained.

An exploration of the rates of evaporation and condensation in this system will help us to better understand the relationship between kinetics and equilibrium. What are the factors that influence the rate of evaporation at the water surface? Certainly the kinetic energy of the liquid molecules is a factor. The higher the energy, the more likely a molecule is to escape the attraction of its near neighbors and enter the gas phase. The average kinetic energy of the water molecules depends on temperature, increasing as temperature increases. A second factor is the surface area, S. The rate of evaporation should be directly proportional to the number of molecules at the surface of the liquid, which is proportional to S. A third factor is the strength of the attractive bonds between water molecules at the surface. This property affects the equilibrium vapor pressure of a liquid, but once we

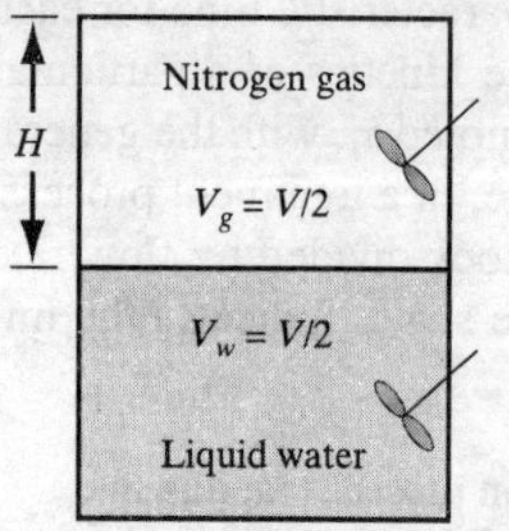

Figure 3.A.3 A sealed jar containing water (H_2O) and nitrogen gas (N_2).

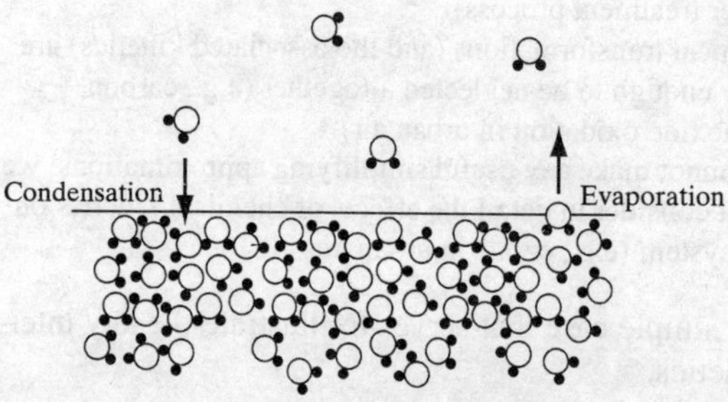

Figure 3.A.4 Evaporation and condensation of water vapor molecules at the liquid-gas interface.

EXHIBIT 3.A.1 *Equilibrium and Kinetics of Phase Change (continued)*

specify that the liquid is water and that the temperature is constant, this property is fixed. In summary, we can say that for a fixed temperature, the rate of evaporation, $r_{\text{evaporation}}$ (with units of molecules per time) is proportional to the gas-water interface area:

$$r_{\text{evaporation}} \propto S \qquad (3.A.24)$$

or

$$r_{\text{evaporation}} = k_1(T_w)S \qquad (3.A.25)$$

where k_1 is a constant that depends on liquid water temperature, T_w.

What can we say about the rate of condensation? A key factor is the rate at which the gaseous water molecules strike the surface. This rate will be proportional to the interfacial surface area, S. It will also be proportional to the gas-phase concentration of water molecules, C_g (units: mol/m^3). Another factor is the mean speed of the water molecules, but this is fixed for a given temperature. Summarizing, the rate of condensation, $r_{\text{condensation}}$, is proportional to both surface area and gas-phase water molecule concentration:

$$r_{\text{condensation}} \propto SC_g \qquad (3.A.26)$$

or

$$r_{\text{condensation}} = k_2(T_g)SC_g \qquad (3.A.27)$$

where T_g is the gas temperature. At equilibrium, the liquid and gas temperatures are equal, $T_w = T_g = T$, and the rates of evaporation and condensation are equal. Therefore,

$$C_g(T) = \frac{k_1(T)}{k_2(T)} = K'(T) \qquad (3.A.28)$$

where K' is a temperature-dependent constant. Thus, the concentration of water molecules in the gas phase at equilibrium depends only on temperature, as we assumed in Example 3.A.3. Recall, from the ideal gas law and from Dalton's law, that the partial pressure of a gas is proportional to the number concentration of gas molecules, given a fixed temperature. Or, more specifically,

$$P_g(T) = \frac{n_g}{V}RT = C_g(T)RT \qquad (3.A.29)$$
$$= K'(T)RT = K(T)$$

On the basis of this argument, we expect the equilibrium partial pressure of pure water (P_g) to be a function only of temperature. This is observed to be true.

Kinetics can be used to describe the time-dependent concentration of water molecules in the gas phase and also to estimate the time required for the system to approach equilibrium. The number of water vapor molecules is equal to the product of the gas volume and the concentration of water vapor molecules, V_gC_g. (We assume that the gas phase is uniformly mixed at all times.) The rate of change in this number is given by the difference between the rates of evaporation and condensation:

$$\frac{d(C_gV_g)}{dt} = r_{\text{evaporation}} - r_{\text{condensation}} \qquad (3.A.30)$$
$$= k_1S - k_2SC_g$$

Since V_g is a constant, it can be taken outside the derivative. Dividing both sides of the equation by V_g (and assuming that the volume of the gas phase can be written as the product HS, where H is the height of the gas-filled portion of the vessel), we obtain

$$\frac{dC_g}{dt} = \frac{k_1}{H} - \frac{k_2}{H}C_g \qquad (3.A.31)$$

with initial condition

$$C_g(0) = 0 \qquad (3.A.32)$$

Differential equations of this form are very frequently encountered in time-dependent environmental systems. The solution method is discussed in detail in Appendix D (§D.1). The solution is given by equation 3.A.33 and is plotted in Figure 3.A.5.

$$C_g(t) = \frac{k_1}{k_2}\left[1 - \exp\left(-\frac{k_2}{H}t\right)\right] \qquad (3.A.33)$$

Note that the steady-state condition from equation 3.A.31 is $d(C_g)/dt = 0$, which leads to the equilibrium result: $C_g = k_1/k_2$. This same condition is obtained from equation 3.A.33 as $t \to \infty$.

The characteristic time required for the water vapor concentration to approach equilibrium can be determined by dividing the stock of water vapor molecules in the system, C_gV_g, by the rate of flow out due to condensation, k_2SC_g. The result is $\tau \sim C_gV_g \times (k_2SC_g)^{-1} = H/k_2$. This is also seen to be the reciprocal of the argument of t in the exponential term in equation 3.A.33.

EXHIBIT 3.A.1 ***Equilibrium and Kinetics of Phase Change (continued)***

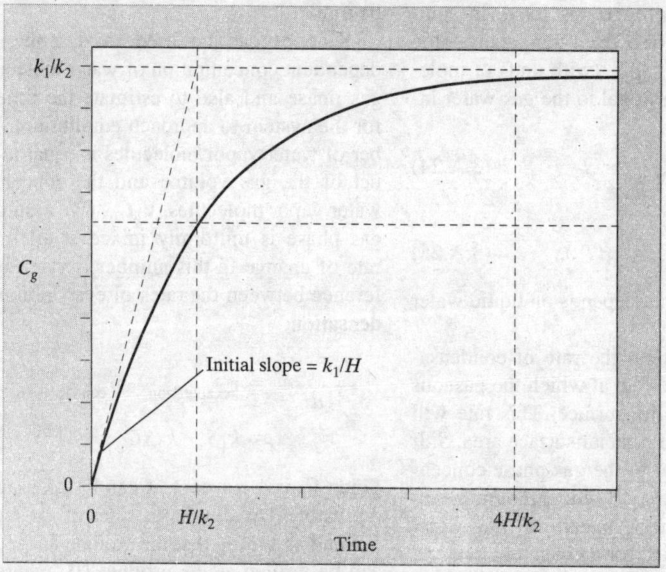

Figure 3.A.5 Time-dependent behavior of the concentration of water vapor molecules in the gas phase of the system depicted in Figure 3.A.3. The steady-state concentration, k_1/k_2, is approached with a characteristic time of H/k_2. The initial rate of increase in vapor concentration is k_1/H and reflects the effects of evaporation alone, since no condensation occurs when $C_g = 0$.

3.B PHASE CHANGES AND PARTITIONING

In this section, we consider transformation processes that involve a change of phase among solid, liquid, and gaseous states. In environmental engineering processes and applications, we must consider many phase-change processes. These include the dissolution of species (whether gas, liquid, or solid) into water, the adsorption of gaseous and liquid species onto solid surfaces, the evaporation of species from liquid to gas, and the volatization of gaseous species from the aqueous phase.

In general, as with other transformation processes, we are concerned with both the equilibrium partitioning of species among states and with the kinetics, or rates, of transformations. However, in many applications, a reasonable approximation for the observed behavior can be made by assuming that equilibrium prevails. Given this fact in combination with the complexity of phase-change kinetics, we will consider only equilibrium behavior here.

3.B.1 Vapor Pressure

The *vapor pressure* of a pure liquid species is the equilibrium partial pressure of the gas molecules of that species above a flat surface of the pure liquid. As discussed in

§3.A, the vapor pressure for a pure liquid species is a function only of temperature. Vapor pressure increases with increasing temperature for all liquids. When the partial pressure of the species is less than the vapor pressure, the liquid will tend to evaporate; this condition is called *subsaturation*. Conversely, under *supersaturation* conditions the partial pressure exceeds the vapor pressure, and there will tend to be net condensation of the gas. When equilibrium is satisfied, we say that the gas phase is *saturated* with the species.

The equilibrium condition between a species in the vapor phase and a flat surface of the pure liquid of that same species can be expressed by

$$P_i^\circ(T) = K_{vp}(T) \tag{3.B.1}$$

where P_i° represents the equilibrium partial pressure of species i and K_{vp} is an equilibrium constant known as the *vapor pressure,* which is a function of temperature.

The most important substances that volatilize and condense in environmental engineering applications are water and volatile organic compounds. The vapor pressure of water as a function of temperature can be determined from the following empirical formula:

$$P_{H_2O}^\circ(T) = P\exp(13.3185a - 1.9760a^2 - 0.6445a^3 - 0.1299a^4) \tag{3.B.2}$$

where $P = 1.01325 \times 10^5$ Pa, $a = 1 - (373.15/T)$, and T is the air temperature (K). This expression is accurate to ±0.1 percent over the temperature range 223 K (–50 °C) $< T <$ 413 K (140 °C) (McRae, 1980). In applying this equation, P should be treated as constant, independent of the atmospheric pressure.

The amount of water vapor in air is commonly expressed as the relative humidity, RH. This parameter is defined by the actual partial pressure of water vapor divided by the saturation vapor pressure:

$$\text{RH} = 100\% \times \frac{P_{H_2O}}{P_{H_2O}^\circ(T)} \tag{3.B.3}$$

When the vapor pressure exceeds the total air pressure applied to the liquid surface, the liquid will boil. The boiling point of water at sea level, where the atmospheric pressure is 1 atm, is 100 °C. At higher elevations, the air pressure is lower, and the boiling point is correspondingly reduced. The water vapor pressure exhibits upward curvature when plotted as a function of temperature (see Figure 3.B.1).

The vapor pressures of organic substances span an enormous range (Table 3.B.1). For compounds with similar chemical structure, species with higher molecular weights tend to have lower vapor pressures. However, the chemical structure is a much stronger factor than the molecular weight, as can be seen in comparing the vapor pressures of naphthalene and chloroform.

In many cases, we are interested in the vapor pressure of an impure liquid. Such a liquid may consist of a single volatile compound with nonvolatile impurities, or it may be a mixture of two or more volatile liquids. In general, the dilution of a liquid with a nonvolatile impurity reduces its vapor pressure. This phenomenon can be understood on the basis of the evaporation and condensation processes described in §3.A: Nonvolatile molecules occupy a portion of the surface, reducing the area from which evaporation takes place, but have less impact on the rate of condensation. For

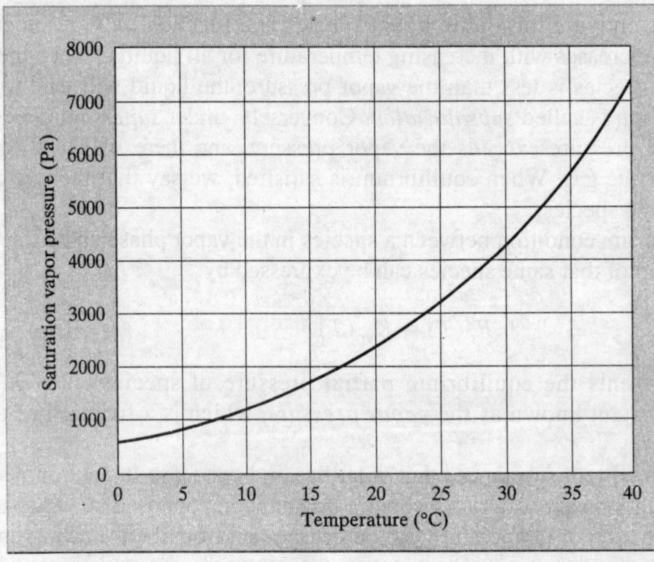

Figure 3.B.1 Saturation vapor pressure of water as a function of temperature.

Table 3.B.1 Vapor Pressure of Some Organic Substances at 25 °C

Species	Chemical formula	Molecular weight (g/mol)	Vapor pressure (Pa)
Acetone	C_3H_6O	58	28,600
Benzene	C_6H_6	78	12,800
Benzo(*a*)pyrene	$C_{20}H_{12}$	252	7×10^{-7}
Chloroform	$CHCl_3$	119.4	26,000
Dieldrin	$C_{12}H_8Cl_6O$	380	7×10^{-4}
Ethylbenzene	C_8H_{10}	106	1,280
Ethylene dibromide	CH_4Br_2	176	270
n-Octane	C_8H_{18}	114	1,890
Naphthalene	$C_{10}H_8$	128	10.6
Phenol	C_6H_6O	94	26
2,3,7,8-TCDD	$C_{12}H_4O_2Cl_4$	322	2×10^{-7}
Tetrachloroethylene	C_2Cl_4	166	2,550
Toluene	C_7H_8	92	3,850
1,1,1-Trichloroethane	$C_2H_3Cl_3$	133.4	16,800
Trichloroethylene	C_2HCl_3	131.4	9,900

Source: Schwarzenbach et al., 1993.

an ideal mixture of a volatile compound with nonvolatile impurities, Raoult's law gives the vapor pressure of the volatile component:

$$P^{\circ}_{i,\text{mix}}(T) = X_i P^{\circ}_i(T) \tag{3.B.4}$$

where X_i is the mole fraction of the volatile component in the mixture.

The vapor pressure of the components in a mixture of volatile substances usually cannot be estimated reliably from simple relationships such as Raoult's law. Instead, empirical data must be used.

3.B.2 Dissolution of Species in Water

Whenever liquid water comes into contact with other materials, there is a tendency for some molecules of these other substances to dissolve into the water. In this section, we consider each of the three phases that may contact water: gases, liquids other than water (and that don't mix with water), and solids.

Partitioning between the Gas Phase and Water: Henry's Law

Consider a sealed vessel that contains liquid water and a gas phase. Gas molecules strike the water surface, and some may become dissolved in the water. As these molecules accumulate in the water, some will escape from the condensed phase and volatilize into the gas. At equilibrium, a balance will be attained such that the rate of dissolution of molecules into the water equals the rate of escape into the gas phase.

In the ideal case, the rate of dissolution is proportional to the rate at which molecules of the gaseous species strike the water surface, and the rate of escape is proportional to the concentration of the species in the aqueous phase. For this situation, we can state that at equilibrium, given a fixed temperature, the species concentration dissolved in water is proportional to the species concentration in the gas phase. This relationship, known as *Henry's law,* is widely used in environmental engineering.

There is no standard form of Henry's law. Many different units can be associated with the aqueous and gas-phase concentrations. Often, one can tell the form of the Henry's law expression from the units associated with the constant; however, some references use a dimensionless form of Henry's law, and in this case the form of the expression cannot be directly inferred from units.

For this text, we will use a single set of units for Henry's law, with the aqueous concentration expressed in moles per liter and the gas-phase quantity expressed as a partial pressure, in atmospheres. Given these units, we present the two complementary forms of Henry's law:

$$C_w = K_{H,g} P_g \tag{3.B.5}$$

or

$$P_g = H_g C_w \tag{3.B.6}$$

where C_w is the equilibrium species concentration in the aqueous phase (M); P_g is the equilibrium partial pressure of the species in the gas phase (atm); $K_{H,g}$ is one form of Henry's law constant, with units of M atm^{-1}; and H_g is a second form of Henry's law constant, with units of atm M^{-1}. Note that the two Henry's law constants are related by $K_{H,g} = H_g^{-1}$.

The Henry's law constants vary widely from one species to another and are also a (weaker) function of temperature. Table 3.B.2 presents Henry's law constants for many species of environmental engineering interest. The dependence of Henry's constant on temperature for oxygen (O_2) is shown in Figure 3.B.2. Examples 3.B.1 and 3.B.2 illustrate how Henry's law is applied.

Table 3.B.2 Henry's Law Constants for Selected Species

Species	Formula	K_H (M atm^{-1})	H_g (atm M^{-1})	Temperature (°C)
Ammonia[a]	NH_3	62	0.016	25
Benzene	C_6H_6	0.18	5.6	20
Benzo(*a*)pyrene	$C_{20}H_{12}$	2040	4.9×10^{-4}	20
Carbon dioxide[a]	CO_2	0.034	29	25
Carbon monoxide	CO	0.0010	1000	20
Chloroform	$CHCl_3$	0.31	3.2	20
Ethylbenzene	C_8H_{10}	0.11	9.1	20
Formaldehyde	HCHO	6300	1.6×10^{-4}	25
Hydrogen sulfide[a]	H_2S	0.115	8.7	20
Methane	CH_4	0.0015	670	20
Naphthalene	$C_{10}H_8$	2.2	0.45	20
Nitric acid[a]	HNO_3	2.1×10^5	4.8×10^{-6}	25
Nitrogen	N_2	0.00067	1500	20
Oxygen	O_2	0.00138	720	20
Phenol	C_6H_6O	2200	4.5×10^{-4}	20
Sulfur dioxide[a]	SO_2	1.24	0.81	25
Tetrachloroethylene	C_2Cl_4	0.083	12	20
Toluene	C_7H_8	0.15	6.7	20
1,1,1-Trichloroethane	$C_2H_3Cl_3$	0.055	18	20
Trichloroethylene	C_2HCl_3	0.11	9.1	20

[a]These species participate in acid-base reactions when dissolved in water. The coefficients listed refer to the solubility of the unreacted species only.

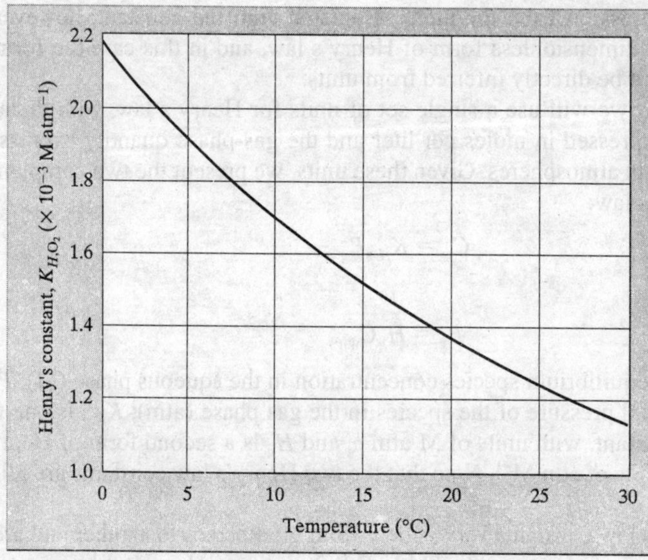

Figure 3.B.2 Henry's law constant for oxygen (O_2) in distilled water as a function of temperature (Whipple and Whipple, 1911).

EXAMPLE 3.B.1 ***Saturation Concentration of Oxygen in Water***

Compute the equilibrium mass concentration of oxygen (O_2) in units of mg/L for water exposed to the atmosphere at sea level at a temperature of 15 °C.

SOLUTION From Figure 3.B.2, we see that at 15 °C, $K_{H,O_2} = 0.00153$ M atm^{-1}. At sea level, the total air pressure is 1 atm. The partial pressure of O_2 is calculated as the product of the mole fraction of oxygen times the total pressure, or 0.209 atm. Thus, from Henry's law, the equilibrium molar concentration of O_2 in water is

$$C_{O_2} = 0.209 \text{ atm} \times 0.00153 \text{ M atm}^{-1} = 0.00032 \text{ M}$$

Multiplying by the molecular weight of oxygen (32 g/mol), we find that the saturation concentration of oxygen at sea level and 15 °C is 0.0102 g/L or 10.2 mg/L.

EXAMPLE 3.B.2 ***Partitioning of Toluene in a Closed System***

A 2 L glass jar is half filled with water and half filled with air at a temperature of 293 K. After 92 mg (=10^{-3} mol) of liquid toluene is added, the jar is sealed. What is the equilibrium concentration of toluene in the water, and what is the equilibrium partial pressure of toluene in the gas phase?

SOLUTION The problem has two unknowns: the quantities of toluene in the aqueous and gas phases, respectively. Henry's law provides one equation linking the unknowns. The second equation is derived from a material balance applied to toluene: The aqueous-phase toluene plus the gas-phase toluene must equal the total added toluene.

Let C_t be the equilibrium molar concentration of toluene in the aqueous phase (M). Let P_t be the equilibrium partial pressure of gaseous toluene (atm). Then the two equations can be written

Henry's law: $\quad C_t = K_{H,t} P_t$

Material balance: $\quad C_t V_w + \dfrac{P_t V_a}{RT} = 10^{-3}$ mol

where $K_{H,t}$ is the Henry's law constant for toluene, V_w and V_a are the water and air volumes in the vessel, R is the gas constant, and T is temperature. Substituting the first equation into the second and replacing the parameters with their values, we find P_t = 5.2×10^{-3} atm and $C_t = 7.8 \times 10^{-4}$ M. In this case, toluene partitions with 78 percent in the aqueous phase and 22 percent in the gas phase.

Solubility of Nonaqueous-Phase Liquids

Some compounds of environmental concern, such as oils, are liquids under environmental conditions and do not mix with water. When present in sufficient quantity, they can form a nonaqueous-phase liquid (NAPL). If a NAPL that is immiscible with water is brought into contact with water, some of the NAPL molecules will dissolve and become distributed throughout the water phase. In environmental engineering, this phenomenon is of particular interest in cases where organic liquids come into contact with water, such as leaks or spills of petroleum products or organic solvents.

Table 3.B.3 Water Solubility of Selected Organic Liquids

Species	Solubility (mg/L)[a]
Benzene	1,780
Benzo(*a*)pyrene	0.0038
Chloroform	8,200
Dieldrin	0.2
Ethylbenzene	152
Ethylene dibromide	4,300
n-Octane[b]	0.72
Naphthalene[b]	31
Phenol	93,000
2,3,7,8-TCDD	0.0002
Tetrachloroethylene	200
Toluene	535
1,1,1–Trichloroethane	4,400
Trichloroethylene	1,100

[a] At 20 °C unless otherwise noted.
[b] At 25 °C.
Source: Sawyer et al., 1994; LaGrega et al., 1994; Schwarzenbach et al., 1993.

Under equilibrium conditions, the quantity of the NAPL that dissolves in water depends on the species and the temperature, but not on the volume of the NAPL. This relationship can be expressed as

$$C_i(T) = K_{ws}(T) \tag{3.B.7}$$

where C_i is the equilibrium concentration of NAPL compound i dissolved in water and K_{ws} is the *water solubility* for the substance. Water solubilities for many species of environmental interest are presented in Table 3.B.3. Example 3.B.3 illustrates the use of these data.

The argument that supports the form of equation 3.B.7 is the same as that used to justify the fact that the equilibrium vapor pressure does not depend on the quantity of condensed-phase species. Molecules of the NAPL move from the nonaqueous phase into the water and from the water back to the nonaqueous phase at rates that are proportional to the interfacial surface area, but independent of the total quantity of NAPL. Thus, the equilibrium state is independent of the quantity of NAPL. It is also independent of the interfacial surface area, a parameter that influences the rate of approach to equilibrium but not the equilibrium condition itself.

EXAMPLE 3.B.3 *Dissolution of a NAPL*

(a) One liter of liquid toluene (density = 0.87 g cm^{-3}) is spilled into 1 m^3 of water. Assuming no loss of toluene or water from the system, what is the steady-state concentration of toluene dissolved in the water?

(b) Repeat (*a*) assuming that the water volume is 10 m^3.

SOLUTION

(a) The solubility of toluene is 535 mg L^{-1} or 535 g m^{-3}. The total amount spilled is 870 g per m^3 of water, which exceeds the solubility. At steady state, equilibrium

EXAMPLE 3.B.3 ***Dissolution of a NAPL (continued)***

will be attained such that the toluene concentration in water will reach saturation, 535 g m^{-3} (= 535 mg/L). A small liquid toluene phase will remain that contains the remaining 335 g.

(**b**) In part (*a*) more than half of the total toluene dissolved into the water. Now, with 10 times as much water, all of the toluene dissolves, so there will be no separate-phase toluene liquid at steady state. The toluene concentration in water will be 870 g/10 m^3 = 87 mg L^{-1}. Note that this concentration is less than the saturation concentration given by $C_{toluene} = K_{ws} = 535$ mg L^{-1}. When the concentration of dissolved species is less than the saturation concentration (equation 3.B.7 and Table 3.B.3), there will be no NAPL at steady state, and the equilibrium equation provides no useful information. Instead, the steady-state concentration is determined by applying a material balance on the dissolved material. Similar situations can arise in the evaporation of a liquid or in the dissolution of solids into water.

Dissolution and Precipitation of Solids

When liquid water contacts solids, some of the solid molecules dissolve into the water. The forward reaction is termed *dissolution*; the reverse process, by which dissolved molecules leave the aqueous solution and form a solid, is called *precipitation*. Dissolution and precipitation are important processes governing the transport and fate of environmental contaminants and are also exploited in environmental treatment processes.

For a general ionic solid of the chemical form A_xB_y, the dissolution/precipitation reaction can be written as follows, where A and B are cations and anions, respectively, and x and y are stoichiometric coefficients:

$$A_xB_y \Leftrightarrow xA + yB \tag{3.B.8}$$

The equilibrium expression for this reaction has this form:

$$[A]^x[B]^y = K_{sp}(T) \tag{3.B.9}$$

The equilibrium constant, K_{sp}, is often called a *solubility product*. If we compare this form of the equilibrium relationship with the general form developed in §3.A.2 (equation 3.A.12), we see that we have the product concentrations raised to the power of their stoichiometric coefficients, but the concentration of reactant is not included in the expression. The rationale for this form is the same one that applies for equilibrium descriptions of vapor pressure and liquid solubility: The quantity of solid material does not influence the amount of dissolved ions in solution. Consider, for example, a system in which a slab of material A_xB_y with thickness H lies beneath a water layer. At equilibrium, the rate of dissolution will be balanced by the rate of precipitation. Both of these processes will occur at the interface between the solid and the liquid. Neither rate depends on the thickness, H, of the solid material, and consequently, the equilibrium condition does not depend on the quantity of the solid.

Table 3.B.4 presents the solubility products for many ionic solids encountered in environmental engineering. Values from two different references are shown to point out the fact that solubility products are not precisely known. In some cases, the two values agree to within about 30 percent; however, in a few cases, disagreement is greater than an order of magnitude. Example 3.B.4 illustrates the use of the solubility product.

Table 3.B.4 Solubility Products for Some Ionic Solids at T = 25 °C

Compound	Equilibrium relationship	K_{sp} [a]	K_{sp} [b]
Aluminum hydroxide	$Al(OH)_3 \Leftrightarrow Al^{3+} + 3\ OH^-$	$1 \times 10^{-32}\ M^4$	$2 \times 10^{-32}\ M^4$
Cadmium hydroxide	$Cd(OH)_2 \Leftrightarrow Cd^{2+} + 2\ OH^-$	$2 \times 10^{-14}\ M^3$	$5.9 \times 10^{-15}\ M^3$
Calcium carbonate	$CaCO_3 \Leftrightarrow Ca^{2+} + CO_3^{2-}$	$5 \times 10^{-9}\ M^2$	$8.7 \times 10^{-9}\ M^2$
Calcium fluoride	$CaF_2 \Leftrightarrow Ca^{2+} + 2\ F^-$	$3 \times 10^{-11}\ M^3$	$4 \times 10^{-11}\ M^3$
Calcium hydroxide	$Ca(OH)_2 \Leftrightarrow Ca^{2+} + 2\ OH^-$	$8 \times 10^{-6}\ M^3$	$5.5 \times 10^{-6}\ M^3$
Calcium phosphate	$Ca_3(PO_4)_2 \Leftrightarrow 3\ Ca^{2+} + 2\ PO_4^{3-}$	$1 \times 10^{-27}\ M^5$	$2.0 \times 10^{-29}\ M^5$
Calcium sulfate	$CaSO_4 \Leftrightarrow Ca^{2+} + SO_4^{2-}$	$2 \times 10^{-5}\ M^2$	$1.9 \times 10^{-4}\ M^2$
Chromium(III) hydroxide	$Cr(OH)_3 \Leftrightarrow Cr^{3+} + 3\ OH^-$	$6 \times 10^{-31}\ M^4$	$6 \times 10^{-31}\ M^4$
Iron(II) hydroxide	$Fe(OH)_2 \Leftrightarrow Fe^{2+} + 2\ OH^-$	$5 \times 10^{-15}\ M^3$	$8 \times 10^{-16}\ M^3$
Iron(III) hydroxide	$Fe(OH)_3 \Leftrightarrow Fe^{3+} + 3\ OH^-$	$6 \times 10^{-38}\ M^4$	$4 \times 10^{-38}\ M^4$
Magnesium carbonate	$MgCO_3 \Leftrightarrow Mg^{2+} + CO_3^{2-}$	$4 \times 10^{-5}\ M^2$	$1 \times 10^{-5}\ M^2$
Magnesium hydroxide	$Mg(OH)_2 \Leftrightarrow Mg^{2+} + 2\ OH^-$	$9 \times 10^{-12}\ M^3$	$1.2 \times 10^{-11}\ M^3$
Nickel hydroxide	$Ni(OH)_2 \Leftrightarrow Ni^{2+} + 2\ OH^-$	$2 \times 10^{-16}\ M^3$	$6.5 \times 10^{-18}\ M^3$

[a] Sawyer et al., 1994, p. 38.
[b] Benefield and Morgan, 1990.

EXAMPLE 3.B.4 ***Ion Concentrations from a Dissolved Solid***

Solid calcium fluoride (CaF_2) is added to pure water so that at equilibrium some solid remains undissolved. Given that the solubility product is $3 \times 10^{-11}\ M^3$, what is the equilibrium concentration of F^- in the water?

SOLUTION Since some solid CaF_2 remains at equilibrium, the solubility product relationship must be satisfied:

$$[Ca^{2+}]\,[F^-]^2 = 3 \times 10^{-11}\ M^3$$

This expression provides one equation, but there are two unknowns. The second equation is obtained either from stoichiometry or electroneutrality; the same mathematical relationship results in either case. From stoichiometry, we know that each Ca^{2+} ion that is dissolved into water is accompanied by two F^- ions. This means that the equilibrium calcium ion concentration must be exactly half the fluoride ion concentration:

$$[Ca^{2+}] = 0.5[F^-]$$

Substituting this expression into the first and solving, we find $[F^-] = 3.9 \times 10^{-4}$ M.

Alternatively, from electroneutrality we know that the cations in solution must balance the anions, so

$$[H^+] + 2[Ca^{2+}] = [OH^-] + [F^-]$$

Since the water was initially pure, and since CaF_2 does not influence the pH, we have $[H^+] = [OH^-]$, and so $[Ca^{2+}] = 0.5[F^-]$, as above.

3.B.3 Sorption

Adsorption is the accumulation of a substance at the interface between two phases, usually a solid and a fluid. In environmental engineering, we are particularly interested in cases in which the fluid phase is either air or water.

Adsorpticn should not be confused with *absorption.* Adsorption is a surface phenomenon in which a chemical species adheres to an interface. Absorption is a bulk phenomenon in which a chemical species becomes distributed throughout a solid or liquid absorbent. Absorption is also used in environmental engineering, but less so than adsorption. Sometimes it is not possible to distinguish between adsorption and absorption. The term *sorption* is commonly used to encompass both phenomena. Here, we emphasize sorption on solids.

Sorption is encountered in diverse environmental engineering situations. Sorption can be applied to transfer contaminants from a fluid (air or water) to the surface of a solid material for subsequent treatment or disposal. It also influences the transport and fate of environmental contaminants and is used in environmental monitoring. For example, in drinking water treatment, sorption is often used for taste and odor control. It is also the best available technique for removing a suite of toxic organic contaminants from water. In air quality engineering, sorption is applied to limit the release of trace organic gases from industrial processes. It is also used to control the release of vapors from automobile gas tanks. Hydrocarbon contaminants in groundwater sorb to soil particles, slowing their movement through the ground. Metals in water sorb to microbial cells in biological treatment systems, enhancing their removal.

The *equilibrium partitioning* of a species between a fluid phase and the sorbed phase is described by semiempirical relationships called "sorption isotherms," or simply *isotherms.* (The word *isotherm* means "uniform temperature"; as with any equilibrium relationship, the partitioning varies with temperature.) Isotherms typically describe the equilibrium relationship between two quantities: (1) the mass of sorbed contaminant per mass of sorbent and (2) the equilibrium concentration of contaminant in the fluid (mass or molar concentration in water, or mass concentration or partial pressure in air).

Isotherms are semiempirical. Their functional forms are based on thermodynamic arguments, but they include data that can be obtained only from experiment. Many isotherm forms have been developed. Three common forms are introduced here:

Linear: $$q_e = K_{\text{ads}} C_e \tag{3.B.10}$$

Langmuir: $$q_e = q_{\max} \frac{b C_e}{1 + b C_e} \tag{3.B.11}$$

Freundlich: $$q_e = K_f C_e^{1/n} \tag{3.B.12}$$

In these equations, C_e is the equilibrium concentration of contaminant molecules in a fluid (typically in moles or milligrams per liter of fluid) and q_e is the equilibrium mass of sorbed molecules per mass of sorbent (typically in moles or milligrams per gram of sorbent).

In a *linear* isotherm, the sorbed mass of contaminant (q_e) increases in direct proportion to the dissolved mass concentration (C_e) with no maximum (see Figure 3.B.3). By contrast, in a *Langmuir* isotherm, the sorbent can become completely saturated, leading to a maximum adsorption capacity, $q_{\max}$. The derivation of this isotherm form assumes a homogeneous adsorbent (see Exhibit 3.B.1). The *Freundlich* isotherm provides an alternative description that applies to cases with surface heterogeneities, which result in a distribution of energies associated with adsorption at surface sites.

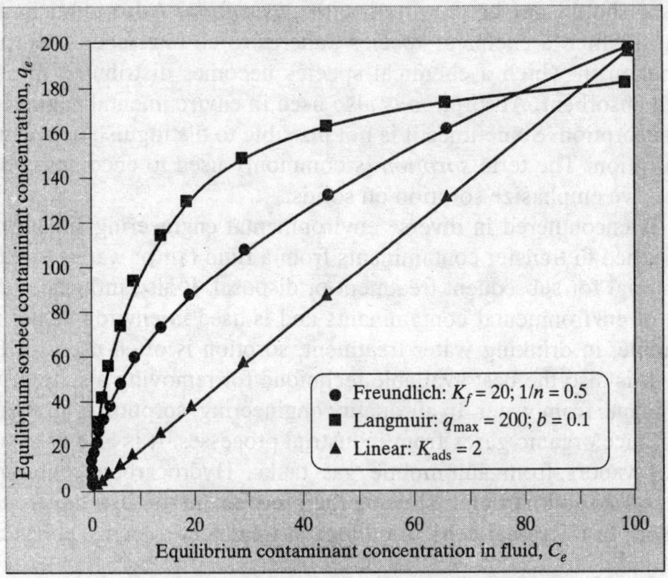

Figure 3.B.3 Shapes of common sorption isotherms.

Typically, one must conduct a set of experiments (a) to test whether a particular functional isotherm form works and (b) to determine the appropriate empirical parameters (K_{ads} for the linear isotherm, q_{max} and b for the Langmuir isotherm, and K_f and $1/n$ for the Freundlich form). Separate experiments may be needed for each combination of pollutant and sorbent. Sample isotherms are shown in Figure 3.B.3. Representative Freundlich isotherm parameters for sorption of organics from water with granular activated carbon are presented in Table 3.B.5.

Equilibrium analysis of systems in which sorption occurs often requires the solution of nonlinear algebraic equations, as illustrated in Example 3.B.5.

Table 3.B.5 Freundlich Sorption Isotherm Parameters for Selected Organic Compounds in Water on Activated Carbon

Species	K_f (mg/g) (L/mg)$^{1/n}$	$1/n$ (–)
Benzene	1.0	1.6
Benzo(*a*)pyrene	34	0.44
Chlordane	190	0.33
Chloroform	2.6	0.73
Dieldrin	606	0.51
Ethylbenzene	175	0.53
Naphthalene	132	0.42
Phenol	21	0.54
Toluene	100	0.45
1,1,1-Trichloroethane	2.5	0.34
Trichloroethylene	28	0.62

Source: Snoeyink, 1990. These data should be used with caution. They apply for a specific sorbent material only.

EXAMPLE 3.B.5 ***Three-Phase Partitioning of Toluene in a Closed System***

Consider the system introduced in Example 3.B.2, in which a 2 L sealed jar, half filled with water and half filled with air, contains 10^{-3} mol of toluene. The equilibrium partitioning is described by Henry's law, so that the aqueous phase contains 0.78×10^{-3} mol and the gas phase contains 0.22×10^{-3} mol. To this system 200 mg of activated carbon is added. Toluene partitioning between the sorbed and aqueous phases is described by the Freundlich isotherm $q_e = 100C_e^{0.45}$, where C_e is the aqueous concentration in mg/L and q_e is the sorbed mass concentration in mg toluene per g activated carbon. What is the new equilibrium concentration of toluene in the water?

SOLUTION The problem has three unknowns: the quantities of toluene in the aqueous, gas, and sorbed phases. Henry's law provides one equation linking the aqueous- and gas-phase concentrations. The Freundlich isotherm links the sorbed- and aqueous-phase concentrations. The third independent equation is derived from a material balance: The sorbed toluene plus the aqueous toluene plus the gaseous toluene must equal 92 mg or 10^{-3} mol.

Let C_t be the molar concentration of toluene in the aqueous phase (M). Let P_t be the partial pressure of gaseous toluene (atm). Let q_e be the sorbed mass of toluene per mass of activated carbon (mg/g). Then the three equations can be written

Henry's law: $C_t = 0.15P_t$

Freundlich isotherm: $q_e = 100 \times (92{,}000C_t)^{0.45} = 1.71 \times 10^4 C_t^{0.45}$

Material balance: $2.17 \times 10^{-6} q_e + C_t + 0.0416P_t = 10^{-3}$

The factor 92,000 in the Freundlich isotherm converts from molar concentration (C_t) to mass concentration in mg/L. The factor 2.17×10^{-6} (= 0.2 g activated carbon divided by 92,000 mg per mole toluene) in the material balance converts from sorbed mass concentration (mg/g) to sorbed moles. The factor 0.0416 in the material-balance expression converts from the toluene partial pressure in atmospheres to gaseous moles, given that the air volume is 1 L and temperature is 293 K.

Use Henry's law and the Freundlich isotherm to eliminate P_t and q_e, respectively, from the material-balance equation, to obtain

$$1.277C_t + 0.0371C_t^{0.45} - 1.0 \times 10^{-3} = 0$$

or

$$C_t + 0.0291C_t^{0.45} - 7.83 \times 10^{-4} = 0$$

Solution techniques for nonlinear algebraic equations such as this are discussed in Appendix D (§D.3). The answer, obtained by an iteration procedure, is $C_t = 0.18 \times 10^{-3}$ M. Comparing this result with the answer to Example 3.B.2 ($C_t = 0.78 \times 10^{-3}$ M), we see that the addition of the activated carbon reduces the aqueous-phase concentration by 77 percent. The same fractional reduction occurs in the gas phase.

Parameter Estimation for Sorption Isotherms

Experiments to test isotherm forms and evaluate parameters can be conducted in a set of well-mixed batch reactors. Each reactor is filled with the fluid of interest, a fixed quantity of the sorbent, and a different quantity of contaminant.

After equilibrium is attained (the time required can be determined from separate kinetic experiments), the concentration of contaminant remaining in solution is measured in each reactor. The mass sorbed is determined from a material-balance equation of this form:

$$q_e = \frac{x}{m} = \frac{M - (C_e V_f)}{m} \tag{3.B.13}$$

where x is the mass of solute sorbed onto the solid (mg), m is the mass of sorbent (g), M is the initial mass of contaminant added (mg), C_e is the equilibrium fluid-phase concentration of contaminant (mg/L), and V_f is the fluid volume (L). The results from each reactor provide one data point defining q_e and C_e. To produce a complete isotherm, data should be generated over a range of equilibrium concentrations that is similar to the expected range in the system of interest. It is important to prevent (or to control and account for) processes that could interfere with sorption measurements such as volatilization, dissolution, and degradation.

The data can then be plotted in linearized forms of the isotherms. The quality of fit to a straight line in the plots determines the most appropriate isotherm form. The slope and intercept of the straight line are then used to determine the isotherm parameters. See Appendix D (§D.5) for information on obtaining the slope and intercept of a best-fit straight line to experimental data by linear regression.

The linear isotherm form requires no rearrangement and can be analyzed directly by plotting q_e against C_e.

The Langmuir isotherm is transformed into a linear form by rearranging it in the following manner:

$$\frac{C_e}{q_e} = \frac{C_e}{q_{\max}} + \frac{1}{bq_{\max}} \tag{3.B.14}$$

The experimental data are plotted with C_e/q_e on the y-axis and C_e on the x-axis. The two unknown parameters can be determined from the slope of the line ($1/q_{\max}$) and the y-intercept ($1/[q_{\max}b]$).

The Freundlich isotherm is converted to linear form by taking the logarithm of equation 3.B.12:

$$\log_{10}(q_e) = \frac{1}{n}\log_{10}(C_e) + \log_{10}(K_f) \tag{3.B.15}$$

A plot of $\log_{10}(q_e)$ against $\log_{10}(C_e)$ will then yield a straight line with a slope of $1/n$ and an intercept of $\log_{10}(K_f)$.

EXHIBIT 3.B.1 *Derivation of the Langmuir Isotherm (Hiemenz, 1986)*

The Langmuir isotherm is the easiest one to derive from basic equilibrium arguments. The main purpose of this derivation is to illustrate the link between sorption isotherms and other equilibrium equations, such as Henry's law.

The derivation of the Langmuir isotherm is based on the following assumptions. (a) The sorbed layer has a thickness of only one molecule. (b) All surface sites have equal sorption affinity for the contaminant relative to the background fluid. (c) The sorbed and aqueous phases are ideal, so that the rates of sorption and desorption are proportional to the aqueous and sorbed concentrations, respectively. We also assume that the contaminant is present in dilute quantities.

A system is depicted in Figure 3.B.4 in which there is a solid and a fluid phase, such as granular activated carbon in water. The fluid (F) contains a dilute concentration of some contaminant (C), such as benzene. There are attractive and repulsive

EXHIBIT 3.B.1 *Derivation of the Langmuir Isotherm (Hiemenz, 1986) (continued)*

Figure 3.B.4 Schematic illustrating single-layer adsorption onto a solid of a contaminant (C) dissolved in a fluid (F).

interactions between the solid and the molecules in the fluid. Only the adsorbed molecules (a single layer) are influenced by these interactions, and the interaction is the same at all surface positions.

Kinetically, the processes of adsorption and desorption can be described in the following terms. An *adsorption* "reaction" occurs if a molecule of the fluid, F, leaves the surface and is replaced by a molecule of contaminant, C. The reverse "reaction," in which an adsorbed C molecule is replaced by a molecule of F, is called *desorption*. The other cases, where F is replaced by F, or C is replaced by C, have no net effect and need not be considered. We can denote the reactions of interest as follows:

$$\text{Adsorption:}\quad C_d + F_s \rightarrow C_s + F_d \qquad (3.B.16)$$

$$\text{Desorption:}\quad C_s + F_d \rightarrow C_d + F_s \qquad (3.B.17)$$

where the subscript denotes the sorbed (*s*) or dissolved (*d*) state of the compound.

By assumption, the rate of the adsorption reaction should be proportional to the concentration of dissolved-phase contaminant times the surface density of sorbed fluid molecules:

$$R_{ads} = k_{ads}[C_d][F_s] \qquad (3.B.18)$$

Likewise, the rate of the desorption reaction is proportional to the amount of sorbed contaminant times the bulk concentration of fluid molecules:

$$R_{des} = k_{des}[C_s][F_d] \qquad (3.B.19)$$

At equilibrium, the two processes will occur at the same rate. Equating R_{ads} and R_{des}, we may write the equilibrium relationship as

$$K' = \frac{k_{ads}}{k_{des}} = \frac{[C_s][F_d]}{[C_d][F_s]} \qquad (3.B.20)$$

where K' is constant at a fixed temperature.

There are assumed to be a fixed number of surface sites, each occupied either by contaminant or fluid molecules. Therefore, the total concentration of surface sites, $[T_s]$, can be expressed as

$$[T_s] = [F_s] + [C_s] \qquad (3.B.21)$$

This equation can be rearranged and substituted into equation 3.B.20 to eliminate $[F_s]$:

$$K' = \frac{[C_s][F_d]}{[C_d]([T_s] - [C_s])} \qquad (3.B.22)$$

Rearranging to solve for $[C_s]$, we obtain

$$[C_s] = [T_s]\frac{\{K'/[F_d]\}[C_d]}{1 + \{K'/[F_d]\}[C_d]} \qquad (3.B.23)$$

Since we have assumed that the contaminant is dilute, $[F_d]$ can be taken as a constant, and so we can designate $\{K'/[F_d]\}$ as the parameter b. Also, the maximum sorption capacity, q_{max}, is proportional to the concentration of surface sites, $[T_s]$, in the same way that the equilibrium amount sorbed, q_e, is proportional to $[C_s]$. (That is, $q_{max}/[T_s]$ is equal to $q_e/[C_s]$, so $[C_s]/[T_s] = q_e/q_{max}$.) Noting that $C_e = [C_d]$, we obtain the conventional form of the Langmuir isotherm

$$q_e = q_{max}\frac{bC_e}{1 + bC_e} \qquad (3.B.24)$$

The Langmuir isotherm is plotted in Figure 3.B.5. We see that for $bC_e \gg 1$, the surface becomes saturated with contaminant and q_e approaches q_{max}. When bC_e is much less than 1, the surface coverage by contaminant is sparse and the Langmuir isotherm reduces to the linear form, $q_e \approx (q_{max}b)C_e$ ($\Rightarrow q_e/q_{max} \approx bC_e$).

EXHIBIT 3.B.1 ***Derivation of the Langmuir Isotherm (Hiemenz, 1986) (continued)***

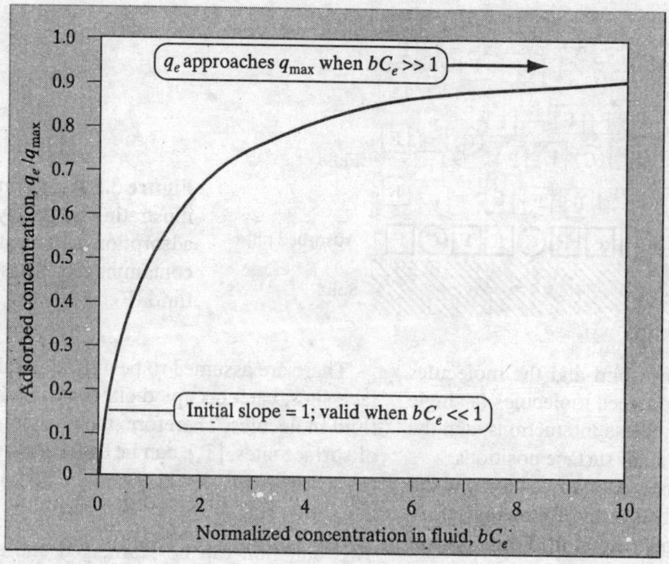

Figure 3.B.5 Langmuir adsorption isotherm.

3.C ACID-BASE REACTIONS

The concentration of hydrogen ions is an important water quality parameter. Because it may span many orders of magnitude, it is commonly measured and reported on a logarithmic basis, using the pH scale, where $\text{pH} = -\log_{10}[H^+]$.

The pH of aqueous solutions is important not only in water quality engineering, but also in air quality engineering and hazardous waste management. In water quality engineering, the pH is controlled to optimize treatment processes and to prevent corrosion in water storage and distribution systems. The pH of rainwater is a major concern both for air quality, because the acids originate as air pollutants, and for water quality, because many potential impacts occur in aquatic environments. Liquid wastes with very high or very low pH values are classified as hazardous, independent of their other attributes.

The pH of a solution is governed through the relative concentrations and strengths of acids and bases. Conversely, acids and bases are those species that influence the pH of aqueous solutions.

Acid-base reactions are generally very fast. For example, under neutral pH conditions, the characteristic time for a hydrogen ion to react with a hydroxide ion is about 70 μs (Laidler, 1987). Because the time scale for acid-base reactions is generally very much faster than time scales of environmental engineering interest, pH predictions are almost always based on the assumption of equilibrium.

This section begins with a review of the definition of acids and bases and proceeds with the simplest acid-base problem: determining the pH of pure water. Next, acids and bases commonly encountered in environmental engineering applications are discussed. Several illustrative examples are then presented that show how acid-base transformation processes arise in environmental engineering applications.

3.C.1 Acid-Base Reactions and the Hydrogen Ion

Hydrogen ions do not exist as free H^+ species in aqueous solutions. Instead, reaction of H^+ with a water molecule to form H_3O^+ is highly favored. The resulting ion may have one or more additional water molecules loosely bound to it. However, for engineering purposes, these details are unimportant. We will consider the free hydrogen ion (H^+), the hydrogen ion bound with a water molecule (H_3O^+), and H_3O^+ with additional water molecules clustered around it to be equivalent and refer to all of these collectively as the hydrogen ion, denoted either H^+ or H_3O^+. The pH is taken to be the base 10 logarithm of the sum of the concentrations of these species. In most environmental engineering applications, errors associated with these simplifications are negligible.

An *acid* is a substance that tends to donate or lose a hydrogen ion, and a *base* is a substance that tends to accept or gain a hydrogen ion. Thus, an acid tends to reduce the pH of an aqueous solution, whereas a base tends to raise the pH. In an acid-base reaction, a proton is transferred from the acid to the base. Consider, for example, hydrochloric acid in water. The acid-base reaction may be written as follows:

$$HCl + H_2O \Leftrightarrow H_3O^+ + Cl^- \tag{3.C.1}$$

In the forward direction, HCl is the acid, donating a hydrogen ion to H_2O, which serves as a base. In the reverse direction, H_3O^+ is the acid and Cl^- is the base. The chloride ion, Cl^-, is called the *conjugate base* of HCl, and H_2O is the conjugate base of H_3O^+.

Note that some species can act as either acids or bases, depending on the circumstances. Water is a good example. In the presence of HCl, it accepts a hydrogen ion, acting as a base. In the presence of a base, such as ammonia, it releases a hydrogen ion, acting as an acid, as follows:

$$NH_3 + H_2O \Leftrightarrow NH_4^+ + OH^- \tag{3.C.2}$$

In the forward reaction, ammonia is a base and water is an acid. The ammonium ion, NH_4^+, is the conjugate acid of ammonia, and the hydroxide ion, OH^-, is the conjugate base of H_2O.

3.C.2 pH of Pure Water

Water undergoes a weak acid-base reaction even in the absence of other species. The reaction can be written as follows:

$$H_2O \Leftrightarrow H^+ + OH^- \tag{3.C.3}$$

According to the general method for writing equilibrium relationships, we expect the following expression to hold:

$$K' = \frac{[H^+][OH^-]}{[H_2O]} \tag{3.C.4}$$

However, in aqueous solutions with dilute levels of impurities, the concentration of H_2O is essentially unchanged by this or any other reaction. The molar concentration of H_2O is 55.5 M, as determined from the density of water (1000 g L^{-1}) divided by its molecular weight (18 g mol^{-1}). So, instead of writing the equilibrium expression with $[H_2O]$ as a variable, we incorporate it into the equilibrium constant, as shown below:

$$K_w = K'[H_2O] = [H^+][OH^-] \tag{3.C.5}$$

K_w is called the equilibrium or dissociation constant for water. As with other equilibrium constants, its value depends on temperature. The following equation may be used to determine $K_w(T)$ (Harned and Owen, 1958):

$$\log_{10}(K_w) = -\frac{4470.99}{T} + 6.0875 - 0.01706T \tag{3.C.6}$$

where T is the temperature in K and K_w has units of M^2. Because of the temperature dependence of K_w, the pH of pure water is 7.0 at only one temperature, T = 25 °C. Example 3.C.1 illustrates this point by showing how to compute the pH of pure water at a different temperature.

EXAMPLE 3.C.1 ***Computing the pH of Pure Water***

What is the pH of pure water at 10 °C?

SOLUTION Pure water contains two species with unknown concentrations, $[H^+]$ and $[OH^-]$. Two independent equations are required to solve for the unknowns. Equilibrium provides one equation. The equilibrium dissociation constant is evaluated from equation 3.C.6, with T = 283.15 K, which yields $K_w = 2.9 \times 10^{-15}$ M^2. Thus, we obtain

$$[H^+][OH^-] = 2.9 \times 10^{-15}\ M^2$$

The second equation is obtained either from stoichiometry or from electroneutrality. Stoichiometry states that the generation of each H^+ ion must be accompanied by the generation of an OH^- ion, so that, in otherwise pure water, their concentrations must be equal. Electroneutrality requires the same condition to maintain charge balance. Substituting $[OH^-] = [H^+]$ into the equilibrium relationship and solving yields

$$[H^+] = [2.9 \times 10^{-15}\ M^2]^{1/2} = 5.4 \times 10^{-8} \Rightarrow \text{pH} = 7.27$$

The pH of cool, pure water is somewhat higher than 7.0.

3.C.3 Strong and Weak Acids

A generic acid is represented by the chemical formula HA, where H represents the hydrogen ion to be donated and A is the conjugate base. In aqueous solution, the general acid dissociation reaction can be written as

$$HA \Leftrightarrow H^+ + A^- \tag{3.C.7}$$

The equilibrium constant for this reaction is written as

$$K_A = \frac{[H^+][A^-]}{[HA]} \tag{3.C.8}$$

One often sees the acid dissociation constant, K_A, reported on a logarithmic scale, where

$$pK_A = -\log_{10}(K_A) \tag{3.C.9}$$

This use is analogous to the definition of pH. The pK_A value of an acid has the interesting property that when $pH = pK_A$ in a solution, exactly half of the acid is dissociated and half is undissociated. This relationship can be demonstrated by taking the logarithm of both sides of equation 3.C.8 and multiplying by −1:

$$\begin{aligned} -\log_{10}(K_A) &= -\log_{10}[H^+] - \log_{10}[A^-] + \log_{10}[HA] \\ \Rightarrow pK_A &= pH - \log_{10}[A^-] + \log_{10}[HA] \end{aligned} \tag{3.C.10}$$

So, if $pH = pK_A$, then $[A^-] = [HA]$. When the pH value is much smaller than the pK_A, the acid is mostly undissociated ($[HA] \gg [A^-]$). When the pH is much larger than the pK_A, the acid is almost completely dissociated ($[A^-] \gg [HA]$). The dissociation constants for many of the acid species encountered in environmental engineering are presented in Table 3.C.1 in terms of their pK_A values.

The *strength* of an acid is quantified by its acid dissociation constant. Strong acids tend to have pK_A values that are small positive numbers (~1 or less) or negative numbers. Strong acids dissociate almost completely in water unless the pH is extraordinarily low. Since they dissociate completely, in a solution that contains only a strong acid the pH is governed by the *amount* of strong acid present per volume of water, but not by the specific pK_A value. Common strong acids include perchloric, hydrochloric, sulfuric, and nitric.

By contrast, *weak* acids only partially dissociate in aqueous solutions at ordinary pH levels. In a solution containing weak acids, the pH is governed both by the amounts *and* by the dissociation constants of the weak acids. Examples of weak acids encountered in environmental engineering include acetic acid, carbonic acid, hydrogen sulfide, and hypochlorous acid.

Table 3.C.1 Acid Dissociation Constants (T = 25 °C)

Species	Chemical formula	pK_A
Perchloric acid	$HClO_4$	−7
Hydrochloric acid	HCl	−3
Sulfuric acid	H_2SO_4	−3
Nitric acid	HNO_3	−1.2
Sulfonic acid	H_2SO_3	1.89
Bisulfate	HSO_4^-	1.92
Phosphoric acid	H_3PO_4	2.12
Acetic acid	CH_3COOH	4.7
Butyric acid	C_3H_7COOH	4.8
Propionic acid	C_2H_5COOH	4.9
Carbonic acid	H_2CO_3	6.35
Hydrogen sulfide	H_2S	7.1
Dihydrogen phosphate	$H_2PO_4^-$	7.21
Bisulfite	HSO_3^-	7.22
Hypochlorous acid	$HOCl$	7.6
Hydrogen cyanide	HCN	9.2
Ammonium	NH_4^+	9.23
Bicarbonate	HCO_3^-	10.33
Hydrogen phosphate	HPO_4^{2-}	12.32
Bisulfide	HS^-	12.9

A monoprotic acid, such as hydrochloric or nitric acid, liberates only a single hydrogen ion per molecule. Polyprotic acids, such as sulfuric or phosphoric acid, can contribute multiple hydrogen ions. Examples 3.C.2 and 3.C.3 demonstrate the difference between strong and weak acids and illustrate how acid-base problems are solved.

EXAMPLE 3.C.2 ***pH of a Solution Containing Strong Acid***

If 10^{-3} mol of a monoprotic strong acid is added to water to make 1 L of solution, what is the equilibrium pH at $T = 25$ °C?

SOLUTION The acid is represented as HA. It dissociates in water to produce H^+ and A^-. Since HA is a strong acid, we can assume it dissociates completely so that there is no HA remaining in the water. (We will check the validity of this assumption at the end.) Since aqueous solutions always contain H^+ and OH^-, there are three species with unknown concentrations: $[H^+]$, $[OH^-]$, and $[A^-]$. Therefore, three independent equations are required to solve the problem.

The electroneutrality relation for this system provides one equation:

$$[H^+] = [OH^-] + [A^-]$$

The dissociation constant for water provides a second equation:

$$[OH^-][H^+] = K_w = 10^{-14}\ M^2$$

The third equation is based on a material balance. From the problem statement, $[A^-] = 10^{-3}$ M if the acid fully dissociates. Combining the three equations results in this expression:

$$[H^+] = \frac{10^{-14}\ M^2}{[H]} + 10^{-3}\ M$$

or, more generally,

$$[H^+] = \frac{K_w}{[H^+]} + C_T$$

where C_T is the concentration of acid initially added to the solution.

This is a second-order equation in $[H^+]$, which can be directly solved by means of the quadratic formula. The answer, $[H^+] = 10^{-3}$ M, or pH = 3, can also be obtained by making a simplifying assumption. Since the water contains a strong acid, we expect $[H^+] \gg [OH^-]$. Therefore, we can neglect $[OH^-]$ in the electroneutrality equation, yielding $[H^+] = [A^-] = 10^{-3}$ M. Indeed, this approximate solution is valid, since $[OH^-] = 10^{-11}\ M \ll [H^+]$.

To check the assumption that the strong acid fully dissociates, we rewrite the equilibrium equation for acid dissociation (3.C.8) as

$$[HA] = \frac{[H^+][A^-]}{K_A} = \frac{10^{-6}\ M^2}{K_A}$$

If the strong acid fully dissociates, then $[HA] \ll [A^-] = 10^{-3}$ M. From the equilibrium equation, we see that as long as $K_A \gg 10^{-3}$ M, or $pK_A < -3$, the condition of full dissociation is a good approximation. This reinforces the point that if $pH \gg pK_A$, then the acid will be fully dissociated.

In summary, we see that the pH of a solution containing a strong acid depends only on the quantity of the acid added and not on its acid dissociation constant.

EXAMPLE 3.C.3 ***pH of a Solution Containing Weak Acid***

If 10^{-3} mol of a monoprotic weak acid is added to water to make 1 L of solution, what is the equilibrium pH at $T = 25$ °C?

SOLUTION As in Example 3.C.2, we represent the acid as HA. It dissociates in water to produce H^+ and A^-. As always in water, the ion OH^- is also present. Since the acid is weak, it will not fully dissociate, so the concentration of the undissociated acid, HA, is a fourth unknown. With four unknowns, we need four independent equations to solve the problem: two equilibrium conditions, the electroneutrality relationship, and a material-balance expression.

Water dissociation: $K_w = [H^+][OH^-]$

Acid dissociation: $K_A = \dfrac{[H^+][A^-]}{[HA]}$

Electroneutrality: $[H^+] = [OH^-] + [A^-]$

Material balance on A: $C_T = [A^-] + [HA]$

In these equations, the parameters K_w and K_A are equilibrium constants and C_T is the total concentration of added acid (10^{-3} M).

To solve this problem, we substitute the other expressions into the electroneutrality equation so that the only unknown is $[H^+]$. So from water dissociation we have

$$[OH^-] = \frac{K_w}{[H^+]}$$

Next, substitute for [HA] from the material-balance expression into the acid dissociation equation and solve for $[A^-]$:

$$[A^-] = \frac{C_T K_A}{K_A + [H^+]}$$

Then substitute into the electroneutrality equation:

$$[H^+] = \frac{K_w}{[H^+]} + \frac{C_T K_A}{K_A + [H^+]}$$

This is a cubic equation for $[H^+]$, which can be solved by methods presented in Appendix D (§D.3), given numerical values for the parameters. It is clear from the form of the equation that in general the pH depends on all three parameters, K_w, C_T, and K_A. In the case of a strong acid, the same general equation would be valid; but $K_A \gg [H^+]$, so $[H^+]$ in the denominator of the right-hand term can be neglected, and the equation simplifies to

$$[H^+] = \frac{K_w}{[H^+]} + C_T$$

as was found in Example 3.C.2.

Returning to the case of the weak acid, we can simplify the problem by application of chemical intuition. As long as the acid dissociates to some extent, we expect that $[H^+] \gg [OH^-]$, and so the latter can be neglected in the electroneutrality

EXAMPLE 3.C.3 ***pH of a Solution Containing Weak Acid (continued)***

relationship. The result is

$$[H^+] = \frac{C_T K_A}{K_A + [H^+]}$$

This is now a quadratic equation whose solution is

$$[H^+] = \tfrac{1}{2}[\sqrt{K_A^2 + 4K_A C_T} - K_A]$$

If $C_T \gg K_A$, this equation can be further simplified to

$$[H^+] = (K_A C_T)^{1/2}$$

or

$$\text{pH} = 0.5\,\text{p}K_A - 0.5\log_{10}(C_T)$$

So, for example, if the acid is acetic with pK_A = 4.7, then the pH of a 10^{-3} M solution is $[0.5 \times 4.7 - 0.5 \times (-3)] = 3.9$. If, on the other hand, the acid is hypochlorous with pK_A = 7.6, then the pH of a 10^{-3} M solution is 5.3. In either case, we see that it is reasonable to neglect $[OH^-]$ in the electroneutrality equation. We also see that the pH of a solution containing a weak acid depends on both the acid's quantity and its dissociation constant. In Example 3.C.2, we saw that a 10^{-3} M solution of a strong acid had a pH of 3. Here we see that the pH of a 10^{-3} M solution of a weak acid is higher, and the weaker the acid, the higher the pH.

3.C.4 Carbonate System

The carbonate system in the natural environment includes these species (see Figure 3.C.1):

CO_2(g)	Carbon dioxide gas
CO_2(aq)	Dissolved carbon dioxide
H_2CO_3(aq)	Carbonic acid (a *diprotic,* weak acid)
HCO_3^-(aq)	Bicarbonate ion
CO_3^{2-}(aq)	Carbonate ion
$CaCO_3$(s)	Calcium carbonate (limestone)

This system is of considerable environmental importance for several reasons:

- It serves to buffer lakes and streams against pH changes due to acidic inputs, such as acid rain.
- It influences the accumulation of CO_2 in the atmosphere from fossil fuel combustion and deforestation, and may therefore affect global climate.
- Species in the carbonate system play a major role in the transport and cycling of carbon throughout the environment since photosynthesizers and other autotrophs obtain carbon for the production of cell mass from carbonate species.

Here, we explore the equilibrium relationships among species in the carbonate system. This exploration illustrates the ways in which Henry's law, acid-base chemistry, and dissolution-precipitation equilibrium combine to influence concentrations and

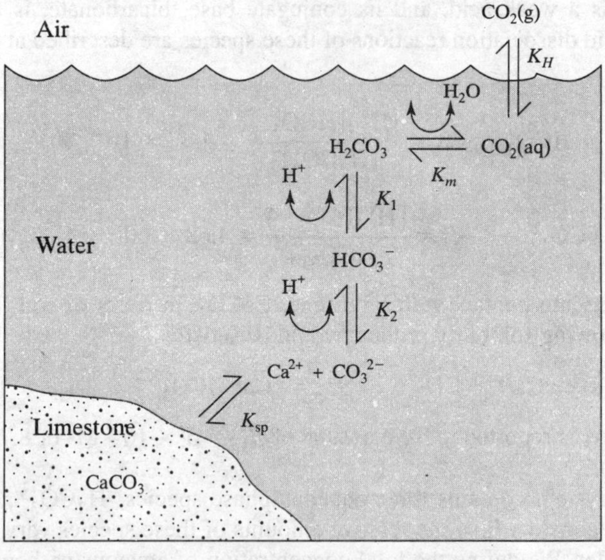

Figure 3.C.1 Species and transformations for the carbonate system in natural air-water-limestone environments.

fates of impurities in air and water. (Additional general information about the carbonate system and its global environmental significance can be found in Erlich et al., 1977; and Berner and Lasaga, 1989.)

Consider a volume of water that is well mixed and exposed to the atmosphere, where the total air pressure is 1 atm and the temperature is 25 °C. The pristine atmosphere contains a mole fraction of 350 ppm of CO_2 (Table 2.B.1). With sufficient exposure time, equilibrium will occur between $CO_2(aq)$ and $CO_2(g)$, as described by Henry's law:

$$[CO_2(aq)] = K_H P_{CO_2} \tag{3.C.11}$$

The Henry's constant for carbon dioxide, K_H, is 0.034 M atm^{-1} (Table 3.B.2), so the equilibrium molar concentration of aqueous CO_2 exposed to clean air at a pressure of 1 atm is $0.034 \times 350 \times 10^{-6}$ M = 12 μM. The dissolved CO_2 reacts rapidly with water to form carbonic acid according to

$$CO_2(aq) + H_2O \Leftrightarrow H_2CO_3 \tag{3.C.12}$$

The equilibrium relationship is described by the following expression, with the water concentration incorporated into the equilibrium constant:

$$\frac{[H_2CO_3]}{[CO_2(aq)]} = K_m = 1.58 \times 10^{-3} \tag{3.C.13}$$

It is difficult to differentiate experimentally between $CO_2(aq)$ and H_2CO_3. For engineering purposes we can combine them into one variable, denoted $H_2CO_3{}^*$ and referred to here as "carbonic acid":

$$[H_2CO_3{}^*] = [CO_2(aq)] + [H_2CO_3] = [CO_2(aq)](1 + K_m) \tag{3.C.14}$$

Since $K_m \ll 1$, $[H_2CO_3{}^*]$ is approximately equal to $[CO_2(aq)]$. At equilibrium, with exposure to a CO_2 partial pressure of 350×10^{-6} atm, $[H_2CO_3{}^*] = 12$ μM.

Carbonic acid is a weak acid, and its conjugate base, bicarbonate, is an even weaker acid. The acid dissociation reactions of these species are described at equilibrium as follows:

$$H_2CO_3^* \Leftrightarrow H^+ + HCO_3^- \qquad K_1 = \frac{[H^+][HCO_3^-]}{[H_2CO_3^*]} = 4.47 \times 10^{-7} \text{ M} \qquad (3.C.15)$$

$$HCO_3^- \Leftrightarrow H^+ + CO_3^{2-} \qquad K_2 = \frac{[H^+][CO_3^{2-}]}{[HCO_3^-]} = 4.68 \times 10^{-11} \text{ M} \qquad (3.C.16)$$

Water may come into contact with limestone, $CaCO_3$, in rocks or soil. If so, at equilibrium, the following solubility product would be satisfied:

$$CaCO_3(s) \Leftrightarrow Ca^{2+} + CO_3^{2-} \qquad K_{sp} = [Ca^{2+}][CO_3^{2-}] \qquad (3.C.17)$$

The solubility product is reported to have a value of $K_{sp} = 5 \times 10^{-9}$ M^2 or 8.7×10^{-9} M^2 (see Table 3.B.4).

The carbonate system contains three aqueous-phase species, $H_2CO_3^*$, HCO_3^-, and CO_3^{2-}. We now consider how the relative amounts of these species vary according to the pH of water. We define the total concentration of aqueous carbonate species by

$$C_{carbonate} = [H_2CO_3^*] + [HCO_3^-] + [CO_3^{2-}] \qquad (3.C.18)$$

The distribution of $C_{carbonate}$ among the three species can be expressed by the following relationships, derived from equation 3.C.18 combined with the equilibrium relationships 3.C.15 and 3.C.16. The left-hand portion of each expression defines α_i as the fraction that species i comprises of the total concentration $C_{carbonate}$:

$$\alpha_{H_2CO_3^*} = \frac{[H_2CO_3^*]}{C_{carbonate}} = \frac{[H^+]^2}{[H^+]^2 + K_1[H^+] + K_1K_2} \qquad (3.C.19)$$

$$\alpha_{HCO_3^-} = \frac{[HCO_3^-]}{C_{carbonate}} = \frac{K_1[H^+]}{[H^+]^2 + K_1[H^+] + K_1K_2} \qquad (3.C.20)$$

$$\alpha_{CO_3^{2-}} = \frac{[CO_3^{2-}]}{C_{carbonate}} = \frac{K_1K_2}{[H^+]^2 + K_1[H^+] + K_1K_2} \qquad (3.C.21)$$

Figure 3.C.2 plots the three fractions versus pH. Note that when pH = pK_1 = 6.35, $[H_2CO_3^*] = [HCO_3^-]$, and when pH = pK_2 = 10.33, $[HCO_3^-] = [CO_3^{2-}]$. This figure also generates the following rules of thumb about the partitioning of aqueous carbonate among the three species:

pH < 4.5	All aqueous carbonate is present as $H_2CO_3^*$.
$4.5 <$ pH < 8.3	$H_2CO_3^*$ and HCO_3^- dominate.
$8.3 <$ pH < 12.3	HCO_3^- and CO_3^{2-} dominate.
pH > 12.3	All aqueous carbonate is present as CO_3^{2-}.

Examples 3.C.4 and 3.C.5 illustrate how acid-base equilibrium problems involving the carbonate system can be formulated and solved. Exhibit 3.C.1 explores the role of the carbonate system in mitigating pH changes that would result from the addition of strong acid to water.

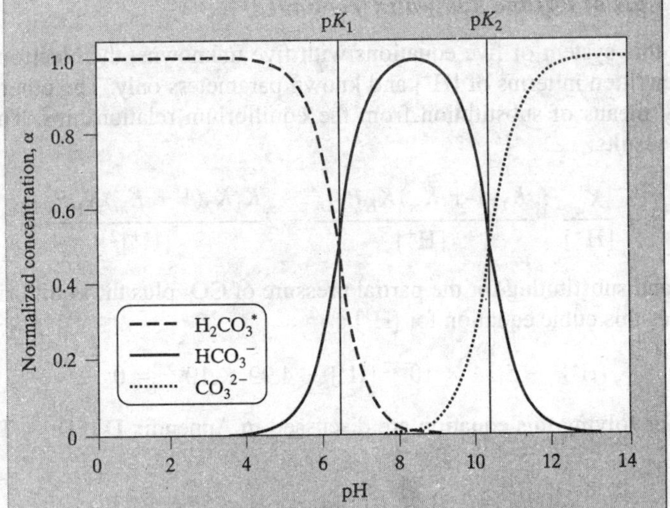

Figure 3.C.2 Distribution of aqueous carbonate species as a function of pH. The vertical axis gives α, the concentration of each species divided by the total concentration of aqueous carbonate ($C_{carbonate}$).

EXAMPLE 3.C.4 *pH of Pristine Rainwater*

What is the pH of rainwater in pristine environments? Assume that the temperature is 25 °C, the total air pressure is 1 atm, and the rainwater is in equilibrium with the atmosphere.

SOLUTION Five species are present in the aqueous phase:

$$H_2CO_3{}^*, HCO_3{}^-, CO_3{}^{2-}, H^+, \text{and } OH^-$$

Five equations are required. One, resulting from Henry's law, determines $[H_2CO_3{}^*]$. Two others derive from the acid dissociation equilibria of carbonic acid and bicarbonate. The fourth and fifth equations describe the dissociation of water and electroneutrality. The equations are

Henry's law: $$[H_2CO_3{}^*] = (1 + K_m)K_H P_{CO_2}$$

Carbonic acid dissociation: $$[HCO_3{}^-] = \frac{K_1[H_2CO_3{}^*]}{[H^+]}$$

Bicarbonate dissociation: $$[CO_3{}^{2-}] = \frac{K_2[HCO_3{}^-]}{[H^+]}$$

Water dissociation: $$[H^+][OH^-] = K_w$$

Electroneutrality: $$[H^+] = [OH^-] + [HCO_3{}^-] + 2[CO_3{}^{2-}]$$

Note the presence of the coefficient, 2, applied to the carbonate concentration in the electroneutrality relationship. It is needed because there are two charges on each carbonate ion.

EXAMPLE 3.C.4 ***pH of Pristine Rainwater (continued)***

To solve this system of five equations with five unknowns, the electroneutrality equation is rewritten in terms of $[H^+]$ and known parameters only. The other ions are eliminated by means of substitution from the equilibrium relationships. This is the equation that results:

$$[H^+] = \frac{K_w}{[H^+]} + \frac{K_1(1 + K_m)K_H P_{CO_2}}{[H^+]} + 2\frac{K_1 K_2(1 + K_m)K_H P_{CO_2}}{[H^+]^2}$$

Rearranging and substituting for the partial pressure of CO_2 plus the equilibrium constants produces this cubic equation for $[H^+]$:

$$[H^+]^3 - 5.34 \times 10^{-12}[H^+] - 4.99 \times 10^{-22} = 0$$

Approaches for solving this equation are discussed in Appendix D (§D.3). The solution is

$$[H^+] = 2.31 \times 10^{-6}\,M \Rightarrow pH = 5.64$$

This analysis shows that when pure water is exposed to the atmosphere, its pH decreases because of the uptake of CO_2 (and, also, the water is no longer pure!). Most rainwater is naturally acidic. Acid rain becomes a problem when strong acid gases are absorbed from air by fog, clouds, and rainwater. Prominent strong acid gases are H_2SO_4 (sulfuric acid) and HNO_3 (nitric acid). The ultimate cause of acid rain is the emission of SO_2 and NO, primarily from combustion. In the atmosphere, these species are converted into sulfuric and nitric acids by chemical reactions (see §7.A.3).

EXAMPLE 3.C.5 ***pH of Water in a Limestone Aquifer***

Initially pure water enters an aquifer lined with limestone, $CaCO_3(s)$. Assume that the water is isolated from the atmosphere at a temperature of 25 °C. Calculate the equilibrium pH.

SOLUTION First we identify the unknown species in the aqueous system. There are six in all: H^+, OH^-, Ca^{2+}, CO_3^{2-}, HCO_3^-, and $H_2CO_3^*$.

As usual, one equation must be specified for each unknown. There are four equilibrium relationships: one for water dissociation, two for acid dissociation (carbonic acid and bicarbonate), and one for the solubility product of limestone. A fifth equation is obtained from electroneutrality. The following argument, based on material balance and stoichiometry, produces the sixth equation.

Without the limestone dissolution reaction, the water would contain only H^+ and OH^-. For each calcium ion that enters the water by dissolution, one carbonate ion is also released. Likewise, for each calcium ion that precipitates from solution back to the solid phase, one carbonate ion is lost. In solution, the carbonate ions can undergo acid-base reactions to become bicarbonate ions or carbonic acid. But these transformations do not cause a net change in the total amount of aqueous carbonate species. Thus, material balance dictates that the sum of the concentrations of aqueous carbonate species must equal the calcium ion concentration.

EXAMPLE 3.C.5 *pH of Water in a Limestone Aquifer (continued)*

In summary, the six equations are

Carbonic acid dissociation: $$[H_2CO_3^*] = \frac{[H^+][HCO_3^-]}{K_1}$$

Bicarbonate dissociation: $$[HCO_3^-] = \frac{[H^+][CO_3^{2-}]}{K_2}$$

Water dissociation: $$[H^+][OH^-] = K_w$$

Electroneutrality: $$2[Ca^{2+}] + [H^+] = [OH^-] + [HCO_3^-] + 2[CO_3^{2-}]$$

Solubility product: $$[Ca^{2+}][CO_3^{2-}] = K_{sp}$$

Material conservation: $$[Ca^{2+}] = [H_2CO_3^*] + [HCO_3^-] + [CO_3^{2-}]$$

Note that the Henry's law relationship used in Example 3.C.4 does not apply here because there is no gas phase.

The solution to this problem follows a procedure similar to but more complex than that used in Example 3.C.4. We again rewrite the electroneutrality relationship in terms of $[H^+]$ and known parameters. The first step is to replace $[OH^-]$ using water dissociation. Then $[Ca^{2+}]$ and $[HCO_3^-]$ are eliminated using the solubility product and bicarbonate dissociation, respectively. The intermediate equation contains $[H^+]$ and $[CO_3^{2-}]$:

$$2\frac{K_{sp}}{[CO_3^{2-}]} + [H^+] = \frac{K_w}{[H^+]} + \frac{[H^+][CO_3^{2-}]}{K_2} + 2[CO_3^{2-}] \quad (*)$$

To eliminate the carbonate ion from this equation, we use the material-conservation expression plus the equilibrium relationships:

$$\frac{K_{sp}}{[CO_3^{2-}]} = \left(\frac{[H^+]^2}{K_1K_2} + \frac{[H^+]}{K_2} + 1\right)[CO_3^{2-}]$$

So

$$[CO_3^{2-}] = K_{sp}^{1/2}\left(\frac{[H^+]^2}{K_1K_2} + \frac{[H^+]}{K_2} + 1\right)^{-1/2}$$

This equation is substituted into (*) to obtain an equation that can be solved for $[H^+]$. See Appendix D (§D.3) for details on the solution method. The answer is $[H^+] = 1.1 \times 10^{-10}$ M, or pH = 9.96.

EXHIBIT 3.C.1 *Alkalinity and pH Buffering*

This discussion has two purposes: (a) to illustrate the significance of alkalinity in buffering water against pH changes, and (b) to provide a sense of the relationship between alkalinity defined by titration and alkalinity defined by the concentrations of key species.

Consider two samples of water contained in closed systems without any gas phase. Sample A consists of pure water to which we add 1 mM of sodium bicarbonate, $NaHCO_3$, which has a very high solubility product and so dissolves completely in water. As we will see below, the pH of this sample rises somewhat above 7.0. Sample B consists of pure water to which we add a small amount of a strong base, NaOH, so that its pH matches that of the first sample. Thus, the samples

EXHIBIT 3.C.1 *Alkalinity and pH Buffering (continued)*

have the same initial pH but very different alkalinities. One means of determining alkalinity for waters containing only carbonate species is through the relationship presented previously (equation 2.C.4):

$$A\ (\text{eq/L}) = [OH^-] + [HCO_3^-] + 2[CO_3^{2-}] - [H^+]$$

A second means of determining alkalinity is through titration. We titrate the solutions by adding strong acid (e.g., HCl) until the pH decreases to 4.5. The alkalinity for each sample is determined as the moles of strong acid protons added per liter of solution. (The reason for choosing 4.5 as the stopping point for titration should be clear by inspection of Figure 3.C.2. The pH of 4.5 is the highest value at which essentially all of the carbonate species are in the form of undissociated carbonic acid. The buffering capacity of carbonate species to resist pH changes is exhausted when the pH reaches 4.5.)

Sample A

Sample A starts with 1 mM of fully dissociated $NaHCO_3$. We must determine the pH of the solution and the concentrations of all of the ions, prior to titration. The concentrations of six species must be evaluated: $[Na^+]$, $[H^+]$, $[OH^-]$, $[H_2CO_3{}^*]$, $[HCO_3^-]$, and $[CO_3^{2-}]$. Therefore, we must have six equations, as written below:

(1) Electroneutrality:

$$[Na^+] + [H^+] = [OH^-] + [HCO_3^-] + 2\,[CO_3^{2-}]$$

(2) Dissociation of water:

$$K_w = [H^+][OH^-] = 10^{-14}\ M^2$$

(3) Sodium conservation:

$$[Na^+] = 10^{-3}\ M$$

(4) Carbonic acid dissociation:

$$K_1 = \frac{[H^+][HCO_3^-]}{[H_2CO_3{}^*]} = 4.47 \times 10^{-7}\ M$$

(5) Bicarbonate dissociation:

$$K_2 = \frac{[H^+][CO_3^{2-}]}{[HCO_3^-]} = 4.68 \times 10^{-11}\ M$$

(6) Carbonate species conservation:

$$10^{-3}\ M = [H_2CO_3{}^*] + [HCO_3^-] + [CO_3^{2-}]$$

The system of equations is solved with the same procedure used for other acid-base problems. The electroneutrality equation is rewritten in terms of $[H^+]$ and known parameters. That equation is then solved by a search procedure (see Appendix D). The results are given below:

$$[H^+] = 5.0 \times 10^{-9}\ M \Rightarrow pH = 8.3$$

$$[HCO_3^-] = 9.8 \times 10^{-4}\ M$$

$$[CO_3^{2-}] = 9.1 \times 10^{-6}\ M$$

$$[OH^-] = 2.0 \times 10^{-6}\ M$$

The alkalinity computed according to equation 2.C.4 is the same as the amount of bicarbonate initially added to the solution:

$$A = [HCO_3^-] + 2[CO_3^{2-}] + [OH^-] - [H^+] = 1.00 \times 10^{-3}\ \text{eq/L}$$

How much strong acid (HCl) must be added to this solution to reduce the pH to 4.5? We solve a new equilibrium problem in which the pH is 4.5, and a seventh species, the chloride ion, is added to the system. We require seven equations. The electroneutrality equation is modified to add chloride, as shown below. Equations (2)–(6) are unchanged. The seventh equation specifies that the final pH is 4.5.

(1) Electroneutrality:

$$[Na^+] + [H^+] = [OH^-] + [HCO_3^-] + 2[CO_3^{2-}] + [Cl^-]$$

(7) pH specification:

$$[H^+] = 3.2 \times 10^{-5}\ M$$

Given the pH, the other equations are easily solved to yield the following results:

$$[OH^-] = K_w/[H^+] = 3.2 \times 10^{-10}\ M$$

$[Na^+] = 10^{-3}$ M (unchanged by addition of acid)

$$[H_2CO_3{}^*] = \frac{10^{-3}\ M}{1 + K_1/[H^+] + (K_1K_2)/[H^+]^2} = 9.86 \times 10^{-4}\ M$$

$$[HCO_3^-] = \frac{K_1[H_2CO_3{}^*]}{[H^+]} = 1.39 \times 10^{-5}\ M$$

EXHIBIT 3.C.1 *Alkalinity and pH Buffering (continued)*

$$[CO_3^{2-}] = \frac{K_2[HCO_3^-]}{[H^+]} = 2.1 \times 10^{-11} \text{ M}$$

$$[Cl^-] = [Na^+] + [H^+] - [HCO_3^-] - [OH^-] - 2[CO_3^{2-}]$$
$$= 1.02 \times 10^{-3} \text{ M}$$

The amount of HCl added is the same as the final concentration of chloride ion. So we needed to add 1.02×10^{-3} eq/L of strong acid. Therefore, according to the titration definition of alkalinity, $A = 1.02 \times 10^{-3}$ eq/L, which differs only slightly (2 percent) from the concentration-based determination.

Sample B

We begin with pure water and want to adjust the pH to 8.3 by the addition of NaOH so that the pH matches that of sample A. What, then, is the composition of the water at pH = 8.3?

$$[H^+] = 10^{-8.3} = 5 \times 10^{-9} \text{ M}$$

$$[OH^-] = K_w/[H^+] = 2 \times 10^{-6} \text{ M}$$

The sodium ion concentration is determined by electroneutrality as

$$[Na^+] = [OH^-] - [H^+] = 2 \times 10^{-6} \text{ M}$$

The alkalinity, as evaluated from the concentration-based equation, is very small:

$$A = [HCO_3^-] + 2[CO_3^{2-}] + [OH^-] - [H^+]$$
$$= 2.0 \times 10^{-6} \text{ eq/L}$$

We now consider how much HCl must be added to reduce the pH to 4.5. The final solution contains four species whose concentrations must be determined: $[Na^+]$, $[H^+]$, $[OH^-]$, $[Cl^-]$. The four relationships are obtained from (a) the target pH, (b) the dissociation of water, (c) the sodium concentration (unchanged by addition of acid), and (d) electroneutrality. Applying these four conditions yields the results

$$[H^+] = 3.2 \times 10^{-5} \text{ M}$$

$$[OH^-] = 3.2 \times 10^{-10} \text{ M}$$

$$[Na^+] = 2 \times 10^{-6} \text{ M}$$

$$[Cl^-] = [H^+] + [Na^+] - [OH^-] = 3.4 \times 10^{-5} \text{ M}$$

Again, the amount of HCl added is the same as the concentration of chloride ion in solution: 34×10^{-6} eq/L. Therefore, by the titration definition, $A = 34 \times 10^{-6}$ eq/L, a different value from that obtained by the concentration-based equation.

Summarizing, we see that sample A, with sodium bicarbonate, has a much higher alkalinity than sample B. This means that sample A is much more effectively buffered against pH changes that result from the addition of strong acid. Based on the titration procedure, the alkalinity of sample A is 30 times greater than the alkalinity of sample B, meaning that 30 times as much acid had to be added to sample A to reduce the pH to 4.5 as was added to sample B. The calculation of alkalinity based on the concentrations of the key constituents worked well for sample A (agreement within 2 percent) but worked only qualitatively for sample B (informing us that alkalinity is low, but not accurately so). In general, the use of equation 2.C.4 for determining alkalinity is accurate only to within hundredths of a meq/L. This is not a practical problem, because the alkalinity of natural and treated waters is typically on the order of meq/L, and the uncertainty in routine environmental measurements is seldom less than 10 percent.

3.D OXIDATION-REDUCTION REACTIONS

In §3.C we considered acid-base reactions, which are characterized by the transfer of a proton from one species to another. Another major class of reactions involves the transfer of *electrons* among species. These are called oxidation-reduction reactions, or *redox* reactions for short.

Redox reactions are encountered in almost every environmental engineering context. Redox reactions are much slower than acid-base reactions, making the subject of kinetics more important and that of chemical equilibrium less important relative to acid-base reactions.

Table 3.D.1 Rules for Determining the Oxidation State of a Chemical Element

1. The oxidation state of an uncharged, uncombined element, whether in atomic or molecular form (as in C, N_2, or O_2), is zero. The oxidation state of an atomic ion is equal to its charge.
2. The oxidation state of a compound is equal to the sum of the oxidation states of the respective atoms within the compound.
3. The oxidation state of a compound is equal to the net electrical charge on the compound. For neutral compounds, the oxidation state is zero. For ions, the oxidation state is equal to the charge of the ion, including the sign.
4. The oxidation state of oxygen in most compounds is –2.[a]
5. The oxidation state of hydrogen in most compounds is +1.[b]

[a]The most common exception occurs in the case of peroxides, such as H_2O_2, in which an oxygen-oxygen bond exists. In this case the oxidation state of oxygen is –1.
[b]When hydrogen is bonded to a hydride (e.g., Li, Na, or K), the oxidation state of H is –1.
Source: Seinfeld and Pandis, 1998; Sienko and Plane, 1961.

The slowness of redox processes poses both challenges and opportunities for environmental engineers. Some waste products can be rendered harmless by subjecting them to oxidation or reduction processes prior to releasing them to the environment. On the other hand, it is generally more challenging to engineer a system to carry out desired oxidation-reduction reactions than it is to change the pH of a solution. Specific examples in which oxidation-reduction reactions are used in engineered treatment systems include catalytic converters for treating automobile exhaust, incineration of hazardous wastes, and processing of wastewater by activated sludge to reduce the concentrations of organic compounds.

3.D.1 Oxidation State

The ability to determine the oxidation state of a chemical element is a useful skill for the study of redox reactions. It is an important step in identifying whether a particular transformation process is a redox reaction.

The oxidation state of an element is the net electrical charge that it possesses (if isolated) or appears to possess (if combined in a compound). The oxidation state of a chemical element is determined by following a set of rules (see Table 3.D.1). Example 3.D.1 demonstrates the use of these rules.

The oxidation state is often denoted using Roman numerals in parentheses following the chemical element. For example, Cr(III) designates chromium in the oxidation state +3, and Cr(VI) represents chromium in the oxidation state +6. This designation does not distinguish whether the elements are free ions or combined in chemical compounds.

EXAMPLE 3.D.1 ***Determining the Oxidation States of Elements in Compounds***

(a) What is the oxidation state of S in sulfuric acid (H_2SO_4)?

(b) What is the oxidation state of C in the carbonate ion (CO_3^{2-})?

SOLUTION

(a) Sulfuric acid is a neutral compound, so its oxidation state, which equals the sum of the oxidation states of the elements, must be zero. Hydrogen has an oxidation state of

EXAMPLE 3.D.1 ***Determining the Oxidation States of Elements in Compounds (continued)***

+1, and oxygen has an oxidation state of –2. Therefore, the oxidation state of sulfur, X_S, must satisfy the algebraic equation

$$2(+1) + X_S + 4(-2) = 0 \quad \Rightarrow \quad X_S = +6$$

The sulfur in sulfuric acid is in oxidation state +6. Sulfur has an oxidation state of +6 whenever it is in the form of sulfate, for example, in the sulfate ion, SO_4^{2-}.

(b) The oxidation state of the carbonate ion is –2. The three oxygen atoms each have an oxidation state of –2. The oxidation state of carbon, X_C, must then satisfy this equation:

$$X_C + 3(-2) = -2 \quad \Rightarrow \quad X_C = +4$$

Note that the oxidation state of C in CO_2, $H_2CO_3^*$, and HCO_3^- is also +4. Carbon dioxide is converted to carbonate ion by dissolving into water and losing two protons. There are no oxidation-reduction reactions involved, so the oxidation state of C remains unchanged.

The nomenclature for redox reactions can be confusing, because the direction of change in the oxidation state is opposite to the direction in which the electrons are transferred. The species that is *oxidized* has its oxidation state *increased* by *losing* one or more electrons from its outer shell. The species that is *reduced* experiences a *decrease* in its oxidation state because it *gains* electron(s). Because free electrons do not exist in ordinary matter, the algebraic sum of the changes in oxidation states in any balanced chemical reaction must be zero. In other words, the oxidation of one species can be accomplished only by a reduction of equal magnitude in another species. Sometimes, however, half-reactions are written in which electrons are consumed or produced. These half-reactions are partial descriptions of an overall redox process and are used to help balance the overall reaction or to evaluate the change in chemical energy.

The existence of oxidation-reduction reactions implies that at least some chemical elements can exist in more than one oxidation state. Examples of environmentally relevant oxidation states for some elements are given in Table 3.D.2.

3.D.2 Corrosion

Corrosion is a process that involves redox reactions and is of great practical concern for the maintenance of society's constructed infrastructure. Corrosion must be controlled to limit damage to metal objects such as ships, bridges, and steel reinforcement rods in concrete. In water and wastewater systems, corrosion must be controlled in pipes and storage tanks.

Undisturbed metals have very low solubility in water. Corrosion occurs when the metal is oxidized, typically into a hydroxide, which has much higher solubility. There are several reasons corrosion is undesirable in water and wastewater systems. The dissolution of metals such as iron and manganese into water supplies can cause taste and color problems. Chemical elements such as lead from old solder joints may leach into potable water and pose a direct health risk. Corrosion also weakens metal pipes, joints, and tanks in water and wastewater systems.

Table 3.D.2 Oxidation States of Some Chemical Elements

Element	Oxidation state	Species	Formula
Sulfur	−2	Hydrogen sulfide	H_2S
	0	Elemental sulfur	S
	+4	Sulfur dioxide	SO_2
	+6	Sulfate ion	SO_4^{2-}
Carbon	−4	Methane	CH_4
	0	Soot, graphite	C
	+2	Carbon monoxide	CO
	+4	Carbon dioxide	CO_2
Nitrogen	−3	Ammonia	NH_3
	0	Nitrogen gas	N_2
	+2	Nitric oxide	NO
	+3	Nitrite ion	NO_2^-
	+4	Nitrogen dioxide	NO_2
	+5	Nitrate ion	NO_3^-
Oxygen	−2	Almost all compounds	
	−1	Hydrogen peroxide	H_2O_2
	0	Oxygen gas	O_2
Hydrogen	0	Hydrogen gas	H_2
	+1	Hydrogen ion	H^+
Chlorine	−1	Chloride ion	Cl^-
	0	Chlorine gas	Cl_2
	+1	Hypochlorous acid	HOCl
	+7	Perchloric acid	$HClO_4$

Corrosion in a Galvanic Cell

Corrosion can be caused by several distinct mechanisms. One type of corrosion can occur when dissimilar conductive materials are connected if both are in contact with water. For example, when copper and iron water pipes are connected, a galvanic cell is established, as depicted in Figure 3.D.1.

A galvanic cell comprises a few key components. There must be two dissimilar metals in contact such that electrical current can flow from one metal to the other. Metals differ in their affinity for electrons. If two different metals are brought into contact, some electrons will flow from the metal with the lower electron affinity to the metal with the higher affinity until a small, counterbalancing electrical potential (on the order of a few volts or less) is established between them. The other key component of a galvanic cell is an electrolytic solution in contact with the two metals that allows ions to be transported from one metal to the other. Water is an electrolytic solution, since, at a minimum, it contains H^+ and OH^- ions. Other ionic impurities enhance the current-carrying capacity of the solution.

For corrosion to proceed in a galvanic cell, there must be an opportunity for oxidation and reduction reactions to occur at the interface between the metals and the solution. When two dissimilar metals are in contact, the more corrodible one (the one with the lower electron affinity) serves as the anode and the less corrodible one as the cathode (see Table 3.D.3). In the common case of contact between copper and iron pipes, copper is the cathode and iron is the anode. Cations, including H^+, migrate toward the cathode. At the cathode, hydrogen is reduced from oxidation state +1 to 0 by

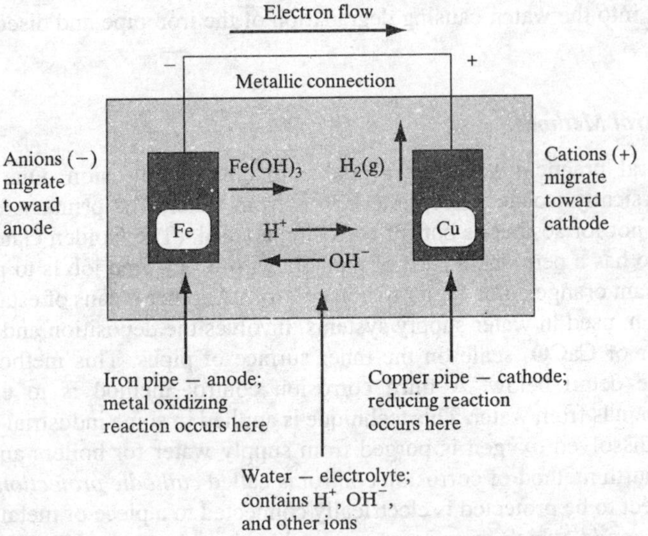

Figure 3.D.1 A galvanic cell may occur in a water supply system when two dissimilar metals, such as iron and copper pipe, are connected. In this case, the iron will corrode.

combining with electrons. Anions, such as OH^-, migrate toward the anode. At the anode, iron is oxidized from 0 to +3, forming iron hydroxide, which can then dissolve into the water. The electrons liberated by the oxidation reaction flow through the metal connection from the iron to the copper, completing the circuit.

The chemical reactions in this galvanic cell are summarized below:

Dissociation of water: $H_2O \rightarrow OH^- + H^+$ (3.D.1)

Oxidation of iron (anode): $Fe \rightarrow Fe^{3+} + 3e^-$ (3.D.2)

Reduction of hydrogen (cathode): $2\,H^+ + 2e^- \rightarrow H_2(g)$ (3.D.3)

Formation of iron hydroxide (rust): $Fe^{3+} + 3\,OH^- \rightarrow Fe(OH)_3$ (3.D.4)

The net effect is obtained by summing the reactions with appropriate stoichiometric factors so that the electrons plus other common terms cancel. If we take reactions 3.D.1–3.D.4 in the relative proportions 6:2:3:2, add them, and cancel the common species from the two sides of the reaction, the overall effect is

$$2\,Fe(s) + 6\,H_2O \rightarrow 2\,Fe(OH)_3(s) + 3\,H_2(g) \quad (3.D.5)$$

So the overall process is a redox reaction in which iron is oxidized from 0 to +3 and hydrogen is reduced from +1 (in water) to 0 (in H_2). Iron and hydrogen participate in this reaction in the molar proportion 1:3. The rust, $Fe(OH)_3$, is moderately soluble and

Table 3.D.3 Metals Listed in Order from Least to Most Corrodible

1. Silver	5. Tin
2. Stainless steel	6. Lead
3. Copper	7. Iron
4. Brass	8. Zinc

so will dissolve into the water, causing degradation of the iron pipe and discoloration of the water.

Corrosion Control Methods

There are several practical measures available to control corrosion. One general means is to physically isolate susceptible metals from water. The primary reason to paint bridges is not for aesthetics but for corrosion control. (The Golden Gate Bridge in San Francisco has a permanent staff of painters whose full-time job is to paint the bridge that brilliant orange color for corrosion control.) Another means of establishing physical isolation, used in water supply systems, involves the deposition and maintenance of a layer of $CaCO_3$ scale on the inner surface of pipes. This method is described in more detail below. A third corrosion control method is to eliminate corrosive compounds from water. This technique is applied in some industrial applications in which dissolved oxygen is purged from supply water for boilers and steam generators. A fourth method of corrosion control is called *cathodic protection*. In this method, the object to be protected is electrically connected to a piece of metal, known as a *sacrificial anode*, which corrodes more readily than the protected metals. The presence of the sacrificial anode allows the other metals in the system to serve as cathodes. Any corrosive oxidation will occur at the anode, which will be gradually consumed. This idea is employed in domestic hot-water heaters, which are internally equipped with a zinc rod that is electrically connected to the iron tank. The zinc gradually corrodes, releasing zinc ions into the water and protecting the iron tank from corrosion. Eventually, the zinc will be completely consumed, and shortly thereafter one can expect the water heater to fail as corrosion begins to attack the tank itself. The principle of cathodic protection can also be used by means of applying an electrical potential between the metals to be protected and a sacrificial anode. For example, in the system depicted in Figure 3.D.1, if a battery is inserted within the metallic connection such that the electron flow is reversed, then the copper pipe will become the anode and the copper will corrode. Cathodic protection is used to protect steel pipelines and ships from corrosion.

Scale Formation and Corrosion Control in Water Supply Systems *Scale* refers to solids that are formed by the precipitation of dissolved ions in water. One form of scale commonly encountered in municipal water supply systems is $CaCO_3$. Water that is in contact with solid $CaCO_3$ will accept dissolved calcium and carbonate ions until the solubility product of the solid is satisfied, that is, $[Ca^{2+}][CO_3^{2-}] = K_{sp}$. If this water is then removed from contact with the solid, pH or temperature changes in the water can shift the equilibrium so that the solubility product is no longer satisfied. For example, if the pH of the water decreases, some of the carbonate will be converted to bicarbonate (see Figure 3.C.2), and $[Ca^{2+}][CO_3^{2-}] < K_{sp}$. On the other hand, if the pH of the water increases, bicarbonate will be converted to carbonate, and $[Ca^{2+}][CO_3^{2-}] > K_{sp}$. To restore equilibrium, solid calcium carbonate will precipitate on surfaces, such as the inner walls of pipes.

Excessive scale formation causes significant problems. In devices such as water heaters or boilers, scale increases the heat transfer resistance of heat exchangers and reduces their thermal efficiency. In pipes, the formation of scale increases the frictional resistance to water flow. In extreme cases, pipes can become plugged.

In principle, scale formation can be eliminated by maintaining conditions in which $[Ca^{2+}][CO_3^{2-}] \ll K_{sp}$. However, the problems of scale formation and corrosion tend to be inversely related. That is, conditions that avoid scale formation tend to be corrosive and vice versa. In practice, one seeks to control treated water to make it slightly scale forming. The goal is to deposit a fine layer of $CaCO_3(s)$ on the inner metal surfaces and maintain this layer to prevent corrosion.

3.D.3 Combustion

Combustion is another system of redox reactions that has great significance in environmental engineering. In typical combustion reactions, fuel is rapidly oxidized, transforming chemical potential energy into thermal energy. Combustion is used as a treatment technology for oxidizing solid, liquid, and gaseous waste products. Combustion is also the single most important source of air pollution. Oxides of sulfur, nitrogen, and carbon are all emitted more from combustion than from any other anthropogenic process. Combustion is a major source of airborne particulate matter and toxic air pollutants.

The enormous air pollution problems associated with combustion can be traced to three causes:

1. Chemical elements other than hydrogen and carbon (and oxygen) may be present in the fuel.
2. Combustion may occur under nonideal conditions, so that the exhaust gases contain incompletely oxidized elements.
3. The high temperature of combustion can lead to partial oxidation of N_2 from air.

Complete combustion of a pure hydrocarbon fuel using O_2 as the oxidizer would produce only carbon dioxide and water vapor as waste products. Increased carbon dioxide emissions would pose the only major environmental concern: global climate change. However, most fuels contain some undesired chemical elements that contribute to air pollutant emissions. Gaseous fuels tend to be cleanest, and liquid fuels commonly have lower levels of impurities than solid fuels.

Combustion as a source of air pollution is discussed extensively in Chapter 7. The application of incineration to treat wastes is considered in Chapter 8. In this section, we discuss the fundamentals of combustion as a redox process.

Characteristics of Fuels and Air

In a generalized combustion redox reaction, carbon in fuels serves as the dominant reduced reactant while the most common oxidized reactant is oxygen (O_2). In most combustion processes, air is the source of oxygen. During combustion, the reduced carbon in the fuel becomes oxidized to form CO_2, while the oxidation state of oxygen is reduced from 0 to –2.

Gaseous fuels are the simplest and their combustion is the best understood, though still complex. The simplest fuel is hydrogen (H_2), which burns according to the overall reaction

$$H_2 + \tfrac{1}{2}O_2 \rightarrow H_2O \tag{3.D.6}$$

In this case, hydrogen is oxidized from 0 to +1 and oxygen is reduced from 0 to –2. This fuel has enormous environmental appeal, since very few air pollutants are generated by hydrogen combustion. However, there are no large natural sources of hydrogen gas. Commonly used gaseous fuels are extracted and processed from petrochemical deposits. A widely burned and relatively clean fuel is natural gas, which is a mixture consisting mostly of low-molecular-weight, straight-chain alkanes: methane (CH_4), ethane (C_2H_6), and propane (C_3H_8). Propane is also widely used as a fuel in small- and medium-scale combustion applications. It is stored in tanks under high pressure as a liquid, but since its vapor pressure is greater than atmospheric pressure, it is easily volatilized for combustion.

The molecular composition of condensed-phase fuels is not easily determined. For both liquids and solids, empirical tests are used to determine the mass fractions of carbon, hydrogen, oxygen, sulfur, and nitrogen in the fuel, plus the mass fraction of noncombustible material, *ash*. Another test is used to characterize the *heating value* of the fuel. The heating value is a measure of the amount of chemical energy that is converted to thermal energy by fuel combustion.

Table 3.D.4 presents the effective elemental compositions and heating values for some common fuels. The elemental compositions and heating values of the solid fuels are only typical values and can vary significantly. To determine the mass fraction of a given element in the fuel from these data, determine the mass of the element per mole of carbon, divide by the effective molecular weight of the fuel, and then multiply by the percentage of fuel that is not ash. So, for example, the mass fraction of nitrogen in the lignite sample is 0.016 ([mol N]/[mol C]) × (14 g)/(mol N) ÷ 16.4 (g/mol C) × 0.751 (non-ash fraction) = 0.010 (1.0 percent).

In the analysis of combustion processes, air is usually represented as a two-component mixture. The active oxidizing agent, oxygen, is taken as one component, present at a mole fraction of 0.209. The second component is taken to be the sum of all the inert species, present at a mole fraction of 0.791, and represented as nitrogen, the dominant constituent. Thus, air is treated as having an effective molecular weight of 28.8 g mol^{-1}, with a molar ratio of nitrogen to oxygen of 3.78:1.

Table 3.D.4 Properties of Fuels

Fuel	Effective chemical formula[a]	Effective molecular weight (g/mol C)	Ash	Heating value ($MJ\ kg^{-1}$)[b]
Hydrogen	H_2	—	—	121
Methane	CH_4	16.0	—	50.2
Propane	C_3H_8	14.7	—	46.3
n-Octane	C_8H_{18}	14.3	—	44.7
Gasoline[c]	$CH_{1.84}O_{0.018}$	14.1	—	42.7
Kerosene	$CH_{1.8}$	13.8	Trace	43.5
Wood, Douglas fir	$CH_{1.45}O_{0.58}N_{0.002}$	22.8	0.8%	19.6
Lignite	$CH_{0.96}O_{0.18}N_{0.016}S_{0.012}$	16.4	24.9%	21.6
Subbituminous coal	$CH_{1.00}O_{0.18}N_{0.008}S_{0.002}$	16.1	3.1%	28.9
Bituminous coal	$CH_{0.79}O_{0.073}N_{0.017}S_{0.008}$	14.5	8.7%	30.5

[a]Dry basis for solid fuels; excludes ash.
[b]Dry basis for solid fuels; lower heating value.
[c]California phase 2 reformulated gasoline, 1996.
Source: Flagan and Seinfeld, 1988; de Nevers, 1995; Kirchstetter et al., 1999.

Combustion Stoichiometry

We may represent a pure hydrocarbon fuel by the chemical formula C_nH_m, where n and m represent stoichiometric coefficients. The complete combustion of a pure hydrocarbon fuel causes the conversion of all carbon to CO_2 and all hydrogen to H_2O.

Fuel and air may be mixed in different relative amounts. If exactly enough air is provided to fully oxidize the fuel without any excess oxygen, then the complete combustion of a pure hydrocarbon fuel using air as the oxidizer can be represented by this overall reaction:

$$C_nH_m + \left(n + \frac{m}{4}\right)(O_2 + 3.78\ N_2) \longrightarrow n\ CO_2 + \frac{m}{2}\ H_2O + 3.78\left(n + \frac{m}{4}\right) N_2 \quad (3.D.7)$$

This situation is called burning with *stoichiometric air.* The requirement of $[n + (m/4)]$ moles of O_2 derives from a balance on oxygen between the reactant and product sides of the expression.

A common specification for combustion conditions is the *fuel-to-air mass ratio.* The stoichiometric fuel-to-air mass ratio,. $(m_f/m_a)_s$, for combustion of a pure hydrocarbon fuel is obtained as the ratio of the mass of fuel to the mass of air on the reactant side of equation 3.D.7:

$$(m_f/m_a)_s = \frac{12n + m}{28.8(n + m/4) \times 4.78} = 0.0291\left(\frac{12n + m}{4n + m}\right) \quad (3.D.8)$$

Here, we have used 28.8 g/mol as the effective molecular weight of air. Combustion of one mole of a C_nH_m fuel requires $[n + (m/4)] \times 4.78$ moles of air. So, for example, the fuel-to-air mass ratio for stoichiometric combustion of methane (CH_4; $n = 1$, $m = 4$) is 0.058.

Often, combustion does not occur with stoichiometric air. The relative amounts of fuel and air for nonstoichiometric combustion may be expressed in terms of the *equivalence ratio*, ϕ, defined by the expression

$$\phi = \frac{(m_f/m_a)}{(m_f/m_a)_s} \quad (3.D.9)$$

where (m_f/m_a) is the fuel-to-air mass ratio as burned. When fuel is burned with excess air relative to the stoichiometric requirement, $\phi < 1$; this condition is called *fuel lean*. When there is insufficient air for complete combustion, $\phi > 1$, and the condition is termed *fuel rich*.

Under fuel-lean conditions, the exhaust gas will contain some unused oxygen. As long as the equivalence ratio, ϕ, is not too small (i.e., not less than about 0.6), complete combustion can occur. The overall reaction for complete combustion of a pure hydrocarbon fuel under fuel-lean conditions can be expressed as follows:

$$C_nH_m + \frac{1}{\phi}\left(n + \frac{m}{4}\right)(O_2 + 3.78\ N_2) \longrightarrow n\ CO_2 + \frac{m}{2}\ H_2O + \frac{3.78}{\phi}\left(n + \frac{m}{4}\right) N_2 + a\ O_2 \quad (3.D.10)$$

where

$$a = \left(\frac{1}{\phi} - 1\right)\left(n + \frac{m}{4}\right) \quad (3.D.11)$$

Example 3.D.2 illustrates the application of fuel-lean combustion stoichiometry. Fuel-rich combustion is generally undesirable from an air pollution or energy-efficiency perspective because there is inadequate oxygen for complete oxidation of the fuel. If a pure hydrocarbon fuel is burned under fuel-rich conditions, the exhaust will contain significant quantities of H_2, CO, and possibly other incompletely oxidized carbonaceous compounds, such as soot (elemental carbon), benzene, formaldehyde, and polycyclic aromatic hydrocarbons. Although stoichiometry is sufficient for estimating the composition of exhaust gas during complete combustion, when combustion is incomplete, additional information is required. Chemical equilibrium and chemical kinetics can provide some insight, but combustion under fuel-rich conditions is sufficiently complex that exhaust gas composition can be reliably determined only by experiment.

The presence of elements other than carbon and hydrogen in the fuel complicates analysis based on stoichiometry. As an approximation, it is reasonable to assume that any sulfur in the fuel is emitted as sulfur dioxide. Depending on the combustion conditions, most nitrogen in the fuel will be emitted as either N_2 or NO. Ash is inert to combustion and may either be removed by a control technique, be deposited at the bottom of the combustion chamber (bottom ash or slag), or be released into the atmosphere as airborne particulate matter (fly ash). Any halogens in the fuel (Cl, F, Br, or I) tend to be emitted as acids, such as HCl.

EXAMPLE 3.D.2 ***Stoichiometry and Combustion***

Lignite is burned in air at an equivalence ratio of 0.95. Assume (a) that the lignite has the chemical composition given in Table 3.D.4, (b) that the fuel nitrogen is all emitted as NO, and (c) that combustion is complete. Determine the exhaust gas composition.

SOLUTION First, the relative amounts of air and fuel must be determined. Under stoichiometric conditions, the following overall reaction applies:

$$CH_{0.96}O_{0.18}N_{0.016}S_{0.012} + x\,(O_2 + 3.78\ N_2)$$
$$\rightarrow CO_2 + 0.48\ H_2O + 0.016\ NO + 0.012\ SO_2 + 3.78x\ N_2$$

The coefficient x is determined from a molar balance on oxygen:

$$0.18 + 2x = 2 + 0.48 + 0.016 + 0.024 \quad \Rightarrow \quad x = 1.17$$

For the specified equivalence ratio, the air is increased by a factor $(1/\phi)$. The overall reaction becomes

$$CH_{0.96}O_{0.18}N_{0.016}S_{0.012} + \frac{1.17}{0.95}(O_2 + 3.78\ N_2)$$
$$\rightarrow CO_2 + 0.48\ H_2O + 0.016\ NO + 0.012\ SO_2 + 4.66\ N_2 + 0.06\ O_2$$

Thus, for each mole of carbon combusted, 6.23 moles of exhaust gas are produced (the sum of the stoichiometric coefficients on the product side of the reaction), with the following mole fractions:

$$Y_{N_2} = \frac{4.66}{6.23} = 0.748$$

$$Y_{CO_2} = 0.161$$

$$Y_{H_2O} = 0.077$$

EXAMPLE 3.D.2 ***Stoichiometry and Combustion (continued)***

$$Y_{O_2} = 9600 \text{ ppm}$$

$$Y_{NO} = 2600 \text{ ppm}$$

$$Y_{SO_2} = 1900 \text{ ppm}$$

3.D.4 Atmospheric Oxidation Processes

Redox reactions in the atmosphere influence the concentrations, fates, and effects of many air pollutants. This section presents some of the main features of atmospheric redox reactions.

The atmosphere is an oxidizing environment since it contains a plentiful supply of oxygen (O_2) in oxidation state 0. Carbon-, nitrogen-, and sulfur-containing compounds are emitted into the atmosphere in reduced or partially oxidized states by both natural and anthropogenic processes. The general trend is for these elements to become more oxidized with time. The reduced C in hydrocarbons is steadily oxidized to form oxygenated organics such as aldehydes, then carbon monoxide, and ultimately CO_2. The nitrogen emitted in the form of NO from combustion processes is oxidized to NO_2 then to nitric acid (HNO_3). Likewise, the sulfur emitted as SO_2 from combustion is oxidized to sulfate (SO_4^{2-}) in the atmosphere. Nitrogen and sulfur emitted in reduced states in compounds such as ammonia (NH_3) and hydrogen sulfide (H_2S) are also oxidized in the atmosphere. Sometimes the oxidized products are more harmful as air pollutants than the precursors. For example, alkanes, which are relatively inert, are oxidized in the atmosphere to form carbonyls, such as formaldehyde and acetone, which are mucous membrane irritants. Nitric oxide is relatively nontoxic, but nitrogen dioxide is a respiratory irritant and nitric acid contributes to acid deposition.

Chemical equilibrium does not adequately describe gas-phase atmospheric chemistry because of kinetic limitations. For example, the equilibrium concentration of carbon monoxide, given mole fractions of O_2 and CO_2 of 0.209 and 350 ppm, respectively, would be zero at ambient temperature. Yet significant CO exists in the atmosphere because emissions from incomplete combustion occur continuously and the rate of oxidation of CO to CO_2 is slow.

On the other hand, equilibrium relationships can be used to estimate the partitioning of a species between the gas phase and liquid water droplets and to describe the acid-base chemistry in fog, clouds, and rain drops. These cases do not involve redox reactions.

Much of the energy needed to initiate atmospheric reactions originates with sunlight. Photolysis reactions generate radicals, which serve as catalysts for atmospheric oxidation processes. Thus, photolysis and radicals are of central importance in air pollution chemistry.

Photolytic Reactions

To initiate a redox reaction, energy must be supplied to break the chemical bonds of the reactants. All of the reactions we have considered so far are classified as *thermal*, initiated by the kinetic energy of the reactant molecules. In *photolytic* reactions, the

Table 3.D.5 Some Photolysis Reactions That Affect Air Pollution in the Troposphere

Reaction	k_{max} (min^{-1})[a]	Comment
$NO_2 + h\nu \rightarrow O\bullet + NO$	0.5	Key step in producing O_3 in urban air
$HNO_2 + h\nu \rightarrow OH\bullet + NO$	0.1	Important as a source of OH•
$H_2O_2 + h\nu \rightarrow 2\ OH\bullet$	0.0005	Hydrogen peroxide photolysis
$NO_3 + h\nu \rightarrow NO + O_2$	1.5	580 nm $< \lambda <$ 630 nm
$NO_3 + h\nu \rightarrow NO_2 + O\bullet$	12	470 nm $< \lambda <$ 635 nm
$HCHO + h\nu \rightarrow HCO\bullet + H\bullet$	0.003	Formaldehyde photolysis
$HCHO + h\nu \rightarrow CO + H_2$	0.005	Formaldehyde photolysis
$CH_3CHO + h\nu \rightarrow CH_3\bullet + HCO\bullet$	0.003	Acetaldehyde photolysis

[a]The reaction rate constants vary with sunlight intensity. These approximate maximum values correspond to midday, clear sky conditions during summer.

energy to break the chemical bonds of a reactant is supplied by absorption of a photon of light.

Gas-phase photolysis causes unimolecular dissociation. A single molecule absorbs a photon and then breaks into two fragments, often highly reactive. In the troposphere, only certain molecules undergo photolytic dissociation. Important examples are summarized in Table 3.D.5. In this table, $h\nu$ represents the energy of a photon, where h is Planck's constant (6.63×10^{-34} J s) and ν is the frequency of light ($= c\lambda^{-1}$, where c is the speed of light, 3.0×10^8 m s^{-1}, and λ is the wavelength). The species labeled with • are radicals.

Radicals are species with an unpaired outer-shell electron, making them highly reactive. Their atmospheric lifetimes are very short, typically μs to s. They are present at very low concentrations, but they are of great importance in atmospheric redox reactions.

Because their generation depends on the photolysis of pollutant molecules, radical concentrations are highest in heavily polluted air during strong sunlight conditions. Under sunny conditions at midday, hydroxyl radical (OH•) mole fractions are reported to range from 0.004–0.04 ppt in unpolluted air to ≥0.4 ppt in heavily polluted air (Finlayson-Pitts and Pitts, 1988). Representative mole fractions of O•, $HO_2\bullet$, and $NO_3\bullet$, in polluted air have been reported as 0.04 ppt, 40 ppt, and 100 ppt, respectively (Seinfeld, 1986; Finlayson-Pitts and Pitts, 1988). Concentrations of these species under relatively clean conditions might be an order of magnitude lower.

The rates of photolytic reactions vary with overall sunlight intensity and wavelength. In the troposphere, photolysis is caused by light of wavelengths 280 nm $< \lambda <$ 730 nm (410 nm $< \lambda <$ 650 nm defines the visible range). Shorter-wavelength light is absorbed by stratospheric ozone and oxygen molecules and does not penetrate to the troposphere. Longer-wavelength light has insufficient energy to break chemical bonds.

Atmospheric Lifetime of Air Pollutants

The atmospheric lifetime of air pollutants varies markedly among species. Radicals typically persist for less than one second. Moderately reactive species such as ozone and nitrogen dioxide have lifetimes of a few minutes during daylight hours. Carbon monoxide has an atmospheric lifetime of weeks, methane lasts many months, and chlorofluorocarbons persist in the atmosphere for decades. The characteristic residence time of pollutants in an urban air basin is on the order of a day before removal by wind. In analyzing the behavior of urban air pollutants, we generally treat the short-lived radicals as being in a steady-state balance. Because of their long lifetimes, carbon monoxide and methane are generally considered nonreactive as impurities in urban air.

Many elementary atmospheric reactions have this form:

$$\text{species} + \text{radical} \rightarrow \text{products} \tag{3.D.12}$$

The rate law for such a reaction can be expressed as

$$R = kY_{\text{species}}Y_{\text{radical}} \tag{3.D.13}$$

where Y_{species} and Y_{radical} are mole fractions of the respective species and k is the reaction rate constant. If we think of the quantity of the species, Y_{species}, as the stock and the reaction rate R as the flow out of the stock, then the characteristic time for this process is

$$\tau_{\text{reaction}} \sim \frac{Y_{\text{species}}}{R} \sim \frac{1}{kY_{\text{radical}}} \tag{3.D.14}$$

The meaning of τ_{reaction} can be stated in this way: It represents a magnitude estimate of the time required for this reaction to consume the current stock of the species, assuming that the radical concentration remains fixed.

Hydroxyl radical attack constitutes the first step in the dominant atmospheric degradation pathway for many species. Table 3.D.6 presents a summary of reaction rate constants for OH• attack of many atmospheric gases. Also indicated in the table are characteristic reaction lifetimes (τ_{reaction}), determined according to equation 3.D.14, with a mole fraction of OH• that corresponds to the upper end of the range reported for clean tropospheric conditions. Under heavily polluted conditions, the characteristic lifetimes would be an order of magnitude shorter. Note that the lifetimes of these species span approximately five orders of magnitude, with isoprene representing a highly reactive species (lifetime ~ 3 h) and the chlorofluorocarbons appearing as essentially non-reactive (lifetime > 50 y).

3.D.5 Microbial Reactions

Microorganisms act as a diverse set of biological catalysts for redox reactions in environmental systems. Microorganisms produce enzymes that speed the rates of reactions that would spontaneously occur much more slowly. The organisms usually extract something useful in the process, either energy or chemical elements needed for growth and reproduction. Chemical species are *biodegraded* or *biotransformed* when they undergo a reaction induced by a microorganism.

Reactions mediated by microorganisms are important in governing the fate of chemicals in the natural environment, especially in aquatic systems, in shallow sediments, and in surface soils. Microorganisms are used in engineered systems to carry out desired reactions, such as in the treatment of municipal wastewater and sludges. Interest has grown in using microorganisms to treat other effluents such as liquid and gaseous hazardous waste streams, and environmental contaminants at or near waste disposal sites (*bioremediation*). These activities build on a long history of microorganism use for producing foods, including wine and beer, cheese, yogurt, and leavened bread.

Table 3.D.6 Reaction Rate Constants and Estimated Atmospheric Lifetimes for Selected Molecules Containing Carbon, Sulfur, or Nitrogen

Species	Formula	Rate constant[a] (ppb^{-1} min^{-1} @ 298 K)	Characteristic lifetime[b] y	d	h
Nitrogen-containing compounds					
Ammonia	NH_3	0.24		70	
Nitrogen dioxide	NO_2	16			30
Sulfur-containing compounds					
Hydrogen sulfide	H_2S	6.9		3	
Sulfur dioxide	SO_2	1.3		10	
C/H/O compounds					
Methane	CH_4	0.01	5		
Ethane	C_2H_6	0.4		40	
n-Octane	C_8H_{18}	13			30
Ethylene	C_2H_4	13			30
Isoprene	C_5H_8	150			3
Benzene	C_6H_6	1.9		9	
Toluene	C_7H_8	9		2	
Methanol	CH_3OH	1.3		10	
Formaldehyde	HCHO	13			30
Carbon monoxide	CO	0.35		50	
Chlorinated hydrocarbons					
Trichlorofluoromethane	CCl_3F	<0.001	>50		
Dichlorodifluoromethane	CCl_2F_2	<0.001	>50		
1,1,1-Trichloroethane	CH_3CCl_3	0.02	2		
Trichloroethylene	C_2HCl_3	3.5		5	
Perchloroethylene	C_2Cl_4	0.3		60	

[a]For reactions with OH•.

[b]Assumes a stable mole fraction of OH• of 4×10^{-5} ppb (corresponding to a concentration of 10^{12} molecules per m^3 at $T = 298$ K, $P = 1$ atm). Results rounded to one or two significant figures.

Source: Finlayson-Pitts and Pitts, 1988.

Microbial Redox Processes

Examples of important redox reactions that are carried out by microorganisms are summarized in this section. In this discussion, the notation $\{CH_2O\}$ is used to denote a fragment of an arbitrary carbohydrate.

Photosynthetic Production of Biomass Photosynthetic microorganisms (algae and some bacteria) carry out photosynthesis reactions. In these reactions, energy-rich carbohydrate molecules are produced by combining carbon dioxide and water, using energy derived from sunlight. Overall, these reactions can be written in the form given below. From a redox perspective, we see that carbon is reduced from oxidation state +4 to 0, while oxygen is oxidized from −2 to 0. Two oxygen atoms are oxidized for each carbon atom that is reduced. Photosynthesis reactions lie at the bottom of many food chains. Light energy is converted and stored as chemical energy that can later be released in separate redox reactions.

$$CO_2 + H_2O \xrightarrow{h\nu} \{CH_2O\} + O_2 \qquad (3.D.15)$$

Aerobic Respiration In the presence of oxygen, microorganisms degrade biomass to form carbon dioxide and water. Chemical energy that is released can be used by the organism. This process is the reverse of photosynthesis: Carbon is oxidized and oxygen is reduced.

$$\{CH_2O\} + O_2 \rightarrow CO_2 + H_2O \qquad (3.D.16)$$

Nitrogen Fixation In the atmosphere, nitrogen is almost entirely in the form of N_2 (oxidation state 0), a very stable molecule. The nitrogen in biological systems is mostly in the form of an amine, $-NH_2$, which is closely related to ammonia (NH_3) and the ammonium ion (NH_4^+). Here, nitrogen is in oxidation state −3. Another common form of nitrogen in water and soils is nitrate (NO_3^-), in which nitrogen is in oxidation state +5. Microorganisms play an essential role in the movement of nitrogen among these oxidation states.

We refer to compounds such as ammonia and nitrate that contain a single nitrogen atom as *fixed nitrogen* species. Certain groups of bacteria are capable of converting gaseous nitrogen to fixed nitrogen, in the form of the ammonium ion. As shown in the following reaction, energy from the oxidation of biomass to CO_2 is used to reduce the nitrogen in N_2 to ammonium. This process is of particular interest in agriculture because fixed nitrogen is a necessary nutrient for plant growth. Bacteria that fix nitrogen live in a symbiotic relationship in the root systems of legumes (beans, peas, etc.). When these plants are grown, soils become enriched in fixed nitrogen rather than depleted. Utilizing legumes in a crop rotation system can reduce the need for chemical fertilizers.

$$3\,\{CH_2O\} + 2\,N_2 + 3\,H_2O + 4\,H^+ \rightarrow 3\,CO_2 + 4\,NH_4^+ \qquad (3.D.17)$$

Nitrification In nitrification, nitrogen in the ammonium ion is oxidized from −3 to +5, with oxygen as the oxidizer. Plants absorb nitrogen more efficiently in the form of nitrate than as ammonium, so this redox reaction can enhance the effectiveness of ammonia-based agricultural fertilizers.

$$NH_4^+ + 2\,O_2 \rightarrow NO_3^- + 2\,H^+ + H_2O \qquad (3.D.18)$$

Nitrate Reduction or Denitrification When oxygen is not available as the oxidizer to degrade biomass, microorganisms can use nitrate as the oxidizer (electron acceptor) instead. Nitrate reduction is used in some wastewater treatment systems to convert fixed nitrogen to N_2 gas, which can then be safely released to the atmosphere. This process is sometimes called *denitrification*, since nitrogen is removed from the aqueous system. Because nitrogen in municipal wastewater begins in a reduced state (−3), the overall treatment process involves two steps: nitrification (according to reaction 3.D.18) in an aerobic reactor, followed by denitrification in an anaerobic reactor. As shown in the reaction below, four nitrogen atoms, being reduced from +5 to 0, can fully oxidize five carbon atoms from 0 to +4. Denitrification is also an important natural process that returns fixed nitrogen to the atmosphere.

$$5\,\{CH_2O\} + 4\,NO_3^- + 4\,H^+ \rightarrow 5\,CO_2 + 7\,H_2O + 2\,N_2 \qquad (3.D.19)$$

Sulfate Reduction Some environments that contain biodegradable materials lack both oxygen and nitrate to serve as the oxidizing agent. In such cases, sulfate may serve that role. As shown in the following reaction, the conversion of one sulfur atom from +6 in sulfate to –2 in hydrogen sulfide oxidizes two carbon atoms from 0 to +4. This reaction can occur in stagnant anaerobic marine sediments that are supplied with decaying biomass, for example, from algae or seaweed accumulation. If the rate of biomass accumulation is high, oxygen can be rapidly depleted from the sediments. Sulfate is plentiful as a dissolved ion in seawater (see Table 2.C.3). The hydrogen sulfide produced in this reaction may be released as a gas, producing an offensive odor downwind.

$$2\,\{CH_2O\} + 2\,H^+ + SO_4^{2-} \rightarrow 2\,CO_2 + 2\,H_2O + H_2S \qquad (3.D.20)$$

Methane Formation (Methanogenesis) In the absence of oxygen, nitrate, and sulfate, biomass can still be converted to carbon dioxide as shown in the following reaction. This methane generation process is exploited in wastewater treatment to convert excess microbiological material to gases, which are more easily handled for disposal. This is an interesting redox reaction, since the two carbon atoms begin in oxidation state zero. One is oxidized to +4 while the other is reduced to –4.

$$2\,\{CH_2O\} \rightarrow CO_2 + CH_4 \qquad (3.D.21)$$

Microbial Kinetics

To engineer treatment processes that use microorganisms to carry out transformation processes, we need tools for predicting system behavior. In general, we want to know how fast microorganisms can degrade contaminants. We also want to know how much and how rapidly other reactants, such as dissolved oxygen, are used in a process. One thing that distinguishes microbial reactions from abiotic chemical transformations is that the microorganisms can themselves grow and reproduce. For example, when carbon in organic molecules is oxidized by bacteria, some of the carbon atoms are fully oxidized to CO_2 (yielding the maximum possible energy), while other carbon atoms are incorporated into new cell mass. As the microorganism mass increases, the degradation rate can also increase. Thus, to predict the rate of contaminant degradation by microorganisms, we must also predict changes in the microbial population.

As presented below, the rates of microbial growth and contaminant degradation are described by kinetic rate equations of a general form. Experiments are conducted to obtain empirical parameters to be used in the kinetic equations. Then, for a specific application, material balance equations are written and solved, taking into account the kinetic behavior of the organisms.

In the equations that follow, we will use the symbol X to represent the mass concentration of active microorganisms in water. The symbol S represents the mass concentration of a material that is being degraded by the microorganisms. In microbiology, this material is referred to as *substrate*. In a common case, S could represent the concentration of biodegradable organic contaminants in water.

Microbial cells reproduce (or grow), and they die (or decay). A balance between the rates of these two processes gives the overall rate of change in the concentration of active microorganisms due to transformation processes. A reasonable model assumes that the rate of each process is proportional to the existing mass concentration of active organisms. So the model equation for the rate of change of microbial cell concentration (X) in a batch reactor has this form:

$$\frac{dX}{dt} = \mu X = \text{cell growth rate} - \text{cell decay rate} = r_g X - k_d X \tag{3.D.22}$$

This equation contains three parameters: μ is the *net specific growth rate of cells*, r_g is the *cell growth rate coefficient*, and k_d is the *cell death rate coefficient*. Each parameter has units of inverse time (typically, d^{-1}).

The cell growth rate depends on the concentration of a limiting substrate, S. The most widely accepted form for describing the dependence of r_g on S is

$$r_g = Y \frac{k_m S}{K_s + S} \tag{3.D.23}$$

In this expression, three new symbols appear. The parameter k_m is called the *maximum specific substrate degradation rate* (with dimensions of mass of substrate per mass of cells per time). The parameter Y is called the *cell yield coefficient*. The dimensions of Y are mass of cells per mass of substrate. The parameter K_s is called the *half-saturation constant* and has the same dimensions as S, that is, mass concentration of substrate. The net specific growth rate of cells can then be written in a form known as the *Monod equation* (sometimes called a saturation reaction):

$$\mu = \frac{1}{X}\frac{dX}{dt} = Y \frac{k_m S}{K_s + S} - k_d \tag{3.D.24}$$

The rate of substrate degradation is related to the cell growth rate through the cell yield coefficient, Y. In fact, the cell yield coefficient represents the mass of cells produced per mass of substrate consumed. Therefore, the rate of change of substrate concentration due to microbial degradation is modeled as

$$\frac{dS}{dt} = -\frac{r_g}{Y} X = -\frac{k_m S}{K_s + S} X \tag{3.D.25}$$

The four equations, 3.D.22–25, are used to describe the kinetic behavior of microorganisms in environmental systems and in engineered treatment processes. Four parameters must be determined experimentally: Y, k_m, K_s, and k_d.

If we consider the functional form of the cell growth rate coefficient, as depicted in Figure 3.D.2, we see that r_g varies as a function of the substrate concentration, S,

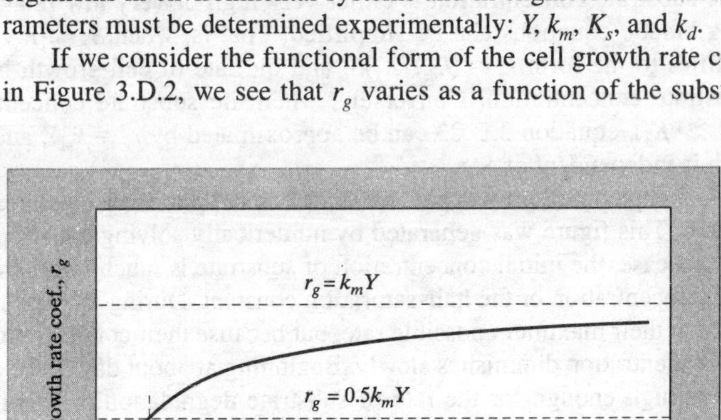

Figure 3.D.2 Dependence of cell growth rate coefficient (r_g) on substrate concentration (S).

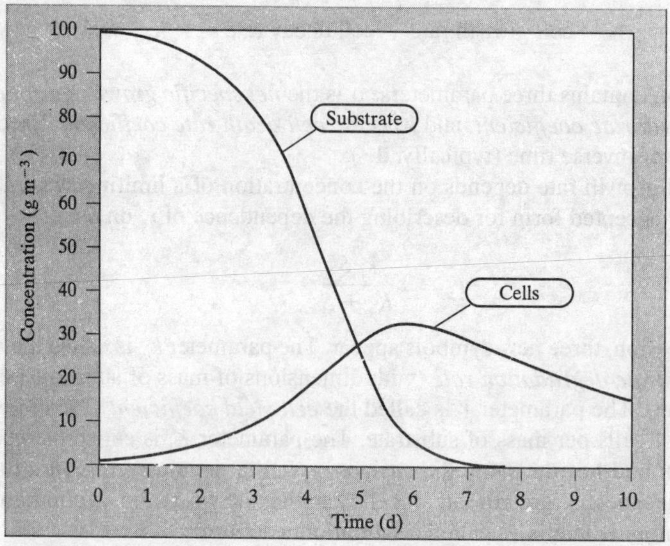

Figure 3.D.3 Predicted cell growth and decay in a batch reactor according to kinetic equations 3.D.22 to 3.D.25. Parameter values are as follows: k_m = 3 (g substrate/g cell) d^{-1}; $Y = 0.5$ (g cell/g substrate); K_s = 50 g substrate m^{-3}; k_d = 0.2 d^{-1}. Initial concentrations are $X(0)$ = 1 g cell m^{-3} and $S(0)$ = 100 g substrate m^{-3}.

such that when the substrate concentration is very large, r_g approaches a maximum value of k_mY. This reflects a condition in which the microorganisms are reproducing at their maximum rate, with unlimited substrate. At the other extreme, when the limiting substrate concentration is very small, the cell growth rate varies strongly with S. The meaning of the half-saturation constant, K_s, is also revealed in this figure. When S is equal to K_s, then the cell growth rate (r_g) is exactly half its maximum value.

When the substrate concentration is either very high or very low relative to K_s, the governing kinetic equations can be simplified. That is, when $S \ll K_s$, equation 3.D.23 simplifies to the form $r_g = (k_mY/K_s)S$, and the rate of cell growth is proportional to substrate concentration. Conversely, when the substrate concentration is very large ($S \gg K_s$), equation 3.D.23 can be approximated by $r_g = k_mY$, and the rate of cell growth is independent of S.

Figure 3.D.3 depicts cell growth and substrate consumption in a batch reactor as a function of time. This figure was generated by numerically solving equations 3.D.22 to 3.D.25. In this case, the initial concentration of substrate is much larger than either the initial cell concentration or the half-saturation constant. During the first few days, cells reproduce at their maximum possible rate, but because their concentration is low, the substrate concentration diminishes slowly. Beginning at about day 3, the cell mass concentration is high enough for the rate of substrate degradation to become rapid. Over the next three days, the cells exhaust the available substrate, and reproduction slows. After day 6, cell decay outpaces cell growth and, as time progresses further, the concentration of active cells decays exponentially toward zero.

Biochemical Oxygen Demand

For water that is in contact with the atmosphere, the most fully oxidized state of common elements is +4 for carbon, +5 for nitrogen and phosphorus, and +6 for sulfur.

Wastewater contains these elements in chemically reduced states, especially carbon in organic compounds. Microorganisms in wastewater use the organic compounds as a source of carbon and chemical energy. Under aerobic conditions, organic carbon is oxidized using dissolved oxygen as the oxidizer. Consequently, the dissolved O_2 content of water becomes depleted. The amount of oxygen required to fully oxidize the compounds in wastewater is referred to as the *theoretical oxygen demand.* It is a measure of the amounts and oxidation states of chemical elements in the water (see §2.C.5). Theoretical oxygen demand is difficult to determine. Instead, in environmental engineering practice, oxygen demand associated with microbiological activity is measured. This parameter is called the *biochemical oxygen demand* (BOD). Although other oxygen demand measurement techniques are available, BOD remains important because it roughly mimics what happens when chemically reduced wastes are discharged to receiving waters such as rivers or lakes.

Two factors are relevant in assessing the oxygen-depleting significance of BOD. The first factor is stoichiometric. We want to know the total amount of oxygen that is required for biodegradation of oxidizable compounds. The second factor is kinetic. We want to know how rapidly oxygen will be consumed in the oxidation process.

The stoichiometric aspect of BOD may be subdivided into two components. That associated with the oxidation of carbon and other reduced elements, excluding nitrogen in ammonia, is termed *carbonaceous* BOD because it mainly reflects the oxidation of carbon. The oxygen required for microbial oxidation of ammonia to nitrate, as described by the nitrification reaction 3.D.18, is called *nitrogenous* BOD. Because reduced carbon is more prevalent in typical wastewater than reduced nitrogen, carbonaceous BOD is generally much higher than nitrogenous BOD. Also, because most microorganisms can oxidize carbon whereas very few oxidize NH_3, the kinetics are faster for oxidation of carbonaceous BOD than for nitrogenous BOD.

Measuring BOD. To understand BOD, one should consider how it is measured. The basic procedure for measuring stoichiometric BOD is simple, consisting of the following steps.

1. Measure the initial dissolved oxygen content of water to be analyzed. Call this DO(0).
2. Fill a 300 mL glass bottle with a sample of the water. Seal the bottle with a stopper.
3. Incubate the water in the dark at 20 °C for 5 days.
4. Measure the dissolved oxygen content of the incubated water. Call this DO_5.
5. Compute the five-day BOD as $BOD_5 = DO(0) - DO_5$, where all have units of mg L^{-1}.

The choice of conditions (300 mL sample and 5 day incubation at 20 °C) has been standardized for ease in comparing results from one laboratory to another. Usually, over 5 days, most of the carbonaceous BOD is expressed but little of the nitrogenous BOD is. If a problem with nitrification is suspected, a specific nitrification inhibitor can be added to the water sample so that only the carbonaceous BOD is determined. The measurement is conducted in the dark to avoid any complications caused by oxygen production through photosynthesis.

Note that BOD is measured by detecting the depletion of dissolved oxygen in the water. This is appropriate since in the reaction

$$BOD + DO \rightarrow \text{oxidized products} \qquad (3.D.26)$$

BOD and DO combine in a 1:1 mass ratio. One must be careful, though, to avoid the pitfall of thinking that the concentration of BOD is directly related to the DO content of the water; these are independent quantities. That is, BOD is a measure of the amount of degradable reduced material, such as organics in the water, while DO is a measure of the water's dissolved oxygen. It is also important to note that BOD includes microorganism cell mass in addition to dissolved organics.

The objective of the standard BOD test is to measure characteristics of the waste products in the water. However, complications may require that the standard test be modified. First, the BOD of the water may be so high that all of the dissolved oxygen is consumed within 5 days. If this happens, $DO_5 \sim 0$, and BOD_5 is unknown. To overcome this difficulty, the water sample can be diluted with BOD-free water. A second problem may arise because of pH changes with time due to the biochemical activity, which may impact microbial degradation kinetics. This problem is addressed by adding a chemical buffer to the solution. Lack of nutrients may also limit microbial degradation of organics. The solution to this problem is to add nitrogen and phosphorus in suitable chemical forms to the test water. Finally, the wastewater may contain too few microorganisms to efficiently oxidize the organics. Seed organisms can be added in this case.

The standard BOD test quantifies a stoichiometric characteristic of BOD. A more general test procedure would involve monitoring the dissolved oxygen content as a function of time throughout the period of the test. The results might look like those presented in Figure 3.D.4.

Figure 3.D.5 shows an idealized representation of the dissolved oxygen content of water in a BOD test. The parameter BOD(t) represents the amount of unexpressed BOD remaining in the water sample at any time t. The total amount of BOD initially in the system is called the *ultimate BOD,* or BOD_u. Note that BOD_u is equal to BOD(0) and also equal to the difference in dissolved oxygen content between the starting condition, DO(0), and the final steady-state condition, DO_u.

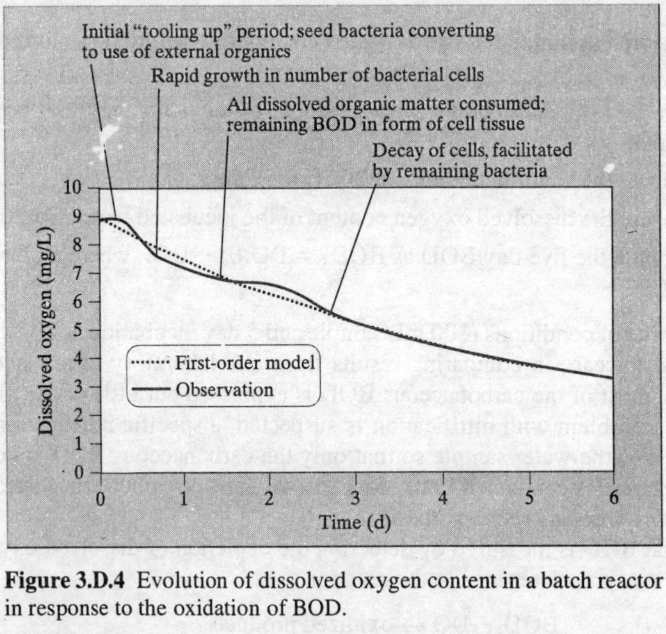

Figure 3.D.4 Evolution of dissolved oxygen content in a batch reactor in response to the oxidation of BOD.

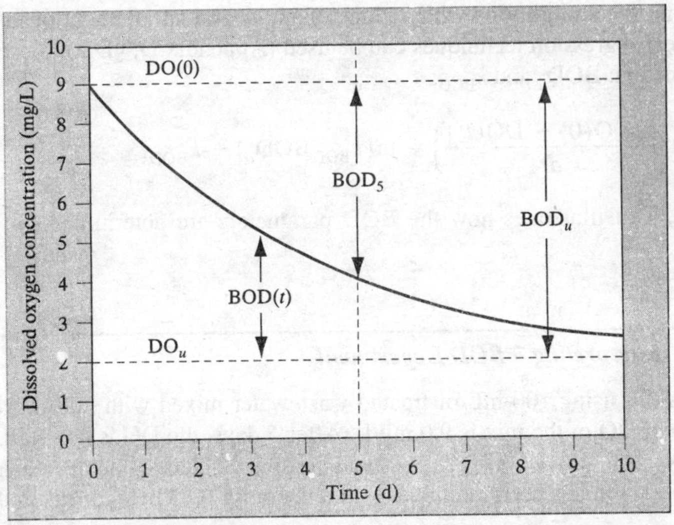

Figure 3.D.5 Idealized decrease in dissolved oxygen concentration in a batch reactor during measurement of BOD.

Be careful not to misunderstand the meaning of BOD_5. This parameter refers to the amount of oxygen demand that is expressed during the first 5 days or the amount of BOD that reacts within 5 days. It is *not* the BOD remaining in the water after 5 days have elapsed. Note that $BOD_u = BOD(5) + BOD_5$.

Although the overall set of reactions is complex, the oxidation of carbonaceous BOD is reasonably well modeled as a first-order process. In a well-stirred batch reactor, the rate of change of BOD with time is described by this expression:

$$\frac{d(\text{BOD})}{dt} = -k_{\text{BOD}}\text{BOD} \tag{3.D.27}$$

where k_{BOD} is the rate constant for BOD oxidation. Given the initial condition $\text{BOD}(0) = \text{BOD}_u$, the BOD content changes with time according to the solution of equation 3.D.27:

$$\text{BOD}(t) = \text{BOD}_u \exp(-k_{\text{BOD}}t) \tag{3.D.28}$$

Without oxygen replenishment from the air, BOD oxidation causes the dissolved oxygen content of the batch reactor to be depleted at the same rate as BOD:

$$\frac{d(\text{DO})}{dt} = -k_{\text{BOD}}\text{BOD} \tag{3.D.29}$$

Given an initial concentration DO(0), equation 3.D.29 can be solved by substituting from 3.D.28 for BOD and integrating, to yield

$$\text{DO}(t) = \text{DO}(0) - \text{BOD}_u[1 - \exp(-k_{\text{BOD}}t)] \tag{3.D.30}$$

In conducting an experiment to determine both k_{BOD} and BOD_u, one could measure DO(0) and DO(t) and determine the unknown parameters by fitting the data to this model. One way to fit the data would be to plot the natural logarithm of the time derivative of (DO(t) – DO(0)) against time. As shown by equation 3.D.31, the data should conform to a straight line with a slope of $-k_{BOD}$ and an intercept of $\ln(k_{BOD} \times \mathrm{BOD_u})$, so linear regression techniques can be used (Appendix D, §D.5) to extract the two key parameters: BOD_u and k_{BOD}.

$$\ln\left(\frac{d[\mathrm{DO}(0) - \mathrm{DO}(t)]}{dt}\right) = \ln(k_{BOD}\mathrm{BOD}_u) - k_{BOD}t \qquad (3.D.31)$$

Example 3.D.3 illustrates how the BOD parameters are determined for an elementary case.

EXAMPLE 3.D.3 ***Interpreting a BOD Experiment***

A BOD test is run using 100 mL of treated wastewater mixed with 200 mL of pure water. The initial DO of the mix is 9.0 mg/L. After 5 days, the DO is 4.0 mg/L. After a long period of time, the DO is 2.0 mg/L and no longer seems to be decreasing. Assume that nitrification has been inhibited so that the only BOD being measured is carbonaceous.

(a) What is the 5-day BOD of the wastewater (mg/L)?

(b) Estimate the ultimate carbonaceous BOD (mg/L).

(c) What is the remaining BOD after 5 days (mg/L)?

(d) Estimate the reaction rate constant, k_{BOD} (d^{-1}).

SOLUTION

(a) The change in dissolved oxygen content in the test sample during the first five days is 9.0 – 4.0 = 5.0 mg/L. Wastewater comprises only one-third of the test sample. To correct for dilution, we multiply by 3, so the BOD_5 content of the wastewater is 15 mg/L.

(b) The ultimate carbonaceous BOD is the difference between the initial and final DO levels, corrected for dilution, so $\mathrm{BOD}_u = (9.0 - 2.0) \times 3 = 21$ mg/L.

(c) $\mathrm{BOD}_u = \mathrm{BOD}_5 + \mathrm{BOD}(5)$, so the BOD remaining at 5 days is $\mathrm{BOD}(5) = 21 - 15 = 6$ mg/L.

(d) The rate constant is estimated from the first-order model, given BOD_u and BOD(5):

$$\mathrm{BOD}(5) = \mathrm{BOD}_u \exp(-k_{BOD} \times 5 \text{ days})$$

Substituting and solving for k_{BOD} yields the result:

$$k_{BOD} = -\frac{1}{5 \text{ days}} \times \ln\left(\frac{6}{21}\right) = 0.25\ \mathrm{d}^{-1}$$

Temperature Dependence of BOD Oxidation Rate The standard BOD test is conducted at 20 °C. Wastewater may be discharged into receiving water with a very different temperature. In Appendix E (§E.4), the following approximate expression is derived for predicting the effect of temperature on kinetic reaction rates:

$$k(T) \approx k(T_{ref})\theta^{(T-T_{ref})} \qquad (3.D.32)$$

In this equation, the rate constant is known at temperature T_{ref} and sought at temperature T. The parameter θ depends on the activation energy of the reaction and other variables. In the present context, it is best to consider θ an empirical parameter for the overall oxidation process.

Typical values of $k_{BOD}(20\ °C)$ for domestic wastewater are 0.2–0.3 d^{-1}. The temperature correction parameter θ for k_{BOD} is commonly assigned a value of 1.047 (Tchobanoglous and Schroeder, 1985). The fact that θ exceeds 1.0 means that BOD oxidation reactions occur more rapidly at higher temperatures. For example, if k_{BOD} is 0.25 d^{-1} at T_{ref} = 20 °C, then the estimated value of k_{BOD} at 10 °C is 0.25 d^{-1} × $1.047^{(10-20)} = 0.16\ d^{-1}$.

REFERENCES

BENEFIELD, L.D., & MORGAN, J.M. 1990. Chemical precipitation. In F.W. Pontius, ed., *Water quality and treatment: A handbook of community water supplies*. 4th ed. McGraw-Hill, New York, Chapter 10.

BERNER, R.A., & LASAGA, A.C. 1989. Modeling the geochemical carbon cycle. *Scientific American*, March, 74–81.

DE NEVERS, N. 1995. *Air pollution control engineering*. McGraw-Hill, New York.

ERLICH, P.R., ERLICH, A.H., & HOLDREN, J.P. 1977. *Ecoscience: Population, resources, environment*. W.H. Freeman, San Francisco.

FINLAYSON-PITTS, B.J., & PITTS, J.N., Jr. 1988. *Atmospheric chemistry: Fundamentals and experimental techniques*. Wiley, New York.

FLAGAN, R.C., & SEINFELD, J.H. 1988. *Fundamentals of air pollution engineering*. Prentice Hall, Englewood Cliffs, NJ.

HARNED, H.S., & OWEN, B.B. 1958. *The physical chemistry of electrolytic solutions*. 3rd ed. Reinhold, New York.

HIEMENZ, P.C. 1986. *Principles of colloid and surface chemistry*. 2nd ed. Marcel Dekker, New York.

KIRCHSTETTER, T.W., SINGER, B.C., HARLEY, R.A., KENDALL, G.R., & TRAVERSE, M. 1999. Impact of California reformulated gasoline on motor vehicle emissions: 1. Mass emission rates. *Environmental Science & Technology*, **33**, 318–328.

LAGREGA, M.D., BUCKINGHAM, P.L., & EVANS, J.C. 1994. *Hazardous waste management*. McGraw-Hill, New York.

LAIDLER, K.J. 1987. *Chemical kinetics*. 3rd ed. Harper & Row, New York.

MCRAE, G.J. 1980. A simple procedure for calculating atmospheric water vapor concentration. *Journal of the Air Pollution Control Association*, **30**, 384.

SANDLER, S.I. 1989. *Chemical and engineering thermodynamics*. 2nd ed. Wiley, New York.

SAWYER, C.N., MCCARTY, P.L., & PARKIN, G.F. 1994. *Chemistry for environmental engineering*. 4th ed. McGraw-Hill, New York.

SCHWARZENBACH, R.P., GSCHWEND, P.M., & IMBODEN, D.M. 1993. *Environmental organic chemistry*. Wiley, New York.

SEINFELD, J.H. 1986. *Atmospheric chemistry and physics of air pollution*. Wiley, New York.

SEINFELD, J.H., & PANDIS, S.N. 1998. *Atmospheric chemistry and physics: From air pollution to climate change*. Wiley, New York.

SIENKO, M.J., & PLANE, R.A. 1961. *Chemistry*. 2nd ed. McGraw-Hill, New York, pp. 107–110.

SNOEYINK, V.L., 1990. Adsorption of organic compounds. In F.W. Pontius, ed., *Water quality and treatment: A handbook of community water supplies*. 4th ed. McGraw-Hill, New York, Chapter 13.

SNOEYINK, V.L., & Jenkins, D. 1980. *Water chemistry*. Wiley, New York.

STUMM, W., & MORGAN, J.J. 1996. *Aquatic chemistry: Chemical equilibria and rates in natural waters*. 3rd ed. Wiley, New York.

TCHOBANOGLOUS, G., & SCHROEDER, E.D. 1985. *Water quality: Characteristics, modeling, modification*. Addison-Wesley, Reading, MA.

WHIPPLE, G.C., & WHIPPLE, M.C. 1911. Solubility of oxygen in sea water. *Journal of the American Chemical Society*, **33**, 362–365.

PROBLEMS

3.1 Concepts in environmental transformations

(a) What is the relationship between chemical equilibrium and chemical kinetics?

(b) What is the relationship between the concepts of *material balance* and *stoichiometry*?

(c) Compare and contrast *microbial oxidation of BOD* versus *hydrocarbon fuel combustion.*

(d) Is rainwater naturally acidic, basic, or neutral? Explain.

(e) What distinguishes a *strong* acid from a *weak* acid?

(f) What does it mean to burn with *stoichiometric air*?

(g) A fuel that contains carbon, hydrogen, and sulfur is burned with sufficient air for complete combustion. What are the most likely chemical forms for C, H, and S in the exhaust?

(h) What thermodynamic property must be fixed for the "equilibrium constant" of a reaction (K) to be constant?

(i) What is the name of the branch of (environmental) chemistry that deals with the rates of reactions?

(j) Between acid-base reactions and oxidation-reduction reactions, which tend to be faster? Why?

(k) A contaminant concentration in a dynamic system is described by the differential equation $dC/dt = S - LC$, with S and L constant and $C(0) = C_0$. What is the characteristic time for the system to approach steady state?

3.2 Water quality in the Mississippi River

Measurements taken of a sample of water from the Mississippi River, near Grand Rapids, Minnesota, yielded the following results (Tchobanoglous and Schroeder, 1985). Note that the error in measuring any single species may be as high as 5 percent.

Species	Mass conc. (g m^{-3})	MW (g mol^{-1})	Molarity (M)
SiO_2	7.5	60	1.25×10^{-4}
Ca^{2+}	31	40	7.75×10^{-4}
Mg^{2+}	13	24.3	5.35×10^{-4}
Na^+	4.7	23	2.04×10^{-4}
K^+	2.3	39.1	5.9×10^{-5}
HCO_3^-	166	61	2.7×10^{-3}
SO_4^{2-}	6.8	98	7.0×10^{-5}
H^+			1.0×10^{-8}

(a) Evaluate the hardness of this sample.

(b) Determine the molar concentration of carbonate (CO_3^{2-}).

(c) Determine the alkalinity of the sample.

(d) Is the electroneutrality relationship satisfied? Demonstrate that your answer is correct.

3.3 Drinking water characteristics

Analysis of treated water from a drinking water supply that originates from runoff of a relatively clean watershed yielded the following species concentrations.

Species	Mass conc. (g m^{-3})
SiO_2	8.3
Ca^{2+}	39
Mg^{2+}	9
CO_3^{2-}	6.4
SO_4^{2-}	22
pH = 9.0	
Total dissolved solids (TDS) = 216 g m^{-3}	

(a) Use electroneutrality to check if any major species are missing from the analysis.
(b) Determine the molar concentration of bicarbonate in this water and recheck the electroneutrality relationship.
(c) Determine the alkalinity of the sample.
(d) Check to see if the solids balance is satisfied, and briefly comment on your finding.

3.4 Air and water
Consider a box with a volume of 1 m^3. The box is initially filled with dry air at a pressure of 1 atm and a temperature of 293 K.
(a) How many moles of air are in the box?
(b) Starting from the condition given in the problem statement, 0.5 mol dry air is replaced with 0.5 mol water. At equilibrium, assuming $T = 293$ K, what is the relative humidity in the box?
(c) Starting from the condition given in the initial problem statement, 2 mol dry air are removed and replaced with 2 mol water. At equilibrium, assuming $T = 293$ K, what is the relative humidity in the box?
(d) What is the total air pressure in the box for condition (*c*)?

3.5 Water vapor in air
(a) Consider a parcel of air at $T = 297$ K, $P = 1$ atm, and relative humidity of 50 percent. What is the mole fraction of water vapor in the air (percent)?
(b) Calculate the mass concentration of water vapor in the atmosphere at 25 °C when the relative humidity is 60 percent and the saturation vapor pressure is 2800 Pa.

3.6 Stormy weather
A warm summer rainstorm has arrived in the middle of the night, causing the nighttime air to become saturated with water vapor. As the sun rises, the air is quickly heated, with the temperature rising from the nighttime value of 20 °C to 25 °C. Assuming that no water has evaporated, what is the relative humidity at the higher temperature? Assume that the air pressure is constant at 1 atm.

3.7 Humidity exercises
(a) Two vessels are partly filled with water, but to different levels. The remainder is initially dry air at 1 atm. The vessels are then sealed. Vessel 1 is maintained at a tem-

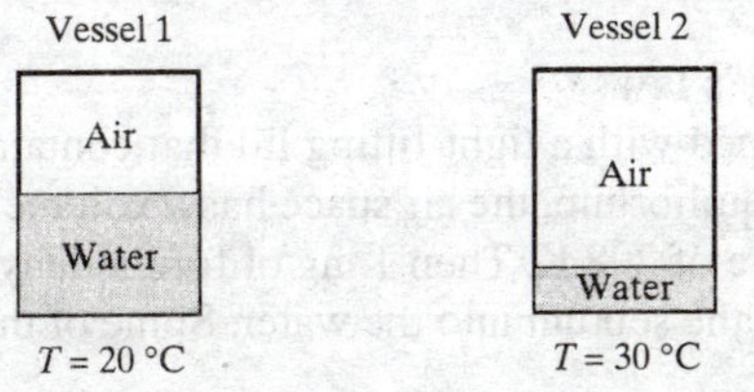

perature of 20 °C, and vessel 2 is kept at a temperature of 30 °C. The air in which vessel has a higher relative humidity at equilibrium? Explain.

(b) Is the following statement true or false? A mixture of two air parcels, each with a relative humidity of less than 100 percent, must also have a relative humidity of less than 100 percent. Justify your answer.

3.8 Mixing parcels of air

As depicted below, two parcels of air are mixed. The properties of the two parcels are indicated in the figure: volume (*V*), temperature (*T*), pressure (*P*), and relative humidity (RH). The volume, temperature, and pressure of the mixture are also specified. What is the relative humidity of the mixture?

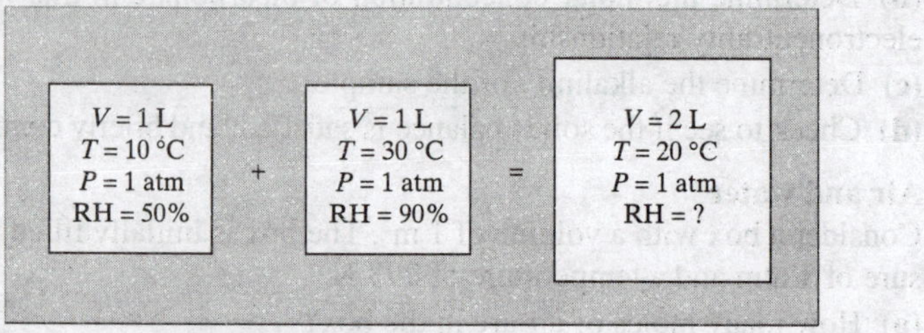

3.9 Evaporation and dissolution

A 1 L jar is initially half filled with pure water and half filled with nitrogen gas (N_2) at $P = 1$ atm, and then sealed. The temperature of the system is maintained at 298 K. Some of the water evaporates into the gas phase. Some of the nitrogen dissolves into the water.

(a) What fraction of the total H_2O in the system is in the gas phase at equilibrium?

(b) What fraction of the total N_2 in the system is in the aqueous phase at equilibrium?

3.10 Nitrogen dissolved in water

(a) Determine the mass concentration of N_2 in water at 20 °C when the water is in equilibrium with the atmosphere.

(b) Compare your result with the equilibrium mass concentration of O_2 in water exposed to the atmosphere: 9.1 mg/L at 20 °C.

3.11 Measuring Henry's law constant for chloroform

The following experiment is conducted to measure the Henry's law constant of chloroform ($CHCl_3$). A vessel with a volume of 2 L is half filled with water and then sealed, so it contains 1 L of water and 1 L of air. One gram of liquid chloroform is injected into the water through a rubber septum. The contents are thoroughly mixed and then allowed to equilibrate. A sample of gas is extracted from the air space above the water. Using a gas chromatograph, it is determined that the mass concentration of chloroform in air is 89 mg/L. Determine the value of the Henry's law constant in units of M atm^{-1}.

3.12 Applying Henry's law

A 2 L jar, equipped with a tight-fitting lid that contains a rubber septum, is half filled with water. At equilibrium, the air space has a volume of 1 L, a total pressure of 1 atm, and a temperature of 298 K. Then 1 mg of formaldehyde (HCHO, MW = 30 g/mol) is injected through the septum into the water. Some of the formaldehyde escapes into the gas phase. Consider the situation when a new equilibrium is established. Determine (*a*) the equilibrium mass concentration of HCHO in water and (*b*) the equilibrium mole fraction of HCHO in air.

3.13 Dissolution of a solid
Begin with pure water. Add 10^{-3} moles of solid $Mg(OH)_2$ (MW = 58 g/mol). The final solution volume is 1 L.
(a) At equilibrium, what is the molar concentration of Mg^{2+}?
(b) What is the equilibrium pH of the solution?
(c) Evaluate the total hardness and the carbonate hardness.

3.14 Solid solubility
Calcium sulfate ($CaSO_4$, solid, solubility product = 2×10^{-5} M^2), is added to initially pure water. If solid $CaSO_4$ remains at equilibrium, what is the equilibrium molarity of calcium ions?

3.15 Is iron sulfide an acid or a base?
A 3 L glass container is partially filled with V_w = 1 L of pure water. The remainder contains clean air, V_a = 2 L (P = 1 atm, T = 25 °C). Solid iron sulfide (FeS) is added to the water in sufficient quantity so that some solid will remain at equilibrium. The container is quickly sealed before any reactions occur. Iron sulfide dissolves to Fe^{2+} and S^{2-}. The sulfide participates in acid-base chemistry, combining with protons to form HS^- and H_2S(aq). The aqueous hydrogen sulfide can escape into the gas phase as H_2S(g).
(a) Is the equilibrium pH of the water higher or lower than 7? Explain.
(b) Write (but do not solve!) a complete set of algebraic equations that could be solved to determine the equilibrium pH of the water.

3.16 Sorption and the Freundlich isotherm
An experiment is conducted to determine the sorption isotherm for benzene in water on activated carbon. To conduct the experiment, three 1 L jars are filled with water plus 1 g of granular activated carbon. Different amounts of benzene are added to each jar. Then the equilibrium concentration of benzene that remains in the water is measured. The experimental data are presented in the table below.

Mass of benzene added (mg)	Equilibrium conc. in water (mg/L)
2.0	1.0
10.0	3.29
50.0	10.0

Use these data to determine the parameters K_f and $1/n$ for a Freundlich isotherm, $q = K_f C_e^{1/n}$, where C_e is the equilibrium concentration of benzene in water and q is the equilibrium mass of sorbed benzene per mass of sorbent. Specify the appropriate units for K_f and $1/n$.

3.17 Activated carbon sorption isotherms
Equilibrium data from batch studies for the sorption of paranitrophenol (PNP, MW = 139 g/mol) from aqueous solution at 25 °C by activated carbon are given below.

PNP sorbed, q (mmol/g)	PNP concentration in solution, C (mM)
0.9	0.006
1.1	0.011
1.4	0.023
1.6	0.045
1.9	0.18
2.1	0.33

(a) Convert the data to units of mg/g for q and mg/L for C.
(b) Determine the coefficients for these data assuming a Freundlich isotherm applies.
(c) Determine the coefficients for these data assuming a Langmuir isotherm applies.
(d) Which isotherm model provides a better fit to the data? Explain.

3.18 Sorbing trichloroethylene on activated carbon

An experiment has been conducted to measure the effectiveness of activated carbon in sorbing trichloroethylene (TCE, C_2HCl_3) from water. The experiments are carried out in 2 L sealed bottles containing 0.5 L of water with 5 g of activated carbon. Different masses of TCE were added to the bottles in each of the experiments. TCE concentrations were measured by taking samples of the gas phase within the sealed bottles. The following data were experimentally determined at 25 °C.

Experiment no.	Amount of TCE added ($\times 10^{-3}$ mol)	TCE in gas phase at equilibrium (atm)
1	24	0.001
2	110	0.02
3	237	0.1

(a) Calculate the equilibrium TCE liquid concentration for each of the experiments (units: mM).
(b) Calculate the moles of TCE sorbed per mass of carbon at equilibrium for each experiment (units: mmol of sorbed TCE per g of activated carbon).
(c) Determine appropriate Freundlich isotherm parameters (K_f and $1/n$) to describe the TCE sorption characteristics of this activated carbon.

3.19 An exercise in determining pH and converting units

(a) For an experimental investigation, a solution with a pH of 4 is required. The solution is to be made by combining pure water with 1.0 M sulfuric acid (H_2SO_4). Determine the volumes of acid and pure water necessary to yield 1 L of solution. (*Notes*: [i] Sulfuric acid will be completely dissociated in water to $SO_4^{\ 2}$; [ii] The molarity of an acid solution refers to the concentration of the principal anion, in this case SO_4^{2-}.)
(b) Determine the normality, the molarity, and the mass concentration of SO_4^{2-} in the final solution.

3.20 A generic acid-base problem

Derive an algebraic equation that relates the concentration of an acid, HA, to the concentration of its conjugate base, A^-. Express your answer in terms of the pK_A value for the acid and the pH of the solution.

3.21 Strong acids and bases

A solution is prepared by dissolving 5 g HNO_3 in pure water to make a total volume of 1 L.

(a) What is the pH of this solution?

(b) What mass of the strong base sodium hydroxide (NaOH) would be required to neutralize the solution?

(c) If the sodium hydroxide is available as a 0.1 M solution, what volume of the basic solution will be needed to neutralize the acid?

3.22 More acid-base chemistry

(a) Hypochlorous acid (HOCl), a weak acid, can be used for disinfecting water. If 1 mg of HOCl is added to pure water to make up 1 L of liquid volume, what is the equilibrium pH?

(b) The pH of the solution in part (*a*) is to be raised to 7.0 by addition of NaOH, a strong base. What mass (g) of NaOH must be added?

3.23 Window cleaner chemistry

One of the active ingredients in window cleaner is ammonium hydroxide (NH_4OH). Consider the following situation: 10^{-3} moles of ammonium hydroxide is added to pure water to form 1 L of solution. The solution is placed into a 1 L glass jar and sealed (without any air space). The ammonium hydroxide fully dissolves to NH_4^+ and OH^-. The liberated ammonium ion is a weak acid ($pK_A = 9.23$). Determine the equilibrium pH of the liquid.

3.24 An equilibrium, acid-base, phase-change problem

A 2 L vessel is half filled with water. The remainder contains air. The vessel is sealed. A small quantity of ammonia, 10^{-3} moles, is injected into the vessel through a rubber septum. The ammonia is volatile. It also participates in an acid-base reaction.

(a) Set up a complete algebraic problem that would permit the pH of the solution to be determined. Specify the unknowns and give one equation for each unknown in the system.

(b) Since ammonia is a basic gas, the algebraic problem can be simplified because $[H^+] \ll [OH^-]$. Another simplification can be made because ammonia is not highly volatile and, therefore, most of the injected ammonia is in the aqueous phase at equilibrium. Apply these simplifications and solve for the pH of the water.

(c) What is the equilibrium partial pressure of ammonia?

3.25 A buffered solution

For this problem, consider a 1 L sealed vessel that is filled with pure water plus 0.0005 mol of sodium carbonate (Na_2CO_3, 53 mg, MW = 106 g/mol). There is no gas phase. The sodium carbonate dissolves completely in water. The equilibrium pH is 10.37. The equilibrium molar concentrations are 237 μM for OH^-, 237 μM for HCO_3^-, and 263 μM for CO_3^{2-}.

(a) What is the molar concentration of Na^+ in the solution?

(b) What is the alkalinity (*A*) of the solution?

(c) Evaluate the total hardness (TH) and carbonate hardness (CH) of the solution.

3.26 Water and limestone

Initially pure water at 25 °C, isolated from the atmosphere, comes into contact with solid $CaCO_3$. For each of the following pairs of species in solution, determine, at equilibrium, which has the higher molar concentration (i.e., mol/L). Explain your reasoning.

(a) Ca^{2+} or CO_3^{2-}
(b) H^+ or OH^-
(c) HCO_3^- or $H_2CO_3^*$

3.27 The pH of bubbly drinks

Pure water is carbonated by means of bubbling gaseous CO_2 through it. Then it is bottled.

(a) If the air space in the bottle above the water contains only CO_2 at a pressure of 2 atm, what is the equilibrium pH of the carbonated water? Assume that $T = 25$ °C.

(b) The bottle is opened, and the water is poured into a glass. Assume that the surrounding atmosphere is unpolluted and $T = 25$ °C. What is the new equilibrium pH? What changes take place in the water during the approach to equilibrium, and what factors govern the rate at which the new equilibrium is approached?

(c) Carry out this experiment: Pour a carbonated beverage into a glass, and then sprinkle some salt into the beverage. Describe what you observe. Does the addition of a small amount of salt influence the equilibrium pH? Does it influence the rate of approach to equilibrium? Explain.

3.28 The pH of seltzer water: An experiment

Consider a new, sealed bottle of seltzer water. The bottle is filled with water except for a small volume of gas that was initially pure CO_2. The pressure in the bottle exceeds 1 atm.

(a) Using only materials that you would find in an ordinary kitchen, describe how you could determine the pH of the seltzer water. (*Hint*: It is permitted to open the bottle.)

(b) Carry out the experiment and report your findings.

3.29 Cows and pH: Basic rain?

Ammonia (NH_3) is a basic gas produced by the decomposition of urea. A major emission source of ammonia is cattle feed lots. Determine the equilibrium pH of initially pure water that is exposed to air containing 25 ppb of ammonia, typical of polluted urban environments. Assume that the temperature is $T = 298$ K and the total air pressure is $P = 1$ atm. Neglect the presence of any other acid/base gases such as CO_2.

3.30 Carbonate system and chemical equilibrium

In a pristine (i.e., pollution-free) area, rainwater falls onto a limestone ($CaCO_3$) outcrop. A small pool collects in a bowl in the rock. Set up an analysis that would permit the determination of equilibrium concentrations of impurities in the water, assuming that the water is in equilibrium both with the solid $CaCO_3$ and with atmospheric CO_2. Neglect any gas-phase species other than CO_2, and assume that only the solid in contact with the water is $CaCO_3$.

(a) List all ionic and nonionic impurities in the water.

(b) List a sufficient number of algebraic equations to solve the problem. Each equation should relate species concentrations to thermodynamic properties and/or to the partial pressure of CO_2. State the scientific principle on which each equation is based.

3.31 Sulfur dioxide and acid rain

Among the reasons that sulfur dioxide is an important air pollutant is that it contributes to acid rain. In analogy with CO_2, SO_2 is a diprotic acid gas. We want to determine the pH of raindrops in equilibrium with polluted air that contains $P_{CO_2} = 350 \times 10^{-6}$ atm and $P_{SO_2} = 3.0 \times 10^{-8}$ atm at 25 °C. (The SO_2 partial pressure corresponds to the primary air quality standard on an annual average basis.)

(a) Write the electroneutrality relationship for ions in the raindrops.
(b) Using appropriate equilibrium relationships, rewrite the electroneutrality relationship in terms of $[H^+]$, P_{CO_2}, P_{SO_2}, and the equilibrium constants.
(c) Solve for the $[H^+]$ concentration of the raindrops. Compare your result with pH = 5.6 for "clean" rain.
(d) Relative to the case of a raindrop in equilibrium with clean tropospheric air ($P_{CO_2} = 350 \times 10^{-6}$ atm and $P_{SO_2} = 0$), will the concentrations of each of the following species in the polluted case be higher, lower, or unchanged? State your reasoning.
(i) $H_2CO_3^*$ **(ii)** HCO_3^- **(iii)** CO_3^{2-} **(iv)** H^+ **(v)** OH^-

3.32 Another acid-base problem involving a phase change
Microorganisms are grown in laboratory reactors to study their effectiveness in destroying hazardous organic compounds. Ammonium chloride (NH_4Cl) is a soluble salt that is often added to the reactor to provide a source of the nutrient nitrogen to the microorganisms. Consider a 2 L closed vessel that is half filled with pure water. A quantity of 500 mg of ammonium chloride is completely dissolved in the water. Some of the ammonium ions dissociate to form ammonia and liberate a proton. Some of the ammonia molecules escape into the gas phase. Calculate the pH of the solution.

3.33 Agricultural runoff
One of the most common types of fertilizers used by farmers is a form of the highly soluble salt ammonium chloride (NH_4Cl). The runoff from a farmer's field carries enough dissolved ammonium chloride into a nearby pond to result in a dissolved NH_4Cl concentration in the lake of 5 mg/L. At the same time, ammonia gas (NH_3) is being produced by the cows that feed around the farmer's lake, resulting in an atmospheric level of 500 ppb above the lake. Assume that the temperature is 298 K and the total air pressure is 1 atm, and neglect any effects of other impurities such as CO_2.
(a) Calculate the pH of the pond water.
(b) Determine whether the lake is an ammonia source or sink under these conditions.

3.34 Spelunking!
You have a couple of days off from school and decide to pursue your favorite sport: spelunking (that is, crawling through caves). So you put on your best Indiana Jones outfit and begin to explore some caves in the foothills of the Sierra Nevada Mountains. As you descend into one especially small and previously undiscovered cave, you begin to smell the rotten-egg odor characteristic of sulfide gas (H_2S). In fact, you notice that many of the rocks around you are glinting with what you recognize to be iron sulfide (FeS) crystals. Suddenly, the rock drops out from under your feet and you fall into a pool of water within a cavern room (see the figure at the top of the next page). The hole you dropped through clogs with debris and closes up after you. The room contains 100 L of air containing hydrogen sulfide gas (H_2S). The water is in contact with rock containing solid iron sulfide, and we know that sulfide is a weak acid governed by the following equilibrium relationships at 25 °C:

$$H_2S(aq) \Leftrightarrow H^+ + HS^- \qquad K_1 = 7.9 \times 10^{-8}\ \text{M}$$

$$HS^- \Leftrightarrow H^+ + S^{2-} \qquad K_2 = 1.2 \times 10^{-13}\ \text{M}$$

$$FeS \Leftrightarrow Fe^{2+} + S^{2-} \qquad K_{sp} = 6.3 \times 10^{-8}\ \text{M}^2$$

$$H_2S(g) \Leftrightarrow H_2S(aq) \qquad K_H = 0.11\ \text{M atm}^{-1}$$

(a) Set up the electroneutrality equation for the water in this system.

(b) Rewrite the equation from (*a*) in terms of known constants, the partial pressure of sulfide gas in the air, and $[H^+]$.

(c) You would like to measure the toxicity of the sulfide gas to determine whether you can expect to escape. Since you happen to be carrying some pH paper, you know that the pH of the water is 8.5. Calculate the partial pressure of sulfide in the air above the pool at equilibrium.

(d) Due to the high pH of the water and the presence of toxic sulfide gases, you must escape this pool immediately. But since you're such a talented environmental-engineering scientist, on your way out you decide to calculate whether the sulfide in the water has all come from dissolving rock or whether some other source of gaseous sulfide must be present. How do you do this, and what do you conclude?

3.35 A corrosion redox problem

Write a balanced chemical reaction for the corrosion of pure zinc metal by dissolved oxygen in water to produce zinc rust, $Zn(OH)_2$.

3.36 Combustion stoichiometry

Use the principle of stoichiometry to determine the amount of oxygen that is required to convert octane (C_8H_{18}) to CO_2 and H_2O.

(a) Write a properly balanced overall reaction.

(b) Calculate the mass of O_2 required per kg of octane.

3.37 Oxygen requirement for organic oxidation

(a) Using the principles of stoichiometry, write an overall reaction for the complete oxidation of acetone (CH_3COCH_3) to carbon dioxide and water.

(b) What mass of oxygen is required to completely oxidize 1 g of acetone?

3.38 By the light of a candle

Candles are made of paraffin, which is a mixture of pure hydrocarbons. For the questions that follow, assume that the effective molecular composition of paraffin is CH_2.

(a) When a candle is burned in air such that paraffin is fully oxidized, what are the chemical products of combustion?

(b) What mass (g) of oxygen (O_2) is required to fully oxidize 1 g of paraffin?

(c) Consider a vessel that has a volume of 0.1 m^3 (= 100 L). The vessel contains dry air at $P = 1$ atm and $T = 298$ K. What is the mass (g) of oxygen (O_2) in the vessel?
(d) An experiment conducted in open air reveals that a burning candle consumes 6 g of paraffin per hour. Use the results of parts (*b*) and (*c*) to estimate how long a candle would burn if placed in a sealed vessel as described in part (*c*).
(e) Give one reason related to the combustion process why the answer to part (*d*) would be considered only a magnitude estimate (i.e., not a precise estimate) of the time required for the candle to become extinguished.

3.39 Redox stoichiometry

Often, wastewater is disinfected with hypochlorous acid (HOCl) before being discharged to the environment. To avoid damaging the receiving waters, the disinfection step is often followed by a "dechlorination" step: Bisulfite (HSO_3^-) is added to the water to react with excess hypochlorous acid. The overall reaction is

$$HOCl + a\, HSO_3^- \rightarrow b\, Cl^- + c\, SO_4^{2-} + d\, X$$

where the letters *a–d* represent stoichiometric coefficients and the symbol X represents a chemical element or compound.
(a) Give the oxidation states of the following elements: Cl in HOCl, Cl^-, S in HSO_3^-, S in SO_4^{2-}.
(b) Determine the stoichiometric coefficients *a–c*.
(c) What is the unknown species, X, and its stoichiometric coefficient *d*?

3.40 Waste incineration

Many organic wastes are treated effectively by incineration, including sulfur-, nitrogen-, and chlorine-containing species. The following reaction describes the combustion of ethyl mercaptan (C_2H_5SH).

$$C_2H_5SH + a\,(O_2 + 3.78\, N_2) \rightarrow b\, CO_2 + c\, H_2O + d\, SO_2 + e\, N_2$$

(a) Determine the values of the stoichiometric coefficients for this reaction.
(b) Calculate the volume of air (measured at $T = 298$ K, $P = 1$ atm) required to combust one mole of ethyl mercaptan.
(c) If 100 mg of liquid ethyl mercaptan were incinerated in a 1 L closed vessel initially containing air at 1 atm and 25 °C, calculate the overall pressure within the vessel after the reaction has ceased and all products have cooled to 25 °C. (*Hint*: At equilibrium the relative humidity cannot exceed 100 percent; excess water would condense to a liquid.)

3.41 Up in smoke

The following reaction describes the incineration of liquid dichloropropane.

$$C_3H_6Cl_2(l) + a\, O_2(g) \rightarrow b\, CO_2(g) + c\, H_2O(l) + d\, HCl$$

(a) Determine the stoichiometric coefficients for this reaction.
(b) What volume of air (measured at $T = 298$ K, $P = 1$ atm) would be required for the stoichiometric combustion of one kilogram of dichloropropane?
(c) After adding 170 mg dichloropropane to a 500 mL vessel containing air at 1 atm and 25 °C, you seal the vessel and incinerate the dichloropropane. Calculate the overall pressure within the vessel after the reaction has ceased and all products have cooled to 25 °C.

(d) Calculate the molar concentration of HCl dissolved in water after the incineration reaction described in (*c*) has proceeded to completion and the exhaust products achieve equilibrium at 25 °C.

3.42 Theoretical oxygen demand

(a) Wastewater from a fruit-canning factory contains 150 mg/L of fructose ($C_6H_{12}O_6$). What is the theoretical oxygen demand of this wastewater (mg/L)?

(b) Determine the theoretical oxygen demand (in mg/L) for water that contains 50 mg/L of acetic acid (CH_3COOH).

3.43 Theoretical oxygen demand for a pure hydrocarbon

Pure hydrocarbons may be represented by the generic molecular formula C_mH_n, where m and n represent integers. Show that the theoretical oxygen demand for converting such a molecule to carbon dioxide and water is as follows:

$$\text{ThOD} = \frac{32m + 8n}{12m + n}$$

where ThOD has units of grams O_2 per gram of hydrocarbon.

3.44 Elementary pollutant dynamics

(a) The concentration, C, of a pollutant species in water is governed by the following differential equation:

$$\frac{dC}{dt} = S - LC$$

$$C(0) = 0.05 \text{ M}$$

where $S = 0.01$ M h^{-1} and $L = 0.5$ h^{-1}. What is the characteristic time for the species concentration to reach steady state?

(b) For this situation, what is the steady-state concentration?

3.45 Elementary reaction kinetics

(a) A species in a batch reactor undergoes first-order decay with a rate constant of 0.3 h^{-1}. How much time must pass before the species concentration is reduced to 10 percent of its initial value?

(b) A species undergoes first-order decay in a batch reactor. Three hours after the reactor is sealed, the species concentration is 20 percent of its initial value. What is the rate constant?

(c) Two reagents, A and B, are placed in a batch reactor where they undergo the following elementary reaction:

$$\text{A} + \text{B} \rightarrow \text{products}$$

The initial concentrations are [A] = 0.1 M and [B] = 0.001 M. The reaction rate constant is 3 M^{-1} h^{-1}. What are the steady-state concentrations of [A] and [B]?

(d) For the conditions in part (*c*), what is the characteristic time to achieve steady state in the reactor?

3.46 BOD evaluation

If the dissolved oxygen concentration measured during a BOD test is 9 mg/L initially, 6 mg/L after 5 days, and 3 mg/L after an indefinitely long period of time, calculate the 10-day BOD.

3.47 Reaction kinetics for BOD

BOD tests were experimentally conducted with a wastewater at two different temperatures, 15 °C and 25 °C, with the following results. Determine the first-order reaction rate for the same wastewater at 10 °C.

Elapsed time (d)	Dissolved oxygen (15 °C)	Dissolved oxygen (25 °C)
0	10.0 mg/L	9.0 mg/L
5	6.0 mg/L	3.7 mg/L
100	3.0 mg/L	2.0 mg/L

3.48 Carbonaceous and nitrogenous oxygen demand

Organic contaminants may be represented by the generic molecular formula $C_mH_nO_pN_r$, where *m, n, p,* and *r* represent integers. The oxidation of an organic may be carried out in two steps, carbonaceous oxidation followed by nitrogenous oxidation.

$$C_mH_nO_pN_r + a\,O_2 \rightarrow b\,CO_2 + c\,H_2O + d\,NH_3$$

$$NH_3 + 2\,O_2 \rightarrow NO_3^- + H_2O + H^+$$

(a) Determine the value of the stoichiometric coefficients *a, b, c,* and *d* in terms of *m, n, p,* and *r.*

(b) Give an expression for the theoretical oxygen demand (ThOD) per gram of organic due to carbonaceous oxidation alone (give answer in units of grams O_2 per gram of organic).

(c) Give an expression for the total ThOD (carbonaceous + nitrogenous) per gram of organic.

(d) Try out your new expressions by computing both the carbonaceous and the nitrogenous ThOD associated with 150 mg/L glycine ($C_2H_5O_2N$).

3.49 Biochemical oxygen demand

An experiment is conducted in a batch reactor to measure the BOD in a water sample. The water sample is placed in the reactor and the dissolved oxygen content (DO) is measured as a function of time. The measurement data are presented below. Use these data to answer the following questions.

Time (d)	DO (mg/L)
0	10
1	8.7
2	7.6
3	6.8
4	6.2
5	5.7
6	5.3
7	5.0
8	4.8
9	4.6
10	4.5
15	4.1
∞	4.0

(a) What is the ultimate BOD of the sample (BOD_u)?

(b) What is the 5-day BOD of the sample (BOD_5)?

(c) What is the reaction rate constant, k, for BOD degradation?

3.50 Elementary reaction kinetics

An experiment is conducted in which species A is placed in a batch reactor with initial concentration A_0. The species undergoes the following elementary reaction, which has rate constant k:

$$A + A \rightarrow \text{products}$$

(a) What is the rate law for this reaction?

(b) What is the rate of change of the concentration of A in the reactor?

(c) What is the characteristic time for the reaction to proceed to completion?

3.51 The mathematics of first-order kinetics

A general solution has been presented to a class of problems that frequently arise in environmental engineering. The mathematical problem can be written as

$$\frac{d[A]}{dt} = k([A]_e - [A]); \qquad [A](t = 0) = [A]_0$$

The solution to this equation is

$$[A](t) = [A]_0 e^{-kt} + [A]_e (1 - e^{-kt})$$

(a) Demonstrate that the given solution satisfies the differential equation and initial condition.

(b) Determine, as a function of k, the time required for $[A](t)$ to reach a value of $([A]_0 + [A]_e)/2$. Compare this result with the characteristic time, $\tau = 1/k$, for this system.

3.52 Elementary kinetics

A closed system contains air pollutant species A, B, and C, which undergo the following elementary reactions. The mole fractions of A and B are fixed at $Y_A = 100$ ppb, $Y_B = 50$ ppb. The initial mole fraction of C is $Y_C(0) = 1$ ppb.

$$A + B \xrightarrow{1} C \qquad k_1 = 2 \times 10^{-4}\ \text{ppb}^{-1}\ \text{min}^{-1}$$

$$C \xrightarrow{2} A + B \qquad k_2 = 0.2\ \text{min}^{-1}$$

(a) Write a differential equation that describes the rate of change of Y_C in terms of rate constants k_1 and k_2 and mole fractions Y_A and Y_B.

(b) Evaluate the steady-state mole fraction Y_C (ppb).

(c) Evaluate the characteristic time required for the system to approach steady state (min).

(d) Sketch Y_C (ppb) versus time (min), providing as much detail as possible (i.e., show the initial condition, steady-state concentration, characteristic time, and general curve shape). (*Hint*: You do not have to solve the differential equation to do this.)

3.53 Determining the rate of a chemical reaction

An experiment is conducted to investigate the rate of ozone disappearance from indoor air in a house. Initially, the windows and doors are opened to allow the indoor ozone concentration to rise to the outdoor level (230 ppb). Then all doors and windows are closed. Assume that the house can be considered sealed at this point. The concentration of ozone in the room air is monitored over time. The results are shown in the following figure.

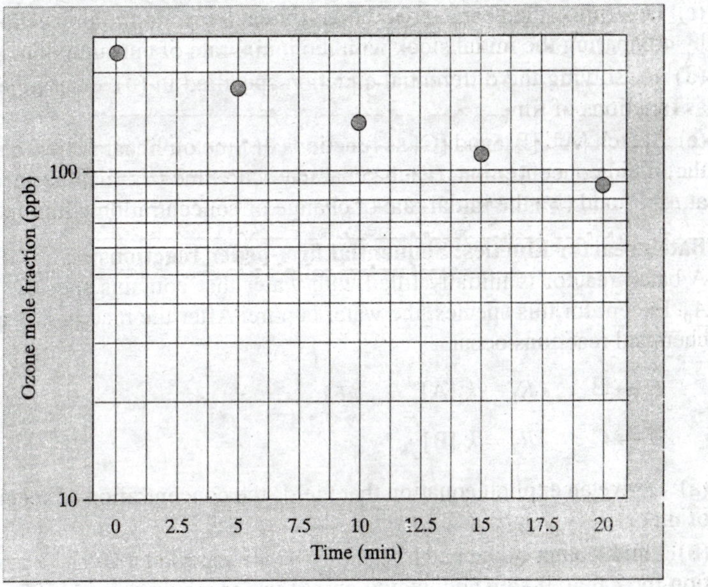

(a) What is the order of the reaction by which ozone disappears?
(b) Use the experimental data to evaluate the rate constant, k, of the reaction. (*Note:* Be sure to specify units.)
(c) Predict the ozone mole fraction at $t = 30$ min.

3.54 Popcorn kinetics: An experiment
The process of popping corn can be described as a kinetic reaction:

unpopped kernel → popped kernel

Devise and conduct an experiment to determine (*a*) the form of the rate law for this reaction and (*b*) the rate constant, k. Describe your experiment and the results in one page. (*Hint*: In formulating a rate law, ignore the initial period required to heat the popcorn kernels to popping temperature.)

3.55 Kinetics of elementary reactions
Consider the following batch reactors and answer parts (*a*)–(*e*) for each.

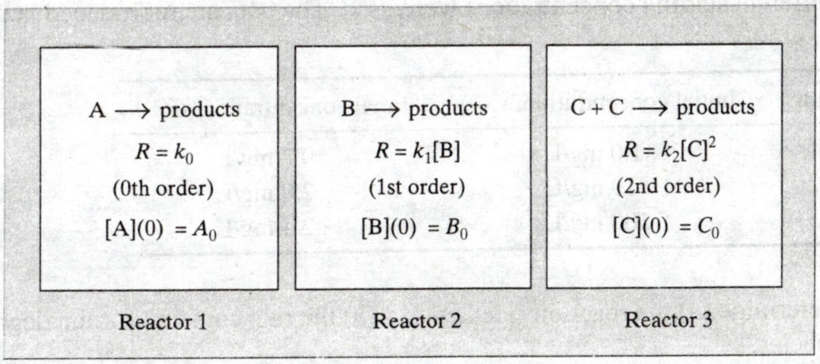

(a) If the units of [A], [B], and [C] are g m^{-3}, what are the units of k_0, k_1, and k_2?
(b) Write a differential equation that gives the rate of change of the species concentration with time for each reactor.
(c) Determine the characteristic time for each reaction to proceed toward completion by comparing the initial stock with the initial rate of consumption.
(d) By solving the differential equation specified in (*b*), determine [A], [B], and [C] as functions of time.
(e) Sketch [A], [B], and [C] as functions of time on linear scales, clearly indicating (i) the initial concentration, (ii) the characteristic time (τ), (iii) the species concentration at $t = \tau$, and (iv) the initial rate of change of concentration with time.

3.56 Batch reactor kinetics: Sequential first-order reactions
A batch reactor is initially filled with water that contains species A at concentration A_0. Except for this species, the water is pure. After the reactor is sealed, the following chemical reactions occur:

$$A \rightarrow B \qquad R_1 = k_1[A]$$

$$B \rightarrow C \qquad R_2 = k_2[B]$$

(a) Derive an explicit equation that yields the concentration of species A as a function of time.
(b) Fluid is kept in the batch reactor for a time period *T*. Write a mathematical condition for *T* that, if satisfied, would permit the approximation $[A](T) \approx A_0$. (*Hint*: $T = 0$ is unnecessarily restrictive.)
Assume for the remainder of the problem that $[A](t) \approx A_0$. Also assume that the reaction rate constants satisfy the condition $k_1 \ll k_2$.
(c) Write a governing equation that describes the rate of change of [B].
(d) Derive an explicit expression for the steady-state concentration of [B].
(e) What is the characteristic time required for the steady-state solution in (*d*) to be attained?
(f) Derive an explicit equation for the time-dependent concentration of [B].

3.57 Batch reactor kinetics: Unknown rate law
An undesired contaminant is present in a fluid. The contaminant is observed to degrade spontaneously, so a suitable treatment technique is simply to detain the fluid for a certain period before use.

To quantify degradation kinetics, a series of experiments was conducted in a well-stirred batch reactor. Each experimental run was conducted for a one-hour period. The contaminant concentration was measured at the beginning and at the end of each run. Different starting concentrations were used. The data are reproduced in the table below.

Run	Initial concentration, $C(0)$	Final concentration, $C(1\ \text{h})$
1	1.0 mg/L	0.7 mg/L
2	10 mg/L	2.0 mg/L
3	100 mg/L	2.4 mg/L

Determine (*a*) the reaction rate law and (*b*) the rate constant for the degradation of this species.

3.58 Kinetics of radioactive decay

A thin layer of a radium salt is deposited onto a metal disk. The layer contains N_{226} atoms of ^{226}Ra. As ^{226}Ra undergoes radioactive decay, the ^{222}Rn atoms that are produced are gaseous and escape from the disk. The disk is placed into a sealed vessel for a period of one year. Initially, there is no ^{222}Rn in the vessel. Within the sealed vessel, the following first-order reactions occur:

$$^{226}\text{Ra} \xrightarrow{1} {}^{222}\text{Rn} + \alpha \qquad k_1 = 1.4 \times 10^{-11}\ \text{s}^{-1}$$

$$^{222}\text{Rn} \xrightarrow{2} {}^{218}\text{Po} + \alpha \qquad k_2 = 2.1 \times 10^{-6}\ \text{s}^{-1}$$

(a) Show that it is reasonable to assume that the ^{226}Ra content of the disk remains constant for the entire year.

(b) Write a material-balance equation that describes the rate of change of the number of ^{222}Rn atoms in the vessel, N_{222}.

(c) Determine the characteristic time for N_{222} to come to steady state in the vessel (assuming that the quantity of ^{226}Ra is fixed).

(d) The "activity" of a radioactive species is given by the product, $I = kN$, where k is the rate constant for radioactive decay and N is the number of atoms of the species in the system. Show that at steady state the activity of ^{222}Rn is the same as the activity of ^{226}Ra, that is, $I_{222} = I_{226}$.

3.59 Kinetics of a third-order reaction

Species A in water is experimentally determined to degrade by a third-order reaction with rate constant k. Consider a batch reactor of volume V filled with water and with an initial concentration of species A given by $[A](0) = A_0$. In answering the following questions, assume that only the third-order reaction occurs.

(a) Complete this differential equation describing the rate of disappearance of A from the reactor:

$$\frac{d[A]}{dt} =$$

(b) If the units of [A] are moles per liter, what are the appropriate units for k?

(c) What is the characteristic time for A to disappear from the reactor?

(d) Solve the governing equation to obtain an explicit equation for $[A](t)$.

3.60 Biotreatment and Henry's law

Methane gas dissolved in water can be used to grow bacteria, which degrade chlorinated solvents.

(a) If each gram of methane can produce 0.7 g of bacteria, calculate the mass of bacteria that could be produced from 1 L of water. The water was initially exposed to 25 percent methane in air at a pressure of 1 atm. After equilibration, it was isolated from further gas-phase contact.

(b) Will there be enough dissolved oxygen in the water for all the methane to be stoichiometrically oxidized? Assume that some of the methane is fully oxidized to H_2O and CO_2. The effective chemical formula for new bacterial cells is $C_5H_7NO_2$. Assume a sufficient quantity of nitrogen is dissolved in the water so that it is not limiting.

3.61 Microbial growth kinetics

In a batch reactor, the growth of microbial cell mass concentration can be described by the relationship

$$\frac{dX}{dt} = r_g X \qquad r_g = \frac{Y k_m S}{K_s + S}$$

where r_g is the cell growth rate coefficient. (In general, one would need to include a decay term in the expression, but for short time periods decay may be safely ignored.) An experiment is established in a batch reactor in which a constant substrate concentration, S, is maintained. Adequate oxygen and nutrients are provided so that they do not limit cell growth. Cell mass concentration is monitored as a function of time.

(a) Define t_d to be the time required for cell mass concentration to double. Derive an expression that relates t_d to the growth rate coefficient, r_g.

(b) An experiment is conducted to measure t_d twice, each time with a different substrate concentration. From the results presented below, determine $(k_m Y)$ and K_s. Be certain to specify appropriate units.

S (g m^{-3})	t_d (h)
10	32
100	8

(c) A third experiment is to be conducted with a substrate concentration of 30 g m^{-3}. Predict the length of time required for the cell mass concentration to increase to 10 times its initial value.

4

Transport Phenomena

Transport phenomena are encountered in almost every aspect of environmental engineering science. In assessing the environmental impacts of waste discharges, we seek to predict the impact of emissions on contaminant concentrations in nearby air and water. Contaminant transport must be understood to evaluate the effect of wastewater discharge to a river on downstream water quality or the effect of an incinerator on downwind air pollutant levels. In waste treatment technologies, contaminant transport within the control device influences the overall efficiency. In many instruments used to measure environmental contamination, performance depends on the effective transport of contaminants from the sampling point to the detector.

The physical scales of concern for contaminant transport span an enormous range. At the low end, we are interested in phenomena that occur over molecular dimensions, such as the transport of a contaminant into the pore of an adsorbent. At the upper extreme, we consider the transport of air and waterborne contaminants over global distances. This range of linear dimensions encompasses approximately 15 orders of magnitude. Not surprisingly, some phenomena that are important at one end of the spectrum are irrelevant at the other.

Transport phenomena also are important in other fields of study, notably chemical and mechanical engineering. In these fields, the subject is often subdivided into fluid mechanics, heat transfer, and mass transfer. In environmental engineering, where the focus is on the movement of contaminants in air and water, elements of all three of these branches are considered. The essential ingredients for understanding contaminant transport processes are knowledge of the basic physical phenomena coupled with the analytical tools of engineering mathematics.

Because of the importance of both molecular and particulate impurities in environmental fluids, the dominant transport processes of both molecules and particles will be discussed in this chapter. We consider both the movement of impurities *with* fluids and their movement *relative to* fluids. In Chapter 5, we will explore how the transformation mechanisms studied in Chapter 3 and the transport mechanisms explored in this chapter are combined to predict system behavior using mathematical models.

4.A BASIC CONCEPTS AND MECHANISMS

4.A.1 Contaminant Flux

Transport of both molecules and particles is commonly quantified in terms of *flux density,* or simply *flux.* Flux is a vector quantity, comprising both a magnitude and a direction. The flux vector points in the direction of net contaminant motion, and the magnitude indicates the rate at which the contaminant is moving.

For example, think of an environmental fluid as depicted in Figure 4.A.1. Imagine a small square frame suspended in this fluid and centered at a point of interest. The frame is oriented so that the transport of contaminants through it is maximized. Then the flux vector points in a direction normal to the frame, aligned with the direction of contaminant transport. The magnitude of the flux vector is the net rate of contaminant transport per unit area of the frame. Since the flux can vary from one position to another, the flux at a point represents the net transport through the frame per area in the limit of the frame becoming infinitesimally small. The common units of flux are mass or moles per area per time. We will use the symbol J to represent flux. An arrow is written above J when we wish to emphasize its vector character. Contributions to flux

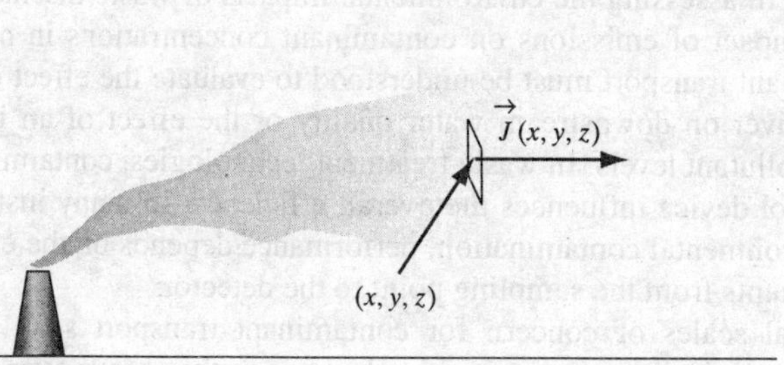

Figure 4.A.1 Flux, $\vec{J}$, is a vector quantity whose value varies with position (x, y, z). The contaminant flux vector points in the direction of transport, and its magnitude is the quantity transported (usually mass or moles) per area per time.

Table 4.A.1 Dominant Mechanisms That Cause Transport of Molecules and Particles in Environmental Fluids

Transport mechanism	Species[a]	Description	1-D flux[b]
Advection	m, p	Movement with fluid	$J_a = CU$
Gravitational settling	p	Transport induced by gravity	$J_g = Cv_t$
Molecular diffusion or Brownian motion[c]	m, p	Transport caused by random thermal motion	$J_d = -D\frac{dC}{dx}$
Turbulent diffusion	m, p	Transport caused by apparently random fluid velocity fluctuations in turbulent flow	$J_t = -\varepsilon_t\frac{dC}{dx}$
Shear-flow dispersion	m, p	Transport caused by nonuniform fluid flow with position	$J_s = -\varepsilon_s\frac{dC}{dx}$
Hydrodynamic dispersion (porous media)	m, p	Transport caused by nonuniform flow through porous material	$J_h = -\varepsilon_h\frac{dC}{dx}$
Electrostatic drift[d]	m, p[e]	Movement of charged species in an electric field	$J_e = Cv_e$
Inertial drift	p	Transport associated with acceleration of a fluid	$J_f = Cv_f$

[a]m = molecule, p = particle.
[b]C = species concentration, U = fluid velocity, v_t = settling velocity, D = molecular or Brownian diffusivity, ε_t = turbulent diffusivity, ε_s = shear-flow dispersivity, ε_h = dispersion coefficient, v_e = electrostatic drift velocity, v_f = inertial drift velocity, J = flux.
[c]Molecular diffusion and Brownian motion apply to molecules and particles, respectively; both occur from the same fundamental process.
[d]See §7.C.
[e]Electrostatic drift applies to charged species only.

may be caused by several mechanisms (Table 4.A.1). A major thrust of this chapter is to explore the basic physical mechanisms that cause contaminant flux.

4.A.2 Advection

Advection is the transport of material caused by the net flow of the fluid in which that material is suspended. Whenever a fluid is in motion, all contaminants in the fluid, including both molecules and particles, are advected along with the fluid.

Figure 4.A.2 shows how an advective flux can be evaluated. Consider a fluid flowing through a tube with a uniform velocity. Assume that the fluid contains a uniform concentration, C, of some contaminant. Focus on the amount of contaminant in a slice of fluid of thickness ΔL. As shown in the upper part of the diagram, at time t, the leading edge of the slice has just reached position x. At some later time, $t + \Delta t$ (see the lower part of the diagram), the trailing edge of the slice passes position x.

The advective flux vector for this contaminant points along the axis of the tube, in the same direction as the velocity vector. The magnitude of the flux vector at position x is obtained by dividing the amount of contaminant that passes position x by the cross-sectional area of the tube (A) and by the time interval required (Δt). The amount of contaminant that passes position x is the contaminant concentration (C) times the volume of the depicted slice ($\Delta L \times A$). Note that the time interval (Δt) is equal to the thickness of the slice divided by the fluid velocity ($\Delta L/U$). So, for this case, the advective flux magnitude, J_a, is the quantity of contaminant passing position x per area

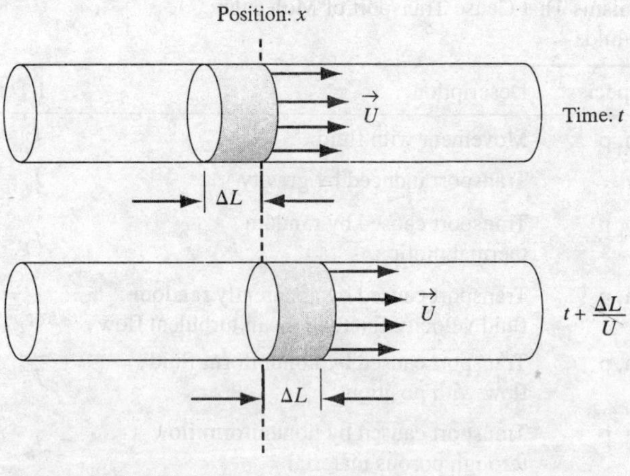

Figure 4.A.2 Advective flux of contaminant through a tube. The two pictures represent the same tube at two points in time, t and $t + \Delta t$, where $\Delta t = \Delta L/U$.

per time:

$$J_a = \frac{C \times \Delta L \times A}{A(\Delta L/U)} = CU \tag{4.A.1}$$

In general, the three-dimensional advective flux vector is the product of the contaminant concentration and the fluid velocity:

$$\vec{J_a}(x, y, z) = C(x, y, z) \times \vec{U}(x, y, z) \tag{4.A.2}$$

4.A.3 Molecular Diffusion

The molecules in air and water are constantly moving. If it could be viewed at a molecular scale, this movement would appear random and chaotic. Molecular movement in gases is particularly frenzied. In air at ordinary environmental conditions, a typical molecule, moving at a speed of ~400 m s^{-1}, collides with other molecules on the order of 10 billion times per second. Each collision involves an exchange of momentum between the participants, causing them to change direction and speed. Every cubic centimeter of air has a phenomenally large number of molecules (~25 million trillion at $T = 298$ K and $P = 1$ atm) participating in this dance. In water, where the molecules are packed a thousand times more densely, molecular motion is less frenetic, but still extremely energetic.

Although the molecular-scale motion seems hopelessly disordered, the macroscopic effects are well understood and predictable. Qualitatively, the random motion of fluid molecules causes a net movement of species from regions of high concentration to regions of low concentration. This phenomenon is known as *molecular diffusion*. The rate of movement depends on the concentration difference, with larger differences leading to higher rates of transport. The rate also varies according to how far the species must travel from high to low concentration: The longer the distance, the lower the flux. The rate of transport varies according to the molecular properties of the species, particularly size and mass, with larger size and larger mass resulting in slower transport. The properties of the fluid itself play a key role: Molecular diffusion is much slower in water than in air because of the much closer packing density of water molecules.

Our discussion is restricted to conditions in which the diffusing species is present at a low mole fraction ($\ll$1), referred to as the *infinite dilution* condition. In most environmental engineering applications, this assumption is appropriate. At high concentrations, diffusion can cause significant net flow of the bulk fluid (Bird et al., 1960; Cussler, 1984).

Fick's Law

To better understand diffusion, it is useful to think about a specific physical system, such as the one depicted in Figure 4.A.3. A glass bulb at 25 °C contains a liquid with a moderate vapor pressure, such as ethylbenzene (1280 Pa at 25 °C; see Table 3.B.1). The bulb is open to the air through a thin cylindrical glass tube. Ethylbenzene molecules evaporate from the liquid, maintaining a partial pressure in the gas phase of the bulb equal to the saturation vapor pressure. Because of their random motion, the ethylbenzene molecules gradually migrate through the tube and escape into the open air. After a short transient period, the net rate of transport of ethylbenzene molecules through the tube will reach a steady value that will be maintained as long as there is liquid ethylbenzene in the bulb. The escape rate of ethylbenzene from the tube varies in inverse proportion to tube length, in proportion to the cross-sectional area of the tube, and in proportion to the partial pressure of the species in the bulb. These characteristics can be expressed quantitatively by the relationship

$$J_d \propto -\frac{\Delta C}{\Delta x} \tag{4.A.3}$$

where J_d is the diffusive flux density (moles per cross-sectional area of the tube per time), ΔC is the change in concentration of ethylbenzene molecules across the tube, and Δx is the tube length. The minus sign appears in this relationship to remind us that the diffusive flux proceeds in the direction of decreasing concentration.

By introducing the proportionality constant, D, this expression can be converted to an equation:

$$J_d = -D\frac{\Delta C}{\Delta x} \tag{4.A.4}$$

The constant, D, is called the *diffusion coefficient*, or *diffusivity*. It is a property of the diffusing species (ethylbenzene in this case), the fluid through which it is diffusing

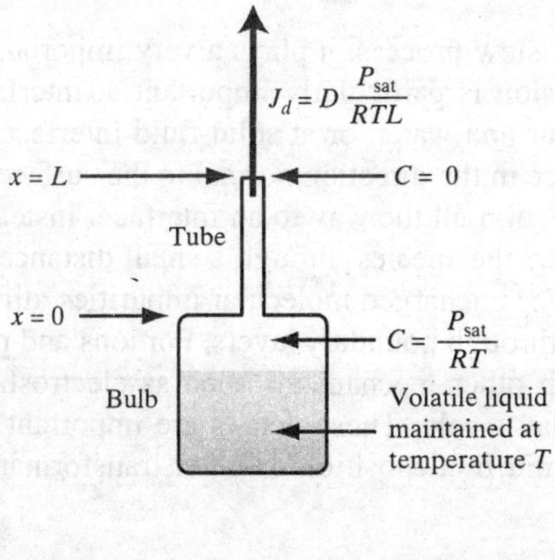

Figure 4.A.3 Apparatus for producing a controlled diffusive flux of a volatile species. P_{sat} is the saturation vapor pressure of the species at temperature T. Provided that the time scale for evaporation and condensation is much more rapid than the time scale for diffusion through the tube, a molar concentration P_{sat}/RT will be attained throughout the bulb. The vapor molecules will diffuse through the tube with a net diffusive flux J_d.

(air), and environmental conditions such as temperature and pressure. The diffusion coefficient has dimensions of length squared per time; typically, D is reported in units of $cm^2\ s^{-1}$. For molecules in air, a characteristic diffusivity is 0.1 $cm^2\ s^{-1}$. In water, molecular diffusivities are on the order of $10^{-5}\ cm^2\ s^{-1}$.

If we treat a fluid as a continuous substance, and take the limit of equation 4.A.4 as the distance Δx becomes infinitesimally small, we can replace $\Delta C/\Delta x$ by the derivative dC/dx and write

$$J_d = -D\frac{dC}{dx} \tag{4.A.5}$$

For applications in three dimensions, the general equation for diffusive flux can be written as follows:

$$\vec{J_d}(x, y, z) = -D\left(\frac{\partial C}{\partial x}, \frac{\partial C}{\partial y}, \frac{\partial C}{\partial z}\right) \tag{4.A.6}$$

In words, equation 4.A.6 states that each component of the flux vector is proportional to the partial derivative of concentration in that coordinate direction, and the proportionality constant is $-D$ for each direction. So diffusive flux occurs in the direction opposite to the concentration gradient at a rate that is proportional to its magnitude. Equations 4.A.5 and 4.A.6 are known as Fick's first law of diffusion, or simply Fick's law. See Exhibit 4.A.1 for more details about the system depicted in Figure 4.A.3.

Significance of Diffusion

Diffusion is a slow transport process. Albert Einstein showed that the characteristic distance a molecule (or a particle) will travel by diffusion in time t is given by

$$x \sim \sqrt{2Dt} \tag{4.A.7}$$

So a gas molecule with a diffusivity of 0.1 $cm^2\ s^{-1}$ can be expected to move a characteristic distance of 5 mm in a second, 3 cm in a minute, 30 cm in an hour, and 1 m in a day. Molecules in water, with diffusivities lower by a factor of 10^{-4}, will travel only 1 percent as far on these time scales.

Another perspective is gained by comparing the simple form of Fick's law (equation 4.A.4) with the simple advective flux expression (equation 4.A.1). If a species concentration diminishes from C to zero over some distance Δx, then diffusion causes transport at a rate equivalent to advection at a velocity $D/\Delta x$. Given the small values of D for air and especially for water, we see that the effective diffusive velocity is very small except when the concentration changes by a large fractional amount over a very small distance.

Although molecular diffusion is a slow process, it plays a very important role in contaminant transport and fate. Diffusion is particularly important at interfaces, for example between two fluids such as air and water, or at solid-fluid interfaces. There can be no fluid advection at an interface in the direction normal to the surface. So impurities cannot be transported by advection all the way to an interface. Instead, some other transport mechanism must convey the species through a small distance, known as a *boundary layer*, to the interface. For uncharged molecular impurities, diffusion is the dominant mechanism of transport through boundary layers. For ions and particles, diffusion always contributes, although other mechanisms such as electrostatic drift and gravitational settling may also play a role. These details are important because impurities may be removed from a fluid by deposition or other transformation processes that occur on surfaces. Often, the rate of transport to the surface governs the overall removal rate.

EXHIBIT 4.A.1 *An Application of Diffusion*

Let's further explore the behavior of the system depicted in Figure 4.A.3. Devices like this are used to release a volatile substance at a constant rate. By diluting the emissions from the top of the tube with a known flow rate of contaminant-free air, one generates an air stream with a constant, known concentration of the volatile substance. This air stream can be used for instrument calibration among other purposes.

When liquid is first placed into the bulb, some time must elapse before the diffusive flux leaving the tube reaches a steady value. An estimate of this time is obtained by rearranging equation 4.A.7 (substituting $\tau_{\text{diffusion}}$ for t and L for x):

$$\tau_{\text{dffusion}} \sim \frac{L^2}{2D} \quad (4.A.8)$$

This expression yields an estimate of the time required for a molecule to diffuse through some distance L. It is a good estimate for the characteristic time required to establish a steady concentration profile throughout the tube length. For typical values in a device of this sort, $L \sim 5$ cm and $D \sim 0.1$ cm^2 s^{-1}, so $\tau_{\text{diffusion}} \sim 2$ min. With conditions held steady for a time $t \gg \tau_{\text{diffusion}}$, the flux will approach a steady value that will be maintained as long as liquid remains in the bulb.

The characteristic time for the liquid to completely evaporate is obtained as the number of moles of liquid in the bulb divided by the molar rate of escape by diffusion:

$$\tau_{\text{evaporation}} \sim \frac{\left(\frac{\rho V}{\text{MW}}\right)}{J_d A} \quad (4.A.9)$$

where ρ is the liquid density, V is the liquid volume, MW is the molecular weight of the diffusing species, and A is the cross-sectional area of the tube. For the case of ethylbenzene, we can estimate an evaporation time, $\tau_{\text{evaporation}} \sim 4 \times 10^7$ s ~ 500 d. We have assumed the following values for the input data: $\rho V = 1$ g, MW $= 106$ g mol^{-1}, $J_d = 7.2 \times 10^{-9}$ mol cm^{-2} s^{-1} ($D = 0.07$ cm^2 s^{-1}, $P_{\text{sat}} = 1280$ Pa, $R = 8.31 \times 10^6$ cm^3 Pa K^{-1} mol^{-1}, $T = 298$ K, and $L = 5$ cm), and $A = 0.031$ cm^2 (0.2 cm inner tube diameter). For times t that satisfy $\tau_{\text{diffusion}} \ll t \ll \tau_{\text{evaporation}}$, the diffusive flux from the tube into the air will be constant. During this time interval, the concentration profile within the tube will also be constant, varying linearly with position x, as shown in Figure 4.A.4.

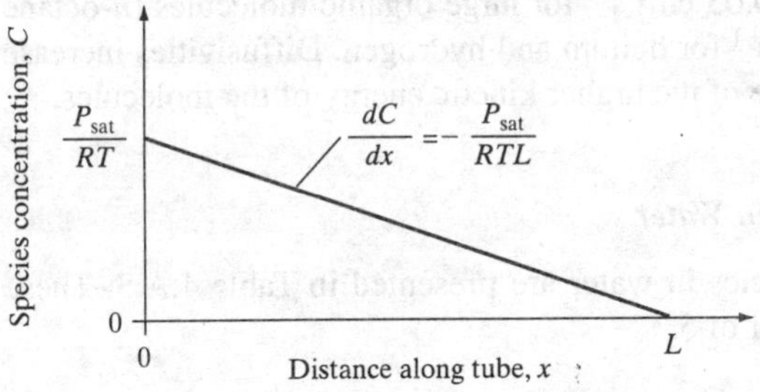

Figure 4.A.4 Steady concentration profile within the tube shown in Figure 4.A.3, valid for times t that satisfy $\tau_{\text{diffusion}} \ll t \ll \tau_{\text{evaporation}}$.

Diffusivity of Molecules in Air

Predictions of diffusive flux depend on the selection of an appropriate diffusion coefficient, D. Measured diffusion coefficients for selected species in air are presented in Table 4.A.2. This table emphasizes species of environmental interest, with air at ordinary environmental temperatures and pressures as the background fluid. Note that there are discrepancies in the data reported from different sources (e.g., ammonia).

Table 4.A.2 Measured Diffusion Coefficients for Species in Air at 1 atm

Species	Formula	T (K)	D ($cm^2\ s^{-1}$)	Reference[a]
Ammonia	NH_3	273	0.198	Bretsznajder, 1971
		273	0.216	McCabe et al., 1993
Benzene	C_6H_6	273	0.077	McCabe et al., 1993
		298.2	0.096	Cussler, 1984
Carbon dioxide	CO_2	273	0.138	McCabe et al., 1993
		317.2	0.177	Cussler, 1984
Ethanol	C_2H_5OH	273	0.102	McCabe et al., 1993; Cussler, 1984
		313	0.147	Reid et al., 1987 (pressure = 1 bar)
Helium	He	276.2	0.624	Cussler, 1984
Hydrogen	H_2	273	0.611	Cussler, 1984; McCabe et al., 1993
Methane	CH_4	273	0.196	Cussler, 1984
Methyl alcohol	CH_3OH	273	0.133	McCabe et al., 1993
n-Octane	C_8H_{18}	273	0.051	McCabe et al., 1993
Naphthalene	$C_{10}H_8$	273	0.051	McCabe et al., 1993
		303	0.087	Reid et al., 1987
Oxygen	O_2	273	0.178	Cussler, 1984; McCabe et al., 1993
Toluene	C_7H_8	273	0.071	McCabe et al., 1993
		299.1	0.086	Cussler, 1984
Water	H_2O	273	0.219	Bretsznajder, 1971; McCabe et al., 1993
		289.1	0.282	Cussler, 1984
		298.2	0.260	Cussler, 1984
		312.6	0.277	Cussler, 1984
		313	0.292	Reid et al., 1987

[a] When two sources disagree by less than 5 percent, only one value is listed.

The data in Table 4.A.2 show that the diffusivity of gases in air varies over about an order of magnitude from about 0.05 $cm^2\ s^{-1}$ for large organic molecules (*n*-octane and naphthalene) to about 0.6 $cm^2\ s^{-1}$ for helium and hydrogen. Diffusivities increase with increasing temperature because of the higher kinetic energy of the molecules.

Diffusivities of Molecular Species in Water

Experimental data for selected species in water are presented in Table 4.A.3. These data exhibit a range of about a factor of 5.

4.A.4 Dispersion

Our everyday experience tells us that impurities released into open air or water do not remain confined at high concentration in a small volume for very long. Cooking odors, incense, and cigarette smoke all become detectable throughout an indoor space within minutes of their release. The visible smoke plume from a fire can be seen to spread significantly as it travels downwind. The rate of contaminant spreading is of substantial interest in environmental engineering. And although fundamentally it occurs as a result of advection and diffusion, the rate of spreading is much more rapid than might be inferred from the discussion so far.

Table 4.A.3 Measured Diffusion Coefficients for Species in Water

Species	Formula	T (K)	D (cm^2 s^{-1})	Reference[a]
Benzene	C_6H_6	298	1.02×10^{-5}	Cussler, 1984
Hydrogen	H_2	298	3.36×10^{-5}	Bretsznajder, 1971
		298	4.50×10^{-5}	Cussler, 1984
Air		298	2.00×10^{-5}	Cussler, 1984
Methane	CH_4	275	0.85×10^{-5}	Reid et al., 1987
Carbon dioxide	CO_2	298	2.00×10^{-5}	Reid et al., 1987
Methanol	CH_3OH	288	1.26×10^{-5}	Reid et al., 1987
		298	0.84×10^{-5}	Cussler, 1984
Ethanol	C_2H_5OH	288	1.00×10^{-5}	Reid et al., 1987
		298	0.84×10^{-5}	Cussler, 1984
Ethylbenzene	C_8H_{10}	293	0.81×10^{-5}	Reid et al., 1987
Oxygen	O_2	298	2.60×10^{-5}	Bretsznajder, 1971
Vinyl chloride	C_2H_3Cl	298	1.34×10^{-5}	Reid et al., 1987

[a]When two sources disagree by less than 5 percent, only one value is listed.

Uniform, steady advection does not cause pollutant spreading, and we have seen that molecular diffusion is a slow process. Consider the following situation (Cussler, 1984). Imagine that a thin stream of smoke particles is released along the western coast of the United States and is steadily advected at a constant, uniform wind speed of 5 m s^{-1} to the eastern coast. Assume that particles have a diameter of 0.1 μm. How much spreading of the plume will occur by diffusion during this transcontinental travel? The time required for the wind to travel that distance is $\sim 10^6$ s or 12 d (5000 km ÷ 5 m s^{-1}). In Example 4.B.3 it will be shown that a 0.1 μm particle has a diffusivity in air of 7×10^{-6} cm^2 s^{-1}. The plume spread caused by diffusion is estimated to be the characteristic distance traveled due to Brownian motion, which, according to equation 4.A.7, is roughly 4 cm. Clearly, this is not an accurate description! It is entirely contrary to our experience to think that we could even detect a smoke plume at a distance of 5000 km from its source. Yet this example accurately applies the tools and concepts we have introduced. What went wrong?

Our error was made in assuming that the wind speed is uniform and steady. It is a property of most fluid flows, including winds and ocean currents, that they are neither constant nor uniform. When fluid velocity varies with time or position, contaminants in the fluid tend to be transported from high concentration to low concentration. The spreading of contaminants by nonuniform flows is called *dispersion.** It is not a fundamentally distinct transport process. Instead, dispersion is caused by nonuniform advection and influenced by diffusion.

Figure 4.A.5 illustrates why it is important to incorporate dispersion into the analysis of environmental transport. Pollutants are shown being released from an elevated stack and blowing downwind. The left-hand figures show that in the case of weak dispersion, the plume spreads slowly. Downwind of the source, the peak concentrations remain very high near the plume centerline and the pollutants do not spread rapidly to the ground. On the right, dispersion is strong. The concentrations within the plume are diminished rapidly by dispersion, but the plume reaches the ground much nearer to the source. This trade-off is characteristic of the effect that

*We will use the term *dispersion* to include both shear-flow dispersion and turbulent diffusion.

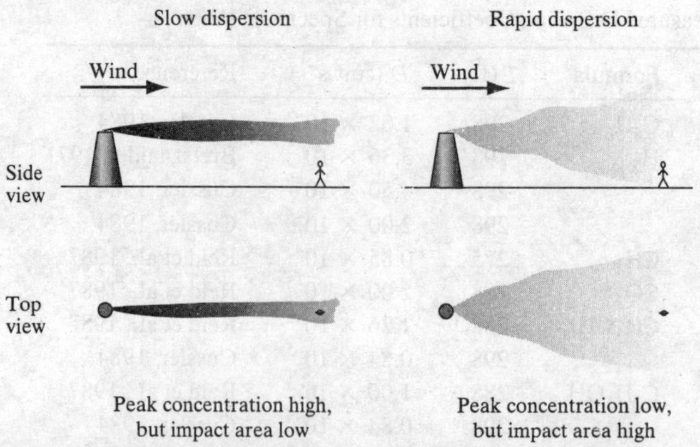

Figure 4.A.5 Effect of dispersion on pollutant concentrations downwind of a localized source. If the release rate of pollutants and mean wind speed are the same in the left- and right-hand frames, then the total flow of pollutants is the same in the two cases. Rapid dispersion leads to smaller peak concentrations but a larger impact area.

dispersion has in all environmental settings: High dispersion rates reduce average concentrations at the expense of increasing the area or duration of impact. A common challenge in environmental transport modeling is to predict concentrations in the region downwind or downstream of a pollution source. To make these predictions in the face of dispersion phenomena, we must solve two problems. First, we must have a model that describes pollutant concentration fields in the presence of dispersion. Second, we must be able to determine how the rate of dispersion varies with environmental conditions.

In complex flow fields, it is impossible (or at least impractical) to describe fluid velocity exactly as a function of space and time. Without this information, dispersion cannot be accurately described in terms of the fundamental mechanisms of advection and molecular diffusion. As an alternative, dispersion can be treated as a random process, analogous to molecular diffusion, by applying Fick's first law of diffusion with the molecular diffusion coefficient replaced by a dispersion coefficient. So in one dimension, we would write

$$J_{\text{dispersion}} = -\varepsilon \frac{dC}{dx} \tag{4.A.10}$$

where ε is a *dispersion coefficient*, obtained through a combination of empirical data and theoretical equations.

Although equation 4.A.10 has the same form as Fick's first law, it is important to bear in mind the great distinction between diffusion and dispersion. Diffusivities are properties of the contaminant and the fluid, depending weakly on environmental conditions (such as temperature) but not at all on fluid flows. Dispersion coefficients, on the other hand, are primarily a function of the fluid flow field. Environmental flows are highly variable and very complex. Dispersion in environmental flows is far less well understood than molecular diffusion. One should not expect a high degree of accuracy from any models that must account for dispersion, especially when applied to environmental transport.

Dispersion phenomena arise in all three branches of environmental engineering discussed in this text. Figure 4.A.5 illustrates one of many instances that arise in air quality engineering. In water quality, the impact of wastewater discharges on contaminant concentrations in rivers, lakes, or oceans depend on dispersion. In hazardous waste management, dispersion controls the movement of contaminants in groundwater.

Several types of dispersion phenomena are encountered in environmental engineering. In this section, two common types are introduced: *shear-flow dispersion* and *turbulent diffusion*.

Shear-Flow Dispersion

In a shear flow, the fluid speed varies with position in a direction that is perpendicular to the fluid velocity. If contaminant concentrations in a shear flow vary in the direction of flow, then dispersion will occur, leading to a net transport of contaminants from regions of high concentration to areas of low concentration.

Shear-flow dispersion is generally important when there is a short-term release of contaminants, such as a spill, in environments in which the velocity gradients are large, such as a river, an estuary influenced by tides, or the near-ground atmosphere. Let's consider a relatively simple example: shear-flow dispersion in fully developed laminar flow through a circular tube or pipe (Figure 4.A.6). The velocity profile of the fluid flow is parabolic because of wall friction. At time $t = 0$, some mass, M, of contaminant is injected into the tube at position $x = 0$ and instantaneously mixed laterally so that its concentration is uniform across the tube. We then observe the contaminant as it is advected past some position, L, downstream. The upper frame in the figure shows the initial condition. The lower frame shows the tube at a subsequent time, $t = L/U_0$. At the later time, the contaminant pulse is centered on downstream position $x = L$, but it has been substantially stretched because those molecules that lie close to the centerline of the tube are advected at a higher velocity than those near the wall. At the same time, contaminant molecules positioned near the leading edge of the pulse and near the center of the tube tend to diffuse toward the walls, reducing their average forward velocity. Conversely, contaminant molecules located near the trailing edge of the pulse along the tube walls tend to diffuse toward the center, increasing their forward velocity. Therefore, surprisingly, the net effect of molecular diffusion is to slow the rate of dispersion, in this case.

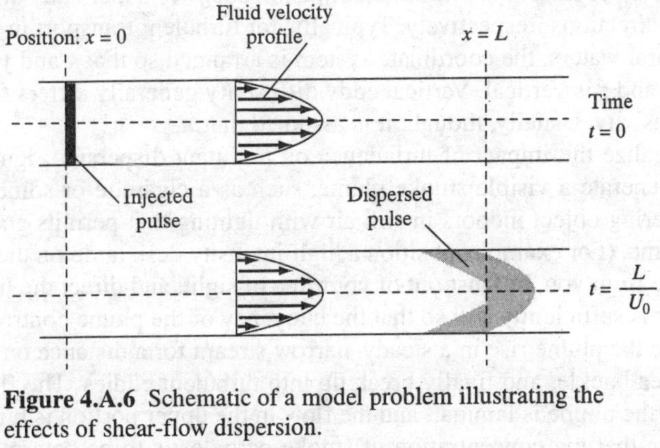

Figure 4.A.6 Schematic of a model problem illustrating the effects of shear-flow dispersion.

For the situation depicted in Figure 4.A.6, the contaminant flux in the direction of flow can be described as the sum of two components: (1) advection, which transports the overall pulse through the tube, and (2) shear-flow dispersion, which causes the pulse to become stretched as it travels. The flux in the direction of flow caused by shear-flow dispersion is written as

$$J_s = -\varepsilon_s \frac{dC(x)}{dx} \tag{4.A.11}$$

where ε_s is the shear-flow dispersivity and $C(x)$ represents the contaminant concentration averaged over a cross-section of the tube at position x.

Shear-flow dispersion is most commonly applied to predict the transport of contaminants in systems in which the flow is confined. In addition to pipe or tube flow, shear-flow dispersion is applied to study pollutant transport in rivers and in estuarine environments influenced by tides.

Turbulent Diffusion

In most systems encountered in environmental engineering, fluid flows are *turbulent* rather than *laminar.* Whereas for laminar flows we can determine the fluid velocity field at each instant and each position, turbulent flows fluctuate in such a manner that predictions are nearly impossible. Instead, we describe turbulent flows in terms of their statistical properties, such as the mean speed and the average size of the fluctuations.

Like molecular diffusion, the apparently random motion that is characteristic of turbulent flows causes a net flux of a contaminant from high to low concentrations. This flux is often described by an equation that is analogous to Fick's law:

$$J_t = -\varepsilon_t \frac{dC}{dx} \tag{4.A.12}$$

where ε_t is the *turbulent diffusion coefficient* and C now denotes the time-averaged value of concentration. For application in three dimensions, this expression can be generalized to vector form:

$$\vec{J_t} = -\left(\varepsilon_{tx}\frac{\partial C}{\partial x}, \varepsilon_{ty}\frac{\partial C}{\partial y}, \varepsilon_{tz}\frac{\partial C}{\partial z}\right) \tag{4.A.13}$$

Here ε_{tx}, ε_{ty}, and ε_{tz} represent the *turbulent diffusion coefficients*, or *eddy diffusivities* in the x, y, and z directions, respectively. Typically, for turbulent transport in the atmosphere or in natural waters, the coordinate system is arranged so that x and y lie in the horizontal plane and z is vertical. Vertical eddy diffusivity generally differs from horizontal eddy diffusivity. Usually, though, it is assumed that $\varepsilon_{tx} = \varepsilon_{ty}$.

To help visualize the impact of turbulence on pollutant dispersion, ignite something that will generate a visible smoke plume, such as a cigarette or some incense. Place the smoldering object indoors in still air with lighting that permits good observation of the plume. (For example, position a high-intensity desk lamp on the opposite side of the plume from you, but just out of your line of sight, and direct the light at the plume.) If the air is sufficiently still so that the buoyancy of the plume controls its motion, you will see the plume rise in a steady, narrow stream for a distance on the order of 10–30 cm, then buckle, and finally break up into turbulent eddies. The flow in the lower portion of the plume is laminar, and the flow in the upper portion is turbulent. If you now imagine that the concentration of smoke particles is to be determined on a

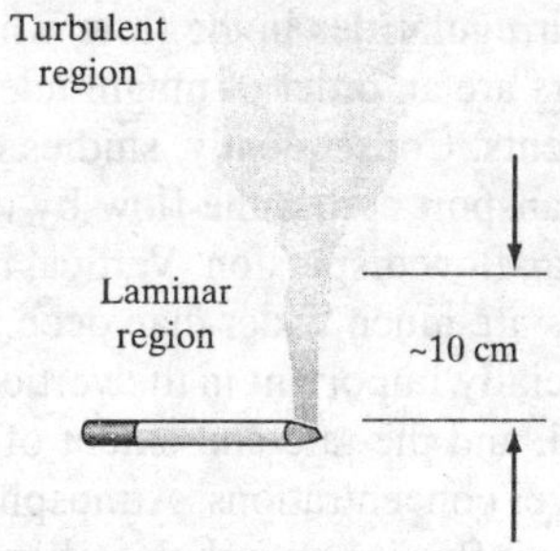

Figure 4.A.7 Schematic representation of time-averaged smoke plume rising from a smoldering cigarette in still air.

time-averaged basis, over a minute or so, you should arrive at a picture like the one shown in Figure 4.A.7. The broadening of the plume in the turbulent region is a direct consequence of turbulent air motion. The mean fluid flow is still upward, but there are fluctuating horizontal components of the flow that cause rapid pollutant dispersion in the horizontal direction. For environmental engineering applications, this is a key property of turbulence: In unbounded flow, turbulent diffusion controls the transport of pollutants in directions normal to the mean flow.

Figure 4.A.8 depicts a plume spreading from a point source discharging either to water or to air. A horizontal slice through the center of the plume is shown from above, and the mean fluid velocity is oriented from left to right. The coordinates are oriented so that x is aligned with the mean velocity and y represents the horizontal distance from the plume centerline. The origin is positioned at the source.

Assume that the rate of pollutant discharge is constant and the mean fluid velocity is independent of position and time. Then Figure 4.A.8 represents the average conditions over some period, such as an hour. On this basis, the plume spreads fairly smoothly and symmetrically from the source. The coordinate axes at right show the time-averaged concentration profile plotted against y at some distance away from the source. The peak concentration occurs at $y = 0$ and gradually approaches zero as the distance from the centerline increases. The spreading of the plume in the y-direction is controlled by turbulent diffusion.

A practical challenge in the use of equation 4.A.13 is to determine the turbulent diffusivities. These parameters vary not only with flow conditions, but also with position and direction. For example, turbulent diffusivities diminish to zero at rigid fluid boundaries, because the fluid velocity itself goes to zero. Correlations exist for

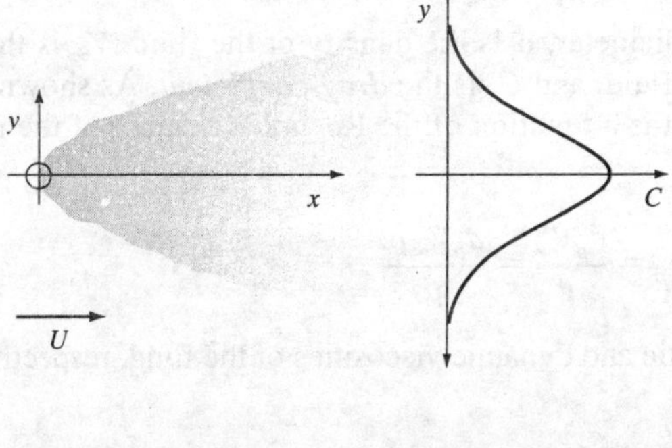

Figure 4.A.8 Plume spreading from a point source. The pollutant discharge rate is constant and the mean fluid velocity is steady and uniform, but the flow is turbulent.

predicting turbulent diffusivities, but these should be used cautiously, as the supporting experimental data are limited.

Turbulent diffusivities are higher in rivers than in regular channels because the irregularities in the river channel create additional irregularities in the flow, which tend to enhance mixing. Turbulent diffusivities in rivers are an order of magnitude smaller than longitudinal (shear-flow) dispersion coefficients. Consequently, studies of transport and mixing in rivers typically account for transport across the flow by turbulent diffusion and along the flow by advection and shear-flow dispersion. Vertical transport is assumed to occur very rapidly, since most rivers are much wider than deep.

In the atmosphere, turbulent diffusion is especially important in the vertical direction. Most pollutants are emitted near the ground, and the rate and extent of vertical turbulent diffusion strongly influences ground-level concentrations. Atmospheric stability, as will be discussed in §7.D, has a very strong influence on turbulent diffusivities, especially in the vertical direction. In unstable conditions, which are characterized by strong heating of the ground by the sun, vertical velocity fluctuations are strong and turbulent diffusivities are high.

Typical turbulent diffusivity values for both rivers and the atmosphere exceed by many orders of magnitude molecular diffusivities of contaminants. This observation reinforces the point that pollutant dispersion away from fluid boundaries is controlled by a combination of shear-flow dispersion and turbulent diffusion, and not by molecular or Brownian motion.

4.B PARTICLE MOTION

We now turn our attention to the movement of particles within fluids by mechanisms other than advection. The transport of particles within fluids is very strongly influenced by their size and mass. Very small particles behave much like molecules, undergoing a diffusive-type process known as *Brownian motion*. For larger particles, transport mechanisms that depend on mass, such as gravitational settling and inertial drift, dominate over diffusive motion.

4.B.1 Drag on Particles

Mechanisms that cause particle movement relative to a fluid are always opposed by drag force. The magnitude of the drag force increases as the particle velocity relative to the fluid increases. The drag force on a spherical particle is usually computed by the expression

$$F_d = \left(\frac{\pi}{4}d_p^2\right)\left(\frac{1}{2}\rho_f V_\infty^2\right)C_d \tag{4.B.1}$$

where d_p represents particle diameter, ρ_f is the density of the fluid, V_∞ is the speed of the particle relative to the fluid, and C_d is the *drag coefficient*. As shown in Figure 4.B.1, the drag coefficient is a function of the Reynolds number of the particle, which is defined by

$$\mathrm{Re}_p = \frac{d_p V_\infty}{\nu} = \frac{d_p V_\infty \rho_f}{\mu} \tag{4.B.2}$$

where ν and μ are the kinematic and dynamic viscosities of the fluid, respectively.

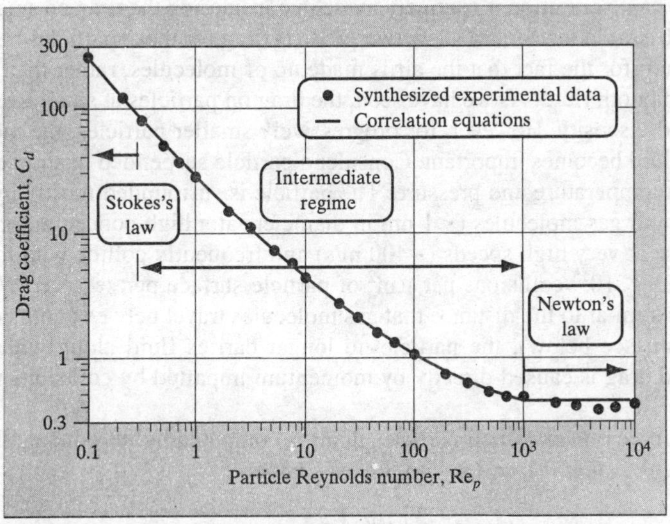

Figure 4.B.1 Drag coefficient as a function of particle Reynolds number for smooth, spherical, nonaccelerating particles in a uniform fluid flow. The experimental data are from Lapple and Shepherd (1940). The correlation equations are described in the text.

Given a particle Reynolds number, we can determine the drag coefficient from the data in the figure, or from the following equations (Perry and Green, 1984):

$$C_d = \frac{24}{\mathrm{Re}_p} \qquad \mathrm{Re}_p < 0.3 \qquad \text{Stokes's law} \tag{4.B.3}$$

$$C_d = \frac{24}{\mathrm{Re}_p}(1 + 0.14\,\mathrm{Re}_p^{0.7}) \qquad 0.3 < \mathrm{Re}_p < 1000 \qquad \text{Intermediate regime} \tag{4.B.4}$$

$$C_d = 0.445 \qquad 1000 < \mathrm{Re}_p < 350{,}000 \qquad \text{Newton's law} \tag{4.B.5}$$

Two intrinsic properties of fluids contribute to drag: viscosity and density. When the particle Reynolds number is low (laminar flow, Stokes's law regime), the drag force depends only on viscosity, as can be seen by substituting equations 4.B.3 and 4.B.2 into 4.B.1:

$$F_d = 3\pi\mu d_p V_\infty \qquad \mathrm{Re}_p < 0.3 \tag{4.B.6}$$

As a particle moves through a stationary fluid, some fluid molecules are drawn along with the particle because of viscous interactions, causing drag. At low Re_p, this contribution dominates.

At high particle Reynolds numbers, the drag force, as described by Newton's law, depends on the density of the fluid, but not on its viscosity.

$$F_d = 0.173\,\rho_f d_p^2 V_\infty^2 \qquad 1000 < \mathrm{Re}_p < 350{,}000 \tag{4.B.7}$$

Work must be done to accelerate molecules out of the way of a particle as it passes through a fluid. This work and the drag that results depend on the mass of the fluid molecules and therefore on the fluid density, but not on the fluid's viscosity. When the particle Reynolds number is large, this contribution to drag is much greater than the viscous contribution.

One important correction is routinely made in considering the drag on small particles (d_p < ~1 μm) in air. Stokes's law overpredicts drag for these particles because it does not account for the fact that the air is made up of molecules, rather than being a continuously smooth fluid. As we have seen, the drag on particles at small Re_p is controlled by fluid viscosity. However, for progressively smaller particles, the molecular nature of the fluid becomes important. Consider a particle suspended in air at ordinary environmental temperature and pressure. The particle is surrounded mostly by empty space. Very small gas molecules (<1 nm in diameter) at a high concentration (2.5×10^{25} m^{-3}) move at very high speeds (~400 m/s) and frequently collide with the particle surface (~3×10^{15} collisions per μm^2 of particle surface per second). When the particle size is similar to the distance that gas molecules travel between collisions (the mean free path; see below), the particle no longer carries fluid along with it as it moves. Instead drag is caused directly by momentum imparted by collisions with the gas molecules.

The drag force on very small particles in air is computed by introducing the *Cunningham slip correction factor,* C_c, into Stokes's law:

$$F_d = \frac{3\pi\mu d_p V_\infty}{C_c} \tag{4.B.8}$$

The correction factor is commonly computed according to the following expression:

$$C_c = 1 + \frac{\lambda_g}{d_p}\left[2.51 + 0.80\exp\left(-\frac{0.55 d_p}{\lambda_g}\right)\right] \tag{4.B.9}$$

The new parameter in this expression, λ_g, is called the *mean free path* and represents the average distance that a gas molecule travels before colliding with another molecule. The mean free path increases in proportion to the absolute temperature and decreases in proportion to pressure. At T = 293 K and P = 1 atm, the mean free path in air is 0.066 μm (Hinds, 1982).

Figure 4.B.2 shows how the slip correction factor varies with particle size in air at standard conditions. For particles larger than about 3 μm, the correction factor is

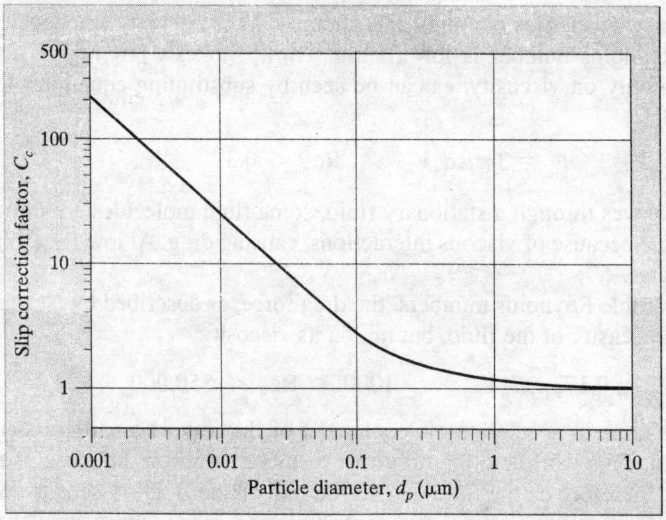

Figure 4.B.2 Slip correction factor for particles in air, assuming that the mean free path is 0.066 μm.

approximately 1 and so can be ignored. However, for very small particles, the slip correction factor can become quite large and must be considered to make even reasonably accurate predictions of airborne particle behavior. Because water molecules are packed so much more closely together than gas molecules, the slip correction factor does not arise in the assessment of particle behavior in water.

Sometimes, in the air pollution literature where Stokes's law holds more often than not, the expression for drag force is written in a different form than equation 4.B.1:

$$F_d = fV_\infty \qquad \text{Re}_p < 0.3 \tag{4.B.10}$$

The parameter f, known as the *friction coefficient*, is equal to $3\pi\mu d_p C_c^{-1}$ (compare with equation 4.B.8).

4.B.2 Gravitational Settling

Particles experience a gravitational force downward that is proportional to their mass. They also experience a buoyancy force upward that is proportional to the mass of fluid that they displace. A particle whose density is greater than that of the surrounding fluid will drift downward relative to the fluid, and a particle whose density is less than the fluid's will drift upward.

Consider a particle suspended in a stationary fluid that has a density greater than the fluid's. If the particle is initially stationary, it will accelerate downward. The downward velocity will be opposed by fluid drag, which increases with increasing velocity. The particle will continue to accelerate until the drag force plus the buoyancy force balances the gravitational force. The particle has then attained its *terminal settling velocity.* (If the particle were less dense than the fluid, the same argument would hold in reverse, with the particle accelerating upward until it reaches its *terminal rise velocity.*) Because the terminal settling velocity is achieved very rapidly for contaminant-sized particles, the details of acceleration and deceleration are of little interest in environmental engineering applications. The effect of gravity on particle motion is usually described only in terms of the terminal velocity.

Gravitational settling is used to remove unwanted particles from water and, to a lesser extent, air. Settling also affects the fate of particles discharged into water or the atmosphere. The rise of bubbles and the fall of liquid water drops through air also are important in certain air and water treatment technologies. The same general approach is used to determine rise and settling velocities for bubbles and water drops as for particulate contaminants.

The force balance in the case of a particle whose density is greater than the fluid density is (see Figure 4.B.3)

$$F_g = F_b + F_d \tag{4.B.11}$$

The gravitational force acting on a spherical particle can be written as

$$F_g = mg = \rho \frac{\pi}{6} d_p^3 g \tag{4.B.12}$$

where g is the acceleration of gravity (980 cm s^{-2}) and ρ is the particle density. The buoyancy force is similar, with the fluid density, ρ_f, replacing the particle density:

$$F_b = \rho_f \frac{\pi}{6} d_p^3 g \tag{4.B.13}$$

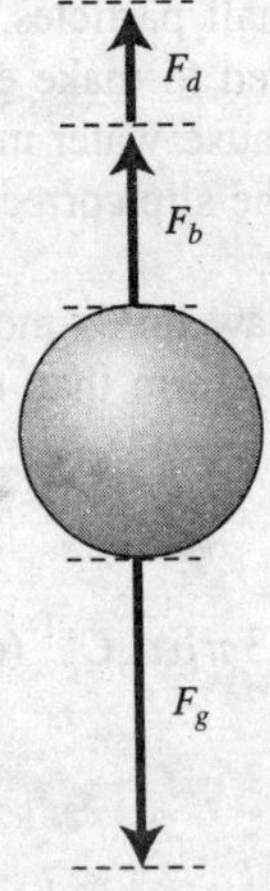

Figure 4.B.3 Force balance on a particle to determine gravitational settling velocity.

Table 4.B.1 Equations and Data for Computing the Terminal Settling Velocity When the Stoke's Law or Newton's Law Expressions for Drag Apply

General information applies in all cases

$g = 980$ cm s^{-2}

- For air (assuming $P = 1$ atm)
  Usually set $\rho_f = 0$ since $\rho_f \ll \rho$
  $\mu = 1.8 \times 10^{-4}$ g cm^{-1} s^{-1} (only a weak function of temperature)
- For water
  $\rho_f = 1$ g cm^{-3}
  $\mu = 0.01$ g cm^{-1} s^{-1} at 20 °C; use Figure 2.A.4 for other temperatures

Stokes's law regime

$$v_t = \left[\frac{C_c g\, d_p^2}{18}\left(\frac{\rho - \rho_f}{\mu}\right)\right] \qquad \text{valid for } \mathrm{Re}_p < 0.3$$

- For air
  Stokes's law applies if $d_p < 53$ µm ($\rho = 1$ g cm^{-3}) and if $d_p < 40$ µm ($\rho = 2.5$ g cm^{-3})
  Use equation 4.B.9 for C_c
- For water
  Stokes's law applies if $d_p < 72$ µm ($\rho = 2.5$ g cm^{-3})
  $C_c = 1$

Newton's law regime

$$v_t = \left[3.00\, g\, d_p\left(\frac{\rho - \rho_f}{\rho_f}\right)\right]^{1/2} \qquad \text{valid for } 1000 < \mathrm{Re}_p < 350{,}000$$

- For air
  Newton's law applies if $d_p > 20$ mm (rare in environmental engineering)
- For water
  Newton's law applies if $d_p > 2.8$ mm ($\rho = 2.5$ g cm^{-3})

The drag force is given by equation 4.B.1. Combining these expressions and rearranging to solve for the particle's terminal settling velocity, v_t, yields

$$v_t = \left[\frac{4gd_p}{3C_d}\left(\frac{\rho - \rho_f}{\rho_f}\right)\right]^{1/2} \qquad \rho > \rho_f \tag{4.B.14}$$

The evaluation of settling velocity using this equation is not trivial because the drag coefficient depends on the particle Reynolds number (equations 4.B.3–5), which, in turn, depends on the settling velocity (equation 4.B.2). In the limits in which Stokes's law or Newton's law holds, the dependence of drag coefficient on velocity is simple enough to permit equation 4.B.14 to be solved explicitly for v_t. Table 4.B.1 provides equations and other information useful for carrying out calculations for these two cases. For the intermediate regime, a nonlinear algebraic equation must be solved (see Appendix D). The data in Figures 4.B.4 and 4.B.5 can be used to obtain a first approximation of v_t, as demonstrated in Example 4.B.1.

The terminal rise velocity of a particle that is less dense than the fluid in which it is suspended is calculated by the same approach. For a rising particle, the force balance becomes $F_g + F_d = F_b$, yielding

$$v_t = \left[\frac{4gd_p}{3C_d}\left(\frac{\rho_f - \rho}{\rho_f}\right)\right]^{1/2} \qquad \rho < \rho_f \tag{4.B.15}$$

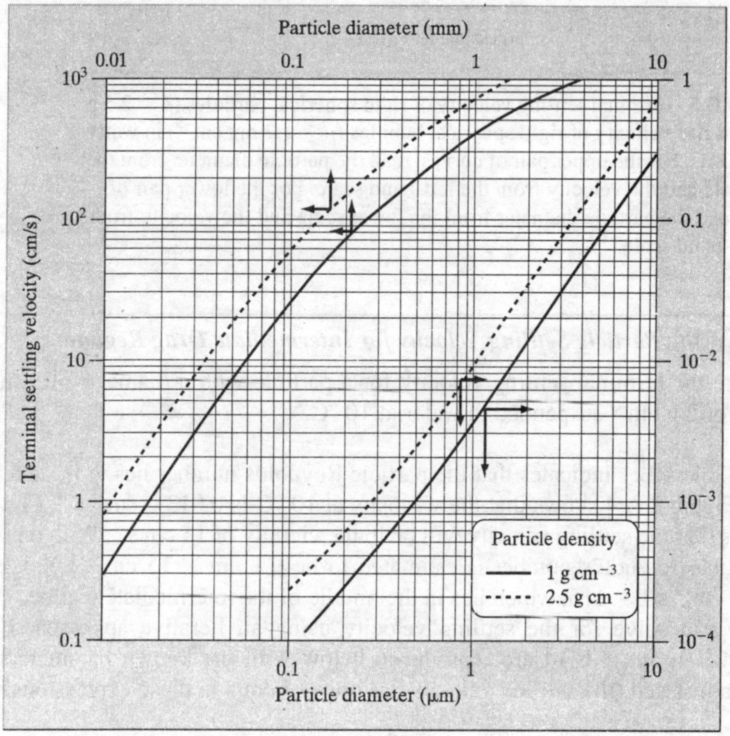

Figure 4.B.4 Terminal settling velocity for rigid spherical particles in air (P = 1 atm, T = 298 K). For the upper pair of curves, read the particle diameter from the upper scale and the settling velocity from the left-hand scale. For the lower pair of curves, read the particle diameter from the lower scale and the settling velocity from the right-hand scale.

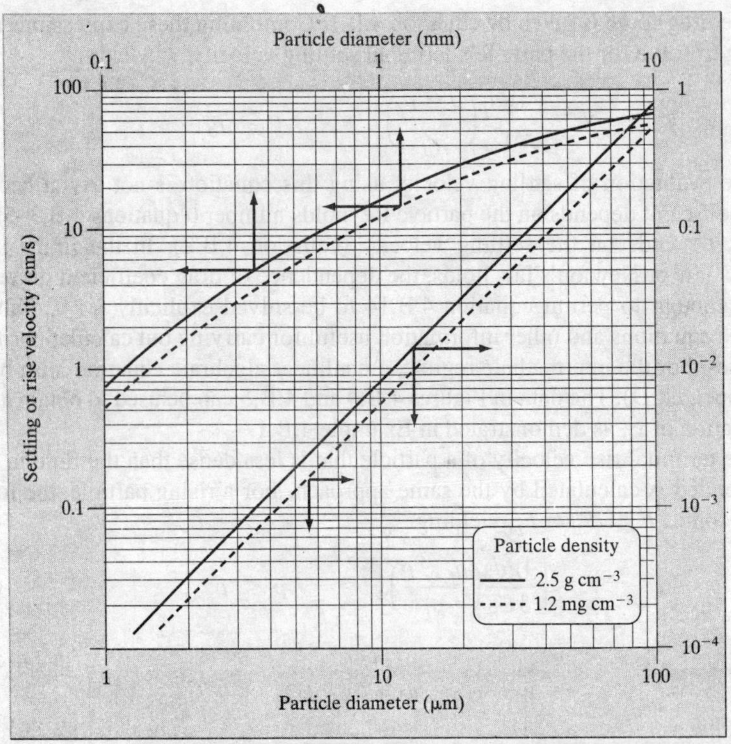

Figure 4.B.5 Terminal settling velocity of rigid spherical particles ($\rho = 2.5$ g cm^{-3}) and rise velocity of rigid spherical bubbles ($\rho = 1.2$ mg cm^{-3}) in water at $T = 20$ °C. For the upper pair of curves, read the particle diameter from the upper scale and the velocity from the left-hand scale. For the lower pair of curves, read the particle diameter from the lower scale and the velocity from the right-hand scale.

EXAMPLE 4.B.1 ***Particle Settling Velocity for Intermediate Drag Regime***

Compute the terminal settling velocity for a particle with a diameter of 1 mm and a density of 2 g cm^{-3} suspended in water at 10 °C.

SOLUTION Table 4.B.1 indicates that the particle Reynolds number lies in the intermediate range. Figure 2.A.4 shows that the viscosity at 10 °C is 0.013 g cm^{-1} s^{-1}. Figure 4.B.5 suggests that the settling velocity will be in the vicinity of 15 cm s^{-1}. With these values, the particle Reynolds number is estimated to be 0.1 cm × 15 cm s^{-1} × 1 g cm^{-3} ÷ 0.013 g cm^{-1} s^{-1} = 115, which lies in the middle of the intermediate regime.

We can solve for the settling velocity using an iterative approach. Equations 4.B.2, 4.B.4, and 4.B.14 are reproduced below with the known parameters for this problem inserted (the settling velocity has units of cm/s in these expressions):

$$\text{Re}_p = 7.69 v_t$$

$$C_d = \frac{24}{\text{Re}_p}(1 + 0.14\,\text{Re}_p^{0.7})$$

$$v_t = 11.4\,C_d^{-1/2}$$

EXAMPLE 4.B.1 ***Particle Settling Velocity for Intermediate Drag Regime (continued)***

Now, we evaluate these equations in series, initially assuming v_t = 15 cm s^{-1} and calculating Re_p, then using that result to calculate C_d, and then using C_d to calculate v_t. We repeat this process until the numbers converge to a stable solution that doesn't change from one cycle to the next. The intermediate results are v_t = 11.3, 10.6, 10.4, 10.4, 10.4 cm s^{-1}. With a reasonable initial guess, a stable answer to three significant figures is achieved after three iterations.

Figure 4.B.4 shows the terminal settling velocity for spherical particles in air as a function of size, according to equation 4.B.14. Figure 4.B.5 shows the settling velocity for particles in water and the rise velocity of bubbles in water as a function of diameter, according to equations 4.B.14 and 4.B.15. Note that for the particle sizes considered, the settling velocity spans an enormous range: seven orders of magnitude in air and six orders of magnitude in water. As a general rule, settling is unimportant for particles smaller than about 1 μm in diameter.

The particle flux caused by gravitational settling is analogous to the contaminant flux caused by advection (equations 4.A.1–2). The magnitude is given by

$$J_g = Cv_t \tag{4.B.16}$$

The vector form of this relationship is

$$\vec{J_g}(x, y, z) = C(x, y, z)\vec{v_t} \tag{4.B.17}$$

where $C(x, y, z)$ is the concentration of particles at the point (x, y, z). The vector points downward, parallel to the gravitational settling velocity, v_t. Example 4.B.2 presents an illustration of how the prediction of particle flux can be applied to a practical environmental question.

EXAMPLE 4.B.2 ***Soiling Due to Particle Deposition***

Studies have shown that a white surface becomes noticeably soiled when 0.2 percent of its area is covered by black particles (i.e., soot) (Hancock et al., 1976). Assume that the same relationship applies for soil dust particles. Estimate the time required for an initially clean, horizontal surface to appear soiled if it is exposed to an atmosphere containing 10 μg m^{-3} of soil-dust particles of diameter 5 μm. Assume that the particles are spherical and have a density of 2.5 g cm^{-3}.

SOLUTION The particle settling velocity is determined by the Stokes's law expression (Table 4.B.1):

$$v_t = \left[\frac{C_c g d_p^2}{18}\left(\frac{\rho - \rho_f}{\mu}\right)\right] = \left[\frac{1.03 \times 980 \times (5 \times 10^{-4})^2}{18}\left(\frac{2.5}{1.8 \times 10^{-4}}\right)\right] = 0.19 \text{ cm s}^{-1}$$

The number concentration of particles is obtained by dividing the mass concentration by the mass per particle:

$$C = \frac{10 \times 10^{-6} \text{ g m}^{-3} \times (10^{-2} \text{ m cm}^{-1})^3}{2.5 \text{g cm}^{-3} \times (\pi/6) \times (5 \times 10^{-4} \text{ cm})^3} = 0.061 \text{ particles cm}^{-3}$$

EXAMPLE 4.B.2 ***Soiling Due to Particle Deposition (continued)***

The flux accumulating onto the surface is given by equation 4.B.16:

$$J_g = Cv_t = 0.061 \text{ particles cm}^{-3} \times 0.19 \text{ cm s}^{-1} = 0.012 \text{ particles cm}^{-2} \text{ s}^{-1}$$

Each particle has a projected area, A_p, equal to that of a circle of diameter 5 μm. The fractional rate R, at which the surface is covered is the product of A_p and J_g:

$$R = \frac{\pi}{4} d_p^2 J_g = \frac{\pi}{4}(5 \times 10^{-4} \text{ cm})^2 \times 0.012 \text{ cm}^{-2} \text{ s}^{-1} = 2.3 \times 10^{-9} \text{s}^{-1}$$

The time required to cover 0.2 percent of the surface is obtained directly as

$$t = \frac{0.002}{R} = 9 \times 10^5 \text{ s} = 10 \text{ days}$$

Note that this is similar to the time scale over which household surfaces become perceptibly dusty.

4.B.3 Brownian Diffusion

In §4.A.3 we discussed molecular diffusion, a transport mechanism caused by the random motion of fluid molecules. An analogous phenomenon applies to particles suspended in a fluid. This transport mechanism is known as Brownian diffusion (or Brownian motion), in recognition of the botanist Robert Brown, who in 1827 reported observing, through a microscope, the wiggling motion of pollen grains suspended in water. Brownian diffusion results from the random collisions of particles suspended in a fluid with surrounding molecules. The overall effect in a fluid with a population of suspended particles is to cause net transport from high concentration to low concentration. The rate of Brownian diffusion is much slower than the rate of molecular diffusion because particles are much larger and much more massive than molecules.

Fick's law effectively describes the flux of particles caused by Brownian diffusion (see equations 4.A.5–6). A key issue in applying Fick's law for particles is the evaluation of the diffusion coefficient, D.

Diffusivity of Particles in Air and Water

The Brownian diffusivity of particles is determined by an equation known as the *Stokes-Einstein relation,*

$$D = \frac{kT}{f} \qquad (4.B.18)$$

where k is Boltzmann's constant (1.38×10^{-16} erg K^{-1}, where the erg is an energy unit equivalent to 1 g cm^2 s^{-2}), T is temperature (K), and f is the friction coefficient computed according to Stokes's law ($3\pi\mu d_p C_c^{-1}$), including the Cunningham slip correction factor if the particle is in air (equations 4.B.9–10). Example 4.B.3 illustrates the application of equation 4.B.18. Figure 4.B.6 shows the Brownian diffusivity of particles in air and water as a function of particle size. As the particles become very small, their diffusivities approach those of large molecules, ~0.1 cm^2 s^{-1} in air and ~10^{-5} cm^2 s^{-1} in water. For large particles, where the slip correction factor approaches 1, the ratio of

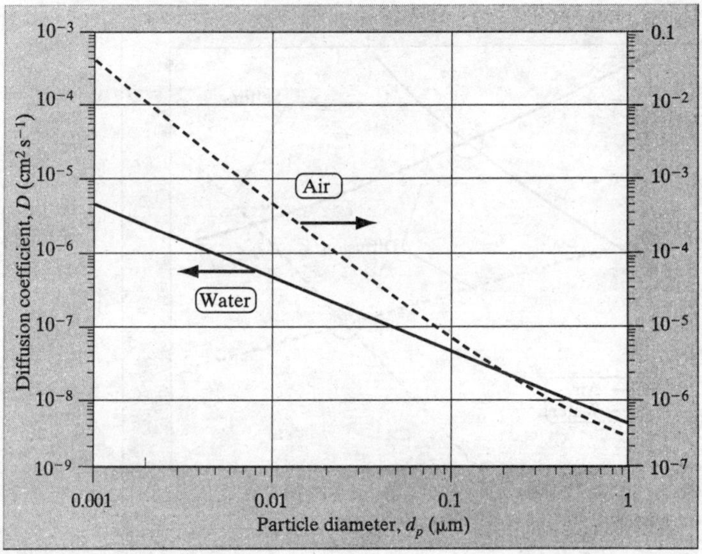

Figure 4.B.6 Brownian diffusivity for particles in air and water; $T = 20$ °C, $P = 1$ atm.

Brownian diffusivity in air to that in water approaches the inverse ratio of the viscosities, approximately 56 at 20 °C.

EXAMPLE 4.B.3 ***Diffusivity of Particles in Air and Water***

Compute the Brownian diffusivity of a 0.1 μm particle in air and in water. Assume $T = 293$ K and $P = 1$ atm.

SOLUTION The slip correction factor for a 0.1 μm particle in air is

$$C_c = 1 + \frac{0.066}{0.1}\left[2.51 + 0.80 \exp\left(-\frac{0.55 \times 0.1}{0.066}\right)\right] = 2.89$$

So the diffusivity of this particle in air is

$$D_{\text{air}} = \frac{1.38 \times 10^{-16} \times 293 \times 2.89}{3\pi(1.8 \times 10^{-4} \times 0.1 \times 10^{-4})} = 6.9 \times 10^{-6} \text{ cm}^2 \text{ s}^{-1}$$

In water, the slip correction factor doesn't apply. The diffusivity is

$$D_{\text{water}} = \frac{1.38 \times 10^{-16} \times 293}{3\pi(0.01 \times 0.1 \times 10^{-4})} = 4.3 \times 10^{-8} \text{ cm}^2 \text{ s}^{-1}$$

Comparing Particle Transport by Settling and Diffusion

It is interesting to compare travel distances for gravitational settling with those for Brownian diffusion. Figure 4.B.7 makes this comparison for particles in air and water. Typical settling and diffusion distances are comparable for particle diameters of

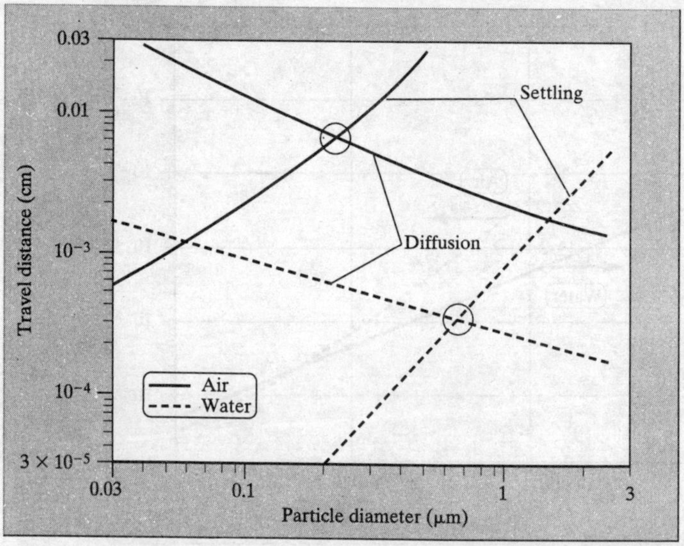

Figure 4.B.7 Distance traveled in 10 s by particles in air or water because of gravitational settling and Brownian diffusion. Particle settling velocities were determined from Stokes's law (Table 4.B.1) with $T = 293$ K and $P = 1$ atm. Diffusion distances were estimated from expression 4.A.7, using the Stokes-Einstein relation to compute diffusivity (equation 4.B.18). Particle density is 2.5 g cm^{-3}.

~0.2 μm for air and ~0.7 μm for water. (The exact crossing point depends on the assumptions about particle density and travel time.) For particles with a diameter of ~1 μm or larger, diffusion is a relatively unimportant transport mechanism in comparison with gravitational settling (and other processes that depend on particle mass or inertia). For particles with a diameter of ~0.1 μm or less, the role of gravitational settling as a transport mechanism is negligible compared with Brownian diffusion.

4.C MASS TRANSFER AT FLUID BOUNDARIES

Several important phenomena occur at the boundaries of fluids. The rate of contaminant transport to and from these boundaries may have a strong influence on contaminant concentration. This influence occurs in both natural and engineered systems. Contaminants can enter or leave a fluid at a boundary, and they may also undergo chemical transformations there. The boundaries of interest include air-water interfaces as well as boundaries between either air or water and (a) soil, (b) other solid surfaces, or (c) nonaqueous liquids such as fuels and solvents.

There is a close relationship between the transport phenomena explored in this section and the transformation processes involving phase change discussed in §3.B. In fact, the distinction between transport and transformation is somewhat blurred here. The kinetics of phase change processes depend, in general, both on the rate of transport to or from the fluid boundary and the kinetics of the transformation processes. Transport and transformation processes occur serially, so the slower process governs the overall rate.

Transport across air-water interfaces is of particular interest in environmental engineering. Several treatment technologies involve transferring pollutants from one phase to the other. For example, scrubbers remove acid gases such as sulfur dioxide and hydrochloric acid from waste air streams by transferring them to water, where they can easily be neutralized. Air strippers remove volatile organic compounds from water. Once the organic molecules are transferred to the gas phase, they can be more easily captured by sorption or destroyed by oxidation processes. In nature, air-water exchanges are also important. The transfer of oxygen from air to water supports aquatic life. The ultimate fate of many air pollutants involves transfer to cloud or rainwater followed by deposition to the earth's surface. The uptake of carbon dioxide by the oceans plays a key role in moderating the impact of fossil fuel combustion on climate.

Transport to a fluid boundary may occur by a variety of mechanisms. Diffusion is always present and often is the rate-limiting process because, at rigid boundaries, flow velocities diminish to zero. Advection usually plays a role in controlling the thickness of the layer through which diffusion must occur. For particles, gravitational settling, electrostatic drift, inertial drift, and other mechanisms may dominate transport near boundaries.

For engineering analysis of transport processes at fluid boundaries, we seek a description that captures the overall effects, that does not violate central principles such as mass conservation, and that is practical to apply. We will employ models that link flux to concentrations and may include an empirical parameter. In the case of turbulent diffusion, we used an equation inspired by Fick's law of diffusion, and the empirical parameter was turbulent diffusivity. For mass transfer at boundaries, similarly inspired flux equations will be written that incorporate a mass-transfer coefficient as the empirical parameter.

4.C.1 Mass-Transfer Coefficient

The net rate of mass flux between a fluid and its boundary is commonly expressed in this form:

$$J_b = k_m(C - C_i) \tag{4.C.1}$$

Here J_b is the net flux to the boundary (amount of species per area per time). When J_b is greater than zero, the flux direction is from the fluid to the boundary; when J_b is less than zero, the flux is directed from the boundary to the fluid. The concentration terms C and C_i, respectively, represent the species concentration in the bulk fluid far from the boundary and the species concentration in the fluid immediately adjacent to the boundary. The other parameter in this equation, k_m, is known as a *mass-transfer coefficient*. The mass-transfer coefficient commonly has units of velocity (length per time).

Figure 4.C.1 shows a simple system for which a mass-transfer coefficient can be directly derived. A glass cylinder is partly filled with a pure volatile liquid (with a sat-

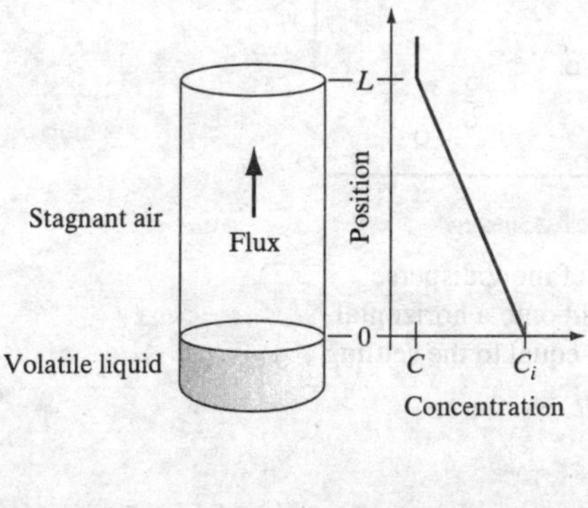

Figure 4.C.1 A simple system in which a mass-transfer coefficient can be directly determined.

uration vapor pressure $\ll$ 1 atm). The remainder of the cylinder contains stagnant air. Molecules of the liquid evaporate, diffuse through the cylinder, and escape from the open top into the free air. Provided that diffusion is not too rapid, the gas-phase concentration of the diffusing species at the liquid-gas interface will be determined directly as $P_{sat}/(RT)$, where P_{sat} is the equilibrium vapor pressure of the liquid. Assume that the concentration at the top of the cylinder is maintained at some value, C, determined by processes occurring in the open air. After an initial transient period, during which the gas-phase concentration profiles may change, a steady-state linear profile will be established, as depicted in Figure 4.C.1. The diffusive flux through the tube, J_d, is given by Fick's law as

$$J_d = D\frac{C_i - C}{L} \tag{4.C.2}$$

The net transport rate from the air to the liquid is equal to the negative of the diffusive flux, J_d. Comparing equation 4.C.2 with equation 4.C.1, we see that the mass-transfer coefficient for this system is

$$k_m = \frac{D}{L} \tag{4.C.3}$$

This is a general result, provided that the concentration profile is steady: For pure diffusion through a stagnant layer, the mass-transfer coefficient is given by the diffusivity divided by the thickness of the layer.

A second case is illustrated in Figure 4.C.2. Particles are suspended in a fluid above a horizontal boundary. The fluid is motionless, the particle concentration is uniform, and the particles migrate only because of gravitational settling. All particles are assumed to have the same settling velocity, v_t. The concentration of particles in the fluid at the interface is taken to be zero. This may seem peculiar, since the particles accumulate on the bottom boundary. However, provided resuspension does not occur (as it will not in a stagnant fluid), once the particles strike the boundary, they are no longer suspended in the fluid, and so it is reasonable to assign $C_i = 0$. The gravitational flux to the surface, J_g, is then given by

$$J_g = v_t C \tag{4.C.4}$$

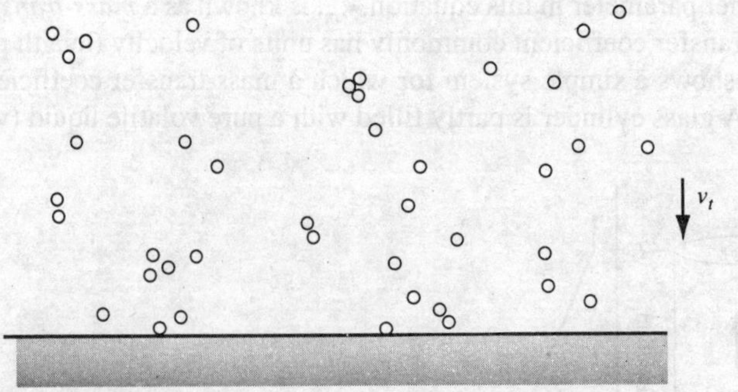

Figure 4.C.2 In a uniform suspension of monodisperse particles settling through a stagnant fluid onto a horizontal surface, the mass-transfer coefficient is equal to the settling velocity.

Comparing this result with equation 4.C.1, we see that for this case the mass-transfer coefficient is equal to the particle settling velocity:

$$k_m = v_t \tag{4.C.5}$$

This too is a general result. Whenever contaminant transport to a surface occurs solely because of a net migration velocity, the mass-transfer coefficient is equal to that velocity.

These examples yield exact expressions for the mass-transfer coefficient. However, in most applications, the use of a mass-transfer coefficient is a simplified description of what is a very complicated set of processes. Therefore, a high degree of precision should not be expected in problems involving mass transfer at boundaries.

Applying equation 4.C.1 wisely requires a sound understanding of the meaning of each of the four variables. First, let's consider the concentration at the interface, C_i. In many cases, C_i can be taken to be zero. This is appropriate when a transformation process that is fast and irreversible occurs at the boundary. Particle deposition by settling is one example where C_i is zero: When a particle strikes a surface with a low incident velocity, it adheres essentially immediately. Chemical transformations can also produce an interface concentration of zero if transport, rather than surface-reaction kinetics, is the rate-limiting step. In some cases, such as that shown in Figure 4.C.1, the interface concentration can be determined by assuming local equilibrium at the boundary.

An important issue arises concerning the species concentration in bulk fluid, C: Where should it be determined? The issue is easily resolved when the fluid is well mixed outside of a thin boundary layer (see Figure 4.C.3). Then C is the concentration anywhere outside the boundary layer. However, some circumstances arise in environmental engineering in which species concentrations vary strongly with position throughout the fluid. In such cases, the use of a mass-transfer coefficient may yield no more than a rough approximation of the mass-transfer rate at surfaces.

A complication arises when the fluid boundary occurs at a surface that has a complex texture. If J_b represents the quantity of contaminant transferred *per area* per time, how do we define the area? Typically, the flux to rough surfaces is determined on the basis of a superficial or apparent area. In other cases, rather than determining the interface area and the mass-transfer coefficient separately, their product is determined and the mass-transfer equation is rewritten in terms of total net transfer of species (flux times area) rather than in terms of flux.

The final issue to address is how an appropriate mass-transfer coefficient is determined for a given situation. There are two main approaches, and both are widely used. For systems with fairly regular geometries, the mass-transfer coefficient is calculated using existing correlations based on theory or experimental data. Some of these correlations are presented below. A second approach, commonly used for more complex systems, is to calculate mass-transfer coefficients based on experiments conducted in laboratory or field settings.

Film Theory

The simplest model system that includes fluid flow divides the fluid into two layers. Adjacent to the surface is a stagnant layer, or film, through which species must diffuse. The concentration profile within the film is assumed to be linear, corresponding to steady diffusive flux. In the second layer, the fluid is well mixed and the species concentration is uniform everywhere.

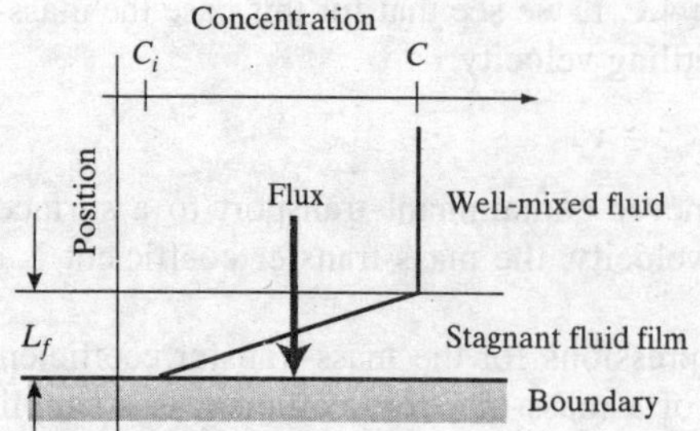

Figure 4.C.3 Schematic of mass transfer to a surface according to film theory.

Mass transfer from the fluid to a surface, according to film theory, is depicted in Figure 4.C.3. This situation is almost identical to that depicted in Figure 4.C.1, and so, in analogy with equation 4.C.3, it should be clear that the mass-transfer coefficient for this case is

$$k_m = \frac{D}{L_f} \tag{4.C.6}$$

where L_f is the film thickness. Film theory suggests that for given flow conditions, the mass-transfer coefficient should scale in direct proportion to species diffusivity ($k_m \propto D$). Experiments consistently demonstrate that, in systems with fluid flow, the mass-transfer coefficient flow increases in proportion to D^α, with $\alpha < 1$. The key weakness of film theory is its dependence on a specific value of the film thickness, L_f. In most circumstances, there is no practical way to determine L_f that is independent of a measurement of mass transfer (or heat transfer) to a surface. Furthermore, the effective thickness of the film through which species must diffuse varies with species diffusivity: A higher diffusion coefficient results in a larger film thickness. Still, although not entirely accurate, film theory provides a helpful conceptual picture of interfacial mass transfer.

Penetration Theory

Recall (equation 4.A.8) that the characteristic time required for a steady-state concentration profile to be established by diffusion is given by $L^2(2D)^{-1}$ where L is the distance over which diffusion occurs. In film theory, the boundary must be in contact with the fluid for a minimum time of $\tau \sim L_f^2(2D)^{-1}$ for the concentration profile to approach the constant-slope condition depicted in Figure 4.C.3. In some circumstances, the contact time between a boundary and a fluid is not long enough for film theory to apply. A second conceptual model, known as penetration theory, has been developed to address this case.

The model is based on transient diffusion in one dimension. At time $t = 0$, the concentration in the fluid is assumed to be uniform everywhere except at the position of the boundary, where it is C_i. Then diffusion to the surface begins and a boundary layer begins to grow (Figure 4.C.4). Fluid motion affects the contact time between the fluid and the boundary.

Analysis of time-dependent diffusion to a flat surface yields the following expression for the mass-transfer coefficient:

$$k_m(t) = \left[\frac{D}{\pi t}\right]^{1/2} \quad \text{instantaneous} \tag{4.C.7}$$

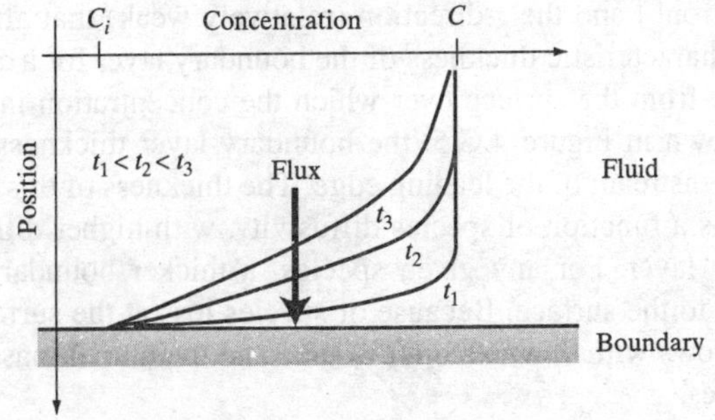

Figure 4.C.4 Schematic of mass transfer to a surface according to penetration theory.

This is the instantaneous mass-transfer coefficient at time t. Because the distance through which species must diffuse increases with time, k_m decreases as t increases. If the contact period is maintained for some interval t^*, the time-averaged mass-transfer coefficient is obtained by integration:

$$\overline{k_m} = \frac{1}{t^*}\int_0^{t^*} k_m(t)\, dt = 2\left[\frac{D}{\pi t^*}\right]^{1/2} \quad \text{average} \tag{4.C.8}$$

From equation 4.C.8, we see that penetration theory predicts that the mass-transfer coefficient should increase in proportion to the square root of species diffusivity. By contrast, we have just seen that film theory predicts that the mass-transfer coefficient is proportional to diffusivity. The reason for the difference lies in the assumptions about the distance through which species must diffuse. In film theory, the film thickness is assumed to be constant and so is independent of D. In penetration theory, the diffusion distance increases with time, and the rate of growth depends on species diffusivity.

Boundary-Layer Theory: Laminar Flow along a Flat Surface

Transport from moving fluids to boundaries is caused by simultaneous advection and diffusion, with advection being stronger far from the boundary and diffusion dominating adjacent to the boundary. Film and penetration theories simplify the analysis by decoupling advection from diffusion. Boundary-layer theory can predict mass-transfer coefficients for simple flows and geometries while simultaneously accounting for the effects of both transport mechanisms.

Let's consider the case of laminar fluid flow at velocity U, parallel to a flat surface (see Figure 4.C.5). The species concentration is C_i at the surface and is C far from the surface. Within the boundary layer there is a concentration gradient, which gives rise to diffusion toward the surface. There are also advective velocity components in

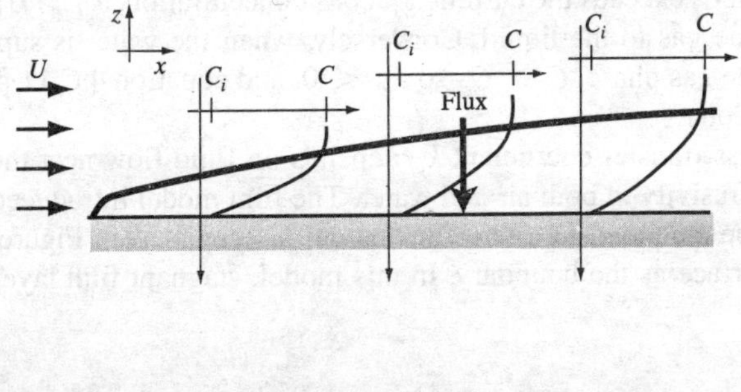

Figure 4.C.5 Schematic of mass transfer to a surface according to boundary-layer theory for the case of a flat surface parallel to a uniform, laminar fluid flow.

the x-direction (relatively strong) and the z-direction (relatively weak) that affect the concentration profile. The characteristic thickness of the boundary layer for a contaminant species is the distance from the surface over which the concentration increases to approximately C. As shown in Figure 4.C.5, the boundary-layer thickness grows with increasing distance downstream of the leading edge. The thickness of this boundary layer at any x-position is a function of species diffusivity, with higher diffusivity causing a thicker boundary layer. For any given species, a thicker boundary layer means slower mass transfer to the surface. Because of species loss at the surface, the boundary-layer thickness grows with downstream distance and the rate of mass transfer correspondingly decreases.

This problem is analyzed using equations that describe the conservation of fluid mass, momentum, and species in the boundary layer (Cussler, 1984; Bejan, 1984). The result is a mass-transfer coefficient calculated at a specific position x downstream of the leading edge of the surface:

$$k_m(x) = 0.323\left(\frac{U}{x}\right)^{1/2} \nu^{-1/6} D^{2/3} \quad \text{local} \tag{4.C.9}$$

where ν is the kinematic fluid viscosity. (Recall that $\nu = \mu/\rho$, where μ is the dynamic fluid viscosity and ρ is the fluid density.) The overall mass-transfer coefficient to the surface is obtained by averaging over the length of the surface:

$$k_m = \frac{1}{L}\int_0^L k_m(x)\, dx = 0.646\left(\frac{U}{L}\right)^{1/2} \nu^{-1/6} D^{2/3} \quad \text{average} \tag{4.C.10}$$

Equation 4.C.10 gives the average mass-transfer coefficient for this flow system for a surface from the leading edge to a distance L downstream. Note that the mass-transfer coefficient, k_m, varies with the diffusion coefficient raised to the 2/3 power. It is generally true that when advection and diffusion are combined, the mass-transfer coefficient increases with diffusivity raised to some power between 0.5 (penetration theory) and 1 (film theory).

4.C.2 Transport across the Air-Water Interface

In environmental engineering applications, the rate of transport of a molecular species across an air-water interface is described by an expression similar to equation 4.C.1:

$$J_{gl} = k_{gl}(C_s - C) \tag{4.C.11}$$

where J_{gl} is the net flux of a species from the gas phase to the liquid phase (mass per interfacial area per time), k_{gl} is a mass-transfer coefficient (length per time), C is the concentration of the species in the bulk liquid phase, and C_s is the saturation (or equilibrium) concentration of the species in the liquid phase that corresponds to the given partial pressure of the species in the gas phase. Typically, C_s is obtained from Henry's law (see §3.B.2). When C_s exceeds the current aqueous concentration, $J_{gl} > 0$ and net transfer occurs from the gas to the liquid. Conversely, when the water is supersaturated with respect to the gas phase, $C > C_s$, so $J_{gl} < 0$, and equation 4.C.11 predicts a net rate of volatilization.

In general, the mass-transfer coefficient k_{gl} depends on fluid flow near the interface and on species diffusivity in both air and water. The film model introduced in the previous section can be extended to a two-film model, as depicted in Figure 4.C.6, with the air-water interface as the boundary. In this model, stagnant film layers exist

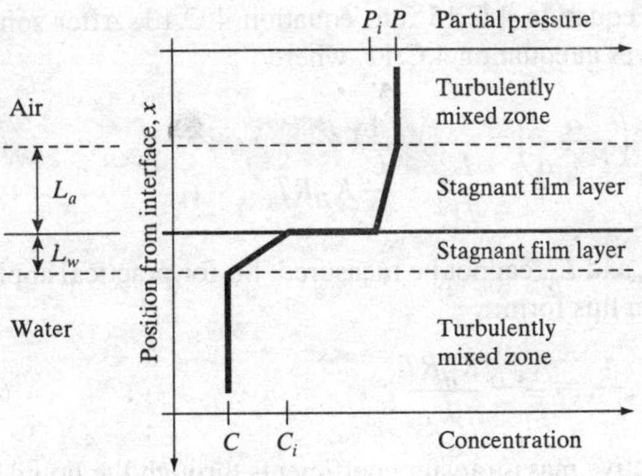

Figure 4.C.6 Schematic of the two-film model for estimating mass transport across an air-water interface.

on either side of the boundary through which species transport occurs only by molecular diffusion. Outside of its stagnant film layer, each fluid is well mixed. Immediately adjacent to the interface, the partial pressure in the gas phase, P_i, is assumed to be in equilibrium with the aqueous concentration, C_i, as described by Henry's law. As in the film model, the concentration profiles in the stagnant film layers are assumed to be linear. Also, because of mass conservation, the flux through the air layer must be equal to the flux through the water layer.

Applying Fick's law, we can write the gas-side flux (from air to the interface) as

$$J_{gl} = D_a \frac{(P - P_i)/RT}{L_a} \tag{4.C.12}$$

where D_a is the species diffusivity through air and L_a is the thickness of the stagnant film layer in the air. Likewise, the liquid-side flux (from the interface into the water) is

$$J_{gl} = D_w \frac{C_i - C}{L_w} \tag{4.C.13}$$

where D_w is the species diffusivity in water and L_w is the film-layer thickness in the water. Since we don't know either P_i or C_i, we would like to eliminate these parameters from the expressions. So far, we have two equations but three unknowns (J_{gl}, C_i, and P_i).

The third equation comes from the equilibrium relationship at the interface:

$$C_i = K_H P_i \tag{4.C.14}$$

where K_H is the Henry's law constant for the species (see Table 3.B.2). Now, we can use algebra to derive an expression for flux that is in the form of equation 4.C.11. Use equation 4.C.14 to replace P_i with C_i/K_H in equation 4.C.12. Then equate the right-hand sides of equations 4.C.12 and 4.C.13 and solve for C_i to obtain

$$C_i = \frac{\alpha C_s + C}{1 + \alpha} \tag{4.C.15}$$

where

$$\alpha = \frac{D_a L_w}{D_w L_a K_H RT} \tag{4.C.16}$$

and

$$C_s = K_H P \tag{4.C.17}$$

Next substitute for C_i from equation 4.C.15 into equation 4.C.13. After some algebraic manipulation, one arrives at equation 4.C.11, where

$$k_{gl} = \frac{D_w}{L_w}\left(\frac{\alpha}{1+\alpha}\right) = \frac{1}{\frac{L_w}{D_w} + \frac{L_a}{D_a}K_H RT} \tag{4.C.18}$$

The film thicknesses, L_w and L_a, cannot be measured. So for practical application, we rewrite equation 4.C.18 in this form:

$$\frac{1}{k_{gl}} = \frac{1}{k_l} + \frac{K_H RT}{k_g} \tag{4.C.19}$$

where k_l and k_g are the respective mass-transfer coefficients through the liquid and gas boundary layers near the interface, each corresponding to the diffusivity divided by the film thickness (D_w/L_w and D_a/L_a, respectively), as in equation 4.C.6. The factor K_H appears with k_g in this expression because k_{gl} is used with the aqueous concentrations to determine flux (equation 4.C.11). The factor RT is needed to convert partial pressure to molar concentration, since we write Henry's law in a way that relates aqueous molar concentration to gaseous partial pressure.

Equations 4.C.11 and 4.C.19 are together sometimes called the *two-resistance model for interfacial mass transfer*. Since species must be transported through fluid on both sides of the interface, the total resistance (k_{gl}^{-1}) is the sum of the resistance in the liquid side of the interface (k_l^{-1}) plus the resistance on the gas side ($K_H RT k_g^{-1}$).

From equation 4.C.19 we see that the relative sizes of the gas and liquid resistances vary according to the Henry's law constant, K_H. The diffusion coefficients, D, of molecular species in a given fluid vary by about one order of magnitude (§4.A.3). The film resistance terms k_l and k_g are expected to vary by no more than D to the first power (§4.C.1). The Henry's law constant, on the other hand, varies by at least eight orders of magnitude among species of interest in environmental engineering (see Table 3.B.2). Therefore, even for fixed flow conditions, the interfacial mass-transfer coefficient, k_{gl}, may vary greatly from one species to another according to the value of the Henry's law constant. This point is illustrated in Figure 4.C.7.

Liss and Slater (1974) reviewed available information on the mass transfer of species between the oceans and the atmosphere and concluded that liquid-side and gas-side mass-transfer coefficients of $k_l = 0.2$ m h^{-1} and $k_g = 30$ m h^{-1} apply for average meteorological and ocean current conditions. These coefficients are suggested to be approximately correct for species with molecular weights between 15 and 65 g/mol. Figure 4.C.7 shows how the overall mass-transfer coefficient varies with the Henry's law constant for molecular species exchange between the atmosphere and the seas. For sparingly soluble gases such as oxygen ($K_H = 0.0014$ M atm^{-1}), the resistance lies entirely on the liquid side and an overall average mass-transfer coefficient of approximately $k_{gl} = 0.2$ m/h applies. On the other hand, the resistance for a highly soluble species such as formaldehyde ($K_H = 6300$ M atm^{-1}) lies entirely on the gas side.

The liquid-side and gas-side mass-transfer coefficients can be combined with values of species diffusivities to estimate effective film thicknesses: $L_a \sim D_a k_g^{-1} \sim 2$ mm and $L_w \sim D_w k_l^{-1} \sim 20$ μm. We see from these estimates that diffusive transport over very small length scales controls the overall rate of interfacial mass transfer. The time scales required for steady-state concentration profiles to be achieved in gas and liquid films of this size are approximately $\tau_g \sim L_a^2(2D_a)^{-1} \sim 0.1$ s and $\tau_l \sim L_w^2(2D_w)^{-1} \sim 0.2$ s.

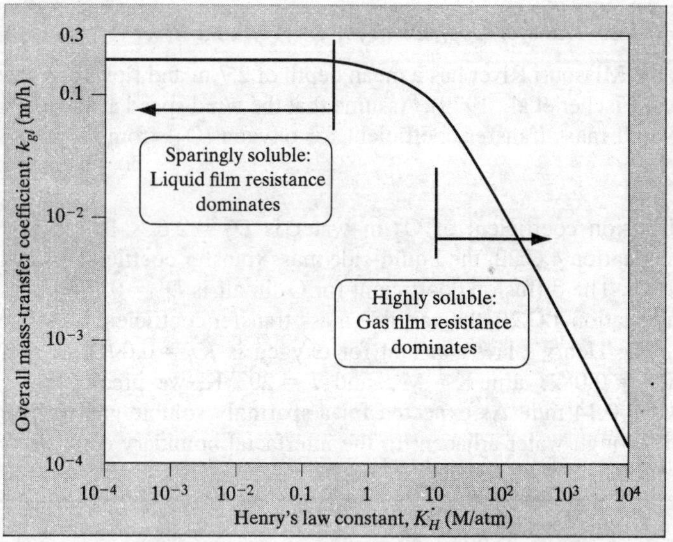

Figure 4.C.7 Dependence of the overall mass-transfer coefficient on the Henry's law constant for average conditions in large bodies of water. The curve traces equation 4.C.19 with $k_l = 0.2$ m h^{-1}, $k_g = 30$ m h^{-1}, and $T = 293$ K (Liss and Slater, 1974).

For natural bodies of water, the following expressions can be applied to estimate gas-side and liquid-side mass-transfer coefficients in relation to environmental and species conditions (Schwarzenbach et al., 1993). For the gas-phase mass-transfer coefficient,

$$k_g = \left[\frac{D_a}{0.26 \text{ cm}^2 \text{ s}^{-1}}\right]^{2/3} (7U_{10} + 11) \tag{4.C.20}$$

For oceans, lakes, and other slowly flowing waters,

$$k_l = \left[\frac{D_w}{2.6 \times 10^{-5} \text{ cm}^2 \text{ s}^{-1}}\right]^{0.57} (0.0014U_{10}^2 + 0.014) \tag{4.C.21}$$

For rivers,

$$k_l = 0.18\left[\frac{D_w}{2.6 \times 10^{-5} \text{ cm}^2 \text{ s}^{-1}}\right]^{0.57} \left(\frac{U_w}{d_w}\right)^{1/2} \tag{4.C.22}$$

In these expressions, U_{10} is the mean wind speed measured at 10 m above the water surface (units are m/s). The term U_w is the mean water velocity in the river (m/s). The mean stream depth is d_w (m). The equations are written so that k_g and k_l have units of m/h. Example 4.C.1 illustrates how these equations are used.

In general, for interfacial transfer of molecular species in any air-water flow system, the overall mass-transfer coefficient for any species can be determined by making measurements of k_{gl} for a minimum of two species with known Henry's law constants, provided one is sparingly soluble and the other is highly soluble. From these measurements, one can determine the values of k_l and k_g. Then the overall mass-transfer coefficient for any other species can be estimated using equation 4.C.19.

EXAMPLE 4.C.1 ***Mass-Transfer Coefficient for Oxygen in a River***

In Nebraska, the Missouri River has a mean depth of 2.7 m and flows at a mean velocity of 1.75 m/s (Fischer et al., 1979). Assume that the wind speed at 10 m is 4 m/s. Estimate the overall mass-transfer coefficient for oxygen (O_2) from the atmosphere to the river.

SOLUTION The diffusion coefficient of O_2 in water is $D_w = 2.6 \times 10^{-5}$ cm^2 s^{-1} (Table 4.A.3). From equation 4.C.22, the liquid-side mass-transfer coefficient is estimated to be $k_l = 0.14$ m/h. The diffusion coefficient for O_2 in air is $D_a = 0.178$ cm^2 s^{-1} (Table 4.A.2). From equation 4.C.20, the gas-side mass-transfer coefficient is estimated to be $k_g = 30$ m/h. The Henry's law constant for oxygen is $K_H = 0.0014$ M atm^{-1} (Table 3.B.2). With $R = 0.0821$ atm K^{-1} M^{-1} and $T = 293$ K, we predict from equation 4.C.19 that $k_{gl} = 0.14$ m/h. As expected for a sparingly soluble gas such as oxygen, mass transfer through water adjacent to the interfacial boundary controls the overall rate of the process.

4.D TRANSPORT IN POROUS MEDIA

Porous materials are solids that contain distributed void spaces. Permeable porous materials contain an interconnected network of voids or pores that permit bulk flow of fluid through the material. Soil is a common example of a permeable porous material. Usually, the pores are highly variable in shape and size, resulting in complex flow channels.

Environmental engineers study the movement of fluids and contaminants through porous media for several reasons. Many treatment technologies for removing pollutants from water and air entail passing the fluid through a porous material. In municipal treatment plants, for example, drinking water is passed through sand filters to remove small, suspended particles. Drinking water is also sometimes treated by passing it through a column of granular activated carbon to remove dissolved organic molecules that are harmful or that may cause taste and odor problems. Air used in industrial processes is often passed through fabric filters to remove suspended particles. Filters of granular activated carbon are also used to remove volatile organic compounds from gas streams. Many hazardous waste treatment technologies also involve passing a fluid through a porous material. Much attention in hazardous waste management focuses on characterizing contaminant migration in subsurface soils, either to predict the threat to water quality or to evaluate a treatment strategy.

The aim of this section is to provide an introduction to transport in porous media, emphasizing the movement of water and air and the contaminants dissolved or suspended within these fluids. The behavior of nonaqueous-phase liquids in subsurface environments is also important in environmental engineering, but the complexities that must be addressed render it beyond the scope of this book. Application of the ideas introduced here for filtering contaminants from water are discussed in Chapter 6. Groundwater contamination is addressed in Chapter 8.

Before proceeding, we must define some basic terms and concepts that arise when dealing with transport through porous materials. One quantitative descriptor of a porous material is its *porosity*, here given the symbol ϕ and defined as

$$\phi = \frac{\text{pore volume}}{\text{total volume}} \tag{4.D.1}$$

The pores may contain both air and water (and, in general, other fluids). We define the *air-filled porosity*, ϕ_a, and the *water-filled porosity*, ϕ_w, in analogy with equation 4.D.1, but with the numerator replaced by the pore volume filled with air or water, respectively. If the air and water are the only fluids contained in the pores, the porosities must satisfy this relationship:

$$\phi = \phi_w + \phi_a \tag{4.D.2}$$

A porous material is *saturated* with a particular fluid if that fluid entirely fills the pores. So, for example, a medium is saturated with water if $\phi_w = \phi$.

Most porous materials encountered in environmental engineering are granular or fibrous. Granular materials, such as soils, typically have porosities in the range 0.3–0.7. The porosities of fibrous materials are usually higher, sometimes as high as 0.99. These materials may have two distinct classes of pores, those that are external to the grains or fibers and those that are internal. Bulk fluid flow occurs only in the external pores, but contaminants can migrate by diffusion into the internal pores and interact with the solid surfaces there. This characteristic is especially important for porous sorbents, such as activated carbon, which have large internal porosities and enormous internal surface areas. However, in this section, we will emphasize fluid and contaminant behavior in the pores that are external to grains and fibers.

Two densities are commonly defined for a porous solid. The *solids density*, ρ_s, represents the mass of solid per volume of solid (often including internal pores). The *bulk density*, ρ_b, represents the mass of solid per total volume. These measures are related by the total porosity:

$$\rho_b = \rho_s(1 - \phi) \tag{4.D.3}$$

The solids density of soil grains is fairly constant at ~2.65 g cm^{-3}. Given a range of porosities of 0.3–0.7, the bulk density would be in the range 0.8–1.9 g cm^{-3}. Soil bulk density varies with grain size, but the dependence is not strong. In soil, as in any granular material, porosity tends to be higher when grain sizes are distributed over a narrow range. With a broad distribution of grain sizes, smaller grains can fill the pores created by larger grains, reducing overall porosity.

Porosity is an area characteristic as well as a volume characteristic of granular materials. Imagine a plane slicing through a porous material. The *area porosity* is the ratio of the area of the plane that intersects pores to the total area that intersects the porous material. Conveniently, if pores are randomly distributed and the number of pores in a plane is large, then the area porosity is equal to the porosity measured on a volume basis. This conclusion is reached by considering a porous material as a collection of infinitesimally thin slices. The volume porosity of the whole is the average of the porosities of the individual slices. If these slices include a random distribution of a large number of pores, then the porosity of each slice will be close to the mean for all slices.

4.D.1 Fluid Flow through Porous Media

During the middle of the nineteenth century, Henri Darcy, a French engineer who was interested in the development of groundwater resources, studied the hydraulics of water flow through a sand column using an apparatus like that shown in Figure 4.D.1. With experiments conducted under steady flow conditions, he found that the following relationship described his results:

$$Q = KA\frac{\Delta h}{L} \tag{4.D.4}$$

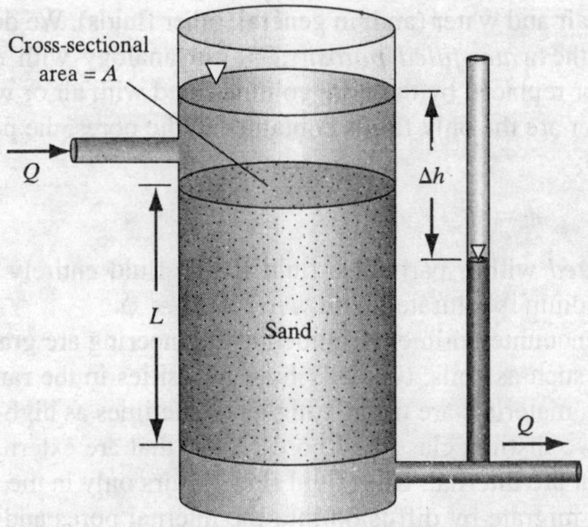

Figure 4.D.1 Apparatus similar to that used by Darcy to study the hydraulics of water flow through sand.

where Q is the volumetric flow rate ($m^3\ s^{-1}$), Δh is the change in fluid head from the inlet to the outlet of the column (m), A is the cross-sectional area of the column (m^2), and L is the length of the column (m). The parameter K, which has units of velocity, was constant for a given sand but varied from one sand sample to another, increasing with increasing grain size.

This relationship has been generalized and is now commonly known as *Darcy's law.* For water flow in one dimension through a water-saturated material, Darcy's law can be written as

$$U = -K\frac{dh}{dl} \tag{4.D.5}$$

where U is called the *Darcy velocity, filtration velocity,* or *superficial flow velocity, K* is called the *hydraulic conductivity;* and dh/dl is the rate of change of pressure head with distance. In relation to Darcy's experiment, U represents Q/A, and the hydraulic gradient, dh/dl, replaces $\Delta h/L$. The filtration velocity (U) should not be confused with the local velocity of water through the pores. Rather, it represents the average volumetric flow of water per unit *total* cross-sectional area of the porous material. Since solids occupy some area, and since water can flow only through pores, the average local velocity of water in the pores must be higher than the filtration velocity. The minus sign in Darcy's law, like the minus sign in Fick's law, reminds us that water flows from high to low pressure head, just as molecules diffuse from high to low concentration. Since head has units of length, dh/dl is dimensionless, and so the hydraulic conductivity must have the same units as U, velocity.

Another form of Darcy's law applies to porous media saturated with any fluid:

$$U = -\frac{k}{\mu}\frac{dP}{dl} \tag{4.D.6}$$

where k is called the *intrinsic permeability,* or simply *permeability,* of the porous material; μ is the viscosity of the fluid; and dP/dl is the derivative of dynamic pressure with respect to distance. This form of Darcy's law assumes that the fluid is incompressible, which, for air, would mean that the pressure drop across the medium must be much smaller than the inlet pressure.

The hydraulic conductivity in Darcy's law (equation 4.D.5) depends primarily on two properties of the system: the viscosity of the fluid and the size of the pores. Fluids that are more viscous flow more slowly, and materials with smaller pores permit less flow. The second form of Darcy's law (equation 4.D.6) separates these two properties. The intrinsic permeability is a function only of the porous material and does not depend on the fluid properties.

Given the units of viscosity (e.g., kg m^{-1} s^{-1}), pressure derivative (kg m^{-2} s^{-2}), and velocity (m s^{-1}), we see that permeability has units of length squared. For a porous medium made up of uniform spherical grains, the permeability increases approximately in proportion to the square of the grain diameter.

The dynamic pressure (P) in equation 4.D.6 is the difference between the total pressure and the hydrostatic pressure. In the absence of motion, fluid pressure must increase with depth to support the mass of the fluid suspended above it against the acceleration of gravity. This hydrostatic change in pressure with height does not induce flow.

The difference between the two forms of Darcy's law lies primarily in the coefficients K and k. By noting that head loss corresponds to dynamic pressure drop according to $\rho_w g \Delta h = \Delta P$, where ρ_w is the density of water and g is gravitational acceleration, we can show that the hydraulic conductivity is related to the intrinsic permeability by

$$K = \frac{k \rho_w g}{\mu_w} \tag{4.D.7}$$

where μ_w is the dynamic viscosity of water.

Intrinsic permeability, k, may be expressed in ordinary area units such as m^2. However, the permeabilities of ordinary materials are so low that a special unit, the darcy, has been defined for permeability, with 1 darcy $= 0.987 \times 10^{-12}$ m^2. Figure 4.D.2 shows representative permeabilities and hydraulic conductivities of soils. Note the enormous range of values between homogeneous clays and coarse sand or gravel.

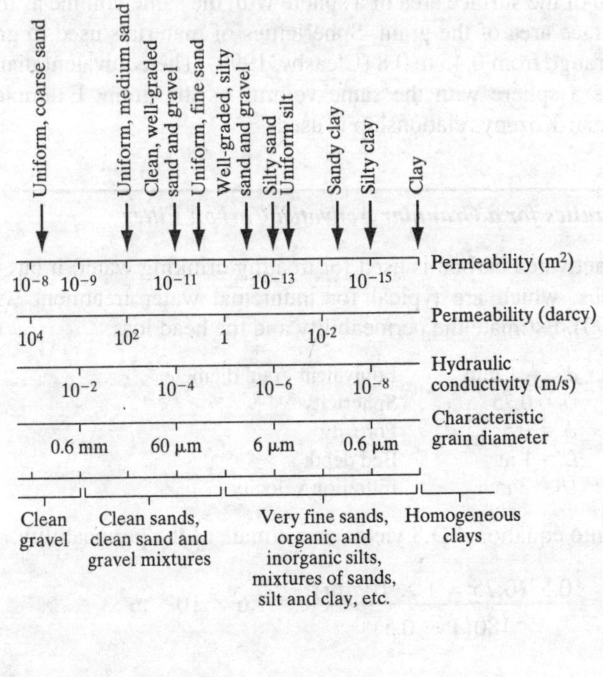

Figure 4.D.2 Permeability and hydraulic conductivity scales, indicating representative values for soils and gravel (Terzaghi and Peck, 1967; Tuma and Abdel-Hady, 1973). Characteristic grain diameters are computed from the Carman-Kozeny equation (4.D.8), assuming $\phi = 0.5$.

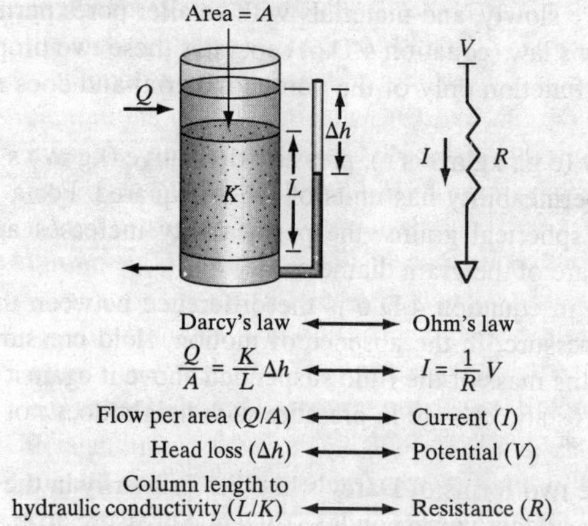

Figure 4.D.3 The analogy between Darcy's law and Ohm's law.

Fluid flow through a sand column is analogous to electrical current flow through a resistor. Figure 4.D.3 makes the comparison, showing that the ratio of the column length to hydraulic conductivity acts like a "resistor" to fluid flow driven by a loss in pressure head.

A widely used expression that relates permeability to grain size is known as the Carman-Kozeny equation. It was derived by analyzing flow through porous media in analogy to flow through a bundle of small capillaries, but with varying size and orientation (Dullien, 1979), giving

$$k = \frac{\phi^3 d_g^2}{180(1-\phi)^2} \tag{4.D.8}$$

where d_g is the grain diameter. For grains that are not spherical, d_g is replaced by $\psi\, d_{eq}$ where ψ is known as the *sphericity* of the grain and d_{eq} is the *equivalent diameter*. The sphericity is the ratio of the surface area of a sphere with the same volume as the grain to the (external) surface area of the grain. Sphericities of materials used in granular-bed filters typically range from 0.45 to 0.8 (Cleasby, 1990). The equivalent diameter is that which produces a sphere with the same volume as the grain. Example 4.D.1 shows how the Carman-Kozeny relationship is used.

EXAMPLE 4.D.1 *Hydraulics for a Granular Activated Carbon Filter*

A filter of granular activated carbon is used for treating drinking water. It has the following characteristics, which are typical for municipal water treatment (Cleasby, 1990; Snoeyink, 1990). Estimate the permeability and the head loss.

$d_{eq} = 1$ mm	Equivalent grain diameter
$\psi = 0.75$	Sphericity
$\phi = 0.5$	Porosity
$L = 1$ m	Bed depth
$U = 3$ m h^{-1}	Filtration velocity

SOLUTION Substitution into equation 4.D.8 yields an estimate of the permeability:

$$k = \frac{0.5^3(0.75 \times 1 \times 10^{-3}\,\text{m})^2}{180(1-0.5)^2} = 1.6 \times 10^{-9}\ \text{m}^2$$

EXAMPLE 4.D.1 ***Hydraulics for a Granular Activated Carbon Filter (continued)***

From equation 4.D.7, the hydraulic conductivity is

$$K = \frac{1.6 \times 10^{-9}\,\text{m}^2 \times 1000\,\text{kg m}^{-3} \times 9.8\,\text{m s}^{-2}}{0.001\,\text{kg m}^{-1}\,\text{s}^{-1}} = 0.015\,\text{m s}^{-1}$$

The head loss is determined from equation 4.D.5:

$$\Delta h = L\frac{dh}{dl} = -L\frac{U}{K}$$

$$= -1\,\text{m} \times \frac{3\,\text{m h}^{-1}}{0.015\,\text{m s}^{-1}} \times \frac{1\,\text{h}}{3600\,\text{s}} = -0.06\,\text{m}$$

The head loss in this case is about 6 cm.

4.D.2 Contaminant Transport in Porous Media

All of the mechanisms that cause contaminant transport in bulk fluids also act on contaminants suspended in fluids in porous media. Contaminants are advected with the fluid flow. Molecules diffuse within the fluid through the pores. Particles settle and undergo Brownian motion. Contaminants sorb onto and within porous media grains. Chemical or biological reactions may cause generation or decay of contaminants. If both air and water are present in the pores, contaminants can partition between the two fluids and migrate within both. Additional transport mechanisms arise, such as surface diffusion, but discussion of these is beyond the scope of this text.

Because of the large surface area of solids in contact with the fluid, contaminant-surface interactions may assume a particularly important role in transport through porous media. Particles may contact and adhere to the solid grains or fibers. Molecular contaminants may sorb onto the solid surface and may undergo chemical reactions there. Our attention in this section is focused mainly on transport mechanisms that do not involve pollutant-surface interactions. The modeling of sorption and transformation reactions in porous media is discussed in §5.B, and applications of particle deposition in porous filters are described in §6.C.2 and §7.C.1.

Diffusion

Molecular diffusion through porous materials affects the migration of gases through soils. In groundwater, diffusion is slow (recall that molecular diffusivities in water are 10,000 times smaller than in air) and therefore of little importance. In treatment technologies, transport is generally dominated by advection because of the need to treat fluids quickly.

In this discussion, we will emphasize gas-phase diffusion. Fick's law relates diffusive flux through a porous medium to the concentration gradient, but requires some clarification and adjustment because of the presence of the solids. For molecular diffusion through a porous solid, Fick's law is written in one dimension as

$$J_d = -D_e\frac{dC}{dx} \tag{4.D.9}$$

For three-dimensional transport applications, Fick's law can be written in vector form:

$$\vec{J_d} = -D_e\left(\frac{\partial C}{\partial x}, \frac{\partial C}{\partial y}, \frac{\partial C}{\partial z}\right) \tag{4.D.10}$$

Here J_d, representing diffusive flux through the porous medium, gives the quantity of contaminant transported per *total* area (not just pore area) per time. The parameter D_e is an effective diffusion coefficient, and C is the species concentration in the pores (per *pore* volume, not per total volume). These equations have the same form as Fick's law for bulk fluids (equations 4.A.5–6), but each of the parameters has an altered meaning.

The effective diffusion coefficient through a porous material is smaller than the corresponding ordinary diffusion coefficient for the given contaminant/fluid combination. For gas molecules in a granular material, such as soil, the following expression is widely used to estimate the effective diffusivity (Millington, 1959):

$$D_e = D\frac{\phi_a^{10/3}}{\phi^2} \tag{4.D.11}$$

In vadose-zone soils (i.e., above the groundwater table), typical values of air-filled and total porosity are $\phi_a = 0.2$ and $\phi = 0.4$. According to equation 4.D.11, the effective diffusivity would be reduced to $0.03D$ in this case. For air-saturated pores, equation 4.D.11 reduces to

$$D_e = D\phi_a^{4/3} \qquad \phi_a = \phi \tag{4.D.12}$$

So with a typical value for dry soil of $\phi_a = 0.4$, $D_e = 0.3D$.

An alternative expression, applied in the case of water- or air-saturated pores, uses an empirical correction factor, the *tortuosity*, T, which accounts for (a) the reduced area through which diffusion can occur and (b) the longer average path a molecule must travel to move a certain distance through the pores. The effective diffusivity is related to the tortuosity by

$$D_e = \frac{D}{T} \tag{4.D.13}$$

Typical values of tortuosity are in the range 2 to 6 (Cussler, 1984), yielding effective diffusivities that are approximately consistent with predictions based on 4.D.12.

Hydrodynamic Dispersion

Hydrodynamic dispersion in porous media is similar to the shear-flow dispersion that occurs in pipes and rivers, as discussed in §4.A.4. The dispersion of a nonreactive groundwater contaminant, or *conserved tracer,* is depicted in Figure 4.D.4. At some time t_1, contamination occurs in the groundwater, as shown in the left-hand portion of the figure. The groundwater flows with a uniform filtration velocity, U, in the $+x$-direction. At some later time, t_2, the center of mass of contamination has been transported downstream to position x_2, given by

$$x_2 = x_1 + (t_2 - t_1)U \tag{4.D.14}$$

In words, the center of mass of contamination moves at the same speed as the fluid. However, the areal extent of contaminated groundwater increases, and typically this increase is much greater than can be explained by molecular diffusion alone. Because

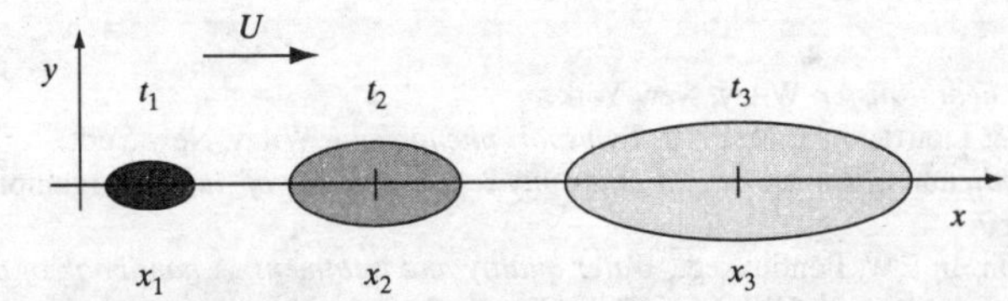

Figure 4.D.4 Transport and dispersion of a fixed quantity of a nonreactive groundwater contaminant. The figure schematically shows the areal extent of contamination at three instants in time, with $t_3 > t_2 > t_1$. The center of mass of contamination is advected in the x-direction at a rate given by the mean velocity, U. Hydrodynamic dispersion causes the contamination to spread more rapidly in the direction of flow than in the transverse direction.

the local velocity through the soil pores is not uniform, some of the contamination travels at an average speed greater than U, whiie other contaminant molecules travel more slowly than U. There is also some spreading in the y-direction, even though the mean velocity in that direction is zero. Since the contaminant is conserved, the greater areal extent of contamination is offset by a lower average concentration. The same characteristics prevail in the interval from t_2 to t_3.

This spreading is not caused by turbulence; it occurs even if the grain Reynolds number is much less than 1, so that the flow is stable and laminar. It is observed to occur in uniform sand columns in the laboratory, although it can be much stronger in the field if the soil contains zones of higher and lower permeability than the average.

Fundamentally, this spreading is caused by the variation of fluid velocities within the pores. At the grain surfaces, the velocities diminish to zero. In the center of a channel between two grains, the local velocity may be much higher than the volume average. Individual contaminant molecules experience independent velocity histories. When averaged over all contaminant molecules, the rate of advection matches that of the fluid, but some individual molecules may travel significantly faster or slower than the mean, causing plume dispersion.

The contaminant flux in the direction of flow caused by hydrodynamic dispersion in porous media is described by an expression analogous to Fick's law in one dimension:

$$J_h = -\varepsilon_h \frac{\partial C}{\partial x} \tag{4.D.15}$$

where ε_h is the *dispersion coefficient* and x is the direction of flow. In this expression, the dispersion coefficient includes molecular diffusion as a limiting condition:

$$\varepsilon_h = D_e + \varepsilon U \tag{4.D.16}$$

Here ε is a characteristic of the porous medium called the *dispersivity* (units of length). As defined above, U is the filtration velocity and D_e is the effective diffusivity. So, the flux J_h represents the sum of transport due to molecular diffusion and due to dispersion caused by nonuniform fluid flow. As with molecular diffusion alone, this flux represents the net rate of contaminant movement per unit *total* cross-sectional area. Similar expressions could be written for the flux components in the transverse direction, but these would require different dispersion coefficients, since it is observed that transverse dispersion is weaker than longitudinal dispersion.

REFERENCES

BEJAN, A. 1984. *Convection heat transfer.* Wiley, New York.

BIRD, R.B., STEWART, W.E., & LIGHTFOOT, E.N. 1960. *Transport phenomena.* Wiley, New York.

BRETSZNAJDER, S., 1971. *Prediction of transport and other physical properties of fluids.* Pergamon Press, Elmsford, NY, Chapter 7.

CLEASBY, J.L. 1990. Filtration. In F.W. Pontius, ed., *Water quality and treatment: A handbook of community water supplies.* 4th ed. McGraw-Hill, New York, Chapter 8.

CUSSLER, E.L. 1984. *Diffusion: Mass transfer in fluid systems.* Cambridge University Press, Cambridge.

DULLIEN, F.A.L. 1979. *Porous media: Fluid transport and pore structure.* Academic Press, New York.

FISCHER, H.B., LIST, E.J., KOH, R.C.Y., IMBERGER, J., & BROOKS, N.H. 1979. *Mixing in inland and coastal waters.* Academic Press, New York.

GEISLING, K.L., TASHIMA, M.K., GIRMAN, J.R., MIKSCH, R.R., & RAPPAPORT, S.M. 1982. A passive sampling device for determining formaldehyde in indoor air. *Environment International,* **8**, 153–158.

HANCOCK, R.P., ESMEN, N.A., & FURBER, C.P. 1976. Visual response to dustiness. *Journal of the Air Pollution Control Association,* **26**, 54–57.

HINDS, W.C. 1982. *Aerosol technology: Properties, behavior, and measurement of airborne particles.* Wiley, New York.

LAPPLE, C.E., & SHEPHERD, C.B. 1940. Calculation of particle trajectories. *Industrial and Engineering Chemistry,* **32**, 605–617.

LISS, P.S., & SLATER, P.G. 1974. Flux of gases across the air-sea interface. *Nature,* **247**, 181–184.

MCCABE, W.L., SMITH, J.C., & HARRIOT, P. 1993. *Unit operations of chemical engineering.* 5th ed. McGraw-Hill, New York.

MILLINGTON, R.J. 1959. Gas diffusion in porous media. *Science,* **130**, 100–102.

PERRY R.H., & GREEN, D.W. (eds.). 1984. *Perry's chemical engineers' handbook.* 6th ed. McGraw-Hill, New York.

REID, R.C., PRAUSNITZ, J.M., & POLING, B.E. 1987. *The properties of gases and liquids.* 4th ed. McGraw-Hill, New York.

SCHWARZENBACH, R.P., GSCHWEND, P.M., & IMBODEN, D.M. 1993. *Environmental organic chemistry.* Wiley, New York, Chapter 10.

SNOEYINK, V.L. 1990. Adsorption of organic compounds. In F.W. Pontius, ed., *Water quality and treatment: A handbook of community water supplies.* 4th ed. McGraw-Hill, New York, Chapter 13.

TERZAGHI, K., & PECK, R.B. 1967. *Soil mechanics in engineering practice.* 2nd ed. Wiley, New York.

TUMA, J.J., & ABDEL-HADY, M. 1973. *Engineering soil mechanics.* Prentice-Hall, Englewood Cliffs, NJ.

PROBLEMS

4.1 Environmental transport: Concepts and exercises

(a) If the concentration of a species is measured in mass per volume, what units are associated with the following parameters?

(i) Concentration gradient

(ii) Flux

(b) What is the difference between the intrinsic permeability of a sand column and its hydraulic conductivity?

(c) Typical molecular diffusivities are 0.1 cm^2 s^{-1} in air and 10^{-5} cm^2 s^{-1} in water. Explain why the diffusion coefficient is so much smaller in water than in air.

(d) Molecular diffusion results from the random motion of molecules in a fluid. Provide an analogous explanation for turbulent diffusion.

(e) We write the two-film model as $J = k_{gl}(C_s - C)$. The mass-transfer coefficient, k_{gl}, can vary markedly from one species to another. What is the dominant species property that controls k_{gl}?

4.2 More on environmental transport processes

(a) Gaseous pollutants are emitted at a constant rate from an industrial stack, as depicted in the figure. Let the x-coordinate be parallel to the wind and the z-coordinate point upward. Name the dominant transport mechanism contributing to flux in each of the x-, y-, and z-coordinate directions within the plume.

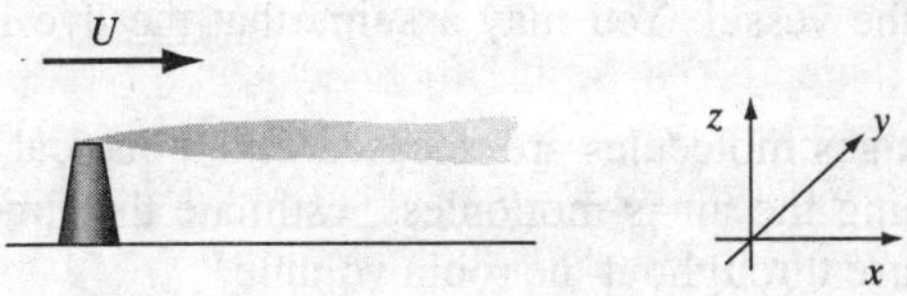

(b) Consider the case of well-developed water flow through a pipe. A short-term pulse of a nonreactive contaminant is injected into the water at the inlet. Sketch the concentration profile, $C(t)$, as a function of time at the outlet, providing as much detail as possible.

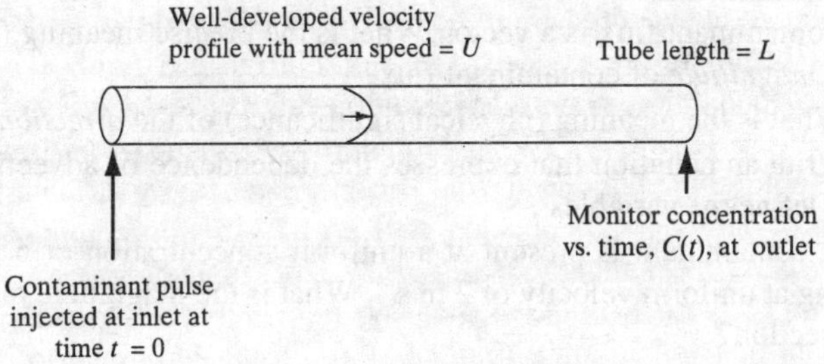

(c) A container is filled with water containing monodisperse particles. Initially the particle concentration is uniform. The fluid is motionless. The particles are large enough so that Brownian motion may be neglected. Sketch the total number of particles suspended in the vessel as a function of time. Label your axes with as much detail as possible. The vessel dimensions are L = length × W = width × H = height, the particle diameter is d_p, the settling velocity is v_t, and the initial particle concentration is C_0.

(d) The following figure shows the idealized concentration profiles for a species near the air-water interface, according to the two-resistance model. Assume that the mole fraction in air is fixed. Sketch the concentration profile at equilibrium.

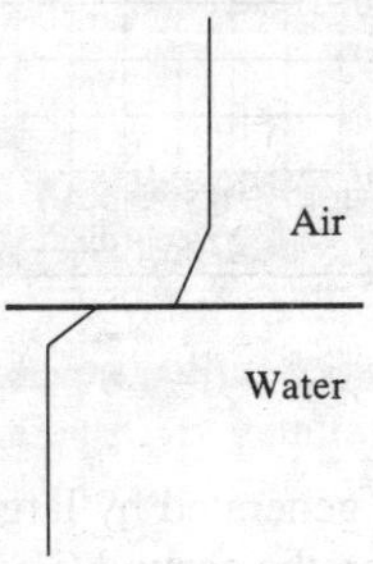

4.3 Magnitude estimates for diffusion

(a) If a fluid flow has a turbulent diffusivity of ε, what is the characteristic distance that a species may be transported by turbulent diffusion in time t?

(b) A cubical vessel, 10 cm on a side, is filled with water. A drop of food coloring is carefully dispensed into the center of the vessel. Assume the water is still (no advection). Estimate the time required for the color to disperse by molecular diffusion throughout the vessel. You may assume that the dye in food coloring is an anionic molecule.

(c) Odorous gas molecules are released into a cubical room that measures 3 m on a side. Assuming the air is motionless, estimate the time required for the odor molecules to diffuse throughout the room volume.

4.4 Comparing transport and reaction time scales

A species diffuses (diffusivity = D) through a stagnant air layer of thickness L. The species also undergoes first-order decay with rate constant k ($k \neq 0$). Write an expression that, if satisfied, would permit decay to be neglected when analyzing transport through the stagnant air layer.

4.5 Contaminant flux

(a) Contaminant flux is a vector. What is the precise meaning (physical significance) of the *magnitude* of contaminant flux?

(b) What is the meaning (physical significance) of the *direction* of the flux vector?

(c) Write an equation that expresses the dependence of advective flux on concentration. Define the variables.

(d) A contaminant is present at a uniform concentration of 5 g m^{-3} in water that is flowing at uniform velocity of 2 m s^{-1}. What is the magnitude of contaminant flux due to advection?

(e) A contaminant is present at a concentration that varies in the x-coordinate direction as shown in the following figure. The contaminant concentration is uniform in the y- and z-coordinate directions. The diffusion coefficient is 0.9 m^2 day^{-1}. What is the flux due to diffusion? Include both magnitude and direction.

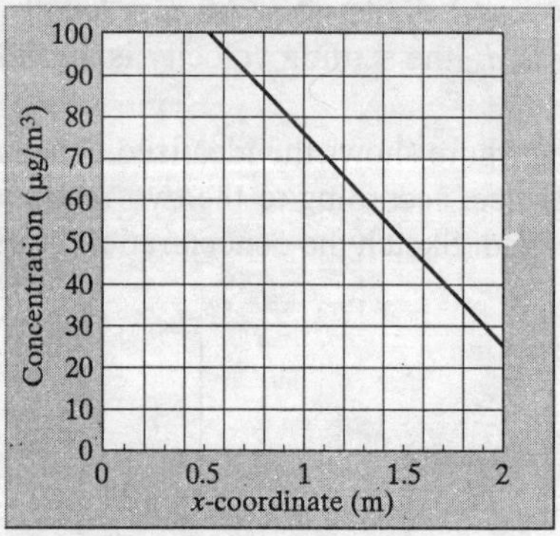

4.6 Estimating plume spread

(a) Commonly, air pollutants generated by large-scale processes are emitted from an elevated stack so that objects on the ground (i.e., people) near the source are protected from exposure to extreme concentrations. Consider air pollutants released from a stack 100 m above the ground. The wind speed is 3 m/s. The turbulent diffusivity in the vertical direction is 0.5 m^2 s^{-1}. Given these data, make a magnitude estimate of the distance downwind where the plume contacts the ground.

(b) Consider the situation depicted below, in which pollutants are emitted from a stack and the plume spreads downwind. The height of the stack is $H = 50$ m. The wind speed is $U = 3$ m s^{-1}. The plume first contacts the ground at a distance $L = 2$ km downwind. Estimate the vertical turbulent diffusivity, ε_{tz}, based on these plume characteristics.

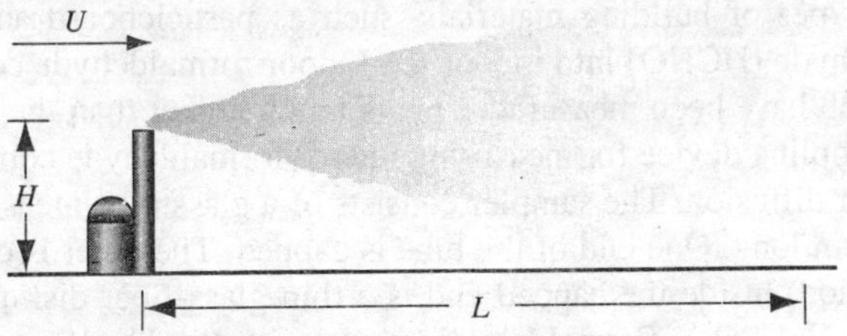

4.7 Drag on a sphere

This problem is concerned with the drag on a spherical particle immersed in a fluid.

(a) The drag force is sometimes expressed by Stokes's law: $F_d = 3\pi\mu d_p V$. Define and give appropriate units for the three parameters on the right-hand side of the expression.

(b) As the parameter d_p or V increases past some limit, Stokes's law begins to fail. What condition must be satisfied for Stokes's law to hold?

(c) Again consider that parameter d_p or V increases past a limit such that Stokes's law begins to fail. Is the drag as predicted by Stokes's law too low or too high? Briefly explain.

4.8 Factors influencing particle settling

(a) Drag on a particle is caused by two intrinsic properties of a fluid: viscosity and density. The particle Reynolds number defines which of these properties is the dominant cause of drag. For each of the two cases below, state which fluid property (or properties) contributes significantly to particle drag.

(i) $0 \le \text{Re} \le 0.3$

(ii) $0.3 \le \text{Re} \le 100$

(b) When analyzing particle settling in environmental engineering, we generally assume that the forces acting on a particle are in balance: drag + buoyancy = gravitational force. Why is this a reasonable approximation?

4.9 Acceleration to the terminal settling velocity

In analyzing particle settling, it is argued that a particle rapidly attains its terminal settling velocity, and therefore particle acceleration need not be considered. Demonstrate the conditions under which this is true. Consider a spherical particle of diameter d and density ρ_p that is initially held at rest in a stagnant fluid of viscosity μ and density ρ_f. (Fluid density is less than particle density.) At time $t = 0$ the particle is released and accelerates downward because of gravity (gravitational acceleration $= g$). Assume that drag is described by Stokes's law, $F_d = 3\pi\mu dV$, where V is the velocity of the particle relative to the fluid.

(a) Derive an expression for the time-dependent settling velocity of the particle.

(b) What condition must be satisfied for the terminal settling velocity to be a valid approximation of the particle's settling velocity?

(c) What is the characteristic time required for a particle of $d_p = 10$ μm ($= 10^{-3}$ cm) and $\rho = 2.5$ g cm^{-3} to approach its terminal (steady) settling velocity in water?

4.10 It's raining gold!

Pure gold has a density of 13.5 g cm^{-3}.

(a) Compute the settling velocity of a 0.1 mm (diameter) gold sphere in water.

(b) Compute the settling velocity of a 10 μm (diameter) gold sphere in air.

Express your answers in units of cm/s.

4.11 Irritating building materials (Geisling et al., 1982)

Certain types of building materials, such as particleboard and plywood, may emit formaldehyde (HCHO) into indoor air. Indoor formaldehyde concentrations in excess of 100 ppb have been measured. This is much higher than the level in polluted urban air. A sampling device for measuring indoor formaldehyde concentrations is based on molecular diffusion. The sampler consists of a glass tube measuring 2.4 cm in diameter by 9 cm long. One end of the tube is capped. The other is opened during the sampling period. Inside the capped end is a thin glass-fiber disk pretreated with sodium bisulfite ($NaHSO_3$). Formaldehyde reacts with the bisulfite, forming a stable solid product on the disk. This reaction maintains the concentration of formaldehyde at zero at the bottom of the sampling tube. The concentration at the open end of the tube is equal to the concentration in the room. A steady-state concentration profile is fairly rapidly established within the tube. A calibration experiment was conducted to study the instrument response. It was observed that the mass of formaldehyde accumulated on the disk was proportional to the average concentration in the room air times the exposure period:

$$M = 0.291\,Ct$$

where M is the mass of formaldehyde accumulated on the disk (in micrograms), C is the average concentration in the room air (in ppm), and t is the sampling period (in hours). From this relationship and the geometry of the sampler, determine the diffusion coefficient of formaldehyde in air.

4.12 The great benzene escape

In studies of air pollutant behavior, it is important to be able to generate a known concentration of contaminant in an air stream. One way to achieve this is through the use of a diffusion source, as depicted below. In this case, pure liquid benzene (C_6H_6) is maintained at a constant temperature of 20 °C in the lower part of the vial. The partial pressure of benzene in the gas phase of the lower part of the vial is maintained at 0.102 atm, in equilibrium with the liquid benzene. Benzene molecules then diffuse through the neck of the vial and escape from the top. By blending the emanations from this source with a known flow rate of air, a controlled, reference concentration of benzene in air may be obtained. The diffusivity of benzene in air is $D = 0.088$ $cm^2\ s^{-1}$. Compute the rate at which benzene escapes from the vial (μg s^{-1}).

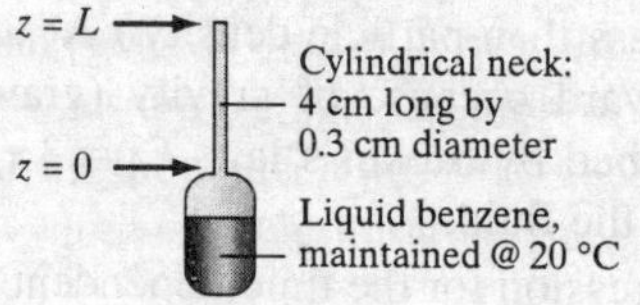

4.13 Contaminant diffusion through a landfill liner

(a) Modern landfills have a membrane liner at the bottom to collect liquid that percolates from above. This liquid, called "leachate," may contain high concentrations of toxic organic compounds. Although the liner is impermeable to flow, the organic contaminants may diffuse through it. Determine the steady-state flux of an organic contaminant from leachate through a membrane landfill liner, using the following data.

$C = 0.8$ kg m^{-3}	Concentration of contaminant in leachate
$L = 1.5 \times 10^{-3}$ m	Thickness of liner
$D = 2 \times 10^{-10}$ m^2 h^{-1}	Diffusivity of contaminant through membrane

(b) Consider the conditions in part (*a*) immediately after a new landfill is constructed. After leachate first comes into contact with the membrane, what is the characteristic time required for the diffusive flux to approach its steady-state value?

4.14 The case of the missing perfume

A bottle of expensive perfume is accidentally left open after its last use. It remains open, unnoticed, as the perfume volatilizes and escapes into the air. How much time elapses before the perfume has completely escaped from the bottle?

Data

$d = 0.3$ cm	Diameter of bottle neck
$L = 1.0$ cm	Length of bottle neck
$M = 0.004$ mol	Perfume in bottle @ $t = 0$
$D = 0.2$ cm^2 s^{-1}	Diffusivity of perfume in air
$V = 3$ cm^3	Volume of perfume bottle
$P_s = 0.02$ atm	Saturation vapor pressure of perfume
$T = 293$ K	Air temperature

4.15 Interfacial mass transfer of oxygen into a lake

Consider a lake with a mean depth of 30 m. The temperature of the lake water is 15 °C. At some time, the lake water contains 3 mg/L of dissolved oxygen. (Incidentally, this is below the level needed to support most game fish.) The lake is at sea level, so the total air pressure is 1 atm.

(a) What is the equilibrium mass concentration of dissolved oxygen in water in the lake?

(b) Compute an interfacial mass-transfer coefficient, k_{gl}, for oxygen transfer from air to water, assuming that the wind speed at 10 m above the lake surface is 3 m/s.

(c) Assuming that no mechanisms change the dissolved oxygen content of the lake water other than interfacial mass transfer, and assuming the lake water mixes rapidly over its full depth, determine a characteristic time for the oxygen content of the lake to be restored to its equilibrium value.

4.16 Filter hydraulics: The Brita® water filter

The Brita® home water purification system is a simple device consisting of two cylindrical water containers, stacked one above the other. The upper chamber is filled with tap water. Under the influence of gravity, the water flows from the upper chamber through a cylindrical filter, packed with specially treated, granular activated carbon, into the lower chamber. The filtered water is poured from the lower chamber for use. The geometry of the system is reflected in the figure. A simple experiment was conducted to characterize the flow through the filter. The upper chamber was filled with water, and water levels in the upper and lower chambers were monitored over time. At one instant, it was noted that the water level in the upper chamber was 14 cm above the water level in the lower chamber. From that point, it required 32 s for the water level in the lower chamber to rise 1 cm.

(a) Determine the hydraulic conductivity, K, of the filter.

(b) One liter of water is to be processed through the filter. The initial head loss across the filter is 14 cm. As the water passes through the filter, the head loss diminishes as the water is transferred from the upper chamber to the lower chamber. The last bit of the 1 L sample of water is driven by a head loss of only 3 cm. How much time is required to process 1 L through the filter under these conditions?

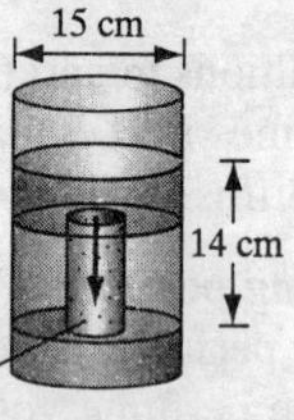

Filter: 5 cm diameter × 10 cm long

5

Transport and Transformation Models

Modeling is a very important activity in environmental engineering. The need for modeling arises from the complex systems and processes that must be addressed. Environmental engineers are frequently challenged to explain why an environmental system behaves as it does, to predict how it will evolve if left undisturbed, or to discern how a system will respond to a change. For example, if subsurface contamination is detected from a leaking chemical storage tank, an environmental engineer may be asked to determine whether it poses a threat to groundwater quality. Or if a new industrial facility is proposed for construction, an environmental engineer's opinion may be sought about the impacts it will have on local air and water quality. Qualitative judgments must be supported by quantitative evidence. It is frequently impossible and seldom practical to address such issues by experiment. Instead, environmental engineers develop and use models to interpret observations or to make predictions.

In the technical literature, the term *model* encompasses physical models in addition to mathematical models. For this text, we will use the term only to describe mathematical models that are based on scientific principles. We focus on models that predict the concentrations of impurities in environmental systems. The models of interest capture the cause-and-effect relationships that determine how concentrations depend on controlling processes, which are usually some combination of pollutant emission and removal rates, plus transport and transformation phenomena.

Environmental modeling can present something of a "chicken and egg" problem. Modeling an environmental system is an effective means of gaining insight about it.

However, some insight is required in advance to produce a sensible model. As introduced in §1.C.8, a preliminary analysis can be conducted using order-of-magnitude estimation techniques. For example, we may have to decide whether a particular transformation process needs to be considered in modeling a specific environmental system. We can estimate a time scale for the transformation to proceed to completion (see §3.A.3) and compare it with the residence time of fluid in the system. If the transformation time scale is of the same order or shorter than the residence time, and if the effect of the transformation process is significant, then that process should be included in the model. If, on the other hand, the transformation time scale is much greater than the residence time, or if the outcome of the completed process is negligible, then the transformation process is nonessential and can safely be ignored. A second-stage effort might involve applying an idealized reactor model, as will be described in §5.A, to approximately describe system behavior. Subsequent efforts, if needed, can be aimed at including more-detailed model descriptions of transport and transformation processes or more-detailed input data. Because of the additional complexity, obtaining model results at these later stages may require much more effort than is needed for the preliminary models.

Many models are commercially or publicly available for use in environmental engineering. For some applications, the use of certain models is prescribed by regulation. Often, however, the user must choose among several competing options. Practicing engineers should consider these questions while evaluating models:

- *Is the model foundation consistent with scientific fundamentals?* For example, is the model formulated so that it satisfies the material-balance principle?
- *Has the model been tested against experimental observations?* Model validation is enormously important in environmental science and engineering. Models that superficially appear reasonable may produce predictions that are grossly wrong. Ideally, the model should have been tested in a circumstance that is similar to the proposed application.
- *Are the input data requirements reasonable?* A concept from computer sciences applies: "Garbage in (yields) garbage out." A model has little practical value if the key input parameters have not or cannot be determined.

In this chapter, two classes of models based upon material balances are described: reactor models and general material-balance models. Reactor models are based on assumptions about species transport that substantially restrict how concentrations may vary in space. By making these assumptions, reactor models offer the advantage of greatly reduced mathematical complexity. Many problems involving reactor models can be solved analytically, and the numerical analysis of those that can't is relatively easy. The discussion in this chapter includes many examples that illustrate potential applications of reactor models for environmental analysis.

Despite the power of reactor models, in some cases more detailed aspects of transport must be considered to accurately model an environmental issue or process. General material-balance models can be used to study these more complex problems. Analytical solutions for these models are possible for only a small fraction of the cases of interest. Commonly, one must either make major simplifying approximations about the system, as is done in applying a reactor model, or use numerical methods to solve the governing equations. Our discussion of general material-balance models emphasizes basic techniques that can be used to achieve analytical solutions. Numerical methods for solving differential equations in environmental engineering is an active area of research that is beyond the scope of this text (Morton, 1996).

5.A REACTOR MODELS

Reactor models are useful for describing a wide variety of environmental systems. They are commonly applied to model transformation processes that occur in fluid-containing vessels, such as those used in water treatment processes. In addition, they are reasonable descriptions of several environmental systems in which the fluid is confined, such as lakes, rivers, and indoor air. Even for systems that they do not describe especially well, magnitude estimates of pollutant concentrations and fates can often be obtained by applying reactor models.

Reactor models reduce the many complex influences of transport processes to the simplest possible description. By making these approximations, the resulting mathematical descriptions of system behavior become relatively easy to solve. Problems described using reactor models often require no more than the solution of algebraic or first-order ordinary differential equations.

We will study the three most widely used models (see Figure 5.A.1). In this discussion, we will assume that the reactors behave according to idealized flow models. Methods for modeling nonideal flow can be found elsewhere (Levenspiel, 1972; Hill, 1977).

A *batch reactor* is a vessel that is loaded with reactants and then sealed, permitting no flow in or out. The contents of the batch reactor are generally mixed. A batch reactor may contain a single fluid or multiple fluids (e.g., air and water). In the idealized mixed batch reactor, the concentration of each species is uniform throughout each fluid in the vessel. Species concentrations within the reactor may change with time because of transformation processes such as chemical reactions or phase changes across a fluid interface. After some reaction period, the vessel is opened and the products are removed. To analyze a batch reactor, we consider only processes that occur within the reactor; there is no transport across its boundaries.

A *completely mixed flow reactor (CMFR)* consists of a fluid container with flow in and flow out. Usually, but not always, the flows are balanced. The contents of the container are thoroughly mixed. In the idealized CMFR, assuming a single fluid phase, mixing occurs so rapidly that the concentration of each species is uniform throughout the vessel. (With multiple fluids, mixing is assumed to be complete and instantaneous for each fluid.) Because of the assumption of perfect mixing, the concentration of a species in the effluent is equal to its concentration throughout the reactor. The analysis of a CMFR is based on a material balance on species within the fluid in the reactor, accounting for processes that occur within the reactor as well as transport into and out of the reactor.

A *plug-flow reactor (PFR)* is conceptually represented as a tube through which fluid flows. In the idealized model, the fluid velocity is uniform over the cross-section of the tube. Unless otherwise stated, we will also assume that the tube has constant cross-section at all axial positions, so that the fluid velocity is constant throughout the reactor. There is no mixing in the axial direction of the tube; transport in this direction

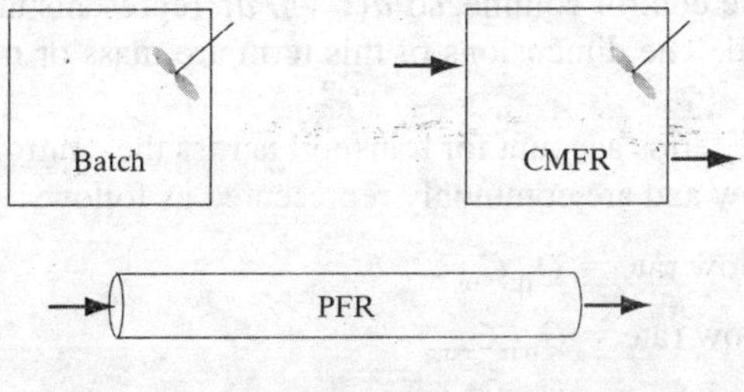

Figure 5.A.1 Schematic representations of three reactor models: batch, completely mixed flow reactor (CMFR), and plug-flow reactor (PFR). A batch reactor is usually mixed, but may be unmixed in some circumstances.

occurs only by advection. There may (or may not) be mixing in the radial direction. The analysis of a PFR involves considering processes that occur as the fluid advects along the axis of the tube.

Throughout our discussion of reactor models, we will make extensive use of two concepts that were introduced at the beginning of the text: material balances (§1.C.2) and characteristic times (§1.C.8).

Reactor Material Balances

The starting point for many problems using reactor models is to write an appropriate material-balance expression. The strategy for constructing a reactor model based on material balances is to choose an appropriate control volume within which any reactions of interest occur. Control volumes can be chosen to include or exclude specific elements of the reactor according to convenience. The overall method for constructing a reactor model is as follows:

1. Draw a simplified schematic of the reactor.
2. Draw a control volume to define the system boundaries for applying a material balance. The most convenient control volume for application of the material-balance equation usually either includes the whole reactor volume or, in the case of a PFR, is a thin volumetric slice through the reactor perpendicular to the direction of flow.
3. Select one or more species of interest and appropriate units for the material balance.
4. List known values such as fluid flows across the control volume boundaries, material concentrations at inlets, initial conditions, and volumes.
5. List expressions for transformation processes that occur within the material-balance control volume.
6. Construct the material-balance equation(s). Usually, a separate equation is needed for each species of interest within each separate fluid contained in the reactor.

Most material-balance equations follow this general form:

$$\text{accumulation rate} = \text{inflow rate} - \text{outflow rate} + \text{net transformation rate} \quad (5.A.1)$$

The accumulation rate represents the change over time in the amount of a species in the fluid within the control volume. A typical expression for the accumulation rate is

$$\text{accumulation rate} = \frac{d(CV)}{dt} \quad (5.A.2)$$

Here, C represents the concentration of the species in units of mass or moles per volume, and V represents the fluid volume. The product CV is the total mass or moles of the fluidborne species within the control volume, so $d(CV)/dt$ represents the rate of change of the amount contained. The dimensions of this term are mass or moles per time.

The inflow and outflow rate terms account for transport across the control volume boundaries due to bulk fluid flow and are commonly represented as follows:

$$\text{inflow rate} = Q_{\text{in}} C_{\text{in}} \quad (5.A.3a)$$

$$\text{outflow rate} = Q_{\text{out}} C_{\text{out}} \quad (5.A.3b)$$

where Q_{in} and Q_{out}, respectively, represent the flow rates of fluid into and out of the control volume in units of volume per time, and C_{in} and C_{out} represent the species concentrations in the respective flow streams. Note that the products QC have units of mass or moles per time.

The net transformation rate represents transport and transformation processes that occur *within* the control volume and that cause the amount of the species to change. Species production or removal by chemical reaction would be included under this heading. It would also include the flux of a species across a fluid interface within the control volume. Such processes can either add the species to or remove the species from the fluid. The associated terms should have appropriate algebraic signs (+ for addition and – for removal). For example, for a species that undergoes first-order decay by a homogeneous chemical reaction, the following term would apply:

$$\text{net transformation rate} = -kCV \qquad \text{first-order decay} \tag{5.A.4}$$

where k is the reaction rate constant and C is the species concentration in the fluid. The minus sign indicates that the reaction removes the species from the control volume. The reaction rate, kC, is multiplied by the fluid volume, V, so that the units match for all terms in the equation: mass or moles per time. More generally, if the species is affected by one or more chemical reactions in the control volume, we write

$$\text{net transformation rate} = rV \qquad \text{general chemical reaction} \tag{5.A.5}$$

where r is the net rate of production (generation minus decay) of species concentration because of chemical reactions (r typically has dimensions of mass or moles per volume per time).

The net transformation rate also accounts for any transport internal to the control volume that alters the amount of the species in the fluid of interest. For example, the effects of a phase change from air to water would be included in this term as follows:

$$\text{net transformation rate} = J_{gl}\,A \qquad \text{flux across fluid interface} \tag{5.A.6}$$

Here J_{gl} represents the species flux (mass or moles per area per time) across the water-air interfacial area, A. One must take care with the sign associated with flux: When a term $J_{gl}A$ is included in a material-balance equation like 5.A.1, that term should be positive when flux adds species to the fluid of interest and negative when flux removes species.

In many situations, the net transformation rate must account for more than one internal transport or transformation process. In such cases, terms should first be determined for each process separately and then summed (with proper signs) to obtain an overall net transformation rate.

Each term in equation 5.A.1 must be expressed in consistent units; typically we use mass per time or moles per time. Extra care should be applied to ensure that the transformation rate is written in correct units, since forms of equations accounting for reaction or flux may vary substantially.

Sometimes, more than one material-balance equation is needed to describe a system. This is the case when two or more reactor volumes are coupled through flow or other transport processes, or when two or more species interact within a single fluid volume. In such cases, separate equations of the form of 5.A.1 should be written for each species in each fluid compartment. Examples are provided later in this chapter.

The resulting mathematical form of equation 5.A.1 varies. For a batch reactor or for a CMFR, a first-order ordinary differential equation describes how the species concentration varies with time. To obtain a time-dependent solution, an initial condition must be specified. In the case of a CMFR, one is often interested only in the steady-state behavior. If so, the accumulation rate is set to zero (since $d[CV]/dt = 0$ in steady state), and an algebraic equation is obtained that describes how concentration varies with conditions. The mathematics associated with the PFR model can be the most complex if a time-dependent solution is sought. Usually, the steady-state behavior is of primary interest. In this case, the mathematical problem becomes a first-order ordinary differential equation that describes how the species concentration changes with distance from the reactor inlet or with time since the fluid entered the reactor. Appendix D contains background information on how the mathematical problems that arise from reactor models can be solved. Solutions to many example problems are presented throughout this section.

Residence Time

An important reactor parameter is the *mean residence time* of fluid molecules, represented by the symbol Θ (see also §1.C.8). In water-based systems, Θ is sometimes called the *hydraulic detention time*. However, the general concept applies for any fluid. In a batch reactor, the residence time is identical for all constituents and is equal to the time that has elapsed since the reactants were added to the reactor. In flow reactors (CMFR and PFR) of volume V (m^3), if the inflow and outflow are equal (Q, m^3 s^{-1}), the mean residence time is given by

$$\Theta = \frac{V}{Q} \tag{5.A.7}$$

In an ideal PFR, each molecule resides in the reactor for an identical amount of time, Θ, during transit from inlet to outlet. An alternative to equation 5.A.7 for a PFR is $\Theta = L/U$, where L is the reactor length and U is the fluid flow velocity. For a CMFR, although the overall average residence time is given by equation 5.A.7, molecules reside in the reactor for different periods. Some that enter the reactor are immediately swept to the outlet and removed. Others remain for a period longer than Θ. These differences in *residence time distribution* are critical in causing different behavior of CMFRs and PFRs, as demonstrated in Example 5.A.1.

EXAMPLE 5.A.1 *Contrasting CMFR and PFR Model Behavior*

Imagine that we have two sets of marbles of distinct design (shown as gray and spotted in Figure 5.A.2). These marbles could represent small parcels or elements of fluid that have different levels of a contaminant associated with them. There are two vessels, each of which holds N marbles. Assume that N is a large number so that it behaves like a continuous variable. One of the vessels, representing a CMFR, is a box that is continuously shaken to mix its contents. The other vessel, representing a PFR, is a tube. Its diameter is only slightly larger than a marble, and so the marbles must proceed through it in single file.

Prior to the experiment each reactor is filled with N gray marbles. Then, at $t = 0$, we begin to add spotted marbles to each reactor at a rate of F marbles per unit time. As each spotted marble is added at the inlet, one marble is removed from the outlet. In the CMFR, since the contents are continuously shaken, each marble in the reactor has an equal likelihood of being removed. In the PFR, the marbles cannot be mixed, and so they are removed in the same order in which they entered the reactor.

EXAMPLE 5.A.1 *Contrasting CMFR and PFR Model Behavior (continued)*

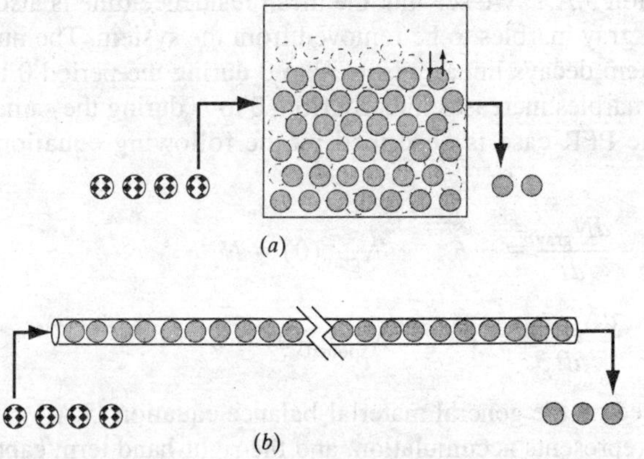

Figure 5.A.2 System illustrating the fluid flow characteristics of two reactor models: (*a*) a CMFR and (*b*) a PFR.

How will the total number of gray and spotted marbles in the two reactors change over time?

SOLUTION Although the total number of marbles in each reactor is constant, N, the numbers of gray and spotted marbles are time dependent and satisfy this "conservation of marbles" equation:

$$N_{\text{gray}}(t) + N_{\text{spotted}}(t) = N$$

The numbers of gray and spotted marbles in each reactor versus time are shown in Figure 5.A.3. First consider the PFR. Each time a spotted marble is added, a gray marble is removed until, finally, the container holds only spotted marbles. Each new marble added to the system spends a fixed amount of time within the PFR, equal to N/F, before being removed at the outlet. This time is analogous to the mean residence time

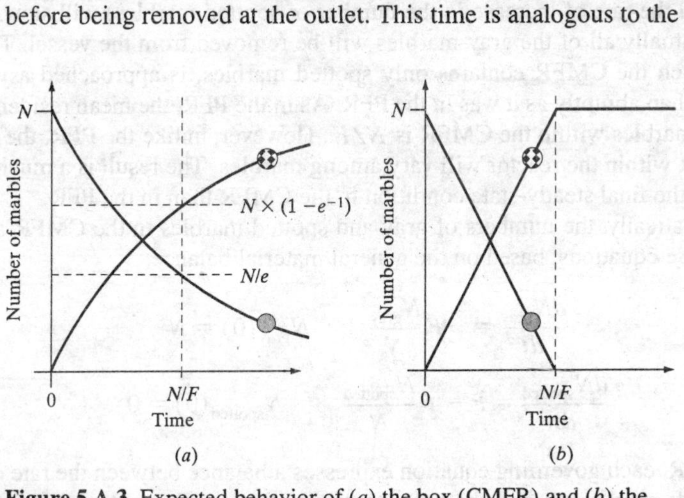

Figure 5.A.3 Expected behavior of (*a*) the box (CMFR) and (*b*) the tube (PFR) in response to a change in the design of marbles supplied at the inlet. The vessels each contain *N* marbles. Initially, all are gray. Beginning at $t = 0$, marbles with a different pattern (spotted) are added to the inlet at rate *F*; marbles are removed from the vessel outlet at this same rate.

EXAMPLE 5.A.1 ***Contrasting CMFR and PFR Model Behavior (continued)***

(Θ) as defined by equation 5.A.7. We see that the mean residence time is also *exactly* the time required for all gray marbles to be removed from the system. The number of gray marbles in the system decays linearly from N to 0 during the period 0 to N/F. The number of spotted marbles increases linearly from 0 to N during the same period.

Mathematically, the PFR case is described by the following equations, valid for $0 \le t \le N/F$:

$$\frac{dN_{\text{gray}}}{dt} = -F \qquad N_{\text{gray}}(0) = N$$

$$\frac{dN_{\text{spotted}}}{dt} = F \qquad N_{\text{spotted}}(0) = 0$$

These equations are based on the general material-balance equation (5.A.1). The left-hand term in each case represents accumulation, and the right-hand term captures the effect of flow out (for gray marbles) and flow in (for spotted marbles). The solutions are obtained by integration, with these results:

$$N_{\text{gray}}(t) = N - Ft \qquad 0 \le t \le N/F$$

$$N_{\text{spotted}}(t) = Ft \qquad 0 \le t \le N/F$$

For $t \ge N/F$, the system is at steady state with $N_{\text{gray}} = 0$ and $N_{\text{spotted}} = N$.

Initially, the behavior of the CMFR is similar. After the first spotted marble is added, the probability of removing a gray marble is extremely high because the box contains N gray marbles and only one spotted marble. So, the initial rates of change of N_{gray} and N_{spotted} are the same in the CMFR as in the PFR. However, as gray marbles are removed from the box and replaced by spotted marbles, the likelihood grows that a spotted marble will be removed instead of a gray marble. Over time, as the proportion of spotted marbles in the vessel increases, the rate of decrease in the number of gray marbles (and the rate of increase in the number of spotted marbles) will slow. Nevertheless, eventually all of the gray marbles will be removed from the vessel. The final state, in which the CMFR contains only spotted marbles, is approached asymptotically rather than abruptly as it was in the PFR. As in the PFR, the mean residence time for spotted marbles within the CMFR is N/F. However, unlike the PFR, the amount of time spent within the reactor will vary among marbles. The result is a much slower approach to the final steady-state condition in the CMFR than in the PFR.

Mathematically, the numbers of gray and spotted marbles in the CMFR are governed by these equations, based on the general material balance:

$$\frac{dN_{\text{gray}}}{dt} = -F\frac{N_{\text{gray}}}{N} \qquad N_{\text{gray}}(0) = N$$

$$\frac{dN_{\text{spotted}}}{dt} = F - F\frac{N_{\text{spotted}}}{N} \qquad N_{\text{spotted}}(0) = 0$$

As in the PFR, each governing equation expresses a balance between the rate of accumulation (left-hand term) and the net rate of addition by flow (flow in minus flow out). For the gray marbles, there is no flow in; for the spotted marbles, the rate of flow in is equal to the total marble flow rate, F. For each type of marble, since marbles are selected randomly for removal, the rate of flow out is proportional to the fraction of marbles of that design in the reactor.

EXAMPLE 5.A.1 ***Contrasting CMFR and PFR Model Behavior (continued)***

The solutions to these equations are (see Appendix D, §D.1)

$$N_{\text{gray}}(t) = N \exp\left(-\frac{F}{N}t\right)$$

$$N_{\text{spotted}}(t) = N\left[1 - \exp\left(-\frac{F}{N}t\right)\right]$$

The exponential functions exhibit the asymptotic behavior shown in Figure 5.A.3. The characteristic time occurs when the argument inside the exponential function is roughly equal to 1: $F\tau/N \sim 1$, or $\tau \sim N/F$.

Comparing the behavior of the gray marbles in these reactor models with chemical kinetics in a batch reactor, we see that flow through a PFR causes behavior similar to a zeroth-order decay reaction, whereas flow through a CMFR produces behavior like a first-order decay reaction (see Example 5.A.2).

5.A.1 Batch Reactor

In a batch reactor, no fluid flow (in or out) occurs during the reaction period. The reactants are initially placed in the reactor. The reactor is then sealed for some reaction period, transformations occur, the reactor is opened, and the products are removed.

Figure 5.A.4 shows that batch reactors may be either mixed or unmixed. Although the mixed condition usually applies, it is sometimes appropriate to use an unmixed model in cases where reactions or phase changes occur at fluid interfaces (see Example 5.A.13).

The material balance equation for a mixed batch reactor assumes this general form:

$$\frac{d(CV)}{dt} = \text{net transformation rate} \tag{5.A.8}$$

where C is the species concentration and V is the fluid volume in the reactor. The net transformation rate term typically has the form of one or more of the equations 5.A.4–6. In the case of chemical reactions within a single fluid, we would write the governing

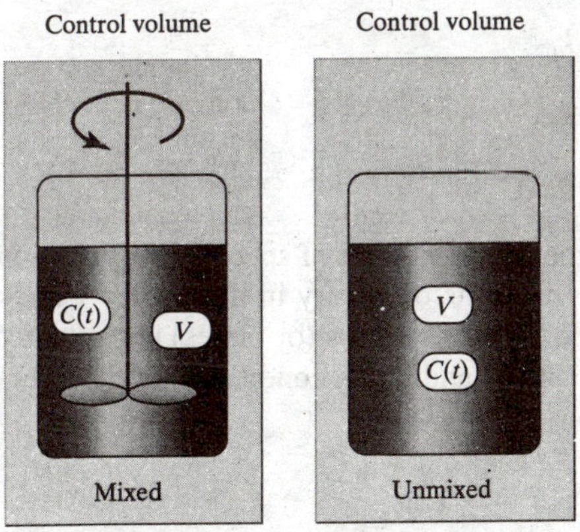

Figure 5.A.4 Schematic of mixed and unmixed batch reactors.

equation as follows:

$$\frac{d(CV)}{dt} = rV \tag{5.A.9}$$

Since there is no fluid flow across the reactor boundaries during operation, V is a constant; it may be taken out of the differential term and then canceled from the equation, leaving

$$\frac{dC}{dt} = r \tag{5.A.10}$$

Example 5.A.2 illustrates the use of this equation. Figure 5.A.5 shows how the species concentration changes with time in a batch reactor in which a decay reaction of zeroth, first, or second order occurs.

If the transformation process involves phase change within the control volume, the material balance can be written in terms of the net flux across the boundary:

$$\frac{d(CV)}{dt} = JA \tag{5.A.11}$$

Here J represents the net flux of species into the fluid (mass per area per time) and A is the interface area over which the flux occurs. Examples 5.A.3 and 5.A.4 illustrate the use of this form of material-balance equation.

For an unmixed reactor, the same material-balance equations apply. However, here C must be interpreted as the spatial average concentration throughout the whole fluid volume in the reactor. Concentration may vary with position, since the reactor contents are unmixed. An example of this sort of problem is presented in Example 5.A.13 (§5.A.4).

EXAMPLE 5.A.2 ***Species Decay in a Batch Reactor as a Function of Reaction Order***

A species is placed in a batch reactor, where it decays by either a zeroth-, first-, or second-order reaction. Derive equations to describe the change in species concentration and characteristic times in each case. Plot the results.

SOLUTION The rates of species decay are given below, where C represents the species concentration and k_i the respective rate constants.

Zeroth order: $r_0 = -k_0$

First order: $r_1 = -k_1 C$

Second order: $r_2 = -2k_2 C^2$

In these expressions, r_i represents the rate of change of species concentration because of the ith-order reaction. Note that the units of k_i vary in these three expressions, so that the decay rate (r_i) is represented in units of mass (or moles) per volume per time. The governing material-balance equations and initial conditions for the three cases are as follows:

Zeroth order: $\frac{d(CV)}{dt} = -r_0 V = -k_0 V \qquad C(t = 0) = C(0)$

EXAMPLE 5.A.2 ***Species Decay in a Batch Reactor as a Function of Reaction Order (continued)***

First order: $\dfrac{d(CV)}{dt} = -r_1 V = -k_1 CV \qquad C(t=0) = C(0)$

Second order: $\dfrac{d(CV)}{dt} = -r_2 V = -2k_2 C^2 V \qquad C(t=0) = C(0)$

Since the batch reactor volumes remain constant throughout the reaction, they cancel from both sides of these equations. The governing equation for the zeroth-order case is valid only for $C(t) > 0$. When $C(t) = 0$, the reaction rate becomes zero and the concentration remains zero indefinitely.

The solutions are obtained by rearranging and directly integrating. The results are presented as equations below and plotted in Figure 5.A.5.

Zeroth order:
$$C(t) = C(0) - k_0 t \qquad t \le C(0)/k_0$$
$$C(t) = 0 \qquad t > C(0)/k_0$$

First order: $C(t) = C(0)\exp(-k_1 t)$

Second order: $C(t) = \dfrac{C(0)}{1 + 2k_2 t C(0)}$

Characteristic times for each case can be calculated by dividing the initial stock by the initial rate of decay as follows:

Zeroth order: $\tau_0 \sim \dfrac{C(0)V}{k_0 V} = \dfrac{C(0)}{k_0}$

First order: $\tau_1 \sim \dfrac{C(0)V}{k_1 C(0) V} = \dfrac{1}{k_1}$

Second order: $\tau_2 \sim \dfrac{C(0)V}{2k_2 C(0)^2 V} = \dfrac{1}{2k_2 C(0)}$

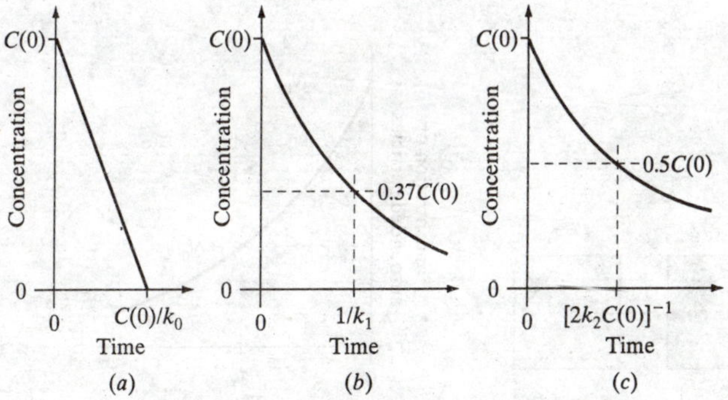

Figure 5.A.5 Change in species concentration as a function of time in a batch reactor in response to (*a*) zeroth-, (*b*) first-, and (*c*) second-order decay. In each case, a characteristic time for the reaction to proceed to completion is indicated on the *x*-axis.

EXAMPLE 5.A.3 ***Particle Settling in a Well-Mixed Batch Reactor***

A mixed batch reactor is initially filled with a uniform suspension of particles that settle under the influence of gravity. Assume that all particles have the same settling velocity. Determine qualitatively and quantitatively the change with time of particle concentration within the reactor. Assume that reactor mixing is sufficient to maintain a uniform particle concentration in the fluid, but not so vigorous as to resuspend particles once they strike the bottom.

SOLUTION Assume that the reactor has a uniform geometry such that its fluid volume can be represented by $V = AH$ where A is the area of a horizontal cross-section and H is the height of fluid in the reactor. Let v_t be the gravitational settling velocity of particles. Then the flux of particles to the bottom is $J = v_t N$ (see Table 4.A.1), where N is the particle number concentration in the well-mixed fluid.

To construct the governing equation, use a material balance of the form given by equation 5.A.11:

$$\frac{d(NV)}{dt} = -v_t NA$$

where the minus sign appears because the net flux of particles to the bottom surface removes particles from the fluid. Since V is constant, we can divide both sides of the equation by V to obtain

$$\frac{dN}{dt} = -\frac{v_t}{H}N \tag{*}$$

The characteristic time for particles to settle through the reactor height is given by the stock (NV) divided by the flux out (NAv_t):

$$\tau_s \sim \frac{NV}{NAv_t} = \frac{H}{v_t}$$

Note that this is equal to the time required for a particle to settle from the top to the bottom of the reactor in the case of motionless fluid.

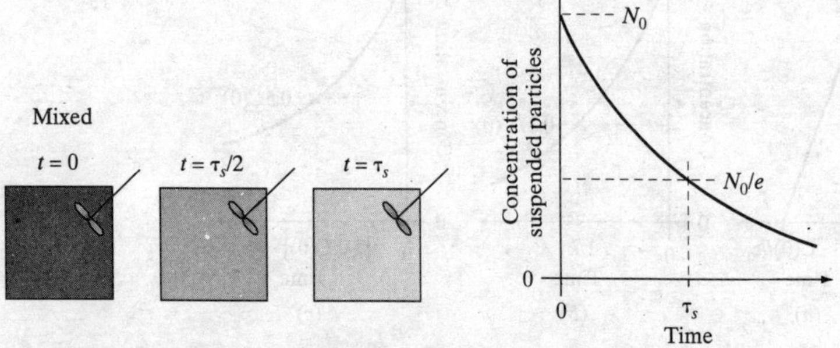

Figure 5.A.6 Change in particle concentration with time in response to gravitational settling in a mixed batch reactor. On the left, the three frames represent successive snapshots of the reactor contents. On the right, the number concentration of particles in the reactor is shown as a function of time.

EXAMPLE 5.A.3 ***Particle Settling in a Well-Mixed Batch Reactor (continued)***

The material-balance equation (*) can be integrated over time and written in terms of τ_s as follows:

$$N(t) = N_0 \exp\left(-\frac{t}{\tau_s}\right)$$

where N_0 is the initial number concentration of particles suspended in the reactor. The particle concentration is spatially uniform (due to mixing) and decays exponentially from the initial value toward zero with a characteristic time τ_s (see Figure 5.A.6).

EXAMPLE 5.A.4 ***Transient Response Involving Interfacial Mass-Transfer***

A batch reactor is partially filled with oxygen-free water. The remainder of the reactor contains air with 21 percent O_2. The air and water are independently well mixed. How does the dissolved oxygen content of the water change with time?

SOLUTION Since oxygen is sparingly soluble, we expect that dissolution will not significantly affect the partial pressure of oxygen in the air. We will initially make the assumption that P_{O_2} is fixed and check the validity of the assumption at the end of the problem.

The first step is to write a material balance on the dissolved oxygen content of the water. Since the net transformation process involves transport of oxygen across a fluid boundary, we use equation 5.A.11 as a starting point. Let A represent the air-water interfacial area, and let H be the average depth of water in the reactor. Noting that $V = AH$ is constant, we can write

$$\frac{d(CV)}{dt} = JA \qquad \Rightarrow \qquad \frac{dC}{dt} = \frac{J}{H}$$

where C is the dissolved oxygen concentration. The interfacial flux is obtained from the two-resistance model (§4.C.2 and equation 4.C.11):

$$J = J_{gl} = k_{gl}(C_s - C)$$

where C_s is the equilibrium concentration of dissolved oxygen in water and k_{gl} is the interfacial mass-transfer coefficient. Substitution yields

$$\frac{dC}{dt} = \frac{k_{gl}C_s}{H} - \frac{k_{gl}}{H}C$$

The initial condition is $C(0) = 0$. Assuming k_{gl}, C_s, and H are constant, this equation can be integrated (see Appendix D, §D.1) to give

$$C(t) = C_s\left[1 - \exp\left(-\frac{k_{gl}}{H}t\right)\right]$$

Figure 5.A.7 shows the exponential growth of the concentration of dissolved oxygen in the water, from zero toward C_s, with a characteristic time of $\tau \sim H/k_{gl}$.

Now let's return to the initial assumption that P_{O_2} is essentially constant. For this to be true, the total quantity of O_2 dissolved in water at equilibrium must be much

EXAMPLE 5.A.4 ***Transient Response Involving Interfacial Mass Transfer (continued)***

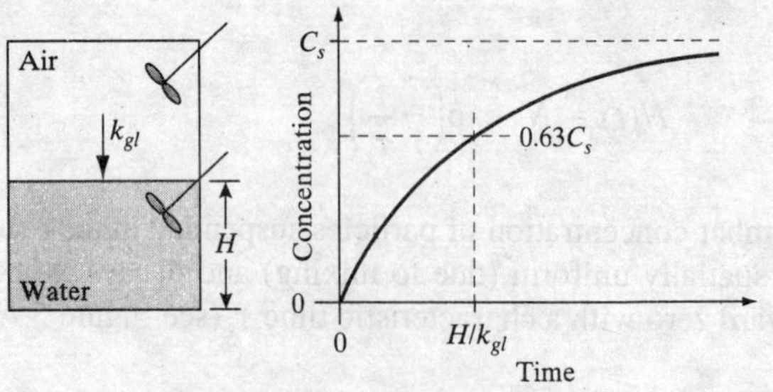

Figure 5.A.7 Kinetics of the dissolution of oxygen into water in a batch reactor. The initial concentration of dissolved oxygen in the water is zero, and C_s is the saturation concentration.

smaller than the initial amount of O_2 in the air; that is,

$$C_s V_w \ll \frac{P_{O_2}}{RT} V_a$$

Here we have used the ideal gas law to convert from partial pressure to molar concentration, and V_w and V_a represent the water- and air-filled volumes of the reactor, respectively.

Rewriting C_s using Henry's law ($C_s = K_H P_{O_2}$; see §3.B.2) and rearranging, we obtain the condition

$$V_w \ll \frac{1}{K_H RT} V_a$$

At 20 °C, $K_H = 0.00138$ M atm^{-1} for O_2 (see Table 3.B.2), so this condition translates to $V_w \ll (3 \times 10^4) V_a$. Therefore, provided there is an appreciable air volume relative to the water volume in the reactor, the original assumption is reasonable.

5.A.2 Completely Mixed Flow Reactor (CMFR)

To model species concentration in a CMFR, we need only modify our treatment of the batch reactor to account for the effects of fluid and species flow into and out of the reactor. It is convenient to choose a control volume that encompasses the entire reactor as shown in Figure 5.A.8. Then fluid flows into the reactor at rate Q_{in} carrying with it species at concentration C_{in}. Fluid flows out of the reactor at rate Q_{out} carrying species at concentration C. According to the central assumption about mixing in this reactor, the concentration at the outlet, C, is identical to the concentration everywhere within the reactor.

For most applications, the fluid flow rates, Q_{out} and Q_{in}, are expressed in units of volume per time. Also, commonly, the fluid flows are balanced ($Q_{in} = Q_{out} = Q$) such that the quantity of fluid in the reactor is constant. We emphasize cases that satisfy these conditions.

To write a material balance in the form of equation 5.A.1, we need to account for the effects of inflow and outflow on accumulation. If the accumulation term is written as $d(CV)/dt$, then it represents the net rate of change of the *total* amount of the species in the reactor. A flow of fluid into the reactor at rate Q (volume per time) carrying

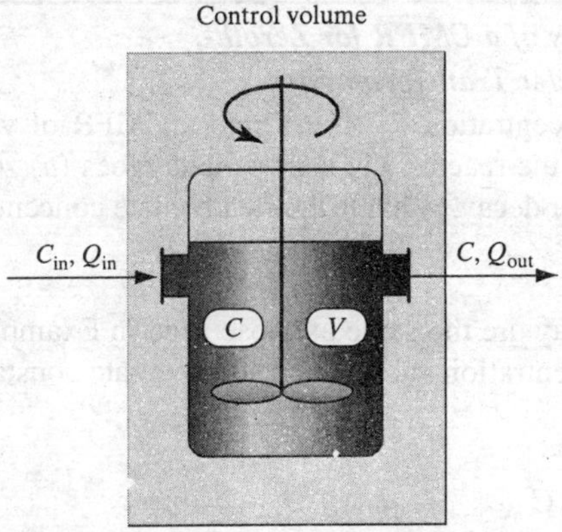

Figure 5.A.8 Schematic of a completely mixed flow reactor (CMFR).

species concentration C_{in} (moles or mass per volume) will add contaminant to the reactor volume at the rate QC_{in}. Likewise, fluid flow out at rate Q carrying concentration C will remove contaminant at the rate QC. So, the basic material-balance equation for a CMFR with balanced fluid flow and a species that may be affected by chemical reactions within the volume is

$$\frac{d(CV)}{dt} = QC_{in} - QC + rV \tag{5.A.12}$$

With balanced flow, V is constant, so the governing equation can be simplified to

$$\frac{dC}{dt} = \frac{1}{\Theta}(C_{in} - C) + r \tag{5.A.13}$$

where $\Theta = V/Q$.

As with the batch reactor, it is sometimes necessary to account for transport of material across a fluid boundary within the control volume. In this case, a term of the form JA is included in the material-balance equation.

For the solution of problems involving a CMFR, choosing a control volume and writing a material balance is always an appropriate first step. Then one considers whether the reactor may be modeled under steady-state conditions. That is, determine whether reactor conditions are unchanging, such that a time-dependent solution is not required. If so, the accumulation term can be eliminated, and the mathematical problem is reduced from a differential equation to an algebraic equation. If reactor conditions are changing and a time-dependent solution is needed, one should next consider whether the material-balance equation can be arranged into the form $dC/dt = S - LC$ with S and L constant. If so, the standard solution discussed in Appendix D (§D.1) can be written immediately. These steps are sufficient to complete the mathematical analysis for most routine applications involving the CMFR model, as demonstrated in Examples 5.A.5–8. Additional examples that illustrate the treatment of some more complex cases are presented in §5.A.4.

Frequently, problems arise in which there is a sudden change in some condition, such as the flow rate or the inlet concentration. These problems are modeled by assuming that the parameters are constant over discrete time intervals with step changes that occur at specific times. The solution for each of the discrete time intervals follows the standard solution. The result at the end of one interval becomes the initial condition for the subsequent interval.

EXAMPLE 5.A.5 ***Steady-State Efficiency of a CMFR for Zeroth-, First-, and Second-Order Transformations***

Fluid that contains contaminant concentration C_{in} flows into a CMFR of volume V with a balanced flow rate Q. Within the reactor, the species undergoes (*a*) zeroth-order, (*b*) first-order, or (*c*) second-order decay. What is the steady-state concentration in the reactor outflow, C?

SOLUTION The three rates of species decay are the same as those used in Example 5.A.2, with C representing the species concentration and k_i the respective rate constants:

Zeroth order: $r_0 = -k_0$

First order: $r_1 = -k_1 C$

Second order: $r_2 = -2k_2 C^2$

(a) Using a control volume that encompasses the entire reactor, the material balance equation for zeroth-order decay is written

$$\frac{d(CV)}{dt} = QC_{in} - QC - k_0 V$$

Since V is constant and $\Theta = V/Q$, we can write

$$\frac{dC}{dt} = \frac{1}{\Theta}(C_{in} - C) - k_0$$

The steady-state outflow concentration, which is independent of the initial condition, is calculated by setting $dC/dt = 0$ and solving for C:

$$C = C_{in} - k_0\Theta \qquad \text{zeroth-order reaction}$$

If the reaction rate (k_0) is sufficiently large, a negative concentration is predicted. Of course, this cannot occur. The failure of the model is the assumption that the reaction proceeds even when the species is completely consumed. When $C = 0$, the reaction rate must go to zero. If $k_0\Theta > C_{in}$, then the steady-state concentration is zero.

(b) For first-order decay, the governing material balance equation becomes

$$\frac{d(CV)}{dt} = QC_{in} - QC - k_1 CV$$

Again, because V is constant, we can simplify the equation to

$$\frac{dC}{dt} = \frac{1}{\Theta}(C_{in} - C) - k_1 C$$

As before, the steady-state solution is obtained by setting $dC/dt = 0$ and solving for C, with the result

$$C = \frac{C_{in}}{1 + k_1\Theta} \qquad \text{first-order reaction}$$

As the mean residence time (Θ) approaches zero, the amount of time available for contaminant decay diminishes, and C approaches C_{in} as expected.

EXAMPLE 5.A.5 ***Steady-State Efficiency of a CMFR for Zeroth-, First-, and Second-Order Transformations (continued)***

(c) For second-order decay, the governing material balance equation becomes

$$\frac{d(CV)}{dt} = QC_{\text{in}} - QC - 2k_2C^2V$$

which can be simplified to

$$\frac{dC}{dt} = \frac{1}{\Theta}(C_{\text{in}} - C) - 2k_2C^2$$

The steady-state solution is obtained by setting $dC/dt = 0$ and applying the algebraic formula for finding the roots of a quadratic equation, yielding

$$C = \frac{(8k_2\Theta C_{\text{in}} + 1)^{1/2} - 1}{4k_2\Theta} \qquad \text{second-order reaction}$$

When the reaction is relatively slow and the mean residence time is short, such that $8k_2\Theta C_{\text{in}} \ll 1$, we can use the approximation $(1 + x)^{1/2} \approx 1 + (1/2)x$, valid for $x \ll 1$, to show that C approaches C_{in}, as expected.

EXAMPLE 5.A.6 ***Flushing a Nonreactive Contaminant from a CMFR***

Consider a CMFR of volume V that initially contains a mass M of nonreactive contaminant. Beginning at time $t = 0$, contaminant-free fluid flows into the reactor at rate Q, and the fluid flow is balanced by equal outflow. How does the concentration vary with time for $t > 0$?

SOLUTION The material-balance equation for the contaminant contains two terms, one accounting for accumulation and one for flow out:

$$\frac{d(CV)}{dt} = -QC$$

or, since V is constant,

$$\frac{dC}{dt} = -\frac{Q}{V}C$$

The initial concentration is obtained by dividing the contaminant mass by the reactor volume:

$$C(0) = \frac{M}{V}$$

Solving the governing equation yields

$$C(t) = \frac{M}{V}\exp\left(-\frac{Q}{V}t\right) = \frac{M}{V}\exp\left(-\frac{t}{\Theta}\right) \qquad (*)$$

Figure 5.A.9 shows a schematic of the system and the time-dependent contaminant concentration. The characteristic time for the concentration to approach steady

EXAMPLE 5.A.6 *Flushing a Nonreactive Contaminant from a CMFR (continued)*

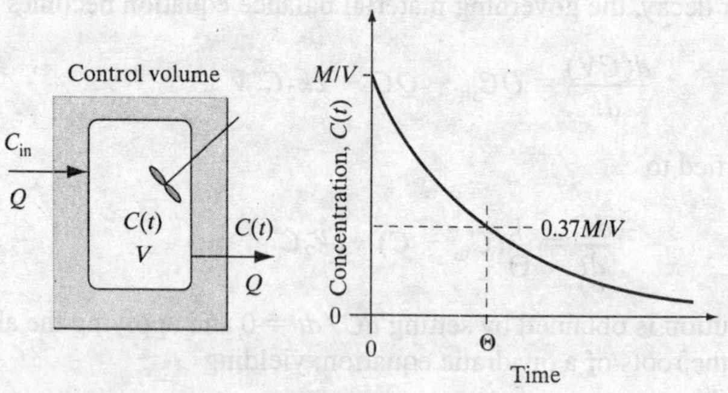

Figure 5.A.9 Schematic of the tracer-decay technique to measure Θ within a CMFR.

state in this CMFR is calculated as the initial stock over the initial flow rate out:

$$\tau_s \sim \frac{M}{QC(0)} = \frac{MV}{QM} = \Theta$$

A method known as the *tracer-decay technique* is used to measure the mean residence time of fluid in a system when either the flow or the volume cannot be easily determined. The technique involves injecting a pulse of a nonreactive tracer into either the inlet or the contained fluid volume, measuring the concentration as a function of time, and then fitting the results to an expression of the following form, obtained by taking the natural logarithm of both sides of equation (*):

$$\ln[C(t)] = \ln[M/V] - \frac{t}{\Theta}$$

A plot of $\ln[C(t)]$ versus time yields a straight line with a negative slope whose reciprocal is the mean residence time.

EXAMPLE 5.A.7 *Response of a CMFR to a Change in Inlet Conditions for a Nonreactive Contaminant*

Again consider a CMFR of volume V with balanced fluid flow rate Q. A nonreactive contaminant is initially present at level C_0 in the reactor. The contaminant concentration in the inlet flow suddenly increases to C_1 at time $t = 0$ and is maintained at this concentration indefinitely. How does the concentration of contaminant in the reactor change with time?

SOLUTION Here the material-balance equation must include both flow terms, accounting for supply and removal:

$$\frac{dC}{dt} = \frac{Q}{V}C_1 - \frac{Q}{V}C \qquad t \geq 0$$

The initial condition is $C(0) = C_0$. This equation fits the standard form $dC/dt = S - LC$ with S and L constant (see Appendix D, §D.1). The solution is the sum of exponential decay from the initial concentration toward zero plus exponential growth from zero toward the steady-state condition $C(\infty) = C_1$. The characteristic time for

EXAMPLE 5.A.7 ***Response of a CMFR to a Change in Inlet Conditions for a Nonreactive Contaminant (continued)***

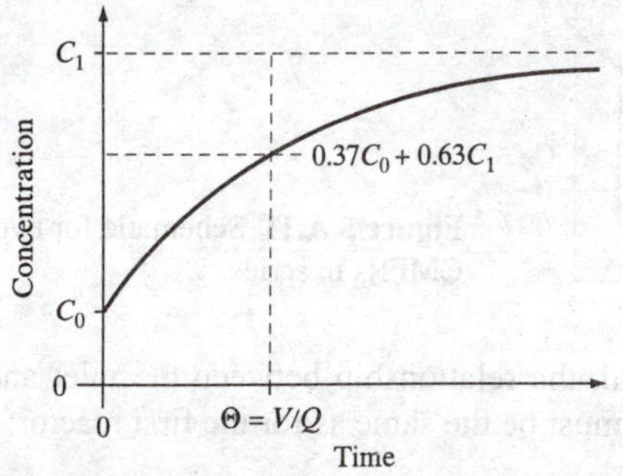

Figure 5.A.10 Response of a CMFR to a sudden change, to C_1, in the concentration of a nonreactive species in the influent stream. The initial concentration in the reactor is C_0.

both components of the solution is given by the mean residence time, $\Theta = V/Q$:

$$C(t) = C_0\left[\exp\left(-\frac{t}{\Theta}\right)\right] + C_1\left[1 - \exp\left(-\frac{t}{\Theta}\right)\right]$$

The result is illustrated in Figure 5.A.10.

EXAMPLE 5.A.8 ***Steady-State Efficiency of Two CMFRs in Series***

A species undergoes first-order decay with a rate constant k.

(a) What mean residence time in a CMFR is required to achieve 90 percent removal (i.e., $C_1/C_{in} = 0.1$, where C_1 is the steady-state outlet concentration for a fixed inlet concentration C_{in})?

(b) If the single reactor is replaced by two CMFRs of equal volume in series, what is the total mean residence time required for 90 percent removal?

SOLUTION

(a) In Example 5.A.5, we found that the steady-state ratio of outlet to inlet concentration for first-order decay in a CMFR is

$$\frac{C_1}{C_{in}} = \frac{1}{1 + k\Theta}$$

To achieve $C_1/C_{in} = 0.1$, the product $k\Theta$ must be 9, so the required residence time is $\Theta = 9/k$.

(b) For a set of two CMFRs in series, the output of the first stage becomes the input for the second, as depicted in Figure 5.A.11. Let Θ^* be the mean residence time in each reactor, so the total residence time in the system is $\Theta = 2\Theta^*$.

From part (*a*), the outlet concentration from the first reactor is

$$\frac{C_1}{C_{in}} = \frac{1}{1 + k\Theta^*}$$

EXAMPLE 5.A.8 ***Steady-State Efficiency of Two CMFRs in Series (continued)***

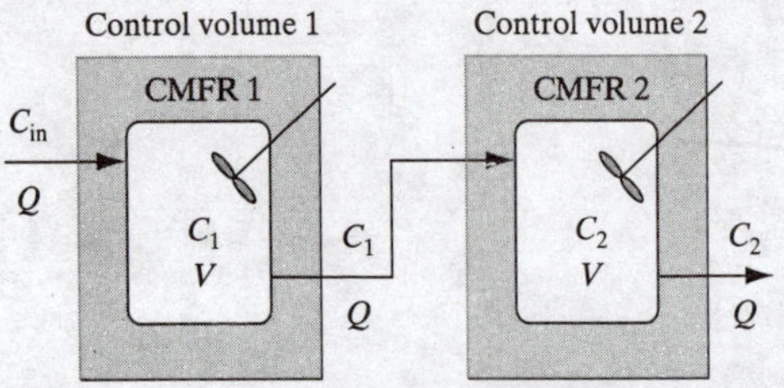

Figure 5.A.11 Schematic for two CMFRs in series.

Because the two reactors are identical, the relationship between the inlet and outlet concentrations for the second reactor must be the same as for the first reactor:

$$\frac{C_2}{C_1} = \frac{1}{1 + k\Theta^*}$$

Multiplying these two equations together yields this result:

$$\frac{C_2}{C_{in}} = \left(\frac{1}{1 + k\Theta^*}\right)^2$$

Setting $C_2/C_{in} = 0.1$ and solving, we find $\Theta^* = 2.16/k$. The total mean residence time is $\Theta = 2\Theta^* = 4.3/k$. Note that this is much less than the residence time required for a single CMFR. Since, for a given flow rate of fluid, reactor volume is proportional to mean residence time, two reactors in series offer a potential cost savings compared with a single reactor by achieving a desired level of transformation in a much smaller volume. This conclusion is valid for any transformation with a reaction order greater than zero. The benefit of reactors in series is greater the higher the reaction order and the higher the overall conversion efficiency desired. Similar benefits can be achieved by using more than two reactors; however, the practical value diminishes because the cost of adding reactors eventually overtakes the savings that result from a smaller total reactor volume.

In the general case, we might consider a fixed reactor volume that is subdivided into n equal-sized CMFRs operated in series. At steady state, if a species undergoes first-order decay, the outlet concentration is related to the inlet concentration by

$$\frac{C_{out}}{C_{in}} = \left(\frac{1}{1 + k\dfrac{V}{nQ}}\right)^n$$

where V/n is the volume of a single reactor. As n becomes very large, the behavior of this system converges to that of a plug-flow reactor, which, as shown in the next section, yields $C_{out}/C_{in} = \exp(-k\Theta)$.

5.A.3 Plug-Flow Reactor (PFR)

In applying a material balance to solve a problem involving a mixed batch reactor or a CMFR, a control volume that encompasses the entire reactor is typically chosen.

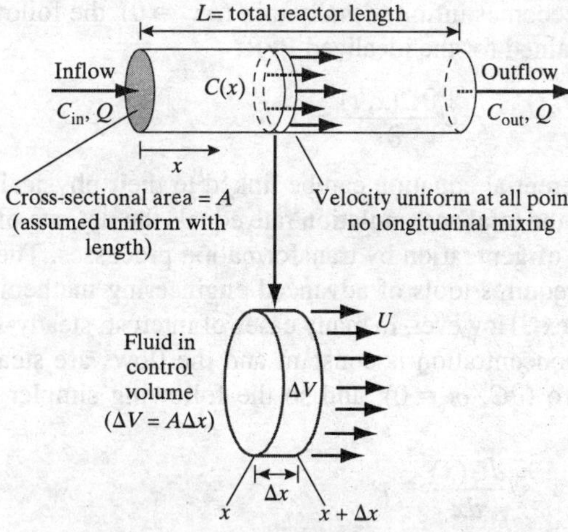

Figure 5.A.12 Schematic of an idealized plug-flow reactor and the control volume used for writing a material balance.

However, if the same type of control volume is chosen for analysis of a plug-flow reactor (PFR), serious mistakes can result, since species concentrations within the reactor are not uniform and the outflow concentration is not easily related to the concentration within the PFR. Instead, it is useful to choose a control volume that encompasses only a thin slice of the reactor volume. Figure 5.A.12 shows an ideal PFR system and a typical control volume. The ideal PFR is a tube with a uniform fluid-filled cross-section of area A, constant flow velocity U along the tube axis, and no mixing in the axial direction. The control volume encompasses a slice of thickness Δx taken at distance x from the reactor inlet. The volume of fluid in the slice is $\Delta V = A\Delta x$.

The material-balance equation for a contaminant that is being transported through the PFR can be written in the form given by equation 5.A.1. We begin by considering transport only by advection and transformation only by a homogeneous chemical reaction. (In later treatments, we will add transport by other mechanisms, such as diffusion and dispersion, and also permit other transformation processes, such as interfacial mass transfer.)

Since the PFR has a uniform cross-sectional area, the fluid flow rate within the PFR can be expressed as $Q = UA$. The contaminant concentration flowing into and out of the control volume will be designated $C(x)$ and $C(x + \Delta x)$, respectively. The material-balance equation can then be written as follows:

$$\frac{\partial (C\Delta V)}{\partial t} = UAC(x) - UAC(x + \Delta x) + r\Delta V \tag{5.A.14}$$

Here the accumulation term on the left represents the rate of change in the quantity of the contaminant in the control-volume fluid, and C represents the average concentration of contaminant in the control volume. Since the contaminant concentration can be a function of both time and distance along the reactor, the accumulation term is written as a partial derivative.

We can replace ΔV with $A\Delta x$ and remove it from the differential since A is constant and Δx does not depend on time. By rearrangement we obtain

$$\frac{\partial C}{\partial t} = \frac{UC(x) - UC(x + \Delta x)}{\Delta x} + r \tag{5.A.15}$$

Taking the limit as the slice becomes infinitesimally thin ($\Delta x \to 0$), the following material-balance equation is obtained for the idealized PFR:

$$\frac{\partial C(x,t)}{\partial t} = -U\frac{\partial C(x,t)}{\partial x} + r \tag{5.A.16}$$

The terms in this partial differential equation can be linked to their physical roots. In words, the equation says that the total accumulation rate equals the net rate of addition by transport plus the net rate of generation by transformation processes. The solution of an equation in this form requires tools of advanced engineering mathematics that are beyond the scope of this text. However, in many cases of interest, steady-state conditions apply: The influent concentration is constant and the flows are steady. Then the accumulation term is zero ($\partial C/\partial t = 0$), and so the following simpler material-balance equation holds:

$$U\frac{dC(x)}{dx} = r \tag{5.A.17}$$

Under steady-state conditions, the concentration is solely a function of position x. We can rewrite equation 5.A.17 in terms of fluid residence time within the reactor by defining Θ_x as the time that an element of fluid at position x has spent within the PFR:

$$\Theta_x = x/U \tag{5.A.18}$$

When x equals L, the total length of the PFR, Θ_x equals Θ, the residence time within the PFR. Since U is constant, we can apply the chain rule of calculus and replace $U\,d/dx$ with $d/d\Theta_x$. Therefore, the steady-state material-balance equation for this idealized PFR can also be written in this form:

$$\frac{dC}{d\Theta_x} = r \tag{5.A.19}$$

Thus, for steady-state conditions, the governing equation for the plug-flow reactor can be written in the same form as that for a batch reactor (compare equation 5.A.10). For the PFR, the time variable Θ_x represents the period during which a particular fluid element has been in the reactor. For the batch reactor, the time variable t represents the period during which all fluid elements have been in the reactor.

Although we invoked a steady-state assumption to derive equation 5.A.19, it is not strictly needed (see Exhibit 5.A.1). A physical analogy can be used to show that plug-flow reactors are closely related to batch reactors. A coordinate transformation shows that the partial-differential material-balance equation (5.A.16) is mathematically equivalent to equation 5.A.19 even for time-varying conditions.

Examples 5.A.9–12 illustrate the application of material balances to model a variety of PFR systems.

EXHIBIT 5.A.1 *Alternative Perspectives on Material Balance in Plug-Flow Reactors*

A PFR can be analyzed as a dynamic system of batch reactors on a conveyor belt (Figure 5.A.13). The reactors are filled one by one and placed on the beginning of the belt (i.e., at the reactor inlet); they advance along the PFR at a steady velocity, U; and when they reach the end of the belt (the reactor outlet), their contents are emptied.

In this representation, the effects of transport are entirely separated from the effects of transformation. Once we have done the mental mapping, we can determine the species concentration at any position and time within the reactor by writing an appropriate material balance for a batch reactor. To do so, we replace the time coordinate with Θ_x,

EXHIBIT 5.A.1 *Alternative Perspectives on Material Balance in Plug-Flow Reactors (continued)*

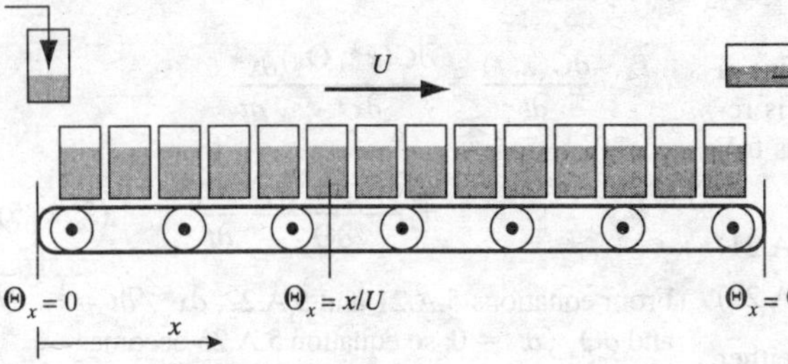

Figure 5.A.13 Representation of a plug-flow reactor as a series of batch reactors on a conveyer belt.

which represents the time interval since the fluid parcel we are considering entered the reactor (or, equivalently, the time since the corresponding batch reactor was placed on the belt). The time coordinate can be translated to a spatial coordinate by the relationship $\Theta_x = x/U$, where x is the distance from the reactor inlet in the direction of flow. When $\Theta_x = \Theta$, $x = L$, and the contents have reached the end of the reactor.

That the batch reactors are placed on a conveyor belt does not in any way affect their contents. So the general governing equation for a batch reactor would apply for each reactor in the system. If the net transformation rate within a batch reactor is rV, then, by analogy with equation 5.A.10, we have

$$\frac{dC}{d\Theta_x} = r \tag{5.A.20}$$

More formally, this result can be derived by means of applying a coordinate transformation to the partial-differential material-balance equation (5.A.16). Instead of writing the equation based on a coordinate system that is fixed in space (x, t), let's permit the coordinate system to flow at the same speed as the reactor fluid. The original and transformed coordinates are depicted in Figure 5.A.14.

A physical analogy may be useful. Think of the water in a river. In the original coordinate system, we sit on the shore and attempt to describe how species concentration changes with time and place as the water flows past. In the transformed coordinates, instead of staying in one place, we walk along the riverbank at the same rate as the water flows downstream. In the original coordinate system, a pulse of reacting contaminant discharged into the river would flow past us as it is transformed, and we would have to account for both transport and transformation to describe the concentration change. In the transformed system, we maintain the pulse adjacent to us as we walk. The

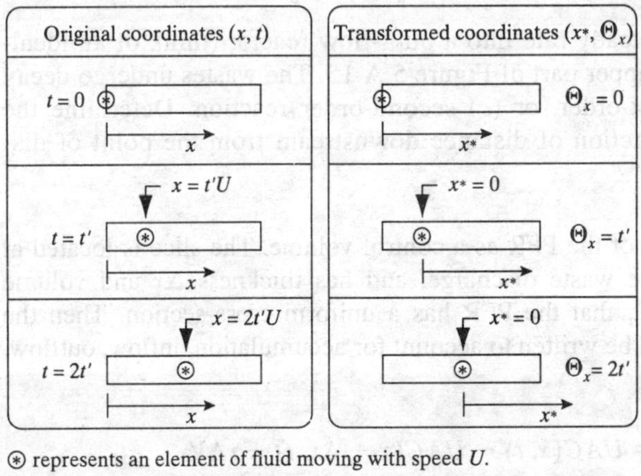

Figure 5.A.14 A plug-flow reactor as viewed from two coordinate systems, (x, t) and (x^*, Θ_x). From top to bottom, successive frames represent snapshots at successive times 0, t', and $2t'$ where t' is arbitrary.

EXHIBIT 5.A.1 *Alternative Perspectives on Material Balance in Plug-Flow Reactors (continued)*

effects of advection are absorbed into the coordinate system rather than having to be incorporated into the mathematics of solving the problem.

To proceed with the derivation, we define a transformed coordinate system (x^*, Θ_x) that is related to the original coordinate system (x, t) as follows:

$$x^* = x - Ut \qquad (5.A.21)$$

$$\Theta_x = x/U \qquad (5.A.22)$$

We can express the species concentration either as a function of the original coordinates, $C(x, t)$, or in terms of the transformed coordinates, $C(x^*, \Theta_x)$. The derivative terms in equation 5.A.16 can now be rewritten as derivatives in the transformed coordinates by applying the chain rule of calculus.

$$\frac{\partial C(x,t)}{\partial x} = \frac{\partial C(x^*, \Theta_x)}{\partial x^*}\frac{\partial x^*}{\partial x} + \frac{\partial C(x^*, \Theta_x)}{\partial \Theta_x}\frac{\partial \Theta_x}{\partial x} \qquad (5.A.23)$$

From equations 5.A.21 and 5.A.22, $\partial x^*/\partial x = 1$ and $\partial\Theta_x/\partial x = 1/U$, so equation 5.A.23 becomes

$$\frac{\partial C(x,t)}{\partial x} = \frac{\partial C(x^*, \Theta_x)}{\partial x^*} + \frac{1}{U}\frac{\partial C(x^*, \Theta_x)}{\partial \Theta_x} \qquad (5.A.24)$$

Applying a similar procedure to the time derivative yields

$$\frac{\partial C(x,t)}{\partial t} = \frac{\partial C(x^*, \Theta_x)}{\partial x^*}\frac{\partial x^*}{\partial t} + \frac{\partial C(x^*, \Theta_x)}{\partial \Theta_x}\frac{\partial \Theta_x}{\partial t} \qquad (5.A.25)$$

From equations 5.A.21 and 5.A.22, $\partial x^*/\partial t = -U$ and $\partial\Theta_x/\partial t = 0$, so equation 5.A.25 becomes

$$\frac{\partial C(x,t)}{\partial t} = -U\frac{\partial C(x^*, \Theta_x)}{\partial x^*} \qquad (5.A.26)$$

Now, substitute equations 5.A.24 and 5.A.26 into equation 5.A.16. We find that the two terms involving derivatives with respect to x^* cancel, leaving the result that concentration no longer depends on the transformed spatial coordinate, x^*, but only on the time coordinate, Θ_x:

$$\frac{dC}{d\Theta_x} = r \qquad (5.A.27)$$

The key point is this: In the transformed frame of reference, the concentration of the reactants varies only according to the chemistry. The effect of flow has been removed, making the mathematical problem of predicting reactor performance much simpler.

EXAMPLE 5.A.9 *Contaminant Decay Downstream of a Steady Discharge*

Wastes are discharged at a steady rate into a plug-flow reactor (think of an idealized river), as shown in the upper part of Figure 5.A.15. The wastes undergo decay by (*a*) zeroth-order, (*b*) first-order, or (*c*) second-order reaction. Determine the waste concentration as a function of distance downstream from the point of discharge.

SOLUTION Choose a vertical slice of the PFR as a control volume. The slice is located at position x downstream of the waste discharge, and has thickness Δx and volume $\Delta V = A\Delta x$. Assume, as usual, that the PFR has a uniform cross-section. Then the material-balance equation can be written to account for accumulation, inflow, outflow, and decay as follows:

$$\Delta V\frac{\partial C(x,t)}{\partial t} = UAC(x,t) - UAC(x+\Delta x, t) + r\Delta V$$

EXAMPLE 5.A.9 ***Contaminant Decay Downstream of a Steady Discharge (continued)***

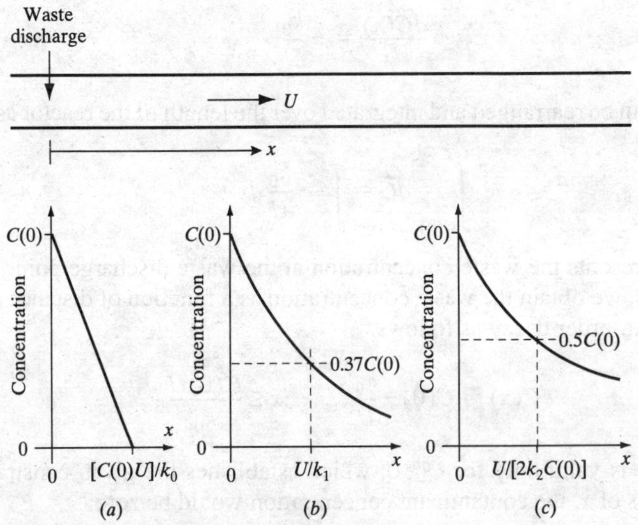

Figure 5.A.15 Concentration of a decaying species versus distance downstream of steady discharge, x, for (a) zeroth-order, (b) first-order, and (c) second-order reaction. The parameter $C(0)$ represents the contaminant concentration in the river immediately downstream of the discharge point, assuming that the waste is well mixed with the river water at that point.

Next divide both sides of the equation by ΔV and substitute $A\Delta x$ in place of ΔV, yielding

$$\frac{\partial C(x, t)}{\partial t} = -U\frac{C(x + \Delta x, t) - C(x, t)}{\Delta x} + r$$

Take the limit as $\Delta x \to 0$ to obtain

$$\frac{\partial C}{\partial t} = -U\frac{\partial C}{\partial x} + r$$

Since the wastes are discharged at a steady rate, we can focus on solving for the concentration profile in steady state. In this case, $\partial C/\partial t = 0$, so C is a function of x only, and we may write

$$U\frac{dC(x)}{dx} = r$$

The three rates for waste decay are the same as those used in Examples 5.A.2 and 5.A.5, with C representing the waste concentration at position x:

Zeroth order:	$r_0 = -k_0$
First order:	$r_1 = -k_1 C$
Second order:	$r_2 = -2k_2 C^2$

EXAMPLE 5.A.9 ***Contaminant Decay Downstream of a Steady Discharge (continued)***

(a) The material balance equation for zeroth-order decay is

$$\frac{dC(x)}{dx} = -\frac{k_0}{U}$$

This equation can be rearranged and integrated over the length of the reactor as follows:

$$\int_{C(0)}^{C} dC = \int_{0}^{x} -\frac{k_0}{U}\, dx$$

where $C(0)$ represents the waste concentration at the waste discharge point. Evaluating the integrals, we obtain the waste concentration as a function of distance along the reactor for zeroth-order decay as follows:

$$C(x) = C(0) - \frac{k_0}{U}x \qquad x \le \frac{C(0)U}{k_0}$$

This expression is valid only for $C \ge 0$, which establishes the given constraint on x. For larger values of x, the contaminant concentration would be zero:

$$C(x) = 0 \qquad x > \frac{C(0)U}{k_0}$$

The results are plotted in frame (a) of the lower part of Figure 5.A.15.

(b) Substitute $r = -k_1 C$ into the material-balance equation, rearrange, and integrate:

$$\int_{C(0)}^{C} \frac{dC}{C} = \int_{0}^{x} -\frac{k_1}{U}\, dx \Rightarrow \ln\left(\frac{C}{C(0)}\right) = -\frac{k_1}{U}x$$

The waste concentration for first-order decay can be written as

$$C(x) = C(0)\exp\left(-\frac{k_1}{U}x\right)$$

This equation is plotted in frame (b) of the lower part of Figure 5.A.15.

(c) Substitute $r = -2k_2C^2$ into the material-balance equation, rearrange, and integrate:

$$\int_{C(0)}^{C} \frac{dC}{C^2} = \int_{0}^{x} -\frac{2k_2}{U}\, dx \quad \Rightarrow \quad \frac{1}{C(0)} - \frac{1}{C} = -\frac{2k_2}{U}x$$

Rearranging, the waste concentration for second-order decay can be expressed as

$$C(x) = \frac{UC(0)}{U + 2k_2C(0)x}$$

Frame (c) of the lower half of Figure 5.A.15 shows this result.

Note that the reaction rate laws, material-balance equations, and results are the same for a plug-flow reactor as for a batch reactor (Example 5.A.2), provided that time, t, is replaced by time since discharge, x/U.

EXAMPLE 5.A.10 ***Comparing CMFR and PFR Performance***

Reactor performance is often characterized by calculating the ratio of the outlet concentration to the inlet concentration under steady-state conditions. Given fixed mean residence times, Θ, compare reactor performance for a CMFR and a PFR for contaminants that undergo zeroth-order, first-order, and second-order decay reactions.

SOLUTION Each of the required results has been calculated in Examples 5.A.5 and 5.A.9. The results are summarized in Table 5.A.1 and Figure 5.A.16.

For the case of a zeroth-order reaction, the reactor configuration does not affect performance: The CMFR and the PFR yield the same results. For all positive reaction orders, though, greater conversion is obtained in a PFR than in a CMFR. The difference in performance is negligible if the overall conversion is small ($C_{out}/C_{in} \sim 1$), but the difference becomes progressively greater as conversion increases ($C_{out}/C_{in} \rightarrow 0$).

Table 5.A.1 Comparison of the Steady-State Performance of CMFRs and PFRs

		C_{out}/C_{in}	
Reaction order	r	CMFR	PFR
Zeroth[a]	$-k_0$	$1 - \dfrac{k_0\Theta}{C_{in}}$	$1 - \dfrac{k_0\Theta}{C_{in}}$
First	$-k_1C$	$\dfrac{1}{1 + k_1\Theta}$	$\exp(-k_1\Theta)$
Second	$-2k_2C^2$	$\dfrac{(8k_2\Theta C_{in} + 1)^{1/2} - 1}{4k_2\Theta C_{in}}$	$\dfrac{1}{1 + 2k_2\Theta C_{in}}$

[a]Expressions are valid provided that $k_0\Theta \le C_{in}$; otherwise, $C_{out} = 0$.

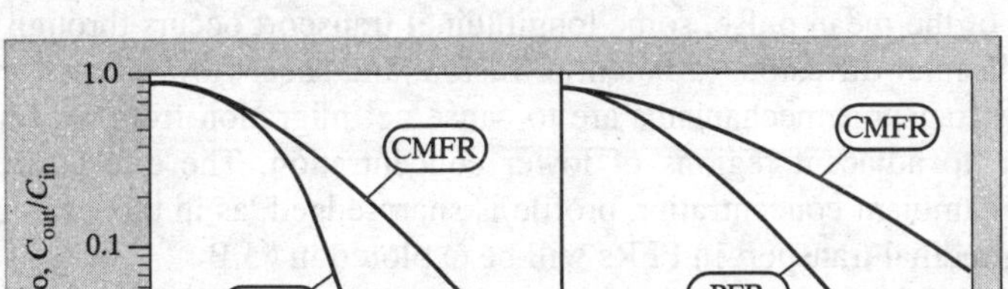

Figure 5.A.16 Steady-state reactor performance, comparing the outlet to inlet concentration ratio for a CMFR and a PFR for a species decaying by a homogeneous reaction of *(a)* first order or *(b)* second order.

EXAMPLE 5.A.11 ***Pulse of Contaminant in a PFR***

This example illustrates the usefulness of the transformed coordinate approach for analyzing time-dependent problems in a PFR.

A pulse of nonreactive contaminant is injected at the inlet of an ideal plug-flow reactor. Describe the concentration as a function of time at the outlet.

SOLUTION Figure 5.A.17 shows a schematic of the PFR and the concentration of the nonreactive contaminant at the inlet (lower left). Qualitatively, since the reactor flow is assumed to be uniform and there is no transport other than advection, and since the contaminant is nonreactive, we expect the pulse to be transmitted through the PFR without distortion. The concentration at the outlet will be the same as the concentration at the inlet, delayed by a time period of $\Theta = L/U$, as shown in the figure (lower right).

Mathematically, this result can be obtained using the governing equation in transformed coordinates as given in equation 5.A.19. For a nonreactive species, $r = 0$, so

$$\frac{dC}{d\Theta_x} = 0$$

The solution to this equation specifies that $C(\Theta_x)$ = constant, meaning that if we track the injected pulse as it flows through the reactor, its shape is unchanged from inlet to outlet.

It is instructive to compare this result with the response of a CMFR to a pulse input, as explored in Example 5.A.6. Mixing in the CMFR causes some of the injected contaminant pulse to appear at the outlet immediately, while other parts of the pulse persist well beyond the mean residence time Θ. In the ideal PFR, on the other hand, the pulse remains entirely unmixed such that all of it is discharged at a precise time, Θ, after it is injected into the reactor.

Of course, in any real reactor, flow does not conform to the ideal PFR. In addition to advection of the mean pulse, some longitudinal transport occurs through a combination of molecular diffusion, turbulent diffusion, and shear-flow dispersion. The effects of these transport mechanisms are to cause net migration from regions of high concentration to adjacent regions of lower concentration. The effects are greatest when the contaminant concentration profile is sharp-edged, as in this example. Modeling of longitudinal transport in PFRs will be explored in §5.B.

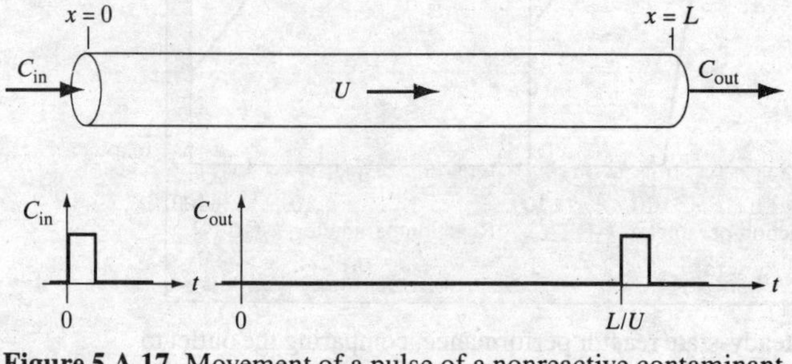

Figure 5.A.17 Movement of a pulse of a nonreactive contaminant through an idealized plug-flow reactor.

EXAMPLE 5.A.12 ***Interfacial Mass Transfer in a Plug-Flow Reactor***

Water flowing in a river is to be modeled as a plug-flow reactor as illustrated in Figure 5.A.18. A volatile organic compound (VOC) is steadily discharged into the river, where it undergoes first-order chemical decay within the water and interfacial mass transfer across the water-air surface. Use the material-balance principle to derive a governing differential equation, and solve the equation to determine the steady-state VOC concentration in the river as a function of distance downstream of the discharge point.

SOLUTION For a control volume, choose a slice of the river of thickness Δx located a distance x downstream of the discharge. The volume of water in this slice is $A\Delta x$, where A is the vertical cross-sectional area of the river. The net flux of VOC from air to water (mass per area per time) is represented by J. The area of the water-air interface in the control volume is $W\Delta x$, where W is the width of the river.

We write a material-balance equation that accounts for the processes affecting VOC transport and transformation in the river water that is contained within the control volume. The VOC is advected along with water into the slice across the upstream face. Likewise, the VOC is advected out of the slice across the downstream face. Interfacial mass transfer causes loss of VOC from the slice across the top surface. Within the slice, the VOC is removed by chemical reaction. The sum of the terms accounting for these four processes, with appropriate algebraic signs, must equal zero since the flow and VOC discharge rate are steady. These statements are translated into the following mathematical expression:

$$UAC(x) - UAC(x + \Delta x) + JW\Delta x + rA\Delta x = 0$$

where, from left to right, the terms account for advection in, advection out, interfacial mass transfer, and chemical reaction. The + sign for interfacial mass transfer is consistent with the presentation of the two-film model, where flux is positive when transfer occurs from air to water (see §4.C.2). We will see, in this case, that $J < 0$ since the flux is actually from water to air.

To proceed, divide both sides of the equation by $A\Delta x$, take the limit as Δx approaches zero, and rearrange to obtain this governing equation:

$$U\frac{dC}{dx} = J\frac{W}{A} + r$$

This equation shows that when interfacial mass transfer occurs in a PFR, its effects add to the net rate of transformation as a term JW/A. Here J is the interfacial flux, W is the width of the interfacial area normal to the direction of flow, and A is the area of a vertical cross-section through which flow occurs. Since $A = WH$, the term W/A can be equated to $1/H$, where H is the mean depth of water in the PFR channel.

Consider the flux as given by the two-film model (equation 4.C.11):

$$J = k_{gl}(C_s - C)$$

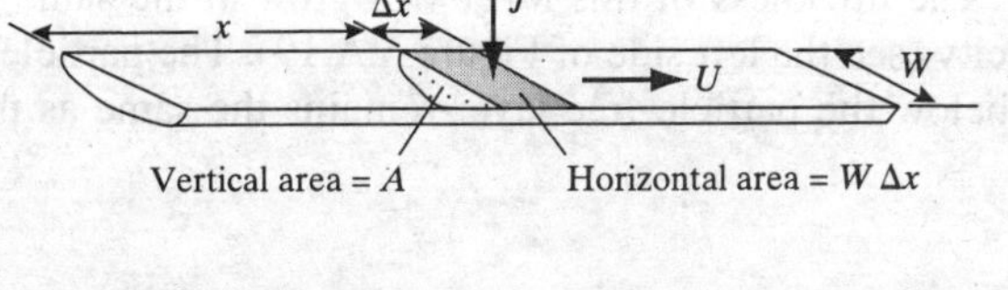

Figure 5.A.18 Schematic of a plug-flow reactor in which interfacial mass transfer occurs.

EXAMPLE 5.A.12 ***Interfacial Mass Transfer in a Plug-Flow Reactor (continued)***

where C_s is the aqueous VOC concentration in equilibrium with the air phase. If the air above the river is free of this contaminant, then $C_s = 0$, making $J = -k_{gl}C$.

If we then assume first-order VOC decay by reaction within the river, with a rate expression given by $r = -k_1C$, we can substitute into the above governing equation to obtain

$$U\frac{dC}{dx} = -k_{gl}\frac{W}{A}C - k_1C$$

The boundary condition is $C = C(0)$ at $x = 0$, where $C(0)$ is the contaminant concentration in the river at the point of discharge ($x = 0$), assuming thorough mixing between the waste stream and the river at that point. The governing equation can be rearranged and integrated along the length of the river:

$$\int_{C(0)}^{C}\frac{dC}{C} = \int_{0}^{x} -\frac{\left(k_{gl}\frac{W}{A} + k_1\right)}{U}dx$$

The result is

$$C(x) = C(0)\exp\left[-\left(k_{gl}\frac{W}{A} + k_1\right)\frac{x}{U}\right]$$

The concentration of VOC decreases exponentially as water moves down the river. The rate of decrease is a function of the mass-transfer coefficient (k_{gl}), the mean depth of the river ($1/H = W/A$), and the chemical decay constant (k_1). This equation can be rewritten with Θ_x substituted for x/U to give C as a function of travel time from the point of discharge.

5.A.4 Advanced Examples

This section contains several examples (Examples 5.A.13–18) that illustrate methods for applying reactor models in situations more complex than those already discussed.

EXAMPLE 5.A.13 ***Particle Settling through Stagnant Fluid***

A fluid-filled reactor initially contains a uniform mixture of suspended particles. The particles settle under the influence of gravity and all have the same settling velocity. In this case, in contrast to Example 5.A.3, the fluid within the reactor is unmixed and assumed to be motionless. Determine qualitatively and quantitatively how the suspension of particles evolves within the container.

SOLUTION For the unmixed reactor, a particle-free layer will appear at the top of the container as settling begins. The thickness of this layer will grow at the same speed as the particle settling velocity (see the left side of Figure 5.A.19). The particle concentration within the zone below the particle-free layer remains the same as the initial concentration.

EXAMPLE 5.A.13 ***Particle Settling through Stagnant Fluid (continued)***

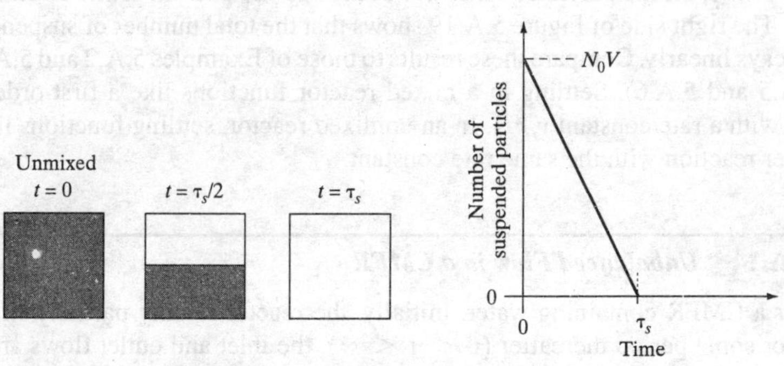

Figure 5.A.19 Evolution of suspended particles in response to gravitational settling in an unmixed batch reactor. On the left, the three frames represent snapshots taken at successive times. On the right, the figure shows that the number of suspended particles decays linearly.

To analyze the problem using a material balance, we equate the rate of accumulation in the reactor fluid to the net flux of particles to the reactor bottom (equation 5.A.11). The particle flux term has the form JA, where, J is the particle flux at the bottom of the container and A is the area of the container bottom. In general, the particle flux due to settling has the form $J = (Nv_t)$ where N is the particle number concentration and v_t is the settling velocity. The relevant particle number concentration is that in the fluid just above the bottom of the container. As long as some particles remain suspended in the container, this concentration is N_0, the initial particle concentration.

With the above assumptions, a material-balance equation can be written as follows:

$$\frac{d(NV)}{dt} = -v_t N_0 A$$

This equation is similar to that derived in the case of the mixed fluid (Example 5.A.3), except that the loss term depends here on N_0 rather than N. In the accumulation term of the governing equation, N represents the number concentration of suspended particles averaged over the entire fluid volume. Since the particle concentration varies with position within the container, it is easier to think about solving the problem in terms of the total number of suspended particles, NV. The initial condition is $NV(t) = N_0V$ at $t = 0$.

The characteristic time for particles to settle through the reactor height is given by the initial stock (N_0V) divided by the initial loss rate by settling (v_tN_0A):

$$\tau_s \sim \frac{N_0 V}{v_t N_0 A} \sim \frac{H}{v_t}$$

Following rearrangement, the material-balance equation can be integrated over time and the solution written as follows:

$$NV(t) = N_0V\left(1 - \frac{v_t}{H}t\right) = N_0V\left(1 - \frac{t}{\tau_s}\right) \qquad \text{valid for } 0 \le t \le \tau_s$$

EXAMPLE 5.A.13 ***Particle Settling through Stagnant Fluid (continued)***

For $t > H/v_t$, all the particles will have settled to the bottom of the container and $NV = 0$. The right side of Figure 5.A.19 shows that the total number of suspended particles decays linearly. Compare these results to those of Examples 5.A.2 and 5.A.3 (Figures 5.A.5 and 5.A.6). Settling in a mixed reactor functions like a first-order decay reaction with a rate constant v_t/H. In an unmixed reactor, settling functions like a zeroth-order reaction with the same rate constant.

EXAMPLE 5.A.14 ***Unbalanced Flow in a CMFR***

Consider a CMFR containing water. Initially, the reactor is only partially filled with water. For some period thereafter ($0 < t < \infty$), the inlet and outlet flows are steady but unequal (Figure 5.A.20). A contaminant species enters the reactor with the inlet flow and decays by a first-order process. Derive a material balance that describes the rate of change of the contaminant concentration in the reactor.

SOLUTION The general material-balance equation for the total contaminant mass in the reactor fluid with unbalanced flow is written

$$\frac{d(CV)}{dt} = Q_{in}C_{in} - Q_{out}C - kVC \qquad C(0) = C_0$$

Unlike cases with balanced flow, the fluid volume is not constant, so we cannot simply take volume out of the derivative. Instead, we write a second material-balance equation for the volume of water in the reactor:

$$\frac{dV}{dt} = Q_{in} - Q_{out} \qquad V(0) = V_0$$

The solution to this equation, given constant Q_{in} and Q_{out}, is obtained by direct integration:

$$V(t) = V_0 + (Q_{in} - Q_{out})t$$

Now, let's rewrite the governing equation for contaminant mass with an expanded accumulation term:

$$C\frac{dV}{dt} + V\frac{dC}{dt} = Q_{in}C_{in} - Q_{out}C - kVC$$

Substitute for dV/dt and V from the material balance on the water and its solution:

$$C(Q_{in} - Q_{out}) + [V_0 + (Q_{in} - Q_{out})t]\frac{dC}{dt} = Q_{in}C_{in} - Q_{out}C$$
$$- k[V_0 + (Q_{in} - Q_{out})t]C$$

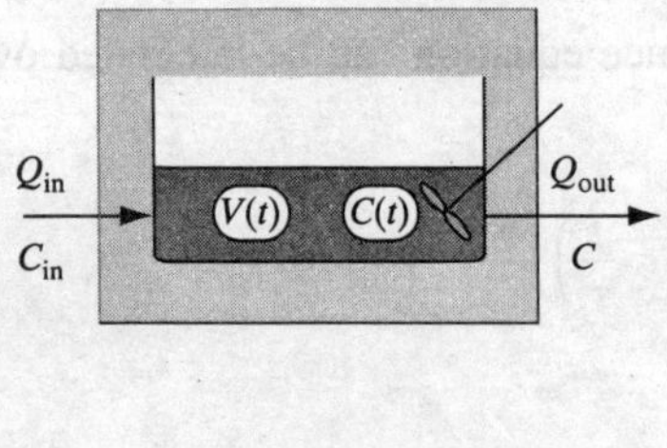

Figure 5.A.20 Schematic of a CMFR with unbalanced fluid flow.

EXAMPLE 5.A.14 *Unbalanced Flow in a CMFR (continued)*

Rearrange to obtain an equation for the rate of change of contaminant concentration in the reactor:

$$\frac{dC}{dt} = \frac{Q_{\text{in}}(C_{\text{in}} - C)}{(Q_{\text{in}} - Q_{\text{out}})t + V_0} - kC \qquad C(0) = C_0$$

Although this equation does not have an analytical solution, it is not difficult to solve it numerically. This example illustrates how to write proper material-balance equations for unconventional situations.

In air-filled reactors, the problem of unbalanced flows is encountered if the temperature or pressure varies significantly in the system. In such cases, in dealing with a gaseous contaminant, it is preferable to write material-balance equations in which the air flow rates are expressed in moles per time and the contaminant concentrations are written as mole fractions. For particulate matter, since the mole fraction is not a convenient unit of measure, it is best to writ · a material balance with flow rates specified in air volume per time but with flows appropriately unbalanced to account for changes in pressure and/or temperature. This issue arises only if the fractional change in pressure or absolute temperature is substantial (e.g., at least 10 percent from inlet to outlet).

EXAMPLE 5.A.15 *Dissolved Oxygen Consumption by BOD in a CMFR*

Water containing biochemical oxygen demand (BOD) and dissolved oxygen (DO) flows into a CMFR. Within the reactor the BOD undergoes first-order decay, consuming DO in the process (cf. §3.D.5). What is the steady-state concentration of DO in the reactor?

SOLUTION Figure 5.A.21 illustrates the situation. We begin by writing a material balance on the dissolved oxygen in the reactor:

$$\frac{d([\text{DO}]V)}{dt} = [\text{DO}]_{\text{in}}Q - [\text{DO}]Q + rV$$

Here r represents the net rate of DO decay in the reactor (mass per volume per time). The transformation process that causes the DO decay (r) is consumption of BOD, as described in reaction 3.D.26 and equation 3.D.29:

$$r = -k[\text{BOD}]$$

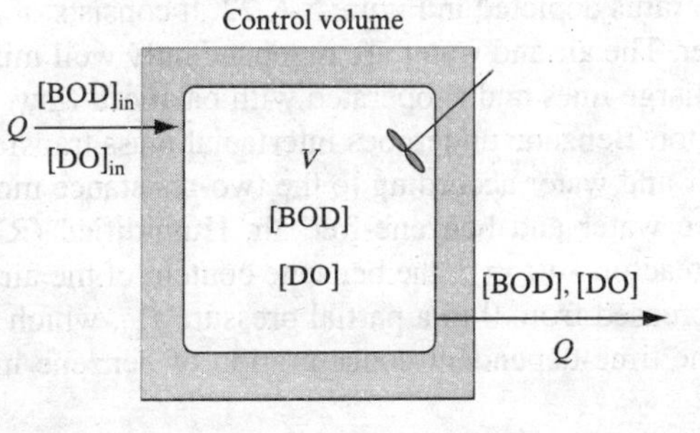

Figure 5.A.21 Schematic of biochemical oxygen demand and dissolved oxygen in a CMFR.

EXAMPLE 5.A.15 ***Dissolved Oxygen Consumption by BOD in a CMFR (continued)***

So,

$$\frac{d([\text{DO}]V)}{dt} = [\text{DO}]_{\text{in}}Q - [\text{DO}]Q - k[\text{BOD}]V$$

Now we have one governing equation with two unknowns, [DO] and [BOD]. To proceed, we write a second material-balance equation, this time applied to BOD in the reactor:

$$\frac{d([\text{BOD}]V)}{dt} = [\text{BOD}]_{\text{in}}Q - [\text{BOD}]Q - k[\text{BOD}]V$$

This is a familiar equation, of the standard form $dC/dt = S - LC$, with S and L constant. The steady-state solution is

$$[\text{BOD}] = [\text{BOD}]_{\text{in}}\frac{1}{1 + k\Theta} \qquad \text{valid for } [\text{DO}] \geq 0$$

where, as usual, $\Theta = V/Q$. The condition $[\text{DO}] \geq 0$ is required because the rate of BOD oxidation must diminish to zero if dissolved oxygen is completely consumed. Now we substitute this result into the material balance equation for DO and solve in steady state to obtain

$$[\text{DO}] = [\text{DO}]_{\text{in}} - \frac{k\Theta}{1 + k\Theta}[\text{BOD}]_{\text{in}} \qquad \text{valid for } [\text{DO}] \geq 0$$

Note that the reaction is zeroth order in [DO], so that if enough BOD is oxidized in the reactor, the dissolved oxygen concentration can be reduced entirely to zero. The reaction then stops.

This example illustrates how problems involving multiple species are addressed. Commonly, one must write a separate material-balance equation for each species and then solve the governing equations simultaneously. In the case of steady-state behavior of a CMFR, the mathematical problem reduces to solving a system of coupled algebraic equations, which are often nonlinear. If one seeks to determine time-dependent concentrations, a system of coupled first-order ordinary differential equations must be solved; often these systems are nonlinear and must be solved numerically. In this example, the problem is mathematically simple, because the equation that governs BOD is independent of the DO concentration (provided it remains above zero).

EXAMPLE 5.A.16 ***Coupled Reactors***

Consider the experimental apparatus depicted in Figure 5.A.22. It consists of a reactor that is partially filled with water. The air and water are independently well mixed. The reactor has air supply and discharge lines and is operated with balanced flow. There is no water flow through the reactor. Benzene undergoes interfacial mass transfer across the interface area A between air and water according to the two-resistance model. Initially, the reactor contains pure water and benzene-free air. Humidified (RH = 100 percent) air flows through the reactor. At $t = 0$, the benzene content of the air flowing into the reactor is suddenly increased from 0 to a partial pressure P_{in}, which is maintained indefinitely. Describe the time-dependent concentration of benzene in air and water.

EXAMPLE 5.A.16 ***Coupled Reactors (continued)***

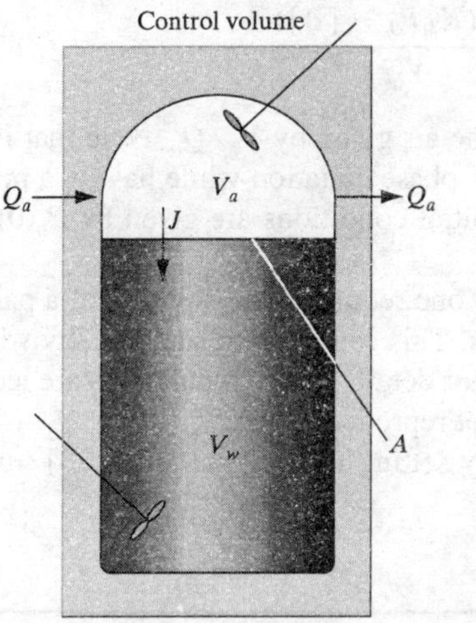

Figure 5.A.22 Schematic of an experimental apparatus in which benzene is transferred from air to water.

SOLUTION We begin by writing two material-balance equations, one for the benzene in the gas phase and a second for the benzene in the aqueous phase. We then solve these equations simultaneously, because they are coupled through the interfacial transfer terms. We will use the partial pressure of benzene in the gas phase, P_b, and the molar concentration in the aqueous phase, [B], as the primary variables. The material balances are written in terms of moles of gaseous and aqueous benzene, respectively, within the two fluid phases in the reactor volume. The partial pressure of benzene in the gas phase is converted to molar concentration through the ideal gas law (P_b/RT equals the number of moles of gaseous benzene per volume of air). The governing equations are

$$\frac{d[V_a(P_b/RT)]}{dt} = Q_a\frac{P_{\text{in}}}{RT} - Q_a\frac{P_b}{RT} - JA$$

and

$$\frac{d(V_w[\text{B}])}{dt} = JA$$

Note that the net rate of interfacial transfer of benzene from air to water (moles per time) is given in terms of the flux (J) times the interfacial area A (§4.C.2 and equation 4.C.11):

$$J = k_{gl}([\text{B}]_s - [\text{B}]) = k_{gl}(K_H P_b - [\text{B}])$$

where k_{gl} is the mass-transfer coefficient across the air-water interface, $[\text{B}]_s$ is the aqueous concentration of benzene in equilibrium with gas-phase benzene at partial pressure P_b, and K_H is the Henry's law constant for benzene. Given this expression, the equations governing the time-dependent levels of benzene in air and water are

$$\frac{dP_b}{dt} = \frac{(P_{\text{in}} - P_b)}{\Theta_a} - \frac{RT}{V_a}k_{gl}A(K_H P_b - [\text{B}])$$

EXAMPLE 5.A.16 ***Coupled Reactors (continued)***

and

$$\frac{d[\mathrm{B}]}{dt} = \frac{k_{gl}A(K_H P_b - [\mathrm{B}])}{V_w}$$

where Θ_a is the mean residence time of the air given by V_a/Q_a. Note that interfacial mass transfer is a net loss term for the gas-phase equation while having a positive effect on the aqueous concentration. The initial conditions are given by $P_b(0) = 0$ and $[\mathrm{B}](0) = 0$.

This problem has been converted into one requiring the solution of a pair of coupled, linear ordinary differential equations. This can be done analytically without too much difficulty (see Appendix D, §D.1.b for details). The calculations are tedious and not particularly illuminating, and so are not reproduced here.

The steady-state results, calculated by setting $dP_b/dt = 0$ and $d[\mathrm{B}]/dt = 0$, are $P_b = P_{\mathrm{in}}$ and $[\mathrm{B}] = K_H P_{\mathrm{in}}$, as expected.

EXAMPLE 5.A.17 ***Characterizing Pollutant Emissions in a CMFR***

A common goal in environmental engineering research and practice is to measure the total amount of a contaminant that is released into an environment through some process or activity. This example demonstrates an experimental approach for evaluating emissions using a CMFR model. The specific data come from a study of the episodic release of air pollutants from dishwashing (potentially significant, but demonstrated to be minor) (Wooley et al., 1990).

A simulated dishwashing activity was conducted in a room-sized test chamber ($V = 20$ m^3). The concentration of ethanol (a constituent of dishwashing detergent that is of concern as a potential contributor to photochemical smog) was measured during and after 20 minutes of dishwashing, with the results shown in Figure 5.A.23. The chamber was ventilated with ethanol-free air at a rate of 0.7 m^3 min^{-1}. Assuming that ethanol is nonreactive, show how the total mass of ethanol emitted by the dishwashing can be determined from these data.

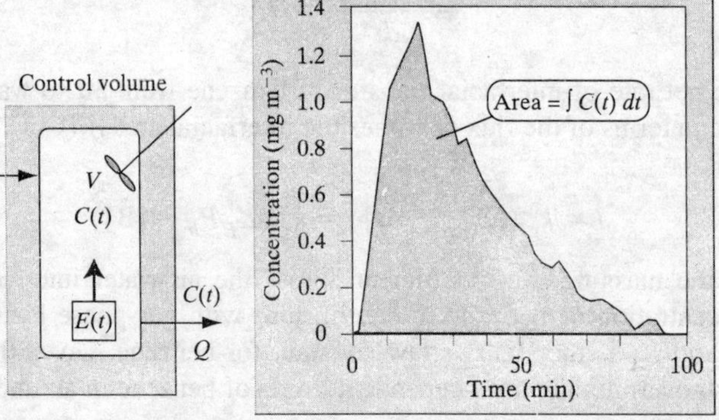

Figure 5.A.23 Ethanol concentration in chamber air resulting from a dishwashing activity conducted during the period 0 to 20 minutes.

EXAMPLE 5.A.17 ***Characterizing Pollutant Emissions in a CMFR (continued)***

SOLUTION We begin by writing a material balance on ethanol in the chamber air. The control volume is drawn around the room, and the only flow of ethanol across the boundary is in the effluent gas. We must include a generation term that accounts for emissions into the chamber air from dishwashing. So,

$$\frac{d(CV)}{dt} = -QC + E(t)$$

where $E(t)$ is the ethanol emission rate in mass per time. In general, the emissions may vary with time in a complex manner whose details are of little interest. What we seek, instead, is the total mass, M, emitted during the entire dishwashing activity. That mass is equal to the integral of the emission rate over time:

$$M = \int E(t)\,dt$$

We can use the material-balance equation to obtain an expression for M in terms of measured parameters. Multiply both sides of the material-balance equation by dt, rearrange, and then integrate to obtain

$$M = \int E(t)\,dt = \int d(CV) + \int QC(t)\,dt$$

Physically, this equation represents a restatement of the material-balance principle: For a nonreactive contaminant, the total mass emitted is the sum of whatever has accumulated in the air (first term on the right) plus whatever has been removed by ventilation (second term).

Integrate, starting from when the emissions begin ($t = 0$) and ending when the chamber no longer contains any ethanol ($t = 100$ min). Since there is no ethanol in the room air at the two integration end points ($t = 0$ and 100 min), there is no net ethanol accumulation, and so the first term on the right equals 0. Then, since the ventilation rate, Q, is a constant, the equation simplifies to

$$M = Q\int C(t)\,dt$$

Given the data in Figure 5.A.23, the integral, which is proportional to the area under the curve, is 47 mg m^{-3} min. Multiplying by the flow rate, we obtain the result we seek: $M = 33$ mg.

As a final note, although we made use of the CMFR model, it happens that the assumption of perfect mixing is not needed in this case. If the concentration is measured in the outlet flow stream, then the principle of material balance can be applied directly: Whatever is emitted must flow past the monitoring point, and so the final expression derived above remains valid, even if the chamber air is not well mixed.

EXAMPLE 5.A.18 ***Plug-Flow Reactor with Recycle***

Dingbat Engineering has a contract to design a new wastewater treatment plant. They studied Example 5.A.10 and know that a PFR works better than a CMFR for treatment

EXAMPLE 5.A.18 ***Plug-Flow Reactor with Recycle (continued)***

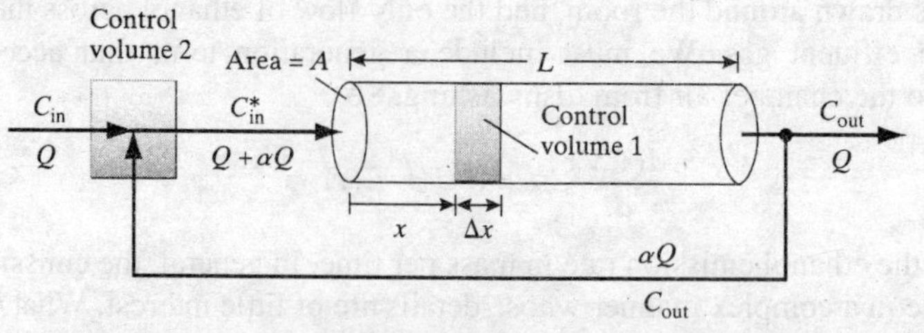

Figure 5.A.24 Schematic of a PFR with recycle.

processes involving first-order decay. The clever engineers at Dingbat reason that they can do even better by recycling a fraction of the outlet flow of the PFR back to the inlet. They think that this will allow the contaminant to react for a longer period and so yield even better conversion efficiency for a fixed reactor volume. Figure 5.A.24 shows a schematic of their design.

Consider the use of this reactor to remove BOD that decays according to

$$r = -kC$$

where C represents the aqueous BOD concentration. In answering the following questions, assume that the flow rates, Q and αQ, and the influent BOD concentration, C_{in}, are constant and that steady-state conditions prevail. A design goal is to minimize the reactor volume such that a fixed fractional removal of BOD is achieved (e.g., 90 percent destruction so $C_{out}/C_{in} = 0.1$).

(a) Derive an expression for BOD in the effluent (C_{out}) in terms of the system parameters (Q, α, A, L, k, and C_{in}) for the limiting case of no recycle (i.e., $\alpha \to 0$).

(b) Derive an expression for C_{out} that is valid for any value of α.

(c) Do the Dingbat engineers have a good idea? In other words, for $\alpha > 0$, does their configuration perform better, worse, or the same relative to a conventional PFR?

SOLUTION

(a) Initially ignoring recycle, we choose a vertical slice of the PFR as a control volume. The slice is located at position x downstream of the inlet, has thickness Δx, and volume $\Delta V = A\Delta x$, where A is the cross-sectional area of the reactor. As usual, we write a material balance that equates the rate of accumulation to the rate of inflow minus the rate of outflow and minus the rate of decay:

$$\Delta V\frac{\partial C(x,t)}{\partial t} = QC(x,t) - QC(x+\Delta x,t) - kC(x,t)\Delta V \qquad \text{control volume 1}$$

where, for small Δx, $C(x, t)$ approximates the average BOD concentration within the control volume. The equation can be rearranged with $A\,\Delta x$ substituted for ΔV to this form:

$$\frac{\partial C(x,t)}{\partial t} = \frac{QC(x,t) - QC(x+\Delta x,t)}{A\,\Delta x} - kC$$

EXAMPLE 5.A.18 ***Plug-Flow Reactor with Recycle (continued)***

Taking the limit as $\Delta x \to 0$ gives us

$$\frac{\partial C}{\partial t} = -\frac{Q}{A}\frac{\partial C}{\partial x} - kC$$

Since the reactor operates at steady state, the BOD concentration is not a function of *t*, but only of *x*, and so we can simplify the equation to this form:

$$\frac{dC(x)}{dx} = -kC\frac{A}{Q}$$

Rearranging and integrating over the length, *L*, of the reactor yields

$$\int_{C_{in}}^{C_{out}} \frac{dC}{C} = \int_0^L -\frac{kA}{Q}dx \quad \Rightarrow \quad \ln\left(\frac{C_{out}}{C_{in}}\right) = -\frac{kA}{Q}L$$

Substituting $\Theta = V/Q = AL/Q$ into the solution and rearranging gives us the following result, which is the expected outcome for first-order decay in a plug-flow reactor:

$$C_{out} = C_{in}\exp(-k\Theta)$$

(b) At first glance, analysis of the system with recycle might appear hopelessly complex. However, the problem can be analyzed without great difficulty by recognizing that the behavior of the reactor itself is unchanged. We apply an additional material-balance relationship to predict the performance of the system with recycle from the result for the system without recycle.

Let's define the concentration entering the reactor as C_{in}^*. With recycle, the fluid flow rate through the reactor is the sum of the inlet flow Q plus the recycle flow αQ. A material balance on a second control volume drawn around the convergence of the inflow and recycle flow can be used to compute the concentration of C_{in}^* as follows:

$$C_{in}Q + C_{out}\alpha Q - C_{in}^*(Q + \alpha Q) = 0$$

or, rearranging,

$$C_{in}^* = \frac{C_{in} + \alpha C_{out}}{1 + \alpha}$$

The hydraulic residence time in the reactor with recycle (Θ^*) is related to the residence time of the reactor without recycle (Θ):

$$\Theta^* = \frac{V}{Q + \alpha Q} = \frac{\Theta}{1 + \alpha}$$

Then, the equation obtained in (*a*) that describes the relationship among C_{out}, C_{in}, and Θ can be modified to obtain C_{out} in terms of C_{in}^*:

$$C_{out} = C_{in}^*\exp(-k\Theta^*) = C_{in}^*\exp\left(-\frac{k\Theta}{1 + \alpha}\right)$$

where $\Theta = LA/Q$, as before. Substituting for C_{in}^* from above and rearranging yields the result we seek: an expression for C_{out} in terms of C_{in}; the recycle ratio, α; plus

EXAMPLE 5.A.18 ***Plug-Flow Reactor with Recycle (continued)***

other primary parameters, k and Θ:

$$C_{out} = C_{in} \frac{\exp\left(-\frac{k\Theta}{1+\alpha}\right)}{1 + \alpha\left[1 - \exp\left(-\frac{k\Theta}{1+\alpha}\right)\right]}$$

(c) Is this a good idea? Comparing the final equations from (*a*) and (*b*), the answer is not immediately apparent. It turns out that this type of recycling is a bad idea in a PFR if the reaction order is greater than zero. That is because the effective residence time within the reactor is smaller with recycle than without (i.e., $\Theta^* < \Theta$). Therefore, in each pass, the transformation reactions have a shorter period of time in which to act upon the contaminant concentration. The benefit of recycling contaminants through the reactor is outweighed by the performance degradation caused by the shorter residence time.

We can demonstrate that for first-order contaminant decay with any specific value of $k\Theta$, C_{out}/C_{in} increases with increasing α, an undesirable result. Furthermore, we can show that as $\alpha \longrightarrow \infty$, the performance of the PFR with recycle approaches the behavior of a CMFR. Intuitively, this makes sense. If the recycle rate is very high, then the transformation within each cycle of the PFR is small, so that the concentration within the reactor approaches that at the outlet, C_{out}. This is just the condition that must be satisfied for the system to perform like a CMFR.

To show that the system behavior converges to the case of a CMFR for large recycle ratios, assume $\alpha \gg 1$, and further assume that α is so large that $k\Theta \ll 1 + \alpha$.

Figure 5.A.25 The effect of the recycle factor, α, on the conversion of a species that undergoes first-order decay with rate constant k in a plug-flow reactor with hydraulic residence time Θ.

EXAMPLE 5.A.18 ***Plug-Flow Reactor with Recycle (continued)***

Then we use the relationship $e^{-x} \approx 1 - x$ for $x \ll 1$ to replace the result from part (*b*) with this approximate relationship:

$$C_{\text{out}} \approx C_{\text{in}} \frac{1 - \dfrac{k\Theta}{1+\alpha}}{1 + \dfrac{\alpha}{1+\alpha} k\Theta} \qquad \text{valid for } \alpha \gg 1 \text{ and } k\Theta \ll 1 + \alpha$$

Since $\alpha \gg 1$ and $k\Theta \ll 1 + \alpha$, the numerator of the fraction is approximately 1, and the denominator is approximately $1 + k\Theta$, so $C_{\text{out}} \approx C_{\text{in}} (1 + k\Theta)^{-1}$, which is precisely the behavior of a CMFR in steady state for a first-order decay reaction (see Example 5.A.5).

Figure 5.A.25 shows how the concentration at the outlet increases with increasing recycle factor α for a first-order decay reaction. The effect is insignificant when the overall conversion is small ($k\Theta \ll 1$) and becomes progressively larger as $k\Theta$ increases.

5.B BEYOND IDEAL REACTORS: GENERAL MATERIAL-BALANCE MODELS

In §5.A we studied reactor models. These models are well suited to describing species behavior when the transport characteristics are simple. For example, a treatment vessel is frequently well described as a CMFR or a PFR. In the natural environment, rivers can often be reasonably described as plug-flow reactors, and small lakes may be well modeled as CMFRs. However, because ideal reactor models are based on simplifying assumptions about transport and mixing, they are ineffective in describing some systems. For example, predicting pollutant concentrations near emission sources requires tools that explicitly account for turbulent diffusion. Reactor models only crudely describe urban air basins, large lakes, or oceans. Additional modeling tools are needed for cases in which the idealized reactor models are inadequate.

In this section we explore an approach that complements reactor modeling. The same fundamental principle of material balance applies. But the simplifying assumptions made for reactor models about transport and mixing are not made here. In this section we derive the fundamental governing equation that expresses species conservation in the presence of transformation and transport processes. We then develop and present solutions to this equation for some commonly encountered cases.

5.B.1 Governing Equation

One-Dimensional System

For simplicity and clarity, we begin by considering a system in which transport occurs only in one coordinate direction (x). The contaminant concentration is assumed to vary only as a function of position in that one direction and, possibly, time (t). In §5.A.3 we derived a governing equation for an ideal plug-flow reactor (equation 5.A.16). In that case, advection was the sole transport mechanism. Here we will consider a similar physical system, but we will allow transport to occur by any mechanism, including molecular and turbulent diffusion.

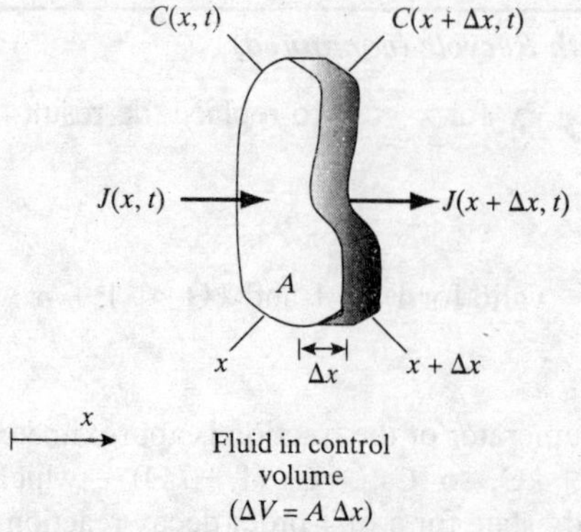

Figure 5.B.1 Schematic for deriving the one-dimensional general material-balance equation. The control volume is located at position x and has width Δx. Contaminant flux occurs only in the $\pm x$-direction. Both the flux and the species concentration are functions of x and t only.

Figure 5.B.1 illustrates a control volume within the fluid environment. We consider a contaminant whose concentration is $C(x, t)$. The contaminant flux, a function of both position and time, is denoted $J(x, t)$. The flux is positive when net transport is in the $+x$-direction and negative when transport is in the $-x$-direction. The control volume in the figure is located at position x, has width Δx, and has cross-sectional area A in the direction normal to the x-coordinate. We consider what happens over a time interval of Δt, beginning at t.

The material-balance principle can be stated in the following words. Over some time interval, the change in the amount of contaminant in the control volume must equal the amount of contaminant that flows into the volume minus the amount of contaminant that flows out of the volume plus the net amount of contaminant produced within the volume. Translating this statement into its mathematical representation, we have

$$C(x, t+\Delta t)A\Delta x - C(x, t)A\Delta x = J(x, t)A\Delta t - J(x+\Delta x, t)A\Delta t + rA\Delta x\Delta t \tag{5.B.1}$$

The product $(A\Delta x)$ is the volume of fluid in the control volume, so the group $(CA\Delta x)$ is the total amount of contaminant in the control volume. Therefore, we can see that the two terms on the left represent the net increase in the amount of contaminant contained in the volume from the beginning (t) to the end $(t + \Delta t)$ of the time interval. On the right-hand side of the equation, the first term expresses the total net flow of contaminant into the volume (flux evaluated at x). Since the flux, J, has units of quantity (mass or moles) per area per time, the group $(JA\,\Delta t)$ represents the total quantity transported during the interval. The second term on the right is the total net flow of contaminant out of the volume (flux evaluated at $x + \Delta x$). This term is subtracted from the equation since, if $J > 0$, it removes mass from the control volume. The final term represents the net effects of transformation within the volume, with r as the net rate of production of contaminant (mass or moles per volume per time).

Note that each term in the equation is proportional to area, so A may be canceled from the equation. Next we divide both sides of the equation by $\Delta x\,\Delta t$ to obtain

$$\frac{[C(x, t+\Delta t) - C(x, t)]}{\Delta t} = \frac{[J(x, t) - J(x+\Delta x, t)]}{\Delta x} + r \tag{5.B.2}$$

We now take the limit as $\Delta x \to 0$ and $\Delta t \to 0$. The difference terms are transformed into partial derivatives, and we obtain the *general material-balance equation for a one-dimensional system*:

$$\frac{\partial C}{\partial t} = -\frac{\partial J}{\partial x} + r \tag{5.B.3}$$

or

$$\frac{\partial C}{\partial t} + \frac{\partial J}{\partial x} - r = 0 \tag{5.B.4}$$

The first term ($\partial C/\partial t$) is called the accumulation term; it accounts for the change in species concentration with time. In a steady-state system, this term vanishes. The second term ($\partial J/\partial x$) accounts for the net effects of transport. Note that if transport flux is constant as a function of position, then $\partial J/\partial x = 0$ and transport does not contribute to accumulation. The third term (r) accounts for the net effects of transformation. For a nonreactive contaminant, $r = 0$.

In the case of an ideal plug-flow reactor, the flux was attributable only to advection and was given by $J = UC$, with U = constant. Substituting this expression into equation 5.B.3 we obtain equation 5.A.16: $\partial C/\partial t = -U\,\partial C/\partial x + r$. Equation 5.B.3 (or 5.B.4) is a more general governing equation that includes the idealized plug-flow reactor as a special case.

Three-Dimensional System

Commonly, we are interested in systems in which species concentrations may vary in all three coordinate directions, and transport may occur in any direction as well. The derivation of a governing material-balance equation in this case follows the same line of reasoning as for the one-dimensional system; however, the mathematical notation is a little more cumbersome.

The same parameters are used in the derivation. However, now all are functions of three coordinate directions (x, y, z) as well as time (t). In addition, the flux, $\vec{J}$, is now a vector quantity with components J_x, J_y, and J_z in the x-, y-, and z-coordinate directions, respectively. The variables are

$C(x, y, z, t)$	Species concentration
$\vec{J}(x, y, z, t)$	Species flux by any transport mechanism, with $\vec{J} = (J_x, J_y, J_z)$
$r(x, y, z, t)$	Net rate of species generation by all transformation processes

The general governing equation for a three-dimensional system is derived by analogy to the one-dimensional derivation. The result is

$$\frac{\partial C}{\partial t} + \left(\frac{\partial J_x}{\partial x} + \frac{\partial J_y}{\partial y} + \frac{\partial J_z}{\partial z}\right) - r = 0 \tag{5.B.5}$$

If we compare this equation with its one-dimensional counterpart (equation 5.B.4), we see that only the transport term has changed. In addition to the x-component of transport, which has become more explicitly specified ($\partial J/\partial x$ in equation 5.B.4 is replaced by $\partial J_x/\partial x$ in equation 5.B.5), new terms have been added to account for the effects of transport in the y- and z-coordinate directions.

It is noteworthy that governing equations similar to equations 5.B.4 and 5.B.5 arise in other fields of study. See Exhibit 5.B.1 for a brief discussion.

EXHIBIT 5.B.1 *Conservation Equations for Other Fluid Properties*

Equations 5.B.4 and 5.B.5 are solved to predict the concentration of species in fluids. Similar governing equations can be derived for many other fluid properties, as shown in Table 5.B.1. In each specific case, the terms that account for production of the property within the control volume (r in the case of species) are specified differently, and this causes the equations to have somewhat different forms. But all of these equations exhibit similar accumulation and transport terms.

The close mathematical relationship among the several transport equations means that solutions developed in one context can often be used for another. Whenever the governing equation and boundary conditions are the same, the mathematical solution of the problem will be the same, regardless of what the symbols represent. This means, for example, that studies of heat transfer are useful for understanding contaminant transport (Carslaw and Jaeger, 1959; Bejan, 1984).

Table 5.B.1 Fluid Properties Measured in Concentration Units for Which a Differential Conservation Equation Applies

Property	Notation[a]	Equation
Mass	ρ	Continuity equation
Momentum	$\rho\vec{U}$	Navier-Stokes equation
Energy	$\rho\hat{c}_v T$	Energy equation
Species	C	Advection-diffusion equation

[a]Definition of symbols: ρ = fluid density, $\vec{U}$ = fluid velocity, $\hat{c}_v$ = heat capacity at constant volume, T = temperature, C = species concentration.

Applying the General Material-Balance Equation

To predict species concentrations, we begin with a general material-balance equation, either equation 5.B.4 (for one-dimensional problems) or equation 5.B.5 (for two- or three-dimensional problems). The next step is to substitute appropriate expressions for reactions and fluxes in terms of species concentration. For this purpose, we use transformation relationships such as those introduced in Chapter 3 and flux expressions such as those described in Chapter 4. We aim to include all effects that are essential and to exclude all unimportant phenomena, so that the model predictions are accurate, yet the mathematical problem is as simple as possible. Deciding which terms to include is an important challenge in environmental modeling.

As an example, let's consider the atmospheric transport and dispersion of a nonreactive ($r = 0$) gaseous pollutant emitted from an elevated point source. Away from boundaries, we expect that the flux will be dominated by a combination of advection and turbulent diffusion. Molecular diffusion can be neglected as insignificant compared with turbulent diffusion. We align the coordinate system such that x points in the direction of the wind, y is the horizontal crosswind direction, and z is vertical. Then the species flux is a vector with these components:

$$\vec{J} = \left(UC - \varepsilon_x \frac{\partial C}{\partial x}, -\varepsilon_y \frac{\partial C}{\partial y}, -\varepsilon_z \frac{\partial C}{\partial z}\right) \tag{5.B.6}$$

where ε_x, ε_y, and ε_z represent turbulent diffusivities in the three coordinate directions.

If the emissions are steady, we might also neglect the turbulent diffusion term in the direction of flow as being a small contributor to flux in comparison with advection. For steady-state behavior, we can assume $\partial C/\partial t = 0$. Applying these simplifications and substituting J into equation 5.B.5 yields the following expression:

$$\frac{\partial(UC)}{\partial x} - \frac{\partial}{\partial y}\left(\varepsilon_y \frac{\partial C}{\partial y}\right) - \frac{\partial}{\partial z}\left(\varepsilon_z \frac{\partial C}{\partial z}\right) = 0 \tag{5.B.7}$$

In atmospheric transport problems, it is reasonable to consider air incompressible, so that $\partial U/\partial x = 0$. The equation then simplifies further to

$$U\frac{\partial C}{\partial x} - \frac{\partial}{\partial y}\left(\varepsilon_y \frac{\partial C}{\partial y}\right) - \frac{\partial}{\partial z}\left(\varepsilon_z \frac{\partial C}{\partial z}\right) = 0 \tag{5.B.8}$$

This is a specific material-balance equation for a nonreactive atmospheric constituent. The first term accounts for the direct effects of wind on advective transport. The other two terms represent the effects of turbulent transport, first in the horizontal, crosswind direction and second in the vertical direction. The fact that the three terms in the equation sum to zero means that advective and turbulent transport must be balanced at each point where the equation applies, so that there is no net accumulation of species.

We must also define boundary conditions and, for time-dependent problems, initial conditions for the system of interest. For each dimension in space and time, there must be a boundary or initial condition for each order of the highest-order derivative. So, for example, equation 5.B.8 requires one boundary condition in the x-direction, since only a first derivative applies with respect to x. Two boundary conditions each are required for the y- and z-directions. Examples of boundary conditions for specific problems are given in §5.B.2.

Solving for Species Concentration

Once the governing equation, boundary conditions, and initial conditions are all defined, we can find the solution for $C(x, y, z, t)$. Sometimes analytical solutions are possible; however, for complex cases, especially those involving two- and three-dimensional systems, no general solution may exist. When analytical solutions are unavailable, the differential material-balance equation can be solved numerically. The essence of the numerical approach is to replace the governing differential equation with a set of algebraic equations, with each equation describing material balance over a small region. The equations are then solved with the help of a computer to obtain information on concentrations as a function of position and time. With the numerical approach, any specific transport/transformation problem can, in principle, be solved. In contrast to analytical solutions, where the concentration is obtained as an explicit function of position and time, numerical solutions yield concentration results that are valid only for the specific simulation conditions. To obtain information for other conditions, the model must be rerun. Numerical modeling of environmental systems continues to grow in importance, assisted by recent rapid growth in computational power.

5.B.2 Approaches for Solving Environmental Transport Problems

In this section we describe some environmental transport problems that can be solved analytically using the general material-balance equation. As we do elsewhere in the text, we rely more on physical reasoning than on formal mathematical analysis. It is especially important for engineers not to lose sight of the physical phenomena represented by mathematical models.

Table 5.B.2 provides an overall summary of the cases considered. For each of the cases the table provides a description of the conditions modeled, the specific governing equation, boundary and initial conditions, and the solution. The following discussion amplifies the summary presented in this table.

Table 5.B.2 Summary of Cases Applying the General Material-Balance Equation

	Case 1
Description	1-D, steady diffusion of nonreactive species through a finite layer
Governing equation	$\frac{d^2C}{dx^2} = 0$
Boundary conditions	$C(0) = C_0 \qquad C(L) = C_1$
Solution	$C(x) = C_0 + (C_1 - C_0)\frac{x}{L}$
	Case 2
Description	1-D, transient diffusion from an instantaneous local source
Governing equation	$\frac{\partial C}{\partial t} - D\frac{\partial^2 C}{\partial x^2} = 0$
Boundary conditions	$C(-\infty, t) = C(\infty, t) = 0$
Initial condition	$C(x, 0) = M_a\,\delta(x)$
Solution	$C(x, t) = \frac{M_a}{(4\pi Dt)^{1/2}} \times \exp\left[-\frac{x^2}{4Dt}\right]$
	Case 3
Description	3-D, turbulent diffusion from an instantaneous point source
Governing equation	$\frac{\partial C}{\partial t} - \varepsilon_x\frac{\partial^2 C}{\partial x^2} - \varepsilon_y\frac{\partial^2 C}{\partial y^2} - \varepsilon_z\frac{\partial^2 C}{\partial z^2} = 0$
Boundary conditions	$C(x, y, z, t) \rightarrow 0$ as $x \rightarrow \pm\infty$, as $y \rightarrow \pm\infty$, or as $z \rightarrow \pm\infty$
Initial condition	$C(x, y, z, 0) = M\,\delta(x)\,\delta(y)\,\delta(z)$
Solution	$C(x, y, z, t) = \frac{M}{(4\pi t)^{3/2}\,(\varepsilon_x\varepsilon_y\varepsilon_z)^{1/2}} \times \exp\left(-\frac{x^2/\varepsilon_x + y^2/\varepsilon_y + z^2/\varepsilon_z}{4t}\right)$
	Case 4
Description	1-D transport by advection and diffusion from a pulsed release
Governing equation	$\frac{\partial C}{\partial t} + U\frac{\partial C}{\partial x} - \varepsilon_x\frac{\partial^2 C}{\partial x^2} = 0$
Boundary conditions	$C(x, t) \rightarrow 0$ as $x \rightarrow \pm\infty$
Initial condition	$C(x, 0) = M_a\,\delta(x)$
Solution	$C(x, t) = \frac{M_a}{(4\pi\varepsilon_x t)^{1/2}} \times \exp\left[-\frac{(x - Ut)^2}{4\varepsilon_x t}\right]$

Table 5.B.2 Summary of Cases Applying the General Material-Balance Equation (continued)

	Case 5
Description	Advection and diffusion in 3-D from a continuous point release
Governing equation	$U\frac{\partial C}{\partial x} - \varepsilon_y \frac{\partial^2 C}{\partial y^2} - \varepsilon_z \frac{\partial^2 C}{\partial z^2} = 0$
Boundary conditions	$C(0, y, z) = \frac{m}{U}\,\delta(y)\,\delta(z)$ $C(x, y, z) \to 0 \quad \text{as } y \to \pm\infty \text{ or as } z \to \pm\infty$
Solution	$C(x, y, z) = \left[\frac{m}{4\pi x(\varepsilon_y \varepsilon_z)^{1/2}}\right] \times \exp\left[-\frac{U}{4x}(y^2/\varepsilon_y + z^2/\varepsilon_z)\right]$
	Case 6
Description	1-D, steady-state advection, dispersion, sorption, and decay of a groundwater contaminant
Governing equation	$U\frac{\partial C}{\partial x} - \varepsilon_h \frac{\partial^2 C}{\partial x^2} + \phi k C = 0$
Boundary conditions	$C(x) \to 0 \text{ as } x \to \infty$ $C(0) = C_0$
Solution	$C(x) = C_0 \exp\left\{\frac{Ux}{2\varepsilon_h}\left[1 - \left(\frac{4\phi k}{U^2/\varepsilon_h} + 1\right)^{1/2}\right]\right\}$

Symbol definitions (dimensions: M = mass or moles, L = length, T = time)

C	Species concentration (M L^{-3})
C_0	Constant concentration at boundary (M L^{-3})
C_1	Constant concentration at boundary (M L^{-3})
D	Molecular diffusivity of species (L^2 T^{-1})
$\delta(x)$	Dirac delta function with respect to x-direction (L^{-1}) [also $\delta(y)$ and $\delta(z)$]
ε_h	Hydrodynamic dispersion coefficient for transport in porous media (L^2 T^{-1})
ε_x	Turbulent diffusivity in the x-direction (L^2 T^{-1}) (similarly, ε_y and ε_z)
k	Rate constant for first-order decay (T^{-1})
L	Distance separating boundaries in Case 1 (L)
M	Species quantity released (M)
M_a	Species quantity released per area in the y-z plane (M L^{-2})
m	Species emission rate for continuous release (M T^{-1})
ϕ	Porosity of soil (–)
t	Time (T)
U	Mean fluid velocity or Darcy velocity (aligned with x-direction) (L T^{-1})
x, y, z	Cartesian coordinates (L)

Modeling Transport by Diffusion Only

We begin by considering cases in which transport occurs only by diffusion. The diffusion coefficient is assumed to be independent of position. We consider cases with either molecular diffusion (or Brownian motion) or turbulent diffusion, but we restrict attention to species that are nonreactive, except, perhaps, at surfaces.

For transport by molecular diffusion or Brownian motion, the flux is related to the concentration gradient by Fick's law (§4.A.3). In one-dimensional systems, we write

$$J = -D\frac{\partial C}{\partial x} \tag{5.B.9}$$

where the partial derivative reminds us that C is a function of both position (x) and time (t). Substituting this relationship into the general one-dimensional governing equation (5.B.4), setting $r = 0$ for a nonreactive species, and applying the fact that D is constant to move it outside of the derivative, we obtain

$$\frac{\partial C}{\partial t} - D\frac{\partial^2 C}{\partial x^2} = 0 \qquad (5.B.10)$$

If steady state can be assumed to apply, then $\partial C/\partial t = 0$, and the diffusion coefficient can be eliminated by division, yielding this simple governing equation:

$$\frac{d^2 C}{dx^2} = 0 \qquad (5.B.11)$$

Case 1 illustrates the use of this specific equation.

For three-dimensional systems, the time-dependent diffusion equation expands to this form:

$$\frac{\partial C}{\partial t} - D\left(\frac{\partial^2 C}{\partial x^2} + \frac{\partial^2 C}{\partial y^2} + \frac{\partial^2 C}{\partial z^2}\right) = 0 \qquad (5.B.12)$$

This expression is known as the *diffusion equation* or *Fick's second law of diffusion*. As in one dimension, a special case arises when steady-state conditions apply:

$$\frac{\partial^2 C}{\partial x^2} + \frac{\partial^2 C}{\partial y^2} + \frac{\partial^2 C}{\partial z^2} = 0 \qquad (5.B.13)$$

If, instead of molecular diffusion, transport is caused by turbulent diffusion, then the molecular diffusion coefficient is replaced by turbulent diffusion coefficients that are functions of coordinate direction, so $J_x = -\varepsilon_x \, \partial C/\partial x$, and so on. For cases in which the turbulent diffusivities are independent of position, the governing material-balance equation can be written as follows:

$$\frac{\partial C}{\partial t} - \left(\varepsilon_x\frac{\partial^2 C}{\partial x^2} + \varepsilon_y\frac{\partial^2 C}{\partial y^2} + \varepsilon_z\frac{\partial^2 C}{\partial z^2}\right) = 0 \qquad (5.B.14)$$

We now consider how to solve these equations for some specific boundary conditions.

Case 1: One-Dimensional, Steady Diffusion of a Nonreactive Species through a Finite Layer Steady diffusion through a layer with fixed concentrations at both boundaries is a simple situation to analyze. Practical examples that are well described by this model include mass transfer to a surface (§4.C.1), interfacial mass transfer (§4.C.2), and diffusive samplers for measuring air pollutant concentrations (Problem 4.11).

The applicable governing equation is 5.B.11. Since this is a second-order differential equation, two boundary conditions are required. These are the fixed concentrations at either edge of the layer ($x = 0$ and $x = L$):

$$C(0) = C_0 \qquad C(L) = C_1 \qquad (5.B.15)$$

Equation 5.B.11 can be directly integrated. Since the second derivative is equal to zero, the first derivative is equal to a constant. Integrating a second time leads to a linear equation:

$$C(x) = A + Bx \tag{5.B.16}$$

where A and B are constants. Substituting the boundary conditions allows us to solve for A and B, yielding

$$C(x) = C_0 + (C_1 - C_0)\frac{x}{L} \tag{5.B.17}$$

It is easily demonstrated that this solution satisfies the boundary conditions and the governing differential equation. In some applications, we are interested in the species flux, which can be derived by substituting equation 5.B.17 into Fick's law:

$$J = -D\frac{dC}{dx} = -D\frac{(C_1 - C_0)}{L} \tag{5.B.18}$$

In this steady-state case, the flux is constant throughout the layer.

Case 2: One-Dimensional Transient Diffusion from an Instantaneous Local Source The next problem we consider is unusually important, not because it directly represents many practical situations, but rather because the resulting solution is a building block from which solutions to many other problems can be constructed. One practical problem to which this description directly applies is the diffusion of an instantaneously injected pulse of nonreactive species through the fluid in a tube, where the concentration is uniform across the tube cross-section but varies along the axis.

Again, we consider stationary fluid, unbounded in the x-direction, and a nonreactive species that is initially absent from the system. A fixed quantity of the species, M_a (mass or moles per cross-sectional area normal to x) is injected at the origin at $t = 0$ and permitted to migrate by diffusion in both the $+x$- and $-x$-directions, but not in the y- and z-directions.

Equation 5.B.10 is the appropriate material-balance equation for the case of one-dimensional transient diffusion. The boundary conditions state that the species concentration must remain zero at an infinite distance from the injection point.

$$C(-\infty, t) = C(\infty, t) = 0 \tag{5.B.19}$$

The initial condition can be written in this form:

$$C(x, 0) = M_a\,\delta(x) \tag{5.B.20}$$

The expression $\delta(x)$ is known as the *Dirac delta function* and is defined by these properties:

$$\delta(x) = 0 \quad \text{for all } x \neq 0 \qquad \text{and} \qquad \int_{-\infty}^{\infty} f(x)\,\delta(x)\,dx = f(0) \tag{5.B.21}$$

What does this definition mean? Qualitatively, the Dirac delta function is a narrow spike. Its value is zero everywhere except when its argument is zero, and then it rises to infinity. The area under the delta function is equal to 1. This means that the product of its infinite height at the origin times the infinitesimal width of its spike equals 1. The dimensions of the delta function are the reciprocal of its argument, so since x has units of length, $\delta(x)$ has units of inverse length. Therefore, equation 5.B.20 states that the initial concentration distribution is zero everywhere except at the origin and that the total amount of contaminant in the system, per cross-sectional area of the tube, is M_a.

A method for solving equation 5.B.10 subject to 5.B.19–20 is discussed in Appendix D (§D.2). Here is the result, which is valid for $t > 0$:

$$C(x, t) = \frac{M_a}{(4\pi Dt)^{1/2}} \times \exp\left(-\frac{x^2}{4Dt}\right) \quad (5.B.22)$$

When species transport occurs by diffusion, solutions to the general material-balance equation frequently have a Gaussian, or bell, shape. Equation 5.B.22 has this property. To see this, divide both sides of the equation by M_a and rewrite the solution with the substitution $(2Dt) = \sigma^2$:

$$\frac{C(x, t)}{M_a} = \frac{1}{\sqrt{2\pi}\,\sigma}\exp\left(-\frac{x^2}{2\sigma^2}\right) \quad (5.B.23)$$

The right-hand side of this equation is the standard form for a Gaussian or normal probability distribution with a mean of zero and a standard deviation of σ. Figure 5.B.2 presents C/M_a versus x for three values of σ. The solution is symmetric about $x = 0$. Since there is no advection, the center of mass of the released species remains fixed at the injection point. Note that as time progresses and σ increases (in proportion to $t^{1/2}$), the peak height diminishes and the plume width spreads. Given the properties of a normal distribution, about two-thirds of the species mass lies in the range $-\sigma < x < \sigma$. About 95 percent is contained between $-2\sigma < x < 2\sigma$. Essentially all of the mass is contained within $\pm 3\sigma$ of the mean. The parameter σ is the characteristic distance that a species travels by diffusion in time t. (See also equation 4.A.7.)

Case 3: Three-Dimensional Turbulent Diffusion from an Instantaneous Point Source For a physical system that exhibits three-dimensional diffusion from a point source, we can think of the sudden discharge of an impurity at mid-depth within a deep lake. The impurity diffuses outward in horizontal and vertical directions. Assuming there is no mean flow, but rather turbulent fluctuations that are uniform throughout

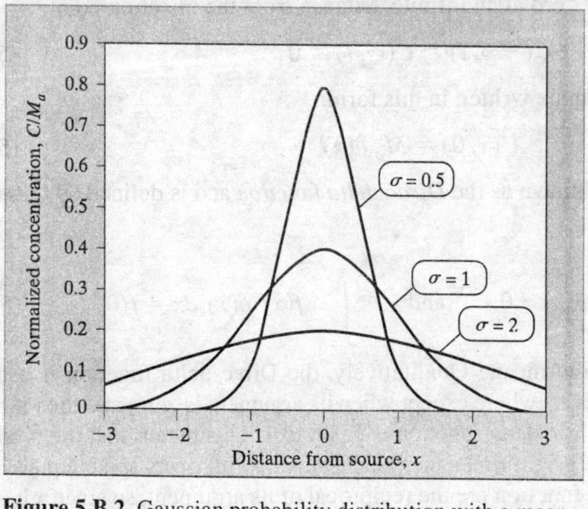

Figure 5.B.2 Gaussian probability distribution with a mean of zero and a standard deviation of σ. For the normalized concentration (C/M_a), x can be interpreted with any length unit; σ must have the same units as x.

the lake, the governing equation is given by equation 5.B.14. The boundary conditions state that the concentration must go to zero at an infinite distance from the origin in the x-, y-, and z-directions. The initial condition says that a fixed quantity of the species is released at the origin at time $t = 0$. Mathematically, these can be written as

$$C(x, y, z, t) \rightarrow 0 \quad \text{as } x \rightarrow \pm\infty, y \rightarrow \pm\infty, z \rightarrow \pm\infty \tag{5.B.24}$$

$$C(x, y, z, 0) = M\ \delta(x)\ \delta(y)\ \delta(z) \tag{5.B.25}$$

The method of separation of variables can be applied to show that the concentration can be written as the product of three solutions to the one-dimensional pulse diffusion problems (see Appendix D, §D.2):

$$C(x, y, z, t) = \frac{M}{(4\pi t)^{3/2}(\varepsilon_x \varepsilon_y \varepsilon_z)^{1/2}} \times \exp\left(-\frac{x^2/\varepsilon_x + y^2/\varepsilon_y + z^2/\varepsilon_z}{4t}\right) \tag{5.B.26}$$

When the turbulent diffusivities are the same in all coordinate directions ($\varepsilon_x = \varepsilon_y = \varepsilon_z = \varepsilon$), the result simplifies to a dependence solely on distance from the origin, $r = (x^2 + y^2 + z^2)^{1/2}$:

$$C(r, t) = \frac{M}{(4\pi\varepsilon t)^{3/2}} \times \exp\left(-\frac{r^2}{4\varepsilon t}\right) \tag{5.B.27}$$

Modeling Transport by Advection and Diffusion

In the previous section, we explored transport by molecular and turbulent diffusion only. The next goal is to extend the transport mechanisms to include advection in a uniform (but possibly turbulent) fluid flow. We begin by considering a localized instantaneous release with one-dimensional and three-dimensional transport (Case 4). Then we analyze the problem of species transport caused by advection and turbulent diffusion from a constant point source of emissions in a three-dimensional system (Case 5).

Case 4: One-Dimensional Transport by Advection and Diffusion from a Pulsed Release Recall that in an ideal PFR, all parts of the fluid exhibit the same net velocity in the direction of flow, so that, according to the model, a sharp-edged pulse of contamination would maintain its shape as it is transported by advection (see Example 5.A.11). However, in any real system, spreading of the pulse will occur because of longitudinal dispersion caused by shear flow and turbulent diffusion.

Let's consider the spreading of a pulse of a nonreactive contaminant in an almost ideal PFR. Assume that the mean fluid velocity is uniform, aligned with the x-coordinate direction, and that the species concentration is uniform in the direction normal to flow (so that there is no net transport except in the x-direction). Contaminant flux in the direction of flow has contributions from both advection and turbulent diffusion:

$$J_x = UC - \varepsilon_x \frac{\partial C}{\partial x} \tag{5.B.28}$$

Substitute this equation into the general one-dimensional transport equation (5.B.4), which, assuming U and ε_x are independent of position and $r = 0$, reduces to this form:

$$\frac{\partial C}{\partial t} + U\frac{\partial C}{\partial x} - \varepsilon_x \frac{\partial^2 C}{\partial x^2} = 0 \tag{5.B.29}$$

The boundary and initial conditions are the same as for transport by diffusion alone (Case 2) and are given by equations 5.B.19–20.

This problem is solved by making a change in coordinates, replacing x by $x^* = x - Ut$. In the new coordinate system (x^*, t), the problem is analogous to the Case 2 situation, with transport by diffusion only. The solution is

$$C(x, t) = \frac{M_a}{(4\pi\varepsilon_x t)^{1/2}} \exp\left[-\frac{(x - Ut)^2}{4\varepsilon_x t}\right] \tag{5.B.30}$$

Compare this result with equation 5.B.22. Note that the equations have the same form. The molecular diffusivity, D, is replaced with the turbulent diffusivity, ε_x. The x-coordinate in 5.B.22 becomes $x^* = x - Ut$ in 5.B.30. Equation 5.B.30 describes a pulse whose center of mass is advected at velocity U in the x-direction. The pulse spreads in $+x$- and $-x$-directions relative to the center of mass because of turbulent diffusion. The concentration profile forms a traveling Gaussian whose peak height diminishes and whose width increases with time, as illustrated for a stationary Gaussian profile in Figure 5.B.2.

The solution to this problem with advection in the x-direction and turbulent diffusion in all three dimensions can be obtained by applying separation of variables (compare with equation 5.B.26):

$$C(x, y, z, t) = \frac{M}{(4\pi t)^{3/2}(\varepsilon_x\varepsilon_y\varepsilon_z)^{1/2}} \times \exp\left[-\frac{\left(\frac{(x - Ut)^2}{\varepsilon_x} + \frac{y^2}{\varepsilon_y} + \frac{z^2}{\varepsilon_z}\right)}{4t}\right] \tag{5.B.31}$$

Case 5: Advection and Diffusion in Three Dimensions from a Continuous Point Release Our treatment so far has emphasized instantaneous emissions. Although this is a reasonable modeling approach for cases involving, for example, accidental spills, in other situations emissions are sustained for extended periods. Of particular interest is the steady-state concentration profile that results from constant emissions at a point source, such as air pollutants emitted from an industrial stack.

The physical situation is depicted in Figure 4.A.8. A contaminant is released at a constant rate m (mass or moles per time) from some fixed point, which is placed at the origin of our coordinate system. The mean fluid velocity (U) is uniform, and we align the coordinates so that the flow is in the $+x$-direction. As before, we assume that the species is nonreactive. Transport occurs by advection in the x-direction and by turbulent diffusion in the y- and z-directions.*

The governing material-balance equation for this case is derived from equation 5.B.5 with $\partial C/\partial t = 0$ (steady state), $r = 0$ (nonreactive species), $J_x = UC$ (advection), $J_y = -\varepsilon_y\, \partial C/\partial y$ (turbulent diffusion), and $J_z = -\varepsilon_z\, \partial C/\partial z$ (turbulent diffusion). Assembling the terms and assuming that U, ε_y, and ε_z are independent of x, y, and z, respectively, we obtain

$$U\frac{\partial C}{\partial x} - \varepsilon_y\frac{\partial^2 C}{\partial y^2} - \varepsilon_z\frac{\partial^2 C}{\partial z^2} = 0 \tag{5.B.32}$$

*Turbulent diffusion also occurs in the x-direction, but for constant emissions its effects on transport are small compared with advection. Consequently, turbulent diffusion in the x-direction is neglected here.

Since the governing equation contains second derivatives with respect to y and z, two boundary conditions are needed for each of these coordinates. These state that the concentration approaches zero as either y or z approaches $+\infty$ or $-\infty$:

$$C(x, y, z) \rightarrow 0 \quad \text{as } y \rightarrow \pm\infty \text{ or } z \rightarrow \pm\infty \tag{5.B.33}$$

In the x-direction, one boundary condition is required. It is

$$C(0, y, z) = \frac{m}{U}\,\delta(y)\,\delta(z) \tag{5.B.34}$$

This condition specifies that at $x = 0$, the concentration is zero everywhere except at the point of release. There, where $x = y = z = 0$, the concentration rises to an infinitely high value, as specified by the delta functions. The coefficient m/U ensures that the species emission rate is m. This problem can be solved by means of mapping it to the case of two-dimensional diffusion from an instantaneous point source. The result is

$$C(x, y, z) = \left(\frac{m}{4\pi x(\varepsilon_y \varepsilon_z)^{1/2}}\right) \exp\left[-\frac{U}{4x}(y^2/\varepsilon_y + z^2/\varepsilon_z)\right] \tag{5.B.35}$$

Sometimes this solution is written as follows, assuming that the turbulent diffusivities are the same in both directions, $\varepsilon_y = \varepsilon_z = \varepsilon$:

$$C(x, y, z) = \frac{m}{4\pi x \varepsilon} \exp\left[-\frac{U(y^2 + z^2)}{4\varepsilon x}\right] \tag{5.B.36}$$

Case 5 is very important as a foundation for the family of expressions called *Gaussian plume models*, which are frequently applied for modeling air pollutant concentrations downwind of localized sources (see §7.D.2).

Accounting for Transformation Processes: First-Order Decay

In developing solutions to the general material-balance equation, we have considered only nonreactive species. In this section, we briefly explore how the same methods can be used to predict the concentrations of reactive species. We restrict our attention to the easiest class of problems: those in which the species undergoes first-order decay, with a rate constant k that is independent of position and time. That is, we consider species for which the net rate of transformation is given by

$$r = -kC \tag{5.B.37}$$

For some situations that we have already considered, it is straightforward to modify the solutions to account for the effects of first-order decay.

First, if the species is emitted instantaneously at some time $t = 0$, then the concentration at some later time t can be determined by first computing the concentration for a nonreactive species and then multiplying that result by a decay factor F_d, given by

$$F_d = e^{-kt} \tag{5.B.38}$$

This approach is valid for some steady-state cases, such as that given in Case 5. In that case, if the emissions occur at $x = 0$, then at any position $x > 0$, the time (t) that species molecules have been in the system is given exactly by the advection travel time:

$$t = \frac{x}{U} \tag{5.B.39}$$

So for these conditions, we can obtain the solution for a species that undergoes first-order decay by multiplying the concentration profile for a nonreactive species (equation 5.B.35) by the appropriate decay factor, $\exp(-kx/U)$:

$$C(x, y, z) = \left(\frac{m}{4\pi x(\varepsilon_y \varepsilon_z)^{1/2}}\right)\exp\left[-\frac{U}{4x}\left(\frac{y^2}{\varepsilon_y} + \frac{z^2}{\varepsilon_z}\right)\right] \times \exp\left(-k\frac{x}{U}\right) \tag{5.B.40}$$

In other cases, one must go back to the general material-balance equation (5.B.3–5), make appropriate substitutions for the transport and transformation terms, and then solve the resulting differential equation. Example 5.B.1 illustrates how these steps can be accomplished.

EXAMPLE 5.B.1 ***Advection, Dispersion, and Decay in One Dimension***

A degradable contaminant is discharged at rate m (mass per time) into a river. Assume that the contaminant rapidly mixes across the width and depth of the river. The contaminant is transported in the downstream direction by advection (velocity U) and dispersion (dispersion coefficient ε). It also undergoes first-order decay (rate constant k). Assume that none of the contaminant is transported upstream of the discharge. What is the contaminant concentration as a function of distance downstream of the discharge?

SOLUTION The governing material-balance equation for this steady-state system is derived from equation 5.B.4. The accumulation term ($\partial C/\partial t$) is set to zero. The flux is attributed to advection plus dispersion ($J = UC - \varepsilon\, dC/dx$, with U and ε assumed to be independent of x). The transformation term accounts for first-order decay ($r = -kC$).

$$U\frac{dC}{dx} - \varepsilon\frac{d^2C}{dx^2} + kC = 0$$

Since the equation contains a second-order derivative in x, we must specify two boundary conditions. The first condition states that if we go infinitely far downstream, the concentration must decay to zero. The second condition states that the total flow of contaminant in the downstream direction at the site of discharge (assumed to be at the origin) must equal the contaminant mass discharge rate. So

$$C(x) \to 0 \quad \text{as } x \to \infty$$

$$AJ = A\left(UC - \varepsilon\frac{dC}{dx}\right) = m \quad \text{at } x = 0$$

Here A represents the vertical cross-sectional area of the river at the point of discharge. Note that the total downstream flux includes contributions from both advection and dispersion.

EXAMPLE 5.B.1 ***Advection, Dispersion, and Decay in One Dimension (continued)***

The general solution of a second-order ordinary differential equation with constant coefficients has the form

$$C(x) = B_1 \exp(r_1 x) + B_2 \exp(r_2 x)$$

where B_1 and B_2 are constants selected to satisfy the boundary conditions, and r_1 and r_2 are roots of the characteristic equation, generated by substituting $C = e^{rx}$ into the differential equation, to obtain

$$-\varepsilon r^2 + Ur + k = 0$$

Applying the quadratic formula, we find

$$r_1 = \frac{U}{2\varepsilon}\left[1 - \left(1 + \frac{4k\varepsilon}{U^2}\right)^{1/2}\right] \qquad r_2 = \frac{U}{2\varepsilon}\left[1 + \left(1 + \frac{4k\varepsilon}{U^2}\right)^{1/2}\right]$$

The first boundary condition (at $x \to \infty$) requires that $B_2 = 0$. Substituting the expression for r_1 into the general solution and then applying the second boundary condition to solve for B_1 leads to the final result, which is valid for $x > 0$:

$$C(x) = \left[\frac{(2m)/(UA)}{\left(\frac{4k\varepsilon}{U^2} + 1\right)^{1/2} + 1}\right] \times \exp\left\{\left(\frac{Ux}{2\varepsilon}\right)\left[1 - \left(\frac{4k\varepsilon}{U^2} + 1\right)^{1/2}\right]\right\}$$

Modeling Transport in Porous Media

Contaminant transport in porous media can be influenced by advection, molecular diffusion, hydrodynamic dispersion, sorption, and transformation processes. Here we will explore methods for incorporating each of these processes into transport models based on the general material-balance equation.

Let's begin with the one-dimensional case. We write the following expression for contaminant flux in the direction of flow, including contributions from both advection and hydrodynamic dispersion (recall from §4.D.2 that the hydrodynamic dispersion term includes molecular diffusion):

$$J = UC - \varepsilon_h \frac{\partial C}{\partial x} \qquad (5.B.41)$$

Here U is the Darcy velocity, that is, the total volumetric flow per unit total cross-sectional area normal to the x-axis ($U = Q/A$; see §4.D.1), and ε_h is the hydrodynamic dispersion coefficient. The flux, J, represents the transported quantity per unit *total* cross-sectional area (not per unit *pore* area) per time.

The next step is to incorporate sorption processes into the governing transport equation. Sorption causes partitioning of the contaminant between solid and fluid (dissolved or gaseous) phases. Hence, the accumulation term that is generally used in the transport equation (e.g., equation 5.B.4) is modified to include sorptive removal from the fluid phase. The material-balance equation is written as follows:

$$\phi \frac{\partial C}{\partial t} + \rho_b \frac{\partial q}{\partial t} + U\frac{\partial C}{\partial x} - \varepsilon_h \frac{\partial^2 C}{\partial x^2} - \phi r = 0 \qquad (5.B.42)$$

where ϕ is the porosity (see §4.D), q is the mass of contaminant sorbed per mass of solid, and ρ_b is the bulk density of the porous medium. Each term in this equation has dimensions of contaminant quantity per unit total volume of the medium (including pores and soil grains) per time. The second term in the equation accounts for the net rate of contaminant removal from the fluid phase due to sorption on the soil grains. To describe the relationship between the sorbed mass concentration, q, and the dissolved mass concentration, C, we make the following substitution:

$$\frac{\partial q}{\partial t} = \frac{\partial q}{\partial C}\frac{\partial C}{\partial t} \tag{5.B.43}$$

If we assume that sorption is reversible and rapid relative to advective and dispersive transport (an approximation that is generally good in groundwater systems but may be questionable in process reactors), we can use the equilibrium partitioning relationship to obtain $\partial q/\partial C$. We assume here that the sorption isotherm is linear (see §3.B.3, equation 3.B.10):

$$q = K_{\text{ads}} C \tag{5.B.44}$$

where K_{ads} is the sorption partition coefficient. Equation 5.B.44 implies that $\partial q/\partial C = K_{\text{ads}}$. Substituting equations 5.B.43–44 into equation 5.B.42 produces the following material-balance equation, which includes advection, hydrodynamic dispersion, sorption, and transformation:

$$\phi\left(1 + K_{\text{ads}}\frac{\rho_b}{\phi}\right)\frac{\partial C}{\partial t} + U\frac{\partial C}{\partial x} - \varepsilon_h\frac{\partial^2 C}{\partial x^2} - \phi r = 0 \tag{5.B.45}$$

It is common to group the variables in the parentheses into a single parameter known as the retardation factor, R (dimensionless):

$$R = \left(1 + K_{\text{ads}}\frac{\rho_b}{\phi}\right) \tag{5.B.46}$$

Physically, the retardation factor represents the ratio of the total contaminant mass in the system to the contaminant mass dissolved in the pore fluid. It is called the *retardation* factor because it has the effect of slowing the overall rate of mass transport in time-dependent situations.

The resulting advection-dispersion-sorption-transformation equation then takes this form:

$$\phi R\frac{\partial C}{\partial t} + U\frac{\partial C}{\partial x} - \varepsilon_h\frac{\partial^2 C}{\partial x^2} - \phi r = 0 \tag{5.B.47}$$

Even for first-order decay, the transformation term can be complex because of the possible interactions of sorption and degradation. For example, it may be that the contaminant decays only when in the dissolved phase. In this case, we would have the expression, $r = -kC$ and the one-dimensional transport of a sorbing contaminant through porous media with first-order decay would then be described by

$$\phi R\frac{\partial C}{\partial t} + U\frac{\partial C}{\partial x} - \varepsilon_h\frac{\partial^2 C}{\partial x^2} + \phi k C = 0 \tag{5.B.48}$$

On the other hand, in the unlikely case that the contaminant undergoes first-order decay with the same rate constant independent of whether it is sorbed or dissolved in the fluid, the reaction term in 5.B.47 would be $r = -kRC$.

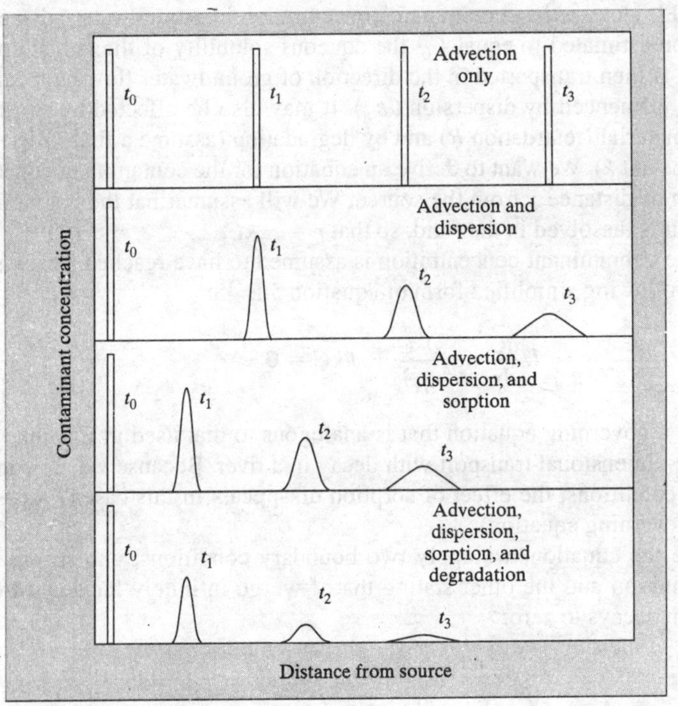

Figure 5.B.3 Movement of a contaminant slug through porous media as influenced by advection, dispersion, sorption, and degradation. In each frame, contaminant concentrations are plotted for four different time periods, $t_3 > t_2 > t_1 > t_0$.

The advection-dispersion-sorption-transformation equation can be expanded to include three-dimensional dispersion as follows:

$$\phi R\frac{\partial C}{\partial t} + U\frac{\partial C}{\partial x} - \varepsilon_x\frac{\partial^2 C}{\partial x^2} - \varepsilon_y\frac{\partial^2 C}{\partial y^2} - \varepsilon_z\frac{\partial^2 C}{\partial z^2} - \phi r = 0 \qquad (5.B.49)$$

where ε_x, ε_y, and ε_z represent the hydrodynamic dispersion coefficients in the x-, y-, and z-directions, respectively.

The effects of advection, dispersion, sorption, and degradation on the time-dependent movement of a slug of contaminant through porous material are illustrated graphically in Figure 5.B.3. Advection moves the contaminant along with the mean fluid flow. Dispersion causes the contaminant to spread from high concentration to low concentration. Sorption slows the rate of transport. Degradation removes contaminant mass from the system.

Case 6: One-Dimensional, Steady-State Advection, Dispersion, Sorption, and Decay of a Groundwater Contaminant The one-dimensional material-balance equation that incorporates advection, dispersion, sorption, and decay can be used to approximately describe the fate of an organic contaminant that has been dissolving for a long period into the groundwater in a homogeneous aquifer. The source of contamination is assumed to be a nonaqueous-phase liquid (NAPL; see §3.B.2) that might have resulted, for example, from a leaking underground storage tank. If groundwater flows past the NAPL slowly, the contaminant concentration in groundwater adjacent to the source can be estimated to equal C_0, the aqueous solubility of the contaminant. The contaminant is then transported in the direction of groundwater flow by means of advection (U), influenced by dispersion (ε_h). It may also be affected by sorption onto the aquifer material (retardation R) and by degradation (assume a first-order reaction, with rate constant k). We want to derive an equation for the contaminant concentration as a function of distance x from the source. We will assume that the contaminant decays only if it is dissolved in the fluid, so that $r = -kC$.

Since the contaminant concentration is assumed to have reached steady state, we can use the following simplified form of equation 5.B.48:

$$U\frac{\partial C}{\partial x} - \varepsilon_h \frac{\partial^2 C}{\partial x^2} + \phi k C = 0 \qquad (5.B.50)$$

The result is a governing equation that is analogous to that used in Example 5.B.1 to describe one-dimensional transport with decay in a river. Because we are considering steady-state conditions, the effect of sorption disappears in this case (R does not appear in the governing equation).

To solve the equation, we apply two boundary conditions, one specified in the problem definition and the other stating that if we go infinitely far downstream, the concentration decays to zero:

$$C(x = 0) = C_0 \qquad (5.B.51)$$

$$C(x) \longrightarrow 0 \quad \text{as } x \longrightarrow \infty \qquad (5.B.52)$$

Figure 5.B.4 Contaminant concentration as a function of distance downstream of a steady-state contaminant source in porous media. The solution for one-dimensional transport, as affected by advection (U), dispersion (ε_h), and degradation (k), is plotted on linear (upper frame) and log-linear (lower frame) coordinates. The soil porosity is ϕ.

The equation can then be solved using the same technique as applied in Example 5.B.1. This is the result:

$$C(x) = C_0 \exp\left\{\frac{Ux}{2\varepsilon_h}\left[1-\left(\frac{4\phi k}{U^2/\varepsilon_h}+1\right)^{1/2}\right]\right\} \quad (5.B.53)$$

At steady state, the contaminant concentration decreases exponentially in the direction of groundwater flow, provided $k > 0$. This result is illustrated in Figure 5.B.4. The individual lines are labeled with different values of the decay constant, k, normalized by $U^2/(\phi\varepsilon_h)$. As the decay constant increases, the concentration decays more rapidly with distance away from the source, as expected.

REFERENCES

BEJAN, A. 1984. *Convection heat transfer*. Wiley, New York.

CARSLAW, H.S., & JAEGER, J.C. 1959. *Conduction of heat in solids*. Oxford University Press, London.

HILL, C.G. 1977. *An introduction to chemical engineering kinetics & reactor design*. Wiley, New York.

LEVENSPIEL, O. 1972. *Chemical reaction engineering*. 2nd ed. Wiley, New York.

MORTON, K.W. 1996. *Numerical solution of convection-diffusion problems*. Chapman and Hall, London.

WOOLEY, J., NAZAROFF, W.W., & HODGSON, A.T. 1990. Release of ethanol to the atmosphere during use of consumer cleaning products. *Journal of the Air and Waste Management Association*, **40**, 1114–1120.

PROBLEMS

5.1 Transport and transformation models

(a) How is a *batch reactor* related to a *plug-flow reactor*?

(b) Consider a system involving transport and transformation processes. How can a comparison between the *characteristic time for reaction*, τ_r, and the *hydraulic residence time*, Θ, be used to simplify analysis? Consider both limits—$\tau_r \ll \Theta$ and $\tau_r \gg \Theta$—as well as the case of $\tau_r \sim \Theta$.

(c) Consider two flow reactors. Reactor A is a PFR with a constant volume of 3 m^3 and a volumetric flow rate of 1.5 $m^3\ h^{-1}$. Reactor B is a CMFR with a constant volume of 1 m^3 and a volumetric flow rate of 0.5 $m^3\ h^{-1}$. Water containing species C enters both reactors with an inlet concentration C_0. Within each reactor, the species undergoes first-order decay with a rate constant k. Which reactor has the longer hydraulic residence time?

5.2 Interpreting a tracer study of a reactor

Tracer studies are often used (1) to determine whether a physical system can be well described by an idealized reactor model and (2) to obtain key parameters of the physical system. A good tracer is chemically inert, nonhazardous, and easily detected. Consider such a test in which 1 g of NaCl was instantaneously released at the inlet of a water-filled system. The outlet concentration of chlorine ion (Cl^-) was measured as a function of time. The results are plotted in the figure at the top of the next page. Use this information to respond to the following questions. (*Note*: The atomic mass of sodium is 23 g/mol and of chlorine is 35.5 g/mol.)

(a) Which reactor model has predicted behavior that is consistent with the measured data?

(b) Determine the volume of the reactor (m^3)

(c) Determine the water flow rate through the reactor ($m^3\ h^{-1}$).

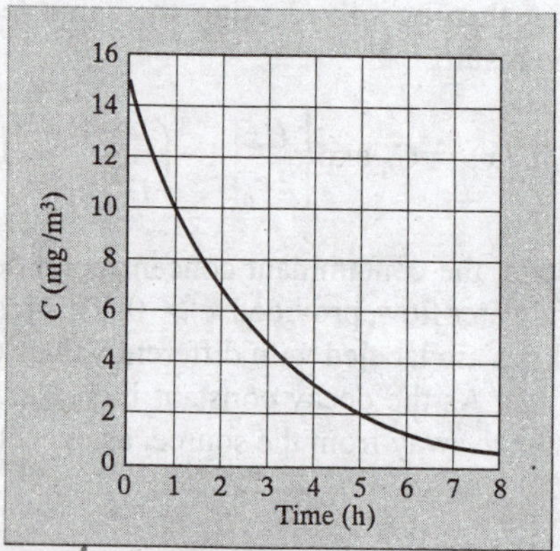

Tracer Expiremets have no reaction.

5.3 Chemical tracers to investigate mixing in reactors
One of the challenges in using reactor models is to determine whether the flow in a particular reactor is well represented by either of the flow models. A common means of characterizing flow in a reactor is to conduct a chemical tracer experiment. Imagine that our system consists of a reactor vessel with single inlet and outlet, as illustrated. The reactor volume is V (m^3). The fluid flow rate into the reactor is equal to the flow rate out of the reactor and has a constant value Q (m^3 s^{-1}). The tracer experiment involves rapidly discharging into the inlet flow stream a mass M (g) of a nonreactive chemical. The rate at which tracer volume is discharged is negligible compared with the total fluid flow rate. The concentration of the chemical is monitored as a function of time at the inlet (C_{in}) and outlet (C_{out}) of the reactor.

Pg 233

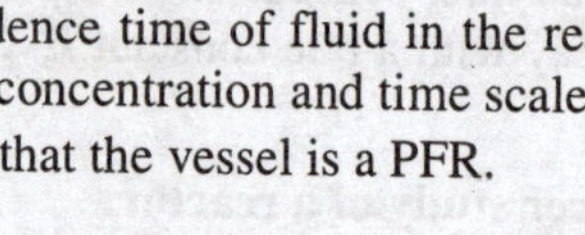

(a) For the inlet concentration profile as shown, sketch the outlet concentration, C_{out}, versus time, assuming that the vessel is a CMFR. Assume that the discharge duration, t_d, is much less than the residence time of fluid in the reactor. In your sketch, be as specific as possible about the concentration and time scales.
(b) Repeat part (*a*) assuming that the vessel is a PFR.

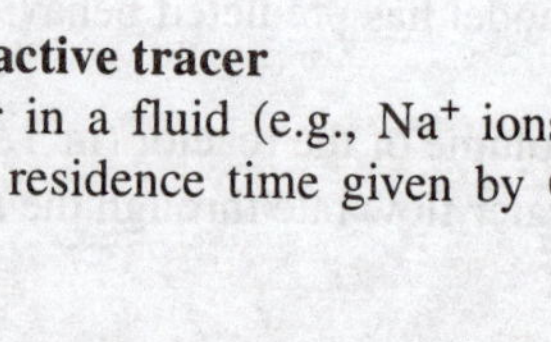

5.4 Reactor response to a nonreactive tracer
Consider a nonreactive tracer in a fluid (e.g., Na^+ ions in water). The fluid flows through a reactor with mean residence time given by Θ. The concentration of the

tracer in the fluid is described as follows

$$C_{in} = 0 \quad \text{for } t < t_1$$

$$C_{in} = C_1 \text{ for } t \geq t_1$$

The time t_1 is greater than zero. At $t = 0$, the tracer concentration within the reactor is zero. For cases (*a*) and (*b*), sketch the tracer concentration at the reactor outlet as a function of time for $t \geq 0$. The concentration and time scales should be properly and completely labeled (i.e., indicate t_1, C_1, Θ, and 0).

(a) Assume that the reactor is an ideal PFR.

(b) Assume that the reactor is an ideal CMFR.

5.5 Transient reactor response with first-order decay

A fluid flows through a reactor with mean residence time given by Θ. The system is illustrated below. The concentration of a contaminant species C in the fluid at the reactor inlet is 0 for all times $t < t_1$. For all times $t \geq t_1$, the concentration at the inlet is maintained at $C_{in} = C_1$. At $t = 0$, the tracer concentration within the reactor is zero. Within the reactor, the species undergoes first-order decay with rate constant k:

$$\text{C} \longrightarrow \text{products} \qquad R = kC$$

For cases (*a*) and (*b*), sketch the time-dependent contaminant concentration at the outlet, $C_{out}(t)$, for $t \geq 0$. The concentration and time scales should be properly and completely labeled.

(a) Assume that the reactor is an ideal PFR.

(b) Assume that the reactor is an ideal CMFR.

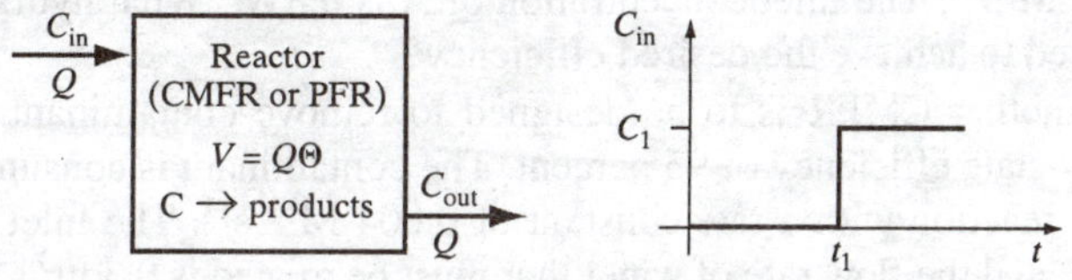

5.6 Lead in lake water; CMFR transient response

Consider the concentration of lead in a lake, using a simplified model. The lake has a river flowing into it and another flowing out of it, and the water level in the lake is constant. The major sources of lead in the lake are direct deposition from the atmosphere and surface runoff carried into the lake with the river. The major removal process is river outflow. Leaded gas was widely used for about 35 years (roughly 1950–1985), leading to a high influx of lead into the lake during those years. Use a CMFR representation of the lake to calculate the growth and decay of lead concentration in the lake.

(a) Derive an analytical expression that describes the lead concentration in the lake water as a function of time beginning in 1950.

(b) Plot the lead concentration versus time (1950–2050) in the lake water.

(c) Assume that the drinking water limit for lead is 50 ppb. During which years, if any, is the water unsuitable for drinking because of the lead concentration?

Data

Lake volume	2×10^9 m^3
River inflow (= river outflow)	1.2×10^8 m^3 y^{-1}
Pb in lake water, 1950	2 ppb
Pb in influent river water, 1950–1985	30 ppb
Pb in influent river water, 1985–2050	5 ppb
Pb deposition from atmosphere:	
1950–1985	3000 kg y^{-1}
1985–2050	300 kg y^{-1}

5.7 Behavior of flow reactors for a zeroth-order reaction

Consider a species that decays according to a zeroth-order reaction. The rate constant is k. A fluid containing the species is continuously supplied to a reactor. The inlet concentration is C_0. The hydraulic detention time of fluid in the reactor is Θ.

(a) For either a PFR or a CMFR (you choose), determine the outlet concentration, assuming that $k\Theta < C_0$.

(b) What is the outlet concentration if $k\Theta > C_0$?

(c) Does the steady-state outlet concentration depend on whether the reactor is a PFR or a CMFR?

5.8 Reactor engineering

Contaminant A is to be treated by chemical reaction in a PFR. If the reaction proceeds at a first-order rate $R = k[A]$, what must the hydraulic detention time, Θ, be to achieve 95 percent removal?

5.9 Effectiveness of a CMFR as a function of reaction order

(a) A CMFR is to be designed to remove contaminant A from water with a steady-state efficiency of 95 percent. The decay reaction is zeroth order, with a rate constant of 0.1 M h^{-1}. The inlet concentration of A is 0.5 M. What hydraulic detention time is required to achieve the desired efficiency?

(b) Another CMFR is to be designed to remove contaminant B from water with a steady-state efficiency of 95 percent. The contaminant is consumed by a second-order decay reaction with a rate constant of 0.004 M^{-1} s^{-1}. The inlet concentration of B is 0.5 M, and the flow rate of water that must be treated is 0.3 m^3 s^{-1}. How large must the reactor be?

5.10 Reactor performance

A continuously mixed flow reactor, as depicted below, is operated in steady state. It has volume V and flow rate Q. We are interested in two species, A and B. They are present in the inlet flow at concentrations A_0 and B_0, respectively. Within the reactor, species A undergoes a first-order decay reaction with rate constant k, producing species B. Determine the concentration of B at the outlet in terms of these parameters.

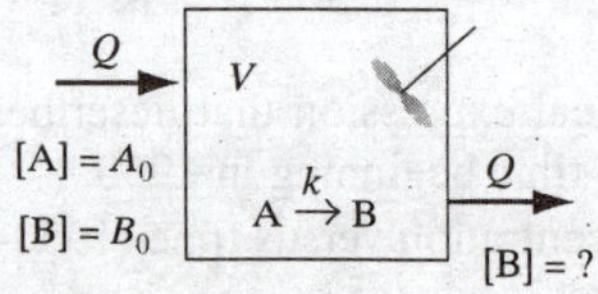

5.11 Reactor performance II

Consider a plug-flow reactor, as depicted below. Species B enters the reactor at concentration B_0. Within the reactor, it undergoes a second-order decay reaction with rate constant k. The reactor has a length L and cross-section A, and the fluid velocity within the reactor is uniformly equal to U. Predict the species concentration, [B], at the outlet of the reactor.

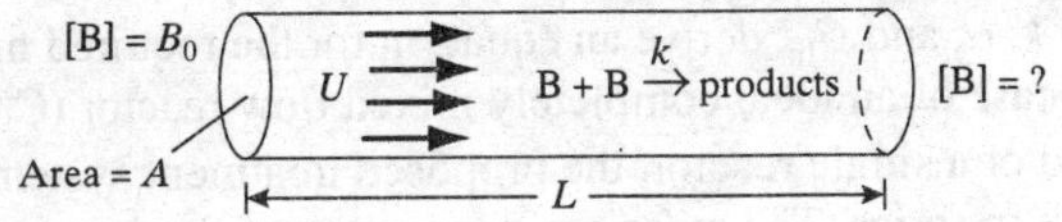

5.12 Reactor performance III

Fluid flows into a reactor of volume V at flow rate Q. At the inlet, the fluid contains species A at concentration A_0 and species B at concentration B_0. Within the reactor, the species undergo a reaction that is first order in both A and B. For parts (*a*) and (*b*) below, write the simplest complete system of equations that can be solved to predict the concentration of A at the outlet of the reactor. (*Hint*: You do not need to *solve* the equation[s].)

(a) Assume the reactor is an ideal PFR.

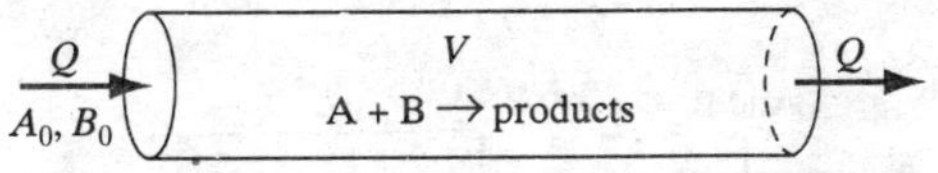

(b) Assume the reactor is an ideal CMFR.

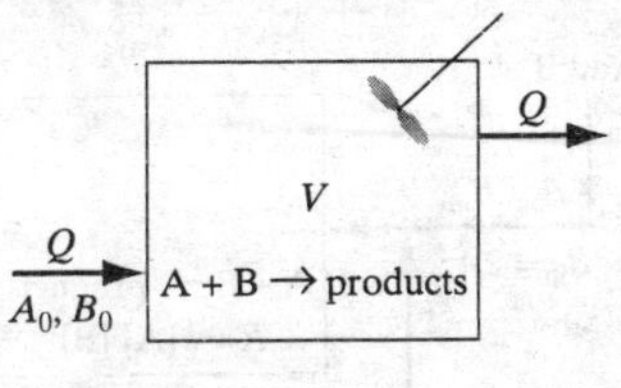

5.13 Conversion in a plug-flow reactor

Consider the system depicted below. Fluid flows with volumetric rate Q through a plug-flow reactor of volume V. The inflow contains species A and B at concentrations A_0 and B_0, respectively. Within the reactor, species A undergoes first-order reaction, producing B according to

$$\mathrm{A} \longrightarrow \mathrm{B} \qquad R = k[\mathrm{A}]$$

Determine the concentrations of [A] and [B] at the outlet of the reactor.

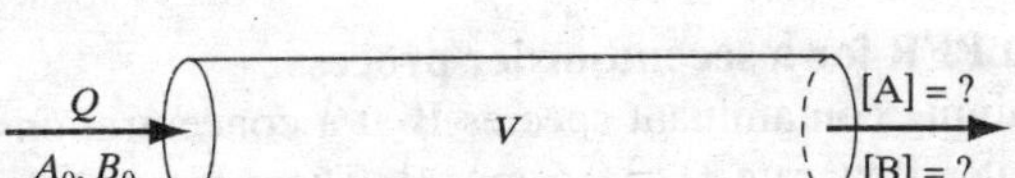

5.14 Reactor design
A species spontaneously decays by a first-order process with a rate constant k. Untreated water contains the species at a concentration C_{in}. It is desired to reduce the concentration to $C_{out} = (1 - \alpha)C_{in}$, where α is the conversion efficiency. (So, for example, if 90 percent control were desired, α would be 0.9 and the outlet concentration would be $C_{out} = 0.1C_{in}$.)

(a) Given k, α, and C_{in}, derive an equation for the required mean fluid residence time, Θ, for the case of an ideal plug-flow reactor (PFR).

(b) Given k, α, and C_{in}, derive an equation for the required mean fluid residence time, Θ, for the case of an ideal completely mixed flow reactor (CMFR).

(c) Instead of a single reactor, the proposed treatment system is to consist of n identical CMFRs in series. The mean residence time of fluid in each reactor is Θ/n, so that the total mean residence time in the system is Θ. Given k, α, C_{in}, and n, derive an equation for Θ.

5.15 Reactor design II
An undesired contaminant, A, can be removed from drinking water by reaction with reagent B that is added to the water. The reaction rate is described by

$$A + B \longrightarrow \text{products} \qquad R = k[A][B]$$

A CMFR is designed to achieve 90 percent removal of A. As shown in the upper schematic, reagent B is added upstream of the reactor such that the inlet concentration of B is twice the inlet concentration of A (i.e., $B_{in} = 2A_{in}$).

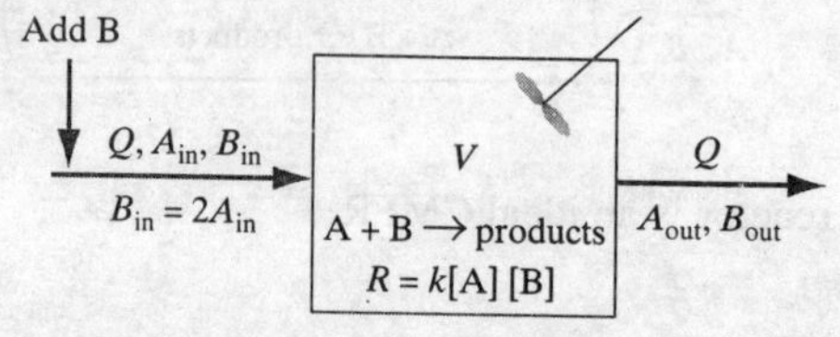

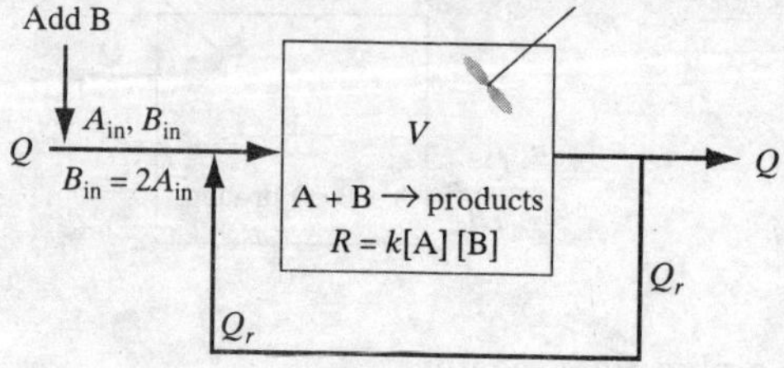

(a) Determine the hydraulic residence time in the reactor required to achieve 90 percent removal of A. The correct answer depends only on some or all of the following parameters: A_{in}, k, Q.

(b) When the reactor is built, it is improperly plumbed so that some flow is recycled, as depicted in the lower schematic. The net flow rate through the reactor system, Q, is unchanged. Will the efficiency of the reactor with recycle flow be higher, lower, or the same as it would be without recycle? Explain.

5.16 Designing a PFR for a second-order process
Water containing contaminant species B at a concentration of 5 g m^{-3} enters a treatment plant at a flow rate $Q = 0.5$ m^3 s^{-1}. There the water is introduced into a plug-

flow reactor in which B is consumed by a second-order process:

$$B + B \longrightarrow \text{products} \qquad R = k[B]^2 \qquad r = -2k[B]^2$$

The rate constant k has a value of 0.04 $m^3\ g^{-1}\ s^{-1}$. How large must the reactor be to achieve 95 percent destruction of B?

5.17 Second-order reaction in a CMFR

Consider a well-mixed reactor as depicted. A species enters the reactor at concentration n_0. Within the reactor, it undergoes an elementary second-order decay reaction with a rate constant k. It is desired to predict the outlet concentration as a function of time, $n(t)$. The analysis is greatly simplified if the reaction may be neglected. Derive a relationship in terms of these parameters that can be used to decide whether the reaction may be ignored: Q, n_0, V, k. (*Hint*: It is known that $k > 0$.)

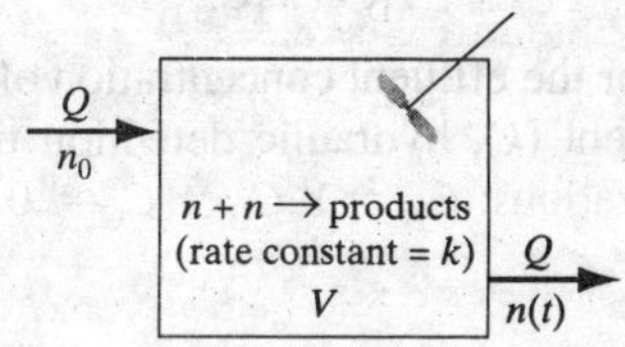

5.18 Sizing a CMFR for a treatment process

A contaminant in water is observed to spontaneously decay.

(a) An experiment is conducted in a mixed batch reactor to investigate the reaction kinetics. The data are reproduced in the following table. Determine the rate law and the rate constant for the reaction.

Time (h)	Conc. (mg/L)
0	60
0.5	44
1	33
1.5	24
2	18
2.5	13

(b) Untreated water contains 50 mg/L of the contaminant. What hydraulic detention time in a CMFR is required to achieve 90 percent removal of the contaminant by decay?

5.19 Water storage as a control technique for radon

Radon from well water may pose a radiological health hazard. One treatment technique is simply to hold the water in storage, allowing the radon to decay. The decay rate is first order:

$$\text{Rn} \longrightarrow \text{products} \qquad k = 0.181\ d^{-1}$$

Consider a situation in which water is delivered from a well to a household at a rate of 1 $m^3\ d^{-1}$. The radon concentration is to be reduced by storage to 5 percent of its level in the untreated supply.

(a) How large must the storage tank be if it acts as a plug-flow reactor (m^3)?

(b) How large must the storage tank be if it acts as a completely mixed flow reactor (m^3)?

(c) If the storage system consists of two identical CMFRs in series, how large must each be to achieve the desired level of control (m^3)?

5.20 Reaction mechanisms and rates

The following set of reactions involving an intermediate and two end products is to be carried out in a completely mixed flow reactor under steady-state conditions.

$$r_A = -k_1 C_A$$

$$r_B = 2k_1 C_A - k_2 C_B - k_3 C_B$$

$$r_C = k_2 C_B$$

$$r_D = k_3 C_B$$

Give expressions for the effluent concentration of each species as a function of the reaction rate coefficient (k), hydraulic detention time (Θ), and the following influent compound concentrations: $C_A = Y$, $C_B = C_C = 0$, and $C_D = Z$.

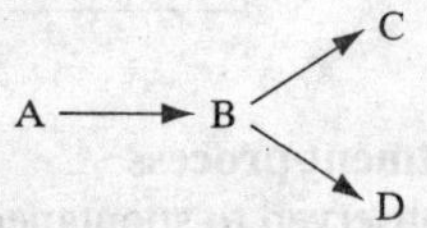

5.21 Comparing reactor efficiencies

Completely mixed flow reactors (CMFRs) and plug-flow reactors (PFRs) may have significantly different removal efficiencies depending on the reaction order. This problem asks you to compare the relative reactor sizes required to ensure 99.5 percent removal of wastewater pathogens. Wastewater from a city, which flows at a constant rate of 10^7 L/d, must be treated for coliform removal before it can be discharged to surface waters. The initial coliform concentration is 250 mg/L and the rate constant for coliform decay is given as $k_i = 2\ d^{-1}\ (mg/L)^{1-i}$, where i = reaction order.

(a) Fill in the table below by calculating the respective CMFR and PFR reactor volumes required for steady-state coliform die-off of 99.5 percent for different reaction orders.

Coliform decay rate	CMFR volume (m^3)	PFR volume (m^3)
Zeroth order: $-k_0$		
First order: $-k_1 C$		
Second order: $-k_2 C^2$		
Third order: $-k_3 C^3$		

(b) Given these results, what generalizations can you make about removal efficiencies of CMFRs versus PFRs?

5.22 More on reactor models and contaminant transport

(a) Describe the central feature(s) that define the fluid flow conditions in PFRs and CMFRs.

(b) An alternative to using reactor models is to describe transport processes using the general material-balance equation. Discuss briefly the important advantages and disadvantages of using reactor models versus the more general governing equation.

(c) Imagine that species A is destroyed chemically by an elementary second-order process:

$$A + A \longrightarrow B$$

If the species is introduced into a PFR at a concentration A_0, determine the hydraulic residence time required for the outlet concentration [A] to be reduced to $0.05A_0$.

(d) Repeat (*c*) for the case of a CMFR.

(e) Discuss briefly the reason(s) why the difference in mixing conditions leads to the difference in the results between (*c*) and (*d*).

5.23 Characteristic time: Transport plus transformation

Consider a well-mixed reactor (CMFR) with a hydraulic detention time $\Theta = V/Q$ (V = reactor volume, Q = fluid flow rate). Within the reactor, species A undergoes first-order decay with a rate constant k. If the inlet concentration of the species is suddenly increased from A_0 to A_1, what is the characteristic time for a new steady-state concentration to be achieved at the outlet of the reactor?

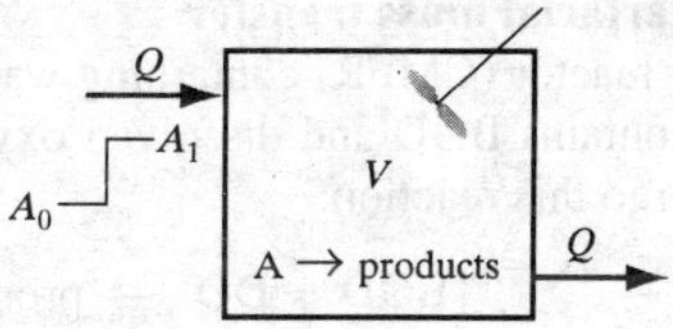

5.24 Radon diffusion through plastic sheeting

Radon is a radioactive air pollutant that is generated in soil and can migrate into buildings. Inhalation of radon's radioactive decay products increases the risk of lung cancer. One of the ideas for preventing radon entry into wood-floored houses is to cover the ground beneath the floor with polyethylene (plastic) sheeting. The rationale is that radon will have to diffuse through the plastic material, and this will greatly reduce the overall entry rate into the house. When this idea was first seriously considered, an experiment was conducted to measure the diffusion coefficient of radon through polyethylene sheeting. The apparatus is depicted below. The central compo-

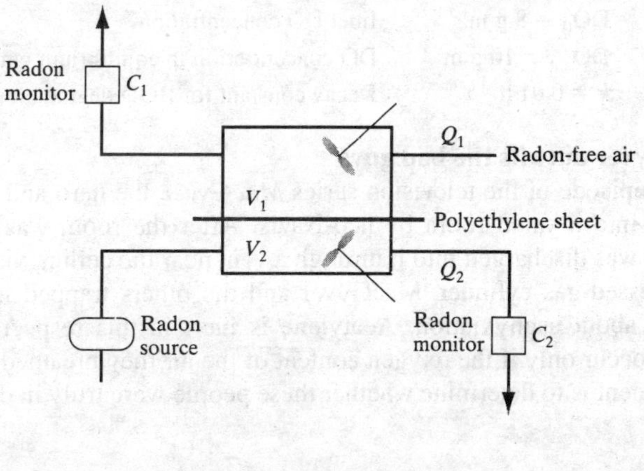

nent is a two-compartment aluminum chamber. The compartments are separated by means of a polyethylene sheet. Radon-rich air flows through one compartment. (The radon source emits radon atoms at a constant rate.) Air that is initially radon free flows through the other compartment. The radon concentration is monitored in each flow stream until steady state is achieved. You are to do some analysis on the results.

Data and Parameters

$Q_1 = 4.0$ cm^3 s^{-1} $V_1 = 750$ cm^3 $C_1 = 1700$ atoms ^{222}Rn cm^{-3}

$Q_2 = 3.0$ cm^3 s^{-1} $V_2 = 350$ cm^3 $C_2 = 3.5 \times 10^6$ atoms ^{222}Rn cm^{-3}

$T = 0.015$ cm = thickness of polyethylene sheet

$A = 150$ cm^2 = exposed area of polyethylene sheet between two compartments

(a) Assuming that Fick's law applies, derive an equation for the diffusion coefficient of radon through polyethylene in terms of measured parameters indicated in the schematic. Using the data indicated, evaluate the diffusion coefficient.
(b) One of the complications with this experiment is that radon undergoes radioactive decay. This is a first-order loss process with a decay constant $\lambda = 2.1 \times 10^{-6}$ s^{-1}. Test whether it is reasonable to neglect radioactive decay in analyzing the data in part (*a*). (*Hint*: There are three transport time scales to consider in the system.)

5.25 DO, BOD, and interfacial mass transfer

A well-mixed flow reactor (CMFR) containing water is open to the atmosphere. The inlet flow stream contains BOD and dissolved oxygen (DO). Within the reactor, the BOD and DO undergo this reaction:

$$\text{BOD} + \text{DO} \longrightarrow \text{products}$$

The reaction is first order in BOD and zeroth order in DO, provided the DO content of the water exceeds 0.1 g m^{-3}. Since the reactor is open to the atmosphere, interfacial mass transfer occurs, replenishing the depleted oxygen.

Solve for the steady-state DO concentration in the reactor.

Data

$V = 100$ m^3	Volume of water in reactor
$Q = 2$ m^3 h^{-1}	Flow rate of water through reactor
$A = 25$ m^2	Air-water contact area
$k_{gl} = 0.2$ m h^{-1}	Interfacial mass-transfer coefficient
$\text{BOD}_0 = 40$ g m^{-3}	Inlet BOD concentration
$\text{DO}_0 = 8$ g m^{-3}	Inlet DO concentration
$\text{DO}_{sat} = 10$ g m^{-3}	DO concentration in equilibrium with atmosphere
$k = 0.01$ h^{-1}	Decay constant for BOD degradation

5.26 MacGyver outwits the bad guys

On an episode of the television series *MacGyver*, the hero and two companions were locked into a small room by neo-Nazis. After the room was closed, acetylene gas (C_2H_2) was discharged into it through a vent near the ceiling via a tube connected to a compressed-gas cylinder. MacGyver and the others trapped in the room were concerned about asphyxiation. Acetylene is inert in this respect, so that asphyxiation would occur only if the oxygen content of the air they breathed became too low. Your assignment is to determine whether these people were truly in danger of asphyxiation.

Assume that the following data and conditions apply:

- The compressed-gas cylinder initially contains 400 mol of pure acetylene.
- The room dimensions are 3 × 3 × 3 m.
- Air pressure in the room is constant at 1 atm.
- Except for flow forced out of the room to balance the injection of acetylene and maintain a constant air pressure, the room is unventilated.
- Temperature is uniform at 298 K everywhere.

(a) MacGyver first recommended that they remain close to the floor of the room. Is this a reasonable suggestion? Why?

(b) Assume that the room behaves as a plug-flow reactor with pure acetylene injected at the ceiling and air leaving the room through a leak near the floor. Describe quantitatively the gaseous constitution of air in the room after the compressed-gas cylinder is empty.

(c) Assume that the room behaves as a completely mixed flow reactor with pure acetylene injected at the ceiling and a mixture of gases leaving the room through a leak near the floor. Describe quantitatively the gaseous constitution of air in the room after the compressed-gas cylinder is empty.

(d) An oxygen content (mole fraction) of 5 percent at a pressure of one atmosphere is required to sustain life. Were MacGyver and his friends in danger of asphyxiation? Explain.

5.27 Reagent requirements in a treatment process

An environmental fluid contains a contaminant A that can be removed by reaction with reagent B according to the following elementary process:

$$\text{A} + \text{B} \longrightarrow \text{inert products}$$

Reagent B is expensive, so we do not wish to use any more of it than necessary. Determine the minimum required inlet concentration of the reagent, $[\text{B}]_{min}$, such that the outlet concentration of the contaminant is a specified fraction of the inlet concentration. Carry out the calculation for two reactor configurations: (*a*) completely mixed flow reactor and (*b*) plug-flow reactor. Express your answers in terms of the following parameters:

f	Fraction of A remaining at the reactor outlet ($[\text{A}]/[\text{A}]_{in}$)
Q	Fluid flow rate [$\text{m}^3\ \text{s}^{-1}$]
V	Reactor volume [m^3]
k	Rate constant [$\text{L mol}^{-1}\ \text{s}^{-1}$]
$[\text{A}]_{in}$	Inlet concentration of contaminant [mol L^{-1}]

(*Note:* For part (*b*) you may write the answer as an implicit equation for $[\text{B}]_{min}$.)
(*Hint:* The following integral may prove useful.)

$$\int \frac{dx}{x(a + bx)} = \frac{1}{a} \ln\left(\frac{x}{a + bx}\right)$$

5.28 Treatment reactor for toluene

You are asked to design a steady-state reactor for the treatment of 100 mg/L toluene in water. Assume that toluene is degraded by a half-order reaction as follows:

$$r_d = -kC^{1/2} \text{ mg L}^{-1}\ \text{d}^{-1}$$

$$k = 0.8\ (\text{mg/L})^{1/2}\ \text{d}^{-1}$$

(a) Calculate the detention time required for 96 percent removal of the toluene in a CMFR.

(b) Calculate the detention time required for 96 percent removal in a PFR.

(c) Derive a differential equation to describe the change in concentration of toluene along a nonideal PFR undergoing turbulent and molecular diffusion under steady-state conditions.

5.29 Ideal reactors and nonideal reactors

A river contaminated with high concentrations of phenol flows at a rate of 5000 m^3/d along a well-characterized and monitored 5 km stretch with a cross-sectional area of 16 m^2. The steady-state concentration of phenol at the start of the stretch is 800 mg/L. The phenol is oxidized as it moves along the river by a first-order reaction; the rate constant is 0.5 d^{-1}.

(a) Determine the concentration at the end of the 5 km assuming no mixing in the forward or reverse direction.

(b) Set up the steady-state differential equation to describe the change in phenol concentration with distance along the river (dC/dx) assuming turbulent diffusivity of 2 m^2/s.

(c) Determine the concentration at the end of the 5 km stretch assuming a turbulent diffusivity of 2 m^2/s.

(d) Determine the concentration at the end of the 5 km assuming complete mixing within the stretch.

(e) Compare the results from (*a*), (*c*), and (*d*) and comment on your findings.

5.30 PCBs in the Hudson River

The Hudson River in New York is contaminated with PCBs (polychlorinated biphenyls). PCBs are hydrophobic organics that sorb strongly to the sediments at the bottom of the river. You are hired to determine the fate of the PCBs in this river. The river can be modeled as a rectangular PFR with water velocity U, depth D, height H, and width W. The PCBs are released into the water from the river sediments at a rate of F_{PCB} (g m^{-2} d^{-1}) starting at $x = 0$ and continuing downstream for an indefinite distance. There is no upstream source of PCBs other than the sediments. In addition to advection, the PCBs undergo horizontal turbulent diffusion (diffusivity $= \varepsilon$) and first-order decay in the water with a rate constant of k_d.

(a) Derive a differential equation describing the steady-state PCB concentration (C) as a function of downstream distance (x).

(b) If we can assume that turbulent diffusion within this river is negligible compared with advection, give an expression for the maximum concentration of PCBs in the water downstream from $x = 0$. (*Hint:* Set $dC/dx = 0$.)

(c) If the PCBs are volatile with a mass-transfer coefficient across the air-water interface of k_{gl} (m/h), and if the atmospheric concentration of the PCBs above the river is zero, derive an expression for the steady-state, maximum downstream concentration of PCBs in the river, again neglecting turbulent diffusion.

5.31 Hydrogen sulfide odors from gold-mining residues

Mercury was commonly used for the separation of gold from ore in the California foothills. For many years, waste mercury was dumped into a small pond of water in gold country near Grass Valley. The anaerobic sediments at the bottom of this lake contain large concentrations of solid mercury sulfide (HgS), which are in equilibrium with water of the pond.

$$HgS \Leftrightarrow Hg^{2+} + S^{2-} \qquad K_{sp} = 4 \times 10^{-23}\ M^2$$

The dissolved mercury ions are absorbed by the local fauna, so that the concentration in the water is maintained at a steady-state level of 10^{-5} M. The sulfide (S^{2-}) reacts with protons in the water to produce hydrogen sulfide gas in the simplified reaction shown below:

$$S^{2-} + 2H^+ \Leftrightarrow H_2S \qquad K_1 = 10^{21.1}\ M^{-2}$$

The H_2S produced in the water is emitted from the pond surface as H_2S gas. The movement of H_2S from the bottom of the pond to the top is driven only by diffusion, with a diffusion coefficient of 1.2×10^{-9} m^2 s^{-1}. A steady wind blows above the pond surface, so the concentration of H_2S gas in the air above the pond is approximately zero. The surface area of the pond is 10 m^2. This pond can be modeled as a one-dimensional steady-state system with the sediments at $z = 0$ m, the pond surface at $z = 2$ m, and no water flow.

(a) Set up a material balance on H_2S within the lake (use C for concentration of H_2S) and derive a differential equation to describe the change in concentration of H_2S with depth (z).

(b) Give the boundary conditions for H_2S within the lake (i.e., the concentrations of H_2S at $z = 0$ and at $z = 2$ m) in terms of the proton concentration [H$^+$] within the water.

(c) Derive an expression to describe the concentration of H_2S as a function of depth [$C(z)$] in terms of [H$^+$].

(d) Given that the rate of H_2S emission into the air is 20 mg d^{-1}, calculate the pH of the water and the concentration of H_2S 0.25 m below the water surface.

5.32 Biodegradation and volatilization

Ethanol (CH_3CH_2OH) is a solvent used in liquid detergents. If this ethanol escapes to the atmosphere, it contributes to photochemical smog. Recent research has shown that little of the ethanol escapes during detergent use, but instead is drained to the sewer. Once in the sewer, the ethanol can have three fates: (1) volatilization to the atmosphere, (2) biochemical oxidation to CO_2 and H_2O, or (3) discharge from the sewer into the wastewater treatment plant unaltered. In this problem, you are to compute the relative importance of these fates for specified conditions. Assume that the flow through the sewer is constant. Use a plug-flow reactor description of the sewer. Assume that the ethanol that remains in the water is biodegraded by a first-order process with rate constant k. Assume that the two-film model describes ethanol release from the sewer to the atmosphere.

Data

$C_s = 0$	Ethanol concentration in sewer water in equilibrium with ethanol in air
$k = 0.15$ d^{-1}	Rate constant for biochemical oxidation of ethanol
$k_{gl} = 0.03$ m d^{-1}	Interfacial mass-transfer coefficient for ethanol in sewer
$W = 1$ m	Width of water channel in sewer
$A = 0.25$ m^2	Vertical cross-sectional area of water flow through sewer
$V = 10$ km d^{-1}	Sewage flow velocity
$L = 10$ km	Distance from point of discharge to wastewater treatment plant

(a) Write a material-balance equation that describes the rate of change of ethanol concentration with residence time following discharge into the sewer: $dC/d\Theta_x$.

(b) Solve the governing equation to obtain $C(\Theta_x)$.

(c) Calculate the fraction of the total ethanol discharge that escapes from the sewer into the atmosphere and the fraction that is degraded.

5.33 Particles in a box

Consider the following situation. An air-filled box of dimensions 1 m × 1 m × 1 m is initially charged with a uniform distribution of spherical particles. Then the box is sealed. The particles are made of a material whose density is 1.5 g cm^{-3}. Their diameter is constant. The initial particle mass concentration is 50 μg/m^{-3}. The particles settle under the influence of gravity, with a velocity of 0.0053 cm s^{-1}, and deposit on the box floor.

(a) What is the characteristic time for particles to be lost from the air in the box by settling?

(b) Assume that the air in the box is motionless. Describe quantitatively and completely the particle concentration versus position in the box 1 h after the box is sealed.

(c) Assume that the air in the box is slowly stirred so that the contents remain well mixed. Assume that the mixing is not strong enough to resuspend particles that deposit on the floor of the box. Describe quantitatively and completely the particle concentration versus position in the box 1 h after the box is sealed.

(d) What is the mass per area of particles deposited on the floor of the box (μg m^{-2}) for cases (*b*) and (*c*)?

(e) A period of time has elapsed equal to 10 times the characteristic time for settling. What is the mass per area of particles deposited on the floor of the box (μg m^{-2})? How does your answer depend on the airflow conditions within the box?

5.34 Particle removal by settling

At the inlet to a completely mixed flow reactor, water contains 150 g m^{-3} of suspended particles, evenly distributed (by mass) among spheres of 1 μm, 10 μm, and 100 μm diameter. The density of each particle is 2.5 g cm^{-3}. The water temperature is 20 °C. (*a*) Compute the number concentration for each particle size in the inlet water. Assuming that the only removal process that influences the particles within the tank is gravitational settling, compute (*b*) the mass concentration and (*c*) the number concentration of particles of each size at the outlet of the reactor. (*d*) Determine for each particle size the rate of mass accumulation per unit area on the bottom of the reactor. The reactor is in the shape of a right-circular cylinder with a vertical axis. Assume steady-state conditions prevail. (*Hint:* Think about deposition on the bottom as a first-order "reaction." What is the "rate" of this reaction?)

Data

H = 5 m is the reactor height.
V = 50 m^3 is the reactor volume.
F = 25 m^3 h^{-1} is the flow rate of water through the reactor.

5.35 Kinetics of a treatment process

You have discovered a series of elementary reactions involving compounds C (contaminant), I (intermediate), and P (product), which may be useful in order to degrade the unwanted contaminant, C. In this series of reactions, one molecule of C degrades to form one molecule of I, which in turn combines with C to produce P as shown below:

$$\mathrm{C} \longrightarrow \mathrm{I} \qquad k_1 = 24\ \mathrm{d}^{-1}$$

$$\mathrm{C} + \mathrm{I} \longrightarrow \mathrm{P} \qquad k_2 = 80\ \mathrm{mM}^{-1}\ \mathrm{d}^{-1}$$

(a) Write differential equations to describe the rate of change of C and I with time in a batch reactor.

(b) You need to design a reactor to treat a constant flow of water initially containing only contaminant C at 5 mM. Set up a material balance for the contaminant in this water flowing through an ideal steady-state CMFR and undergoing the above reactions.

(c) Given that the water flow is 10 L/s and the concentration of combined contaminant and intermediate (C + I) in the effluent must be ≤0.5 mM, evaluate whether a CMFR with a 0.5-day detention time will be sufficient for the treatment of this water.

5.36 Modeling urban air toxics

The goal of this problem is to develop a model that can be used to estimate concentrations of a toxic air pollutant within an air basin. Assume that the species is emitted uniformly throughout the air basin (e.g., benzene emitted from motor vehicles). For this model, we treat the air as well mixed in the vertical direction. The wind is assumed to blow at a constant speed U from west to east. Turbulent diffusion and dispersion in the horizontal plane are neglected.

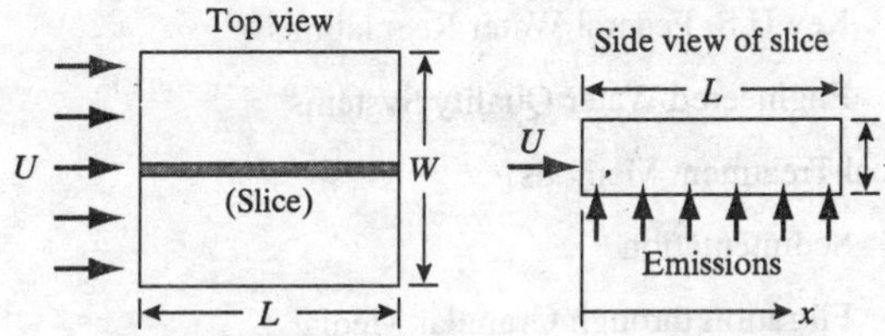

We focus on a slice of the air basin as depicted in the figure. Upwind of the air basin, the pollutant concentration is zero. As air moves through the basin, pollutants are emitted into the air parcel from below. The pollutant emission rate is uniform over the whole basin and constant with respect to time. Within the air parcel, the pollutant species can undergo degradation by first-order reaction. The following parameters apply:

L	Length of air basin in the wind direction (m)
W	Width of air basin normal to the wind (m)
H	Mixing height of air basin (m)
E	Total emission rate of pollutant into the air basin (g s^{-1})
U	Wind speed across air basin (m s^{-1})
k	Degradation constant for pollutant accounting for homogeneous chemistry (s^{-1})

(a) Develop a governing equation for $dC(x)/dx$, where C is the concentration of the air toxic species (g m^{-3}) and x is the coordinate direction parallel to the wind.

(b) Solve the governing equation to obtain $C(x)$.

(c) What characteristic length is required for the concentration C to become independent of x?

(d) The model can be simplified if some of the processes can be neglected. Consider the case of homogeneous degradation. Assume that the degradation rate is finite but small. Specify a mathematical condition that, if satisfied, would allow degradation to be neglected.

6

Water Quality Engineering

6.A THE NATURE OF WATER QUALITY PROBLEMS

Worldwide per-capita water consumption continues to increase while the overall supply of freshwater remains fixed (Table 6.A.1). This condition necessitates new strategies for conserving and reusing water. Indeed, increasing human intervention in the hydrologic cycle is causing overall declines in the quality of freshwater available for

Table 6.A.1 Distribution of Fresh and Saline Water on Earth

Site	Volume (km^3)
Oceans	1,350,000,000
Polar ice caps and glaciers	29,000,000
Groundwater	8,300,000
Freshwater lakes	125,000
Saline lakes and inland seas	104,000
Soil and subsoil water	67,000
Atmospheric water vapor	13,000
Living biomass	3,000
Stream channels	1,000

Source: Harte, 1988.

consumption. To ensure that a safe, adequate, and sustained water supply is available in the future, three important objectives have been recommended (AEEP, 1998):

1. Expand our knowledge of the toxicological and environmental impacts of chemical and microbiological water impurities.
2. Develop improved capabilities for repurifying previously used freshwater in an efficient and economically effective manner.
3. Develop and apply improved water conservation techniques to decrease agricultural, industrial, and domestic water use.

These three objectives are interrelated. That is, a greater understanding of the toxicological and environmental impacts of impurities will facilitate development of new repurification technologies, which can, in turn, lead to better water conservation. For example, it is now understood that certain organic contaminants in water may act as endocrine disrupters, mimicking the activity of hormones in humans and wildlife. If we can determine the types and amounts of chemicals that cause such effects, we can modify practices that contribute to the release of these chemicals. We can also develop improved water and wastewater treatment systems to ensure adequate removal of these compounds. Increased understanding of the specific nature of water quality impacts on human and ecosystem health will lead to greater confidence in conservation techniques that rely on water reuse. Water conservation techniques that rely on modified agricultural practices, redesigned industrial processes, and altered domestic behavior are also important.

An important goal for future environmental engineering practice is to shift to a proactive stance of anticipating environmental quality problems before they occur. In addition to responding to damage caused by pollution, it is important to avoid problems by anticipating and limiting undesirable environmental releases. The use of methyl *tert*-butyl ether (MTBE) as a gasoline additive provides a recent example of the problems that can occur because of a failure to predict adverse environmental consequences of a large change in practice (see Exhibit 6.A.1).

EXHIBIT 6.A.1 *The Case of MTBE*

Methyl *tert*-butyl ether (MTBE) is an oxygen-containing organic molecule (see Figure 6.A.1). In the 1970s MTBE was introduced and used to a limited extent as a gasoline additive to increase fuel octane ratings. In the 1990s the Clean Air Act amendments mandated that oxygenated compounds

EXHIBIT 6.A.1 *The Case of MTBE (continued)*

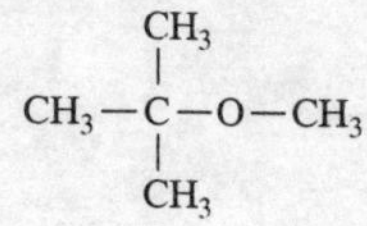

Figure 6.A.1 Chemical structure of methyl *tert*-butyl ether (MTBE).

be added to gasoline in many parts of the United States to reduce atmospheric emissions of carbon monoxide and volatile organic compounds from motor vehicles. Fuel oxygenates tend to shift the fuel-air equivalence ratio, ϕ, toward a more fuel-lean condition (see §3.D.3), thereby reducing emissions of the products of incomplete combustion. This mandate led to much wider use of MTBE in gasoline. Indeed, by the late 1990s, approximately 30 percent of gasoline sold in the United States contained an average of 11 percent MTBE by volume (Andrews, 1998). What was not recognized at the time, however, was that MTBE is highly soluble in water and not as easily biodegraded as other components of gasoline. Consequently, MTBE from leaking underground storage tanks and incidental gasoline spills entered many groundwater and surface-water sources. In addition, gasoline leaks and spills from watercraft engines caused formerly pristine lakes to become contaminated with measurable concentrations of MTBE. Since little is known about the health effects of MTBE, and since the taste and odor thresholds of this compound are extremely low, water supplies with very low concentrations of MTBE are considered unsuitable for potable use. To halt the further degradation of water supplies, the state of California decided in late 1999 to phase out the use of MTBE in gasoline by 2003. In early 2000, the USEPA announced a plan to ban or phase out the use of MTBE as a fuel additive nationwide. Other, more benign oxygenates, such as ethanol, are likely to replace MTBE. In summary, a control measure that was initially introduced to improve air quality resulted in unacceptable impairment of water quality. If the environmental fate and transport issues associated with MTBE had been thoroughly evaluated before it was put into widespread use, then much cost and water quality degradation would have been avoided.

The major sources of water pollution include point sources, such as domestic sewage and industrial waste, and nonpoint sources, such as agricultural and urban runoff. These sources can be responsible for discharging a range of pollutant types to water bodies (Table 6.A.2). The occurrence and effects of specific pollutants on natural and engineered water bodies will be discussed in the following sections.

Table 6.A.2 Major Pollution Sources for Fresh and Saline Waters

	Point sources		Nonpoint sources	
Pollutant class	Municipal sewage	Industrial waste	Agricultural runoff	Urban runoff
BOD	✓	✓	✓	✓
Nutrients	✓	✓	✓	✓
Pathogens	✓	✓	✓	✓
Suspended solids	✓	✓	✓	✓
Salts		✓	✓	✓
Toxic metals		✓		✓
Toxic organics		✓	✓	✓
Heat		✓		

Source: Davis and Cornwell, 1991.

6.A.1 Rivers and Streams

The characteristics of a river or stream influence the effect of water pollutant discharges on those receiving waters. For example, a large, slow-moving river such as the Mississippi is able to absorb a much greater loading of contaminants without significant adverse downstream effects than a small, babbling brook. Water volume and flow, depth, sediment type, surrounding vegetation and land use, and even climate affect pollutant impacts.

Water quality monitoring has been conducted for 53 percent of the 1.3 million miles (2.1 million km) of perennial rivers and streams in the United States (USEPA, 1996a). The results indicate that 64 percent had good water quality capable of supporting diverse communities of fish, plants, and aquatic insects, as well as the array of human activities assigned to each river by the respective states. On the other hand, 36 percent of the monitored reaches had impaired water quality that limited its ability to support one or more of the designated uses (Table 6.A.3), with aquatic life support as the most severely impacted characteristic.

The four most prevalent water quality problems impacting rivers and streams in the United States are siltation, nutrients, pathogens, and oxygen-depleting substances (Table 6.A.4). Siltation, the accumulation of small soil particles on riverbeds, suffocates fish eggs and destroys aquatic insect habitats, damaging the food web that supports fish and other wildlife. Siltation can occur as a result of agriculture, urban runoff, construction, and forestry. Nutrient pollution generally refers to elevated quantities of nitrogen and phosphorus in the water. Excessive nutrients cause increased plant and algae growth, resulting in depleted populations of fish and other desirable aquatic species. Municipal and industrial wastewater discharges and runoff from agricultural lands, forestry operations, and urban areas are major nutrient sources. Pathogen contamination of surface waters can cause significant human health problems, from simple skin rashes to acute gastroenteritis. Common sources of pathogens include inadequately treated municipal wastewater, agricultural and urban runoff, and wildlife fecal material. The discharge of oxygen-demanding material into rivers and streams causes a depletion of dissolved oxygen downstream. Oxygen depletion can cause problems for aquatic life and severely impact the natural biota of a river or stream. Sources of oxygen-demanding material include municipal and industrial wastewater, as well as agricultural and urban runoff.

Table 6.A.3 Summary of Water Quality in U.S. Rivers and Streams

Designated use	Length surveyed ($\times 10^5$ mi)	Suitability for designated use (%)[a] FS	S/T	PS	NS	NA
Aquatic life support	6.4	60	8	23	8	<1
Fish consumption	3.8	84	1	14	2	<1
Primary contact: swimming	4.3	76	3	10	10	<1
Secondary contact: boating	2.6	78	2	16	4	<1
Drinking water supply	1.9	79	5	10	6	<1
Agriculture	3.1	93	<1	3	3	<1

[a]Percentage of surveyed length of rivers and streams that support designated use: FS = fully supporting, S/T = supporting/threatened, PS = partially supporting, NS = not supporting, NA = not attainable.
Source: USEPA, 1996a.

Table 6.A.4 Major Pollutants and Processes That Impair River and Stream Water Quality in the United States

Pollutant or process	Length impaired ($\times$1000 mi)[a]	Proportion impaired[b]
Siltation	127	18%
Nutrients	98	14
Pathogens	80	12
Oxygen-depleting substances	73	10
Pesticides	52	7
Other habitat alterations	47	7
Suspended solids	45	7
Metals	40	6

[a]A single segment of river or stream may be impacted by more than one pollutant or process.
[b]Percentage of monitored river and stream length whose water quality is impaired by the indicated pollutant or process.
Source: USEPA, 1996a.

Dissolved Oxygen (DO) Sag Curve

Municipal wastewaters and many industrial wastewaters contain high concentrations of organic material. When biodegradable organics are discharged into a river, aerobic biodegradation of the organics consumes dissolved oxygen (see §3.D.5). The concentration of dissolved oxygen downstream of a wastewater discharge is controlled primarily by the competing processes of oxygen consumption due to biodegradation of organic pollutants and oxygen replenishment due to aeration at the water surface. Consequently, the concentration of oxygen downstream of a large pollutant discharge tends to decrease for a distance before recovering, as depicted in Figure 6.A.2.

Key issues associated with oxygen depletion include the minimum oxygen concentration that occurs downstream and the distance to that minimum. Streeter and Phelps (1925) developed a model to predict these values for an idealized river. This

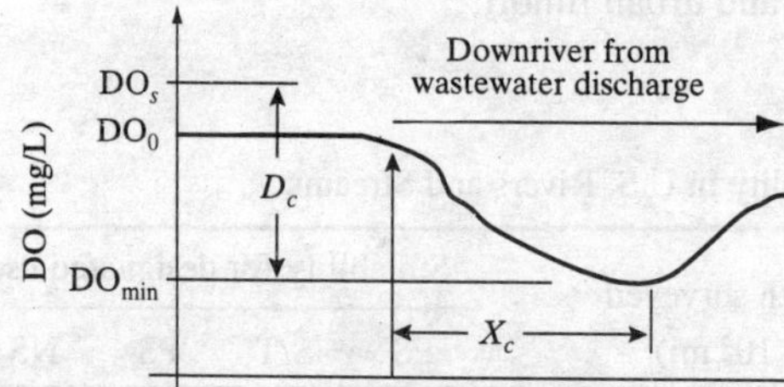

Figure 6.A.2 Idealized depiction of the dissolved oxygen (DO) concentration downstream of a wastewater discharge. The minimum oxygen concentration is related to the critical oxygen deficit, D_c, by $\text{DO}_{min} = \text{DO}_s - D_c$, where DO$_s$ is the dissolved oxygen content of the water at equilibrium with the atmosphere. The distance downstream of the discharge at which D_c occurs is denoted as X_c, the critical distance.

model provides a nice illustration of the integration of transport and transformation processes in a reactor model.

In the model, the consumption of oxygen is proportional to the biochemical oxygen demand (BOD) (see equation 3.D.29):

$$\left(\frac{d(\mathrm{DO})}{dt}\right)_{\text{reaction}} = r_{\mathrm{O}_2} = -k_1\,\mathrm{BOD} \tag{6.A.1}$$

This expression describes the change in dissolved oxygen content, DO, per time caused by biochemical reactions alone. Here BOD represents the current concentration of biodegradable organics (measured in terms of the biochemical oxygen demand) and k_1 is the reaction rate constant.

As shown in Figure 6.A.2, we will let x be the downstream coordinate, with the origin occurring at the wastewater discharge point. If we assume instantaneous and complete mixing of the wastewater and the river at the point of discharge, then species concentrations immediately downstream of the origin are the flow-weighted averages of the concentrations in the river and wastewater:

$$\mathrm{DO}(0) = \frac{\mathrm{DO}_r \times Q_r + \mathrm{DO}_w \times Q_w}{Q_r + Q_w} \tag{6.A.2}$$

$$\mathrm{BOD}(0) = \frac{\mathrm{BOD}_r \times Q_r + \mathrm{BOD}_w \times Q_w}{Q_r + Q_w} \tag{6.A.3}$$

In these equations, the subscripts r and w refer to the river and wastewater, respectively, upstream of the discharge point. Flow rates are indicated by Q. Downstream of the discharge point, the total flow rate is $Q_r + Q_w$.

BOD decays exponentially as the water flows downstream (see equation 3.D.28):

$$\mathrm{BOD}(t) = \mathrm{BOD}(0)e^{-k_1 t} \tag{6.A.4}$$

where the time coordinate, t, can be replaced by the transit time for water to flow a distance x downstream:

$$t = \frac{x}{U} \tag{6.A.5}$$

Here U is the flow velocity downstream of the discharge, given by $(Q_r + Q_w)/A$, where A is the vertical cross-sectional area of the river. Substituting equation 6.A.5 into 6.A.4, and the result into 6.A.1, we get

$$r_{\mathrm{O}_2} = -k_1\,\mathrm{BOD}(0)e^{-k_1 x/U} \tag{6.A.6}$$

Oxygen is supplied to the river by reaeration from the atmosphere. The reaeration rate is expressed using the two-film model (see §4.C.2):

$$J_{gl} = k_{gl}(\mathrm{DO}_s - \mathrm{DO}) \tag{6.A.7}$$

Here J_{gl} is the flux of oxygen across the air-water interface (mass per area per time), k_{gl} is the mass transfer coefficient (length per time), and DO_s is the saturation concentration of dissolved oxygen in water exposed to air.

We can now derive the final form of the model by assuming that the river acts as an ideal plug-flow reactor. We write a steady-state material balance on dissolved oxygen in a slice of the river, as depicted in Figure 6.A.3:

$$UA\,\mathrm{DO}(x) - UA\,\mathrm{DO}(x + \Delta x) + J_{gl}W\,\Delta x + r_{\mathrm{O}_2}A\,\Delta x = 0 \tag{6.A.8}$$

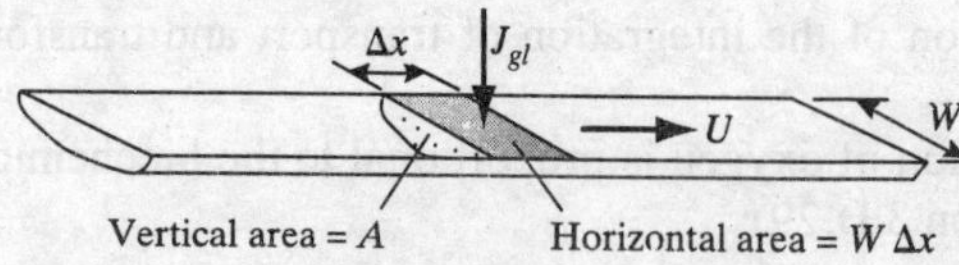

Figure 6.A.3 Control volume for material balance on dissolved oxygen in an idealized river, modeled as a steady-state plug-flow reactor.

From left to right, the terms of equation 6.A.8 describe the rates of advection in, advection out, reaeration, and consumption due to BOD. Each term has dimensions of mass (of oxygen) per time.

Next, we divide both sides of the equation by $A\,\Delta x$, take the limit as Δx approaches zero, and rearrange to obtain this governing differential equation for dissolved oxygen:

$$U\frac{d(\mathrm{DO})}{dx} = J_{gl}\frac{W}{A} + r_{O_2} \tag{6.A.9}$$

Substituting from equations 6.A.6 and 6.A.7, we obtain

$$U\frac{d(\mathrm{DO})}{dx} = k_{gl}\frac{W}{A}(\mathrm{DO}_s - \mathrm{DO}) - k_1\,\mathrm{BOD}(0)e^{-k_1x/U} \tag{6.A.10}$$

To simplify the form of the governing equation, let's define two new parameters: the oxygen deficit $D = \mathrm{DO}_s - \mathrm{DO}$, and the reaeration constant (units: inverse time) $k_2 = k_{gl}W/A$. Since DO_s is not a function of x, $dD = -d(\mathrm{DO})$. Substituting, we can rewrite equation 6.A.10 as

$$U\frac{dD}{dx} = -k_2D + k_1\,\mathrm{BOD}(0)e^{-k_1x/U} \tag{6.A.11}$$

Given the boundary condition that at $x = 0$, the oxygen deficit is $D_0 = \mathrm{DO}_s - \mathrm{DO}(0)$, we can integrate equation 6.A.11 to obtain an expression for the dissolved oxygen deficit at any distance x downstream of the pollutant discharge. The result is

$$D(x) = \frac{k_1\mathrm{BOD}(0)}{k_2 - k_1}\left(e^{-k_1x/U} - e^{-k_2x/U}\right) + D_0e^{-k_2x/U} \tag{6.A.12}$$

To calculate the minimum dissolved oxygen concentration (or maximum oxygen deficit), we can set the derivative of the deficit equal to zero ($dD/dx = 0$) and solve equation 6.A.11. This yields the critical deficit in terms of the critical distance downstream, X_c:

$$D_c = \frac{k_1}{k_2}\,\mathrm{BOD}(0)e^{-k_1X_c/U} \tag{6.A.13}$$

The critical distance is obtained by differentiating equation 6.A.12 with respect to x, setting the result equal to zero, and solving for $x = X_c$:

$$X_c = \frac{U}{k_2 - k_1}\ln\left[\frac{k_2}{k_1}\left(1 - \frac{D_0(k_2 - k_1)}{k_1\mathrm{BOD}(0)}\right)\right] \tag{6.A.14}$$

Example 6.A.1 illustrates the application of these equations.

EXAMPLE 6.A.1 ***Protecting Largemouth Bass Downstream of a Wastewater Discharge***

The lovely town of River City is situated along a scenic stretch of the Sweet-N-Pure River. The city is seeking a waiver from the federal government to permit it to discharge

EXAMPLE 6.A.1 ***Protecting Largemouth Bass Downstream of a Wastewater Discharge (continued)***

its partially treated wastewater into the Sweet-N-Pure River. The regulators will give permission for the discharge if the city can show that the largemouth bass population in the river is sufficiently protected. This species of bass requires a minimum of 3 mg/L of dissolved oxygen to survive. Given the following information about the wastewater discharge and the river, estimate the minimum downstream dissolved oxygen concentration to determine whether or not the city should be compelled to upgrade its treatment system.

Data

Maximum steady-state waste discharge	$Q_w = 0.5\ \text{m}^3/\text{s}$
Minimum steady-state river flow	$Q_r = 9\ \text{m}^3/\text{s}$
Average cross-sectional river area	$A = 20\ \text{m}^2$
DO content in the wastewater	$\text{DO}_w = 0.2$ mg/L
DO content of the upstream river	$\text{DO}_r = 8$ mg/L
DO saturation concentration at 20 °C	$\text{DO}_s = 9.0$ mg/L
BOD concentration in the wastewater	$\text{BOD}_w = 80$ mg/L
BOD concentration in the upstream river	$\text{BOD}_r = 0.1$ mg/L
BOD decay constant	$k_1 = 0.2\ \text{d}^{-1}$
Reaeration constant for the river	$k_2 = 0.4\ \text{d}^{-1}$

SOLUTION First, calculate the concentration of DO and BOD in the mixture of river and wastewater:

$$\text{DO}(0) = \frac{(8\ \text{mg/L} \times 9\ \text{m}^3/\text{s}) + (0.2\ \text{mg/L} \times 0.5\ \text{m}^3/\text{s})}{9\ \text{m}^3/\text{s} + 0.5\ \text{m}^3/\text{s}} = 7.6\ \text{mg/L}$$

$$\text{BOD}(0) = \frac{(0.1\ \text{mg/L} \times 9\ \text{m}^3/\text{s}) + (80\ \text{mg/L} \times 0.5\ \text{m}^3/\text{s})}{9\ \text{m}^3/\text{s} + 0.5\ \text{m}^3/\text{s}} = 4.3\ \text{mg/L}$$

The oxygen deficit at the point of discharge is given by

$$D_0 = \text{DO}_s - \text{DO} = 9\ \text{mg/L} - 7.6\ \text{mg/L} = 1.4\ \text{mg/L}$$

The river velocity downstream of the discharge, U, can be calculated as follows:

$$U = (Q_r + Q_w)/A = (9.5\ \text{m}^3/\text{s})/20\ \text{m}^2 = 0.48\ \text{m s}^{-1} = 41\ \text{km d}^{-1}$$

We can then calculate the critical oxygen deficit and the distance downstream to that deficit using equations 6.A.13 and 6.A.14:

$$X_c = \frac{41\ \text{km d}^{-1}}{0.4\ \text{d}^{-1} - 0.2\ \text{d}^{-1}} \ln\left[\frac{0.4\ \text{d}^{-1}}{0.2\ \text{d}^{-1}}\left(1 - \frac{(1.4\ \text{mg/L})(0.4\ \text{d}^{-1} - 0.2\ \text{d}^{-1})}{(0.2\ \text{d}^{-1})(4.3\ \text{mg/L})}\right)\right] = 61\ \text{km}$$

$$D_c = \frac{0.2\ \text{d}^{-1}}{0.4\ \text{d}^{-1}}(4.3\ \text{mg/L}) \exp(0.4\ \text{d}^{-1} \times 61\ \text{km}/41\ \text{km d}^{-1}) = 3.9\ \text{mg/L}$$

From these model calculations, the minimum concentration of dissolved oxygen in the river would be $\text{DO}_s - D_c = 9 - 3.9 = 5.1$ mg/L. The minimum would occur 61 km downstream of the wastewater discharge. Therefore, River City should be able to get its waiver without upgrading its wastewater treatment system.

6.A.2 Lakes and Reservoirs

The characteristics of a lake are defined primarily by its morphology (shape, depth, volume), geology, light, temperature, and wind mixing. These physical characteristics, along with the biological productivity of the lake, control the water quality (Horne and Goldman, 1994). Lake productivity is controlled by the amounts of nutrients available to support algae, plant, and animal life within the lake. Lakes with small amounts of available nutrients are called *oligotrophic* and are generally characterized by clear water with little oxygen demand. Because of this, oligotrophic lakes generally have high oxygen concentrations at all depths and are often capable of supporting cold-water fish that require high oxygen levels in the lower depths. *Eutrophic* lakes, on the other hand, have high levels of nutrients that result in enhanced algal growth, turbid water, high oxygen demand, and, consequently, low dissolved oxygen concentrations in the lower depths. Although eutrophication of lakes may occur as a natural aging process due to the deposition of silts and other deposits over geologic time, human activities can markedly speed the process by discharging high concentrations of nutrients and oxygen-demanding wastes into lakes.

The water quality and subsurface ecology of lakes and reservoirs can be significantly affected by the transport of oxygen, nutrients, organic matter, and microorganisms from the surface to deep waters. The temperature profile within the water body is pivotal in controlling this transport. Thermally stratified water bodies can be characterized in terms of three zones: the near-surface, warm-water layer called the *epilimnion*; the deep, cold-water layer called the *hypolimnion*; and the *thermocline** region connecting them, in which the temperature changes rapidly with depth (Figure 6.A.4[*a*]). Lakes and reservoirs often take on this type of thermal stratification in the summer, when intense energy from the sun and high air temperatures heat the epilimnion waters sufficiently to produce a warm, buoyant, stable layer. The greater density of the cooler waters in the hypolimnion adds to the stability of the stratification. In typical clear lake water, 50 percent of the incident sunlight is absorbed in the upper 2 m, and very little light energy penetrates 10 m below the surface (Wetzel, 1975). In turbid

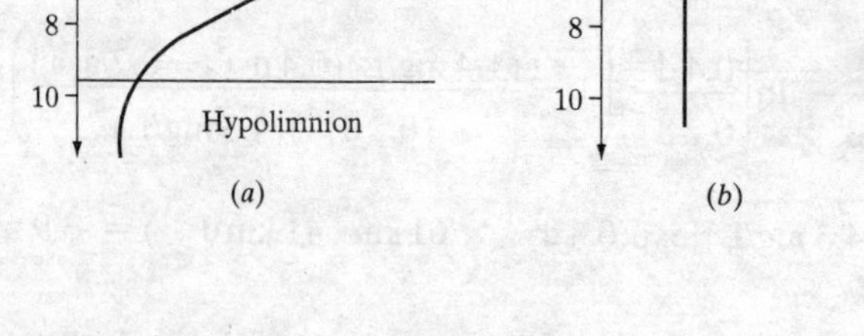

Figure 6.A.4 Temperature profile in a stably stratified lake: (*a*) summer conditions and (*b*) winter conditions.

*These terms can be most easily remembered by connecting them to their Greek roots. The prefixes *epi* and *hypo* mean "outer" and "beneath," respectively. (Think of "epilogue" and "hypodermic.") *Limnion* refers to lakes: "Limnology" is the study of lakes and other freshwater aquatic systems. *Thermocline* combines two terms: *Thermo* refers to temperature (as in "thermometer"), and *cline* refers to slope (as in "incline").

waters, the absorption of light energy is even more concentrated near the upper surface. Mixing caused by wind action tends to cause the epilimnion to be a well-mixed layer, although in periods of weak winds a stable temperature profile can be established within this layer.

The thermocline occurs at a depth that is near the lower limit of wind-induced mixing and sunlight-generated heat penetration. As long as the stratification remains, the hypolimnion is effectively isolated from the surface, separated from the epilimnion by a strongly stable thermocline. Little sunlight or heat penetrates to the hypolimnion, and mass transfer between the epilimnion and hypolimnion is slow.

As summer wanes, air temperature and sun intensity decrease, the water in the epilimnion cools, and its density increases. Eventually, when the epilimnion cools to a temperature close to that of the hypolimnion, surface winds can cause the layers to mix, destroying the thermal stratification. This phenomenon is called a *turnover* of the lake and can be important in transporting dissolved oxygen from the surface to deep waters and nutrients from deep to shallow waters.

Although lakes can remain well mixed throughout the winter in warm climates, in cold climates a new stable stratification may be established by the formation of ice at the lake surface (Figure 6.A.4[*b*]). In this case, the ice thermally insulates the lake and physically isolates the water from the mixing effects of winds, resulting in a weak but persistent stratification. In the spring the ice melts and the surface waters warm, gradually reestablishing the summer stratification.

Except when the surface is frozen, wind plays an important role in the mixing of water in lakes. As wind blows across the surface, frictional forces cause the surface waters to flow in the same direction as the wind. The water cannot accumulate on the downwind shore, so a return flow, in the direction opposite the wind, is established beneath the surface. For thermally stratified lakes in which the wind-induced flows are insufficient to overcome the stable stratification across the thermocline, the return flow can occur in the lower portion of the epilimnion (Figure 6.A.5). A weaker mixing cell can then be established in the hypolimnion by movement of water at the thermocline. Otherwise, the flow cycle may penetrate through the full depth of the water, causing mixing throughout the lake.

Figure 6.A.6 shows measurement data from a lake to illustrate some of these points. Temperature profiles as a function of depth are plotted at four times during a year for Lake Tahoe on the California-Nevada border (Goldman and Carter, 1965). The February profile shows an unstratified lake, with uniformly cool waters throughout. The profile in May shows moderate stratification with the approach of summer. At summer's end in September, the lake is at its warmest, and a clear delineation can be seen at 25 m between the epilimnion and the thermocline. The relatively thick epilimnion in this case results from two factors. The extreme clarity of the lake permits sunlight to penetrate and heat the surface water through many meters. In addition, the

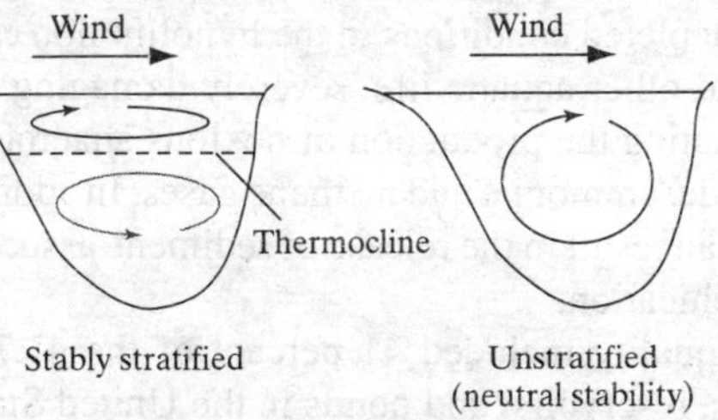

Figure 6.A.5 Wind-induced mixing in a thermally stratified lake compared with an unstratified lake.

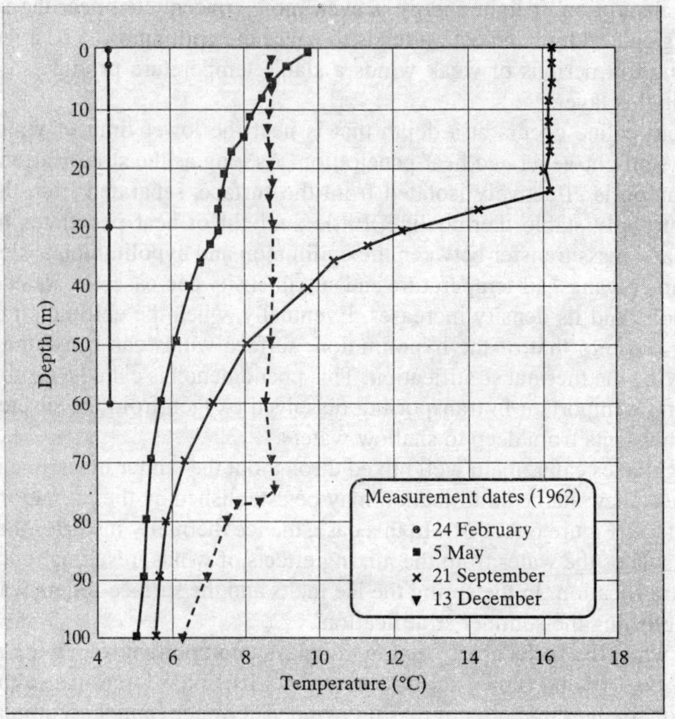

Figure 6.A.6 Measured temperature profiles in Lake Tahoe, California and Nevada. (*Source*: Goldman, C.R., & Carter, R.C. 1965. An investigation by rapid carbon-14 bioassay of factors affecting the cultural eutrophication of Lake Tahoe, California–Nevada. *Journal of the Water Pollution Control Federation*, **37**, 1044–1059. Copyright © Water Environment Federation. Reprinted with permission.)

lake's large size allows winds to induce strong flows at the surface, which causes mixing of the heated water to greater depths. The boundary between the thermocline and the hypolimnion is not sharp but occurs in the vicinity of 60 m depth. In the December profile, the lake is nearly unstratified, with a small thermocline at 75 m. The investigators attribute this deep thermocline position to a strong storm that had occurred in October.

In thermally stratified lakes, transport of oxygen from the surface to deep waters can be hindered. Consequently, if there is significant oxygen demand in the hypolimnion, anaerobic conditions in that layer may result. Oxygen demand in the depths of lakes and reservoirs can occur as a result of particulate organic matter settling from the epilimnion when sufficient nutrients are available to support significant photosynthetic algal growth (eutrophic conditions). Organic matter from pollutant discharges may cause similar effects. Oxygen-depleted conditions in the hypolimnion can result in the reduction of cold-water fish and other aquatic life, severely damaging the food chain within the ecosystem and promoting the production of noxious anaerobic degradation products such as hydrogen sulfide, ammonia, and methane gases. In addition, anaerobic conditions in the hypolimnion can result in the release of sediment-associated nutrients, accelerating the cycle of eutrophication.

A recent survey of water quality included 41 percent of the 41.7 million acres (16.9 million hectares) of lakes, reservoirs, and ponds in the United States (excluding

Table 6.A.5 Summary of Water Quality in U.S. Lakes, Reservoirs, and Ponds

Designated use	Area surveyed ($\times 10^6$ acres)	Suitability for designated use (%)[a]				
		FS	S/T	PS	NS	NA
Aquatic life support	14.2	55	14	25	6	<1
Fish consumption	10.9	60	5	32	3	<1
Primary contact: swimming	15.4	63	12	21	4	<1
Secondary contact: boating	8.3	62	13	23	2	<1
Drinking water supply	8.5	81	10	7	1	0
Agriculture	4.0	84	5	10	1	0

[a]Percentage of surveyed acreage of lakes, reservoirs, and ponds that support designated use: FS = fully supporting, S/T = supporting/threatened, PS = partially supporting, NS = not supporting, NA = not attainable.

Source: USEPA, 1996a.

the Great Lakes).* The results indicate that 60 percent of the water bodies were of sufficient quality to support all designated uses such as aquatic life support, swimming, fishing, and water supply, while 40 percent failed to support one or more designated uses (Table 6.A.5).

The pollutants that most commonly impair water quality in freshwater bodies in the United States are nutrients and metals (Table 6.A.6). Nutrients, primarily nitrogen and phosphorus, increase lake productivity and can hasten eutrophication as described above. Significant nutrient sources include agricultural runoff, industrial and municipal wastewater discharges, and atmospheric deposition. Most reports of metal contamination in lakes are a result of the detection of mercury in fish tissue. Although it is difficult to detect trace concentrations of mercury in water, the bioaccumulative nature of this toxic metal causes it to become concentrated in fish. The major source of mercury contamination in lakes is thought to be atmospheric transport and deposition from electrical power stations.

Table 6.A.6 Major Pollutants and Processes That Impair Water Quality in Lakes, Reservoirs, and Ponds in the United States

Pollutant or process	Area impaired ($\times 10^6$ acres)	Proportion impaired[a]
Nutrients	3.3	20%
Metals	3.3	20
Siltation	1.6	10
Oxygen-depleting substances	1.4	8
Noxious aquatic plants	1.0	6
Suspended solids	0.9	5
Total toxics	0.9	5

[a]Percentage of monitored area of lakes, reservoirs, and ponds whose water quality is impaired because of the indicated pollutant or process.

Source: USEPA, 1996a.

*Perhaps for simplicity, the USEPA reports lake, reservoir, and pond sizes in acres, a unit of area, rather than in terms of volume.

6.A.3 Groundwater

Water that percolates into the ground from rainfall, snowmelt, rivers, streams, lakes, and ponds passes through a partially air-filled region called the *unsaturated* or *vadose* zone and enters a water-filled or *saturated* zone that is referred to as the *aquifer* (see Figure 6.A.7). The top of the saturated zone is the *water table*, and the region immediately above the water table, referred to as the *capillary fringe*, contains water that rises into the solid soil matrix by means of capillary action. Water within the saturated zone can be extracted by pumping from wells. Water within the capillary fringe is generally unavailable for withdrawal.

It is estimated that of the 2–4 percent of water on earth that is freshwater (rather than saltwater), three-fourths is frozen in polar ice and glaciers. Of the remaining quarter, groundwater represents more than 98 percent. However, not all groundwater is available for use since the majority is in aquifers that are too deep to economically exploit.

The major use of groundwater in the United States is for agricultural irrigation (Figure 6.A.8); in addition, 50 percent of the population, including 95 percent of the rural populace, rely on groundwater as their primary drinking water source.

The amount of water that is available for pumping from any given aquifer is controlled by many factors, including the aquifer thickness, the permeability of the soil or rock matrix, and the presence of impermeable confining layers. When water is pumped from an aquifer at rates that exceed the rate of resupply by infiltration, several outcomes are possible. The well may run dry, necessitating relocation to a deeper or

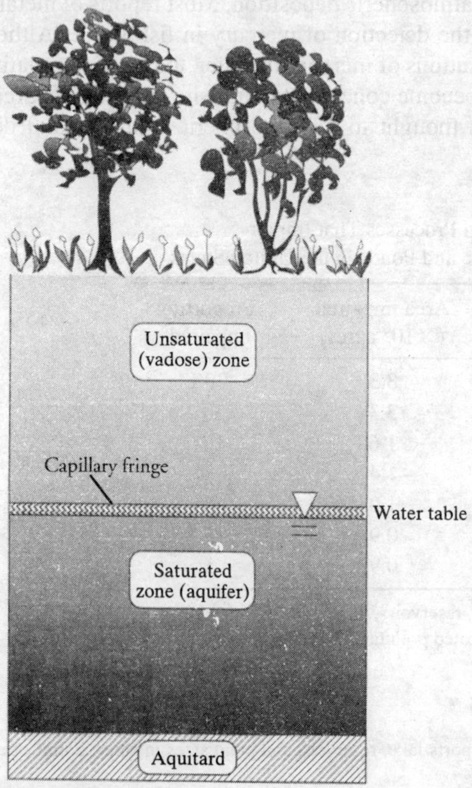

Figure 6.A.7 Schematic of a subsurface aquifer.

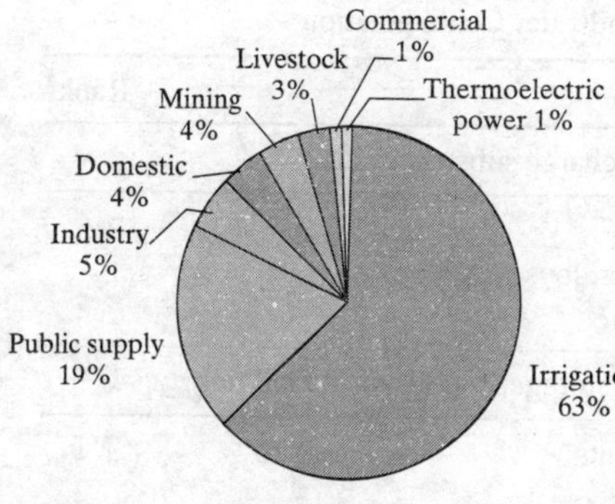

Figure 6.A.8 Groundwater uses in the United States (USEPA, 1996a).

more productive aquifer. In addition, the aquifer formation may partially collapse, causing land subsidence. This has been the case in the Central Valley of California, where land elevations have dropped more than 10 m in the past several decades due to excessive withdrawals from the aquifer for agricultural irrigation. When groundwater is extracted along ocean coasts, saltwater may infiltrate the aquifer in place of the withdrawn freshwater in a process called saltwater intrusion. In this case, the progressively decreasing quality of the withdrawn well water generally results in discontinued or reduced use of the affected wells.

Although the solid matrix through which groundwater flows can act as a filter, effectively removing particulate matter and some other types of contamination from the produced water, increasing amounts of groundwater contamination by dissolved constituents have been detected in the past decades. Since groundwater demand is expected to increase significantly in the future as urban populations continue to grow, groundwater contamination, which often occurs in heavily populated areas, represents a significant threat to water supplies. There are many potential sources of groundwater contamination, including landfills, septic tanks, above- or belowground storage tanks, irrigation flows, and urban runoff (Table 6.A.7). In the United States, the predominant source of groundwater contamination is hazardous waste releases, which are discussed in detail in Chapter 8. Common groundwater contaminants include inorganic compounds such as nitrate, chromium, and lead, and organic compounds such as gasoline constituents and chlorinated solvents. Recently emerging groundwater concerns include the fuel additive MTBE, discussed in the beginning of this chapter, and perchlorate.

For decades the U.S. Air Force has used ammonium perchlorate (NH_4ClO_4) as an oxidizer in solid rocket propellants. Perchlorate interferes with hormone production in the thyroid when present at very low concentrations, making it a serious human health threat. The extensive use of rocket propellants and inadvertent releases during their production, storage, and utilization have resulted in the widespread distribution of the highly soluble perchlorate in aquifers and drinking water sources. Since perchlorate is recalcitrant to chemical reduction, biological methods for its removal from groundwater are currently being sought. Several methods for groundwater remediation are described in §8.E.3.

Organic groundwater contaminants often exist in the form of nonaqueous-phase liquids (NAPLs), also known as "free product." To understand what a NAPL is, think about a bottle filled half with water and half with oil. When the bottle is vigorously shaken, the two liquids briefly mix, but when shaking is discontinued, the liquids quickly separate back into two discrete phases. Although both the oil and water are liquids, they are not miscible, and we therefore refer to the oil as a NAPL. In addition,

Table 6.A.7 Common Sources of Groundwater Contamination

Contamination source	Rank[a]
Sources designed to discharge substances	
Septic tanks and cesspools	4
Injection wells of hazardous and non-hazardous waste	
Land application of wastewater and sludge	
Unplanned releases from treatment, storage, and disposal facilities	
Hazardous waste landfills or impoundments	1
Sanitary landfills	2
Aboveground and underground storage tanks and pipes	5
Waste piles	
Releases that arise as a consequence of other activities	
Pesticide or fertilizer applications	3
Animal waste disposal	6
Mine drainage	7
Oil drilling	8
Saltwater intrusion	9
Irrigation return flow	10
Urban runoff	
Atmospheric deposition	
Sources designed to transport substances	
Pipelines	
Trucks, trains, ships	

[a]Rankings indicate estimated order of importance for impacting groundwater in the United States.

Source: Adapted from USEPA, 1994.

the oil is less dense than the water, as demonstrated by the fact that it floats on the water in the bottle. Therefore, we refer to oil as a light NAPL, or LNAPL. Other LNAPLs include petroleum hydrocarbon fuels such as gasoline, diesel, and jet fuel. Some compounds form dense NAPLs, or DNAPLs, which are denser than water. Examples include many chlorinated solvents and PCBs. When NAPLs are released into the environment, they are drawn into the subsurface by gravity and capillary action within the porous media. NAPLs can also be transported into the subsurface by the downward movement of water from precipitation events. LNAPLs tend to migrate downward through the unsaturated subsurface until they reach an impermeable layer or the water table, where they spread laterally (Figure 6.A.9). DNAPLs released to the subsurface tend to migrate downward through both the unsaturated and saturated zones until they reach an impermeable barrier such as a clay or consolidated rock layer.

In the subsurface, NAPL contamination dissolves into groundwater, volatilizes into pore gases, and sorbs onto solids. Since the partitioning of NAPL contamination into other phases is generally a slow process that is limited by mass-transfer kinetics, NAPLs can be long-term sources of groundwater contamination. Movement of NAPLs in the subsurface is governed by a combination of gravitational and capillary forces, described by models that are beyond the scope of this text. The movement of

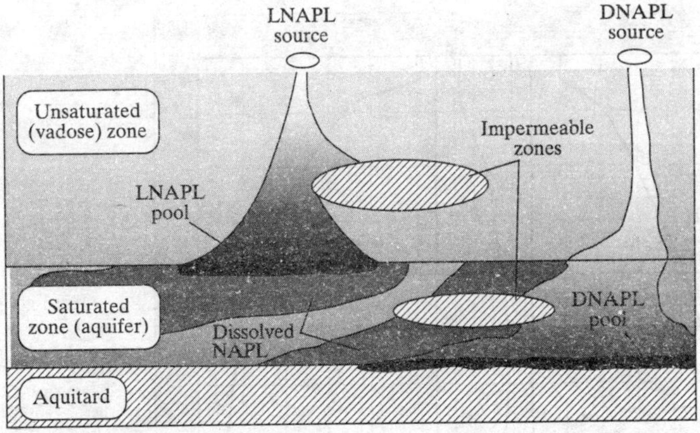

Figure 6.A.9 Idealized migration patterns of LNAPL and DNAPL releases into the subsurface.

dissolved contaminants in groundwater can be modeled by accounting for advection, dispersion, sorption, and decay as described in §4.D.2 and §5.B.2. Methods for remediation of groundwater NAPL contamination are discussed in §8.E.

Although most groundwater contamination of concern is due to human activity, there are also natural groundwater contaminants, such as magnesium, chloride, boron, selenium, and arsenic, which dissolve from aquifer geological formations. In some cases, natural groundwater contamination can be severe. For example, millions of adults and children in Bangladesh have become acutely or chronically ill within the last decade due to arsenic poisoning from groundwater withdrawn from the prevailing geological formations in parts of that country that are naturally high in arsenic salts. This occurrence has been described as one of the biggest mass poisonings in history. Ironically and tragically, the wells that supply this groundwater were recently dug (within the past 20 years) to address the widespread occurrence in Bangladesh of disease caused by microbial pathogens in low-quality surface-water supplies.

6.A.4 Oceans and Estuaries

Oceans cover three-quarters of the earth's surface and contain water that is far too saline for human consumption. The less saline (brackish) coastal waters that are found where rivers meet the oceans are called *estuaries* and are generally subject to tidal action. The high dissolved-solids concentrations of estuarine waters typically make them impractical for use as a water supply. However, estuarine waters in bays and tidal rivers, along with adjacent wetlands, are extremely productive habitats that support large populations of commercial fish, shellfish, and wildlife.

In the oceans, both temperature and salinity affect mixing. Temperature profiles are controlled by means of the same processes that operate in lakes. Salinity in surface layers is affected by the relative amounts of evaporation and precipitation. Regions with excess evaporation tend to have higher salinities, and regions with excess precipitation tend to have lower salinities. Because oceans are so much larger than lakes, they are less prone to seasonal changes in temperature profiles. In the open oceans at the mid-latitudes, temperature and salinity profiles are relatively stable over long periods, with small perturbations (2–8 °C) occurring every 3 to 4 years in a phenomenon referred to as "El Niño."

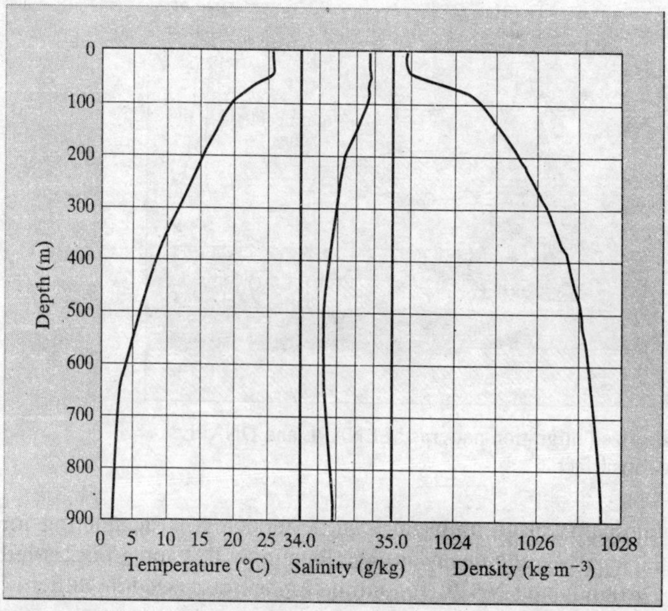

Figure 6.A.10 Temperature and salinity profiles measured in the mid-Pacific (Gregg, 1973). Below 4 km, the Pacific has a stable temperature of 1 °C and salinity of 34.7 g/kg. The density was computed from the temperature and salinity measurements using the approach and data presented by Fischer et al. (1979). The temperature and salinity profiles are reproduced with permission of M.C. Gregg.

Figure 6.A.10 shows measured temperature and salinity as a function of depth in the mid-Pacific. The water density, computed from the temperature and salinity, shows that the surface layer is well mixed down to about 50 m, below which a thick thermocline is present that extends down to about 800 m below the surface. In the open ocean, temperature variations are much greater than salinity variations and therefore have a much more significant impact on water density. The density profile approximately mirrors the temperature profile.

Wind-induced mixing of the warm, shallow surface layer of the ocean causes nutrients to be transported from the cooler, deep waters, providing sustenance for phytoplankton and supporting healthy fish populations. During El Niño years, however, the thickness of the warm surface layer can be increased significantly so that wind-induced mixing is not sufficient to cause effective upwelling of nutrients, with the result that phytoplankton and fish populations can be decimated.

Although salinity has only a modest influence on mixing in the open oceans, it is a dominant contributor to mixing in estuaries. The salinity in an estuary ranges from effectively zero to the value of undiluted ocean water, ~35 g kg^{-1}. The corresponding range of water densities, from 1000 to ~1025 kg m^{-3} is roughly five times larger than the range caused by typical environmental temperature differences. Consequently, salinity profiles act more prominently than temperature profiles in estuarine stratification.

The water flow characteristics in an estuary are subject to two predominant influences: freshwater flow from the upstream river and tidal flow from the ocean. The tidal flow oscillates twice per day, as the tides rise and fall. The river flow is usually much steadier, varying on a time scale of weeks or longer.

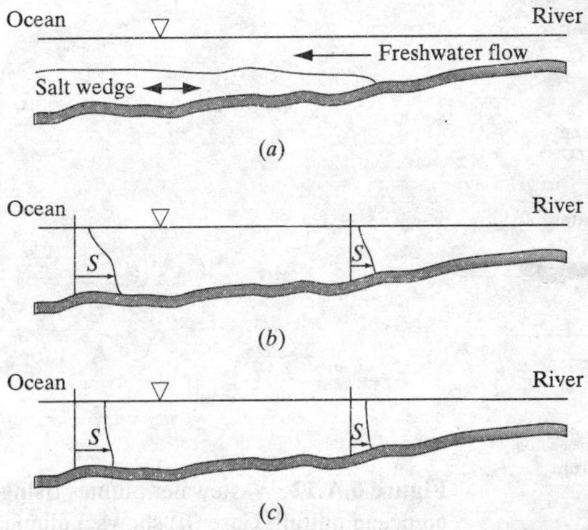

Figure 6.A.11 Stratification in an estuary caused by variable salinity. Case (*a*) shows a fully stratified estuary, case (*b*) shows a partially stratified system, and case (*c*) depicts a well-mixed, less stratified system. *S* represents the salinity (mass of salts per mass of water). (Reproduced with permission of Academic Press, Inc., and N.H. Brooks from H.B. Fischer et al. *Mixing in Inland and Coastal Waters*, Academic Press, Orlando [1979].)

If the flow velocities are low, so that the turbulence within the mixing waters is modest, the freshwater will tend to flow over the top of the saltwater, leading to strongly stable density stratification, as shown in Figure 6.A.11(*a*). Here the salinity (*S*) varies with depth from zero at the surface to ~35 g kg^{-1} at the bottom. At the other extreme, if the flows are intense and turbulence is high, strong vertical mixing can occur such that the salinity varies only weakly in the vertical direction (Figure 6.A.11[*c*]). Partial stratification, the intermediate case, is depicted in Figure 6.A.11(*b*).

Since estuarine and ocean waters are generally too high in dissolved solids to serve as sources of water for drinking or agricultural irrigation, the protected designated uses for these waters include support of aquatic life, harvestable fish and shellfish, swimming, and boating. A very important water quality issue in estuaries is excess nutrients such as nitrogen and phosphorus. As in lakes, nutrient discharges from human activities can promote enhanced growth of algae in estuaries, resulting in eutrophic conditions that can stifle fish growth and damage shellfish habitats. In the United States in the mid-1990s, 22 percent of surveyed estuarine waters were significantly impacted by excess nutrients (USEPA, 1996a). Excess bacterial concentrations and toxic organics are also important estuarine pollutants. In ocean waters, a commonly reported water quality issue is high bacterial concentrations. While it is recognized that pathogenic bacteria and viruses do not generally pose a threat to aquatic life, shellfish in particular can accumulate pathogens, causing disease when harvested and consumed. Therefore, the harvest and sale of shellfish from estuaries and ocean waters with excess bacteria are strictly regulated. The major sources of pollutants in estuaries include industrial discharges, urban runoff, and municipal point sources, in that order of importance; in ocean waters, urban runoff, septic systems, and municipal point sources are the major causes of contamination.

Ocean Outfalls

The predominant municipal point sources that discharge to oceans are outfalls, long pipes that typically extend for miles along the ocean floor, discharging wastewater into deep ocean waters. It is common practice for coastal cities to dispose of their treated municipal wastewater through such devices.

The major function of ocean outfalls is to minimize the potential for bacterial contamination on beaches due to wastewater discharge. They are designed to achieve

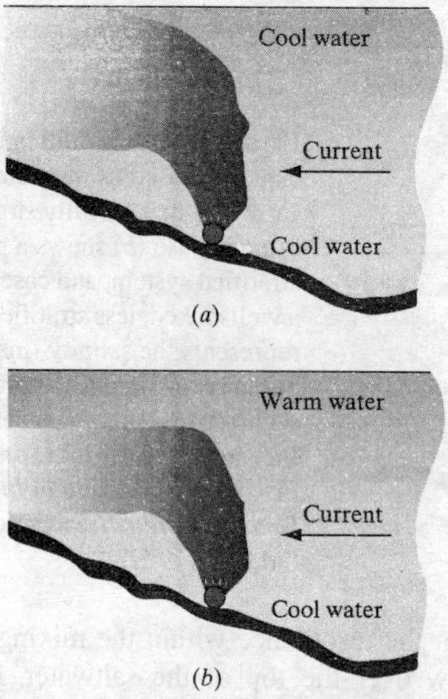

Figure 6.A.12 Wastewater plumes rising from an ocean outfall. Case (*a*) shows a plume ascending to the surface due to lack of ocean stratification or adequate mixing, and case (*b*) shows the more desirable trapped plume.

this goal based on three characteristics: distance, depth, and dilution. The farther offshore an outfall extends, the longer it takes for contaminants to be transported back to the beach. Therefore, extended distances result in increased transport times, which provide a greater opportunity for pathogens and contaminants to decay prior to human exposure. The offshore length of municipal outfalls in the United States is typically greater than several miles. The depth at which the discharge occurs also greatly affects potential beach pollution, since winds and currents facilitate contaminant transport in surface waters. Because the density of seawater is approximately 1030 kg m^{-3} while the density of wastewater is about 1000 kg m^{-3}, wastewater is buoyant in the ocean and tends to rise to the surface (see Figure 6.A.12). To minimize the potential for beach pollution, the discharged water must be sufficiently well mixed with the receiving seawater to ensure effective dilution and buoyancy control. If the buoyant wastewater can be sufficiently mixed with the deep, cold seawater prior to its ascension to the surface, then the mixture will be more dense than the surface water, and the diluted wastewater can be trapped below the surface (Figure 6.A.12[*b*]). This is advantageous because contaminants that make it to the surface are more easily transported to the shore.

Although outfalls assist in eliminating bacterial contamination of beaches, problems may be encountered in ocean bottom sediments near outfalls, including regions of low dissolved oxygen, high concentrations of hazardous organic chemicals such as DDT and PCBs, and toxic metals such as Pb, Cu, and Zn. These problems result from sludge particles present in the discharged waste. These particles coagulate in the seawater and settle near the outfalls. Wastewater sludge particles are generally characterized by high BOD and may also have high concentrations of toxic organics and metals. Past observations of fish with tumors and fin rot in the vicinity of ocean outfalls suggest that discharge of inadequately treated wastewater can result in significant habitat and fisheries damage. Currently in the United States, wastewater must be treated to significantly reduce BOD and particulate levels prior to discharge to prevent such problems.

6.B OVERVIEW OF WATER QUALITY REGULATIONS AND TREATMENT SYSTEMS

Historically, the design and operation of treatment systems applied to water supply and wastewater were the main activities of the branch of civil engineering known as sanitary engineering. The emergence of environmental engineering, with broader goals in water quality engineering as well as air quality and hazardous waste emphases, has not diminished the importance of water and wastewater treatment. Access to clean drinking water and satisfactory sanitation facilities remains an elusive goal for a large portion of the world's population. In industrialized countries, the public continues to demand water quality that contributes as little as possible to the risk of disease from exposure to pathogens and toxic chemicals. This demand is embodied in tough standards that are precipitating extensive changes in the practice of water and wastewater treatment engineering. Meanwhile, continued industrial and population growth is threatening water availability as well as water quality in many areas. The desire to treat poor-quality (e.g., brackish) water for use or wastewater for reuse continues to grow.

This section begins with an introduction to the most important regulations governing water and wastewater treatment in the United States. The remainder of this section presents an overview of technologies used for water and wastewater treatment, emphasizing municipal systems. In subsequent sections we will examine details of a number of specific treatment techniques. It is not our goal to provide a complete discussion of water and wastewater treatment. Rather, our central aim is to demonstrate how the principles developed in the first five chapters of this text can be applied to understanding the design and functioning of the most common treatment techniques.

Regardless of the specific process, the purpose of water and wastewater treatment is either to *separate* an impurity from the liquid stream or to *transform* the impurity from an undesirable form to a more desirable form. Thus, the themes of transport and transformation, emphasized in the first half of the text, reemerge here. If the treatment technique functions by separation, then the issue of how the removed material is subsequently treated or disposed of must also be considered.

Treatment processes for controlling water quality are generally applied at two points in the hydrologic cycle. *Water treatment* is applied to water taken from the environment before it is delivered to a user. *Wastewater treatment* is applied to water that has been used, before it is returned to the environment. Of course, the treatment objectives for water and wastewater differ. In addition, the treatment requirements vary according to both the input (the origin of the water/wastewater) and the output (the intended use of the water or the characteristics of the environment into which the wastewater will be discharged). For example, the amount and type of processing required for drinking water treatment vary according to the water source (see Table 6.B.1). In comparison with river water downstream of metropolitan centers (for example, the Mississippi River at New Orleans), groundwater may have high concentrations of dissolved minerals but low risk of contamination by pathogenic microorganisms. Water treatment also varies according to the intended use: The requirements for municipal drinking water are not the same as those for industrial use. Similarly, wastewater from a refinery or a manufacturing facility will have different characteristics and therefore will require different treatment techniques from those used for municipal wastewater. Whether the treated wastewater will be discharged to the ocean, to a lake, or to a river also impacts the types and levels of contamination permitted in the treated wastewater.

Table 6.B.1 General Characteristics of Water Sources

Source	Attributes
River	Water quality variable over periods of hours to days; susceptible to accidental spills; may have high turbidity and organic content; expect high turbidity following spring thaw and heavy rains; generally low mineral content
Lake/reservoir	Water quality seasonally variable; may have high turbidity and organic content; generally low mineral content; summer stratification can produce anoxic conditions at depth, causing high levels of dissolved iron and manganese and/or hydrogen sulfide
Groundwater	Water quality constant over time but variable with position; usually excellent quality for turbidity, microbial content, and organics; may have high levels of dissolved minerals, including hardness ions, iron, and manganese

Source: Hamann et al., 1990.

6.B.1 Key U.S. Federal Water Regulations

Outbreaks of acute waterborne diseases such as cholera and typhoid fever in the nineteenth and early twentieth centuries prompted the development of widespread water and wastewater treatment systems to improve water quality. Today, the regulations that govern water treatment are based on both health concerns and aesthetics. Water-related health concerns include both acute effects such as gastroenteritis caused by pathogenic microorganisms and chronic effects such as carcinogenicity caused by disinfection by-products. Aesthetic concerns for drinking water include hardness, color, turbidity, taste, and odor.

Waterways, such as the Great Lakes and the Cuyahoga and Chicago Rivers, were severely degraded during the 1960s and 1970s because of unchecked pollutant releases. This situation led to the passage in the United States of two important pieces of water quality legislation, the Clean Water Act (CWA) in 1972 and the Safe Drinking Water Act (SDWA) in 1974. Although the goal of these two acts is the same—to ensure the availability of high-quality water for beneficial uses—the approaches are quite different.

The CWA regulates the discharge of pollutants to rivers, lakes, estuaries, and wetlands by means of wastewater discharge permits and effluent standards. The main objective of the CWA is to protect the nation's surface water and groundwater at a level that can be considered "swimmable and fishable"; however, the ultimate goal of the CWA is to eliminate the discharge of pollutants into navigable waters. The CWA established the National Pollution Discharge Elimination System (NPDES), administered by the individual states, as a mechanism by which the quantity and quality of wastewater effluents are regulated for each individual discharger. In this manner, a particular water body receives a level of protection that is appropriate for its specific ecological condition, water use, and pollutant loading. For example, discharges to a pristine lake that is used as a drinking water source are regulated much more stringently than those to a heavily impacted river used as a major shipping lane. In addition, NPDES requirements may include "best available technology" (BAT) standards, such as the requirement of biological treatment for municipal wastewater. Specific contaminant limitations can be mandated along with BATs, for instance, biological treatment for organic removal combined with a 0.5 mg/L maximum concentration of phosphorus. Alternatively, multiple BATs may be combined, such as biological treatment for organic removal and precipitation for phosphorus removal. Wastewater regulations in the past focused mainly on mechanisms for maintaining sufficient dissolved oxygen in receiving waters by requiring biological treatment to limit BOD releases. More recently, additional constituents such as nutrients (nitrogen and phosphorus), suspended solids, pathogens, and toxic contaminants, such as metals, pesticides, and specific organics, have been regulated.

The SDWA establishes water quality standards for all public water systems that serve an average of 25 or more people daily. These standards include the national primary drinking water standards and the national secondary drinking water standards (see Appendix F). Primary standards are based on the protection of human health and include the enforceable maximum contaminant levels (MCLs) and the nonenforceable maximum contaminant level goals (MCLGs). MCLs are the maximum permissible levels of contaminants in water delivered to any user, while the MCLGs are the maximum levels of contaminants in drinking water for which no known or anticipated adverse health effects have been documented. In some cases, MCLs are replaced by treatment techniques (TTs), procedures that public water systems must employ to ensure contaminant control. Treatment techniques are applicable for lead and copper, organic flocculant aids, and pathogens. The treatment techniques for pathogens include the Surface Water Treatment Rule (SWTR) and the coliform standard. The SWTR requires that surface water, and groundwater that is under the direct influence of surface water, undergo some combination of disinfection and filtration to achieve the following criteria: (1) A minimum of 99.9 percent of *Giardia lamblia* and 99.99 percent of viruses are to be killed or inactivated. (2) The turbidity must be maintained at less than 5 nephelometric turbidity units (NTU) at all times and less than 1 NTU in 95 percent of daily samples. (3) The total heterotrophic bacterial level, as determined by plate count, must be less than 500 colonies per milliliter. The coliform standard requires that no more than 5 percent of water samples test positive for coliforms in a month, and of those positives, that there be no samples that test positive for *fecal* coliforms.

Secondary standards are nonenforceable guidelines that are based on potential adverse cosmetic effects, such as skin discoloration and laundry staining, or aesthetic effects, such as taste, odor, and color. These standards apply to specific water contaminants such as aluminum, copper, and iron, as well as general water characteristics such as odor, color, corrosivity, and total dissolved solids.

6.B.2 Engineered Water Quality Systems

A primary role of environmental engineers concerned with water and wastewater treatment can be described in this way:

1. Given certain characteristics of a water supply and certain water quality objectives, design, construct, and then operate a reliable and cost-effective water treatment plant.
2. Given certain characteristics of wastewater and certain standards governing the manner of discharge of this wastewater to the environment, design, construct, and then operate a reliable and cost-effective wastewater treatment plant.

The history of water and wastewater treatment was briefly summarized in §1.B.1. Many treatment technologies mimic or exploit natural processes. Engineered systems are generally optimized to function more efficiently than their natural counterparts. However, as often happens with a technological discipline that is driven by urgent human needs, the *science* of water and wastewater treatment engineering has lagged behind the *practice*. Although great strides have been made during the last few decades, there remains a lack of comprehensive fundamental understanding of some aspects of treatment technology. What understanding we do have is important to master because it can lead to improvements in practice. Much of the practice of water quality engineering is based on the following characteristics:

1. Use proven designs. If a technique works in one place, it commonly (but not always) will work elsewhere.
2. Conduct bench and pilot testing of proposed treatment processes. Test methods at one or more small scales before going to full-scale operation.
3. Use theory to guide the design and interpretation of bench and pilot tests. Apply the insight gained from prior practice, experiment, and theory as a basis for design.

Water Treatment

In 1996 there were over 170,000 public drinking water systems providing water to over 250 million people in the United States (USEPA, 1996b). Approximately 61 percent of the U.S. population were served by the 8 percent of those systems that rely on surface-water sources while the rest relied on groundwater sources. The per-capita public water withdrawal in the United States is estimated at 570 L d^{-1}. Although the corresponding worldwide number is much lower, it has increased approximately eightfold over the past century (Gleick, 1998).

A typical municipal water treatment plant for a surface-water supply is shown schematically in Figure 6.B.1. Water from the source is delivered to the treatment plant, typically through a pipe or open aqueduct. At the inlet to the plant, the water passes through coarse screens to remove any large objects such as logs, fish, or turtles. A pump may be used to lift the water to the highest elevation in the plant; flow through the rest of the plant can then be achieved by gravity. The total water throughput is measured with a meter near the inlet.

The function of the aeration stage is both to saturate the water with oxygen and to remove gases that may be objectionable (such as hydrogen sulfide or volatile organics) or that may interfere with other treatment stages (such as carbon dioxide, which increases the lime requirement for softening). The goal in the design of an aerator is to provide a large interfacial contact area between the air and water phases. A common aeration technique in water treatment entails spraying the water into the air and allowing it to fall, as in a fountain (Cornwell, 1990).

Next, chemicals may be added and rapidly mixed with the water to aid in subsequent treatment steps. The chemicals that may be used at this stage include alum and ferric chloride, which cause particles to coagulate so that they become easier to remove in the following stages (see §6.D.2), and lime, which causes chemical precipitation of unwanted ions (see §6.D.4). In addition, powdered activated carbon may be added here to remove taste- and odor-causing molecules and volatile organic compounds by sorption (see §6.D.3). If coagulants or precipitants are added, the next stage typically is a flocculator. Here the water is gently mixed to promote collisions between particles; this allows them to adhere, forming larger particles.

In the sedimentation stage, coarse particles are separated from the water by gravitational settling (see §6.C.1). Since sedimentation is effective only for coarse particles, the water may also be passed through a sand filter to ensure removal of fine particles (see §6.C.2). If powdered activated carbon was added in a previous stage, it is removed from the water at this stage. As an alternative to the addition of powdered

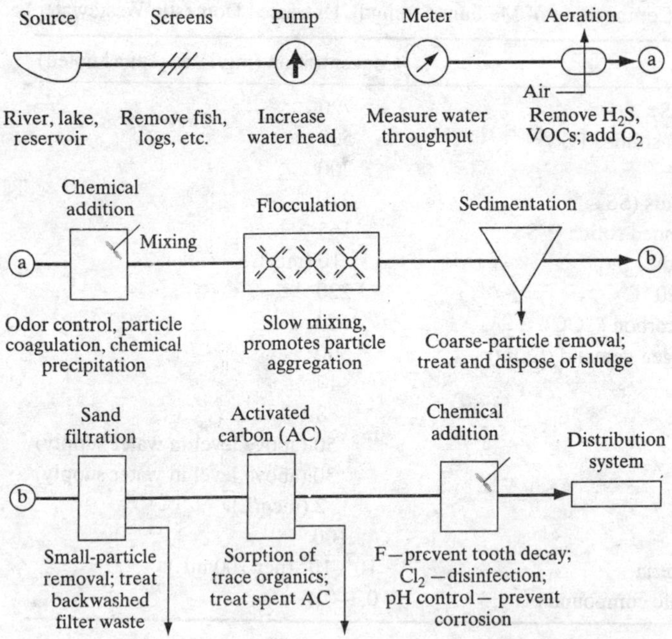

Figure 6.B.1 Schematic of a typical municipal water treatment system.

activated carbon, the water may be passed through a fixed bed of granular activated carbon. As an alternative to chemical precipitation, an ion exchange column may be used to remove undesired ions from the water (see §6.D.4).

Finally, the treated water typically passes through one more stage of chemical addition before entering the delivery system. At this point, a disinfectant is added to the water to kill or inactivate pathogenic microorganisms (see §6.D.1). To help prevent tooth decay, fluoride may also be added to the water. Also, to control precipitation and corrosion in the water distribution network, chemicals may be added to adjust the pH so that the water is slightly scale forming (see §3.D.2).

Wastewater Treatment

Municipal wastewater has four primary components: domestic wastewater, industrial wastewater, infiltration and inflow to the sewage system, and storm runoff. Domestic wastewater is generated from households at 60–85 percent of the water usage rate, since some water is lost to outdoor uses such as landscape irrigation, car washing, and pool filling (Metcalf & Eddy, 1991). The composition of typical domestic wastewater is summarized in Table 6.B.2. The quantity and quality of industrial wastewater is highly specific to the industry. In many cases, these waste streams must be pretreated on site to be acceptable for discharge to the sewer. The other sources of wastewater often contribute significantly to the volume of wastewater that must be managed, but generally do not contribute much to the total load of contaminants.

The processes that may be found in a municipal wastewater treatment system are shown in Figure 6.B.2. Water entering the plant from the sewer first passes through screens to remove large objects such as logs, rags, and toy boats. After grinding coarse solid objects that are small enough to pass through the screens, the water may

Table 6.B.2 Composition of Medium-Strength, Untreated Domestic Wastewater

Contaminant	Concentration (mg/L, except as noted)
Total solids (TS)	720
Total dissolved solids (TDS)	500
Volatile TDS	200
Suspended solids (SS)	220
Volatile suspended solids (VSS)	165
Settleable solids	10 (mL/L)
BOD (5-day, 20 °C)	220
Total organic carbon (TOC)	160
Chemical oxygen demand (COD)	500
Nitrogen	40
Phosphorus	8
Chlorides	50 (above level in water supply)
Sulfates	30 (above level in water supply)
Alkalinity	2 (meq/L)
Grease	100
Coliform bacteria	10^7–10^8 (per 100 mL)
Volatile organic compounds	0.1–0.4

Source: Metcalf & Eddy, 1991.

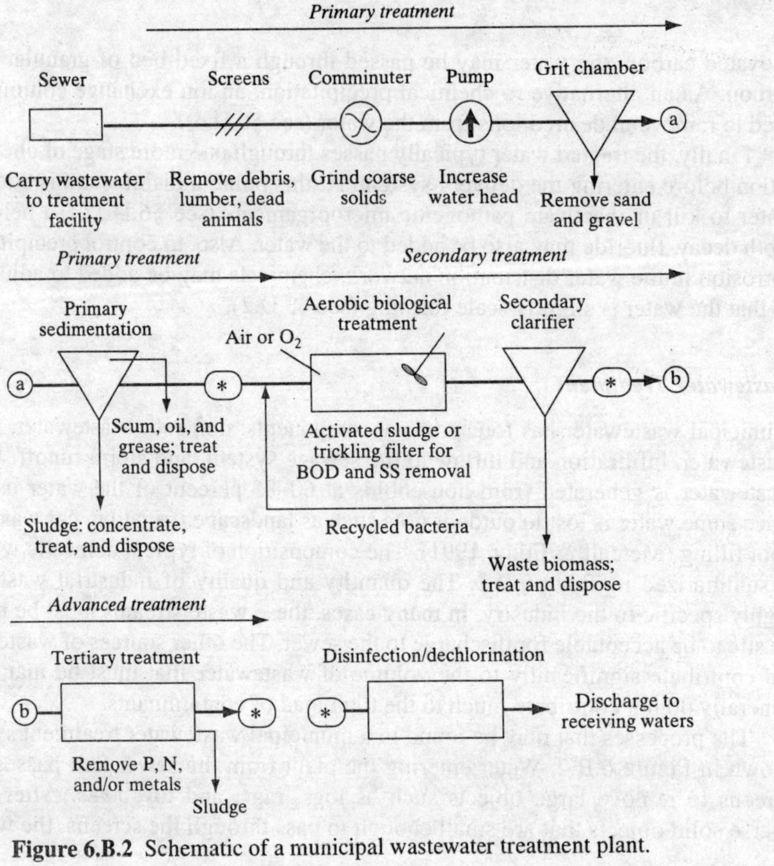

Figure 6.B.2 Schematic of a municipal wastewater treatment plant.

be lifted with a pump to a grit chamber, the first of two stages that use gravity to separate materials from the wastewater according to density. In the grit chamber, the wastewater residence time is short (on the order of minutes), and so only the largest solids such as gravel and coarse sand are removed. In the second stage, the primary sedimentation basin, the residence time is much longer. Here other coarse solid particles are removed from the bottom of the unit while materials such as oil and grease are removed by skimming the upper surface. The treatment train to this point is considered *primary* treatment. Until recently, primary treatment was sufficient if the wastewater was to be discharged to the ocean. However, USEPA regulations now require at least secondary treatment of all municipal wastewater before discharge. The main objective of *secondary* treatment is to reduce the concentration of biochemical oxygen demand (BOD) (see §3.D.5) in the wastewater. This objective is usually achieved by means of a biological treatment process, such as activated sludge, trickling filters, or oxidation ponds (see §6.E). In addition to removing BOD, secondary treatment also removes some of the suspended solids that pass through primary sedimentation. After secondary treatment, wastewater typically has low concentrations of BOD and particulate matter, but may still have high concentrations of nutrients such as nitrogen and phosphorus, heavy metals, or refractory organics. If the receiving water is sensitive to these impacts, a *tertiary* or *advanced* treatment stage may be required. Tertiary processes may include (a) biological treatment for nitrogen or phosphorus removal, (b) chemical precipitation for phosphorus or metals removal, (c) single or multimedia filters that may contain activated carbon for additional solids and refractory organics removal, and (d) air stripping for ammonia removal. Finally, it may be necessary to disinfect wastewater to inactivate pathogens prior to discharge (see §6.D.1). When chlorine is used as the disinfectant, the high concentration required to adequately destroy pathogens sometimes necessitates dechlorination.

In the United States, there are approximately 16,000 municipal wastewater treatment facilities, with a total capacity of approximately 10^{11} m^3 d^{-1} and an average per-capita capacity of 440 L d^{-1} (USEPA, 1998). More than 98 percent of these plants currently apply at least secondary treatment, and many include advanced treatment stages.

The remaining sections of this chapter discuss the principles and practice of specific water and wastewater treatment processes. The processes are organized into three groups. In *physical* processes (§6.C), physical forces are applied to separate impurities from the water. In *chemical and physicochemical* methods (§6.D), impurities are transformed or removed by the promotion of chemical reactions. In *biological* treatment (§6.E), impurities are transformed through oxidation-reduction reactions carried out by microorganisms. The boundaries separating these three groups are blurred. Some treatment processes, such as adsorption and coagulation-flocculation, operate through a mixture of chemical and physical phenomena.

6.C PHYSICAL TREATMENT METHODS

6.C.1 Sedimentation

In sedimentation basins, also known as *settling tanks* or *clarifiers*, the force of gravity is used to separate suspended particles from water. In the basic design, water moves slowly through a tank in the horizontal direction. Suspended material rises or falls depending on whether its density is less than or greater than that of water. Sedimentation is applied in both water and wastewater treatment. In both cases, coarse solids that settle to the bottom of the tank are removed. In wastewater treatment, greases, oils, and other materials that float are skimmed from the water surface.

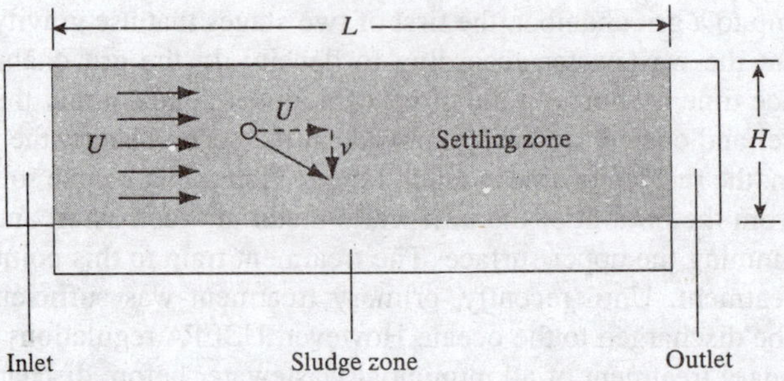

Figure 6.C.1 Vertical cross-section of an idealized rectangular sedimentation basin.

Figure 6.C.1 is a schematic diagram of a rectangular sedimentation basin. Water enters on the left, in the inlet section, and then flows slowly through the settling zone (shaded) toward the outlet, on the right. Ideally, the water flow is laminar and uniform throughout the settling zone. Suspended particles are transported horizontally by means of advection and vertically by means of gravitational settling. Although we consider only the removal of settling particles in this discussion, the same approach can be applied to materials that float.

For simplicity, we assume that the particles have settling velocities that are constant and stable. Then each particle in the basin will follow a diagonal trajectory across the settling zone that is defined by the horizontal fluid velocity, U, and vertical settling velocity, v. Any particle that moves from the settling zone into the sludge zone is captured and removed from the water, while any particle that enters the outlet zone escapes into the effluent.

We define the removal efficiency, η, as the fraction of entering particles that are removed from the water by settling to the sludge zone. As we would expect, the efficiency varies strongly with particle size. If we define H and L as the height and length of the settling zone, respectively, then the transit time for water through the settling zone is given by L/U. Particles with sufficient settling velocity (v) to descend through the full height of the settling zone, H, will be removed 100 percent from the water. This situation is depicted in Figure 6.C.2(*a*).

If the settling velocity is lower, so that the distance settled during the period L/U is less than H, a particle that enters at the top of the settling zone will not settle into the sludge zone, but rather will escape to the outlet. In this case, a critical trajectory from height H' to the far corner of the sludge zone can be defined, as depicted in Figure 6.C.2(*b*). All particles that enter the tank at a height below H' will be captured, while none of the particles that enter the tank above H' will be removed. If we assume that entering particles are evenly distributed along the inlet height, then the efficiency is $\eta = H'/H$. Furthermore, from simple geometry, we can see that $H'/v = L/U$. This expression can be rearranged to give the settling distance (H') of any particle as a function of its settling velocity (v) and the reactor configuration (L/U). We can then use the design parameters of a particular settling basin to define a *critical settling velocity*, v_c, for particle removal in that system:

$$v_c = \frac{UH}{L} \tag{6.C.1}$$

Here v_c is the minimum particle settling velocity required for 100 percent particle removal. Particles that settle faster than the critical velocity ($v > v_c$) will be completely

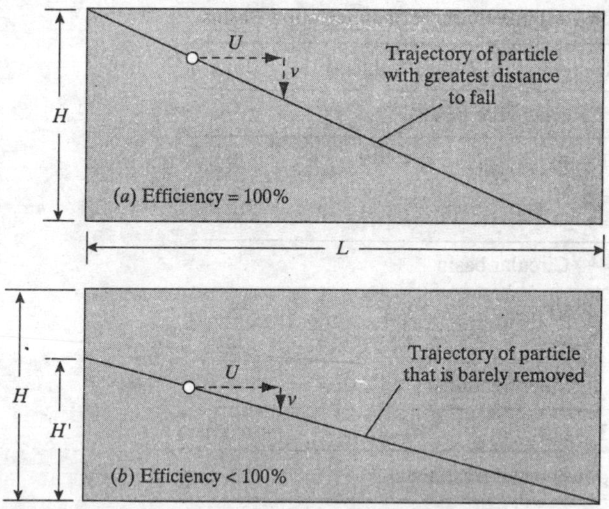

Figure 6.C.2 Schematic of particle trajectories in a settling zone of an idealized rectangular sedimentation basin. In (*a*) the settling velocity is sufficiently high for all particles to be removed; in (*b*) the settling velocity is lower, and less than 100 percent of the particles are captured.

removed since the time required to settle through the entire settling zone is less than the time required to flow from the inlet to the outlet ($H/v < L/U$). On the other hand, for particles that settle more slowly than the critical velocity ($v < v_c$), the efficiency is given by

$$\eta = \frac{H'}{H} = \frac{Lv}{UH} = \frac{v}{v_c} \qquad (v \le v_c) \tag{6.C.2}$$

Alternatively, if we define the horizontal surface area of a rectangular tank as $A_s = LW$ and the volumetric flow rate of water through the tank as $Q = UHW$, the critical settling velocity can be written as

$$v_c = \frac{UH}{L} \times \frac{W}{W} = \frac{Q}{A_s} \tag{6.C.3}$$

The ratio Q/A_s is known as the *overflow rate* and is equal to the critical settling velocity v_c.

Sedimentation basins are often designed to be circular rather than rectangular. In this case, water is introduced at the central axis and flows radially outward. The horizontal water velocity decreases with radial distance, so that the trajectories traced by particles are no longer straight lines. However, even in this design, the critical settling velocity is equal to the overflow rate, $v_c = Q/A_s$, and the efficiency is calculated in the same manner as for a rectangular basin.

Table 6.C.1 shows typical design values for sedimentation basins in water and wastewater treatment. As with many treatment process designs, overflow rates are based on a balance between efficiency and cost. Typically, an engineer must design for a specified flow rate, Q. Increasing the surface area (A_s) produces lower overflow rates and critical settling velocities that correspond to smaller particles, yielding greater particle removal rates. However, sedimentation basins represent a large fraction of the

Table 6.C.1 Common Design Values for Sedimentation Basins

Parameter	Range	Typical value(s)	Units
	Rectangular basin		
Length	15–90	25–40	m
Depth	3–5	3.5	m
Width	3–24	6–10	m
	Circular basin		
Diameter	4–60	12–45	m
Depth	3–5	4.5	m
	Water treatment		
Overflow rate	35–110	40–80	m d^{-1}
	Wastewater treatment		
Overflow rate	10–60	16–40	m d^{-1}

Source: Tchobanoglous and Schroeder, 1985.

land area in a typical treatment plant, and increasing A_s entails greater land and construction costs.

Figure 6.C.3 and Example 6.C.1 show that only coarse particles can be efficiently removed by sedimentation basins. Figure 6.C.3 illustrates the relationship between overflow rate and particle size. The range of typical overflow rates reported in Table 6.C.1 (16–80 m d^{-1} = 0.02–0.09 cm s^{-1}) corresponds to particle diameters in the range 15–32 μm. Smaller particles may pass through the sedimentation basin without being captured. In water treatment, coagulation and flocculation can be applied to re-

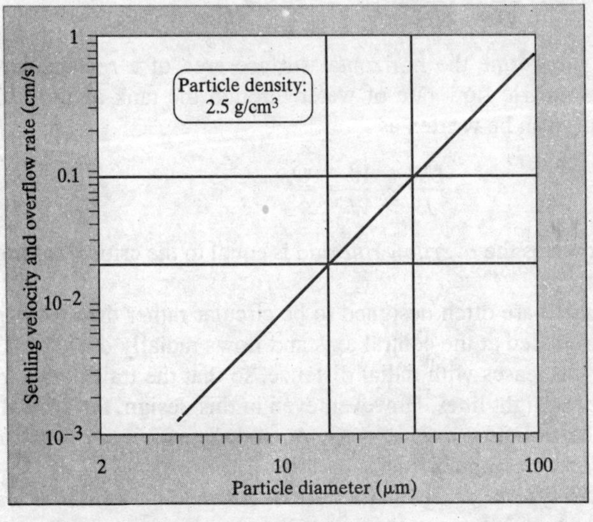

Figure 6.C.3 Settling velocity of mineral particles (spherical shape with density 2.5 g/cm^3) as a function of particle diameter. The horizontal lines indicate the typical range of overflow rates for sedimentation basins in water and wastewater treatment. The vertical lines indicate the smallest particle sizes that would be removed with 100 percent efficiency at these overflow rates.

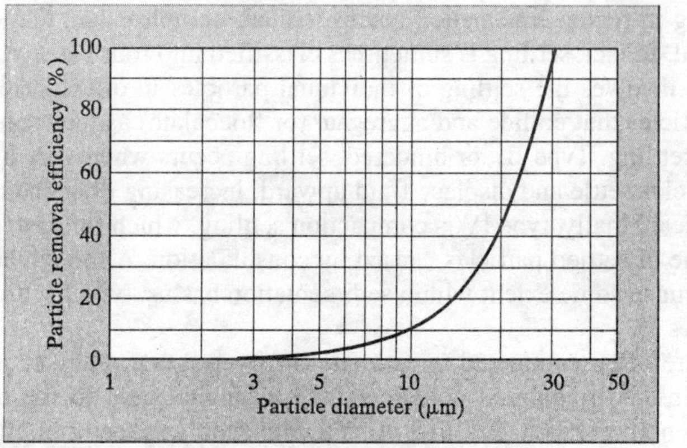

Figure 6.C.4 Theoretical particle removal efficiency for the sedimentation basin considered in Example 6.C.1.

duce the concentration of small particles, improving the particle collection efficiency in the sedimentation basin. Figure 6.C.4 shows that the collection efficiency diminishes rapidly with decreasing particle size once the critical settling velocity is reached. In Example 6.C.1, particles with diameters larger than 32 μm are collected with 100 percent efficiency while 10 μm particles are collected with only 10 percent efficiency. Since the settling velocity for a 10 μm particle is only 10 percent of the settling velocity for a 32 μm particle, removing all 10 μm particles by sedimentation requires 10 times as much surface area as needed to remove all 32 μm particles.

It is impractical to design a sedimentation basin to efficiently remove particles less than about 10 μm in diameter; consequently, sedimentation is not effective in reducing the waterborne concentrations of potentially pathogenic bacteria and viruses and is relatively ineffective at improving the clarity of water. A second particle removal stage, filtration, is commonly required for water treatment applications.

EXAMPLE 6.C.1 ***Sedimentation Basin Performance***

A sedimentation basin in a local water treatment plant is designed to handle a water flow rate of 0.2 m^3 s^{-1}. The basin is rectangular, with a length of 32 m, a width of 8 m, and a depth of 4 m. For particles with a density of 2.5 g cm^{-3}, plot the theoretical removal efficiency of the sedimentation basin as a function of particle diameter.

SOLUTION The overflow rate is

$$\frac{Q}{A_s} = \frac{0.2 \text{ m}^3 \text{ s}^{-1}}{8 \text{ m} \times 32 \text{ m}} = 7.8 \times 10^{-4} \text{ m s}^{-1} = 0.078 \text{ cm s}^{-1}$$

From Figure 6.C.3, it appears that particles of diameter greater than about 30 μm have settling velocities greater than v_c and are removed with 100 percent efficiency. For smaller particles, the settling velocity is computed using the Stokes's law expression from Table 4.B.1 (assuming T = 20 °C):

$$v = \left[\frac{g d_p^2}{18}\left(\frac{\rho - \rho_f}{\mu}\right)\right]$$

The removal efficiency for each particle diameter is then computed using equation 6.C.2 and is plotted in Figure 6.C.4.

Particle settling in real sedimentation basins is more complex than the idealized case just considered. In fact, settling is sometimes classified into four categories. Type I, considered here, involves the settling of individual particles in dilute suspensions. The settling of particles that collide and aggregate (or flocculate) as they settle is described as type II settling. Type III, or hindered, settling occurs when very high concentrations of particles settle and displace fluid upward, increasing drag and reducing net settling velocities. Finally, type IV is compaction settling, which refers to the slow reduction in volume of settled particles caused by consolidation. Although each type of settling may occur to some extent within sedimentation basins, type I settling commonly predominates.

Experiments are often conducted to estimate the overall efficiency of proposed sedimentation basins. A cylindrical container with a height equal to the proposed depth of the sedimentation basin (i.e., 3–5 m) and with evenly spaced sampling ports is used. After the cylinder is filled with a sample of the water to be treated, samples are extracted from each sampling port at prescribed time intervals. All samples collected at a given time are mixed together and analyzed for turbidity or suspended-solids content. The average particle concentration in the mixed sample corresponds to that at the outlet of a sedimentation basin with a hydraulic detention time equal to the interval between filling the cylinder and sampling.

The efficient operation of a sedimentation tank depends on ideal fluid flow. The fluid velocity should be uniform and laminar with no dead zones, no short-circuiting, and no vertical mixing. The inlet and outlet regions are designed with baffles to promote uniform laminar flow. In reality, ideal flow is never achieved. Uneven flow may be caused by several mechanisms: thermal stratification, wind, salinity stratification, and kinetic energy from the inlet stream.

As a rule of thumb, water and wastewater treatment plants are always designed with at least two sedimentation basins operating in parallel (as opposed to a single large unit) so that repair and maintenance operations can take place without requiring that the whole plant be taken off-line. For wastewater treatment, such redundancy is legally required.

All settling tanks have some mechanism for continuous or batch removal of settled sludge. One approach for sludge removal is to have a bottom surface that slopes slightly downward toward a drain, which is typically positioned near the inlet. A set of scrapers mounted on a chain drive sweeps the accumulated sediments toward the drain. The sludge can then be removed by pumping it from the tank.

Sedimentation tanks in wastewater treatment plants also commonly have a mechanism for removing materials that float by skimming them from the surface. Flotation is sometimes promoted by injecting air bubbles into the wastewater near the inlet to assist in the removal of solids and greases. The process of dissolved air flotation is discussed in the context of hazardous waste treatment in §8.C.1.

Sludge treatment and disposal are major issues in both water and wastewater treatment plants. Sludges are generally thickened, dewatered, and disposed of in a practical manner. For sludge management, some options are anaerobic digestion (see §6.E.3), composting, land application, landfilling, and incineration. Deep-ocean dumping was once practiced but generally is no longer permitted.

6.C.2 Filtration through Granular Media

In water treatment plants, filtration is applied after sedimentation to "polish" the water, that is, to remove fine suspended particles. As demonstrated in the previous section, sedimentation basins are not capable of removing free suspended particles smaller than about 10 μm. Smaller particles must be removed to produce water that is aesthetically acceptable and safe for drinking. Particles comparable in size to the wavelength of light are the most efficient (per mass) at scattering light. These particles, therefore, are responsible for much of the turbidity of water. Since the wavelength of visible light is ~400–700 nm, or 0.4–0.7 μm, particles such as clays, with diameters around a micrometer, must be removed from water to ensure its clarity. Furthermore, and more important, pathogens, such as bacteria and viruses are often smaller than 10 μm (see Figure 2.C.3) and are not effectively removed by sedimentation tanks. By physically removing pathogens from water, filtration can augment chemical disinfection (see §6.D.1) for pathogen treatment. Filtration is also needed when chemical precipitation is applied to remove unwanted ions from water (see §6.D.4).

In the United States, filtration is almost always applied to treat drinking water that originates from surface sources, as required by the EPA's surface-water treatment rules. On the other hand, filtration may not be needed to treat groundwater since the aquifer itself acts much like a deep-bed filter.

In municipal water treatment plants, filtration is accomplished by passing water through a fine granular medium, such as sand. Elements in the design of a conventional *rapid-sand filter* are shown in Figure 6.C.5. The filter is housed in a concrete box, open at the top, with dimensions of up to 100 $m^2 \times 3$ m high. In ordinary operation, water flows in through valve 1 and out through valve 2 while valves 3 and 4 are closed. As the water passes through a layer of sand with thickness L, particles are captured by mechanisms that are discussed below. The fine pores of the sand create resistance to flow, so a head loss, h, occurs across the filter. The sand itself is supported by an underdrain system; one such system consists of a horizontal network of perforated pipes placed in a layer of gravel. The gravel is graded, with relatively coarse grains at the bottom and fine grains at the top, so that water flow is not restricted and sand does not escape from the filter.

As the filter operates, the pores become clogged with trapped particles, so the filter must be cleaned periodically using a process known as *backwashing*. In this process, valves 1 and 2 are closed while valves 3 and 4 are opened. Water is forced upward through the filter with sufficient velocity to suspend the sand grains. The high water velocity causes pore spaces in the sand to expand and the grains to collide, freeing the filtered particles for discharge to waste through the washwater trough and valve 4. From there, the washwater may be discharged to a sewer for treatment in a

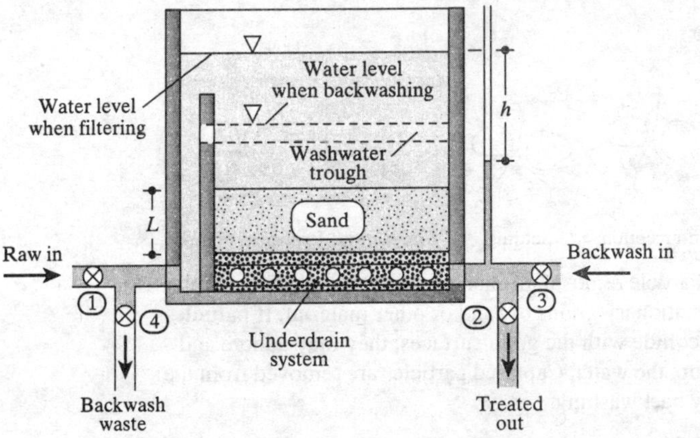

Figure 6.C.5 Deep-bed sand filter for removing fine particles from water.

wastewater treatment plant, or it may be otherwise treated to remove the waste particles. In a conventional rapid-sand filter, the upward water velocity during backwashing is approximately 30 m h^{-1}.

Two key elements govern the design and function of a granular filter: hydraulics and particle capture mechanisms. As the grain size becomes smaller, the pores between grains also become smaller and the particle capture efficiency tends to improve. However, smaller grain sizes also increase the resistance of the filter bed to water flow. The selection of granular material and depth of the filter bed is based, in part, on the necessary compromise between filter efficiency and hydraulic throughput.

Particle capture within sand filters relies on transport leading to contact between the particle and a filter grain, and on subsequent adherence of the particle to the grain surface. Although the details of particle filtration in porous media are understood qualitatively, they are not sufficiently quantifiable to serve as a basis for design. Five processes that contribute to particle capture are illustrated in Figure 6.C.6. The mechanism of *straining* occurs when a particle is too large to pass through a pore and so its progress is mechanically blocked. Straining is most effective against relatively large particles. The mechanism of *interception* occurs when the streamline of a particle causes it to strike a grain. A third mechanism, *settling*, occurs when a particle settles under the influence of gravity onto the upper surface of a grain. The fourth mechanism, *impaction*, occurs when the particle's inertia is sufficiently large that its path deviates from a streamline as fluid flows around a grain, resulting in a collision. For very small particles, random particle movements caused by *Brownian motion* may result in collisions with filter grains. Particle adhesion following contact is enhanced if a coagulant is added to the water upstream of the filter (see §6.D.2). The particle removal performance of candidate filter designs is commonly tested using bench-scale and pilot plant experiments.

In contrast to particle capture, the hydraulic aspects of granular filters can be easily quantified. The underlying principles are discussed in §4.D.1. Many specific equations have been developed to relate water flow through a filter to the head loss, h, and the filter properties (Tchobanoglous and Schroeder, 1985). The equations all reduce to Darcy's law (equation 4.D.5) in the limit of low Reynolds numbers. They differ in the details: how hydraulic conductivity is related to filter properties, and how flow is related to head loss at Reynolds numbers above the Darcy limit. One example is the Carman-Kozeny equation (Cleasby, 1990; see also §4.D.1):

$$\frac{h}{L} = \frac{\kappa\mu}{\rho g}\frac{(1-\phi)^2}{\phi^3}S^2\frac{Q}{A_s} \tag{6.C.4}$$

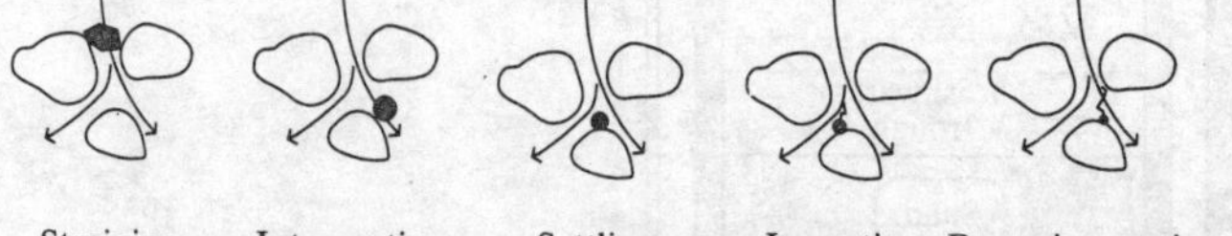

Figure 6.C.6 Particle removal mechanisms in a granular filter. Water flows through stationary grains of sand or other material. If particles are induced to collide with the grain surfaces, they may adhere and be separated from the water. Captured particles are removed from the granular bed by backwashing.

where κ is the Kozeny constant, a dimensionless variable commonly taken to be 5 for water filters; μ is the dynamic viscosity of the fluid; ρ is the fluid density; g is gravitational acceleration; ϕ is the porosity of the filter bed; S is the specific surface area of the filter grains (surface area per volume); Q is the volumetric flow rate of water through the filter; and A_s is the horizontal surface area of the filter bed. The specific surface area of a filter grain is $6/d$ for a smooth sphere of diameter d. For uniform granular materials in water filters, S is commonly estimated by $S = 6(\psi d_{eq})^{-1}$, where ψ is the sphericity and d_{eq} is the diameter of a sphere with the same volume as the grain. The ratio Q/A_s represents the superficial or Darcy velocity, U. Equation 6.C.4 is based on Darcy's law and is valid for Re < 6, where Re $= d_{eq}(Q/A_s)\rho\mu^{-1}$. Example 6.C.2 illustrates the use of this equation with representative values for municipal water treatment.

During filter operation, two parameters are measured to monitor performance. The head loss across the filter is measured to monitor the decline in permeability. The turbidity of the effluent is measured to monitor particle removal efficiency. Because small particles are so efficient at scattering light, turbidity measurements taken with a light scattering nephelometer (see §2.C.8) can be used to determine the filtration effectiveness. Turbidity is most commonly reported in terms of "nephelometric turbidity units" (NTU). The objective for drinking water is 0.5–1 NTU.

EXAMPLE 6.C.2 ***Sizing a Deep-Bed Sand Filter***

A water treatment plant is being designed to supply 1 $m^3\ s^{-1}$ of water for a nearby community. If a sand filter is used, calculate the minimum surface area of the filter necessary to provide treated water at this rate. Assume that the following data apply:

Porosity	$\phi = 0.4$
Head loss across filter	$h = 1$ m
Sand grain diameter	$d_{eq} = 0.5$ mm
Length of filter bed	$L = 0.75$ m
Sphericity of sand	$\psi = 0.8$

SOLUTION We apply equation 6.C.4 to determine the Darcy velocity, Q/A_s. Since Q is given in the problem statement, we can then evaluate A_s as Q/U. The calculation proceeds as follows, with all dimensional parameters expressed in units of m, kg, and s.

$$\frac{Q}{A_s} = \frac{h}{L}\frac{\rho g}{\kappa\mu}\frac{\phi^3}{(1-\phi)^2}\left(\frac{\psi d_{eq}}{6}\right)^2$$

$$= \frac{1}{0.75}\frac{998 \times 9.8}{5 \times 1 \times 10^{-3}}\frac{0.4^3}{(1-0.4)^2}\left(\frac{0.8 \times 0.5 \times 10^{-3}}{6}\right)^2$$

$$= 2.1 \times 10^{-3}\ \text{m s}^{-1} = 7.4\ \text{m h}^{-1} = 180\ \text{m}^3\ \text{m}^{-2}\ \text{d}^{-1}$$

$$A_s = \frac{1\ \text{m}^3\ \text{s}^{-1}}{2.1 \times 10^{-3}\ \text{m s}^{-1}} = 480\ \text{m}^2$$

Since individual filter boxes are limited to $A_s \sim 100\ m^2$, and since one filter box is cycled off for backwashing at any time, approximately six filters are needed to provide the total required supply flow rate.

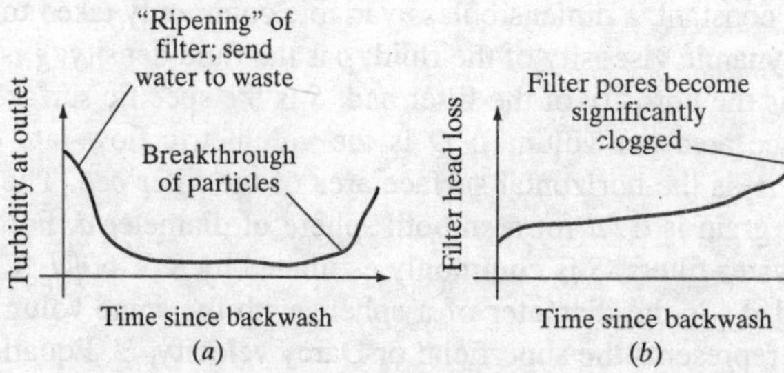

Figure 6.C.7 Evolution of performance parameters during operation of a rapid-sand filter: (*a*) turbidity of filtered water and (*b*) head loss across filter, assuming constant water throughput.

Figure 6.C.7 shows how the performance of a filter evolves following the end of a backwash cycle. Initially, the particle level in the outlet water is high. Following backwash, as the filter grains settle, large pores become clogged and residual particles from the backwash operation are released. The initial outlet flow is discarded until the turbidity diminishes to an acceptable level. The filter then yields good particle removal performance for some interval until increased head loss or particle breakthrough is observed. Particle accumulation decreases permeability within the filter, resulting in increased localized flow velocities, which in turn can cause loosely adhered particles to be released. When either the turbidity rises to an unacceptable level or the hydraulic resistance becomes too large, the filter is cleaned by backwashing. The interval between backwashing episodes ranges from approximately 10 hours to several days. Typically, backwashing requires about 10 minutes, and the quantity of water wasted is a small but nonnegligible percentage of the volume successfully treated during a filter cycle.

One goal in developing granular filters has been to increase throughput (Q/A_s) without reducing filter efficiency, so that the land and capital costs of constructing filters are minimized. The first granular water filters, introduced at the beginning of the nineteenth century, were "slow-sand" filters. Grain diameters were small (0.2–0.4 mm), so hydraulic resistance was high and water velocities were low (Q/A_s = 0.1–0.3 m h^{-1}). In the slow-sand filter, most particles were retained in the top few centimeters of the sand layer, and a layer of microorganisms would grow on the surface of the filter due to the low water flow. Pore sizes at the top of the sand bed were small, and permeability was low. After a period of 20 to 60 days, the layer of particle and microbial buildup at the sand surface (called *schmutzdeck*) was scraped off and discarded. Slow-sand filters have the advantages of simple design and low operator skill requirements, making them attractive for use in developing countries. However, they have the disadvantages of large land requirements and low fluid throughputs, limiting their usefulness in urban areas. Now, particularly in large installations, deep-bed filters (also called rapid-sand filters) are more commonly employed. The grains are larger (0.4–1 mm diameter), the fluid velocities are higher (2–10 m h^{-1}), particles are removed throughout the bed rather than just on the surface, and much less land area is required to process a given flow rate of water.

To further improve performance, some deep-bed filters are designed with two or even three layers of distinct granular materials (see Figure 6.C.8). Table 6.C.2 summarizes the characteristics of granular materials used in water filters. Relative to a single layer of sand, dual- or multimedia filters can increase the extent to which

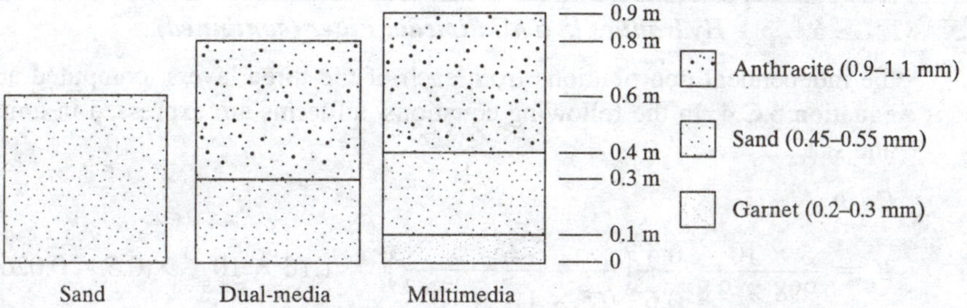

Figure 6.C.8 Filter media commonly used in municipal water treatment plants.

Table 6.C.2 Properties of Granular Materials Used in Water Filters

Parameter	Silica sand	Anthracite	Garnet
Grain diameter, d_{eq} (mm)	0.45–0.55	0.9–1.1	0.2–0.3
Grain density (g cm^{-3})	2.65	1.45–1.73	3.6–4.2
Sphericity, ψ	0.7–0.8	0.46–0.60	0.6
Porosity, ϕ	0.42–0.47	0.56–0.60	0.45–0.55

Source: Cleasby, 1990.

particle capture occurs within the entire depth of the filter bed. The following paragraph explains why.

With a single layer of grains that are not perfectly uniform, the backwashing cycle causes size sorting, with the larger particles migrating to the bottom of the filter bed and the smaller particles moving to the top. Consequently, the smallest pore sizes occur in the upper part of the bed, and particle filtration is most effective there. Particles that penetrate the upper portion may not be captured at all because they tend to encounter larger pores as they move through the filter. With the dual- or multimedia systems, the water flows through successively finer materials. Particles that penetrate the first layer can be captured in the second or third.

The key to the stability of these layered beds lies in the density differences among the materials. The finest material has the highest density and remains at the bottom, even through backwashing. The most consistently observed practical benefit of multilayer filter beds is a longer interval between backwashings. By using a greater fraction of the bed to capture particles, pores are blocked more slowly and the head loss can remain acceptably low for a longer interval. Example 6.C.3 shows how to assess head loss through a multimedia filter.

EXAMPLE 6.C.3 ***Hydraulics in a Multimedia Filter***

A water treatment plant uses a multimedia filter. The filter has three layers: anthracite coal (effective grain diameter, sphericity, and porosity of $d_c = 1.1$ mm, $\psi_c = 0.5$, and $\phi_c = 0.6$, respectively), sand ($d_s = 0.5$ mm, $\psi_s = 0.7$, $\phi_s = 0.45$), and garnet ($d_g = 0.3$ mm, $\psi_g = 0.6$, and $\phi_g = 0.5$). The entire filter is 0.9 m thick with 0.5 m, 0.3 m, and 0.1 m layers of coal, sand, and garnet, respectively. Calculate the clean-bed head loss in this filter at a loading rate of 100 m^3 m^{-2} d^{-1}.

SOLUTION By the principle of mass conservation, the filter loading rate (Darcy velocity) must be the same through each layer. The total head loss is then obtained as the sum of

EXAMPLE 6.C.3 ***Hydraulics in a Multimedia Filter (continued)***

the independent contributions from each of the three layers, computed according to equation 6.C.4. In the following equations, all terms are expressed in units of m, kg, and s.

Coal:

$$h_c = \frac{5 \times 10^{-3}}{998 \times 9.8} \frac{0.4^2}{0.6^3}\left(\frac{6}{0.5 \times 1.1 \times 10^{-3}}\right)^2 \times 1.16 \times 10^{-3} \times 0.5 = 0.026 \text{ m}$$

Sand:

$$h_s = \frac{5 \times 10^{-3}}{998 \times 9.8} \frac{0.55^2}{0.45^3}\left(\frac{6}{0.7 \times 0.5 \times 10^{-3}}\right)^2 \times 1.16 \times 10^{-3} \times 0.3 = 0.17 \text{ m}$$

Garnet:

$$h_g = \frac{5 \times 10^{-3}}{998 \times 9.8} \frac{0.5^2}{0.5^3}\left(\frac{6}{0.6 \times 0.3 \times 10^{-3}}\right)^2 \times 1.16 \times 10^{-3} \times 0.1 = 0.13 \text{ m}$$

The total head loss is $h = h_c + h_s + h_g = 0.33$ m. This is a little less than the head loss (0.35 m) in a conventional rapid-sand filter that uses the same sand material and has a bed depth of 0.6 m.

6.C.3 Membrane Separation Processes

Membranes, thin layers of material that permit the transmission of water at a different rate relative to certain impurities, can be used to remove contaminants from water. Membrane technologies are used in several environmental engineering applications. In water quality engineering, reverse-osmosis membranes are used for desalination of seawater or brackish water. Other applications include softening of water, removal of dissolved organics, preparation of high-purity water for industrial processes, preparation of bottled drinking water, and treatment of wastewater, such as agricultural drainage, for reuse.

The use of membranes for water and wastewater treatment tends to be restricted more by economic than by technical considerations. Membrane water treatment methods are likely to become more important in the future for several reasons:

- Membrane manufacturing processes are not yet mature; expected advances include improvement of membrane treatment efficiencies and lower manufacturing costs.
- Membrane separation processes are already economically competitive with alternative technologies for removing ions from water, either for desalination (conversion of seawater or brackish water to potable water) or for demineralization (preparation of high-purity water from potable water). The number of ion removal applications will increase rapidly as population and water demand continue to grow.
- Stricter environmental regulations on drinking water quality will create pressure to expand the use of membrane technologies. In particular, membranes can be effective in controlling turbidity, pathogens, and disinfection by-products in treated drinking water.

Membrane technologies are also applied in air quality engineering (e.g., for separating volatile organic gases from waste air streams) and in hazardous waste management (e.g,. to recover valuable resources from waste streams, such as metals from electroplating operations).

Membrane separation processes are commonly divided into several groups according to both the type of membrane and the application. The following presentation is organized according to these groupings.

Reverse Osmosis

Reverse osmosis (RO), also known as *membrane hyperfiltration*, is the most widely applied membrane technology in water quality engineering. The primary application is desalination of seawater and brackish water to produce drinking water. As of the late 1980s, there were approximately 4000 land-based RO plants (each with a capacity of at least 25,000 gallons per day, or about 1 L s^{-1}), with a total installed capacity of about 10^9 gal d^{-1} (44 m^3 s^{-1}) (Wiesner et al., 1992).

Reverse-osmosis membranes are asymmetric, consisting of a thin, dense skin, typically 0.1–1 μm thick, supported by a porous substructure with a thickness of 20–100 μm. The skin functions to separate water from impurities while the substructure provides mechanical strength. The detailed mechanisms by which RO membranes function are not well understood. The challenge is to explain how water molecules can be transmitted while ions, being of similar size, are rejected. One theory holds that a strong attraction between the skin and water molecules allows them to pass through the membrane while ions are excluded. A second theory posits that water sorbed to the membrane surface flows by convection through the very fine pores (Wiesner et al., 1992).

The overall functioning of an RO process is illustrated in Figure 6.C.9, which shows a vessel divided into two parts. These compartments are separated by an ideal RO membrane, which permits passage of water but not salt ions. Freshwater is placed on one side of the membrane, saltwater on the other. Both halves of the vessel are open to the atmosphere and initially have the same water level. Because the concentration of water molecules is higher on the freshwater side, there will be a net migration, known as *osmosis*, through the membrane from the fresh toward the saline side. In *reverse osmosis*, the natural tendency of water to flow from fresh to saline is overcome by applying a large pressure, P_a, to the saline water. The net flux of water across the membrane depends on the applied pressure, as described by this equation:

$$J_w = k_p(P_a - \pi) \qquad (6.C.5)$$

where k_p reflects the permeability of the membrane and π is known as the *osmotic pressure*. It is clear from equation 6.C.5 that the osmotic pressure is equal to the

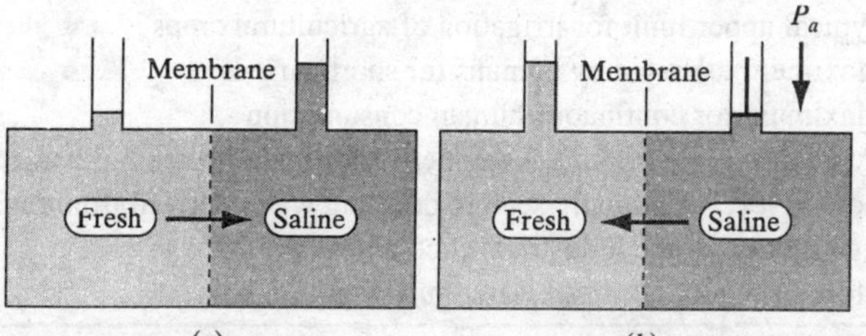

Figure 6.C.9 Illustration of (*a*) osmosis and (*b*) reverse osmosis.

applied pressure at which there is no net water migration across the membrane. The osmotic pressure is a function of the difference in salinity across the membrane. Seawater, with a total dissolved solids (TDS) content of 35 g/L, generates an osmotic pressure relative to freshwater of 23 atm; for brackish water with a TDS of 5 g/L, the osmotic pressure is 3 atm (Crossley, 1983). In typical applications, the applied pressure exceeds the osmotic pressure by 20–50 atm.

Two types of membranes are commonly used in RO units. One type, introduced about 1960, is made from sheets of cellulose acetate. These sheets may be wound into a spiral so that a large membrane area can be maintained in a small volume. The second type of RO unit uses fine, hollow fibers of an aromatic polyamide with the skin on the inside and the supporting substructure on the outside. In operation, RO systems are configured such that the membrane skin is exposed to the raw water at high pressure. The treated water is collected as it passes through the membrane. Concentrated brine remains on the high-pressure side of the membrane and must be discarded. For desalination of seawater, only about 25–30 percent of the inlet water is collected as product while 70–75 percent of the feed water is discharged as waste (Crossley, 1983).

Typical water fluxes (J_w) in commercial RO units are 0.4–0.8 $m^3\ m^{-2}\ d^{-1}$ (10–20 gal $ft^{-2}\ d^{-1}$) for cellulose acetate membranes and 0.04–0.12 $m^3\ m^{-2}\ d^{-1}$ (1–3 gal $ft^{-2}\ d^{-1}$) for aromatic polyamide fibers (McCabe et al., 1993). To serve the entire needs of a city of 100,000 inhabitants, the water supply capacity must be about 0.7 $m^3\ s^{-1}$(16 million gallons per day) (Tchobanoglous and Schroeder, 1985). An RO system of this scale based on cellulose acetate would require a total membrane area on the order of 100,000 m^2. The largest RO plant (as of the mid-1980s), in Yuma, Arizona, has a capacity of 3.2 $m^3\ s^{-1}$ (72 million gallons per day). It is used to treat agricultural drainage water, reducing its total dissolved solids level from 3000 to 285 mg/L, so that it can be returned to the Colorado River (Applegate, 1984).

In addition to water flux, the performance of an RO membrane is described by the salt rejection ratio, R, defined as

$$R = \frac{C_{\text{raw}} - C_{\text{treated}}}{C_{\text{raw}}} \tag{6.C.6}$$

where C_{raw} is the concentration of total dissolved solids (TDS) in the untreated feed water and C_{treated} is the TDS concentration in the product water. As indicated by the data in Table 6.C.3, the conversion of seawater to potable water requires a salt rejection ratio of approximately 99 percent. Membranes used for desalting brackish water do not require such a high salt rejection ratio.

Table 6.C.3 Levels of Total Dissolved Solids (TDS) in Different Waters

TDS level (ppm)	Description
35,000	Seawater
>2000	Brackish water
2000	Typical upper limit for irrigation of agricultural crops
1000	Maximum tolerated by humans for short periods
500	Maximum for continuous human consumption
100–150	Typical level in drinking water derived from collected rainwater
30	Lowest level commonly seen in drinking water derived from rainwater
5	Required by some industrial processes
<1	Most stringent industrial water purity requirement

Source: Silver, 1983; Spiegler and El-Sayed, 1994.

The cost (as of 1991) for large-scale production of freshwater by RO, per cubic meter of product, was about \$1.50 for seawater applications and about \$0.40–0.70 for brackish waters with TDS in the range 2000–6000 ppm (Spiegler and El-Sayed, 1994). By comparison, in 1990 irrigation water was delivered to agricultural users in the Central Valley of California for about \$0.01 per m^3 and to residential consumers in San Francisco for about \$0.20 per m^3.

In addition to removing ions from water, reverse osmosis membranes are very effective in reducing concentrations of particles, microorganisms, and organic molecules. However, to reduce fouling and extend membrane service life, water supplied to RO units is usually pretreated to reduce concentrations of contaminants that can be easily removed by other technologies.

The major alternative to RO for producing potable water from saltwater is thermal distillation (i.e., evaporation/condensation cycles). Relative to this method, RO offers compelling energy advantages. The theoretical thermodynamic minimum energy required to extract pure water from seawater is 2.8 kJ per kg of extracted water. Because of the high heat of vaporization of water, practical systems using thermal distillation require 240–280 kJ/kg. For reverse osmosis, energy is mainly used to compress untreated water to high pressure. The energy requirement for practical RO systems is about 45 kJ/kg (Silver, 1983). Thus, since RO membranes became available in the 1960s, RO has steadily displaced thermal distillation in most situations for purifying saltwater.

Electrodialysis

Like reverse osmosis, electrodialysis (ED) is applied to remove ions from water. However, whereas RO is based on a membrane that allows water to pass while preventing ion transit, membranes used in ED permit ions to pass rather than water.

Figure 6.C.10 shows the configuration of an ED unit for water treatment. Two distinct types of membranes are employed. Those labeled A in the figure permit the passage of anions while rejecting cations. Those labeled C permit only the passage of cations. The A and C types of membranes are alternated in an ED stack. Electrodes are placed parallel to these membranes, at the outer sides of the stack, and during operation a DC voltage is applied. For the arrangement shown in the figure, as ions are

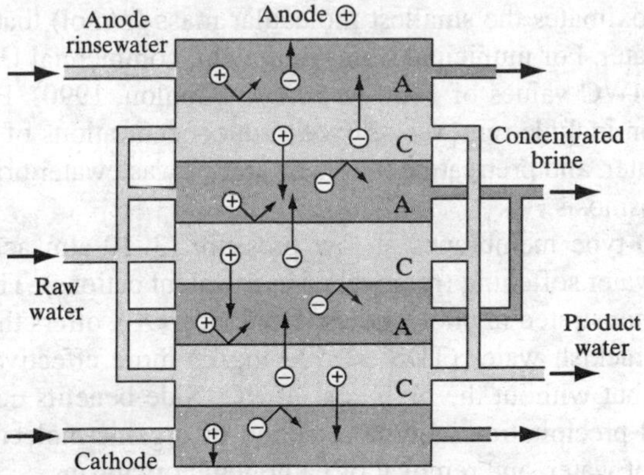

Figure 6.C.10 Schematic of an electrodialysis unit for removing ions from water.

transported from left to right by water flow, they also migrate up (anions) or down (cations). If an anion encounters an A membrane, it passes to the next cell, but if it encounters a C membrane, it is confined. The opposite conditions apply for cations. In this manner, ions are removed from every alternate cell (clear cells) in the stack and concentrated in the adjacent (shaded) cell. In practical applications, tens to hundreds of membranes are placed between each pair of electrodes (Spiegler and El-Sayed, 1994).

In comparison with reverse osmosis, ED systems operate with low pressure differences across the membranes. Since separation is accomplished by the application of an electrostatic force to the ions, the energy requirements are directly related to the salinity of the raw water. Consequently, ED is economically attractive as an alternative to RO for desalination of brackish waters, but not for desalination of seawater. Worldwide, there are several hundred ED plants in operation for generating potable water (Wiesner et al., 1992). Other applications include preparation of high-purity water for industrial purposes and the extraction of salt from seawater (McRae, 1983).

To prevent the accumulation of salts at membrane and electrode surfaces, some systems are designed to regularly switch the direction of the electric field, typically a few times per hour. These *electrodialysis reversal* (EDR) systems must be equipped with appropriate valves. The flow channels that deliver product water with the electric field operating in one direction will deliver concentrated brine when the field direction is changed.

Microfiltration, Ultrafiltration, and Nanofiltration

Membranes used for microfiltration, ultrafiltration, and nanofiltration processes are analogous to RO membranes in that they permit passage of water while retaining impurities. However, they are more permeable than RO membranes, which means that (a) they permit greater passage of impurities, especially ions, and (b) they operate at much lower applied pressures.

In microfiltration and ultrafiltration (UF), water penetrates by convection through distinct membrane pores, and the impurities are retained by mechanisms similar to those operating in granular filters (see Figure 6.C.6). Microfiltration membranes have pore dimensions of about a micrometer and are designed only to remove particles. Ultrafilters have much smaller pores that are capable of capturing large organic molecules but not ions. Ultrafilters are characterized in terms of their molecular-weight cutoff (MWC), which approximates the smallest molecular mass (g/mol) that can be efficiently removed from water. For municipal water treatment, commercial UF membranes are available with MWC values of 1000 to 50,000 (Conlon, 1990). Potential applications of ultrafiltration include removal of excessive concentrations of organic molecules from drinking water, and pretreatment of seawater or wastewater prior to final treatment with reverse osmosis.

Nanofiltration uses RO-type membranes at low pressure (3–10 atm across the membrane) to accomplish water softening (removal of multivalent cations). This technology, which has been demonstrated in pilot studies (Conlon, 1990), offers the possibility of treating slightly brackish water (TDS < 500 mg/L) more effectively than conventional technologies, but without the high cost of RO. Side benefits compared with conventional chemical precipitation include rejection of organic molecules that contribute to discoloration of water, and removal of pathogenic organisms.

Membrane technologies other than reverse osmosis and electrodialysis have not yet been widely adopted in water and wastewater treatment practice. As of 1992, worldwide, there were fewer than 30 plants in operation using ultrafiltration or microfiltration. There were no full-scale systems operating in the United States (Wiesner et al., 1992).

6.D CHEMICAL AND PHYSICOCHEMICAL TREATMENT METHODS

6.D.1 Disinfection

The central aim of disinfection is to limit the risk of disease transmission associated with potable water and wastewater. Physical treatment processes that separate fine particles from water, such as granular and membrane filtration (§6.C.2–3), contribute to this goal. However, most disinfection is achieved by treating water or wastewater to kill or inactivate microorganisms. This section explores disinfection processes, emphasizing chlorination, which has long been the dominant technology for disinfecting drinking water in the United States (Haas et al., 1992).

Microorganisms in water can be inactivated by either physical or chemical treatment methods. Two common physical methods are heating water (boiling) and irradiating water with ultraviolet light. Boiling is energy intensive and expensive, so its use is reserved for emergencies. Ultraviolet light is an attractive alternative to chemical treatment methods in some circumstances. Both of these techniques work by damaging molecules in the microorganisms. Chemical disinfection methods operate by adding a strong oxidant to the water. The oxidant acts by either (1) damaging molecular bonds in the cell wall, leading to disintegration of the cell, or (2) diffusing into the cell and interfering with biochemical activity or damaging cellular DNA. The oxidants used for disinfection include chlorine, chlorine dioxide, chloramines, and ozone.

There is no ideal disinfectant. Table 6.D.1 summarizes the characteristics of dominant drinking water disinfectants. In addition to germicidal potency, perhaps the two most important characteristics of disinfectants are the presence or absence of a residual and the production of harmful disinfection by-products. In municipal water treatment, it is desirable to have some residual disinfectant activity in the water when it leaves the treatment plant. The presence of this residual provides a measure of protection against contamination as water travels through the distribution system. The most important disadvantage of ozone and ultraviolet light as drinking water disinfectants is their lack of a residual.

Disinfection by-products are created when disinfectants interact with organics in the water. In the past few decades, the generation of trihalomethanes by chlorination

Table 6.D.1 Summary of Disinfectant Properties for Drinking Water

Disinfectant	Solubility	Residual[a]	Germicidal potency	Cost rank[b]	Harmful by-products
Chlorine	High	Yes	Very good (pH < 7)	1	Yes
Chloramines	High	Yes	Fair	2	Maybe not
Chlorine dioxide	High	Yes	Good	4	Yes
Ozone	Limited	No	Best	3	Yes, but limited
Ultraviolet light	n/a	No	Fair	?	None known

[a]Does the use of this disinfectant leave a residual product that can protect the water from microbial contamination in the distribution system?
[b]Lowest cost = 1, highest cost = 4.
Source: Selleck, 1989.

of drinking water has become a concern (see Appendix C, §C.3). Current efforts aim to reduce the generation of harmful by-products without compromising the primary goal of microbial disinfection.

Disinfection with Chlorine

In the United States, the use of chlorine to disinfect municipal drinking water supplies began in the early twentieth century. Its introduction, which required the support of a court decision in response to a lawsuit, was accompanied by dramatic reductions in the incidence of typhoid and other waterborne diseases (Sawyer et al., 1994).

The active agent in chlorination is the chlorine in hypochlorous acid (HOCl). It is supplied to water through one of three means. The least expensive method, used in many large systems, involves the application of pure chlorine gas (Cl_2) from compressed-gas cylinders. There are significant safety concerns associated with this approach, as chlorine gas is highly toxic and must be managed with care. An alternative approach, somewhat more expensive but much safer, involves the use of bleach (sodium hypochlorite, NaOCl), which is added to water as a liquid. The third method is to add calcium hypochlorite, $Ca(OCl)_2$, a solid, to the water. This is the most expensive of the three methods but is appropriate for some low-demand systems. (Calcium hypochlorite is commonly used to chlorinate water in swimming pools.)

Chlorination of drinking water is accomplished by adding the chemical to the water and holding it for a minimum time period to allow contact between the oxidant and microorganisms. Typically, disinfection systems are designed so that the product of the contact time and the residual concentration exceeds a prescribed minimum, as discussed below. In current U.S. practice, median contact time is 45 minutes and the median residual chlorine concentration (measured at the nearest consumer) is 1.1 mg/L (Haas et al., 1992).

Wastewater is also commonly disinfected with chlorine. In both water and wastewater treatment, some of the chlorine is consumed in oxidizing chemical impurities rather than acting on microorganisms. Because of the higher level of impurities, the chemical chlorine demand is higher for wastewater than for drinking water. Typical chlorine doses for wastewater treatment are 40–60 g m^{-3} compared with 2–5 g m^{-3} for water treatment (Tchobanoglous and Schroeder, 1985).

The design of an effective chlorine disinfection process depends on basic knowledge of (a) the chemistry of chlorine in water and (b) disinfection kinetics.

Chemistry of Chlorine in Pure Water Chlorine gas readily dissolves in water. The equilibrium partitioning between gas-phase and dissolved-phase species is described by Henry's law (§3.B.2) (Haas, 1990):

$$Cl_2(g) \Leftrightarrow Cl_2(aq) \qquad K_H = \frac{[Cl_2(aq)]}{p_{Cl_2}} = 0.062 \text{ M atm}^{-1} @ 298 \text{ K} \qquad (6.D.1)$$

Aqueous Cl_2 reacts rapidly with water, with a characteristic reaction time of 10^{-2} s (Aieta and Roberts, 1986), to form hypochlorous acid (HOCl) according to the following equilibrium relationship (Stumm and Morgan, 1996):

$$Cl_2(aq) + H_2O \Leftrightarrow HOCl + H^+ + Cl^-$$

$$K_1 = \frac{[HOCl][H^+][Cl^-]}{[Cl_2(aq)]} = 5 \times 10^{-4} \text{ M}^2 @ 293 \text{ K} \qquad (6.D.2)$$

The chlorine atoms in Cl_2 are in oxidation state zero. In reaction 6.D.2, chlorine in hypochlorous acid has an oxidation state of +1 and the chloride ion is reduced to −1. This type of process, in which a single element is both oxidized and reduced, is called a *disproportionation* reaction. The oxidized chlorine in hypochlorous acid is the active disinfectant. Chloride ion is produced as a harmless by-product.

Thermodynamics strongly favor the forward direction of reactions 6.D.1 and 6.D.2. Consequently, it is a reasonable approximation that each mole of chlorine gas (Cl_2) delivered to the water yields one mole of hypochlorous acid. However, hypochlorous acid is a weak acid and so may dissociate to form the hypochlorite ion, OCl^-:

$$HOCl \Leftrightarrow H^+ + OCl^- \qquad K_2 = \frac{[H^+][OCl^-]}{[HOCl]} = 2.6 \times 10^{-8}\ M\ @\ 293\ K \qquad (6.D.3)$$

The hypochlorite ion (OCl^-) is not nearly as good a disinfectant as HOCl. For example, the germicidal potency of OCl^- for coliform bacteria is a factor of 80 lower than the potency of HOCl (Chang, 1971). Consequently, for chlorination to work efficiently, the pH must be controlled to maintain a high ratio of HOCl to OCl^-. From equation 6.D.3, we see that to ensure $[HOCl] \gg [OCl^-]$, it is necessary that $[H^+] \gg 2.6 \times 10^{-8}$ M, which implies that the pH must be less than approximately 7 (see Figure 6.D.1).

The alternative reagents used for chlorine disinfection undergo an initial dissociation step that generates hypochlorite ion.

$$NaOCl \rightarrow Na^+ + OCl^- \qquad (6.D.4)$$

$$Ca(OCl)_2 \rightarrow Ca^{2+} + 2\,(OCl^-) \qquad (6.D.5)$$

In each case, the forward reaction is strongly favored. Once hypochlorite is present, it can be easily converted to hypochlorous acid by maintaining an acidic pH. Note that while the addition of chlorine gas acidifies the water, addition of sodium hypochlorite or calcium hypochlorite tends to increase the pH (since the hypochlorite ion is a base). In general, pH control is an important factor in chlorine disinfection.

The sum of hypochlorite ion and hypochlorous acid concentrations is referred to as *free available chlorine,* or simply *free chlorine.* To quantify this, free chlorine is

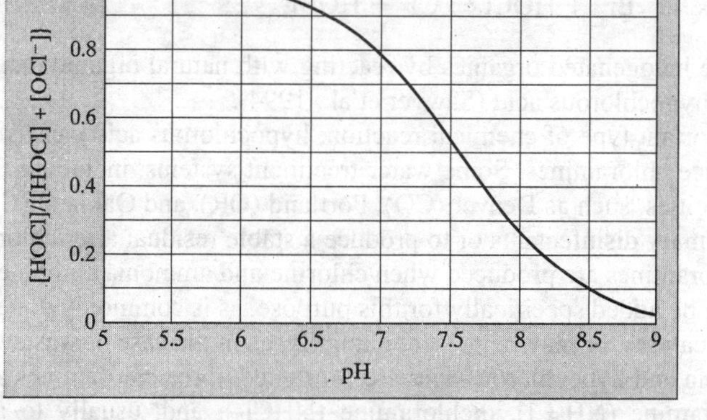

Figure 6.D.1 The fraction of hypochlorous species ($HOCl + OCl^-$) that is present as undissociated hypochlorous acid (HOCl), versus pH.

commonly expressed in terms of the mass of chlorine gas (Cl_2) per unit volume of water that would produce the given level. Thus, since the molecular weight of Cl_2 is 70.9 g/mol, a free available chlorine level of 1 mg/L implies that $[HOCl] + [OCl^-] = 10^{-3}$ g $L^{-1} \times (70.9 \text{ g/mol})^{-1} = 14\ \mu M$. To achieve a free available chlorine level of 1 mg/L, the required doses of the three main reagents per liter of water are similar: 1 mg for Cl_2, 1.05 mg for NaOCl, and 1.013 mg for $Ca(OCl)_2$. To obtain these numbers, we have taken into account that 1 mol of free available chlorine requires either 1 mol of Cl_2, 1 mol of NaOCl, or 0.5 mol of $Ca(OCl)_2$.

Reactions of Hypochlorous Acid with Chemical Impurities in Water Hypochlorous acid is a good disinfectant because it is moderately reactive toward microorganisms. However, it also reacts with chemical impurities that may be present in water. There are two major outcomes of such reactions: (1) The required dose of chlorine to achieve disinfection is increased, and (2) the chemical characteristics of the water are altered. Some of the chemical reactions have a beneficial effect on water quality; others are detrimental.

One class of chemical reactions involving hypochlorous acid is oxidation-reduction processes. In these, the chlorine atom is reduced from oxidation state +1 in hypochlorous acid (or hypochlorite) to −1, as the chloride ion. Reduced species that tend to be oxidized by chlorination include the sulfur in hydrogen sulfide and metal ions such as Fe^{2+}:

$$H_2S + HOCl \rightarrow S + H_2O + H^+ + Cl^- \tag{6.D.6}$$

$$2\,Fe^{2+} + HOCl + 5\,H_2O \rightarrow 2\,Fe(OH)_3(s) + 5\,H^+ + Cl^- \tag{6.D.7}$$

Even though they consume disinfectant, these reactions are beneficial because they convert undesirable compounds to better forms. Hydrogen sulfide is toxic and smells bad. Iron in the +2 oxidation state is soluble and imparts a metallic taste to water. Elemental sulfur and iron in the +3 oxidation state are relatively insoluble and can be easily removed by filtration.

Undesirable chemical reactions involving hypochlorous acid are the production of low-molecular-weight chlorinated and brominated organic compounds, such as the trihalomethanes. Hypochlorous acid can react with the natural organic matter present in untreated waters, especially surface waters, to form chlorinated organics. Also, if the water contains bromine, hypobromous acid, a compound that is more reactive than hypochlorous acid, may be formed:

$$Br^- + HOCl \Leftrightarrow Cl^- + HOBr \tag{6.D.8}$$

HOBr can produce halogenated organics by reacting with natural organic matter in a manner similar to hypochlorous acid (Sawyer et al., 1994).

In a third important type of chemical reaction, hypochlorous acid can react with ammonia to produce chloramines. Some water treatment systems, including those in several large U.S. cities, such as Denver (CO), Portland (OR), and Oakland (CA), use chloramines as primary disinfectants or to produce a stable residual after chlorination (Haas, 1990). Chloramines are produced when chlorine and ammonia combine in water. Ammonia may be added specifically for this purpose, as is commonly done in water treatment applications, or may be present naturally, as in the case of wastewater.

When ammonia and hypochlorous acid are combined, three chloramines are produced: monochloramine (NH_2Cl), dichloramine ($NHCl_2$), and, usually to a much smaller extent, trichloramine (NCl_3).

$$NH_3 + HOCl \rightarrow H_2O + NH_2Cl \tag{6.D.9}$$

$$NH_2Cl + HOCl \rightarrow H_2O + NHCl_2 \tag{6.D.10}$$

$$NHCl_2 + HOCl \rightarrow H_2O + NCl_3 \tag{6.D.11}$$

The chlorine atoms in the chloramine molecules have an oxidation state of +1, the same as in hypochlorous acid. This means that they are available as oxidants and take on two electrons per atom when being reduced to Cl^-. However, the potency of chloramines as disinfectants is similar to that of the hypochlorite ion, that is, much less than that of hypochlorous acid. There are two primary advantages of using chloramines in water disinfection: (1) The production of trihalomethanes is decreased, and (2) the residual disinfectant has a longer persistence in the water supply system.

To characterize the overall disinfectant characteristics of a system, we define two more terms. *Combined chlorine* refers to the total amount of chlorine present in chloramines. As with free chlorine, it is usually measured in mass concentration units, based on the amount of chlorine gas that would produce the same total oxidation potential, rather than on the basis of the chloramine molecules themselves. The following relationship is used:

$$\text{combined chlorine (g/L)} = 70.9\,[NH_2Cl] + 142\,[NHCl_2] + 213\,[NCl_3] \tag{6.D.12}$$

where the chloramine concentrations are expressed in units of moles per liter. The coefficients represent the product of the molecular weight of Cl_2 (70.9 g/mol) and the relative number of reducible Cl atoms per chloramine molecule compared with the single reducible atom per Cl_2.

The *chlorine residual* refers to the sum of the combined and free chlorine, with both expressed on the same mass concentration basis. Figure 6.D.2, known as a *breakpoint curve*, shows how the chlorine residual varies as chlorine is added to water. If the water is free of chlorine demand (i.e., reactions that consume chlorine or form combined chlorine), then the chlorine residual is all free chlorine and is exactly equal to the chlorine dose. On the other hand, if the water contains ammonia, then combined chlorine is formed, as in zone 1 in Figure 6.D.2. When the molar ratio of

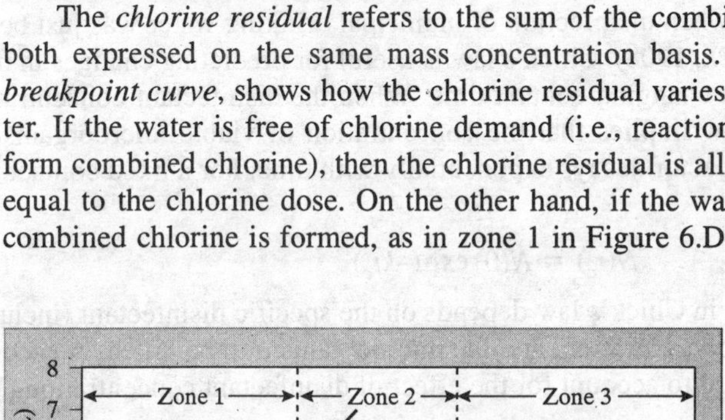

Figure 6.D.2 Breakpoint chlorination curve, for an initial ammonia nitrogen concentration of 1 mg/L. (Reprinted from *Water Chlorination Principles and Practices* [Manual M20], by permission. Copyright © 1973, American Water Works Association.)

added chlorine to nitrogen in ammonia increases beyond approximately 1.0, the combined chlorine concentration actually decreases with increasing chlorine dose, as occurs in zone 2. In this zone, the nitrogen in chloramines is being oxidized to N_2 gas, among other possible products. This chemistry can be deliberately used to remove ammonia nitrogen from water (Pressley et al., 1972).

At the position called the *breakpoint*, the chlorine residual reaches a local minimum. As chlorine is added beyond this point, the residual increases (zone 3), but now in the form of free chlorine rather than as combined chlorine. Many disinfection systems are based on providing sufficient chlorine dose to go beyond the breakpoint.

Disinfection Kinetics (for Drinking Water)

In a disinfection process, the desired reaction might be written in this form:

$$\text{live microorganism} + \text{disinfectant} \rightarrow \text{dead microorganism} \tag{6.D.13}$$

From the principles of chemical kinetics (§3.A.3), one expects the rate of this reaction to depend both on the concentration of the live microorganisms and the concentration of the disinfectant. Usually, however, the rate law is written as

$$\left(\frac{dN}{dt}\right)_{\text{disinfection}} = -kN \tag{6.D.14}$$

where N is the number concentration of viable (live) organisms and k is the reaction rate constant. This relationship is known as *Chick's law*, after the scientist who proposed it at the time when disinfection of municipal drinking water was just beginning to be practiced (Chick, 1908). Chick's law is useful for predicting changes in inactivation percentage with changing contact time. When the disinfectant concentration (C) is stable, Chick's law implies that the concentration of viable microorganisms in a batch or plug-flow reactor decays exponentially with time. For a fixed contact time, t_c, we have

$$N(t_c) = N(0)\exp(-kt_c) \tag{6.D.15}$$

The rate constant, k, in Chick's law depends on the specific disinfectant (including its chemical form), its concentration, and the microorganism to be killed. A second relationship was proposed to account for the effect of disinfectant concentration (Watson, 1908):

$$C^n t_c = \alpha \tag{6.D.16}$$

Here C is the disinfectant concentration, and n is an empirical constant. For a fixed level of inactivation (i.e., a fixed value of the ratio $N(t_c)/N(0)$), α is a constant. This second relationship is known as *Watson's law*. Together, the Chick and Watson relationships guide disinfection practice. Federal regulations for drinking water disinfection in the United States specify a product of disinfectant concentration multiplied by contact time, Ct_c, that must be maintained. Tables of critical Ct_c values to achieve fixed disinfection goals (i.e., a fixed ratio of $N(t_c)/N(0)$) have been published for specific combinations of pathogens and disinfectants (Haas, 1999). Implicit in this approach is the assumption that $n = 1$ in Watson's law. For more information about the relationship between Chick's law and Watson's law, see Exhibit 6.D.1. Examples 6.D.1 and 6.D.2 illustrate how to solve basic problems in disinfection kinetics.

EXHIBIT 6.D.1 *A Closer Look at Disinfection Kinetics*

Disinfection is achieved by maintaining water in contact with a disinfectant concentration, C, for a minimum period, t_c. We assume that Chick's law and Watson's law both apply.

The disinfection goal can be specified in terms of a certain percentage of inactivation of microorganisms, which is equivalent to $\{1-[N(t_c)/N(0)]\} \times 100$ percent, where $N(t)$ is the number concentration of live microorganisms at time t (see equation 6.D.15). Inactivation is also expressed in terms of R-log removal, where $R = -\log_{10}[N(t_c)/N(0)]$. For example, 99.9 percent inactivation implies $N(t_c) = 0.001N(0)$, which corresponds to 3-log removal ($R = 3$).

For a specified disinfection goal, by Chick's law we have

$$\frac{N(t_c)}{N(0)} = 10^{-R} = \exp(-kt_c)$$

Substituting from Watson's law for t_c, we can manipulate this equation to show that

$$k = k'C^n = \left(\frac{R\ln(10)}{\alpha}\right)C^n = \frac{C^n}{\alpha}\ln\left[\frac{N(0)}{N(t_c)}\right]$$

This relationship indicates how the reaction rate constant, k, varies with disinfectant concentration, C. For ideal behavior, α is proportional to R and $n = 1$. In this case, the reaction rate constant is proportional to the disinfectant concentration.

EXAMPLE 6.D.1 *Effect of pH on Disinfection*

Water is to be disinfected by the addition of chlorine gas (Cl_2). From experiments, it is determined that at pH = 7, 99.99 percent kill (4-log removal) is achieved with $Ct_c = 20$ min g m^{-3}, where C is the free available chlorine concentration. Predict the product Ct_c that is required to attain 4-log removal at pH = 8. Assume that the only effect of changing pH is to alter the balance between HOCl and OCl^-. Also assume that OCl^- is completely ineffective as a disinfectant. Assume that $n = 1$ in Watson's law.

SOLUTION The easiest way to approach this problem is to determine the relative amounts of [HOCl] in water at pH = 7 and pH = 8 assuming a constant concentration of free available chlorine. The ratio [HOCl]/C is determined as follows:

$$[\text{HOCl}] + [\text{OCl}^-] = \frac{C}{\text{MW}_{\text{Cl}_2}}$$

where $\text{MW}_{\text{Cl}_2} = 70.9$ g/mol is the molecular weight of chlorine gas. Equation 6.D.3 yields

$$[\text{OCl}^-] = \frac{K_2[\text{HOCl}]}{[\text{H}^+]}$$

Combining the two equations leads to this result:

$$[\text{HOCl}] = \frac{C}{\text{MW}_{\text{Cl}_2}} \times \left(\frac{[\text{H}^+]}{[\text{H}^+] + K_2}\right)$$

At pH = 7, the term in brackets equals 0.79, whereas at pH = 8 it has a value of 0.28. Thus, for a fixed free available chlorine concentration, C, the concentration of effective disinfectant (HOCl) is reduced by a factor 2.8 (0.79/0.28) by a change of pH from 7 to 8. Since Ct_c must be increased by the same factor, we predict $Ct_c = 2.8 \times 20 = 56$ min g m^{-3} at pH = 8.

EXAMPLE 6.D.2 ***Disinfection Kinetics***

A series of experiments was conducted in a batch reactor to test the disinfection by chlorine of the protozoan *Giardia lamblia*. In each experimental run, the concentration of free available chlorine was varied and the time required to kill 99 percent of the organisms was determined. The results showed that a Ct_c value of 100 min mg L^{-1} was sufficient to achieve 2-log removal.

(a) In Chick's law, what value of k corresponds to a chlorine concentration of 1 mg/L?

(b) At a chlorine dose of 1 mg/L and a contact time of 10 minutes, what fraction of the *G. lamblia* organisms would be killed?

(c) At a chlorine dose of 10 mg/L and a contact time of 5 minutes, estimate the fraction of the *G. lamblia* organisms that would be killed.

SOLUTION

(a) Since Ct_c = 100 min mg L^{-1} for 2-log removal, and given that C = 1 mg/L, we must have t_c = 100 min. We can use Chick's law (equation 6.D.15) to write the following expression:

$$\frac{N(100 \text{ min})}{N(0)} = \exp(-k \times 100 \text{ min}) = 0.01 \qquad \Rightarrow \qquad k = 0.046 \text{ min}^{-1}$$

(b) Again we apply Chick's law, now using the rate constant determined in part (*a*). We find that 63 percent of the *G. lamblia* organisms remain viable, or 37 percent are killed with 10 min of exposure to 1 mg/L of free available chlorine.

$$N(10 \text{ min}) = N(0) \exp(-0.46) \Rightarrow \frac{N(10 \text{ min})}{N(0)} = 0.63$$

(c) If we assume that n = 1 in Watson's law, we can assume that Ct_c is constant for a given disinfection goal. Then, using the result from (*b*) of a 37 percent inactivation for a C of 1 mg/L, we can set Ct_c for the two doses equal as follows:

$$Ct_c = 1 \text{ mg/L } (10 \text{ min}) = 10 \text{ mg/L } (X \text{ min})$$

The time required to achieve 37 percent inactivation at a C of 10 mg/L is X = 1 min. We can now apply Chick's law to evaluate k for this dose:

$$\frac{N(1 \text{ min})}{N(0)} = \exp(-k \times 1 \text{ min}) = 0.63 \qquad \Rightarrow \qquad k = 0.46 \text{ min}^{-1}$$

Comparing this result with part (*a*), we see that a 10× increase in disinfectant concentration is estimated to produce a 10× increase in the disinfection rate constant. Finally, this rate constant can be used in Chick's law to evaluate the *G. lamblia* inactivation at a chlorine dose of 10 mg/L and a contact time of 5 min:

$$N(5 \text{ min}) = N(0) \exp(-0.46 \times 5 \text{ min}) \qquad \Rightarrow \qquad \frac{N(5 \text{ min})}{N(0)} = 0.10$$

We see that 90 percent of the *G. lamblia* organisms are inactivated with a Ct_c value of 50 min mg L^{-1}. In this case each unit of log removal requires a Ct_c increment of 50 min mg L^{-1}. The constant α in equation 6.D.16 has a value of 50R min mg L^{-1}, and $k = 0.046C$ (min^{-1}), where C has units of mg L^{-1}. This example illustrates an important point: *G. lamblia* is fairly resistant to disinfection with chlorine (Hoff and Akin, 1986).

6.D.2 Coagulation and Flocculation

Coagulation and flocculation are typically combined in a two-stage process applied in drinking water treatment to assist in the removal of small, suspended particles. Coagulation and flocculation are transformation processes. The goal is to cause small particles to combine into larger aggregates that can be more readily separated from water by sedimentation and filtration.

As will be discussed in more detail below, there is a natural tendency for clays and other colloidal particles (diameter range 0.001–1 μm) to have negative electrical charge on their surface. In freshwater, electrostatic interactions between the particles and the ions that surround them prevent the particles from colliding to form aggregates. Given their very small size, these particles may remain suspended in water for long periods. (For example, a 1 μm mineral particle, with a settling velocity of 10^{-4} cm s^{-1}, would require about a month to settle through a distance of 3 m.) Suspensions of colloidal particles are called *stable* if their surface properties prevent them from aggregating and they therefore remain suspended in the water column for a long period. The processes of coagulation and flocculation destabilize particles and promote collisions so that particles that may otherwise remain suspended for months in freshwater are formed into aggregates and can be separated from water in an hour or less (Amirtharajah and O'Melia, 1990).

In the *coagulation* stage of the process, a chemical reagent is added and rapidly mixed into the water to destabilize the colloidal particles. In the *flocculation* stage, the water is slowly stirred to promote collisions between particles. The aggregates formed in this process are known as *flocs*.

When coagulation and flocculation are applied for the treatment of drinking water from surface sources, the process is placed ahead of the sedimentation basin. The flocs are then removed by sedimentation or, if they pass through this stage, by filtration. If the source water is relatively free of particles, a sedimentation basin may not be needed. In this case, a coagulant may be added upstream of a rapid-sand filter to improve its performance. Raw water with turbidity of 10 NTU applied to a well-designed filter will be clarified only to 5 NTU without coagulants, but can be controlled to 0.2 NTU with optimal coagulation and flocculation (ASCE/AWWA, 1990).

In addition to improving efficiency for removing inert particles, coagulation and flocculation have been found to improve the removal of many other undesirable impurities from water. Microorganisms are more efficiently removed from water by sand filters when preceded by coagulant addition. Coagulation and flocculation also help in the removal of natural organic matter—humic and fulvic acids—present in the form of macromolecules or fine colloidal particles. If not removed, these materials can cause color and contribute to the formation of undesirable disinfection by-products in drinking water. Other materials, such as toxic metals, synthetic organic molecules, iron, and manganese, may be sorbed to fine particles and removed with those particles.

To understand how coagulation and flocculation function, we need to consider two major aspects of coagulation: collision-inducing mechanisms and particle interactions. For particles to collide, they must move relative to one another. Such movement may originate from one or more of four main processes, as illustrated in Figure 6.D.3. In *differential settling*, a large particle settles more rapidly than a smaller particle, overtaking it. *Brownian motion* induces collisions due to random particle motion. *Laminar* and *turbulent shear* cause particles on different fluid streamlines to move toward one another because of a velocity gradient in the fluid. It is the rate of shear-induced collisions that one seeks to enhance in a flocculator.

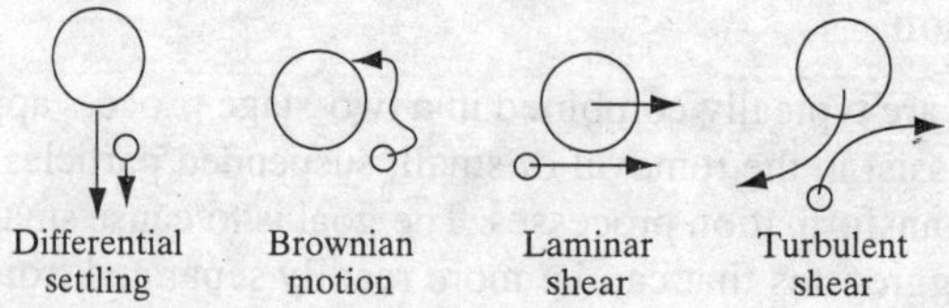

Figure 6.D.3 Major mechanisms that cause collisions between particles suspended in water.

Collisions between colloidal particles would generally cause the particles to adhere to one another. The nonspecific van der Waals forces that exist between molecules in solids often cause sufficient adhesion between particles. Furthermore, even in the absence of a flocculator, colloidal particles would tend to collide frequently enough to aggregate rapidly in freshwater were it not for electrostatic repulsion. This repulsion, caused by surface charge on particles, is strong enough to make colloidal suspensions stable in freshwater.

To understand the electrostatic repulsion between particles, we must explore why particles have surface charge. There are many types of colloidal particles in water, and the mechanisms that induce surface charge vary among them. To keep the discussion relatively simple, we will focus on the illustrative case of clay particles, which are a major cause of turbidity in water.

Clays are small soil particles that are formed chemically by dissolution and precipitation processes acting on larger soil particles and rocks. Most clay grains are made of aluminum oxides and silicon oxides, organized into alternating sheets. Clays are capable of absorbing large quantities of water between the sheets, which causes swelling.

When suspended in freshwater, clay particles tend to have a negative surface charge due to three dominant mechanisms (Stumm and Morgan, 1996): (1) acid-base reactions caused by hydroxyl groups at the surface that donate protons at pH above 4, (2) isomorphic substitution caused by the replacement of cations with high oxidation state (e.g., Si(IV)) by those with lower oxidation state (e.g., Al(III); see Figure 6.D.4), and (3) ion adsorption (see §3.B.3).

In a solution that contains negatively charged particles, one might expect that direct electrostatic repulsion would cause the particles to repel each other, leading to a stable suspension. The reality is somewhat more complex. As we know, water contains both cations and anions. The cations are attracted by the negative surface charge of colloidal particles, so that clouds of positive charge surround these particles. The electrostatic attraction between the negative surface charge and the positive ions is offset by the electrostatic repulsion among the cations themselves. The size of the cationic clouds surrounding colloidal particles depends on the ionic strength of the solution. In freshwater with a low ionic strength, the cloud of positive charge is relatively

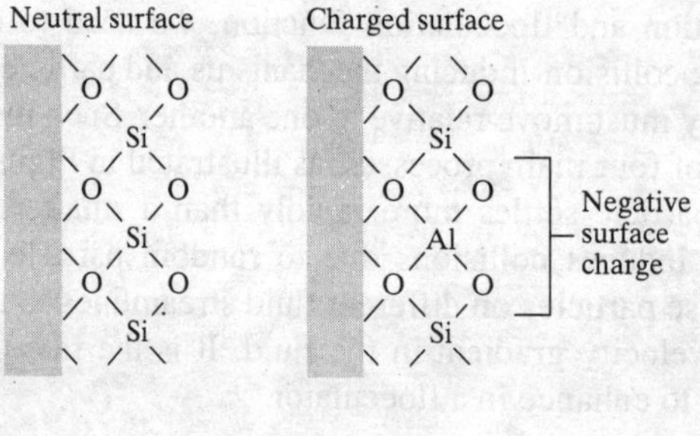

Figure 6.D.4 The isomorphic substitution of a cation with a lower oxidation state (in this case, Al(III) replaces Si(IV)) produces a negative surface charge in clay particles (Amirtharajah and O'Melia, 1990). The same process can occur when a divalent cation (e.g., Ca^{2+} or Fe^{2+}) substitutes for Al(III).

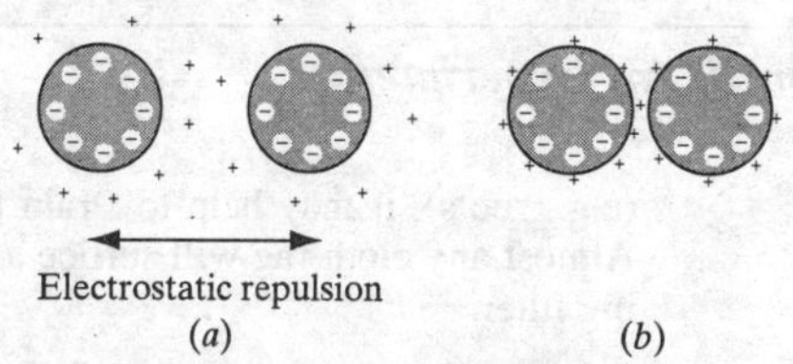

Figure 6.D.5 Schematic representation of negatively charged particles surrounded by cation clouds. In (*a*) the ionic strength is low and the cation clouds are diffuse. Electrostatic repulsion between the cationic clouds prevents particles from colliding. In (*b*) the ionic strength is high allowing compression of the cationic clouds to the immediate vicinity of the particle surfaces, thereby permitting particles to collide.

diffuse because fewer ions are available in solution to be attracted to the particle (Figure 6.D.5[*a*]). Particles are prevented from colliding with one another by electrostatic repulsion between their diffuse cation clouds. Thus, a suspension of clay particles in freshwater is stable. In seawater, on the other hand, where the ionic strength is quite high, sufficient cations to neutralize a particle's charge can be attracted to the immediate vicinity of the particle surface (Figure 6.D.5[*b*]). In this case, with small separation between the surface charge and the counteracting cations, particles appear almost electrically neutral, and they can move close enough together to collide.

The ionic cloud plus the charged particle surface is called an *electrical double layer.* The reduction in cloud size that occurs when the ionic strength increases is called *double-layer compression*. A natural example of the effect of double-layer compression on the stability of colloidal particles occurs where high-flow rivers discharge into the ocean (Figure 6.D.6). Fine particles that are suspended and transported great distances in freshwater rapidly coagulate and settle when mixed with saltwater. In aerial photographs of the world's largest rivers, turbid freshwater can be seen extending great distances from the river's mouth and a fairly sharp line is exhibited, demarcating the boundary between saltwater and freshwater. Sailors have used this feature to identify freshwater while traveling near continental shores. This phenomenon also contributes to the formation of deltas. An experiment that can be conducted with ordinary materials dramatically demonstrates these effects (see Exhibit 6.D.2).

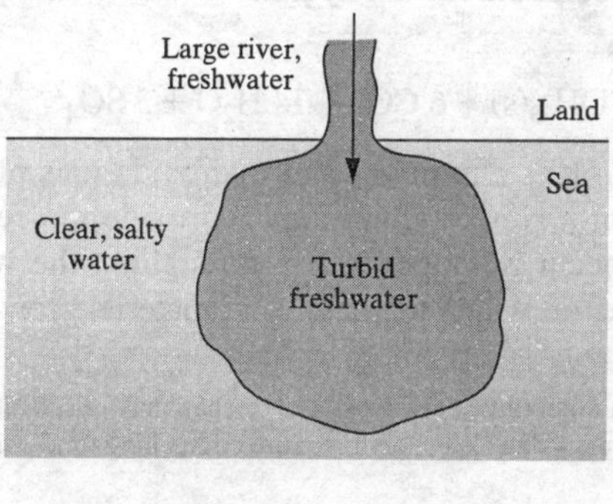

Figure 6.D.6 Stable particle suspensions in freshwater become unstable when mixed with seawater because of double-layer compression.

EXHIBIT 6.D.2 *How Deltas Are Formed: An Experiment*

You will need the following materials:

- Soil, roughly 10 g (high clay content is best)
- Two jars with tight-fitting lids
- Two additional water containers of any type (e.g., rinsed milk cartons)
- Freshwater, roughly 2 L (tap water is fine)
- Clean seawater, roughly 250 mL*

Once you have collected the materials, follow this procedure:

1. Make a stable suspension of clay in water. Mix the soil with 200 mL freshwater in one of the containers. Eliminate the particles that settle rapidly by decanting the murky supernatant into the second container, and then rinsing and discarding the particles from the bottom of the first. Repeat this process a few times until it appears that most of the particles will remain in suspension for many minutes. If your soil has many items that float (e.g., roots), it may help to strain the water. Almost any cloth rag will suffice as a straining filter.
2. Prepare the receiving waters. Label the two jars "freshwater" and "saltwater," respectively. Fill each jar to 50 percent of its capacity with the appropriate, clear water.
3. Add equal amounts of the clay suspension to each jar, seal, and mix thoroughly. Place the jars where they can be observed without being disturbed. Check the turbidity in the water in the two jars at intervals after mixing. The turbidity can best be seen by shining a light at the jars at an angle of 90 degrees from the viewing position. Initially, look at them 15, 30, and 60 minutes after mixing. Then observe them again many hours and a day later. You should see a dramatic difference emerge in the clarity of the two samples.

Although double-layer compression is commonly observed in nature, it is not the dominant mechanism employed for destabilizing particles in water treatment applications. In fact, coagulants applied in water treatment operate through four mechanisms: double-layer compression, charge neutralization through adsorption of oppositely charged ions, interparticle bridging, and precipitate enmeshment. Of these four, double-layer compression is perhaps the least important. Interparticle bridging refers to a condition in which a coagulant forms a polymer chain that becomes attached to two or more particles. In this way, the particles are bound together even though they may never come into contact. Enmeshment occurs when particles are trapped within a colloidal floc either as it forms or as it subsequently settles through the water (Amirtharajah and O'Melia, 1990).

The most widely used coagulant in drinking water treatment is alum—that is, hydrated aluminum sulfate, $Al_2(SO_4)_3 \cdot 14\,H_2O$—which is added to water either in dry or liquid form. The typical dose is about 10 mg/L (ASCE/AWWA, 1990), which corresponds to 33 μM (0.9 mg/L) of aluminum. If the water has sufficient alkalinity, alum undergoes a reaction like this to form solid aluminum hydroxide (Davis and Cornwell, 1998):

$$Al_2(SO_4)_3 \cdot 14\,H_2O + 6\,HCO_3^- \rightarrow 2\,Al(OH)_3(s) + 6\,CO_2 + 14\,H_2O + 3\,SO_4^{2-} \quad (6.D.17)$$

In reality, the solid that is formed involves aluminum and hydroxide ions plus water molecules, and is more complex than a simple aluminum hydroxide molecule. The precipitation process also does not occur homogeneously throughout the water. Instead, existing particles serve as nuclei on which the new solid material grows.

*Water from the ocean is best; however, if you do not live near the coast, you can make a substitute by dissolving 35 g of table salt (NaCl) per liter of water. The experiment is more compelling (and collecting the materials more fun) if you use real seawater.

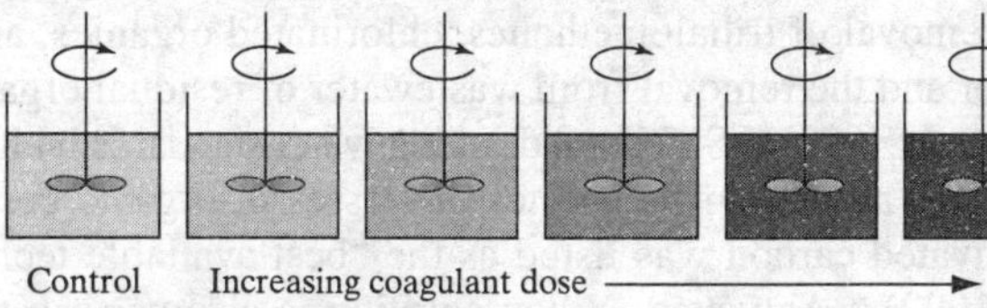

Figure 6.D.7 Schematic representation of the jar test used to manage the coagulation process in water treatment.

Instead of alum, coagulants based on inorganic iron ions, such as ferric chloride, $FeCl_3$, or ferric sulfate, $Fe_2(SO_4)_3$, are sometimes used. In this case, solid ferric hydroxide, $Fe(OH)_3$, is formed instead of aluminum hydroxide. In addition to the main coagulant, some processes add a synthetic organic polymer, known as a *polyelectrolyte*, as a coagulant aid. The use of a coagulant aid can reduce the required coagulant dose and may yield larger, tougher flocs through promotion of interparticle bridging.

The coagulant addition step occurs in a rapid-mix reactor. To optimize performance of the coagulant, the pH may be manipulated, either by adding sulfuric acid (to lower pH), or lime (CaO) or soda ash (Na_2CO_3) (to raise pH). The optimum pH is determined by a jar test, as described below. Typically, for alum it is in the range 5.5–7.5; with ferric chloride the optimum pH may lie in the range 5–8.5 (ASCE/AWWA, 1990).

After a brief period of vigorous mixing, the water is transferred to a flocculator (unless in-line filtration is being used), where it undergoes slow stirring for a period of 20–40 minutes to promote floc growth. The goal of the flocculator is to maximize the rate of interparticle contact without stirring so vigorously that flocs are broken. Flocculators are designed to achieve mixing either hydraulically or mechanically, in the latter case using either mixing blades or paddles. The flocculated particles may be smaller than 100 μm or may be as large as 0.1–3 mm (ASCE/AWWA, 1990).

Coagulant types and doses are generally estimated using a procedure known as a *jar test*. With one simple test apparatus, the performance of different coagulants and potential coagulant aids can be compared, while the optimum pH, flocculator mixing rate, and detention time can be chosen (ASCE/AWWA, 1990). The test can be repeated at intervals ranging from many hours to a few days, and the results can be used to obtain the best performance in response to changing water conditions.

The basic procedure involves executing a series of batch experiments that mimics the full-scale process. For example, to test for proper coagulant dose, samples of untreated water are collected in a series of jars to which different coagulant doses are added (Figure 6.D.7). The coagulant is first mixed vigorously for a brief period (~1 minute), and then the water is stirred slowly for 15–20 minutes to represent the flocculation stage. The mixer is stopped and the flocs and other particles are allowed to settle for an additional 30 minutes. Finally, the turbidity of the supernatant fluid is determined. The best conditions among those tested are those that produce the lowest residual turbidity.

6.D.3 Sorption of Organic Molecules

Sorption, using activated carbon, is applied in water and wastewater treatment to remove undesired organic molecules. Historically, the main purpose for using activated carbon in water quality engineering was to remove compounds that cause taste and odor problems in drinking water. More recently, other applications have emerged, including the removal of trihalomethanes, chlorinated organics, and pesticides from drinking water and the removal from wastewater of residual organics that cannot be biodegraded. As of 1992, U.S. federal drinking water standards included specific concentration limits for more than 50 species or classes of organic compounds. Sorption on granular activated carbon was listed as the "best available technology" (BAT) for 52 (Pontius, 1992).* Activated carbon sorption is also used in the treatment of hazardous wastes and in air pollution control.

Activated carbon is a relatively inexpensive material with an enormous specific surface area, typically about 1000 m^2 g^{-1}. To put this in everyday terms, packed into a small handful of activated carbon is a total area greater than that of a football field. By contrast, soil minerals might have 0.1–10 m^2 per gram and glass beads (1 mm diameter) would have only 0.004 m^2 g^{-1} of area. The enormous surface area of activated carbon means that the use of small amounts can be sufficient to capture significant quantities of pollutants. The surface area in activated carbon is distributed throughout the grains in pores as small as 1 nm.

Activated carbon is produced in a two-stage process from some carbonaceous material, such as wood, bark, lignite, coal, bone, solid petroleum residues, coconut husks, or nut shells. The raw material is pyrolyzed, that is, heated to a high temperature ($\leq$700 °C) in the absence of air. This step causes all of the organic molecules to be volatilized. Then the residual material is "activated" by heating to 800–900 °C in the presence of a weak oxidizing agent such as steam or carbon dioxide. What remains is a highly porous structure consisting almost entirely of carbon. Properties of activated carbons vary because of differences in source materials and processing.

Another important property of activated carbon is its nonpolar nature. Water is a highly polar molecule, while many organic molecules are nonpolar. With respect to polarity, "like attracts like," so nonpolar organics in water are attracted to the surface of activated carbon. On the other hand, charged species are not effectively captured by activated carbon. Even among organic molecules, the sorption affinity of activated carbon varies broadly. For example, organics with the following characteristics are generally more effectively sorbed: nonpolar, low aqueous solubility, high molecular weight, nonionic, branched chain, and unsaturated.

For water treatment, activated carbon is used in one of two forms: powdered (PAC) or granular (GAC). The grain size of PAC is 5–100 μm while GAC grains are typically 0.6–0.9 mm in diameter (ASCE/AWWA, 1990). When PAC is used, it is added to water as a slurry upstream of the granular filter. The dose is typically less than 25–50 mg/L but might be as high as 100 mg/L for some applications (Snoeyink, 1990). The PAC laden with sorbed organic molecules is then removed from water and discarded, either in the sedimentation basin or the granular filter. Powdered activated carbon is widely used for taste and odor control. Surveys of water utilities in the United States in the 1970s and 1980s showed that 25–30 percent used PAC.

Granular activated carbon is typically used in a fixed bed, like a granular sand filter (see §6.C.2). In fact, GAC is sometimes used in place of anthracite in a dual-bed filter or even in place of sand as the only filter medium. Alternatively, a separate contactor is provided downstream of filtration. The useful life of a GAC bed might range from a few months to several years. Spent GAC can be regenerated by means of heating or it may be discarded. In some applications involving volatile organic compound control, the pollutant is first transferred from the aqueous to the gas phase by aeration

*For 29 compounds it was the only BAT listed; for the other 23, it was listed along with packed-tower aeration.

(see §8.C.1), and then the gas stream is passed through a GAC bed (Crittenden et al., 1988).

The performance of activated carbon for removing contaminants from water depends on both equilibrium and kinetic characteristics of the system. Equilibrium is quantified by a sorption isotherm (§3.B.3), which specifies the equilibrium relationship between the aqueous- and sorbed-phase concentrations of the contaminant (see, e.g., Table 3.B.5). The kinetics of sorption dictate how rapidly equilibrium conditions are approached and are a function of mass transfer rates. Contaminant sorption on a grain of activated carbon can be described in four sequential steps:

1. Transport from the bulk fluid to the vicinity of the grain
2. Transport across a thin boundary layer adjacent to the grain
3. Transport into the pores of the grain to a free sorption site
4. Sorption of the contaminant

The first and fourth steps are usually rapid, so the overall rate of sorption is governed by the rate of the second and third steps: The time required for contaminants to diffuse throughout a spherical grain varies with the square of particle radius. Consequently, the time required to approach equilibrium is expected to be 200 times less for a 50 μm PAC particle than for a 0.75 mm GAC grain.

One usually neglects sorption kinetics in the design and operation of a treatment process that utilizes PAC because of the complexity involved in quantifying these rates. Instead, one seeks to ensure that the contact time is long enough, say at least 30 minutes, to approach equilibrium. Assuming that equilibrium prevails, the relationship between the dose of PAC applied and the removal of single contaminants can be addressed using isotherm data and material balances, as demonstrated in Example 6.D.3.

EXAMPLE 6.D.3 *Use of Powdered Activated Carbon for Organics Control*

(a) Untreated drinking water is found to contain 20 μg/L of geosmin, which imparts an undesirable earthy-musty odor to the water. What fractional removal can be achieved by the addition of 10 mg/L of PAC? Assume that geosmin sorption on PAC is described by a Freundlich isotherm, $q_e = K_f C_e^{1/n}$. Here q_e represents the mass (μg) of sorbed geosmin per milligram of PAC, C_e represents the mass concentration of dissolved geosmin in water (μg/L), $K_f = 0.5$ (μg/mg) (L/μg)$^{1/n}$ and $1/n = 1.08$.

(b) A water supply is contaminated with 1 mg/L of 1,1,1-trichloroethane (TCA). What dose of powdered activated carbon must be added to reduce the concentration to the 0.2 mg/L standard? Again assume that a Freundlich isotherm applies, with $K_f = 2.5$ (mg/g) (L/mg)$^{1/n}$ and $n = 2.94$.

SOLUTION

(a) In this problem, both the final concentration of dissolved geosmin and the amount of sorbed geosmin in the treated water are unknown, so two equations are needed. One is the Freundlich isotherm. The other is based on mass conservation of geosmin. Let x be the mass of sorbed geosmin per volume of water (μg/L). Then we can write

$$x + C_e = 20 \ \mu\text{g/L}$$

$$q_e = \frac{x}{10 \text{ mg/L}} = 0.5 C_e^{1.08}$$

EXAMPLE 6.D.3 ***Use of Powdered Activated Carbon for Organics Control (continued)***

Substitute the second equation into the first and solve to find $C_e = 3.1$ μg/L. (See Appendix D, §D.3.) Thus, since 3.1 is 16 percent of 20, 84 percent of the geosmin can be removed from the water by using this dose of PAC.

(**b**) The initial concentration of TCA in water is $C = 1$ mg/L, and the desired final state is $C_e = 0.2$ mg/L. By material balance, this means that we must add enough PAC to sorb 0.8 mg of TCA for every liter of water. As in (*a*), define x (= 0.8 mg/L) as the quantity of TCA sorbed per volume of water. Let m be the mass of PAC added per volume of water. Then

$$q_e m = x = 0.8 \text{ mg/L}$$

But q_e must also satisfy the adsorption isotherm,

$$q_e = \frac{x}{m} = K_f C_e^{1/n}$$

Rearranging and substituting, we find

$$m = \frac{x}{K_f C_e^{1/n}} = \frac{0.8 \text{ mg/L}}{2.5 \text{ mg/g} \times 0.2^{0.34}} = 0.55 \text{ g L}^{-1}$$

So a dose of 550 mg of PAC per liter of water is required to achieve the desired removal. This is an impractically large dose. We see that TCA is only weakly sorbed by activated carbon and that PAC is not likely to be a practical control method.

Next, let's explore how a bed of GAC functions in water treatment by considering a specific experimental situation. A column with uniform cross-sectional area A and length L is filled with GAC. Water flows through the column at rate Q (volume per time). Initially, the water is free of organic contaminants. At time $t = 0$, the concentration of a single contaminant at the inlet is suddenly increased to C^* and maintained at that value throughout the rest of the experiment. The concentration of the contaminant in water at the outlet of the column, C_{out}, is continuously measured. Throughout the experiment, water can be sampled and analyzed for contaminant concentration from positions along the length of the column.

Figure 6.D.8 illustrates how the contaminant concentration varies with position at some specific time during the experiment. The column is divided into three zones. In the *saturated zone*, the GAC is at equilibrium with the inlet contaminant concentration. No net transfer of contaminant from water to GAC occurs here; instead, the contaminant flows through this region to the *mass-transfer zone*. In the mass-transfer zone, the GAC has less sorbed contaminant than it would at equilibrium. So, in this region, contaminant is transferred from water to the GAC grains. At the downstream edge of the mass-transfer zone, the contaminant has been completely sorbed, and the concentration in water has dropped to zero. The remainder of the column is the *unexposed zone*, where the GAC has no sorbed contaminant.

As the experiment continues, the saturated zone grows and the unexposed zone shrinks as more contaminant is sorbed by the GAC. The length of the mass-transfer zone is a function of the sorption kinetics, with slower kinetics producing a longer zone. Under typical conditions in which sorption is favorable, the shape of the concentration profile in the mass-transfer zone does not change once it is initially established, but rather is advected at a constant, slow velocity toward the outlet (McCabe et al., 1993).

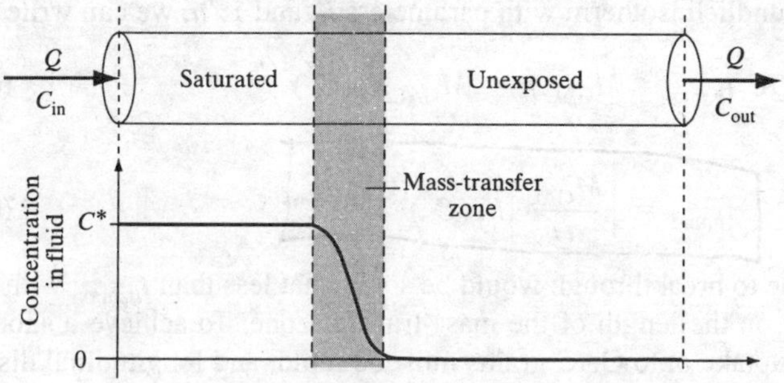

Figure 6.D.8 A concentration profile for a sorbing contaminant in water flowing through a column of granular activated carbon.

Figure 6.D.9 shows how the contaminant concentration at the outlet varies as a function of time. An important parameter is the *breakthrough time*, t_b, defined as the period from the beginning of contaminant exposure until the outlet concentration rises to some prescribed fraction of C^*, typically 5–10 percent (LaGrega et al., 1994). The breakthrough time represents the period in which a column can be operated before the GAC must be regenerated or replaced. The value of t_b is a function of the inlet concentration, the flow rate through the column, the equilibrium sorption capacity of the activated carbon for the contaminant, and the quantity of activated carbon in the column. It is also a function of the length of the mass-transfer zone. If the mass-transfer zone is short (rapid kinetics), then the outlet concentration rises above zero only when the bed is almost fully saturated. When the mass-transfer zone is long (slow kinetics), breakthrough occurs well before the GAC grains in the mass-transfer zone are in equilibrium with the inlet concentration.

An upper bound on the breakthrough time, $t_{b,\max}$, can be estimated using a material balance by assuming that equilibrium has been reached throughout the bed. (This assumption is equivalent to assuming that the mass-transfer zone has zero length.) In this case, since no contaminant escapes the column prior to breakthrough, the mass of contaminant that enters the column during the exposure period ($QC^*t_{b,\max}$) must equal the mass that is sorbed onto the GAC. This mass can be calculated by multiplying the mass of GAC in the column, M_{GAC}, by q_e, the mass of contaminant sorbed per mass of GAC at equilibrium with the inlet concentration. Assuming that the equilibrium is

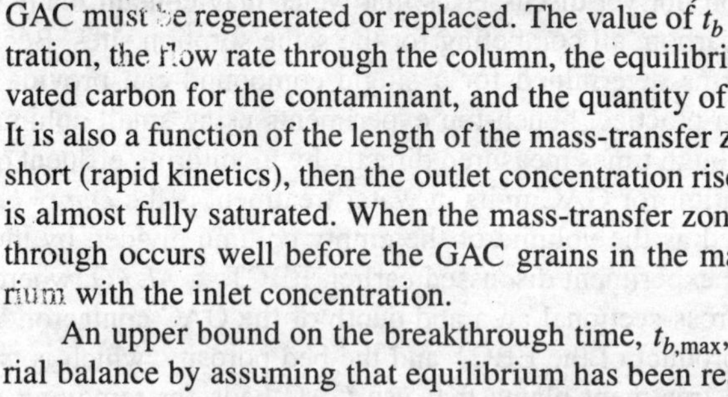

Figure 6.D.9 Inlet concentration and breakthrough curves, showing how the contaminant concentration at the outlet of a GAC column (right) evolves after the inlet concentration (left) is changed from zero to C^* at time $t = 0$. The shaded area is proportional to the mass of contaminant that is sorbed by the GAC column when it has reached equilibrium.

described by a Freundlich isotherm with parameters K_f and $1/n$, we can write

$$QC^* t_{b,\max} = M_{\mathrm{GAC}} q_e = M_{\mathrm{GAC}} K_f (C^*)^{1/n} \quad (6.D.18)$$

or

$$t_{b,\max} = \left[\frac{M_{\mathrm{GAC}}}{Q}\right] K_f (C^*)^{(1-n)/n} \quad (6.D.19)$$

The actual time to breakthrough would be somewhat less than $t_{b,\max}$, with the difference depending on the length of the mass-transfer zone. To achieve a short mass-transfer zone, the uptake onto GAC grains must be rapid, and longitudinal dispersion must be weak (McCabe et al., 1993). Mass transfer within grains can be a particularly important factor. Since the time required for a grain to become saturated with a contaminant varies as the square of the grain diameter, it is desirable to use the smallest GAC grains possible. A trade-off must be made against hydraulic head loss. As grain size becomes smaller, head loss increases. Another design trade-off involves bed length: A longer bed is associated with higher head loss but allows longer operation before regeneration is needed. In some systems, multiple columns are used in series. This design requires a greater capital investment (for additional valves and plumbing) but permits greater GAC utilization since the first bed can be completely spent before it is taken off-line for regeneration (Snoeyink, 1990).

A major complication not yet discussed is that water may contain many species that sorb onto activated carbon, all competing for the same sorption sites. Because of this, the sorption isotherms determined for a single compound can provide only a rough guide to design. In practice, bench-top experiments using small columns may be conducted or breakthrough times measured directly by monitoring effluent quality.

A common specification for GAC units in water treatment is the *empty-bed contact time* (EBCT), defined as the volume of the empty column divided by the water flow rate. In terms of the experiment discussed earlier, EBCT $= AL/Q$, where A and L are, respectively, the cross-sectional area and depth of the GAC contactor. The actual contact time is the product of the EBCT and the bed porosity, which is typically 0.4–0.5. A survey of 47 treatment plants that use GAC beds for removing organic compounds from water found median values of 10 min for EBCT and 1 m for bed depth (Roberts and Summers, 1982).

Granular activated carbon offers several advantages over PAC. Lower effluent contaminant concentrations can be practically achieved with GAC because the contaminant is reduced to progressively lower concentrations in the mass-transfer and unexposed zones. Consequently, GAC may be suitable for removing trihalomethanes and other volatile organic compounds from water for which the required PAC doses would be impractically large. The use of a GAC bed also provides a stable form of protection against short-term contaminant releases into the water supply. If treatment with activated carbon is needed on a continuous basis, GAC is generally less expensive than PAC. On the other hand, PAC offers considerable flexibility and can be the method of choice if treatment is needed only intermittently.

6.D.4 Chemical Precipitation and Ion Exchange

Chemical precipitation and ion exchange are two techniques that are applied to remove undesired ions from water. For some applications, these methods are attractive alternatives to membrane technologies such as reverse osmosis, nanofiltration, and electrodialysis (§6.C.3). Chemical precipitation and ion exchange are widely used for water softening, that is, the removal of excess concentrations of multivalent cations from water, especially Ca^{2+} and Mg^{2+} (§2.C.4). Both technologies can also be applied for the removal of other undesirable inorganic cations such as barium, cadmium, chromium, nickel, and radium. Although they are based on fundamentally different chemical mechanisms, the two processes are discussed together because of the large overlap of their application areas. We emphasize water softening because this is one of the most common uses of chemical precipitation and ion exchange. This application also illustrates many of the important issues that must be considered when using these methods.

Chemical Precipitation

The objective in chemical precipitation is to convert dissolved ions to solids that can be removed by sedimentation. To achieve this objective, one must first identify an appropriate solid that can be formed from each ion of concern. To be efficient, the solid must have a low solubility product and the reagent should be inexpensive and innocuous.

Water softening by chemical precipitation is generally achieved using the *lime-soda process*. The target precipitates in this process are calcium carbonate ($CaCO_3$) to remove Ca^{2+} and magnesium hydroxide ($Mg(OH)_2$) to remove Mg^{2+}. The precipitation-dissolution relationships for these compounds are (see §3.B.2 for background)

$$CaCO_3(s) \Leftrightarrow Ca^{2+} + CO_3^{2-} \qquad K_{sp1} = 8.7 \times 10^{-9}\,M^2 \tag{6.D.20}$$

$$Mg(OH)_2(s) \Leftrightarrow Mg^{2+} + 2\,OH^- \qquad K_{sp2} = 1.2 \times 10^{-11}\,M^3 \tag{6.D.21}$$

Recall that the solubility product is satisfied when the dissolved ions are in equilibrium with a solid. From reactions 6.D.20 and 6.D.21, the concentrations of the principal hardness ions ($[Ca^{2+}]$ and $[Mg^{2+}]$) that can exist at equilibrium in water are given by the relationships

$$[Ca^{2+}] \leq \frac{K_{sp1}}{[CO_3^{2-}]} \tag{6.D.22}$$

$$[Mg^{2+}] \leq \frac{K_{sp2}}{[OH^-]^2} \tag{6.D.23}$$

These expressions show that by increasing the concentrations of appropriate anions, in this case carbonate (CO_3^{2-}) and hydroxide (OH^-), we can reduce the concentration of the undesired cations by forcing them to form solid products. Carbonate and hydroxide satisfy the other conditions of being relatively inexpensive and posing no hazard if properly managed.

The name *lime-soda process* derives from the raw chemicals that are applied. Lime (CaO) is typically used to raise the pH by increasing $[OH^-]$. The lime undergoes a treatment step in advance of application in which it is converted to calcium hydroxide:

$$CaO + H_2O \rightarrow Ca(OH)_2 \tag{6.D.24}$$

Sodium carbonate (Na_2CO_3), also known as *soda ash* or simply *soda,* is added to increase the concentration of carbonate in the water. Adequate carbonate may already be available in the water in the form of bicarbonate (HCO_3^-). If so, raising the pH by

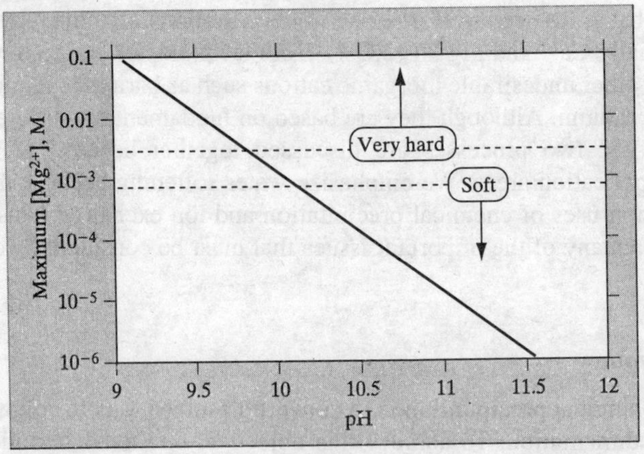

Figure 6.D.10 Concentration of magnesium ions in equilibrium with solid $Mg(OH)_2$ as a function of pH according to reaction 6.D.21. The horizontal lines labeled "very hard" and "soft" correspond to total hardness due to $[Mg^{2+}]$ alone of 6 meq/L and 1.5 meq/L (or 300 and 75 mg/L as $CaCO_3$), respectively.

adding calcium hydroxide will convert the bicarbonate to carbonate (see §3.C.4 and especially Figure 3.C.2), which can then precipitate with calcium. Both calcium hydroxide and sodium carbonate are highly soluble in water and therefore can be assumed to fully dissociate into their ionic constituents—Ca^{2+} and OH^- in one case, and Na^+ and CO_3^{2-} in the other.

Figures 6.D.10 and 6.D.11 illustrate the effect of increasing pH on the equilibrium concentrations of $[Mg^{2+}]$ and $[Ca^{2+}]$. The effect is particularly dramatic for magnesium. Since the solubility product of magnesium hydroxide depends on the

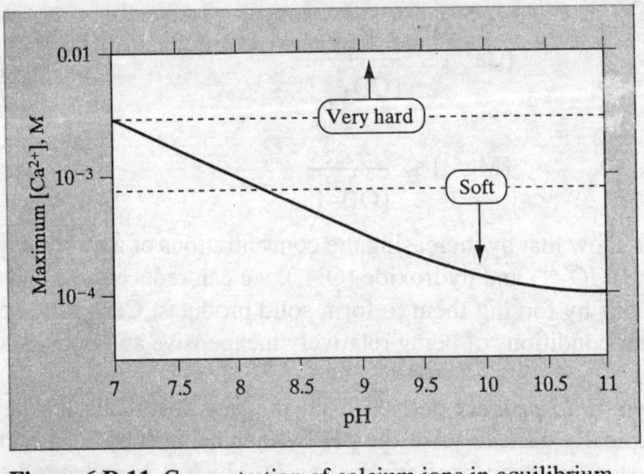

Figure 6.D.11 Concentration of calcium ions in equilibrium with solid $CaCO_3$ as a function of pH, according to reaction 6.D.20. The alkalinity of the untreated water is assumed to be equal to the total hardness, which is assumed to be associated entirely with $[Ca^{2+}]$. The pH change is assumed to occur without addition of $[Ca^{2+}]$. The horizontal lines labeled "very hard" and "soft" correspond to total hardness of 6 meq/L and 1.5 meq/L (or 300 and 75 mg/L as $CaCO_3$), respectively.

square of the hydroxide concentration, increasing the pH by one unit reduces the maximum concentration of Mg^{2+} that can be present at equilibrium by a factor of 100. As the pH increases above 10, the maximum potential contribution of magnesium to hardness becomes insignificant. In practice, when magnesium removal is required, the pH is raised to approximately 11 to ensure as much removal as possible.

The effect of increasing pH on the equilibrium concentration of the calcium ion is more complex since the concentration of carbonate itself varies with pH and with the formation of the calcium carbonate solid. Figure 6.D.11 is plotted with these assumptions: (1) all hardness is caused by $[Ca^{2+}]$, (2) alkalinity is equal to the calcium hardness ($[Ca^{2+}] = [CO_3^{2-}] + 1/2\,[HCO_3^-]$), and (3) the pH is raised without increasing $[Ca^{2+}]$. We can use these assumptions along with equation 6.D.22 to show that the calcium ion concentration in equilibrium with solid $CaCO_3$ is given by

$$[Ca^{2+}] = \left[K_{sp1}\left(1 + \frac{[H^+]}{2K_2}\right)\right]^{1/2} \tag{6.D.25}$$

where $K_2 = 4.68 \times 10^{-11}$ M is the acid dissociation constant for bicarbonate (see reaction 3.C.16). For these conditions, $[Ca^{2+}]$ decreases by a factor of $10^{1/2}$ for each unit of pH increase when pH $\ll$ pK_2 (= 10.3) and reaches a constant value of $(K_{sp1})^{1/2} = 0.09$ mM for pH $\gg$ pK_2. The target pH value for $CaCO_3$ precipitation by lime-soda softening is in the range 10.0–10.6 (Benefield and Morgan, 1990).

You might find it puzzling that calcium hydroxide is used as a reagent in water softening. After all, the goal is to *remove* hardness ions from water. How can it be that adding a calcium-containing reagent reduces hardness? The answer to this paradox is that calcium hydroxide's primary role is to increase pH, and it does this more substantially than it increases $[Ca^{2+}]$. Each added calcium ion releases two hydroxide ions, which in turn can convert two bicarbonate ions into carbonate ions. Each carbonate ion can then combine with calcium to form solid calcium carbonate. Thus, two calcium ions can be removed for each one added, decreasing the hardness of water.

Alternative reagents can be used to raise the pH of water and promote precipitation reactions. In fact, some systems use sodium hydroxide (NaOH), also known as caustic soda. Sodium carbonate could also be used alone, as it too increases the pH of water. However, both of these reagents are more expensive than lime, and as a result most softening at water treatment plants is based on the lime-soda system.

To determine the reagent requirements for lime-soda softening, one must know the alkalinity, the carbonic acid content, the total hardness, and the noncarbonate hardness of the untreated water (§2.C.4), and one must also determine whether magnesium is to be removed. Table 6.D.2 summarizes the reagent requirements based on the component reactions shown stoichiometrically in Table 6.D.3. Example 6.D.4 illustrates the use of these relationships. As shown in reaction 3, Table 6.D.3, additional reagent must be added to eliminate carbonic acid from the water. Recall that CO_2 is an acid gas (§3.C.4) that forms carbonic acid as it dissolves in water, and then dissociates

Table 6.D.2 Reagent Requirements for Lime-Soda Softening[a]

Condition	$Ca(OH)_2$ (eq/L)[b,c]	Na_2CO_3 (eq/L)[c]
ALK $\geq$ TH	TH + $2[Mg^{2+}] + 2[H_2CO_3^*]$ + 0.00125	None
ALK $<$ TH	ALK + $2[Mg^{2+}] + 2[H_2CO_3^*]$ + 0.00125	NCH

[a]ALK = alkalinity, TH = total hardness, NCH = noncarbonate hardness, each expressed in eq/L.
[b]If Mg^{2+} removal is not required, these requirements can be reduced by 0.00125 eq/L.
[c]Each mole of these reagents contains two equivalents.

Table 6.D.3 Stoichiometry of Reactions in the Lime-Soda Softening Process

(1) Addition of calcium hydroxide to precipitate $CaCO_3(s)$

$$Ca^{2+} + 2HCO_3^- + Ca(OH)_2 \rightarrow 2CaCO_3(s) + 2H_2O$$

(2) Addition of calcium hydroxide to precipitate $Mg(OH)_2(s)$

$$Mg^{2+} + 2HCO_3^- + 2Ca(OH)_2 \rightarrow 2CaCO_3(s) + Mg(OH)_2(s) + 2H_2O$$

(3) Addition of calcium hydroxide to eliminate carbonic acid (needed to raise the pH)

$$H_2CO_3^* + Ca(OH)_2 \rightarrow CaCO_3(s) + 2H_2O$$

(4) Addition of calcium hydroxide to raise the pH to 11 (needed to precipitate $Mg(OH)_2$)

$$pH = 11 \Rightarrow [OH^-] = 10^{-3}\ M = 1.0\ meq/L, \text{ unreacted;}$$

as rule of thumb, add 1.25 meq/L to raise pH if Mg^{2+} removal is required

(5) Addition of sodium carbonate to precipitate Ca^{2+} if alkalinity of raw water is insufficient

$$Ca^{2+} + Na_2CO_3 \rightarrow CaCO_3(s) + 2Na^+$$

into bicarbonate and carbonate ions, releasing H^+. To raise the pH of water, the protons from these dissociation reactions must be consumed by adding additional hydroxide ions. One equivalent of lime is sufficient to consume one equivalent of carbonic acid.

It has been suggested that ideal water for municipal purposes has a total hardness of 1.5–1.7 meq/L, with a magnesium hardness no greater than 0.8 meq/L (ASCE/AWWA, 1990). Lime-soda softening can achieve no lower than about 1 meq/L total hardness. The contribution of Ca^{2+} to hardness following lime-soda softening is typically 0.6–1 meq/L. Magnesium's contribution can be reduced to about 0.2–0.4 meq/L, but at a penalty of somewhat increased Ca^{2+} levels, relative to the 0.6–1 meq/L range, because of the excess lime added to raise the pH (Benefield and Morgan, 1990).

EXAMPLE 6.D.4 *Reagent Requirements for Lime-Soda Softening*

The ion content and pH of water samples from two sources were measured and are reported in the table below. Determine the appropriate reagent requirements for lime-soda softening of each sample.

Parameter	Sample A	Sample B
Ca^{2+} (mg/L)	100	60
Mg^{2+} (mg/L)	10	40
Na^+ (mg/L)	20	30
HCO_3^- (mg/L)	240	150
SO_4^{2-} (mg/L)	125	230
Cl^- (mg/L)	10	12
pH	6.5	7.5
TH (meq/L)	5.8	6.3
NCH (meq/L)	1.9	3.8
ALK (meq/L)	3.9	2.5

EXAMPLE 6.D.4 ***Reagent Requirements for Lime-Soda Softening (continued)***

SOLUTION On the basis of total hardness, we would characterize both samples as very hard, so lime-soda softening is warranted. Let us next consider whether magnesium removal is warranted in the two cases. For sample A, the contribution of magnesium to hardness is (10/24.3) × 2 = 0.8 meq/L. At this level, there would be little benefit from further reduction, and so we might choose to remove only Ca^{2+}. For sample B, the magnesium contribution to hardness is four times larger, about half the total, and definitely worth reducing.

The other parameter necessary to specify reagent requirements and not given in the problem statement is the concentration of carbonic acid. This can be determined from the pH and the bicarbonate concentration, assuming acid-base equilibrium (see reaction 3.C.15):

$$[H_2CO_3^*] = \frac{10^{-pH}[HCO_3^-]}{K_1} \qquad K_1 = 4.47 \times 10^{-7}\ M$$

So, for sample A, $[H_2CO_3^*] = 2.78$ mM, and for sample B, $[H_2CO_3^*] = 0.17$ mM.

In both cases, alkalinity is less than total hardness, so some sodium carbonate will be required. Applying the expressions in Table 6.D.2, we obtain

Sample A:

$$Ca(OH)_2 = 3.9 + 2\frac{10}{24.3} + 2 \times 2.78 = 10.3\ \text{meq/L}\ (= 380\ \text{mg/L})$$

$$Na_2CO_3 = 1.9\ \text{meq/L}\ (= 100\ \text{mg/L})$$

Sample B:

$$Ca(OH)_2 = 2.5 + 2\frac{40}{24.3} + 2 \times 0.17 + 1.25 = 7.4\ \text{meq/L}\ (= 270\ \text{mg/L})$$

$$Na_2CO_3 = 3.8\ \text{meq/L}\ (= 200\ \text{mg/L})$$

The final step in the lime-soda softening process is called *recarbonation*. The solid flocs that are produced in the precipitation process are never completely removed by sedimentation. Furthermore, the pH of 10–11 established for precipitation is too high for drinking water. Recarbonation, which occurs after the sedimentation basin but before the granular filter, serves two purposes: (1) to redissolve the remaining suspended precipitates of $CaCO_3$ and $Mg(OH)_2$, and (2) to reduce the pH to approximately 8.5–9.

Recarbonation involves the addition of carbon dioxide to the water. As we know, CO_2 is an acid gas. When it dissolves in water it forms carbonic acid, which dissociates, releasing protons and decreasing the water's pH (§3.C.4). Decreased pH promotes dissolution of both solid $CaCO_3$ and $Mg(OH)_2$ by reversing the reactions used to form the precipitates.

Two methods are commonly used to add CO_2 to water. In one, natural gas (mainly CH_4) is burned with air in submerged burners to generate carbon dioxide, which dissolves as it bubbles through the water. Alternatively, pure CO_2 is discharged from pressurized storage tanks directly into the water through diffusers. The underwater burner system is capable of transferring CO_2 to water with about 75 percent efficiency. The pure CO_2 system typically achieves 80–85 percent transfer efficiency (ASCE/AWWA, 1990).

The lime-soda process contributes to the removal of other undesired cations. The mechanisms are similar: formation of solid precipitates of the cation with hydroxide or carbonate followed by separation from water by sedimentation. Lime-soda softening is rated as either good (removal efficiency in the range 60–90 percent) or excellent (efficiency greater than 90 percent) for the following inorganic species (Hamann et al., 1990): arsenic (oxidation state +5), barium, cadmium, chromium (+3), copper, iron, lead, manganese, nickel, radium, uranium, and zinc.

Ion Exchange

As the name suggests, ion exchange functions by removing some ions from water in exchange for others. A net benefit can be realized if the removed ions are more harmful, more hazardous, or less desirable than the replacements.

Ion exchange occurs on the surface of certain solids. Zeolites, natural clay minerals, were the first materials used for ion exchange in water treatment. Most applications now use synthetic materials such as resin beads made from polystyrene polymers, cross-linked with divinylbenzene molecules. The ion exchange properties of these resins are a result of surface molecules added during manufacture. Resins used to exchange cations contain the conjugate base of a strong acid (such as sulfonate, SO_3^-) or of a weak acid (such as carboxylate, COO^-). Resins used to exchange anions have amine surface groups such as $N^+(CH_3)_3$ or $N(CH_3)_2$ (Clifford, 1990). Relative to zeolites, resins offer the following advantages: high chemical and physical stability, uniform size and composition, high exchange capacity, fast kinetics, high reversibility, and durability.

Water softening is an important technological application of ion exchange. Because the process functions in a simple manner, this application is an attractive alternative to chemical precipitation for small municipal water utilities and for residential water softeners. Ion exchange is not typically used for softening in large water treatment plants because it is more expensive than the lime-soda process.

Besides hardness ions, ion exchange can also be used to remove other unwanted cations from solution. Along with chemical precipitation, it is listed as one of the best available technologies for removing excess levels of barium, cadmium, chromium, nickel, and radium from drinking water (Pontius, 1992). It is also a suitable method for treating some undesirable anions, such as cyanide (CN^-), fluoride (F^-), nitrate (NO_3^-), and sulfate (SO_4^{2-}).

Consider the case of calcium removal using a strong-acid cation exchanger. If the active ion exchange sites are sulfonates initially charged with sodium, the exchange reaction may be written in this form:

$$2 \equiv SO_3Na + Ca^{2+} \Leftrightarrow (\equiv SO_3)_2Ca + 2\, Na^+ \qquad (6.D.26)$$

where $\equiv SO_3$ represents a sulfonate group attached to the resin surface. In the forward direction, two surface groups release their attached sodium ions into solution in favor of binding one calcium ion. When unspent resin is exposed to ordinary water, the forward direction of this reaction is strongly favored. Many other cations will also be preferentially bound on the surface in exchange for sodium.

To be treated by ion exchange, water is passed through a fixed bed of granular material, similar to the way granular activated carbon is used (§6.D.3). The ion exchange process continues until most of the surface sites are occupied by exchanged ions, provided that the kinetics are not limiting. At this point, the material has no further capacity to capture ions, and the cation concentrations at the outlet will rise to the levels at the inlet. To function again, the ion exchange material must be regenerated. For a cation exchange resin, this can be accomplished by passing concentrated NaCl brine through the contactor. The very high concentration of sodium ions forces reaction 6.D.26 to the left so that surface-bound multivalent cations are released back into solution in favor of the sodium ions. The highly concentrated waste brine is then disposed of. Once the regeneration process is complete, water treatment can again begin. In typical applications, a volume of water equal to 300–60,000 times the volume of the empty contactor bed can be processed before it is necessary to regenerate the resin (Clifford, 1990). Regeneration requires 1–5 bed volumes of brine plus 2–20 bed volumes of clear rinsewater, which must also be wasted. Disposal of the concentrated brine and the increase in the concentration of Na^+ in treated water are the two most important problems associated with ion exchange (see Example 6.D.5).

The operating capacity of ion exchange resins is typically on the order of 1 eq per L of bed volume (including voids between grains) (Clifford, 1990). This means that in a typical application, in which water is to be softened by 4 meq/L, each liter of bed volume can remove the hardness ions from 250 L of water before regeneration is required. Bed depths are commonly in the range of 0.7–3.7 m and are operated with an empty-bed contact time (empty-bed volume divided by water flow rate) of 1.5–7.5 min.

EXAMPLE 6.D.5 ***Impact of Ion Exchange Water Softening on Sodium Levels***

Sodium intake is linked to high blood pressure and inner ear problems for some members of the public. Food is the major source of Na^+ (typical intake 1.1–3.3 g d^{-1}). Nevertheless, the American Heart Association recommends that drinking water contain less than 20 mg L^{-1} (Tate and Arnold, 1990).

Consider water that is to be softened using ion exchange with sodium as the mobile ion. What increase in Na^+ concentration would be caused by decreasing the hardness from 6 to 1.5 meq/L? (This change would be from the lower limit of "very hard" to the upper limit of "soft"; see Table 2.C.4.)

SOLUTION Removing 4.5 meq/L of hardness requires the release of 4.5 meq/L (= 4.5 mM) of Na^+, as can be seen from reaction 6.D.26. Since the atomic mass of sodium is 23 g/mol, this process adds approximately 100 mg/L of sodium to water. This level is high enough to be of serious concern for people who are on a sodium-restricted diet.

Ion exchange can also be applied to deionize water for industrial applications (Applebaum, 1968). In this case, cationic exchange and anionic exchange resin beds are used in series. Instead of regenerating with salts, a strong acid is used to regenerate the cation exchanger so that H^+ is the ion released into solution on exchange. Similarly, a strong base is used to regenerate the anionic resin, so that OH^- is the exchanged ion.

The disposal of concentrated brines is a significant concern that constrains the use of ion exchange technology. When used for water softening and deionization, brine can often be discharged into the sewer. For applications in coastal areas, discharge to the oceans may be possible. However, if ion exchange is used for removing toxic materials, these options may not be available. Possible alternative disposal methods include storage in evaporation lagoons, recovery for use, and disposal at hazardous waste sites (ASCE/AWWA, 1990).

6.E BIOLOGICAL WASTEWATER TREATMENT

Wastewater commonly contains high concentrations of biodegradable organic matter. If this organic material is discharged to a lake, river, or even the ocean, microorganisms use it as a food source. If the discharge is large, problems occur (see §6.A). The dissolved oxygen content of water is depleted as microorganisms oxidize the organic wastes. Higher forms of life, such as fish, have difficulty surviving in oxygen-depleted water. In sediments near the discharge of inadequately treated wastewater, where dissolved oxygen may be completely lacking, sulfur in the wastes can be reduced to hydrogen sulfide and organic sulfur compounds, which then escape into the air, producing unpleasant odors.

Standard practice for municipal wastewater treatment now includes a minimum of two stages. In the first stage, known as *primary treatment*, wastewater is passed through a sedimentation basin to remove solids and oil/grease that can either settle or rise fairly rapidly. In the second stage, *secondary treatment*, the main objective is to remove biodegradable organic material (also known as biochemical oxygen demand, BOD), much of which is dissolved in the water or present in small particles. This goal is accomplished by biological treatment. The same fundamental microbial degradation processes that occur in natural systems are used in engineered systems. However, conditions are optimized in engineered systems to enhance reaction rates so that organic removal that would take many days in nature can be accomplished within a few hours. In addition to reducing BOD levels, biological treatment can significantly reduce the concentration of suspended solids in wastewater. Table 6.E.1 illustrates this point with data for a typical municipal plant. Overall, both suspended solids (SS) and BOD are removed from the wastewater with better than 90 percent efficiency. About 40 percent of the total suspended solids removal (80 mg/L out of 205) and about 60 percent of the BOD removal (120 mg/L out of 200) occurs in the activated sludge unit (secondary treatment).

Several specific biological treatment methods have been devised to reduce the BOD content in wastewater. Among the high-technology options are activated sludge, trickling filters, and rotating biological contactors. These options are commonly used for large facilities in urban parts of industrialized countries. Lower-technology options include wastewater lagoons, land application of wastewater, and septic tank/leach fields. These are more commonly applied in rural areas and in less developed countries. Here we will consider two methods with long histories: activated sludge and trickling filter processes, each still in common use throughout the United States and other industrialized countries. We will also explore anaerobic digestion for the treatment of sludge that is generated by primary and secondary wastewater treatment.

Table 6.E.1 Concentration of Contaminants in a Typical Wastewater Treatment System

Location	SS (mg/L)	BOD_5 (mg/L)
Influent to plant	220	220
Effluent from primary sedimentation	95	140
Effluent from activated sludge unit	15	20

Source: Metcalf & Eddy, 1991.

6.E.1 Activated Sludge

In the activated sludge process, organic matter is biodegraded by bringing it into contact with a high concentration of microorganisms (the activated sludge) in the presence of dissolved oxygen. Activated sludge is characterized as a *suspended* growth process because the microorganisms are suspended in water within the reactor. The first activated sludge unit was built in Worcester, England, in 1916. During the following decades, large units for municipal wastewater treatment were constructed in a number of U.S. locations (Ganczarczyk, 1983).

Aerobic, heterotrophic bacteria (see §2.C.9) do most of the work in the activated sludge process. Some of the organic matter is oxidized to carbon dioxide to generate energy for the metabolic needs of the organisms. Much of the remaining organic material is incorporated into new microbial cells. These two fates can be termed *oxidation* and *synthesis*, respectively. The overall reaction can be represented in this form:

$$\text{organics} + O_2 \xrightarrow{\text{cells}} CO_2 + \text{cells} + \cdots \tag{6.E.1}$$

where "cells" refers to the bacteria and other microorganisms in the activated sludge. The rate of cell growth is roughly half the rate of BOD consumption when both are expressed on a mass basis.

The basic process configuration contains two units, as depicted in Figure 6.E.1. The removal of BOD is carried out in the first unit (reactor). Air or pure oxygen is supplied to the reactor to maintain an adequate supply of dissolved oxygen for the aerobic biological reactions. The water leaving the reactor contains a high concentration of microorganisms and a low concentration of dissolved BOD. The cell separator is a sedimentation basin that separates the treated wastewater from the bacterial flocs. The wastewater effluent can be discharged or, if necessary, subjected to further treatment. The concentrated slurry of microorganisms extracted from the bottom of the separator is split into two streams. One stream is returned to the inlet of the reactor, while the other is removed from the system and processed as wasted sludge.

The cell separator serves three essential functions in the activated sludge system. First, by removing the microorganisms from the treated wastewater, the discharged BOD is greatly reduced. The microbial cells themselves constitute a source of BOD since they consist of biodegradable organics, and their discharge would have a similar detrimental impact on water quality. The second role of the cell separator is to recycle cells, increasing their concentration in the reactor. Cell recycle allows the microorganisms to have a mean residence time in the reactor that is much longer than the hydraulic detention time. This allows the incoming BOD load to be rapidly consumed. The

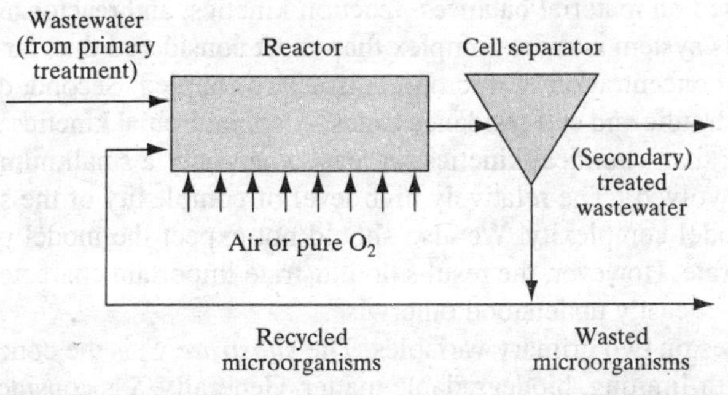

Figure 6.E.1 Typical configuration of an activated sludge unit for wastewater treatment.

third function of the cell separator is to provide a waste sludge stream with a relatively high concentration of microorganisms and low water content. Extracting and wasting solids from an activated sludge system is necessary to remove nondegradable organics and inactive cells from the system. Reduced waste-sludge water content means that a smaller volume of waste sludge must subsequently be treated.

The microbial ecology in an activated sludge system is complex. Many aspects of the operation of this treatment process are aimed at encouraging the growth and reproduction of beneficial microorganisms and discouraging the growth of cells that would disrupt the process. A key feature of a properly functioning activated sludge system is a population of microorganisms that settles rapidly in the cell separator and forms compact sludge. In general, individual bacteria do not settle rapidly enough to be recycled by a typical cell separator. However, some bacteria grow in flocs that contain many cells. Although individual bacterial cells have dimensions of a few micrometers, floc size is generally 50–500 μm (Horan, 1990). The rapid settling of microbial flocs in the cell separator allows them to be recycled, creating a selective pressure that favors reproduction of flocculating cells and discourages reproduction of discrete suspended bacteria.

When an activated sludge process is functioning well, the flocs are relatively dense and settle effectively. An important failure mode of the process involves excessive reproduction of filamentous bacteria. Flocs with high concentrations of filamentous bacteria are bulky and loosely packed. These flocs do not settle well and may overflow into the treated effluent. Much attention in the practical management of activated sludge processes is devoted to preventing or correcting this problem, which is known as *sludge bulking* (Jenkins et al., 1993).

The reactor in an activated sludge system may be configured to operate either as a completely mixed flow reactor (CMFR) or as a plug-flow reactor (PFR). An intermediate option involves the use of a series of well-mixed reactors. PFRs perform better but are more susceptible to disturbances from variable influent conditions, such as the short-term release of a chemical that is toxic to bacteria. As a rule of thumb, PFRs are used for constant-load systems whereas CMFRs are applied in cases with variable loads. The performance of a PFR system cannot be modeled simply, unless the assumption is made that microbial growth during a single pass of water through the reactor is negligible. The performance of a CMFR system is simpler and captures the key features of the activated sludge process, so we will focus on this case.

A quantitative analysis of the activated sludge process can achieve three distinct objectives: (1) to predict the effluent BOD concentration and how it varies with operating conditions, (2) to determine the wastage rate of sludge, and (3) to determine how much oxygen must be supplied to the reactor.

The analysis is based on material balances, reaction kinetics, and reactor models. For several reasons, this system is more complex than most considered thus far. First, the BOD level and the concentration of microorganisms are coupled. Second, cell recycle decouples the hydraulic and cell residence times. Also, microbial kinetics are inherently more complex than chemical kinetics (at least when only a small number of chemical species are involved). The relatively high level of complexity of the system results in increased model complexity. We also should not expect the model predictions to be highly accurate. However, the results do illustrate important characteristics of the system that are not easily understood otherwise.

The analysis focuses on two primary variables. The *substrate*, S, is the concentration of dissolved, growth-limiting, biodegradable matter. Generally, S is considered to be the 5-day biochemical oxygen demand, BOD_5, (or sometimes the ultimate BOD)

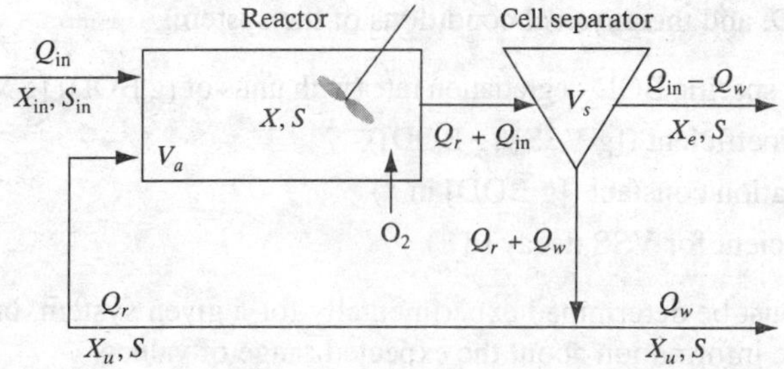

Figure 6.E.2 Schematic of the activated sludge system with major variables defined for analysis.

of the wastewater and is measured in units of mg/L. The second parameter, X, also measured in mg/L, is the concentration of microbial cells, often measured as volatile suspended solids (VSS or, sometimes, as total suspended solids, SS) in the water (see §2.C.8). Although VSS measurements include much more than microbial cells, VSS is relatively easy to measure and, in the activated sludge reactor, varies more or less in proportion to the microbial level.

Figure 6.E.2 shows the activated sludge system, as we will analyze it, with variables labeled. The flow rates are denoted Q, the volumes V, and the suspended solids and dissolved BOD levels X and S, respectively. The subscript "in" denotes inlet conditions, e is the effluent, r stands for return or recycle, u means underdrain, w means waste, a indicates aeration tank, and s stands for separator. In Figure 6.E.2, material balances have already been applied on water, assuming that steady-state conditions prevail. The three fundamental flow rates are Q_{in}, the flow into the system, Q_w, the waste sludge flow rate, and Q_r, the recycle flow rate. All other flows are derived from these three.

The kinetics of microbial metabolism and reproduction in an activated sludge reactor are complex because of the mixed microbial population and the chemically diverse substrate. However, reasonable success has been found in representing this complex system by expressions similar to the Monod equation (see §3.D.5):

$$\left(\frac{dX}{dt}\right)_{\text{reaction}} = r_g X - k_d X \tag{6.E.2}$$

and

$$\left(\frac{dS}{dt}\right)_{\text{reaction}} = -\frac{r_g}{Y} X \tag{6.E.3}$$

with

$$r_g = \frac{k_m S Y}{K_s + S} \tag{6.E.4}$$

where r_g is the cell growth rate coefficient (with units of inverse time). The other new parameters are defined below. Equations 6.E.2 and 6.E.3 express the rate of change of VSS and BOD_5 in the activated sludge reactor as a consequence of microbial reactions alone. The first term, on the right side of equation 6.E.2, captures cell growth. The second term reflects the fact that cells eventually die and cell mass decays. The rate expression for microbial growth, r_g, varies with substrate concentration as described by equation 6.E.4. Four kinetic parameters depend on the mix of microorganisms, the composition of BOD, and the physical conditions of the system:

k_m = maximum specific BOD degradation rate (with units of [g BOD] [g VSS]$^{-1}$ d^{-1})
Y = cell yield coefficient ([g VSS]/[g BOD])
K_s = half-saturation constant ([g BOD] m^{-3})
k_d = rate coefficient for VSS decay (d^{-1})

These parameters must be determined experimentally for a given system, but past experience can provide information about the expected range of values.

To keep the mathematics as simple as possible, we make several assumptions. First, we only consider steady-state conditions. Second, we assume that the microbial reactions described by equations 6.E.2–4 occur only in the reactor. Reactions in the cell separator are of lesser importance because the volume is typically smaller ($V_s < V_a$) and the oxygen needed to promote aerobic reactions is not provided there. Third, we assume that the cell separator functions perfectly, so that (a) no biomass escapes into the effluent ($X_e = 0$) and (b) no separation of dissolved BOD occurs ($S_e = S_u = S$, as shown in Figure 6.E.2). Finally, we assume that the inlet concentration of VSS is negligible ($X_{in} = 0$).

The two primary unknowns in the system are X and S. The two primary equations that link these unknowns to each other and to the system parameters are the respective material balances applied to the reactor:

$$\frac{d(SV_a)}{dt} = S_{in}Q_{in} + SQ_r - S(Q_{in} + Q_r) + V_a\left(\frac{dS}{dt}\right)_{reaction} \tag{6.E.5}$$

$$\frac{d(XV_a)}{dt} = X_uQ_r - X(Q_r + Q_{in}) + V_a\left(\frac{dX}{dt}\right)_{reaction} \tag{6.E.6}$$

We substitute for the reaction terms from equations 6.E.2–4, set the accumulation terms to zero because of the steady-state assumption, and rearrange to obtain these equations:

$$X = \left(\frac{S_{in} - S}{\Theta}\right)\left(\frac{K_s + S}{k_mS}\right) \tag{6.E.7}$$

$$X_uQ_r = X\left[Q_{in} + Q_r + k_dV_a - V_a\left(\frac{k_mSY}{K_s + S}\right)\right] \tag{6.E.8}$$

In equation 6.E.7, we have introduced the hydraulic residence time,

$$\Theta = \frac{V_a}{Q_{in}} \tag{6.E.9}$$

Next, we apply a material balance to VSS as it moves through the cell separator (see Figure 6.E.3). At steady state, the flow of VSS into the cell separator balances the flow out:

$$X(Q_r + Q_{in}) = X_e(Q_{in} - Q_w) + X_u(Q_r + Q_w) \tag{6.E.10}$$

Using the idealization that $X_e = 0$, we can rearrange this equation to yield an expression for X_u in terms of X:

$$X_u = \left(\frac{Q_{in} + Q_r}{Q_w + Q_r}\right)X \tag{6.E.11}$$

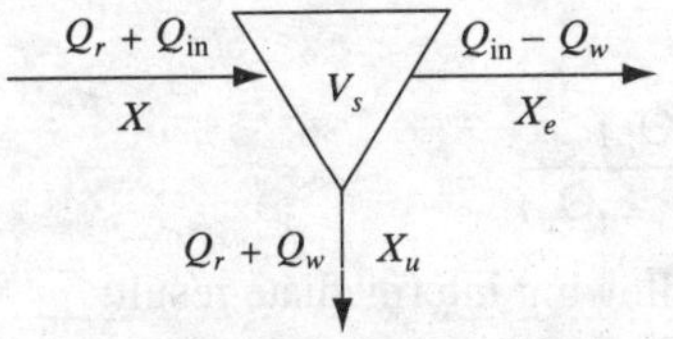

Figure 6.E.3 Schematic of the cell separator used to derive a relationship between X_u and X.

It is convenient here to define the *mean cell residence time* (MCRT), denoted by Θ_c, as the ratio of the cell mass in the reactor to the cell wastage rate:

$$\Theta_c = \frac{V_a X}{Q_w X_u} = \frac{V_a}{Q_w}\left(\frac{Q_w + Q_r}{Q_{in} + Q_r}\right) \qquad (6.E.12)$$

Here we have used equation 6.E.11 to give the MCRT in terms of only flows and volumes. Under typical operating conditions, Q_w is much less than Q_r, so equation 6.E.12 can be approximated by

$$\Theta_c = \frac{\Theta}{W}\left[\frac{R}{1 + R}\right] \qquad (6.E.13)$$

where $W = Q_w / Q_{in}$ is the wastage ratio and $R = Q_r / Q_{in}$ is the recycle ratio. In a conventional activated sludge process, the recycle ratio is in the range 0.25–1, the hydraulic detention time, Θ, is 3–5 h, and the mean cell residence time is 5–15 days (Table 6.E.2), so the wastage ratio must lie in the range 0.2–2 percent. Equation 6.E.13 shows how the MCRT can be controlled by means of manipulating the flow rates, and especially the wastage ratio, W.

Now, if we substitute equation 6.E.11 into the left-hand side of equation 6.E.8, we find that all terms in the equation are proportional to X, so, provided $X \neq 0$, we may divide the equation by X and rearrange to solve for S. After some algebra, and using

Table 6.E.2 Parameter Values for Conventional Activated Sludge Systems Using a Completely Mixed Flow Reactor

Parameter	Typical range	Typical value
Microbial parameters		
k_m (mg BOD_5 per mg VSS per day)	2–10	5
K_s (mg BOD_5 per L)	25–100	60
Y (mg VSS per mg BOD_5)	0.4–0.8	0.6
k_d (d^{-1})	0.025–0.075	0.06
Operational parameters		
Mean cell residence time, Θ_c (d)	5–15	
F/M (mg BOD_5 per mg VSS per day)	0.2–0.6	
Loading (kg BOD_5 m^{-3} d^{-1})	0.8–1.9	
Total suspended solids (kg m^{-3})	2.5–6.5	
Hydraulic detention time, Θ (h)	3–5	
Recycle ratio, $R = Q_r / Q_{in}$	0.25–1	
BOD removal efficiency (%)	85–95	

Source: Metcalf & Eddy, 1991.

equation 6.E.12, we arrive at this result:

$$S = \frac{K_s(1 + k_d\Theta_c)}{k_mY\Theta_c - (1 + k_d\Theta_c)} \tag{6.E.14}$$

In deriving equation 6.E.14, we obtain the following intermediate result:

$$\frac{k_mYS}{K_s + S} = \frac{1}{\Theta_c} + k_d \quad \Rightarrow \quad \frac{K_s + S}{k_mS} = \frac{Y\Theta_c}{1 + k_d\Theta_c} \tag{6.E.15}$$

By substituting this relationship into equation 6.E.7, we obtain the following expression for evaluating the VSS level in the reactor:

$$X = \left(\frac{S_{\text{in}} - S}{\Theta}\right)\left(\frac{Y\Theta_c}{1 + k_d\Theta_c}\right) \tag{6.E.16}$$

So, in summary, we have applied the principle of material balance along with a model of microbial reactions based on Monod kinetics to derive equations 6.E.14 and 6.E.16.

Now, to predict the values of the two system variables, X and S, we need input data for microbial kinetics: k_m, Y, K_s, and k_d. We also need to know the flows and volumes in the system: Q_r, Q_w, Q_{in}, and V_a. For any given reactor volume and set of flow rates, we can use equations 6.E.9 and 6.E.12 (or 6.E.13) to obtain the hydraulic detention time, Θ, and the MCRT, Θ_c. The microbial kinetic parameters either are determined from separate experiments or are assumed from literature values such as those given in Table 6.E.2.

Example 6.E.1 illustrates the analysis results using typical parameter values. Several noteworthy points emerge. First, the VSS level in the reactor is quite high. At a few grams per liter (or kg per m^3), the VSS level in the reactor is roughly 10 times the level of suspended solids in untreated sewage. However, the solids content still represents significantly less than 1 percent of the total mass in the reactor, since the density of water is 1000 kg m^{-3}. Second, the predicted BOD level in the reactor (and, according to the model, in the treated effluent) does not depend strongly on the mean cell residence time. In all cases shown in Figure 6.E.4, the outlet concentration of S would be less than 15 mg/L, which represents better than 90 percent removal from the inlet

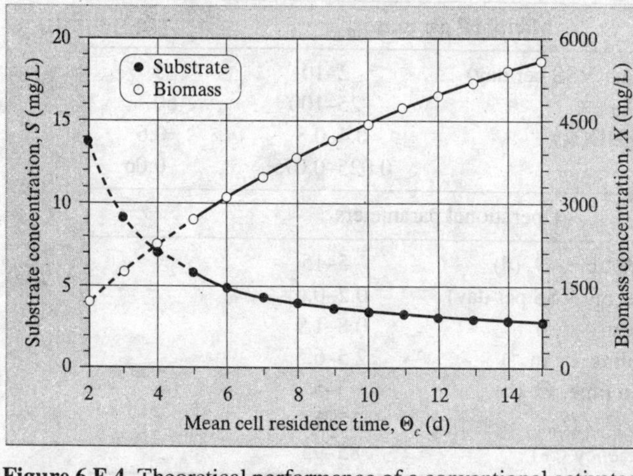

Figure 6.E.4 Theoretical performance of a conventional activated sludge unit with a completely mixed flow reactor.

condition of 200 mg/L. Table 6.E.2 shows that observed removal efficiencies for BOD in activated sludge units are good, 85–95 percent, but not quite as high in practice as predicted by this model. Some of the discrepancy is attributable to microorganisms that escape into the outlet stream. Although assumed to be zero here, a more typical value of X_e is 5–35 mg/L (Tchobanoglous and Schroeder, 1985).

EXAMPLE 6.E.1 ***Predicted Performance of a Conventional Activated Sludge Unit***

Considering typical process parameter values, evaluate the effectiveness of an ideal activated sludge unit for removing $S_{in} = 200$ g m^{-3} of BOD. Plot X and S as a function of the mean cell residence time Θ_c, given a hydraulic residence time of $\Theta = 4$ hours. For kinetic parameters, use the typical values given in Table 6.E.2.

SOLUTION Apply equations 6.E.14 and 6.E.16. The results are plotted in Figure 6.E.4.

Two other parameters listed in Table 6.E.2 are important. The *food to microorganism ratio*, F/M, is a commonly reported characteristic of the activated sludge process. It is defined as the rate of BOD supply to the reactor (g BOD_5/d) divided by the quantity of biomass in the reactor (g VSS). It can be evaluated as

$$\text{F/M} = \frac{Q_{in}S_{in}}{V_a X} = \left(\frac{S_{in}}{S_{in} - S}\right)\left(\frac{1 + k_d\Theta_c}{Y\Theta_c}\right) \tag{6.E.17}$$

where the first equality represents the definition of F/M and the second has been obtained by substitution from equation 6.E.16. Using typical values of 0.9 for BOD removal efficiency, 0.06 d^{-1} for k_d and 0.6 g VSS per g BOD_5 for Y, equation 6.E.17 predicts that F/M varies between 0.23 and 0.48 g BOD_5 per g VSS per day for 5 d $< \Theta_c <$ 15 d.

The *loading* provides a basis for rapidly estimating the required volume of an activated sludge reactor. The per-capita contribution to BOD_5 in domestic sewage in the United States is found to be in the range 60–110 g d^{-1} (Metcalf & Eddy, 1991).* Combining this range of values with the range of loading rates reported in Table 6.E.2 leads to the result $V_a = 0.03 - 0.14$ m^3 person^{-1}, for systems that treat mainly domestic sewage.

The cell wastage rate is given by $Q_w X_u$. Using equations 6.E.11 and 6.E.16, we have

$$Q_w X_u = Q_w\left(\frac{Q_{in} + Q_r}{Q_w + Q_r}\right)\frac{(S_{in} - S)Y}{(1 + k_d\Theta_c)}\left(\frac{\Theta_c}{\Theta}\right) \tag{6.E.18}$$

Using equations 6.E.9 and 6.E.12 to rewrite Θ_c/Θ in terms of flow rates, this expression simplifies to

$$Q_w X_u = Q_{in}(S_{in} - S)\left(\frac{Y}{1 + k_d\Theta_c}\right) \tag{6.E.19}$$

The product $Q_{in}(S_{in} - S)Y$ represents the rate at which BOD is converted to biomass. The denominator corrects for loss of biomass in the reactor due to microbial decay.

The oxygen demand in the reactor is estimated by applying a material balance on oxygen around the entire process (see Figure 6.E.5). The difference in oxygen demand

*This value is exclusive of the contribution of ground kitchen wastes. Widespread use of in-sink garbage disposals increases the total to 80–120 g per capita per day.

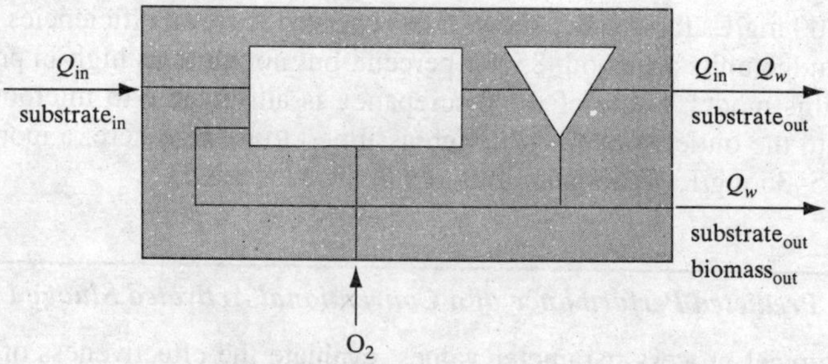

Figure 6.E.5 Control volume applied to determine the oxygen demand for an activated sludge process.

between the flows in and the flows out must be balanced by oxygen added to the reactor. We may write this balance as

$$R_{O_2} = \alpha(S_{in} - S)Q_{in} - \beta Q_w X_u \tag{6.E.20}$$

where R_{O_2} represents the total oxygen demand (mass per time). The first term on the right represents the oxygen demand that would be required if all BOD entering the reactor were completely oxidized. The second term on the right corrects for the oxygen demand that is not exerted because some BOD is only partially oxidized to produce microorganisms. When the substrate S is measured as 5-day BOD, the factor α represents the ratio of the ultimate BOD to the 5-day BOD, for which an appropriate value is 1.47 (corresponding to a first-order rate constant for BOD oxidation of 0.23 d^{-1}; see §3.D.5). The factor β represents the mass of oxygen required per mass of cell for full oxidation of microbial cells. It has a value of 1.42 g O_2 (g cell)$^{-1}$, as derived from stoichiometry, based on the following reaction:

$$C_5H_7O_2N + 5\,O_2 \rightarrow 5\,CO_2 + 2\,H_2O + NH_3 \tag{6.E.21}$$

Here $C_5H_7O_2N$ represents an approximate average microbial composition. As shown by this reaction, 5 mol of oxygen is required for complete oxidation of every five atoms of carbon in bacterial cells. In terms of mass, $5 \times 32 = 160$ g of O_2 is required for every $5 \times 12 + 7 + 32 + 14 = 113$ g of biomass, or 1.42 g O_2 per g VSS. Equation 6.E.20 may be divided by the reactor volume (using equation 6.E.12) to yield the following equation. This equation provides an estimate of the required mass of oxygen per volume of reactor per time, r_{O_2}:

$$r_{O_2} = 1.47\,\frac{S_{in} - S}{\Theta} - 1.42\,\frac{X}{\Theta_c} \tag{6.E.22}$$

Using the parameters and variables defined in Example 6.E.1, the oxygen requirement ranges from 0.95 kg m^{-3} d^{-1} for $\Theta_c = 5$ days to 1.2 kg m^{-3} d^{-1} for $\Theta_c = 15$ days.

Finally, we should make a few more observations about the modeling results. Equation 6.E.14, which predicts the concentration of dissolved BOD_5, has two peculiar characteristics. First, it is entirely independent of the inlet concentration, S_{in}. We find that the discharged contaminant level depends on the four microbial kinetic parameters and on the mean cell residence time, an operational parameter that depends on flow rates and the reactor volume, but *not* on the inlet BOD level. This surprising result is a consequence of the coupling between the amount of BOD converted and the biomass concentration in the reactor. Increased BOD at the inlet causes the steady-state concentration of microbial cells in the reactor to increase (see equation 6.E.16), which means that the BOD is degraded more rapidly, so that the effluent BOD concentration is unaffected. The net effect is that the effluent BOD is a function of the MCRT (Θ_c) rather than influent BOD.

The second peculiarity in equation 6.E.14 is subtler. The denominator is expressed as the difference between two terms, and the nature of these terms is such that the denominator can become zero, or even negative depending on the value of Θ_c. Consequently, under some conditions, predictions of $S > S_{\text{in}}$ or $S < 0$ may result, neither of which is physically meaningful. Where does the analysis fail? The key is contained in a condition stated just before equation 6.E.14: "provided $X \neq 0$." If the mean cell residence time is made too short, a condition known as *cell washout* occurs. Under this condition, cells do not reproduce as rapidly as they are removed from the system via the waste line. The proper steady-state solution in this case is $X = 0$ and $S = S_{\text{in}}$. In other words, the activated sludge unit completely fails to function. Equation 6.E.16 remains valid, but equation 6.E.14 is invalid in this case, as an assumption used to derive it is violated. If we apply the "typical values" for microbial parameters in Table 6.E.2 to equation 6.E.14 and solve for a typical value of $S = S_{\text{in}} = 200$ mg/L, we find that cell washout occurs when $\Theta_c \leq 0.45$ d. For an activated sludge unit to function properly, one must operate it with Θ_c much greater than the value at which cell washout would occur.

6.E.2 Trickling Filters

Like the activated sludge process, trickling filters are designed to bring BOD-laden wastewater into contact with a large quantity of microorganisms under aerobic conditions. However, in contrast to activated sludge units, where microorganisms are suspended in flocs within the reactor, in a trickling filter the microbial growth is attached to a solid substrate.

As with the activated sludge process, there is a long history in the use of trickling filters for wastewater treatment. The technology was introduced in England in 1893 (Metcalf & Eddy, 1991), and the same basic process is still employed. Trickling filters are easy to operate, are fairly reliable, and exhibit good removal efficiencies for BOD_5 and suspended solids.

Figure 6.E.6 shows the configuration of a typical trickling filter. A circular cylinder of height H and diameter D is filled with an inert material—stone, plastic, or wood—on which microorganism growth is promoted. Following primary treatment, wastewater is sprayed over the top of the unit and allowed to trickle down to the bottom, where it is collected. The water distribution system consists most commonly of a set of nozzles attached to arms that rotate horizontally about the central (vertical)

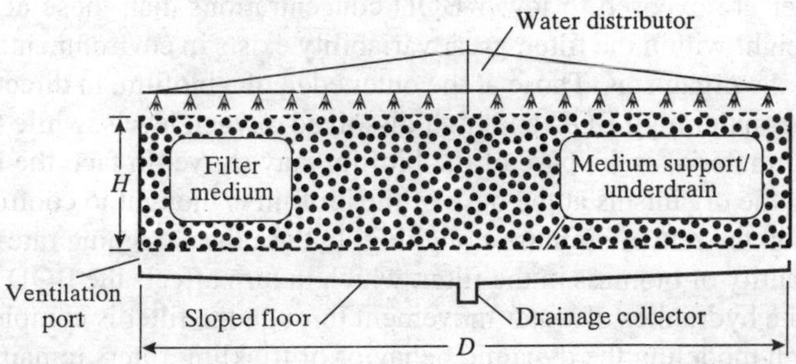

Figure 6.E.6 Schematic of a trickling filter.

axis of the filter. The nozzles are arranged so that the water is distributed uniformly over the open top surface of the filter. The hydraulic loading rate is an important design parameter. It is defined as the volume rate of wastewater (volume per time) per land surface area of the filter ($A = 0.25\pi D^2$ for the case depicted in Figure 6.E.6).

The flow of air needed to maintain aerobic conditions occurs through the pores of the filter medium and through ventilation ports positioned along the bottom of the unit. Air may be supplied mechanically, but most often it is provided by natural convection, caused by differences in temperature between the ambient air and the air in the filter. The underdrain provides mechanical support for the filter materials while permitting water and sloughed microbial biofilms to flow down into a collection system.

That this unit is called a "filter" is potentially misleading. The mechanism of operation is nothing like those used to capture particles in a granular filter (§6.C.2). Instead, as the water flows in thin streams over the solid surfaces, attached microbial biofilms extract the organic material and use it for metabolism and cell synthesis. Under steady-state operation, the microbial biofilm grows to a certain thickness and then is released or sloughed from the surface. As with the activated sludge process, a secondary clarifier must be placed downstream of the trickling filter to remove the sloughed biomass from the treated effluent.

Until the mid-1960s, most trickling filters used rock or slag from industrial furnaces as the filter medium. Since then, plastics have been available as an alternative packing material, and now both rock and plastic are widely used. The individual stones in rock filters range in diameter from 2.5 to 10 cm. Their high density limits the height of a rock filter bed to about 2.5 m because of structural considerations. Plastic packing materials are available in two main forms. In one type, corrugated sheets are bonded together into rectangular modules that are stacked in a regular pattern. The second type of system consists of small molded units that are randomly packed. The plastic materials offer three advantages over rocks. First, they are much less dense, so that taller filters may be used—up to 12 m. Second, the plastic packings have much higher porosity than stone, typically 95 percent compared with 50 percent. For reasons summarized below, this permits a higher loading rate of BOD onto the filter. Third, the plastics have a higher specific surface area than stone, 80–200 $m^2\ m^{-3}$ as compared with 40–70 $m^2\ m^{-3}$. In combination, these factors reduce both the land area and the total filter volume required for a trickling filter using plastic packing material. However, rocks are still used in many applications, because they are less expensive than plastic media.

The process dynamics in a trickling filter are considerably more complex than for an activated sludge system. The biomass is heterogeneous. Microorganisms at the bottom of the filter are exposed to lower BOD concentrations than those at the top. Even at a given height within the filter, great variability exists in environmental conditions surrounding the organisms. Those at the outer edge of a biofilm, in direct contact with the water, encounter high BOD and high dissolved oxygen levels, while those attached to the filter material at the base of the biofilm may starve. In fact, the low supply rate of food to the organisms at the base of the biofilm is thought to control, along with hydraulic drag, the rate of biomass sloughing. In turn, the sloughing rate strongly influences the quantity of biomass in the filter, which in turn affects the BOD removal efficiency. Even the hydraulics of water movement through the filter is complex. Consequently, although modeling the dynamic behavior of trickling filters remains an active area of research, design is guided largely by past practice, augmented by pilot

Table 6.E.3 Design and Operational Data for Trickling Filters

Parameter	Low rate	Intermediate	High rate	Super-high rate
Medium	Rock, slag	Rock, slag	Rock	Plastic
Specific surface area (m^2 m^{-3})	40–70	40–70	40–70	80–200
Porosity (m^3 m^{-3})	0.4–0.6	0.4–0.6	0.4–0.6	0.90–0.97
Mass density of medium[a] (kg m^{-3})	800–1500	800–1500	800–1500	30–100
Hydraulic loading,[b] Q/A (m^3 m^{-2} d^{-1})	0.5–3	3–10	8–40	10–70
BOD_5 loading (kg m^{-3} d^{-1})	0.1–0.4	0.2–0.5	0.5–1	0.5–1.5
Depth, L (m)	1–2.5	1–2.5	1–2.5	3–12
Recirculation ratio, R	0	0–1	1–2	1–2
BOD_5 removal efficiency (%)	80–90	50–70	65–85	65–80

[a]Volume includes voids.

[b]Including recycled flow.

Source: Metcalf & Eddy, 1991; Tchobanoglous and Schroeder, 1985.

tests, which are interpreted using model equations based on greatly simplified descriptions of process dynamics.

Table 6.E.3 presents a summary of the characteristics of four classes of trickling filters. The standard, low-rate filter can achieve a BOD_5 removal efficiency of 80–90 percent, only slightly inferior to the performance of a much more operationally demanding activated sludge unit. The primary disadvantage of the trickling filter is its size, which is much larger than that required for an activated sludge unit. The BOD_5 loading rate for a standard trickling filter is 0.1–0.4 kg m^{-3} d^{-1}, a factor of 5–8 lower than the loading rate for conventional activated sludge units (compare Tables 6.E.2 and 6.E.3). Therefore, for a given wastewater load, the standard trickling filter must be 5–8 times larger (in volume) than an activated sludge reactor. Given a per-capita contribution to BOD_5 in domestic sewage of approximately 100 g d^{-1}, the volume of a low-rate trickling filter for treating domestic sewage must be in the range 0.25–1 m^3 per person served. The maximum dimensions of a standard filter with a circular plan are $H = 2.5$ m and $D = 60$ m, so the limit on population served by a single unit is 7500–30,000. Tchobanoglous and Schroeder (1985) suggest that the standard trickling filter becomes uneconomical when total flow exceeds about 4000 m^3 d^{-1}, which corresponds to the rate of domestic sewage generation by a population of approximately 20,000.

One of the failure modes of a trickling filter, known as *ponding*, occurs when biomass growth is so rapid that the pores of the filter become plugged and wastewater pools at the top of the filter bed. This phenomenon constrains the BOD_5 loading rate of the standard, low-rate filter, since BOD loading controls the biomass growth rate within the filter. When biofilm growth outpaces the sloughing process in these filters, the pores in the filter become filled with biomass, and ponding occurs. In the intermediate and high-rate processes, treated effluent is recycled through the filter. The higher hydraulic loading rate produces a stronger shear force on the biomass, increasing the rate of sloughing and allowing higher BOD loading. However, the BOD removal efficiency of these units is not as good as for the low-rate filters because the filter configuration moves from approximately plug-flow toward a well-mixed reactor. With plastic media, a much higher BOD_5 loading can be used because the higher porosity of the plastics means that oxygen can be delivered at higher rates and thicker biofilms can be grown without the risk of ponding.

6.E.3 Anaerobic Digestion of Wastewater Sludges

In primary and secondary wastewater treatment, a significant fraction of the removed BOD is extracted as sludge. This concentrated liquid waste stream typically must be further treated before it can be safely disposed of. One of the most widely employed sludge treatment technologies is anaerobic digestion. Here a large fraction of the organic material in the sludge is converted to carbon dioxide and methane by microorganisms that act in the absence of oxygen. This process reduces the volume of sludge that must be ultimately disposed of, improves the degree to which the solids in the sludge can be separated from water, and greatly reduces the tendency of the sludge to further decompose after disposal.

Like the aerobic processes discussed already, the primary goal of anaerobic digestion is to treat carbonaceous material to reduce its impact on the environment. Anaerobic treatment is effectively applied to wastes that have a high concentration of organic material. In comparison with aerobic treatment, it is much slower. However, it converts a larger fraction of organic carbon to gases, thereby producing less residual sludge (only about 10 percent of that produced by aerobic treatment). Also, a main end product of anaerobic digestion is methane, which can be used as a fuel to operate equipment for heating and pumping requirements at the wastewater treatment plant.

Besides its use in the stabilization of primary and secondary sludge, anaerobic digestion is also used in place of aerobic methods for the treatment of some high-strength industrial organic wastes. If the wastes are suitable as a substrate for generating methane and if the BOD level exceeds about 1500 mg/L, anaerobic digestion may be the most economical wastewater treatment method (Sawyer et al., 1994). Anaerobic treatment eliminates the need for aeration, which represents a major part of the operating cost of aerobic biological treatment. Some hazardous organic wastes that are resistant to microbial degradation under aerobic conditions, such as chlorinated solvents, can be biologically treated under anaerobic conditions. Here we focus on the treatment of municipal wastewater sludge using anaerobic digestion.

The sludges produced by primary and secondary wastewater treatment contain most of the offensive materials that were present in the raw sewage. Primary treatment is purely a separation process that removes solids, greases, and oils in concentrated streams from the wastewater. In secondary treatment, transformations occur, but only about half of the incoming organic carbon is converted to gas and lost from the system; the remainder is converted to microbial mass that is wasted as sludge. Some characteristics of wastewater sludge streams are summarized in Table 6.E.4. Note that the solids fraction of sludges remains relatively small. More than 90 percent of sludge

Table 6.E.4 Characteristics of Municipal Wastewater Sludges

Parameter	Primary sludge	Activated sludge
Total dry solids (TS, %)	2–8	0.83–1.16
Volatile solids (% of TS)	60–80	59–88
Nitrogen, N (% of TS)	1.5–4	2.4–5
Phosphorus, P_2O_5 (% of TS)	0.8–2.8	2.8–11
Potassium, K_2O (% of TS)	0–1	0.5–0.7
pH	5.0–8.0	6.5–8.0
Alkalinity (meq/L)	10–30	12–220
Organic acids (mM)	3–30	18–28

Source: Metcalf & Eddy, 1991.

is water. A large fraction of the solids are volatile, indicating that they are principally of an organic nature and therefore may be subject to chemical and biological degradation if left untreated. Wastewater sludges also contain significant quantities of the three main elements used in agricultural fertilizers: nitrogen, phosphorus, and potassium. One of the important end uses of digested sludge is as a soil amendment.

What the table does not show is that wastewater sludges typically contain pathogenic microorganisms, posing a distinct health risk, and they can putrefy if not properly treated, resulting in odor problems. Consequently, the treatment and ultimate disposal of wastewater sludges is an important aspect of the overall wastewater treatment process. In fact, it has been estimated that in the United Kingdom, about 40 percent of the total wastewater treatment cost is associated with the treatment and disposal of sludges (Forster, 1985).

The treatment of wastewater sludges consists of two main phases. In the first phase, the goal is to separate the water from the sludge in processes known as *thickening* and *dewatering*. These steps decrease the volume of material that must ultimately be disposed of. The second phase is known as *sludge stabilization*. There are three primary goals of sludge stabilization: to reduce the level of pathogens in the residual solids; to eliminate offensive odors; and to inhibit, reduce, or eliminate the potential for putrefaction (Metcalf & Eddy, 1991). Sludge stabilization significantly increases the options available for the ultimate disposal of wastewater solids.

Anaerobic digestion of municipal wastewater sludges has been widely practiced since the early 1900s and is the most widely used sludge stabilization method. Overall, it converts about 40–60 percent of the organic solids in sludges to methane and carbon dioxide. The residual organic matter is chemically stable, nearly odorless, and contains significantly reduced levels of pathogens. The resulting suspended solids are also more easily separated from water relative to untreated sludge or aerobically treated sludge.

Figure 6.E.7 shows one of the configurations used for anaerobic sludge digestion. The first of the two tanks is always sealed to maintain anaerobic conditions; the second may be either sealed for additional gas collection or open to the atmosphere. The input to the digester is typically a mixture of settled materials from primary sedimentation, wasted microorganisms from secondary treatment, and the surface scums from both primary and secondary clarifiers. These sludges may be added to the reactor continuously or intermittently. The first-stage reactor is heated and mixed to speed the

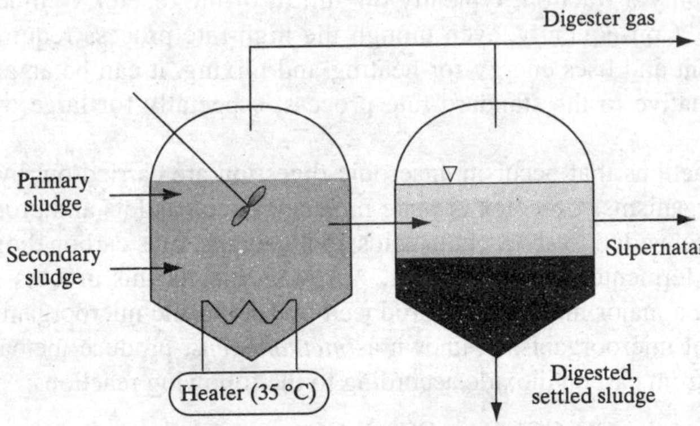

Figure 6.E.7 Schematic of a two-stage, high-rate anaerobic sludge digester (Metcalf & Eddy, 1991).

rate of biological conversion. After a typical residence time of 10–20 days, the mixed, digested sludge passes to the second reactor. Here it is allowed to undergo further digestion, but without mixing or heating. Settled sludge is removed from the second reactor, either incrementally or continuously. The removed sludge is then dewatered and disposed of. The supernatant fluid from a digester is typically highly contaminated, with about 1000 mg/L of BOD and 2000 mg/L of suspended solids. It is recycled to the beginning of the wastewater treatment plant.

The typical composition of the digester gas is 60–65 percent CH_4, 30–35 percent CO_2, plus small quantities of H_2, N_2, H_2S, and H_2O. The heating value of methane is 36.5 MJ m^{-3} (at 1 atm, 273 K), which corresponds to 820 kJ mol^{-1}. Overall, the digester gas has a heating value of 22–24 MJ m^{-3}. Typical gas production is about 1 m^3 (44.6 mol) per kg of volatile suspended solids digested. This quantity of production can be understood stoichiometrically. Assume that the VSS consists mainly of bacterial cells with an effective molecular composition of $C_5H_7O_2N$. Then the mass fraction of carbon in biodegradable VSS is $(5 \times 12) \div 113 = 0.53$. Digestion of 1 kg of VSS should then liberate 530 g = 44 mol of C, which corresponds, approximately, to the number of moles of CH_4 plus CO_2 produced. On a per-capita basis, a digester that treats the sludge from primary and secondary municipal wastewater produces about 28 L of gas per day!

In addition to maintaining an acceptable temperature (32–35 °C), the process requirements for a high-rate anaerobic digester include maintaining the pH between 6.5 and 8.5, ensuring the absence of molecular oxygen, promoting good mixing, and operating with an adequate solids retention time. In contrast to the activated sludge process, in which the microorganisms are recycled, the solids retention time in a sludge digester is the same as the hydraulic detention time. The typical design is sized so that the loading rate of volatile solids is in the range 1.6–4.8 kg m^{-3} d^{-1}. Alternatively, reactors can be roughly sized on a population basis, with high-rate units that treat combined primary and secondary sludge requiring a volume of 70–120 L $person^{-1}$. Reactor tanks are typically cylindrical, with a diameter in the range 6–38 m and a sidewall water depth of from 7.5 m to 14 m or more (Metcalf & Eddy, 1991).

An alternative to the high-rate digester shown in Figure 6.E.7 is the standard-rate process consisting of a single unheated and unmixed reactor. Because of the lower temperature, a detention time of 30–60 days is required to complete the digestion process. Because the reactor is unmixed, stratification occurs. From top to bottom are the gas zone, scum layer, supernatant layer, actively digesting sludge, and digested sludge. Only a fraction, typically one-third, of the reactor volume can be filled with sludge. Consequently, even though the high-rate process requires more complex equipment and uses energy for heating and mixing, it can be an attractive low-volume alternative to the standard-rate process, especially for large treatment facilities.

The major reactions that occur in anaerobic digestion are carried out by several groups of microorganisms. Complex organic molecules such as fats and proteins are decomposed and converted to short-chain acids and alcohols, plus carbon dioxide and hydrogen gas, by fermenters and acetogens. A key species in this process is acetic acid (CH_3COOH), a major intermediate produced by acetogenic microorganisms. An additional group of microorganisms, known as *methanogens*, produce methane from the acetic acid or from carbon dioxide according to the following reactions:

$$CH_3COOH \rightarrow CH_4 + CO_2 \qquad (6.E.23)$$

$$CO_2 + 4\,H_2 \rightarrow CH_4 + 2\,H_2O \qquad (6.E.24)$$

Methanogens and acetogens have a symbiotic relationship. The methanogens need the by-products of the acetogens as their fuel, and the acetogens benefit when the methanogens decompose their waste products.

The methanogens generally function more slowly than the acetogens, so they constitute the rate-limiting step in the degradation process. For a digester to function properly, a dynamic balance must be maintained so that both bacterial populations prosper. The maintenance of proper pH is key. The products of acetogenesis lower the pH in the digester. If the pH drops below about 6.5, the activity of the methanogens begins to slow. Additional acid production further reduces the pH, ultimately terminating methanogenesis and causing the digester to turn *sour*. A useful indicator of digester health is the level of volatile organic acids, such as acetic acid, in the digester liquid. A value of 50–250 mg/L (as acetic acid) is indicative of proper functioning. Upsets can cause the level to rise to thousands of mg/L (Sawyer et al., 1994).

The pH can be maintained in the appropriate range by means of controlling bicarbonate alkalinity. In its normal, healthy state, a digester will have an alkalinity in the range 20–100 meq/L. A rise in the CO_2 level of the digester gas is another early sign of trouble. Upsets in sludge digesters can be caused by anything that inhibits the metabolism of the microbial community, including excessive levels of toxic metals and sulfides, dissolved oxygen, overloading or underloading of the reactor with organic material, and poor mixing.

Figure 6.E.8 summarizes the movement of organic carbon through the major stages of a typical wastewater treatment plant. Of the 3.3 moles of organic C per person per day that enters the wastewater system, about 60 percent is converted to gas (36 percent CO_2 and 24 percent CH_4), 6 percent escapes in the treated wastewater, and the remaining one-third must be disposed of as digested and settled sludge. The digested sludge is typically a liquid slurry with a solids content in the range of 2.5–7 percent (mass of dry solids per total mass) (Metcalf & Eddy, 1991). After digestion, and depending on the ultimate fate of the residuals, the sludge may be subjected to further treatment, including dewatering and disinfection.

The ultimate disposal of the residual sludges from wastewater treatment is problematic. The major options are addition to the soil as an amendment, disposal in a

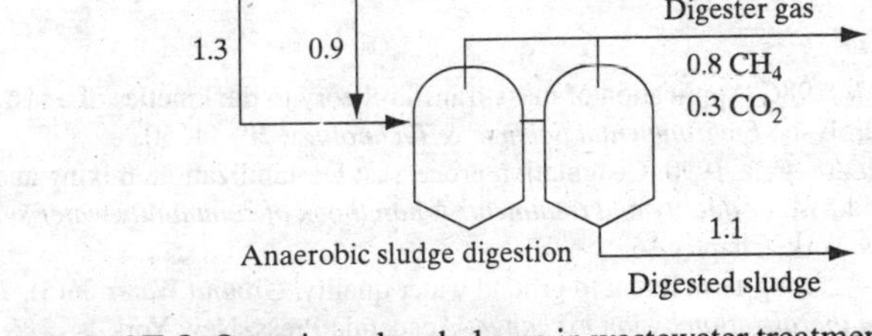

Figure 6.E.8 Organic carbon balance in wastewater treatment. The numbers are estimates of the molar flow of organic carbon (moles C per capita per day) for domestic sewage. (Estimates are based on data presented in Metcalf & Eddy, 1991.)

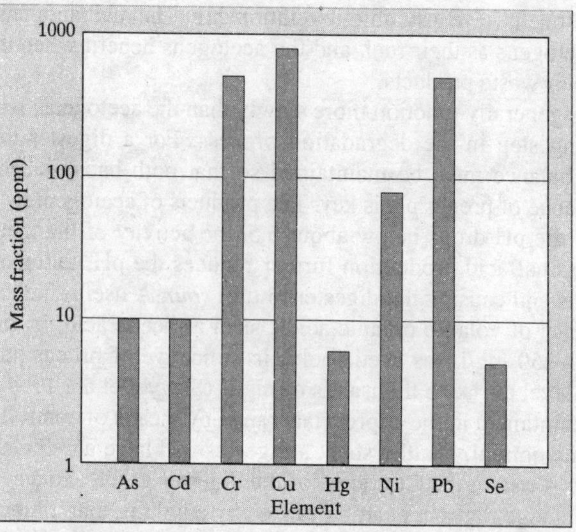

Figure 6.E.9 Median content of toxic metals in wastewater sludge (mass fraction of dry solids) (USEPA, 1984, as cited in Metcalf & Eddy, 1991).

landfill, disposal in the oceans, and incineration. Sludge is moderately rich in nutrients and can be beneficial for plant growth if applied to soil. Unfortunately, sludge effectively concentrates heavy metals from wastewater (see Figure 6.E.9), limiting its usefulness as a soil amendment. Better control of industrial discharges of metals to municipal treatment plants may improve the usefulness of digested wastewater sludge for soil application. As for the other options, landfill capacity is constrained in many parts of the United States and other industrialized nations, and is a problem for all aspects of solid waste disposal. Ocean disposal is banned. Incineration raises air pollution concerns and faces stiff public opposition in many areas. Furthermore, incineration only reduces the volume of sludge residuals; it does not eliminate the need for ultimate disposal of some solid residuals. More stringent requirements recently imposed for wastewater treatment will only exacerbate the problem of sludge management, because many advanced wastewater treatment processes generate substantial sludge waste streams. Sludge treatment and disposal will no doubt remain a challenging problem in the future.

REFERENCES

AIETA, E.M., & ROBERTS, P.V. 1986. Application of mass-transfer theory to the kinetics of a fast gas-liquid reaction: Chlorine hydrolysis. *Environmental Science & Technology*, **20**, 44–50.

AMIRTHARAJAH, A., & O'MELIA, C.R. 1990. Coagulation processes: Destabilization, mixing and flocculation. in F.W. Pontius, ed., *Water quality and treatment: A handbook of community water supplies.* 4th ed. McGraw-Hill, New York, Chapter 16.

ANDREWS, C. 1998. MTBE—A long-term threat to ground water quality. *Ground Water* **36**(5), 705–706.

APPLEBAUM, S.B. 1968. *Demineralization by ion exchange*. Academic Press, New York.

APPLEGATE, L.E. 1984. Membrane separation processes. *Chemical Engineering*, **91**(12), 64–89.

AEEP. 1998. *Research frontiers for environmental engineers and scientists*. Report of a workshop conducted by the Association of Environmental Engineering Professors. Monterey, CA.

ASCE/AWWA. 1990. *Water treatment plant design*. 2nd ed. McGraw-Hill, New York, Chapters 5, 10, and 15.

AWWA. 1973. Manual M20. *Water chlorination principles and practices*. American Water Works Association, New York, p. 21.

BENEFIELD, L.D., & MORGAN, J.M. 1990. Chemical precipitation. In F.W. Pontius, ed., *Water quality and treatment: A handbook of community water supplies*. 4th ed. McGraw-Hill, New York, Chapter 10.

CHANG, S.L. 1971. Modern concept of disinfection. *ASCE Journal Sanitary Engineering Division* **97**, 689–707.

CHICK, H. 1908. An investigation of the laws of disinfection. *Journal of Hygiene*, **8**, 92–158.

CLEASBY, J.L. 1990. Filtration. In F.W. Pontius, ed., *Water quality and treatment: A handbook of community water supplies*. 4th ed. McGraw-Hill, New York, Chapter 8.

CLIFFORD, D.A. 1990. Ion exchange and inorganic adsorption. In F.W. Pontius, ed., *Water quality and treatment: A handbook of community water supplies*. 4th ed. McGraw-Hill, New York, Chapter 9.

CONLON, W.J. 1990. Membrane processes. In F.W. Pontius, ed., *Water quality and treatment: A handbook of community water supplies*. 4th ed. McGraw-Hill, New York, Chapter 11.

CORNWELL, D.A. 1990. Air stripping and aeration. In F.W. Pontius, ed., *Water quality and treatment: A handbook of community water supplies*, 4th ed. McGraw-Hill, New York, Chapter 5.

CRITTENDEN, J.C., CORTRIGHT, R.D., RICK, B., TANG, S.R., & PERRAM, D. 1988. Using GAC to remove VOCs from air stripper off-gas. *Journal of the American Water Works Association*, **80**(5), 73–84.

CROSSLEY, I.A. 1983. Desalination by reverse osmosis. In A. Porteous, ed., *Desalination technology: Developments and practice*. Applied Science Publishers, London, pp. 205–248.

DAVIS, M.L., & CORNWELL, D.A. 1991. *Introduction to environmental engineering*. 2nd ed. McGraw-Hill, New York.

DAVIS, M.L., & CORNWELL, D.A. 1998. *Introduction to environmental engineering*. 3rd ed. WCB McGraw-Hill, Boston.

FISCHER, H.B., LIST, E.J., KOH, R.C.Y., IMBERGER, J., & BROOKS, N.H. 1979. *Mixing in inland and coastal waters*. Academic Press, Orlando, FL.

FORSTER, C.F. 1985. *Biotechnology and wastewater treatment*. Cambridge University Press, Cambridge.

GANCZARCZYK, J.J. 1983. *Activated sludge process: Theory and practice*. Marcel Dekker, New York.

GLEICK, P.H. 1998. *The world's water:1998–1999*. Island Press, Washington, DC.

GOLDMAN, C.R., & CARTER, R.C. 1965. An investigation by rapid carbon-14 bioassay of factors affecting the cultural eutrophication of Lake Tahoe, California–Nevada. *Journal of the Water Pollution Control Federation*, **37**, 1044–1059.

GREGG, M.C. 1973. The microstructure of the ocean. *Scientific American*, **228**(2), 65, February.

HAAS, C.N. 1990. Disinfection. In F.W. Pontius, ed., *Water quality and treatment: A handbook of community water supplies*. 4th ed. McGraw-Hill, New York, Chapter 14.

HAAS, C.N. 1999. Disinfection. In R.D. Letterman, ed., *Water quality and treatment: A handbook of community water supplies*. 5th ed. McGraw-Hill, New York, Chapter 14.

HAAS, C.N., JACANGELO, J.G., BISHOP, M.M., CAMERON, C.D., et al. 1992. Survey of water utility disinfection practices. *Journal of the American Water Works Association*, **84**(9), 121–128.

HAMANN, C.L., MCEWEN, J.B., & MYERS, A.G. 1990. Guide to selection of water treatment processes. In F.W. Pontius, ed., *Water quality and treatment: A handbook of community water supplies*. 4th ed. McGraw-Hill, New York, Chapter 3.

HARTE, J. 1988. *Consider a spherical cow*. University Science Books, Mill Valley, CA.

HOFF, J.C., & AKIN, E.W. 1986. Microbial resistance to disinfectants: Mechanisms and significance. *Environmental Health Perspectives*, **69**, 7–13.

HORAN, N.J. 1990. *Biological wastewater treatment systems: Theory and operation*. Wiley, New York.

HORNE, A.J., & GOLDMAN, C.R. 1994. *Limnology*. 2nd ed. McGraw-Hill, New York.

JENKINS, D., RICHARD, M.G., & DAIGGER, G.T. 1993. *Manual on the causes and control of activated sludge bulking and foaming*. 2nd ed. Lewis, Boca Raton, FL.

LAGREGA, M.D., BUCKINGHAM, P.L., & EVANS, J.C. 1994. *Hazardous waste management*. McGraw-Hill, New York, §9.3.

MCCABE, W.L., SMITH, J.C., & HARRIOT, P. 1993. *Unit operations of chemical engineering*. 5th ed. McGraw-Hill, New York, Chapters 25–26.

MCRAE, W.A. 1983. Electrodialysis. In A. Porteous, ed., *Desalination technology: Developments and practice*. Applied Science Publishers, London, pp. 249–264.

METCALF & EDDY. 1991. *Wastewater engineering: Treatment, disposal, and reuse*. 3rd ed. McGraw-Hill, New York.

PONTIUS, F.W. 1992. A current look at the federal drinking water regulations. *Journal of the American Water Works Association*, **84**(3), 36–50.

PRESSLEY, T.A., BISHOP, D.F., & ROAN, S.G. 1972. Ammonia-nitrogen removal by breakpoint chlorination. *Environmental Science & Technology*, **6**, 622–628.

ROBERTS, P.V., & SUMMERS, R.S. 1982. Performance of granular activated carbon for total organic-carbon removal. *Journal of the American Water Works Association*, **74**(2), 113–118.

SAWYER, C.N., MCCARTY, P.L., & PARKIN, G.F. 1994. *Chemistry for environmental engineering*. 4th ed. McGraw-Hill, New York.

SCHWARZENBACH, R.P., GSCHWEND, P.M., & IMBODEN, D.M. 1993. *Environmental organic chemistry*. Wiley, New York.

SELLECK, R.E. 1989. University of California, Berkeley. Personal communication.

SILVER, R.S. 1983. Seawater desalination. In A. Porteous, ed., *Desalination technology: Developments and practice*. Applied Science Publishers, London, pp. 1–30.

SNOEYINK, V.L. 1990. Adsorption of organic compounds. In F.W. Pontius, ed., *Water quality and treatment: A handbook of community water supplies*. 4th ed. McGraw-Hill, New York, Chapter 13.

SPIEGLER, K.S., & EL-SAYED, Y.M. 1994. *A desalination primer*. Balaban Desalination Publications, Santa Maria Imbaro, Italy.

STREETER, H.W., & PHELPS, E.B. 1925. A study of the pollution and natural purification of the Ohio River. Public Health Bulletin No. 146. U.S. Public Service, Washington, DC.

STUMM, W., & MORGAN, J.J. 1996. *Aquatic chemistry: Chemical equilibria and rates in natural waters*. 3rd ed. Wiley, New York.

TATE, C.H., & ARNOLD, K.F. 1990. Health and aesthetic aspects of water quality. In F.W. Pontius, ed., *Water quality and treatment: A handbook of community water supplies*. 4th ed. McGraw-Hill, New York, Chapter 2.

TCHOBANOGLOUS, G., & SCHROEDER, E.D. 1985. *Water quality: Characteristics, modeling, modification*. Addison-Wesley, Reading, MA.

USEPA. 1984. *Environmental regulations and technology: Use and disposal of municipal wastewater sludge*. EPA 625/10-84-003.

USEPA. 1994. *Groundwater and wellhead protection handbook*. EPA/625/R–94/001, Washington, DC.

USEPA. 1996a. National Water Quality Inventory—1996 Report to Congress. http://www.epa.gov/305b/.

USEPA. 1996b. Providing safe drinking water in America: 1996 national public water system annual compliance report and update on implementation of the 1996 Safe Drinking Water Act amendments. http://www.epa.gov/ogwdw/pwsinv96.html.

USEPA. 1998. *Water pollution control: 25 years of progress and challenges for the new millennium*, EPA/833/F–98/00, Washington, DC.

WATSON, H.E. 1908. A note on the variation of the rate of disinfection with change in the concentration of the disinfectant. *Journal of Hygiene*, **8**, 536–542.

WETZEL, R.G. 1975. *Limnology*. Saunders College Publishing, Philadelphia.

WIESNER, M.R., CLARK, M.M., JACANGELO, J.G., LYKINS, B.W., et al. 1992. Committee report—Membrane processes in potable water treatment. *Journal of the American Water Works Association*, **84**(1), 59–67.

PROBLEMS

6.1 Issues in water and wastewater treatment

(a) Before sewers were developed, what was done with human waste generated in cities?

(b) A municipal water treatment plant that obtains its source water from a river typically has a sedimentation basin followed by a deep-bed sand filter. Why are both units needed?

(c) Why is pH control important when disinfecting with chlorine?

(d) What does it mean to *destabilize* a suspension of fine particles?

(e) Name two common methods for carrying out secondary treatment in a municipal wastewater treatment plant.

(f) What health concern is causing some municipal water treatment agencies to convert from chlorine to chloramines for drinking water disinfection?

(g) For what specific purpose is reverse osmosis used in water treatment?

(h) What happens to powdered activated carbon after it is added as a reagent in drinking water treatment?
(i) What reagents are added to water to achieve chemical precipitation?
(j) What is the significance of the *overflow rate* in a sedimentation basin?
(k) For what purpose is an anaerobic digester used in wastewater treatment?

6.2 Cultural eutrophication
(a) Define cultural eutrophication.
(b) Explain how the runoff of water from agricultural land can contribute to the eutrophication of lakes.
(c) How can cultural eutrophication be controlled?

6.3 Toxic metals and the role of pH
One of the major concerns associated with acid precipitation is the mobilization of toxic metals, such as aluminum. Aluminum in water is toxic to fish at levels as low as 0.1 mg/L. The California standard for aluminum in drinking water is 1 mg/L. Consider water in contact with gibbsite ($Al_2O_3{\cdot}3H_2O$), a common clay mineral. The equilibrium dissolution of gibbsite may be described as follows:

$$\tfrac{1}{2}\,Al_2O_3{\cdot}3H_2O(s) \Leftrightarrow Al^{3+} + 3\,OH^- \qquad K_{sp} = 1 \times 10^{-34}\ M^4\ (25\ °C)$$

(a) At what pH would the concentration of Al^{3+} in equilibrium with gibbsite be 0.1 mg/L?
(b) Repeat part (*a*) for an Al^{3+} concentration of 1 mg/L.

6.4 "Your water and your health"
An advertising flyer from the local water district gives the following data concerning the tap water that they deliver to customers (concentrations of minor ions such as iron and manganese are omitted). No information is given about other ions, but you may assume that the other dominant species are HCO_3^- and CO_3^{2-}.

Cations (mg/L)		Anions (mg/L)	
Calcium (Ca^{2+})	8.6	Chloride (Cl^-)	3
Magnesium (Mg^{2+})	0.77	Sulfate (SO_4^{2-})	3.4
Potassium (K^+)	0.64	Fluoride (F^-)	0.84
Sodium (Na^+)	3.7		
pH	8.46		

(a) What is the equilibrium ratio of $[HCO_3^-]$ to $[CO_3^{2-}]$ in this water?
(b) Write an electroneutrality relationship for this water.
(c) Use the electroneutrality relationship and the data given to determine the molar concentration of HCO_3^-.
(d) Evaluate each of the following properties of the water:
 (i) Total hardness (meq/L)
 (ii) Carbonate hardness (meq/L)
 (iii) Ionic strength (M)
 (iv) Alkalinity (meq/L)

6.5 Water quality measures applied to Perrier

The following information was printed on a can of Perrier to indicate its contents:

Concentration (mg/L)			
Ca^{++}	148	HCO_3^-	388
Mg^{++}	3.7	SO_4^{2-}	36.6
Na^{++}	8.9	Cl^-	21
K^+	0.6	NO_3^-	17

(a) One of the ions is incorrectly listed. Which one? What is the correct form?

(b) Evaluate the alkalinity of the water (meq/L), assuming that the pH is less than 9.

(c) Evaluate the total hardness of the water (meq/L). Determine the carbonate hardness and noncarbonate hardness. How would this water be classified with respect to hardness?

(d) The printed information is insufficient to conclusively determine the pH of the water. However, we can obtain an upper bound for the pH. Given a solubility product for $CaCO_3(s)$ of 5×10^{-9} M^2, determine the maximum concentration of CO_3^{2-} (M) that this water can contain. Then use the acid dissociation constant for $HCO_3^- \Leftrightarrow H^+ + CO_3^{2-}$, $K_2 = 4.7 \times 10^{-11}$ M, to determine the maximum pH. (*Note*: A higher pH than this maximum would lead to precipitation of $CaCO_3(s)$ in the water, which would not be great for sales of Perrier!)

6.6 Fundamentals of water and wastewater treatment

(a) A water treatment plant requires three processes for particle removal: coagulation-flocculation, filtration, and sedimentation. In which order would you place these processes? Defend your answer.

(b) What is the main function of an activated sludge treatment unit and how does it work?

(c) What are the specific functions of lime (CaO or $Ca(OH)_2$) and soda ash (Na_2CO_3) when these reagents are used for water softening?

(d) What are the key advantages and disadvantages of using ozone versus chlorine for disinfection?

(e) Why might activated carbon be used in a water treatment plant?

(f) What are disinfection by-products and how are they formed?

6.7 Overall design of a municipal water treatment plant

The following treatment processes are commonly used in municipal drinking water treatment:

- Coagulant addition–flocculation tank
- Deep-bed sand filter
- Disinfectant addition–contact basin
- Granular activated carbon sorption unit
- Sedimentation basin

(a) State the purpose and means of functioning of each unit.

(b) If each of these processes were to be used in a single treatment plant, specify the order in which they would be applied. Justify your answer.

6.8 Turbidity control
The source water for a conventional municipal drinking water treatment plant contains excessive turbidity caused by a stable suspension of colloidal particles.
(a) Specify the treatment step or steps needed *only* for the purpose of reducing turbidity.
(b) State the specific purpose of each treatment step listed in (*a*).
(c) Briefly describe the chemical or physical processes in each treatment step that contribute to turbidity reduction.

6.9 Chemicals used in water and wastewater treatment
Each of the following chemicals may be used in a municipal water treatment or wastewater treatment plant. State the purpose for which each is used.

(a) $FeCl_3$	(b) $Ca(OH)_2$	(c) $Cl_2(g)$	(d) NaF	(e) O_3
(f) NaOCl	(g) Na_2CO_3	(h) NH_3	(i) CO_2	(j) O_2

6.10 Wastewater treatment system
(a) Sketch a typical treatment process scheme for municipal wastewater that will be discharged into a local river.
(b) Briefly describe the major function and mechanism of each of the treatment processes.

6.11 Technologies for treating water and wastewater
Name a treatment technology that is commonly used to remove or transform each of the following species from water or wastewater in municipal treatment plants.
(a) Pathogens from either water or wastewater
(b) Calcium ions from drinking water
(c) Undesired dissolved organic molecules from drinking water
(d) Undesired dissolved organic molecules from wastewater
(e) Suspended particles >30 μm in diameter from either wastewater or drinking water
(f) Suspended particles in the diameter range 5–20 μm from drinking water
(g) Hydrogen sulfide from drinking water

6.12 Wastewater treatment
In a municipal wastewater treatment plant, what is the principal objective of each of the following stages of treatment? How is this objective achieved?
(a) Primary treatment
(b) Secondary treatment
(c) Tertiary treatment

6.13 Fundamental aspects of wastewater treatment
(a) Describe the purpose and method of operation of a primary sedimentation basin.
(b) What is the main function of an activated sludge treatment unit and how does it work?
(c) What are the main by-products generated in the activated sludge process?

6.14 Reactor design for a treatment process
You are assigned to carry out a preliminary design for a water treatment process. Some species of concern, denoted A, is known to degrade in water. But you do not know the rate law or reaction rate constant. You conduct a series of batch experiments. In each case, you establish an initial concentration of 50 mg/L of species A in water in

a sealed vessel. After a certain period, you open the vessel and measure the remaining concentration of A. The results of your experiment are tabulated below.

Experiment	Duration of experiment	Final concentration of A (mg/L)
1	1 hour	35.0
2	3 hours	17.2
3	6 hours	5.9

(a) What is the reaction order?

(b) What is the rate constant?

(c) You are to design a reactor to handle a constant flow rate of 0.3 m^3 s^{-1} of water containing 50 mg/L of A. If the reactor is configured as a CMFR, how large must it be to achieve 90 percent removal of A?

(d) Repeat (*c*) for the case of a plug-flow reactor.

(e) Repeat (*c*) for two CMFRs in series. Assume the reactors have equal volumes.

6.15 Total particle mass removal in sedimentation tanks

A sedimentation tank has the characteristics listed below. Assume that the influent (untreated) water contains spherical particles in discrete sizes with the mass concentrations listed.

(a) Calculate the total removal efficiency for particle mass (i.e., mass removed per mass entering) in this sedimentation tank.

(b) What is the total rate of particle mass removed by the sedimentation tank (tonnes per day)?

Data

Overflow rate	$Q/A_s = 48$ m d^{-1}
Height of settling zone	$H = 3.5$ m
Width of settling zone	$W = 10$ m
Length of settling zone	$L = 40$ m
Particle density	$\rho = 2.65 \times 10^3$ kg m^{-3}

Influent Particle Mass Concentration

$d_p = 2$ μm	$M_2 = 100$ g m^{-3}
$d_p = 5$ μm	$M_5 = 150$ g m^{-3}
$d_p = 10$ μm	$M_{10} = 200$ g m^{-3}
$d_p = 20$ μm	$M_{20} = 175$ g m^{-3}
$d_p = 50$ μm	$M_{50} = 125$ g m^{-3}

6.16 Grit chamber effectiveness

At a municipal wastewater treatment plant, there are grit chambers located upstream of the primary treatment facilities. The approximate dimensions of each grit chamber are 18 m (long) by 3.6 m (wide) by 2.7 m (deep). At a typical design load, the hydraulic detention time is 1.2 minutes. Determine and plot the expected particle removal efficiency in one of these grit chambers as a function of particle diameter. Carry out the analysis for two idealized reactor models: (*a*) a CMFR and (*b*) a PFR. Assume that the suspended solids are spherical with a grain density of 2.65 g cm^{-3}. How does the answer to this problem help to understand the design and function of a grit chamber?

6.17 Primary sedimentation tanks

A wastewater treatment facility has 16 primary sedimentation tanks that operate in parallel. Each tank has the following dimensions:

$$L = 55\,\text{m (length)}$$
$$W = 10.7\,\text{m (width)}$$
$$H = 3.0\,\text{m (depth)}$$

The tanks each have a design capacity of 0.8 $m^3\ s^{-1}$ for wastewater. Answer the following questions, assuming that the system is operating at 50 percent of design capacity.

(a) What is the overflow rate ($m\ h^{-1}$)?

(b) What is the hydraulic detention time (h)?

(c) For spherical particles with a density of sand (2.65 g cm^{-3}), what is the smallest particle that is removed with 100 percent efficiency?

(d) If the average solids removal efficiency is 50 percent and the solids content of untreated wastewater is 500 g m^{-3}, what is the total rate (for all 16 tanks) at which sludge mass is collected (metric tons of solids per day)?

6.18 A sedimentation basin design problem

Consider an ideal rectangular sedimentation tank for removing spherical particles from water. The fluid velocity is uniform and horizontal throughout the settling region.

(a) For the data given below, calculate the minimum length of the tank (m) that would remove 40 μm diameter particles with 100 percent efficiency.

(b) Using your answer to part (*a*), calculate and plot the removal efficiency for particle diameters ranging from 1 μm to 50 μm in diameter. Use a logarithmic scale for particle diameter.

Data

Height of settling zone	$H = 3.5$ m
Width of settling zone	$W = 6$ m
Water flow rate through tank	$Q = 0.2\ m^3\ s^{-1}$
Particle density	$\rho = 2.65 \times 10^3\ kg\ m^{-3}$

6.19 Settling velocities of particles

Water flowing into a rectangular plug-flow reactor contains 180 g m^{-3} of suspended spherical particles with density of 2.2 g cm^{-3}, evenly distributed (by mass) among 10 μm, 100 μm, and 1000 μm diameters. The water is 20 °C and flows at a rate of 20 m^3 h^{-1}. The height of the rectangular reactor is 4 m, and the width is 3 m.

(a) Compute the number concentration (per m^3) for each particle size in the *inlet* water.

(b) Calculate the settling velocity for each of the particle sizes.

(c) If we assume that any particles that settle to the bottom of the reactor before exiting the reactor are removed, calculate the reactor length that is required to ensure 100 percent particle removal.

6.20 Application of a sedimentation basin for coal cleaning

A large portion of the electricity generation in the United States is based on coal combustion. A widely used technology is pulverized coal combustion, in which the coal is ground to a powder, typically 40–80 μm diameter, and then mixed with air before being injected through burners into a furnace. One of the environmental problems associated with coal combustion is the generation of fly ash. Coal cleaning can be used to reduce the generation of fly ash. Impurity particles have a different density than the coal particles, and this can be used as a basis for separation. In this problem, you are to consider the potential application of a settling tank or sedimentation basin for separating coal from mineral impurities. Assume that we have a mixture of coal and soil particles in water. For simplicity, assume that each of the particles is spherical with a 50 μm diameter. The density of the coal particles is 1.3 g cm^{-3}. The density of the soil particles is 2.6 g cm^{-3}.

(a) Compute the settling velocities of the coal and soil particles in water.

(b) In a sedimentation basin, what overflow rate would be required to achieve 90 percent removal of the soil particles?

(c) At the overflow rate obtained in part (*b*), determine the fraction of coal particles entering the basin that would be lost due to settling into the sludge zone.

6.21 Preliminary sizing of a sedimentation basin

You would like to construct a settling basin to remove 85 percent of the particles in the influent water. A characterization of the influent particles indicates that they are evenly divided (by mass) between 20 μm and 15 μm diameter spherical particles with a density of 2.4 g/cm^3.

(a) Calculate the settling velocities of the particles within the settler (cm/s).

(b) Calculate the critical settling velocity that is required to achieve the design goal.

(c) Given that the settling reactor can be no more than 5 m wide and 10 m long, what influent flow would this unit be able to treat within the given constraints (m^3/d)?

6.22 Filtration velocity through a dual-media filter

A dual-media filter with the properties shown below is employed at a water treatment plant. Estimate the filter throughput (Q/A_s, in units of $m^3\ m^{-2}\ d^{-1}$). Assume that the total head loss through the filter is 0.5 m. The temperature is 20 °C.

Parameter	Anthracite	Sand
Depth, L (m)	0.5	0.3
Grain diameter, d_{eq} (mm)	1.0	0.5
Porosity, ϕ	0.6	0.45
Sphericity, ψ	0.5	0.75

6.23 Sorption for controlling dissolved organics

Groundwater contaminated with trichloroethylene (TCE) is to be treated by sorption on powdered activated carbon (PAC). An experimental study on the sorptive capacity of the PAC yields data that are well fit by a Freundlich isotherm:

$$q_e = 129 C_e^{0.73}$$

where q_e = mass of sorbed TCE per mass of PAC (mg/g) and C_e = equilibrium concentration of TCE remaining in the aqueous phase (mg/L).

(a) If the TCE concentration in untreated water is 1 mg/L, what mass of PAC must be used per volume of water ([mg PAC]/[L H_2O]) to reduce the TCE concentration to the allowed maximum contaminant level of 0.005 mg/L?

(b) If the amount of PAC applied is only half that indicated in part (*a*), what is the equilibrium aqueous concentration of TCE?

6.24 Sorption by activated carbon

In drinking water treatment systems, powdered activated carbon (PAC) is commonly used to control taste and odor. One of the molecules responsible for odors in drinking water is geosmin. To determine the required dose of PAC, a treatment plant operator conducted a sorption experiment. Plotted in the figure are the experimental results. The vertical axis represents q_e, the mass of sorbed geosmin per mass of PAC. The horizontal axis represents C_e, the mass concentration of geosmin that remains dissolved in water at equilibrium. The straight-line fit to q_e versus C_e on a log-log plot indicates that the data are well described by a Freundlich isotherm: $q_e = K_f C_e^{1/n}$.

(a) From the data in the figure, determine K_f and n. If C_e has units μg/L and q_e has units mg/g, what are the units associated with K_f and n?

(b) If a sample of untreated water contains 10 μg/L of geosmin, how much PAC must be added (mg/L) to reduce the dissolved geosmin concentration to 0.5 μg/L?

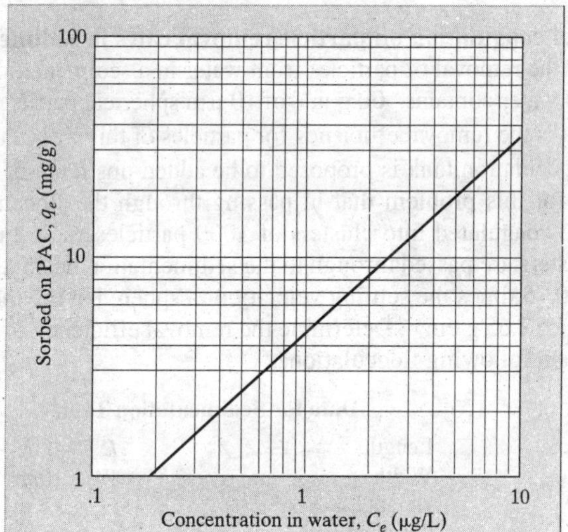

6.25 Sorption for pesticide removal from drinking water

A municipal drinking water treatment plant discovers that it has an unacceptably high level of the pesticide carbofuran and must develop a treatment technique to reduce the concentration to satisfy the relevant water quality standard. The treatment method of choice is sorption on activated carbon. Your firm gets the design job, and you are assigned to do the preliminary calculations. Research has revealed that the sorption of carbofuran onto activated carbon from water follows this isotherm:

$$q_e = 266 C_e^{0.41}$$

where C_e (mg/L) is the equilibrium concentration of the species in water and q_e (mg/g) is the mass of sorbed carbofuran per mass of activated carbon.

The carbofuran concentration in untreated water is 1 mg/L. The drinking water standard is 0.04 mg/L. You must design to treat a steady water flow rate of 0.4 m^3 s^{-1}.

(a) One possible approach is to mix powdered activated carbon (PAC) with the water, wait for equilibrium to be attained, then remove the PAC. At what rate must PAC be added (g s^{-1}) to meet the proposed standard?

(b) An alternative configuration is a packed column of granular activated carbon, similar to a deep-bed sand filter. Given the following parameters, specify the cross-sectional area (in the direction normal to flow) and depth of the column needed to treat the water.

$K = 0.02$ m s^{-1}	Hydraulic conductivity through GAC bed
$h = 0.5$ m	Permissible head loss across GAC bed
$t_c = 600$ s	Contact time for water in GAC column (to reach equilibrium)

(c) If the bulk dry density of the GAC used in (*b*) is 500 kg m^{-3}, estimate the upper-bound, maximum time, $t_{b,\max}$, that the column could be operated before break-through.

(d) At time $t = t_{b,\max}/2$, sketch the concentration of carbofuran in the water within the column as a function of length along the column. It is important only that you show proper inlet and outlet concentrations as well as the correct general shape of the curve.

6.26 Impact of coagulation on particle removal rates in sedimentation tanks

Consider the removal of particles from water in a sedimentation basin. Assume that the untreated water contains 500 g m^{-3} of 10 μm spherical particles of density 2.65 g cm^{-3}.

(a) What is the removal efficiency for particles of this size in the sedimentation basin?

(b) A flocculation tank is proposed to be added upstream of the sedimentation basin. Assume for this problem that in passing through the flocculation tank, all particles would be coagulated into clusters of 1000 particles each. Further assume that when these clusters are passed through to the sedimentation basin, they settle with a velocity equal to 0.75 times the settling velocity of a sphere having the mass of the cluster and a density of 2.65 g cm^{-3}. Determine the removal efficiency of particles in the sedimentation basin following flocculation.

Data for Sedimentation Tank

Length	$L = 40$ m
Width	$W = 10$ m
Height	$H = 3.5$ m
Overflow rate	$Q/A_s = 50$ m/d

6.27 Sedimentation basin efficiency

Water is treated in a rectangular sedimentation basin. The basin is operated with a laminar flow velocity of $U = 450$ m d^{-1}. Its dimensions are length $L = 30$ m, width $W = 8$ m, and depth $H = 4$ m. The influent contains spherical particles with sizes distributed as given below. The particles have a solid density of 2400 kg/m^3.

Particle diameter (μm)	Number concentration (m^{-3})
2	1.3×10^{12}
8	3.9×10^{10}
16	1.5×10^{10}

(a) Calculate the mass concentration of suspended solids in the influent (g/m^3).

(b) Calculate the settling velocities of the particles (cm/s).

(c) Calculate the fraction of particle mass removed.

(d) Calculate the fraction of particle number removed.

6.28 Softening limestone-hardened water

You have been asked to design a process to soften water that comes from a reservoir in contact with a limestone formation. The water is in equilibrium both with the solid limestone ($CaCO_3$) and with the atmosphere.

(a) Sketch the carbonate system in the reservoir complete with each carbonate species and governing equilibrium constants.

(b) Write an electroneutrality relationship for the ions in the reservoir.

(c) Using appropriate equilibrium relationships, rewrite the electroneutrality relationship in terms of $[H^+]$ and the equilibrium constants.

(d) The pH of the reservoir is 8.3. Explain why this is higher than the pH of rainwater.

(e) Calculate the alkalinity, the total hardness, and the carbonate hardness of the water.

(f) Calculate the moles of lime and soda ash that would be required to soften this water.

6.29 Water softening and the hardness of water

(a) Data on major cations in drinking water supplied by a local utility are listed below. Determine the total hardness of this water in meq/L.

(b) Would softening by the lime-soda process improve the quality of this water? Explain.

Ion	Mass concentration (mg/L)
Al^{3+}	0.10
Fe^{3+}	0.02
Ca^{2+}	9.8
Mg^{2+}	1.8
Na^+	4.6

6.30 Water softening: Reagent needs and sludge formation

Water containing 150 g m^{-3} of Ca^{2+} and 75 g m^{-3} of HCO_3^- alkalinity is to be softened using the lime-soda process.

(a) For a water flow rate of 1 m^3 s^{-1}, calculate the daily chemical requirements of lime ($Ca(OH)_2$) and soda (Na_2CO_3) (metric tons per day).

(b) Calculate the rate of generation of sludge solids mass (metric tons per day).

(c) Calculate the net change in total dissolved solids (TDS, g m^{-3}) caused by softening.

6.31 Precipitation as a treatment method for barium

Barium (Ba) is a toxic metal that may be found at trace levels in drinking water. The national primary drinking water regulations limit Ba in municipal drinking water to 2 mg/L. The drinking water in many communities in Illinois, Kentucky, Pennsylvania, and New Mexico can contain barium at five times the standard. Barium can be removed from drinking water by precipitation as barium carbonate, $BaCO_3$. Consider the following situation. Untreated water contains 10 mg/L of Ba^{2+}, 100 mg/L of Ca^{2+}, and 350 mg/L of HCO_3^-, and is at a pH of 7.0. (Other ions may be present but are of no consequence here.) Sodium hydroxide (NaOH, a strong base) is to be added to the water to raise the pH and cause $BaCO_3$ to precipitate. Some of the calcium may precipitate as $CaCO_3$ as well. The solubility product for $BaCO_3$ is $K_{sp1} = 8.1 \times 10^{-9}$ M^2. Assume that the solubility product for $CaCO_3$ is $K_{sp2} = 8.7 \times 10^{-9}$ M^2.

(a) Determine the total hardness, carbonate hardness, and noncarbonate hardness of the untreated water (meq/L).

(b) What is the minimum carbonate concentration (M) for which the equilibrium barium concentration satisfies the standard?

(c) At the carbonate concentration (M) determined in part (*b*), how much calcium will remain in solution at equilibrium?

(d) Determine how much NaOH must be added to the water (g/m^3) such that the equilibrium barium concentration would satisfy the standard.

6.32 Water softening by the lime-soda process

Water with the following properties is to be softened with the lime-soda process:

Species	Concentration (g m^{-3})
Ca^{2+}	0
Mg^{2+}	120
$H_2CO_3{}^*$	4.8
$HCO_3{}^-$	300
$CO_3{}^{2-}$	1.4
$SO_4{}^{2-}$	236

The pH of the untreated water is 8.

(a) Determine the total hardness, carbonate hardness, and noncarbonate hardness of this water.

(b) Determine the reagent requirements. What is the mass concentration (g m^{-3}) of soda ash (Na_2CO_3) and slaked lime ($Ca(OH)_2$) that must be added to raise the pH to 11 and fully soften the water?

(c) Assuming that all of the Mg^{2+} precipitates as $Mg(OH)_2$ and that all of the Ca^{2+} (from lime addition) precipitates as $CaCO_3$, what is the sludge production rate in kilograms of sludge per cubic meter of treated water?

(d) In fact, not all of the magnesium will precipitate. Given a solubility product for $Mg(OH)_2$ of 1.2×10^{-11} M^3 at 25 °C, what fraction of the Mg^{2+} initially present will remain in solution at equilibrium at pH = 11?

(e) After precipitation and settling to remove hardness, the pH of the water is lowered from 11 to 7 by the addition of gaseous carbon dioxide. Consider a sample of water in which the only impurities are magnesium (Mg^{2+}, at the level determined in part [*d*]), species in the carbonate system ($H_2CO_3{}^*$, $HCO_3{}^-$, and $CO_3{}^{2-}$), the hydroxide ion (OH^-), and protons (H^+). How much gaseous CO_2 (in g CO_2 per m^3 of treated water) must be added to lower the pH from 11 to 7?

6.33 Drinking water treatment by chemical precipitation

(a) Cadmium is a toxic metal. It can be removed from water by chemical precipitation of solid cadmium hydroxide, $Cd(OH)_2$. The drinking water standard for cadmium is 5 μg/L (0.005 mg/L). The solubility product of $Cd(OH)_2$ is 2×10^{-14} M^3. What is the minimum pH necessary to reduce the dissolved cadmium ion concentration to the standard?

(b) How much $Ca(OH)_2$ must be added to initially pure water to raise the pH to 11? Express your answer as a mass concentration.

(c) In the use of chemical precipitation for water softening, the pH is typically raised by adding $Ca(OH)_2$. There is a certain irony in the use of this reagent. What is this irony?

(d) The difficulty identified in (*c*) would not occur if NaOH were used to raise the pH instead of $Ca(OH)_2$. Why is calcium hydroxide preferred?

(e) After chemical precipitation of drinking water is completed, the pH of the treated water must be reduced from 10–11 to near neutral. What reagent is most commonly used to reduce the pH?

6.34 Chemical precipitation at a metal plating shop

A small facility, the Nickel 'n' Dime Shop, provides a nickel plating service to the metal casting industry. Their wastewater has elevated concentrations of Ni^{2+}. The local wastewater utility requires that they reduce the Ni^{2+} level in their wastewater from the current level of 2 mg L^{-1} to 0.1 mg L^{-1}. You are hired as a consultant to provide an assessment of the possibility of using chemical precipitation to treat the wastewater. Preliminary investigation reveals that it might be practical to form solid $Ni(OH)_2$ by adding $Ca(OH)_2$ to raise the pH. The solubility product of $Ni(OH)_2$ is $K_{sp1} = 6.5 \times 10^{-18}$ M^3.

(a) What is the minimum pH required to achieve the discharge standard for Ni^{2+}?

(b) How much $Ca(OH)_2$ (mg/L) would you have to add to initially *pure* water to raise the pH to the level determined in part (*a*)?

(c) If the initial wastewater conditions were as given below, how much $Ca(OH)_2$ (mg/L) would you have to add to raise the pH to the level needed to precipitate Ni^{2+} such that the equilibrium concentration was 0.1 mg/L? (*Hint*: The potential precipitation of $CaCO_3$ and the buffering against pH change by alkalinity should be considered.)

initial pH = 7.0

initial alkalinity = 3.0 meq/L

initial concentration of Ni^{2+} = 2.0 mg/L

6.35 Disinfection with chlorine

An experiment reveals that a concentration of 0.10 g m^{-3} of free available chlorine yields a 99 percent kill of bacteria in 8 minutes. What contact time is required to achieve 99.9 percent kill at a free available chlorine concentration of 0.05 g m^{-3}? Assume that Chick's law and Watson's law hold with $n = 1$.

6.36 Chemistry of a disinfectant: Household bleach

The label on a bleach bottle states that it contains 5.25 percent (by mass) sodium hypochlorite (NaOCl). Assume that bleach is manufactured by adding this chemical to otherwise pure water. Take the density of bleach to be the same as that of pure water, 1.0 g cm^{-3}. Assume that NaOCl dissociates completely.

(a) What is the pH of bleach?

(b) Bleach can be used as an emergency disinfectant—for example, for water after an earthquake. Assume that the target dose of free residual chlorine ($HOCl + OCl^-$) to be added is 50 g m^{-3}. How much bleach (cm^3) would you add to 1 L of water to disinfect it?

(c) If too much bleach is added to water, it is possible to dechlorinate the water by adding sulfur dioxide (SO_2). The sulfur is oxidized from aqueous sulfonic acid to sulfate (SO_4^{2-}), while the chlorine is reduced from the hypochlorite ion (OCl^- or HOCl) to chloride ion (Cl^-). Write a stoichiometrically balanced reaction that shows how this dechlorination reaction proceeds.

(d) Of course, most people don't have SO_2 gas cylinders lying around in their homes. An alternative dechlorination agent is ammonia (NH_3). The nitrogen in ammonia is oxidized to either N_2O or N_2. Write separate, stoichiometrically balanced reactions for dechlorination by ammonia for each of these end products.

6.37 Chlorine chemistry

Chlorine gas is bubbled into water to disinfect it by killing pathogens.

(a) To improve the effectiveness of chlorine as a disinfectant, water should be maintained at acidic pH. Find the partial pressure of $Cl_2(g)$ that is required to bring pure water to a pH of 5 at equilibrium.

(b) What percentage of the added chlorine would be in the form of HOCl at this pH?

6.38 Disinfection kinetics for chlorine

The following data were obtained from batch disinfection studies of the bacteria *Streptococcus*. Given a contact time of 15 minutes between chlorine and the water, estimate the concentration of free available chlorine necessary to kill 99.99 percent of the bacteria.

	Survival (%)	
Free available chlorine g m^{-3}	1-min contact time	5-min contact time
0.05	77.4	27.8
0.10	60.0	7.8

6.39 Disinfection stoichiometry

Chick's law suggests that the rate of inactivation is first order with respect to microorganisms and zeroth order with respect to the disinfectant. Use a stoichiometric argument to demonstrate that this assumption is reasonable. The typical bacterial number concentration in clean, natural water is 10^4 per liter, and the typical free-residual chlorine concentration used for disinfection is 20 μM. The typical dimension for a coliform bacterium is 2 μm, and the elemental composition is $C_5H_7O_2N$.

6.40 More issues in drinking water disinfection

(a) Chlorine is a common drinking water disinfectant. What is the most favorable chemical form of chlorine in water when it is to be used as a disinfectant?

(b) Experiments reveal that the inactivation of *Giardia* cysts by chloramines is first order with a rate constant that is proportional to the total chlorine residual (TCR). At a TCR of 2 mg/L in a batch reactor, 98 percent inactivation is achieved after a contact time of 300 min. Estimate the inactivation rate constant at a concentration of 1 mg/L.

(c) Ozone may be used as a disinfectant as an alternative to either chlorine or chloramines. Ozone has one major disadvantage relative to these other chemical disinfectants. Name it.

6.41 Chlorine disinfection and pH

Dissolved chlorine rapidly hydrolyzes in water to form hypochlorous acid. In turn, hypochlorous acid dissociates in water to form OCl^-, a much less effective disinfectant.

(a) What percentage of added chlorine would be in the form of the most active oxidizer (HOCl) at pH 6.5? At pH 7.5?

(b) Given that an HOCl concentration of 0.03 mM has been shown to result in a 95 percent kill in 10 minutes, how much time would be required for a 98 percent kill with a total chlorine dose of 0.5 mM at a pH of 6.5? At a pH of 7.5? (*Hint*: Assume that the water contains only cells and the added chlorine, and that only HOCl is an effective disinfectant.)

(c) What percent kill would be achieved with a 20-minute exposure time at the dose and at the two values of pH given in part (*b*)?

6.42 Disinfection kinetics and reactor design

The following figure shows contact time as a function of disinfectant concentration required to achieve 99 percent inactivation of *E. coli* by free chlorine (Haas, 1990). These experiments were conducted in a batch reactor.

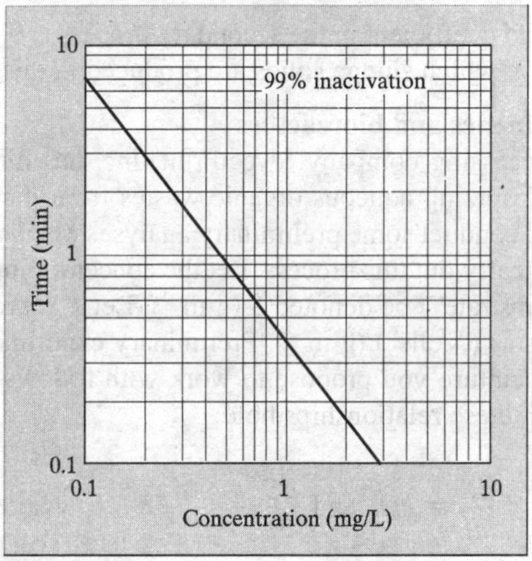

(a) Consider a contact basin for disinfection that performs as a perfect plug-flow reactor. The hydraulic residence time is 6 min. Specify the free chlorine concentration required to achieve 99.9 percent inactivation of *E. coli* within the contact basin.

(b) An alternative to a plug-flow reactor is a series of well-mixed reactors. Consider a design in which three equal-volume reactors are placed in series; in each reactor the hydraulic residence time is 2 min. Specify the free chlorine concentration required to achieve 99.9 percent inactivation of *E. coli* for this configuration.

6.43 Drinking water disinfection

An experiment is conducted in a batch reactor to test the effectiveness of hypochlorous acid for killing the fecal coliforms *E. coli*. It is found that 99 percent inactivation is achieved in 2 min. with a disinfectant concentration of 0.3 mg/L. In a separate experiment, it is found that 99 percent inactivation is achieved in 0.1 min. of contact with a disinfectant concentration of 3 mg/L. Assume that disinfection kinetics are described by Chick's law and Watson's law.

(a) From the experimental data, determine the values of k' and n for the relationship $k = k'C^n$.

(b) Specify the hydraulic detention time, Θ, for a PFR designed to achieve 99.9 percent removal of *E. coli* with a disinfectant concentration of 1 mg/L.

(c) Specify the hydraulic detention time, Θ, for a CMFR designed to achieve 99.9 percent removal of *E. coli* with a disinfectant concentration of 1 mg/L.

6.44 Sludge production in a wastewater treatment system

A municipal wastewater treatment plant is equipped with primary and secondary facilities. The following information is available on the effectiveness of the total treatment system.

Total suspended solids (mg/L)	
Plant influent	250
Final effluent	20
Oxygen demand (mg/L)	
Plant influent	340
Final effluent	21
Wastewater flow rate	3.2 $m^3\ s^{-1}$

If the cell yield coefficient in the secondary process is 0.6 g VSS (g BOD)$^{-1}$, estimate the total rate at which sludge solids are produced in this plant (metric tons per day).

6.45 Microbial kinetics and bioreactors

The food-processing company SweetFruit, Inc. has hired you to design a treatment process for oxidizing aqueous organic wastes from their canning plant. In this problem, you will conduct some preliminary analyses associated with the design of a batch bioreactor to carry out this process. Let the concentration of the biodegradable organic material ("substrate") be denoted S (g m^{-3}). Let X represent the concentration of live microorganisms ("cells") (g m^{-3}). Preliminary experiments reveal that the mixed microorganism culture you propose to work with follows Monod kinetics. That is, in a batch reactor, these relationships hold:

$$\frac{dX}{dt} = r_g X \text{ and } \frac{dS}{dt} = -\frac{r_g}{Y}X \qquad \text{where } r_g = \frac{k_m S}{S + K_s}Y$$

The parameter values for this system are as follows:

$Y = 0.4$ (g cell per g substrate)
$k_m = 3.0$ (g substrate per g cell) per day
$K_s = 25$ (g substrate per m^3)

Over the time scale of interest for this system, cell death can be neglected.

(a) Consider a batch reactor in which the initial cell mass concentration is much greater than the substrate concentration. Furthermore, the initial substrate concentration is much less than the half-saturation constant. For the specific case $X(t = 0) = 200$ g m^{-3} and $S(t = 0) = 5$ g m^{-3}, determine the processing time needed to reduce the substrate concentration to 10 percent of its initial value (i.e., to 0.5 g m^{-3}). (*Hint*: It will facilitate your analysis if you start by *demonstrating* that $X(t) \sim$ constant. [*Note*: "Demonstrating" is not equivalent to "assuming" or "asserting."])

(b) Consider a batch reactor in which the initial substrate concentration is much greater than the cell mass concentration. For the specific case $X(t = 0) = 100$ g m^{-3} and $S(t = 0) = 1000$ g m^{-3}, determine the processing time needed to reduce the substrate concentration to 10 percent of its initial value (i.e., to 100 g m^{-3}). (*Hint*: It is not reasonable to assume that $X =$ constant in this case.)

(c) Consider operation of the bioreactor as a CMFR. A steady volumetric flow rate of $Q = 500$ m^3 d^{-1} must be processed. The inlet concentration of cells is $X_0 = 50$ g m^{-3}. The inlet substrate concentration is $S_0 = 200$ g m^{-3}. What volume, V, must the reactor have to achieve 90 percent removal of S (i.e., S at the outlet $= 20$ g m^{-3})?

(d) Considering the situation in part (*c*), what configuration changes could be employed to reduce the volume of the reactor?

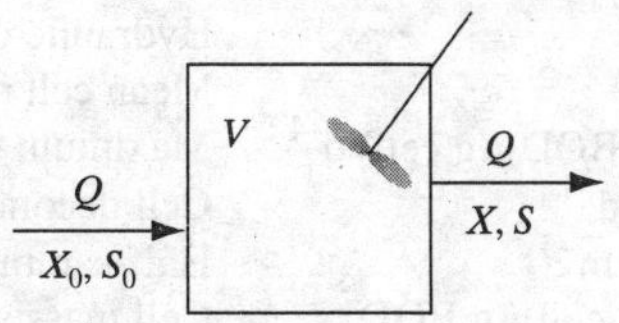

6.46 Performance parameters for an activated sludge unit

To predict the performance of an activated sludge unit, one must have data on microorganism properties. To obtain these data, a laboratory experiment is conducted using a completely mixed flow reactor (CMFR) without recycle. For a given inlet concentration of BOD (S_0), the steady-state outlet concentration of BOD (S) and cell mass (X) are measured at two values of the hydraulic residence time ($\Theta = V/Q$).

(a) Given the hypothetical experimental data, compute the biomass parameters k_m, k_d, K_s, and Y.

(b) What outlet concentrations of BOD (S) and cell biomass (X) would result if the hydraulic residence time were 0.5 days?

Run	S_0 (g m^{-3})	X_0 (g m^{-3})	Θ (d)	S (g m^{-3})	X (g m^{-3})
1	250	0	3.0	10	120
2	250	0	1.0	45	110

6.47 Secondary (biological) treatment of wastewater

(a) What is the main purpose of secondary treatment of wastewater?

(b) Consider a conventional activated sludge process. What are the fates for the organic carbon that enters the reactor?

(c) In a typical wastewater treatment plant, the sludge from both primary sedimentation and secondary treatment is sent to a sludge digester for further processing. Name two benefits that result from sludge digestion.

(d) Digested sludge is rich in nutrients and has a good texture for use as a soil amendment. However, one major health concern limits the use of digested sludge as a soil amendment. What is this concern?

(e) The phenomenon known as *cell washout* can occur in an activated sludge process. Describe (qualitatively) the circumstances in which cell washout might occur.

6.48 Activated sludge operations

An activated sludge unit is being used to treat 1500 m^3/d of wastewater containing a BOD concentration of 250 g m^{-3}. Measured parameters for this particular unit are as follows: $Y = 0.6$ (g cells)/(g BOD), $k_m = 2.0$ (g BOD)/(g cells) d^{-1}, $k_d = 0.1$ d^{-1}, $K_s = 4$ (g BOD) m^{-3}. Make the typical assumptions for activated sludge modeling.

(a) Calculate the mean cell residence time (units: d) and CMFR volume (units: m^3) required for 99 percent removal of BOD for a system involving no cell recycle.

(b) Repeat these calculations for a system that recycles 50 percent of the cells in 10 percent of the influent flow. Assume the cell separator works perfectly.

6.49 Activated sludge operations II

Wastewater from a primary treatment tank is to undergo secondary treatment in an activated sludge unit. The following data and operational parameters are specified:

$\Theta = 6$ h	Hydraulic detention time
$\Theta_c = 8$ d	Mean cell residence time
$k_m = 3$ (g BOD)/(g cell) d^{-1}	Maximum specific BOD degradation rate
$k_d = 0.05$ d^{-1}	Cell decomposition rate
$K_s = 30$ g m^{-3}	Half-saturation constant for cell growth
$Y = 0.4$ (g cell)/(g BOD)	Cell mass yield coefficient
$S_0 = 150$ g m^{-3}	BOD content at inlet of reactor

Two identical units are designed to operate in parallel and to process a combined wastewater flow rate of 0.5 m^3 s^{-1}. Assume that the units operate ideally.

(a) How large must each reactor be (m^3)?

(b) What is the rate at which cell mass must be discharged from the units in steady-state operation (metric tons per day)?

6.50 Two-chamber bioreactor for wastewater treatment

Researchers have proposed a two-compartment reactor for the biological treatment of volatile organics in wastewater. The reactor is composed of two plug-flow reactors stacked together and separated by a semipermeable membrane, as represented below.

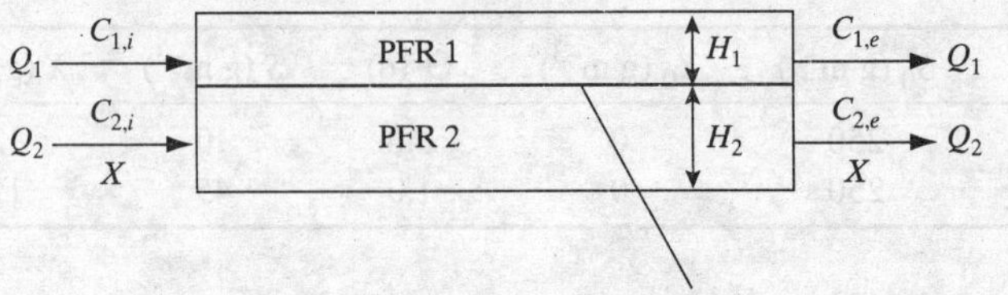

The two reactors are rectangular, with width W, length L, and heights H_1 and H_2 respectively. Assume that fluid in each PFR remains well mixed in the direction normal to flow, but that there is no mixing in the direction of flow (i.e., they are ideal PFRs). Water containing 100 mg/L of a volatile contaminant (inlet concentration = $C_{1,i}$) flows at a steady rate through PFR 1 such that the hydraulic residence time is 12 hours. Water containing active microbial cells (concentration = X) and no influent contaminant ($C_{2,i} = 0$) flows at a steady rate through PFR 2 with a detention time of 24 hours. The contaminant diffuses through the membrane from PFR 1 to PFR 2 due to a concentration gradient and is degraded within PFR 2 by the microorganisms. The flux of contaminant through the membrane is well described by the two-film model. The reasonable assumption that the contaminant concentration in PFR 2 is much lower than that within PFR 1 makes the following simplification possible:

$$J = K(C_1 - C_2) \approx KC_1$$

where $K = 0.15$ m d^{-1}. Contaminant degradation within PFR 2 is first order with respect to both contaminant concentration and cell concentration, with a rate constant $k = 2$ L mg^{-1} d^{-1}. Assume that the active cell mass is conserved within PFR 2, that is, the cells neither reproduce nor die during the 24-hour detention time in the reactor.

(a) Derive a material balance on the contaminant within PFR 1.

(b) Calculate the required H_1 for 99 percent removal within PFR 1.

(c) Using a material balance on the contaminant within PFR 2, derive an expression for the change of contaminant concentration over distance (dC_2/dx) in terms of C_1 and C_2, X, W, U, H_2, and the rate constants.

(d) The PFR configuration described above uses cocurrent flow (flows through the reactors are in the same direction) rather than countercurrent flow. Would cocurrent or countercurrent flow result in superior contaminant removal for the given conditions? Why?

6.51 Controlling VOC emissions from wastewater treatment

In recent years, wastewater treatment plants have come under scrutiny as sources of toxic air pollutants. One approach to reducing emissions involves covering the treatment vessels and passing exhaust air through a control device such as an activated-carbon adsorption unit.

Pictured below is a schematic representation of a covered wastewater treatment process with ventilated air space. The lower portion of the vessel is filled with water (volume = V_w, m^3), and the upper portion contains air (volume = V_a, m^3). Assume that each portion of the vessel may be represented as a CMFR. Water and air flow through the vessel at steady rates, Q_w and Q_a (m^3 s^{-1}), respectively. Consider a specific volatile compound. Prior to $t = 0$, assume that the compound is absent from the system. At $t = 0$, the compound suddenly appears in the incoming water stream at concentration C^* (mol m^{-3}) and is maintained at that concentration indefinitely. The partial pressure of the compound in the inlet air, P_{in}, is always zero. However, within the vessel, the compound escapes from the water to the air as described by the two-film model, $J = K(C_w - C_s)$, where J is a flux density (mol m^{-2} s^{-1}), K is a mass-transfer coefficient (m s^{-1}), and C_w is the compound concentration in the water (mol m^{-3}). In this model, C_s is the concentration in water that would be present in equilibrium with the partial pressure P_a of the compound in air. Assume that C_s and P_a are related by Henry's law, $C_s = K_H P_a$, where P_a has units of atm and K_H has units of atm^{-1} mol m^{-3}. The area across which interfacial mass transfer occurs is A (m^2). Neglect any chemical or biochemical reactions other than volatilization within the reactor.

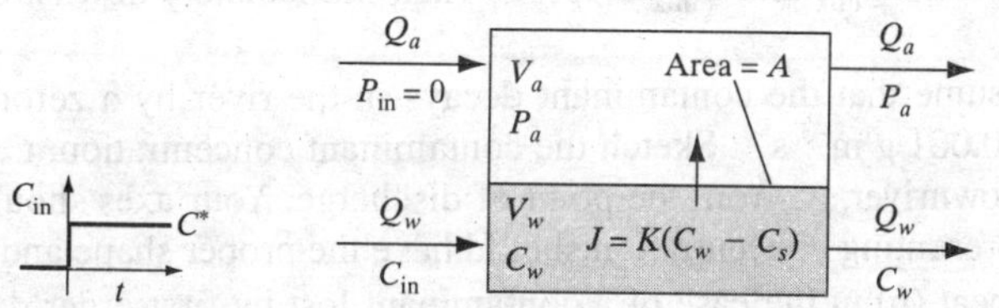

(a) The evolution of C_w and P_a over time may be determined by solving a pair of first-order ordinary differential equations that account for material conservation. Write the governing equations and specify appropriate initial conditions.

(b) Solve the governing equations in the limit $t \to \infty$ to determine steady-state values of C_w and P_a.

(c) What is the characteristic time needed for this system to approach steady state following the change in input conditions at $t = 0$?

6.52 Biochemical oxygen demand

A test of the biochemical oxygen demand (BOD) of domestic wastewater reveals that at 25 °C, (1) the ultimate carbonaceous BOD is 250 mg L^{-1} and (2) the first-order rate constant for oxidation of the carbonaceous BOD is 0.10 d^{-1}. The waste is to be discharged to receiving water at 15 °C. Calculate the time after discharge at which 50 percent of the oxygen demand associated with the waste is consumed. Assume a temperature coefficient of $\theta = 1.047$ applies.

6.53 Wastewater discharge into a river
The city of Leeds, England (pop. 700,000), discharges about 1.4 $m^3\ s^{-1}$ to the Aire River. Without the wastewater discharge, the river flow rate would be 12.7 $m^3\ s^{-1}$. Assuming that the following data apply, determine the concentrations of BOD and DO in the river-wastewater mix just below the point of discharge.

Parameter	River	Wastewater
Biochemical oxygen demand	$BOD_r = 3$ mg/L	$BOD_w = 30$ mg/L
Dissolved oxygen	$DO_r = 8.5$ mg/L	$DO_w = 2$ mg/L

6.54 Contaminant decay in a river
The figure below shows wastewater discharge into a river. The symbols Q represent volumetric flow rates of water. The symbols C represent contaminant concentrations. The cross-sectional area of the river in a vertical plane is uniform at A. The subscripts r, w, and "mix" refer to the uncontaminated river, the waste stream, and the mixed river plus waste stream, respectively. Assume that the river is well modeled as a plug-flow reactor and that the flows and waste contaminant concentration are steady.

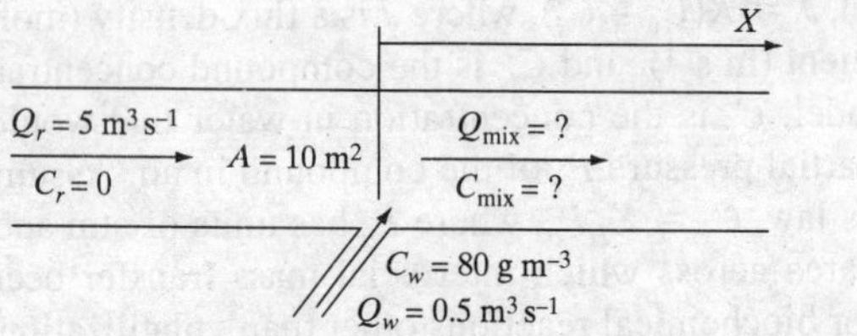

(a) Evaluate Q_{mix} and C_{mix} at $X = 0$ (i.e., immediately downstream of the waste discharge).
(b) Assume that the contaminant decays in the river by a zeroth-order reaction at a rate of 0.001 $g\ m^{-3}\ s^{-1}$. Sketch the contaminant concentration, C_{mix}, as a function distance downriver, X, from the point of discharge. Your axes should be labeled and the line representing concentration should have the proper shape and intercept(s).
(c) Repeat (*b*) in the case of a contaminant lost by first-order decay with a rate constant of 0.002 s^{-1}.

6.55 The impact of wastewater discharge on a river
Wastewater containing BOD = 50 mg/L is discharged at a rate of 1.5 $m^3\ s^{-1}$ into a river that flows at a rate of 8 $m^3\ s^{-1}$. Upstream of the discharge, the river is saturated with dissolved oxygen (10.8 mg/L). The first-order rate constant for BOD in the wastewater is 0.25 d^{-1} at 20 °C, and the temperature of the river is 12 °C. The reaeration rate constant for the river is 0.3 d^{-1}. Assume that the river contains no BOD upstream of the wastewater discharge.
(a) Evaluate the BOD of the river-wastewater mix (mg/L).
(b) Evaluate the first-order rate constant for BOD degradation in the river (d^{-1}). Assume that $\theta = 1.047$ and that the river temperature is not changed by the wastewater discharge.

(c) How much time is required following discharge for half of the BOD content of the river-wastewater mix to be consumed (d)?

(d) Determine the critical oxygen deficit in the river downstream of the discharge (mg/L).

6.56 Wastewater discharge into a first-class fishing river

You've just been hired as a consultant to the American Fly-Fishing Association (AFFA) to analyze the effects of a wastewater discharge on the dissolved oxygen content of Bitesabound River. You secure the following information from local water quality officials: The wastewater flows at a rate of 1.2 $m^3\ s^{-1}$ into the Bitesabound River, which, upstream of the discharge, flows at a rate of 10 $m^3\ s^{-1}$. The temperature of the river is 15 °C. Upstream of the discharge there is 2.3 mg L^{-1} of BOD and 9 mg L^{-1} of dissolved oxygen. (The saturation concentration of DO in water at 15 °C is 10 mg L^{-1}.) The reaeration rate constant for the river is 0.4 d^{-1} (not class 5 rapids, but great for trout!). You begin your analysis by experimentally determining the BOD of the wastewater (at 20 °C). The best data come from a test in which 10 percent wastewater is diluted with 90 percent pure water. The dissolved oxygen concentration in these samples is 9.0 mg L^{-1} initially, 6.0 mg L^{-1} after 5 days, and 2.5 mg L^{-1} after an indefinitely long period.

(a) Determine the BOD of the wastewater (mg L^{-1}) and the first-order decay rate (d^{-1}).

(b) Assume that the predominant contaminant in the wastewater is bacteria, with a chemical composition of $C_5H_7O_2N$. Calculate the theoretical percentage of dissolved oxygen that would be taken up by carbonaceous versus nitrogenous oxygen demand, given that (1) in carbonaceous BOD nitrogen is converted to ammonia (NH_3) and (2) the nitrogenous BOD accounts for the oxidation of ammonia to nitrate (NO_3^-).

(c) Calculate the BOD of the river water immediately after it has mixed with the wastewater discharge.

(d) Evaluate the first-order rate constant for BOD degradation in the river (d^{-1}). Assume that $\theta = 1.047$ and that the temperature of the river is not changed by the wastewater discharge.

(e) How much time (d) is required following discharge for half of the BOD content of the river-wastewater mix to be consumed?

(f) Determine whether your favorite species of game fish—rainbow trout, which requires 6 mg L^{-1} dissolved oxygen—would be able to thrive in all reaches of this river.

6.57 Deep-ocean outfalls and trapped plumes

Water density in the ocean is a function of temperature and salinity. Describe how these features factor into the design of deep-ocean outfalls to maintain a trapped subsurface plume.

6.58 Dissolved oxygen in a goldfish bowl

Consider a simple fish bowl that contains one goldfish, as depicted in the figure at the top of page 384. Changes in the dissolved oxygen (DO) concentration in the water are governed by three processes: (1) the fish consumes oxygen directly, (2) the fish excretes material that exerts biochemical oxygen demand as microorganisms further oxidize it, and (3) interfacial mass transfer adds DO to the water from the air. Assume that the BOD is oxidized according to the usual first-order rate model. Assume that the two-film model applies for interfacial mass transfer. In answering the questions posed, assume that each of these processes occurs at a steady rate.

(a) What is the steady-state BOD concentration in the fish bowl?

(b) What is the characteristic time for the BOD concentration to reach steady state?

(c) What is the steady-state concentration of dissolved oxygen in the fish bowl?

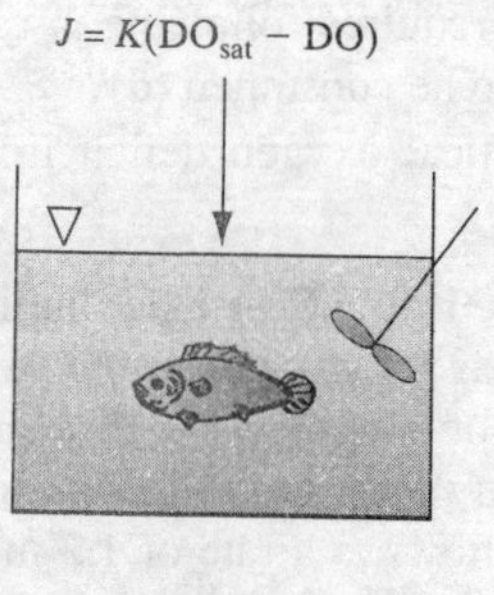

Data

Water depth	$H = 25$ cm $= 0.25$ m
Interfacial surface area	$A = 120$ cm^2 $= 0.012$ m^2
Volume of fish bowl	$V = 3$ L $= 0.003$ m^3 $= 3000$ cm^3
DO consumption by fish	$L_{fish} = 2.0$ mg d^{-1}
BOD excretion by fish	$E_{fish} = 1.35$ mg d^{-1}
Rate constant for BOD exertion	$k = 0.15$ d^{-1}
Saturation concentration of DO	$DO_{sat} = 10$ g m^{-3} $= 10$ mg/L
Interfacial mass-transfer coefficient for oxygen	$K = 8$ cm d^{-1} $= 0.08$ m d^{-1}

(d) Suggest an effective means to increase the DO concentration in the fish bowl and explain briefly how it works. (*Hint*: Removing the fish is not an acceptable solution!)

6.59 Assessment of a toxic waste spill

In the dark of night, toxic polluters secretly dump 2000 kg (about ten 55-gallon drums) of waste toluene into Lake Purewater. As it is a windy autumn night, the toluene mixes rapidly throughout the small lake. After the discharge, toluene escapes from the lake water via a river that is fed by the lake and by volatilization.

Data

$A = 2 \times 10^5$ m^2	Lake surface area
$D = 10$ m	Average lake depth
$Q = 5$ m^3 s^{-1}	River flow rate through lake
$K = 0.13$ m h^{-1}	Interfacial mass-transfer coefficient
$C_s = 0$	Equilibrium toluene concentration in lake

(a) Use the principle of material balance to write a differential equation that describes the toluene concentration versus time in the lake after the secret discharge. Assume that the lake is well represented by a CMFR model and that interfacial mass-transfer flux is well described by the two-film model, $J = K(C_s - C)$.

(b) If the time of discharge is designated as $t = 0$, then the toluene concentration versus time in the lake is described by the following equation: $C(t) = C_0 e^{-\alpha t}$. Evaluate α, given the data above.

(c) According to our model, all of the contamination will eventually escape the lake either through volatilization or in the river discharge. What fraction of the contamination escapes the lake by volatilization?

6.60 Growth and decay of algae

Lovetofishya Lake has a serious problem with algae because of the influx of phosphorus carried by the local river. The lake contains a water volume V and an algae concentration A (mg/L). The river flowing into the lake, at a rate Q_r, carries a concentration of algae of A_0 (mg/L) into the lake. Water leaves the lake via a river that also flows at rate Q_r (evaporation is negligible). Algae grow in the lake due to photosynthesis according to the following rate equation: $r_g = k_g A$. Algae decay in the lake according to the following rate expression: $r_d = -k_d A$.

(a) Assuming that the lake is completely mixed and that the lake volume and reaction rates are constant over time, derive an equation to describe the accumulation of algae in the lake over time.

(b) Calculate the steady-state concentration of algae in the lake, given the following data: $Q_r = 100\ m^3/d$, $A_0 = 10$ mg/L, $V = 1000\ m^3$, $k_g = 0.95\ d^{-1}$, $k_d = 0.90\ d^{-1}$.

(c) Compare the algae concentration in the effluent of the completely mixed lake to that of an identical lake that acts as a plug-flow reactor.

6.61 Analyzing the effects of sewage discharge into a lake

A small city (population 10,000) discharges its untreated sewage into a nearby shallow freshwater lake. A river flows into the lake and another flows out of it such that the water volume remains constant. The winds are strong enough to keep the lake water well mixed. Determine the dissolved oxygen content of the lake under steady-state conditions (mg L^{-1}). Represent the lake as a CMFR. Use the first-order decay model for biochemical oxygen demand (BOD) and the two-film model for oxygen transfer from the atmosphere. Ignore any effect of aquatic life on the dissolved oxygen balance.

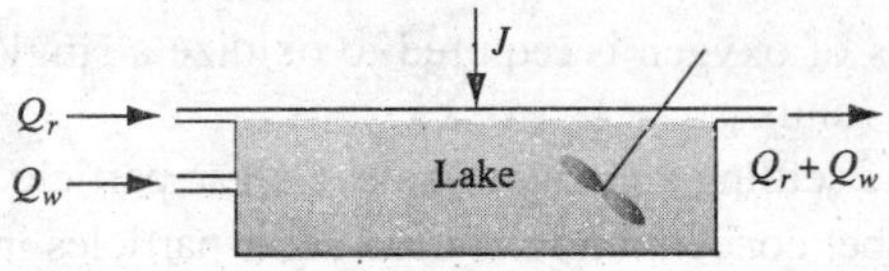

Data

- Lake

  Volume: $V = 2 \times 10^7\ m^3$ Average depth: $d = 10$ m Temperature: $T = 20$ °C

- River (incoming)

  Flow rate: $Q_r = 0.6\ m^3\ s^{-1}$ Dissolved oxygen: $DO_r = 9.0$ mg/L

- Oxygen transfer from atmosphere to lake

  Transfer coefficient: $K = 4 \times 10^{-5}$ cm s^{-1}
  Saturation concentration, dissolved oxygen: $DO_{sat} = 9.1$ mg/L

- Waste

  Flow rate: $Q_w = 0.075\ m^3\ s^{-1}$ BOD content: BOD = 120 mg/L
  BOD decay rate: $k = 0.2\ d^{-1}$ Dissolved oxygen: $DO_w = 8$ mg/L

6.62 Oxygen depletion in the hypolimnion of a lake

Consider the depletion of dissolved oxygen in the hypolimnion of a stratified lake. Assume that the epilimnion and the hypolimnion can each be represented as a CMFR. Turbulent diffusion and particle settling across the thermocline are the only means of coupling the CMFRs. The oxygen content of the epilimnion is maintained at the saturation level. In the hypolimnion, oxygen is depleted because microorganisms oxidize the particles of dead algae that settle into the hypolimnion from above. Oxygen is restored to the hypolimnion by diffusion across the thermocline. The following differential equation describes the rate of change of dissolved oxygen in the hypolimnion:

$$\frac{d(\text{DO})}{dt} = -\alpha \frac{A}{V} J_m + K \frac{A}{V}(\text{DO}_{\text{sat}} - \text{DO})$$

where A/V represents the ratio of the upper surface of the hypolimnion to its volume (m^{-1}), $\alpha = 1.42$ is the mass of dissolved oxygen required for oxidization per mass of algae, J_m is the mass flux density of algae particles across the thermocline (g m^{-2} s^{-1}), DO_{sat} is the dissolved oxygen content of the water at saturation, and K is a mass-transfer coefficient for the diffusion of oxygen through the thermocline. This equation is only valid for DO > 0, where DO is the dissolved oxygen content of water in the hypolimnion. Assume that the dead algae particles are spherical with a diameter of 25 μm, have a density of 1.5 g cm^{-3}, and have an effective molar composition of $C_5H_7O_2N$. Assume that $A/V = 0.1$ m^{-1}.

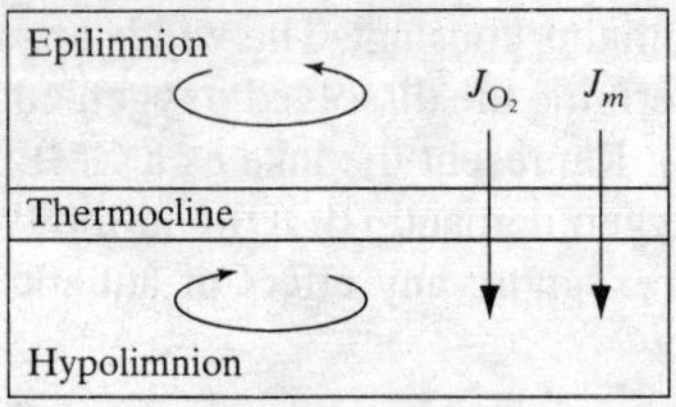

(a) What mass of oxygen is required to oxidize a single dead algae particle to CO_2, H_2O, and NH_3 (units g)?
(b) What is the settling velocity of a dead algae particle (m/s)?
(c) If the number concentration of dead algae particles in the epilimnion remains constant at 10 cm^{-3}, what is the value of J_m (g m^{-2} s^{-1})?
(d) If the turbulent diffusivity of O_2 in the thermocline is 1×10^{-5} m^2 s^{-1} and the thickness of the thermocline is 5 m, what is the value of K (m s^{-1})?
(e) Given the results of (*c*) and (*d*), assuming $\text{DO} = \text{DO}_{\text{sat}} = 12.5$ mg/L at $t = 0$, determine the time required for the dissolved oxygen content of the hypolimnion to decrease to 5 mg/L (d)?

6.63 Contamination of river water from sediments

Consider the RunsThruIt River that has a uniform rectangular channel of width W (units m). The water depth in the channel is D (units m). Assume that the water flow along the channel is uniform at a velocity U (units m h^{-1}).

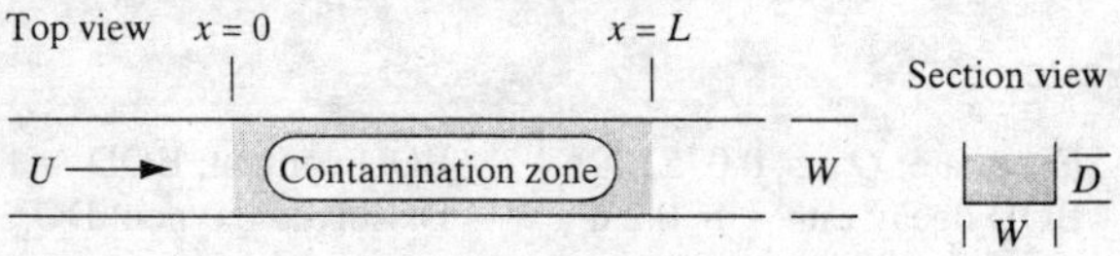

Because of improper past disposal practices, sediments in one stretch of the river (length L, units m) are contaminated. The contaminated sediments are uniformly distributed along the bottom of the river (but not on the sides) throughout the contamination zone. The contaminant is constantly emitted from these sediments into the river water. The emission rate is E (units g m^{-2} h^{-1}). Within the water, the contaminant rapidly mixes vertically so that the concentration can be modeled as uniform in any plane normal to the flow direction. Within the water, the contaminant undergoes first-order decay with a rate constant k (units h^{-1}). In addition, the contaminant is volatile. Its escape from the water surface to the atmosphere follows the two-film model with interfacial mass-transfer coefficient K (units m h^{-1}). Let x be the coordinate in the direction of flow and define $x = 0$ at the leading edge of the contamination. Upstream of the contaminated sediments, the contaminant is absent from the water. The concentration of the contaminant in air is negligible.

Derive an equation that predicts the contaminant concentration in the water at position $x = L$. Your analysis should take into account all of the processes mentioned in the problem statement: advection, emission, reaction, and volatilization. Specify your answer as a concentration C (units g m^{-3}) in terms of some combination of the parameters given: U, W, D, L, K, k, and E.

7

Air Quality Engineering

This chapter addresses some major aspects of air quality engineering, stressing scientific and technical perspectives. The chapter is divided into four sections. The first presents an overview of many current air pollution problems. The second section discusses air pollutant emissions, emphasizing combustion, which is the most important human activity that contributes to air pollution. The third section describes some key technologies for controlling emissions of pollutants from sources. The final section presents models that are used to predict air pollutant concentrations.

7.A NATURE OF AIR POLLUTION PROBLEMS

Air quality engineers apply science and technology to prevent or limit adverse effects of air pollution on human health, human welfare and ecosystems. The problems of interest occur over a wide range of spatial and time scales, as summarized in Table 7.A.1.

Table 7.A.1 Time and Space Scales of Air Pollution Problems

Environment	Length scale	Time scale[a]	Examples
Indoor air	10 m	1 h	Radon, environmental tobacco smoke
Industrial plumes	1 km	10 min	Toxic organics, metals
Urban air basin	100 km	1 day	Ozone, carbon monoxide, particles
Regional	1000 km	1 week	Acid deposition
Global	20,000 km	Decades	Climate change, stratospheric ozone depletion

[a]Characteristic residence time of a nonreactive pollutant in the indicated environment.

Regardless of their scale, air pollution problems have a common structure, as depicted in Figure 7.A.1. The key initial step in any air pollution problem is the emission of pollutants. Transport, transformation, and removal processes, combined with emissions, affect the concentrations of pollutants. Excessive pollutant concentrations cause adverse effects.

Except for global air quality issues, the time scales associated with air pollution problems are relatively short. Consequently, air quality control measures generally aim to reduce emissions from sources, rather than to increase the rate of removal of pollutants from the environment.

For pollutants that may cause health effects, two strategies have emerged for air quality control. One is based on limiting airborne concentrations to a level at which no adverse health effects are expected. Currently in the United States, this approach is applied for six *criteria pollutants*, as discussed in §7.A.1. In cases where the concentrations exceed standards, plans must be developed and implemented to reduce emissions to a level at which the concentration standards are met. The second strategy is based on reducing emissions to the lowest practical level, given the available technology. This strategy is applied to *hazardous air pollutants* (HAPs), as discussed in §7.A.2. Activities that emit substantial quantities of HAPs are subject to regulations that require them to apply the *best available control technology* (BACT) to keep emissions as low as possible. For other air quality problems, control strategies are less well developed.

In pursuit of the broad goal of avoiding the adverse effects of air pollution in a cost-effective manner, air quality engineers are involved in a wide range of activities, including (1) creating technical knowledge, (2) designing and applying air pollution control techniques, and (3) developing air pollution control strategies. Activities in the first category focus on building an understanding of the cause-effect relationships in air pollution problems. The ultimate goal is to understand the relationship between emissions and effects so that we can predict the impacts that would result from control measures in advance of their application. Air pollution control techniques include the prevention of pollution by product modification or process change and the design, construction, and operation of control devices.

Although air pollution problems share the general structure shown in Figure 7.A.1, they also constitute a diverse set of issues that defy simple categorization. The remainder of this section is organized into six topics that encompass most of the air pollution problems facing society. The first two topics, *criteria pollutants* and *hazardous air pollutants,* are defined as much by regulatory criteria as by scientific or technical characteristics. The air pollution problems described in the subsequent four sections (acid deposition, photochemical smog, indoor air quality, and global change)

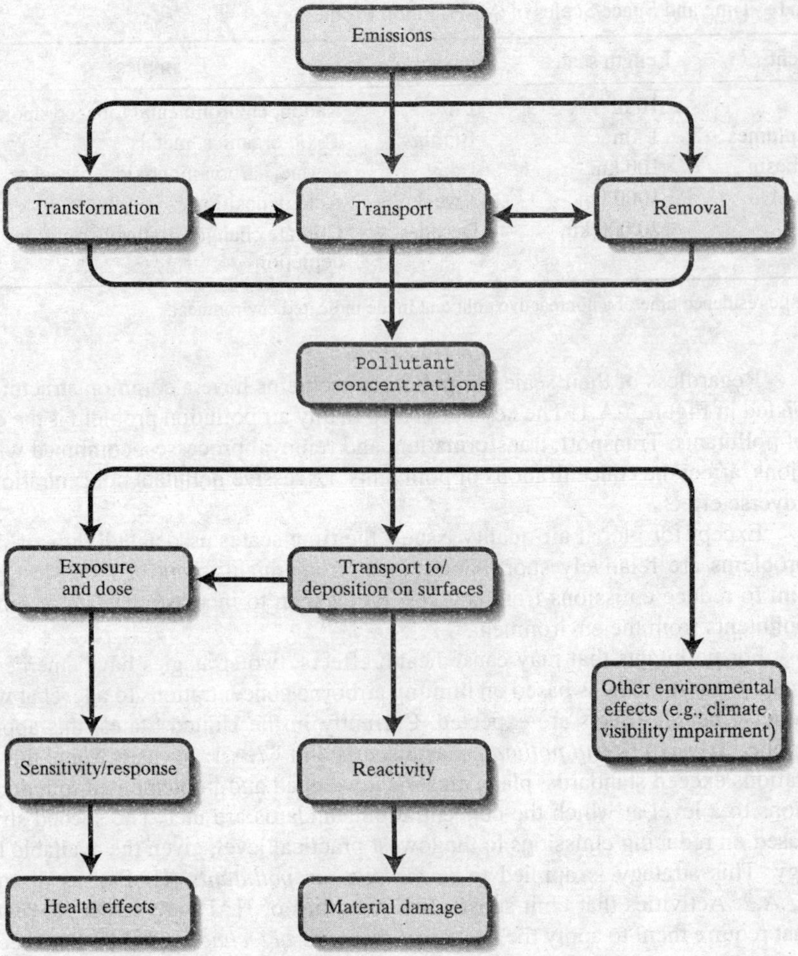

Figure 7.A.1 Framework for understanding air pollution problems.

are only partially under regulatory control, and so the organization here is based on technical aspects of these topics.

7.A.1 Criteria Pollutants

Beginning with the 1970 amendments to the Clean Air Act, the USEPA has been empowered to establish concentration-based air quality standards. *Primary* standards are set at levels necessary to protect public health, including sensitive individuals. *Secondary* standards are designed to protect the public welfare and the environment from known or anticipated adverse effects of pollutants. By law, these standards, known as *national ambient air quality standards* (NAAQS), are set without considering the cost or practicality of attainment.

It is important to distinguish between *primary* and *secondary* pollutants.* Primary pollutants are emitted directly from sources. Secondary pollutants are formed from precursor species by means of transformation processes in the atmosphere.

* Do not confuse primary and secondary pollutants with primary and secondary standards!

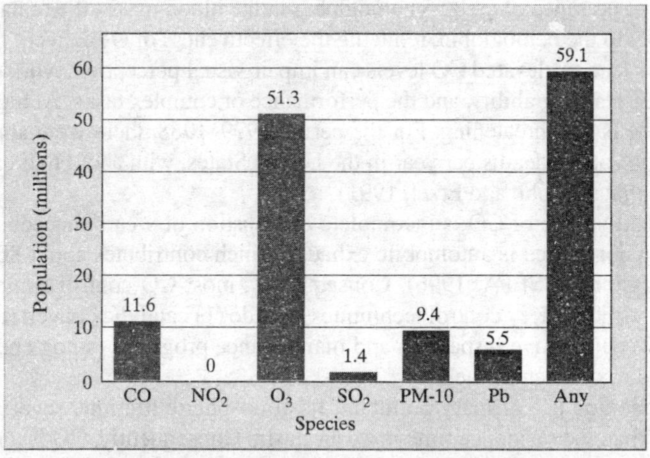

Figure 7.A.2 United States population living in counties that did not meet the NAAQS for air pollutants (population data from 1990 and air pollution data from 1993) (USEPA, 1994). The ozone standard in effect at the time was 120 ppb for a 1-hour averaging period.

Primary pollutants are controlled by reducing their emissions into the atmosphere. To control the concentration of secondary pollutants, one must reduce the emissions of precursors.

The USEPA has established NAAQS for six species, known as criteria pollutants. States and local air districts are required to monitor concentrations of criteria pollutants and, in cases where they do not meet the standards, to prepare and file plans (state implementation plans, or SIPs) to make progress toward meeting the goals. States are allowed to set more stringent standards. Areas that meet the NAAQS are called *attainment areas.* For a nonattainment area, the only practical means of achieving the NAAQS is to reduce emissions.

The six criteria pollutants are carbon monoxide (CO), nitrogen dioxide (NO_2), ozone (O_3), sulfur dioxide (SO_2), particulate matter (PM), and lead (Pb). These six pollutants are not the only ones that have adverse health effects. But, more so than other air pollutants, they possess the following features: (1) They are very common air pollutants, with long histories of human exposure. (2) They each appear to have a threshold below which no adverse effects occur. (3) Minor effects are reversible. (4) There is a significant body of clinical and epidemiological data showing adverse human health effects at commonly encountered levels.

The current standards for the criteria pollutants are summarized in Appendix F, Table F.3. To a significant degree, the United States is not in compliance with these standards. As shown in Figure 7.A.2, as of 1993 about 25 percent of the U.S. population lived in areas that violated one or more of the standards. The ozone standard is the most commonly violated. The following paragraphs describe the health effects and also briefly summarize sources, atmospheric dynamics, and control opportunities for the six criteria pollutants.

Carbon Monoxide (CO)

CO binds reversibly with hemoglobin in red blood cells, impairing the oxygen-carrying capacity of the blood. The affinity of hemoglobin for CO is about 250 times as high as for oxygen. Also, the presence of carboxyhemoglobin in the blood makes the remaining O_2 bind more tightly to the hemoglobin, reducing the effectiveness of O_2 delivery to tissues and organs. Exposure to elevated CO levels can impair visual perception, work capacity, manual dexterity, learning ability, and the performance of complex tasks. At high levels, carbon monoxide is life threatening. For the period 1979–1988, there were about 1000 accidental CO poisoning deaths per year in the United States, with about half caused by motor vehicle exhaust (Cobb and Etzel, 1991).

The dominant source of CO is incomplete combustion of a carbonaceous fuel. In ambient air, a major source is automobile exhaust, which contributes about 80 percent of the total emissions (USEPA, 1996). Consequently, most CO control strategies are aimed at motor vehicles. Key control techniques include (1) catalytic converters to treat auto exhaust, (2) automotive inspection and maintenance programs (smog check), and (3) reformulated (oxygenated) fuels.

Carbon monoxide is a primary pollutant. Its atmospheric lifetime, several weeks, is much longer than the residence time in an air basin. Consequently, CO is a nonreactive species in urban air, and atmospheric concentrations vary in direct proportion to emissions. Carbon monoxide does not react rapidly with surfaces and has low water solubility. Once it is released to the atmosphere, its main fate is oxidation (by reaction with OH•) to carbon dioxide.

Nitrogen Dioxide (NO_2)

NO_2 is a respiratory irritant. Inhalation exposure can lower resistance to respiratory infections. Chronic exposure to elevated levels may cause increased incidence of acute respiratory disease in children. In addition to its adverse respiratory impact, NO_2 is a major factor in several other air pollution problems, including ozone formation, visibility impairment, and acid deposition. However, its health effects are the sole basis for the primary standard.

NO_2 is principally a secondary pollutant, formed by the oxidation of nitric oxide (NO) in the atmosphere. Nitric oxide is a combustion by-product. The sum of NO and NO_2 (plus other, typically minor, oxides of nitrogen) is termed NO_x. In the atmosphere, NO is oxidized to NO_2 by reaction with ozone and with radicals that contain an O_2 group (peroxy radicals). The air pollution chemistry of nitrogen oxides in the atmosphere is summarized in §7.A.3–4.

Industry and electric utilities account for about 47 percent of total NO_x emissions in the United States, and motor vehicles contribute another 45 percent. Key control techniques include combustion manipulation to promote burning at a lower flame temperature and the use of catalytic converters in automobile exhaust systems (in which NO is reduced to N_2). Emission characteristics and control techniques are summarized in §7.B.2.

Three separate sections of the 1990 Clean Air Act amendments (CAAA) contain provisions that affect NO_x emissions: those related to ambient air quality, mobile source control, and acid deposition (Lee, 1991). The annual NO_x emission threshold that triggers regulatory action was significantly reduced. Although all parts of the United States are currently in compliance with the NO_2 standard, emission controls are still required because NO_x is a precursor to ozone and to nitric acid. All regulated NO_x sources are required to apply "reasonably achievable control technology" (RACT) to reduce emissions as much as possible. Offset requirements that require new sources to reduce existing emissions were made more stringent. For new automobiles and light trucks, emission standards were tightened, and a provision was included for allowing the USEPA to require an even tougher standard to be met in 2003 if needed. Finally, the 1990 CAAA required significant reduction of NO_x emissions from fossil fuel–burning power plants to be achieved by the year 2000.

Ozone (O_3)

Ozone is a strong chemical oxidant. When inhaled, it damages lung tissue and sensitizes the lung to other irritants. Ozone exposure may cause chest pain, coughing, and nausea. Chronic exposure to low levels of ozone may cause permanent structural damage to the lungs and may accelerate lung aging. Ozone also causes damage to agricultural crops and forest trees.

Ozone is strictly a secondary pollutant, produced by photochemical reactions in the atmosphere. The precursors that lead to ozone formation are nitrogen oxides and volatile organic compounds (VOCs). VOC emissions in the United States originate from a variety of human activities and natural sources, including motor vehicle exhaust, fuel and solvent evaporation, industrial processes, and emissions from vegetation. Ozone control necessitates reducing emissions of NO_x and VOCs. VOC emission control can be achieved by product substitution or reformulation, sorption (on activated carbon), improved vapor containment, catalytic oxidation, thermal oxidation, and a few other processes. One of the challenges in the development of ozone control strategies is the complex, nonlinear dependence of peak ozone concentrations on NO_x and VOC emissions (see §7.A.4).

Ozone has a short atmospheric lifetime, especially in polluted air. It reacts rapidly with NO and at significant rates with alkenes (hydrocarbons with a double C-C bond) and NO_2. It also readily decomposes by reaction at surfaces. In each of these processes, ozone is an oxidizing agent. The key consequences of its high reactivity are that (1) ozone does not persist for long periods in urban air basins and (2) ozone is not transported over long distances. Thus, ground-level ozone concentrations typically exhibit a strong peak during the daylight hours, dropping to near zero overnight. Downwind of urban areas, ozone concentrations may be elevated because of the transport of precursor pollutants to that area.

Ozone is an essential component of the stratosphere, protecting the biosphere from the harmful effects of ultraviolet radiation from the sun. However, its presence in the lower troposphere is another matter. Ozone has been, by far, the most troublesome of the criteria pollutants. For more than 20 years, the U.S. federal primary standard for ozone was set at 120 ppb, based on a 1-hour averaging period. In 1997 the standard was revised to a lower concentration with a longer averaging time: 80 ppb based on an 8-hour average. As of 1993, about 20 percent of the U.S. population lived in counties where the ozone standard was violated, and this percentage is expected to increase under the new standard. Despite large-scale efforts, only moderate progress has been made in improving urban ozone levels in the United States over the previous decades (National Research Council, 1991). The worst situation has been in the Los Angeles air basin, where the 1-hour, 120 ppb NAAQS was exceeded for about 100 days per year during the mid-1990s, with peak levels more than twice the standard. Urban ozone is also a very serious problem internationally. For example, in Mexico City, peak hourly-averaged ozone concentrations are as high as 450 ppb.

Sulfur Dioxide (SO_2)

SO_2 is a respiratory irritant that can have adverse effects on breathing, contribute to respiratory illness, alter the defense mechanisms of the lungs, and aggravate existing pulmonary and cardiovascular disease. Sulfur dioxide can also damage crops and trees. It is the most important precursor for acid deposition.

Sulfur dioxide is produced by the combustion of fuels that contain sulfur and by some industrial processes. Important sources of SO_2 emissions in the United States are coal and oil combustion, steel mills, oil refineries, pulp and paper mills, and nonferrous smelters. Stationary-source fuel combustion for electricity generation and for industrial processes contributes 85 percent of the total emissions in the United States. SO_2 emissions can be controlled by reducing the S content of fuel or by treating the flue gas to remove SO_2.

Sulfur dioxide is a primary pollutant. It may deposit to the earth's surface either as a gaseous molecule or dissolved in rain or fog droplets. The more important fate, however, is atmospheric oxidation to sulfuric acid and subsequent deposition. The oxidation process has a time scale of several days and may occur either in the gas phase or in liquid water droplets.

SO_2 sources tend to be large, widely dispersed, and remote from urban areas. Consequently, most monitoring stations, which are located in urban areas, do not show violations of the NAAQS standard. However, on the basis of modeling studies of plumes in the vicinity of major sources, 47 locations are designated as nonattainment areas for SO_2. Large sources of sulfur dioxide are also coming under regulatory control because of acid deposition (see §7.A.3).

Particulate Matter (PM_{10} and $PM_{2.5}$)

Two standards apply to particulate matter in outdoor air. The PM_{10} standard limits the mass concentration of airborne particles whose aerodynamic diameter is less than 10 μm. A newly proposed $PM_{2.5}$ standard applies to particles with aerodynamic diameter smaller than 2.5 μm.* (The aerodynamic diameter is the diameter of a spherical particle with the density of water that exhibits the same inertial transport properties as the particle in question.) Different standards for PM_{10} and $PM_{2.5}$ acknowledge the different sources, composition, atmospheric fate, and respiratory deposition patterns of *coarse* (diameter $\geq$ 2 μm) and *fine* (diameter $\leq$ 2 μm) particles.

The primary health concern with particulate matter is impairment of the respiratory system. Particle exposure may have adverse effects on breathing, may aggravate existing respiratory and cardiovascular disease (e.g., asthma), may impair the body's defense mechanisms against foreign materials, may damage lung tissue, and may contribute to premature mortality. Recent epidemiological research indicates a surprisingly strong correlation between particulate matter levels in ambient air and mortality (Pope et al., 1995). Suspended particulate matter is also the major contributor to atmospheric visibility impairment.

Airborne particle behavior is a sensitive function of particle size. Mass and inertia control the behavior of coarse particles. They have an atmospheric lifetime of hours to days and are deposited by gravitational settling and inertial impaction. *Ultrafine* particles ($\leq$0.1 μm in diameter) may be lost by diffusional deposition onto surfaces, including coagulation with other airborne particles. Their atmospheric persistence is in the range of a few days to a few weeks. The intermediate particle size range (0.1–2 μm diameter) is known as the *accumulation mode*. These particles are too small to settle, too light to impact, and too large to diffuse. Their atmospheric lifetime is on the order of weeks. This means that they can be effectively transported over regional distances before depositing.

* At the time of this writing, implementation of the $PM_{2.5}$ standard is being delayed by a court decision in response to a lawsuit.

Particulate matter is a mixed primary and secondary pollutant. Coarse particles are mainly primary pollutants, with dominant sources that include fugitive dust from roads, construction, and agricultural emissions. Wind erosion is another important source, especially during dry years. Industrial emissions are significant sources of both coarse and fine particles. A major transportation source of fine particulate matter is diesel soot. Residential wood smoke is a notable source of fine particles wherever this is a common source of home heating. Secondary processes contribute to fine particulate mass, including oxidation of hydrocarbons into low-vapor-pressure organics, and the formation of aerosol salts such as ammonium nitrate and ammonium sulfate. Particle control techniques applied at the stack are based on a variety of filtration mechanisms, including electrostatic precipitation, mechanical filtration, and inertial impaction, as discussed in §7.C.1. Distributed and natural sources are controlled through changes in management practices.

Lead (Pb)

Lead is regulated both as a criteria pollutant and as a hazardous air pollutant. A summary of lead as an environmental pollutant was presented in §2.C.4.

Since it is a chemical element, lead is strictly a primary pollutant. In the atmosphere, lead is associated with suspended particulate matter, mainly in the fine mode. Lead in fine particles has a long atmospheric residence time and deposits broadly over the globe. Atmospheric lead problems in urban areas of the United States have largely been eliminated because of the removal of lead from gasoline.Thirteen locations are designated by the USEPA as nonattainment areas, mainly due to large, inadequately controlled industrial sources (e.g., lead smelters, battery manufacturing, and waste incineration).

7.A.2 Hazardous Air Pollutants

Hazardous air pollutants (HAPs) are species that are known or suspected carcinogens, or that have been shown to cause other serious health effects, such as reproductive problems or birth defects. HAPs include metal compounds, hydrocarbons, halogenated organics, and pesticides (see Table F.4 in Appendix F). Several features differentiate HAPs from criteria air pollutants, as summarized in Table 7.A.2; but ultimately, the distinction reflects legislative and regulatory considerations more than toxicological properties of the pollutants.

Characterization and control of HAPs focus on major emission sources, which are usually associated with industrial processes. The USEPA is charged with defining the *maximum achievable control technology* (MACT) for each major industrial activity or

Table 7.A.2 Characteristics of Hazardous Air Pollutants and Criteria Pollutants

Characteristic	Hazardous air pollutants	Criteria pollutants
Number of species	Hundreds	Six
Ambient concentration data	Sparse	Extensive
Ambient concentrations	Often below detection limit	Relatively easily measured
Scale of problems	Localized (plume)	Dispersed (urban airshed or regional)
History of control effort	Limited, recent (years)	Extensive (decades)

Source: USEPA, 1994.

process that emits HAPs. Any source that emits (after control) at least 10 tons per year of any of the regulated HAPs or at least 25 tons per year of any combination of HAPs is required to apply the MACT to reduce emissions. However, not all exposure to HAPs results from large-scale industrial processes. For example, benzene is a significant constituent of gasoline, and a major source of exposure is evaporation of fuel. Benzene is also produced by incomplete combustion. Smoking tobacco is the dominant cause of human exposure to benzene.

In practice, industrial emissions of HAPs are regulated through a permitting process. In California, new and modified sources of HAPs are required to conduct a health risk assessment as part of the permitting process. The assessment involves combining the results from three sequential steps. First, the rate of emissions from the facility is estimated (see §7.B.1). Then, given the emissions estimate, Gaussian plume models are combined with local meteorological data to estimate exposure levels (see §7.D.2). Usually, the exposure is predicted for a hypothetical maximally exposed individual (MEI) who lives somewhere outside the facility boundaries. Given this exposure, the adverse health impact is estimated using toxicological data. For carcinogens, this step involves multiplying estimated exposure by a unit risk factor (see Table 7.A.3 and Example 7.A.1). For noncancer risk, the hazard index is computed by dividing the expected exposure by an acceptable, no-effect exposure level. The levels of cancer and noncancer risk are used to decide whether the permit should be granted and, if so, what level of control must be applied.

Table 7.A.3 Unit Risk Factors for Inhalation of Carcinogenic Hazardous Air Pollutants[a]

Compound	Unit risk[b] $(\mu g\ m^{-3})^{-1}$
Dioxins	38
Chromium VI	0.14
Cadmium	4.2×10^{-3}
Inorganic arsenic	3.3×10^{-3}
Nickel	2.6×10^{-4}
Ethylene oxide	8.8×10^{-5}
Vinyl chloride	7.8×10^{-5}
Ethylene dibromide	7.1×10^{-5}
Carbon tetrachloride	4.2×10^{-5}
Benzene	2.9×10^{-5}
Ethylene dichloride	2.2×10^{-5}
Perchloroethylene	8.0×10^{-6}
Formaldehyde	6.0×10^{-6}
Chloroform	5.3×10^{-6}
Trichloroethylene	2.0×10^{-6}
Methylene chloride	1.0×10^{-6}
Asbestos[c]	1.9×10^{-4}

[a]Approved by the California Air Resources Board Scientific Review Panel (1984–1992).
[b]The unit risk factor multiplied by the lifetime average exposure concentration yields the excess lifetime risk of cancer incidence due to exposure.
[c]Unit risk per 100 fibers/m^3.

EXAMPLE 7.A.1 ***Estimating Cancer Risk Associated with Benzene Exposure***

The estimated mean inhalation exposure concentration for the U.S. public to benzene is 15 $\mu g\ m^{-3}$ (Wallace, 1989). Estimate the lifetime cancer risk associated with this level of exposure.

SOLUTION The unit risk factor for benzene is 2.9×10^{-5} $(\mu g\ m^{-3})^{-1}$ (Table 7.A.3). Multiplying the mean exposure concentration by the unit risk factor yields the estimated risk: 4.4×10^{-4}. This result implies that for 1 million people exposed at average levels over a lifetime, 440 cases of benzene-caused cancer are predicted to occur. Wallace points out that about half of the total exposure is caused by smoking tobacco.

7.A.3 Acid Deposition

Some compounds emitted into the atmosphere can be converted to acidic species. The deposition of excessive amounts of these species can cause economic and ecosystem damage. Most of the acid deposition research and legislation in the United States has focused on sulfuric acid. At the heart of the issue are the extensive emissions of sulfur dioxide from the combustion of coal for electricity generation in the Midwest, which leads to deposition of sulfuric acid in the Northeast. However, species other than sulfuric acid also influence atmospheric acidity, as summarized in Table 7.A.4.

The atmospheric dynamics of the two most important species that contribute to acid deposition are shown schematically in Figure 7.A.3. Sulfur contained in fuel is emitted as SO_2 during combustion. In the atmosphere, the SO_2 is oxidized to sulfuric acid. The atmospheric oxidation step can occur either in the gas phase, initiated by hydroxyl radicals, or within water droplets (e.g., clouds), initiated by one of several atmospheric oxidants. The processes that lead to the formation of nitric acid are more complex. A major pathway begins with the emission of nitric oxide from combustion.

Table 7.A.4 Species That Influence the pH of Water in the Atmosphere

Species	Notes
CO_2	Carbon dioxide is mostly natural, although heavy fossil fuel use and changing land use practices are increasing the atmospheric abundance. Slightly soluble, CO_2 combines with water to form the weak, diprotic carbonic acid (H_2CO_3) that acidifies rain and cloud drops. In pristine environments, the pH of rainwater due to CO_2 is 5.6. (See §3.C.4.)
HNO_3	Nitric acid originates mostly as NO from high-temperature combustion. HNO_3 is a highly soluble strong acid. It is the dominant acidic air pollutant in the western United States.
H_2SO_4	Sulfuric acid originates from combustion of sulfur-containing fuels (mainly coal) and from some industrial processes. It is a highly soluble, strong, diprotic acid and is the dominant acid species in eastern North America and in Europe.
Organic acids	Organic acids are formed by photochemical transformation processes acting on hydrocarbon precursors. These species may be important in urban areas. They tend to have lower solubility in water and to have less acidic strength than the inorganic strong acids.
NH_3	Ammonia is the primary basic gas in the atmosphere. It has moderate solubility and moderate strength as a base. Major sources include animal waste and microbial degradation of soil humus.
Soil dust	Wind-blown soil dust tends to be alkaline and so has some acid-neutralizing capacity. It may have a significant influence in nonurban areas of the western United States.

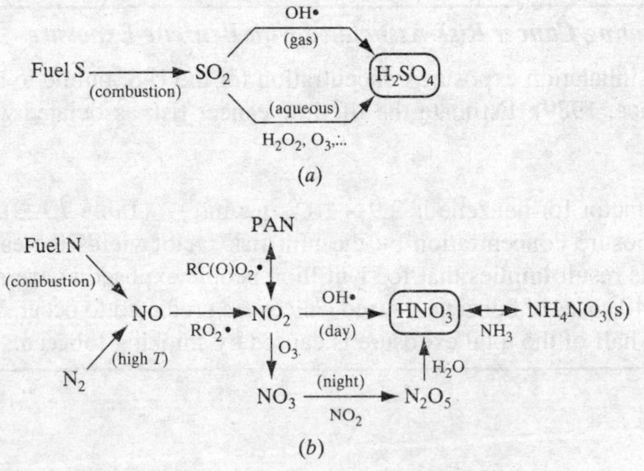

Figure 7.A.3 Transformation dynamics that produce atmospheric acids from (*a*) sulfur oxides and (*b*) nitrogen oxides. The arrows are labeled with the major reactant that is involved in the transformation process. R denotes a hydrocarbon fragment such as CH_3.

Table 7.A.5 Reaction Rate Constants ($T = 298$ K) for Production of Sulfuric and Nitric Acid by Hydroxyl Radicals

Reaction	Rate constant (ppb^{-1} min^{-1})
$NO_2 + OH\bullet \rightarrow HNO_3$	16
$SO_2 + OH\bullet\ (+ O_2 + H_2O) \rightarrow H_2SO_4 + HO_2\bullet$	1.6

In the atmosphere, NO is oxidized to NO_2 by peroxy radicals ($HO_2\bullet$ and $RO_2\bullet$, where R denotes a hydrocarbon fragment such as CH_3). Then NO_2 can be oxidized by the hydroxyl radical ($OH\bullet$) to nitric acid. Rate constants for the production of nitric and sulfuric acid by gas-phase reaction with $OH\bullet$ are presented in Table 7.A.5. Example 7.A.2 illustrates their use.

EXAMPLE 7.A.2 *Production of Nitric Acid in Urban Air*

During daylight hours, in polluted urban air, typical mole fractions of nitrogen dioxide and hydroxyl radicals are 50 ppb and 0.4 ppt, respectively.

(**a**) Determine the rate of production of nitric acid by reaction between NO_2 and $OH\bullet$.

(**b**) Assuming that hydroxyl radicals remain at a constant level, what is the characteristic time for NO_2 to be oxidized to nitric acid?

SOLUTION

(**a**) Table 7.A.5 shows the relevant reaction. The rate of production of nitric acid is given by

$$\frac{dY_{HNO_3}}{dt} = kY_{NO_2}Y_{OH\bullet} = 16 \times 50 \times 0.4 \times 10^{-3} \text{ ppb min}^{-1} = 0.32 \text{ ppb min}^{-1}$$

EXAMPLE 7.A.2 ***Production of Nitric Acid in Urban Air (continued)***

(b) The characteristic time for NO_2 to be converted to nitric acid is

$$\tau \sim \frac{1}{kY_{OH\bullet}} = \frac{1}{16 \times 0.4 \times 10^{-3}\ \text{min}^{-1}} = 160\ \text{min}$$

In polluted skies with strong sunshine, nitrogen dioxide is converted to nitric acid over a time scale of just a few hours. If it remains in the gas phase, the nitric acid will deposit rapidly on the earth's surface, rather than traveling far from the emission source. Oxidation of SO_2 to sulfuric acid occurs more slowly (Table 7.A.5 shows that the rate constant for reaction with OH• is an order of magnitude less for SO_2 than for NO_2), so sulfuric acid can deposit a long distance away from the SO_2 emission sources.

Acid deposition contributes to many environmental problems. Material damage concerns include corrosion, paint deterioration, and degradation of buildings and monuments. Vegetation is potentially at risk. As yet, there is no compelling documentation of damage to agricultural crops, but some high-elevation forests are exhibiting damage from acidic cloud water. The disruption of freshwater aquatic ecosystems is also a serious concern. At issue here are both the direct effects of lower pH on aquatic life and the indirect effects of enhanced dissolution of toxic metals, such as aluminum and arsenic, from rock and soil. Closely related to the acid deposition problem is visibility impairment. Sulfate, originating from SO_2 emissions, forms a substantial fraction of the mass of fine particles throughout the northeastern United States and is responsible for a large portion of the regional haze that persists throughout most of the summer. Finally, there is concern about possible human health effects. "Although substantial uncertainty exists with the present state of knowledge, the body of data raises concern that acidic aerosols, alone or in concert with other pollutants, may be contributing to health effects in exposed populations" (Irving, 1991). Note that although both SO_2 and NO_2 are criteria pollutants, the NAAQS for these species do not take into account damage that may be caused by sulfuric acid and nitric acid.

Because of the presence of carbon dioxide in the atmosphere, rainfall tends to have a moderately acidic pH, typically 5.6, even in pristine environments (see Example 3.C.4). This acidity plays an important role in hydrogeochemical processes, such as the weathering of stone. The average pH of rainfall in some northeastern U.S. locations is now in the low 4s. Water has 40 times more H^+ at a pH of 4.0 than at a pH of 5.6.

Acid deposition is not directly regulated. Instead, as in the case of hazardous air pollutants, acid deposition is controlled indirectly, by regulating emissions. This approach can lead to an overall reduction of impact but may not prevent damage in any specific area.

A major research program known as the National Acid Precipitation Assessment Program (NAPAP) was conducted throughout the 1980s to study the causes, effects, and potential for control of acid deposition in the United States. The 1990 Clean Air Act amendments require a reduction of 10 million tons per year of SO_2 emissions from electric utilities. (This is about a 50 percent reduction relative to the emissions originally projected for the year 2000.) Emissions of NO_x from large-scale combustion sources are to be reduced by 2 million tons per year.

7.A.4 Photochemical Smog

During moderate to severe episodes of photochemical smog, the sky is no longer blue. Instead, visibility is obscured by an orange-brown haze. Air assaults the senses in other ways, too. One's eyes water. The air has a slightly acrid taste and smell. The American Lung Association estimates that the combined effects of urban air pollution account for 400,000 asthma attacks, 1 million respiratory problems (many in children), and 15,000 premature deaths among the elderly annually. Even healthy people can experience respiratory discomfort, a tightness of the chest that makes it difficult to breathe deeply. In addition, materials are damaged by exposure to photochemical smog: Agricultural crops are harmed, rubber cracks, and some pigments fade.

A remarkable characteristic of photochemical smog is that the pollutants that cause the adverse effects are generally *not* the pollutants that were emitted from sources. Central to this system are transformation processes—chemical kinetic reactions and phase changes—that produce secondary pollutants whose effects are worse than the primary emissions.

The starting ingredients in the photochemical smog system are nitrogen oxides (mainly NO), volatile organic compounds (mainly hydrocarbons), and sunlight. Given sufficient quantities of these ingredients and enough time, significant levels are produced of secondary pollutants such as ozone, aldehydes, organic acids, nitrogen dioxide, nitric acid, and peroxy acetyl nitrate (PAN). Some of these pollutants—ozone and PAN in particular—are strong chemical oxidants. Others, such as the aldehydes, irritate mucous membranes. The sky becomes discolored because NO_2 effectively absorbs blue light from the solar spectrum, allowing only the longer-wavelength components of sunlight to pass. The haze results because some of the chemical reactions produce species that have low vapor pressures and condense into fine particles that effectively scatter light.

In current practice, control of photochemical smog focuses on reducing urban and regional ozone concentrations. Standards do not exist for many of the other pollutants in the photochemical smog system.

How is ozone created from NO_x and hydrocarbon emissions? To begin to answer this question, we consider a set of three reactions called the primary photolytic cycle:

$$NO_2 + h\nu \rightarrow NO + O\bullet \quad k_{1,\max} = 0.5 \text{ min}^{-1} \tag{7.A.1}$$

$$O\bullet + O_2 \rightarrow O_3 \quad k_2(298 \text{ K}) = 21.8 \text{ ppm}^{-1} \text{ min}^{-1} \tag{7.A.2}$$

$$NO + O_3 \rightarrow NO_2 + O_2 \quad k_3(298 \text{ K}) = 26.6 \text{ ppm}^{-1} \text{ min}^{-1} \tag{7.A.3}$$

In the first reaction, nitrogen dioxide absorbs a photon of light ($h\nu$ denotes the energy of the photon) and dissociates into nitric oxide and a free oxygen atom (radical). In the second reaction, the oxygen radical combines with an oxygen molecule to form an ozone molecule.* In the third reaction, ozone oxidizes nitric oxide to nitrogen dioxide and is converted back to molecular oxygen in the process.

If these three reactions occur at equal rates, there will be no change in the levels of any of the constituents. Nevertheless, these reactions play a major role in governing the daily rise and fall of ozone levels in the troposphere. Reaction 7.A.1 is the primary source

*Reaction 7.A.2 belongs to a class known as "recombination" reactions. These reactions exhibit a rate constant that depends on pressure because they need the participation of a third molecule. When O• and O_2 combine, the resulting ozone molecule has excess internal energy. The molecule can be stabilized only by collision with a third molecule that carries away this excess energy in the form of kinetic energy. In air, the third molecule is likely to be either N_2 or O_2, since those are the most prevalent. Without this stabilizing collision, the excited ozone molecule may decompose back into the original reactants. Sometimes the reaction is written with an explicit reference to this third body, denoted "M": $O\bullet + O_2 + M \rightarrow O_3 + M$.

of O• radicals. Reaction 7.A.2 is the only significant means by which ozone is formed in the troposphere, and reaction 7.A.3 is an important removal mechanism for ozone.

During daylight hours, these reactions operate on fast time scales. For example, the rate constant for reaction 7.A.1, which depends on sunlight intensity, can be as high as 0.5 min^{-1}, resulting in a characteristic time for NO_2 photolysis of only 2 min. Reaction 7.A.2 is extremely fast: O• is a highly reactive radical and O_2 is plentiful in the atmosphere. Given that the mole fraction of oxygen in air is 0.21, or 210,000 ppm, the characteristic lifetime of the O radical is only 13 μs! Insight into the formation of ozone can be gained by analyzing reactions 7.A.1–3 under the assumption that they operate in a steady-state balance so that the rate of each reaction is the same:

$$k_1 Y_{NO_2} = k_2 Y_{O\bullet} Y_{O_2} = k_3 Y_{NO} Y_{O_3} \tag{7.A.4}$$

Here, Y_i represents the mole fraction of species i. An expression known as the *photo-stationary-state relation* is derived by means of equating the left-hand and right-hand terms of equation 7.A.4 and solving for the ozone level:

$$Y_{O_3} = \frac{k_1 Y_{NO_2}}{k_3 Y_{NO}} \tag{7.A.5}$$

Two important points emerge from equation 7.A.5. First, the level of ozone, Y_{O_3}, depends directly on sunlight intensity through the photolytic rate constant, k_1. Therefore, ozone concentrations are expected to be highest during sunny midday hours and to decline to near zero overnight. Second, the ozone level depends on the relative amounts of NO_2 and NO. A higher NO_2/NO ratio results in a higher ozone concentration, and vice versa.

These features can be seen in the monitoring data presented in Figure 7.A.4. Here the ozone level exhibits a typical pattern of very low values during the night and early morning hours, rising to a strong peak of 170 ppb at about 3 P.M., and then declining rapidly to near zero as the sun sets. When O_3 levels are high, the sunlight intensity is high and the NO_2/NO ratio is large.

To complete this exploration, we must determine how NO_2 is formed, since most NO_x emissions are in the form of NO. Reaction 7.A.3 shows one method; however, it cannot account for the net accumulation of ozone, since each NO to NO_2 conversion by this path also consumes one molecule of ozone.

Peroxy radicals, HO_2• and RO_2•, convert NO to NO_2. These species are formed by the photolysis of volatile organic compounds and by conversion of other radicals. Consider, for example, the photolysis of formaldehyde. There are two outcomes, one of which involves the production of new radicals (see Table 3.D.5):

$$HCHO + h\nu \rightarrow H\bullet + HCO\bullet \qquad k_{max} = 0.003\ min^{-1} \tag{7.A.6}$$

The two radicals each react quickly with molecular oxygen to form hydroperoxy radicals:

$$H\bullet + O_2 \rightarrow HO_2\bullet \tag{7.A.7}$$

$$HCO\bullet + O_2 \rightarrow HO_2\bullet + CO \tag{7.A.8}$$

The rates of reactions 7.A.7 and 7.A.8 are very rapid. Therefore, the net overall effect of this reaction pathway for the photolysis of formaldehyde is the production of two hydroperoxy radicals:

$$HCHO + h\nu (+ 2\ O_2) \rightarrow 2\ HO_2\bullet + CO \qquad k_{4,max} = 0.003\ min^{-1} \tag{7.A.9}$$

The hydroperoxy radicals then oxidize NO to NO_2:

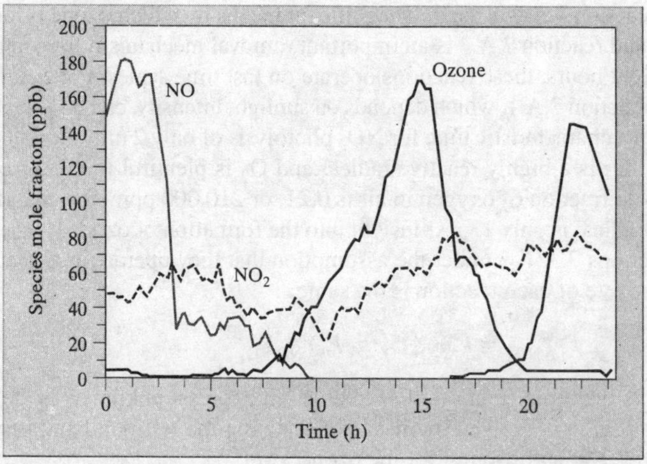

Figure 7.A.4 Ozone, nitric oxide (NO), and nitrogen dioxide (NO_2) mole fractions for a day in the Los Angeles air basin (November 4, 1984). The maximum 8-hour average ozone concentration is 97 ppb, which exceeds the 80 ppb standard. Time 0 corresponds to midnight.

$$NO + HO_2\bullet \rightarrow NO_2 + OH\bullet \qquad k_5(298\ K) = 12{,}000\ ppm^{-1}\ min^{-1} \qquad (7.A.10)$$

The hydroxyl radical (OH•) can be reconverted to hydroperoxy by several reactions, such as 7.A.11 followed by 7.A.7.

$$CO + OH\bullet \rightarrow CO_2 + H\bullet \qquad k_6(298\ K) = 440\ ppm^{-1}\ min^{-1} \qquad (7.A.11)$$

Organics that photolyze, such as formaldehyde, may be emitted directly from sources. They may also be formed in the atmosphere from hydrocarbon precursors. Radicals such as OH• can initiate an attack on a hydrocarbon species, ultimately converting it to an oxygenated compound, such as an aldehyde. The aldehyde can then photolyze, increasing the pool of radicals available for oxidation processes such as the conversion of NO to NO_2 (7.A.10) or CO to CO_2 (7.A.11), as well as the oxidation of other hydrocarbons.

In summary, then, tropospheric ozone is formed when a mixture of organic gases and nitrogen oxides is exposed to sunlight. The sunlight causes nitrogen dioxide to dissociate, liberating an oxygen radical that combines with an oxygen molecule to produce ozone. Sunlight also causes certain organic gases, such as aldehydes, to dissociate, producing peroxy radicals that convert nitric oxide to nitrogen dioxide. This step increases the rate at which ozone is produced by NO_2 photolysis and decreases the rate at which it is consumed by reaction with NO. The net effect is to increase the atmospheric concentration of ozone.

When ozone levels in the troposphere are too high, the only practical means of reducing them is to reduce the emissions of the precursor pollutants. A key question is, *How much reduction in nitrogen oxide and organic emissions must be achieved to control ozone?*

Unfortunately, since ozone formation is complex, there is no simple way to answer this question. One approach is based on predictions from sophisticated mathematical models that combine the modeling framework introduced in §5.B with information on emissions, transport, transformation, and removal mechanisms.

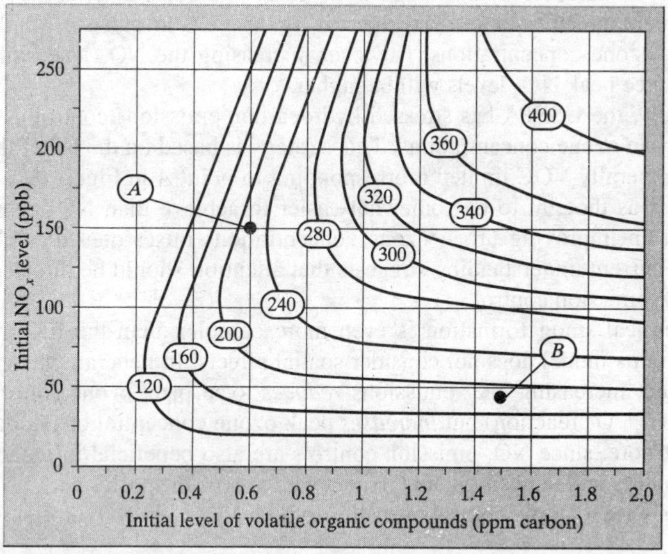

Figure 7.A.5 EKMA (empirical kinetic modeling approach) predictions of peak ozone mole fractions in ppb (labels on curved lines) as a function of initial nitrogen oxide (NO_x) levels and volatile organic compound (VOC) levels in a parcel of air (USEPA, 1981; Seinfeld, 1989). The initial VOC level represents a weighted sum of all VOC mole fractions, where the weighting factor is equal to the number of carbon atoms per molecule. Points *A* and *B* are discussed in the text.

A simpler model, emphasizing the chemical kinetics of ozone formation, illustrates some of the key features. Figure 7.A.5 shows model predictions of peak ozone mole fractions (in ppb) in an air parcel as a function of the initial concentrations of nitrogen oxides and volatile organic compounds (VOCs). The figure shows, for example, that a peak ozone concentration of 160 ppb can be generated from an initial VOC level of 1 ppm carbon in combination with an initial NO_x level of ~35 ppb, or from an initial mix of 0.4 ppm carbon of VOCs and ~110 ppb of NO_x.

A key point revealed by this figure is that the system is *highly nonlinear*. The peak ozone level in an air parcel does not increase in direct proportion to the initial amounts of the precursors. In fact, under the circumstances that correspond to point *A* in the figure, *decreasing* the initial NO_x level leads to *higher* peak ozone levels. On the other hand, at point *B*, peak ozone concentrations decrease sharply with a decrease in initial NO_x levels.

With such complex behavior, the optimal control strategy for reducing peak ozone concentrations clearly must depend on the current circumstances in an air basin. For example, at point *A*, ozone control will probably be more effective if the strategy focuses on reducing VOC emissions. Conversely, at point *B*, ozone control should be achieved by stressing NO_x emission reductions.

Why does the system behave this way? At point *A*, ozone accumulation is limited by the availability of radicals to promote the conversion of NO to NO_2. At this position, increasing the level of VOCs increases the NO_2/NO ratio, thereby increasing the ozone concentration (equation 7.A.5). Conversely, increasing the initial NO_x level decreases the NO_2/NO ratio since NO_x emissions are primarily in the form of NO.

At point *B* there are more than enough hydrocarbon radicals to effectively convert each NO molecule to NO_2. In this case, increasing the VOC level has negligible influence on peak ozone concentrations. However, increasing the NO_x level will lead to more ozone since peak NO_2 levels will be higher.

Historically, the USEPA has stressed hydrocarbon emissions control as a means to improve urban ozone concentrations. This strategy is based on the belief that urban systems are generally VOC limited (corresponding to point *A* in Figure 7.A.5). Also, VOC control was thought to be somewhat easier to achieve than NO_x control. The slow progress in improving urban ozone concentrations raises questions about this strategy. Our current understanding suggests that attention should be directed at both VOC and NO_x emission control.

Photochemical smog formation is even more complex than the EKMA results suggest, since this model does not consider spatial effects. In general, when a system is VOC limited, increasing NO_x emissions *reduces* local peak ozone concentrations (due to the $NO + O_3$ reaction) but *increases* peak ozone concentrations in downwind areas. Furthermore, since NO_x emission controls are also beneficial in reducing visibility impairment, acid deposition, and irritancy effects from species such as NO_2 and PAN, it seems wise to apply control measures to both VOC and NO_x emissions.

7.A.5 Indoor Air Quality

Indoor air pollution started with the use of fires in caves, but prehistoric peoples had many more pressing concerns than that health hazard. The earliest systematic studies of indoor air, conducted in the 1920s and 1930s, focused on determining how much ventilation was needed to maintain a proper mix of the metabolic gases—oxygen and carbon dioxide—and to control body odors in buildings. During the 1960s and early 1970s, research began to evaluate the dependence of indoor air pollutant concentrations on outdoor levels of industrial and urban air pollutants such as sulfur oxides, particulate matter, ozone, and nitrogen oxides. Beginning in the mid-1970s, concentrations of several classes of air pollutants were found to be commonly higher indoors than outdoors. This finding, which applies to combustion by-products, volatile organic compounds, radon, and bioaerosols, has spurred substantial interest in indoor air quality as an important environmental issue.

Environmental engineering is largely devoted to protecting human health. To achieve this goal, one cannot ignore indoor air. As illustrated in Figure 7.A.6, in the United States (and presumably in many other developed countries) about 75 percent of the mass of environmental fluids that enters one's body on a daily basis is indoor air. Even if drinking water and outdoor air were restored to and maintained in a pristine state, human pollutant exposures could still be substantial if indoor air pollutant levels were high.

Concerns about indoor air quality focus on health risks such as disease, as well as other aspects of human well-being, such as odors and irritancy or allergic reactions to indoor air pollutants. For example, much attention is now being given to the "sick-building syndrome" and its effects on workplace productivity.

In addition, indoor air quality poses material damage concerns. Many of society's most precious possessions—works of art and cultural artifacts, for example—are kept indoors and may be damaged by exposure to air pollutants. Electronic equipment failures may also be caused by the deposition of indoor air pollutants onto circuits or switches.

Many investigations of indoor air quality utilize the continuously mixed flow reactor (CMFR) introduced in Chapter 5 (see Figure 7.A.7). Applying the principle of

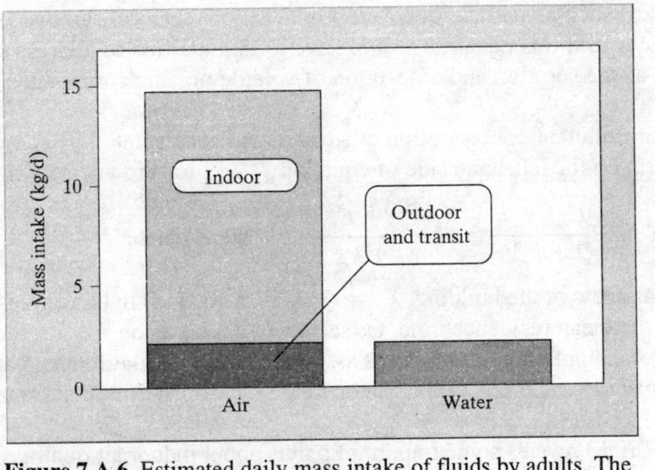

Figure 7.A.6 Estimated daily mass intake of fluids by adults. The estimate assumes ingestion of 2 L water and inhalation of 12 m^3 air per day, which represents the mean lifetime average for men and women (Layton, 1993). The distribution between indoor and outdoor air intakes is based on a survey of Californian activity patterns (Jenkins et al., 1992), which shows that the average time spent outdoors is 6 percent of the day, with 87 percent spent indoors and 7 percent spent in transit.

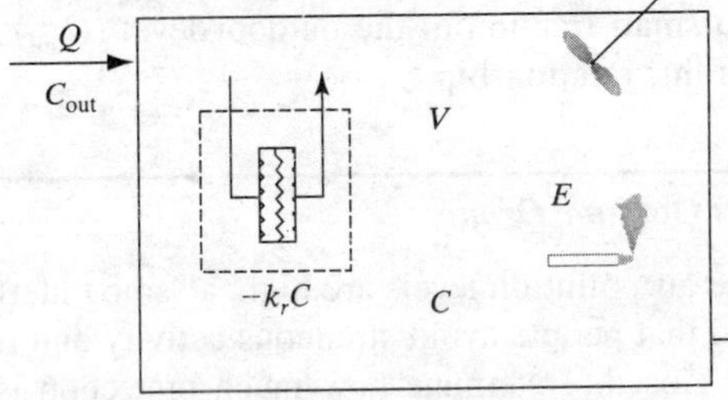

Figure 7.A.7 Schematic of an indoor air quality model based on a CMFR. See text for definition of symbols.

material balance, we can write the following governing equation for the concentration, C, of a pollutant in the room air:

$$\frac{d(CV)}{dt} = QC_{\text{out}} + E - QC - k_r CV \tag{7.A.12}$$

The term on the left represents the net accumulation of the pollutant in indoor air. The terms on the right represent pollutant supply from outdoor air by ventilation (QC_{out}), pollutant generation by direct indoor emissions (E), pollutant removal by ventilation (QC), and pollutant removal by all other mechanisms (k_rCV), such as air treatment or reaction with indoor surfaces. In equation 7.A.12, the airflow rates into and out of the building are assumed to be equal. Because of mass conservation, the mass flow rate of air in must approximately equal the mass flow rate out. The volume flow rates, Q, must also be equal if the indoor and outdoor air temperatures are the same.* Equation 7.A.12 assumes that pollutant removal is a first-order process with rate constant k_r.

*By the ideal gas law ($PV = nRT$), the indoor-outdoor temperature difference must be 30 K for the ventilation flow rates to differ by 10 percent.

Many processes, such as particle deposition onto indoor surfaces, are well described by this treatment. But this equation would have to be modified to address some phenomena, such as the sorption and desorption of volatile organic compounds on indoor surfaces.

The indoor pollutant concentration at steady state is calculated by setting the accumulation term on the left-hand side of equation 7.A.12 to zero and solving for C:

$$C = \frac{QC_{out} + E}{Q + k_r V} = \frac{\lambda_v C_{out} + E/V}{\lambda_v + k_r} \qquad \text{steady state} \qquad (7.A.13)$$

The air-exchange rate of the building, $\lambda_v = Q/V$, is introduced in this expression. It is the inverse of the mean residence time, Θ (see §5.A). In equation 7.A.13, the numerator is the sum of the pollutant supply rates associated with ventilation from outdoor air and indoor emissions, and the denominator is the sum of the pollutant removal rate coefficients.

Equation 7.A.13 reveals some important points about indoor air quality, especially when limiting conditions are considered. For example, in the absence of indoor emissions, the steady-state indoor concentration reduces to a fraction of the outdoor level:

$$\frac{C}{C_{out}} = \frac{\lambda_v}{\lambda_v + k_r} \qquad \text{no indoor emissions} \qquad (7.A.14)$$

The relationship between indoor and outdoor concentrations in this case is governed by the relative magnitude of the ventilation rate, λ_v, and the rate of removal by any other mechanism, k_r. If ventilation dominates ($\lambda_v \gg k_r$), then C/C_{out} approaches 1. On the other hand, if removal by other means dominates ventilation ($k_r \gg \lambda_v$), then the indoor concentration (C) is a small fraction of the outdoor level (C_{out}). Example 7.A.3 illustrates an application of this relationship.

EXAMPLE 7.A.3 ***Protection against Outdoor Ozone***

In urban areas, at a time when the air pollution levels are high, a "smog alert" may be issued. Officials then recommend that people avoid strenuous activity and remain indoors with doors and windows closed. Determine how much protection is afforded against ozone by remaining indoors. Consider the case of a small residence ($V = 150$ m^3). Even with the doors and windows closed, we can assume that $Q = 2$ m^3 min^{-1} of air leaks into the house from outdoors through cracks. Ozone reacts with indoor surfaces; this process can be represented as a first-order chemical reaction that consumes ozone with a rate constant of $k_r = 1.5$ h^{-1} (Weschler et al., 1989).

(a) Compute the indoor/outdoor ratio of ozone under steady-state conditions.

(b) Comment on the effectiveness of remaining indoors as a means of protecting oneself against ozone exposure.

SOLUTION

(a) The steady-state, indoor/outdoor ozone concentration ratio is given by equation 7.A.14. Substituting the data given in the problem statement, we find that $C/C_{out} =$ 0.35.

(b) The indoor concentration of ozone is reduced to about a third of the outdoor value under these conditions. Being indoors offers substantial, but not perfect protection, against ozone in outdoor air.

Another interesting point relates to the significance of indoor emissions. From equation 7.A.13, we see that if $E \gg QC_{out}$, then the indoor concentration does not depend significantly on the outdoor concentration, but rather is approximately given by the relationship

$$C = \frac{E/V}{\lambda_v + k_r} \quad \text{strong indoor emissions} \tag{7.A.15}$$

Equations 7.A.14–15 demonstrate an important point. Increased ventilation increases the concentration of pollutants that are of outdoor origin but decreases the indoor concentrations of pollutants that are emitted directly indoors. With strong indoor emissions and no removal mechanisms other than ventilation, the steady-state indoor pollutant concentration is inversely proportional to the ventilation rate, $C = E/Q$.

Thus, ventilation is a very important parameter in indoor air quality studies. It not only strongly influences indoor pollutant concentrations but also is associated with a large energy demand. That is, when the outdoor weather is not comfortable, then the air supplied for ventilation must be heated or cooled, and sometimes humidified or dehumidified.

There are three modes of building ventilation. *Infiltration* refers to the leakage of air through cracks and other unintentional leaks in the shell of the building, driven by small indoor-outdoor pressure differences. *Natural ventilation* is the flow of air, driven by the same forces, through designed openings, such as windows and doors. *Mechanical ventilation* refers to the supply of air by means of ducts and fans. In the United States, most residential buildings and small commercial buildings are ventilated by a combination of infiltration and natural ventilation. Large commercial buildings usually have mechanical ventilation systems.

Typical infiltration rates for single-family dwellings, expressed as air exchange rates ($\lambda_v = Q/V$), are in the range 0.2–2 h^{-1}. When windows are opened, ventilation rates can increase by an order of magnitude, especially if the wind is blowing. Mechanical ventilation rates are usually specified as a certain minimum volume flow rate of outside air per building occupant; current standards are in the range 8–30 L s^{-1} person^{-1}, depending on how the space is used (ASHRAE, 1990).

Apart from ventilation, the other dominant factor controlling indoor air quality is indoor emissions. The indoor air quality problems of greatest concern are those for which strong indoor sources exist such that the indoor concentrations exceed those outdoors. Three pollutant classes with significant indoor sources are briefly discussed next.

Combustion By-products

It is well known that combustion is an important source of outdoor air pollution. Any combustion activities that occur indoors can contribute to indoor air quality problems. The most common activities are cooking, heating, smoking, and the use of internal combustion engines (i.e., from the operation of motor vehicles inside buildings or in attached garages). Other indoor combustion sources include candles, kerosene lamps, incense, and mosquito coils. Pollutant species of concern include carbon monoxide, nitrogen oxides, sulfur oxides, organic compounds, and particulate matter, including soot.

Indoor air quality problems caused by combustion may be explored through the use of the CMFR model depicted in Figure 7.A.7. Example 7.A.4 uses this approach to illustrate how great an impact cigarette smoking can have on indoor particle levels.

One especially large combustion-related indoor air quality problem is caused by the use of biomass cookstoves in developing countries. Throughout much of the world, food is cooked indoors over stoves in which fuels, such as wood, dung, and crop residues, are burned under poor combustion conditions. The resulting concentrations of emitted pollutants can be staggeringly high. For example, Smith et al. (1983) have pointed out that "the woman cooks are inhaling as much [benzo(*a*)pyrene] as if they smoked 20 packs of cigarettes per day." Benzo(*a*)pyrene is a known carcinogen. Worldwide, an enormous number of people, perhaps half a billion, are exposed to biomass cookstove emissions. Smith et al. suggest that the best solution to this problem is to convert to cleaner cooking fuels.

EXAMPLE 7.A.4 *Particulate Matter from Tobacco Smoke*

A single cigarette is smoked over a 6-minute period in a studio apartment (V = 100 m^3), releasing 9 mg of particulate matter ($PM_{2.5}$). The major removal mechanism is ventilation ($\lambda_v = 0.5\ h^{-1}$). Predict the concentration of environmental tobacco smoke (ETS) particles as a function of time.

SOLUTION Write and solve the differential material-balance equation for two discrete periods: t = 0–6 min, while emissions are occurring, and $t > 6$ min, after emissions are finished. During the period of smoking, the material-balance equation is (from equation 7.A.12)

$$\frac{dC}{dt} = E/V - \lambda_v C \qquad 0 < t < 6 \text{ min}$$

where $E = 9 \text{ mg} \div 6 \text{ min} = 1.5 \text{ mg min}^{-1} = 1500\ \mu\text{g min}^{-1}$. The initial condition is not specified, so we will assume $C(0) = 0$. The solution is

$$C(t) = (E/Q)[1 - \exp(-\lambda_v t)] \qquad 0 < t < 6 \text{ min}$$

where $Q = \lambda_v V = 50\ m^3\ h^{-1} = 0.83\ m^3\ \text{min}^{-1}$, so $E/Q = 1800\ \mu\text{g m}^{-3}$. At t = 6 min, the concentration is $C(6 \text{ min}) = 88\ \mu\text{g m}^{-3}$. After the cigarette is smoked, the

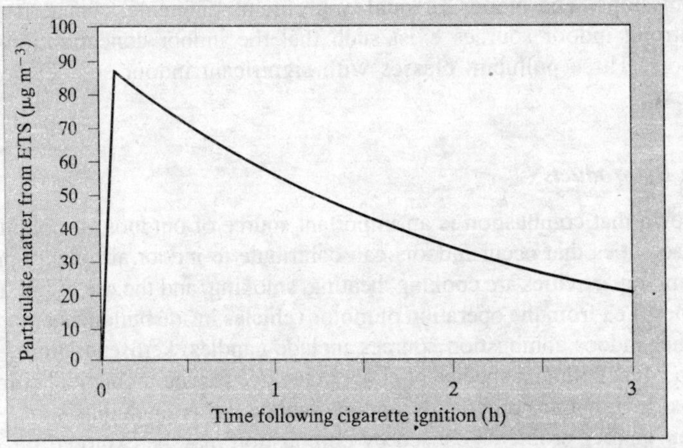

Figure 7.A.8 Predicted concentration of airborne particulate matter, C, caused by smoking a cigarette in a studio apartment.

EXAMPLE 7.A.4 ***Particulate Matter from Tobacco Smoke (continued)***

concentration decays because of removal by ventilation:

$$\frac{dC}{dt} = -\lambda_v C \qquad t > 6\,\text{min}$$

where $C(6\text{ min}) = 88\ \mu\text{g m}^{-3}$. The solution is

$$C(t) = 88\ \mu\text{g m}^{-3}\ \{\exp[-\lambda_v(t - 6\text{ min})]\} \qquad t > 6\text{ min}$$

The particle concentration is plotted versus time in Figure 7.A.8. Note that the federal NAAQS for $PM_{2.5}$ in outdoor air is 65 μg m^{-3} on a 24 h basis. In this indoor environment, tobacco smoke particles alone would cause that level to be exceeded on a 24 h average basis if nine or more cigarettes were smoked indoors daily.

Volatile Organic Compounds

Volatile organic compounds (VOCs) include thousands of individual species whose impact ranges from innocuous to highly hazardous. In outdoor air, these species are of concern in part because they are precursors to photochemical smog. In indoor air, the main concerns are their impact on the health and well-being of building occupants. Adverse effects include disturbances associated with short-term exposure, such as odor and irritation, as well as problems that may result from long-term exposure, such as cancer.

Many VOCs are present in indoor air at levels that, on average, are higher than outdoor air concentrations (see Figure 7.A.9). The sources of these VOCs are

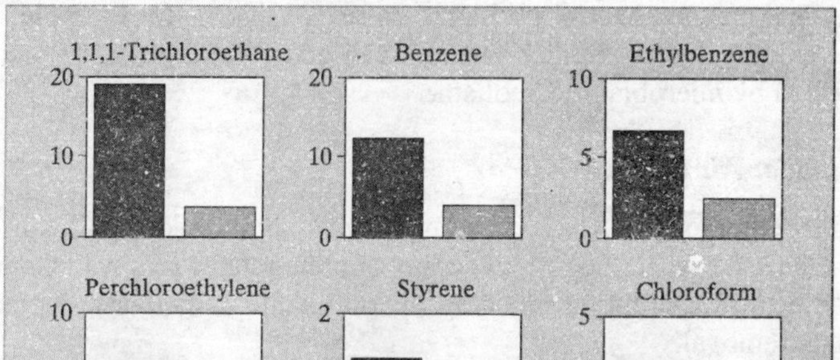

Figure 7.A.9 Geometric mean concentrations of selected air pollutants (μg m^{-3}) measured in residences ($n = 384$) and outdoors ($n = 86$) overnight in the cities of Elizabeth and Bayonne, NJ, September–November 1981 (Wallace et al., 1987).

numerous, including building materials, furnishings, dry-cleaned clothes, cigarettes, gasoline vapors, cleansers, pesticides, moth crystals, and tap water. VOC concentrations vary markedly from one building to another, by as much as a factor of 100 or more. Indoor emissions are the most important determinant of whether an individual building will have high VOC levels or not. Concentrations of individual VOCs are seldom high enough to trigger a direct toxicological concern because of short-term exposure, but little is known about the health implications of exposure to a mixture of VOCs.

Microbial Contaminants and Other Bioaerosols

Several infectious diseases are transmitted by inhalation of bacteria or viruses. For example, new tuberculosis infections can occur when someone inhales a bacterium that was released by a person with an active case of the disease. Tuberculosis is an enormous public health problem worldwide: It is estimated that 1.7 billion people, or 30 percent of the world's population, are infected, and 3 million deaths are caused each year by tuberculosis, mostly in countries without modern health care systems. Other diseases that are transmitted by airborne pathways include chicken pox, measles, and Legionnaires' disease. It is also believed that the airborne route contributes to the transmission of influenza.

Other airborne particles of biological origin can cause allergic response and trigger asthma attacks. Allergic agents include pollens, animal dander (bits of skin, fur, and saliva shed by house pets), and dust mites. Dust mites are barely visible creatures (roughly 300 μm in dimension) that live in places such as carpeting and bedsheets and feed on human skin flakes. Their fecal matter and body fragments, which are highly allergenic, can become airborne by mechanical processes such as walking along a carpet.

Molds and mildew also create problems in indoor air. These fungi grow readily on moist surfaces and may cause odor problems. In severe cases—for example, in the aftermath of a building flood—moldy surfaces can generate large emissions of toxic chemicals created by microbial metabolism.

Controlling Indoor Air Quality

As suggested by equation 7.A.13, it is useful to think about indoor pollutant levels as reflecting a balance between the rate of supply and the rate of removal. Control measures can then be aimed either at reducing the pollutant sources or at increasing the rate of pollutant removal.

The indoor air quality problems of greatest concern are those caused by indoor emissions. Limiting these emissions is a powerful method of controlling indoor concentrations. This objective may be achieved by modifying behaviors (e.g., smoking only outdoors) or by modifying products (e.g., using consumer products that have lower emission rates of toxic compounds). Ventilation always serves as one means of removing pollutants caused by indoor emissions, so increasing the ventilation rate serves to reduce indoor concentrations. However, in practice, only moderate improvements in indoor air quality can be achieved by increasing ventilation because of the cost and discomfort associated with high ventilation rates when weather conditions are unfavorable. Ventilation can be augmented by direct air-cleaning techniques. For example, indoor air can be recycled through particle filters or gas sorption units (typically based on activated carbon) to reduce concentrations. However, like ventilation, the effectiveness of this approach is often limited to modest or moderate improvements.

7.A.6 Global Change

Evidence accumulated over the past few decades demonstrates that human activities may perturb the atmosphere on a global scale. Attention focuses on two classes of disturbance: changes to climate and stratospheric ozone depletion. The most astonishing feature of these two issues is that human activities have reached a scale at which the dynamics of the entire earth are affected! This sobering fact should provide strong motivation for giving greater consideration to mankind's impact on the environment.

Climate Change

The earth's climate is governed by the complex interplay of many factors, including incoming solar radiation, evaporation and condensation of water, radiant heat transfer from earth's surface to space, and the rotation of the earth about its axis. Of central importance is the role that the atmosphere plays in the earth's energy budget. To gain some understanding of the problem of climate change, we must consider how energy flows through and interacts with the atmosphere.

Almost all of the incoming energy to the earth's surface comes from the sun in the form of electromagnetic radiation. The sun is extremely hot, about 6000 K at its surface, so much of the energy it emits is found in short-wavelength radiation (see Figure 7.A.10). Averaged over the area of the entire earth's surface, the energy from the sun that reaches the outer edge of the earth's atmosphere is about 338 W m^{-2} (see Example 7.A.5).

A fraction of the energy incident at the top of the earth's atmosphere is not absorbed but rather is reflected or otherwise scattered back into space. That fraction, known as the *albedo*, has a value of 0.31 when averaged over the entire earth-atmosphere system. Thus, the net rate at which earth absorbs energy from the sun averages to approximately 233 W m^{-2}.

The earth exists in a state of approximate thermal balance. The energy absorbed from the sun is balanced by energy lost from the earth to space, which also occurs in the form of electromagnetic radiation. Because the earth is much cooler than the sun, this radiation occurs at longer wavelengths, centered at about 10 μm, in contrast to the peak at 0.4–0.5 μm for solar radiation.

The measured average temperature of the earth's surface is 288 K. Applying the Stefan-Boltzmann law (see Example 7.A.5), we predict that the earth should emit electromagnetic radiation at a rate of 389 W m^{-2}, substantially higher than the rate of energy absorption from the sun. To reconcile this discrepancy in the energy balance, we must consider how the atmosphere interacts with electromagnetic radiation.

The atmosphere contains gases and particles that both absorb and scatter light. Absorption represents a net energy conversion process, with light energy being converted to thermal energy (that is, increasing temperature). Overall, the atmosphere absorbs energy relatively effectively in the ultraviolet and infrared portions of the electromagnetic spectrum, but it is largely transparent to visible light. Atmospheric gases absorb much of the infrared radiation that is emitted from the earth's surface. The atmosphere warms as a result and reemits radiation, both toward the earth and in the direction of space. The net rate of release of energy from the earth-atmosphere system, 233 W m^{-2}, balances the energy absorbed from the sun and corresponds to the blackbody emissions from a system at 254 K.

Figure 7.A.10(*b*) shows the absorption spectrum of the atmosphere. Molecular oxygen absorbs light in the far ultraviolet. Stratospheric ozone is responsible for

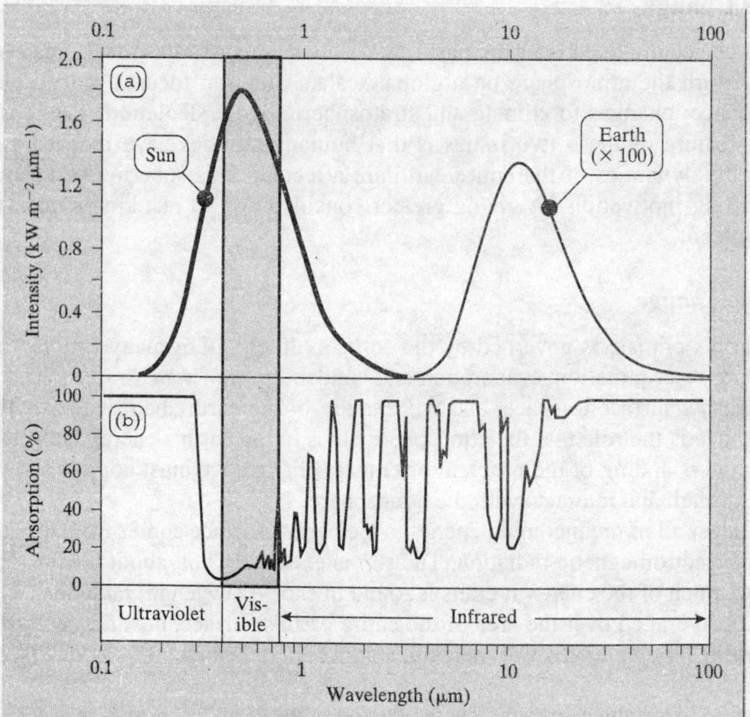

Figure 7.A.10 (*a*) The electromagnetic energy spectrum measured outside the earth's atmosphere. The radiation from the sun corresponds approximately to theoretical emissions from a blackbody at 5783 K. Earth's spectrum, magnified by a factor of 100, corresponds well to a blackbody at 254 K. (*b*) The absorption spectrum of the atmosphere, showing the fraction of light of a given wavelength that is absorbed during one pass from the top of the atmosphere to sea level, with a clear sky and the sun at about 40° above the horizon. (From *Atmospheric Radiation: Theoretical Basis,* Second Edition, by R.M. Goody and Y. L. Yung, copyright © 1995 by Oxford University Press, Inc. Used by permission of Oxford University Press, Inc.)

absorbing other portions of the ultraviolet spectrum, particularly UV-C (0.20–0.28 μm) and UV-B (0.28–0.32 μm). In the infrared portion of the spectrum, carbon dioxide and water vapor are the strongest light absorbers. Other gases also contribute, including ozone, methane, nitrous oxide, and chlorofluorocarbons.

EXAMPLE 7.A.5 *Energy from the Sun*

A result from theoretical physics, known as the *Stefan-Boltzmann law,* states that the total power radiated by a blackbody at temperature T (K) per unit surface area is given by

$$E = \sigma T^4$$

where $\sigma = 5.67 \times 10^{-8}$ W m^{-2} K^{-4}. Assuming that the sun is a perfect blackbody at $T = 5783$ K, estimate the average incidence of solar energy to the top of the earth's atmosphere.

EXAMPLE 7.A.5 ***Energy from the Sun (continued)***

Data

$d_s = 1.39 \times 10^9$ m	Diameter of the sun
$d_e = 1.27 \times 10^7$ m	Diameter of the earth
$R_{es} = 1.49 \times 10^{11}$ m	Mean earth-to-sun distance

SOLUTION Applying the Stefan-Boltzmann law, we estimate the energy from the sun to be $E_s = 6.3 \times 10^7$ W m^{-2}, determined at the surface of the sun. This energy radiates uniformly in all directions at a rate that is constant for small time intervals. By energy conservation, the total rate of energy emission from the surface of the sun ($\pi d_s^2 E_s$, since πd^2 is the surface area of a sphere of diameter d) must equal the rate of energy flow through a sphere of radius R_{es} centered on the sun (see Figure 7.A.11):

$$\pi(2R_{es})^2 E_e = \pi d_s^2 E_s$$

Solving yields $E_e = 1380$ W m^{-2} as the estimated energy flux density at the top of the earth's atmosphere through a plane normal to the sun. The measured energy flux density from the sun through a normal plane at the top of the earth's atmosphere, known as the *solar constant*, is 1353 W m^{-2}, only a few percentage points lower than our estimate.

The surface area of a sphere is equal to four times its cross-sectional area. Therefore, the incident solar energy flux to the top of the earth's atmosphere, averaged over the entire earth's surface, is $1353 \div 4 = 338$ W m^{-2}.

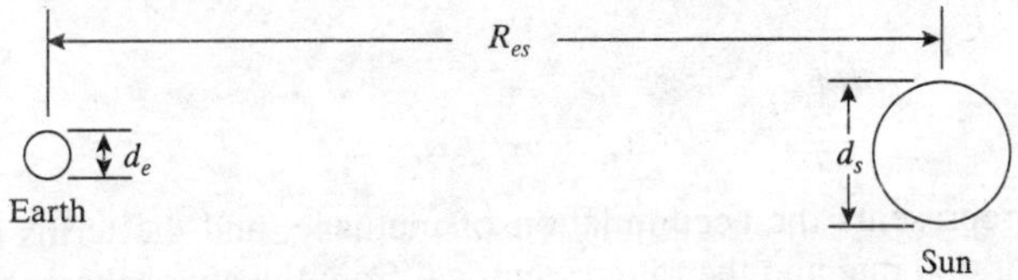

Figure 7.A.11 The geometry of the earth-sun system (not to scale).

The main reason for scientists' concern about climate change is that the atmospheric abundance of some species that absorb infrared radiation, known as *greenhouse gases*, is increasing (see Table 7.A.6 and Example 7.A.6). Incoming (visible)

Table 7.A.6 Characteristics of Some Atmospheric Gases That Influence Climate

Species	Abundance (ppm)	Rate of increase[a] (% y^{-1})	Lifetime (y)	Radiative forcing[b] (W m^{-2})	GWP[c] (–)
CO_2	358	0.4	50–200	1.56	1
CH_4	1.72	0.6	12	0.47	21
N_2O	0.31	0.25	120	0.14	310

[a]Average annual growth rate in atmospheric concentration for the period 1984–1994.

[b]Radiative forcing represents an estimate of the perturbation in the earth-atmosphere energy balance based on changes in the atmospheric abundance of a gas from preindustrial times (ca. 1750) to the present. The estimated preindustrial abundances are 280 ppm for CO_2, 0.70 ppm for CH_4, and 0.275 ppm for N_2O.

[c]GWP (global warming potential) is a weighting factor that estimates the global warming impact of the species relative to CO_2 on a constant mass emissions basis. Data here are for a 100-year period. The value of 21 for methane implies that release of 1 kg of methane today will have as much impact on climate over the next 100 years as release of 21 kg of carbon dioxide.

Source: IPCC, 1996.

radiation is largely unaffected by these gases while outgoing (infrared) radiation is absorbed. According to a simple energy balance model, to achieve a new steady-state balance, the temperature of the earth-atmosphere system would have to rise.

EXAMPLE 7.A.6 ***Methane Accumulation in the Atmosphere***

The current global average mole fraction of methane in the atmosphere is 1.72 ppm and is increasing at an average rate of about 0.6 percent per year. Gas-phase degradation by reaction with OH• radicals is the main removal mechanism. The average first-order loss rate for methane from the atmosphere is approximately $0.1\ y^{-1}$.

(a) Given that the atmosphere has a total mass of 5.1×10^{18} kg, what is the rate of accumulation of methane mass (grams per year) in the atmosphere?

(b) What mass emission rate of methane (grams per year) is needed to sustain the measured concentration and its growth rate?

SOLUTION

(a) The molecular weight of air is approximately 29 g mol^{-1}. We can estimate that the atmosphere contains 5.1×10^{21} g $\div$ 29 g mol^{-1} = 1.8×10^{20} mol of air. Multiplying by the mole fraction of methane and its molecular weight (16 g mol^{-1}), we estimate that the mass of methane in the atmosphere is 4.8×10^{15} g. The observed accumulation rate of 0.6 percent per year implies that methane is added to the atmosphere at a net rate of 2.9×10^{13} g y^{-1}, or 29 million metric tons per year.

(b) This material-balance equation applies to the total mass of methane in the atmosphere:

$$\frac{dM_{CH_4}}{dt} = E_{CH_4} - k_r M_{CH_4}$$

The term on the left represents the accumulation of methane, and the terms on the right represent the emission rate and the rate of removal from the atmosphere, respectively. The accumulation term is equal to 0.6 percent per year of the atmospheric mass of methane, and the removal term is equal to 10 percent per year ($k_r = 0.1\ y^{-1}$). Therefore, the total annual emission rate must be 10.6 percent of the methane mass: 5.1×10^{14} g y^{-1}, or 510 million metric tons per year.

Comment These numbers agree to within about 20 percent or better of the estimates published by the Intergovernmental Panel on Climate Change (IPCC, 1996).

The earth's climate is very complex and is subject to significant natural variation. These characteristics make it difficult to discern whether the climate has changed as a consequence of the increased abundance of greenhouse gases. The best current scientific estimate is that the global mean surface temperature has increased by 0.3–0.6 °C during the past century (IPCC, 1996). For a mid-range estimate of how emissions will evolve, the global mean surface temperature is expected to increase by 1–3.5 °C between 1990 and 2100. Because of thermal expansion and melting of the polar ice caps, the mean sea level is predicted to rise 15–95 cm over this period. It is expected that the occurrence of extremely hot days will increase (and, conversely, the frequency of extremely cold days will diminish). Increased surface temperatures are also expected to drive the hydrologic cycle more vigorously by increasing the rate of evaporation, producing more frequent and severe storms and changing the patterns of flood and drought. Rapid local changes in climate might have serious consequences for some terrestrial ecosystems where native plants are sensitive to perturbations in moisture or temperature conditions.

The most important greenhouse gas is carbon dioxide. This species plays a central role in the natural cycling of carbon through the environment. By photosynthesis, CO_2 is taken up from the atmosphere by terrestrial plants. When plant material decays, some of the carbon is released back to the atmosphere, either in the form of CO_2 or as methane, and some remains stored as organic matter in the soil. Carbon dioxide dissolves in water, forming carbonic acid, bicarbonate, and carbonate (see §3.C.4). Marine organisms extract carbon from one of these forms to create new organic and inorganic (i.e., $CaCO_3$) mass.

The anthropogenic perturbations of the natural CO_2 cycle are primarily a result of fossil fuel combustion and the clearing of tropical forests. By burning coal and petroleum products, we release into the atmosphere carbon that had been sequestered by plants in earlier geologic times. As tropical forests are cleared, the rate of terrestrial uptake of carbon by plants is reduced and much of the previously sequestered carbon is released to the atmosphere. The total estimated anthropogenic emission of carbon dioxide to the atmosphere is $(7.1 \pm 1.1) \times 10^{15}$ g C y^{-1} or 2.6×10^{16} g CO_2 y^{-1} (IPCC, 1996). Of this total, about half (47 percent) accumulates in the atmosphere, 28 percent is estimated to be taken up by the oceans, 7 percent goes into the regrowth of Northern Hemisphere forests, and the fate of the remaining 18 percent is unknown. The atmospheric persistence of CO_2 is long, on the order of a century.

Efforts to limit human-induced climate change will probably focus on reducing the anthropogenic emissions of carbon dioxide. However, this objective will not be easily achieved. Most air quality problems result from the emissions of undesired, minor by-products of human activities. In contrast, carbon dioxide is the *major* by-product of burning carbonaceous fuels. The mass fraction of carbon in fossil fuels is large. Because of the added mass of oxygen, the mass of carbon dioxide produced by fossil fuel combustion actually *exceeds* the mass of fuel burned. Global emissions of CO_2 from human activities are staggeringly high, equivalent to about a metric ton (10^6 g) of carbon per year for every person on the planet.

A major means of reducing the emissions of CO_2 is to lower the rate of fossil fuel use by improving end-use energy efficiency and by conversion to alternative energy sources. Other possibilities are being explored, such as capturing CO_2 in combustion flue gas for disposal into the deep ocean. Overall, it appears that CO_2 emissions cannot be reduced without major economic investment and, probably, significant lifestyle changes.

Stratospheric Ozone Depletion

Ultraviolet radiation is harmful to life. It can damage molecular bonds, leading to a range of undesirable effects in biological organisms, such as skin cancer in humans. At the earth's surface, we are substantially protected from the sun's ultraviolet radiation by the presence of O_2 and O_3 in the stratosphere. Molecular oxygen (O_2) efficiently absorbs radiation at wavelengths less than 0.24 μm; ozone is an efficient absorber of light with wavelengths of 0.24–0.32 μm.

Ozone in the stratosphere is maintained in a dynamic balance with a short lifetime. Changes in the rate of production or destruction of ozone can influence stratospheric concentrations. About 90 percent of all atmospheric ozone is in the stratosphere, with peak mole fractions of about 10 ppm occurring at altitudes of 15 km (over the poles) to 25 km (over the equator) (Rowland, 1990).

Stratospheric ozone is produced by photolysis of molecular oxygen, followed by reaction of the oxygen radical with a second oxygen molecule (Abbatt and Molina, 1993):

$$O_2 + h\nu \rightarrow O\bullet + O\bullet \quad (7.A.16)$$

$$O\bullet + O_2 \rightarrow O_3 \quad (7.A.17)$$

Stratospheric ozone can be consumed by photolysis and by reaction with oxygen radicals:

$$O_3 + h\nu \rightarrow O_2 + O\bullet \quad (7.A.18)$$

$$O\bullet + O_3 \rightarrow 2\,O_2 \quad (7.A.19)$$

Stratospheric ozone can also be consumed by a series of chemical reactions of this general form:

$$O_3 + X \rightarrow O_2 + XO \quad (7.A.20)$$

$$O\bullet + XO \rightarrow O_2 + X \quad (7.A.21)$$

In these reactions, X denotes a molecule or atom that is capable of being oxidized by ozone and of being regenerated by reaction with the oxygen radical. The net effect of reactions 7.A.20–21 is the same as for reactions 7.A.18–19: ozone and O• are consumed to produce two oxygen molecules. A balance between the rates of production (reactions 7.A.16–17) and destruction (reactions 7.A.18–21) determines the concentration of stratospheric ozone.

The species X serves as a catalyst for degradation of stratospheric ozone. The main species that can act as X are nitric oxide (NO), chlorine (Cl), bromine (Br), and the hydroxyl radical (OH•). Each of these species is present in the stratosphere as a result of natural sources. For example, nitrous oxide (N_2O), which is a natural by-product of denitrification reactions carried out by microorganisms in the oceans and in soils, is stable in the troposphere. Nitrous oxide can be transported to the stratosphere where, in the presence of ultraviolet light, it can be photochemically converted to nitric oxide (Seinfeld and Pandis, 1998). Likewise, methyl chloride (CH_3Cl) and methyl bromide (CH_3Br), emitted naturally from the oceans, can be transported to the stratosphere, where UV photochemical reactions liberate the chlorine and bromine atoms. Once started, the ozone-consuming reactions continue until the catalyst is scavenged from the atmosphere—for example, in the form of an acid species such as HNO_3 or HCl. A single chlorine atom in the stratosphere can destroy 10^5 molecules of ozone (Rowland, 1990).

The stratospheric concentrations of ozone scavengers have been significantly increased by human activities. Increasing the concentration of these scavengers leads to an increase in the rate of destruction of stratospheric ozone and a decrease in its concentration. Consequently, the degree of protection that the stratosphere provides by absorbing ultraviolet light is degraded.

Over the past few decades, considerable attention has focused on the role of chlorofluorocarbons (CFCs) as causes of stratospheric ozone depletion (Rowland, 1990). These species, which contain only carbon, fluorine, and chlorine atoms, were first synthesized in the 1920s. They were widely used as the working fluid in compression refrigeration units, as commonly found in air conditioners and food refrigerators. They were also used to manufacture plastic materials such as styrofoam, incorporated as a propellant in aerosol sprays, and used as industrial solvents. Characteristics of the three most widely used CFCs are summarized in Table 7.A.7.

In the troposphere, CFCs are extremely stable, with estimated atmospheric lifetimes on the order of a century. After mixing rapidly through the troposphere, the compounds are slowly transported into the stratosphere. There, intense UV radiation can

Table 7.A.7 Characteristics of Common Chlorofluorocarbons

Species	Formula	Mole fraction[a] (ppt)	Emissions[a] ($\times 10^9$ g y^{-1})	Lifetime (y)	Applications
CFC-11	CCl_3F	220	350	60	Foaming agent for open-cell plastic foams, aerosol propellant
CFC-12	CCl_2F_2	375	450	120	Home refrigerators, auto air conditioners, aerosol propellant
CFC-113	CCl_2FCClF_2	30	150	90	Cleaning solvent in microelectronics manufacturing

[a]Atmospheric abundance and emission estimates for 1985 (WMO, 1990).
Source: Rowland, 1990; WMO, 1990.

cause photodegradation, liberating the chlorine atoms and triggering the ozone-destroying catalytic cycle shown by reactions 7.A.20–21. Because of the combination of their industrial usefulness, nontoxic properties, and atmospheric stability, the atmospheric concentrations of CFCs grew rapidly between 1960 and 1990 (see Figure 7.A.12).

Molina and Rowland (1974) originally advanced the theory that CFCs could cause stratospheric ozone depletion. Subsequent experimental investigations have substantiated this concern. Above Antarctica, massive springtime ozone destruction has been observed for several successive years. Less dramatic but still detectable ozone depletion has been detected above midlatitudes in the Northern Hemisphere.

Through a combination of industry and government efforts that included substantial international cooperation, remarkable progress has been made toward eliminating the production and use of chlorofluorocarbons (Benedick, 1991). The use of CFCs as

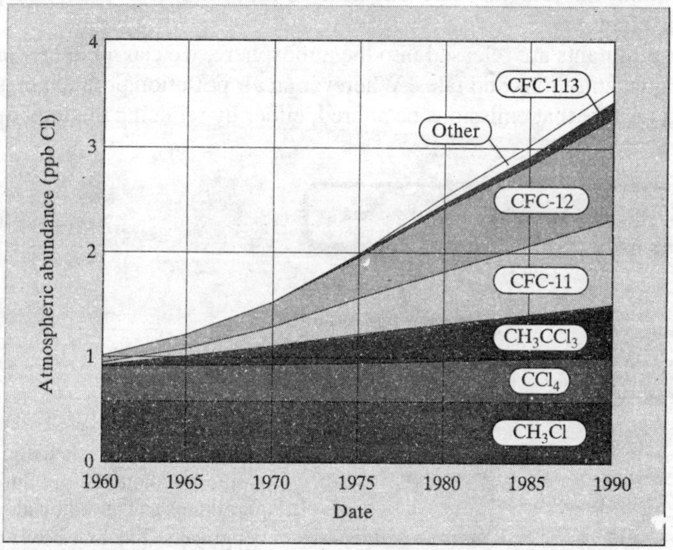

Figure 7.A.12 Mole fraction of some important chlorine-containing molecules in the atmosphere, expressed in moles of Cl per 10^9 moles of air (WMO, 1990). The potential for ozone depletion is greatest for the compounds that do not contain hydrogen (i.e., CCl_4 and the CFCs) because they are not degraded in the troposphere.

an aerosol spray propellant was banned in the United States, Canada, and some northern European countries around 1980. In 1987, 23 nations signed the Montreal Protocol, agreeing to reduce the use of key CFCs by 50 percent by 1999. Amendments to the Montreal Protocol were signed in London (1990) and in Copenhagen (1992) that increased the number of signatory countries, accelerated the phase-out schedule, and expanded it to include other ozone-depleting industrial chemicals. In 1996, the United States ceased production and import of CFCs and related chemicals.

Global reductions in CFC emissions are being achieved by several methods (Rowland, 1990). For aerosol sprays, alternative propellants are being used as well as mechanical push-pumps (e.g., window cleaner) and non-aerosol delivery systems (e.g., roll-on deodorant). For refrigeration units, alternative working fluids are being developed. One option has been the replacement of CFCs by HCFCs (molecules containing hydrogen in addition to carbon, fluorine, and chlorine) and HFCs (molecules composed of hydrogen, fluorine, and carbon). The presence of a hydrogen atom in these molecules makes them subject to attack by OH• radicals in the troposphere, thereby reducing their transport into the stratosphere. For example, HCFC-22 (CF_2HCl), widely used as the refrigerant in home air conditioners, has an atmospheric lifetime of 15 y and has only about 5 percent of the ozone depletion potential of CFC-11 (WMO, 1990). The HFCs, such as HFC-134A (CH_2FCF_3), contain no chlorine and so have very low ozone-reducing potential. However, even with rapid elimination of the production of CFCs, because of their long persistence and the large quantities that are in current use, the impact of these chemicals on stratospheric ozone levels will continue to be experienced for decades.

7.B AIR POLLUTANT EMISSIONS AND CONTROLS

Figure 7.B.1 illustrates the structure of a typical air pollution problem. Pollutants are emitted from sources. Transport, transformation, and removal processes act on the emitted pollutants to establish airborne concentrations. These concentrations may cause adverse effects.

Once air pollutants are released into the atmosphere, we can no longer practically control their concentrations and fates. Wherever an air pollution problem exists, effective solutions require that emissions be altered, either by reducing the total quantity of

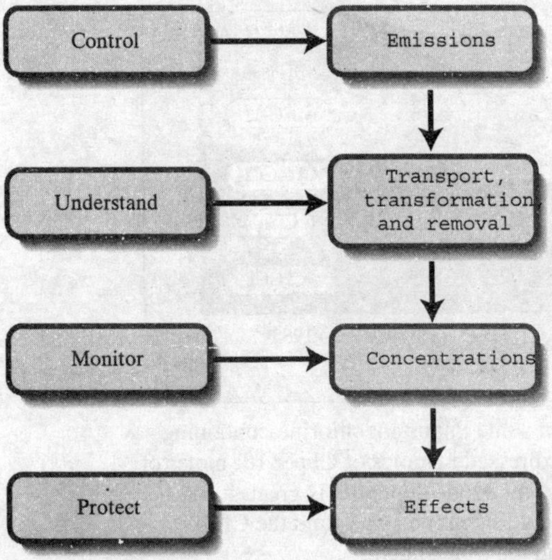

Figure 7.B.1 A schematic representation of air pollution problems as viewed by air quality engineers. The four boxes on the right represent the natural system. The boxes on the left indicate the activities and objectives of air quality engineers in interacting with this system.

pollutants released to the atmosphere, by changing the chemical or physical form of the releases, or by altering the manner of release. Even in indoor air, where increasing removal rates is sometimes practical (e.g., by increasing ventilation or adding filtration), often the most efficient control method is to reduce emissions. Thus, air quality engineers must understand how to quantify and control pollutant emissions.

Furthermore, policy decisions related to air pollution are often based on pollutant emissions. For example, air basins that are not in compliance with national ambient air quality standards are required to develop plans to make progress toward attainment. A key element in these plans is the identification and quantification of pollutant emission sources in the air basin. This information is known as an *emission inventory.* Emission inventories are produced by combining information on the intensity of activities with *emission factors* that specify the quantity of a pollutant emitted per unit of activity. Whether at an urban, regional, or larger scale, developing and maintaining an accurate emission inventory is essential in the design of rational air pollution control programs.

7.B.1 Characterizing Emissions

Air pollution sources are many and diverse. Table 7.B.1 and Figure 7.B.2 show the overall U.S. emissions inventory for several major air pollutants. Stationary fuel com-

Table 7.B.1 Estimated Emissions of Air Pollutants from Major Source Categories for the United States in 1995

	Emissions (10^{12} g y^{-1})				
Source category	CO	NO_x[a]	VOC[b]	SO_2	PM_{10}[c]
Fuel combustion					
Electric utilities	0.3	5.7	0.03	10.9	0.2
Industrial	0.6	2.8	0.12	2.8	0.2
Other	2.7	0.6	0.5	0.5	0.4
Chemical and allied product mfg.	2.0	0.3	1.5	0.4	0.06
Metals processing	2.0	0.08	0.07	0.7	0.13
Petroleum and related industries	0.3		0.6	0.3	0.02
Other industrial processes	0.7	0.3	0.4	0.4	0.4
Solvent utilization	0.002	0.003	5.8	0.001	0.002
Storage and transport	0.06	0.003	1.6	0.005	0.05
Waste disposal and recycling	1.6	0.08	2.2	0.03	0.2
Highway vehicles	53.2	6.9	5.5	0.3	0.3
Off-highway vehicles	14.2	2.7	2.0	0.3	0.4
Miscellaneous	5.9	0.2	0.4	0.007	34.4
Natural sources					2.0
Total	83.6	19.8	20.8	16.6	38.7

[a]Nitrogen oxide emissions are reported as if they were all in the form of NO_2. So, for example, the total *mass* emission rate of 19.8 × 10^{12} g y^{-1} of NO_x implies a *molar* emission rate of 4.3 × 10^{11} mol y^{-1} since the molecular weight of NO_2 is 46 g mol^{-1}.

[b]This column reports emissions of reactive volatile organic compounds that contribute to photochemical smog; being essentially nonreactive on a time scale of days, methane is excluded. Biogenic emissions of reactive VOCs are not included in this table. They are estimated to be of similar magnitude on a nationwide basis as the anthropogenic emissions.

[c]The dominant emissions of PM_{10}, in the category "miscellaneous," are from fugitive dust suspended mainly from travel on unpaved and paved roads, construction, agricultural tilling, and wind erosion.

Source: USEPA, 1996.

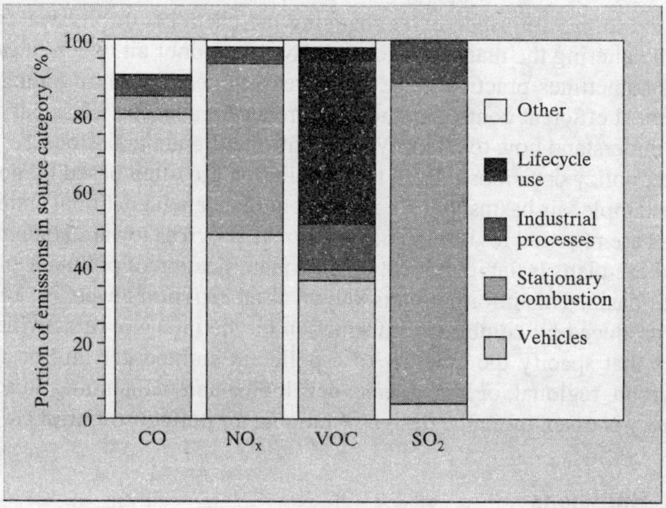

Figure 7.B.2 Air pollutant emissions contributed by broad source categories for the United States, 1995 (USEPA, 1996). Life cycle use for volatile organic compounds (VOCs) includes solvent utilization, storage and transport, and waste disposal and recycling.

bustion and motor vehicles are the major contributors to carbon monoxide, nitrogen oxides, and sulfur dioxide emissions. These two source categories are discussed in §7.B.2–3.

The data in Table 7.B.1 suggest that particulate matter emissions from the "miscellaneous" category are so large—89 percent of the total—that only this source deserves attention. However, it is important to recognize that particulate matter is a complex pollutant class and that health and other environmental concerns do not scale in direct proportion to the magnitude of the emissions from different sources. Table 7.B.2 provides some additional information on the characteristics of airborne particles from different sources. The contributions from mechanical generation—mainly fugitive dust from construction activities and from roadways, crop activities in agriculture and forestry, and wind erosion—are very large contributors to the total PM_{10} emissions. However, these mechanical generation processes mainly add coarse-mode soil dust to the air. Transport distances are not large; visibility impairment is much less (per mass) than for fine particles; and chemically, soil dust is not especially potent as a respiratory health hazard.

As shown in Figure 7.B.3, although mechanical generation processes dominate, combustion-related emissions of particulate matter are substantial (see §7.B.2 for dis-

Table 7.B.2 Major Source Categories of Airborne Particulate Matter

Category	Examples
Mechanical generation	Sea salt, wind-blown dust, tire wear, road dust
Combustion generation	Soot from diesel and gasoline engine exhaust; ash from industrial and power-plant boilers consuming fuel oil, natural gas, or coal; forest fires
Process generation	Fumes from welding, paint spray
Secondary gas-to-particle formation	Sulfates, nitrates, organics, ammonium, water

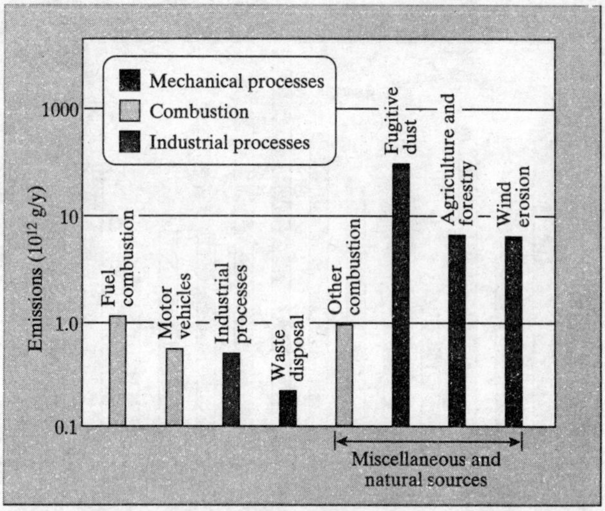

Figure 7.B.3 U.S. nationwide emission estimates for particulate matter smaller than 10 μm in diameter (PM_{10}). Data are for 1993, except for wind erosion, which represents the annual average for 1985–1993 (USEPA, 1994).

cussion). Particle emissions are also substantial from industrial processes, such as steel manufacturing, coal processing, cement production, aluminum processing, and paper production.

Secondary particle formation in the atmosphere is of considerable interest. Gaseous pollutants can change phase to add to the particle mass. For example, emissions of sulfur dioxide, nitrogen oxides, and ammonia in the gas phase can produce ammonium nitrate and ammonium sulfate in the particle phase. Similarly, oxidation of organic gases can produce species with low vapor pressure that contribute to the total mass of organic particles. Although not included in emissions inventories, secondary particle formation in polluted urban air may be of a similar magnitude as primary particle emissions as a contributor to fine particle mass.

In addition to emissions related to the criteria pollutants, there is considerable concern about the release of hazardous air pollutants, especially from industrial processes. Figure 7.B.4 presents data from the U.S. Toxics Release Inventory (TRI). The species shown in the figure are those on the list of hazardous air pollutants for which reported emissions from industrial plants and facilities are the largest. To provide a sense of perspective, note that the maximum emission rate for a single species (9×10^4 Mg y^{-1} for methanol and toluene) corresponds to about 1 g per person per day averaged over the entire U.S. population. Since the TRI is based on industrial self-reporting, some possible emission sources are not included, so it is likely that total nationwide emissions exceed the values shown in this figure.

Measuring Air Pollutant Emissions

Air pollutants are released into the atmosphere from many distinct activities and processes. The nature of these processes influences the methods that can be used to quantify emissions. In this subsection, we'll consider some examples of how air pollutant emissions are determined.

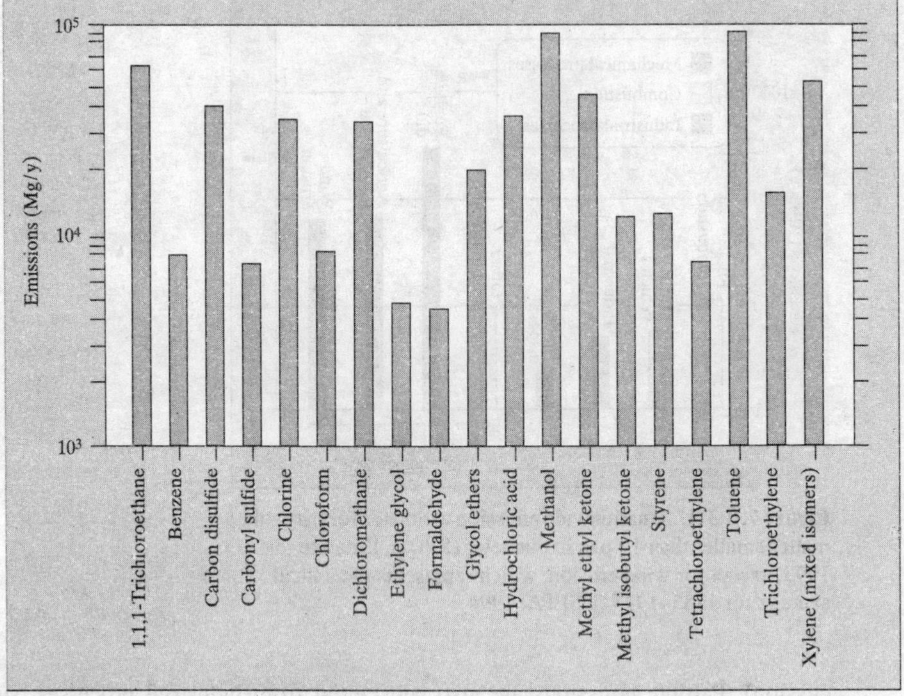

Figure 7.B.4 Nationwide emission estimates of some hazardous air pollutants in the United States, as reported in the Toxics Release Inventory. Included are all species with emissions >4500 Mg/y (USEPA, 1993).

Stack Emissions In large-scale industrial processes, air pollutants are emitted from a flue or stack. The principle of material balance can be applied to determine the mass emission rate, E (g s^{-1}). If the pollutant species is well mixed within the flue, then measure the species concentration in the flue gas, C (g m^{-3}), and multiply it by the rate at which flue gas is emitted, Q (m^3 s^{-1}), to yield

$$E = QC \tag{7.B.1}$$

Automobile Tailpipe Emissions One method of measuring emissions in automobile exhaust is a variant of the stack emissions procedure. The pollutant concentration in a sample of vehicle exhaust is measured and multiplied by the exhaust flow rate to give the mass emission rate, as shown in equation 7.B.1.

This procedure is easy to apply when a vehicle is idling, but not when it is moving. Motor vehicle emissions depend significantly on driving conditions. To test emissions under more realistic conditions, dynamometers are used in which the drive wheels of a vehicle are placed on a roller so that the vehicle remains stationary even with the transmission engaged. Standard driving cycles have been developed to mimic patterns of acceleration, deceleration, and high- and low-speed cruising as might typically be encountered in highway and surface-street driving. Pollutant emissions are measured by collecting and analyzing exhaust gas while the car is "driven" on the dynamometer.

An important alternative measurement method has been developed and applied in recent years (Bishop and Stedman, 1990). This method is based on optical remote sensing and can be used to measure emissions of some species, such as CO, as vehi-

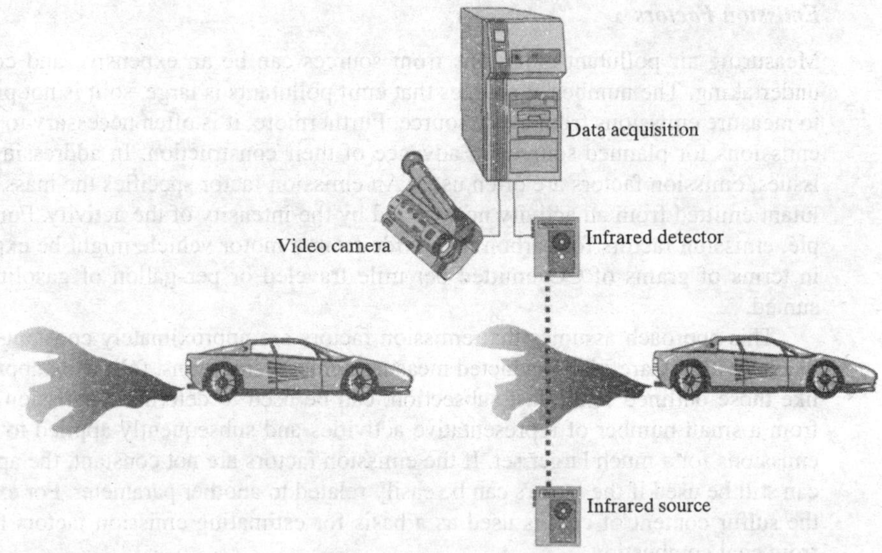

Figure 7.B.5 System for on-road measurement of CO emissions from motor vehicles. Infrared absorption is used to measure the relative amounts of CO_2, CO, and hydrocarbons in the exhaust plume. The video camera records the license plate so that the vehicle model and age can later be determined from registration records.

cles are driving past a fixed sampling station (see Figure 7.B.5). The relative amounts of CO, CO_2, and hydrocarbons (HC) in the exhaust gas are determined by measuring the absorption of infrared light of specific wavelengths. Almost all of the carbon in the exhaust originates from carbon in the fuel and is emitted as CO, CO_2, or HC. By measuring the amount of CO relative to the total amount of carbon-containing species in the exhaust plume, the emission rate of CO per mass of carbon in consumed fuel can be directly determined. Combining this result with information on fuel composition and fuel economy (km traveled per L of fuel) allows the CO emission rate to be determined in grams of pollutant emitted per km traveled.

Emissions from Small-Scale Activities or Processes Sometimes pollutants are not emitted from a stack or a tailpipe, but rather from some process or activity (e.g., welding) or from a component in a larger system, such as a leaky valve. When the release occurs on a sufficiently small physical scale, it is possible to measure emissions by capturing them within a ventilated enclosure, measuring the species concentration in the exhaust of the enclosure, and applying a material-balance equation. An example of this approach was presented in §5.A.4 (Example 5.A.17).

Determining Emissions from Overall Material Balance In some situations, the overall mass of emitted pollutants can be determined without direct measurement from basic knowledge about the process and products. For example, when an organic solvent is used in an unconfined area, it is reasonable to assume that the entire mass of applied solvent will become a gaseous emission. To determine the mass emissions of volatile organic compounds (VOCs) from painting, one can multiply the volume of paint used by the mass concentration of VOCs in the paint. A similar idea, employing the principle of stoichiometry, is applied in estimating emissions of some species, such as CO_2 and SO_2, from combustion systems, as described in §7.B.2.

Emission Factors

Measuring air pollutant emissions from sources can be an expensive and complex undertaking. The number of sources that emit pollutants is large, so it is not practical to measure emissions from every source. Furthermore, it is often necessary to predict emissions for planned sources in advance of their construction. In addressing these issues, emission factors are often used. An emission factor specifies the mass of pollutant emitted from an activity normalized by the intensity of the activity. For example, emission factors for carbon monoxide from a motor vehicle might be expressed in terms of grams of CO emitted per mile traveled or per gallon of gasoline consumed.

This approach assumes that emission factors are approximately constant among like activities. Carefully conducted measurements of emissions, following approaches like those outlined in the last subsection, can be used to determine emission factors from a small number of representative activities and subsequently applied to predict emissions for a much larger set. If the emission factors are not constant, the approach can still be used if the factors can be easily related to another parameter. For example, the sulfur content of coal is used as a basis for estimating emission factors for SO_2 from coal combustion.

The U.S. Environmental Protection Agency maintains the standard reference for emission factors data in a series of documents known as AP-42, which are available electronically (www.epa.gov) as well as in paper form (USEPA, 1985). As an example, Table 7.B.3 presents emission factors for three fuels that are used in utility boilers to generate electricity. Example 7.B.1 illustrates the use of these data.

Table 7.B.3 Emission Factors and Heating Value Data for Selected Fuels Consumed in Large Boilers Used for Electricity Production

	Fuel		
Parameter	Bituminous coal[a] (pulverized, 1.8% S)	Residual fuel oil (no. 6, 2% S)	Natural gas[b]
Heating value[c]	24.2 kJ/g	41.7 MJ/L	38.3 MJ/m^3
	Emission factor[d]		
Particles	31 kg/Mg	2.9 kg/m^3	16–80 kg/10^6 m^3
SO_2	35 kg/Mg	38 kg/m^3	9.6 kg/10^6 m^3
NO_x (as NO_2)[e]	10.5 kg/Mg	8 kg/m^3	8800 kg/10^6 m^3
CO[f]	0.3 kg/Mg	0.6 kg/m^3	640 kg/10^6 m^3
Nonmethane organics	0.04 kg/Mg	0.09 kg/m^3	23 kg/10^6 m^3
Methane	0.015 kg/Mg	0.03 kg/m^3	4.8 kg/10^6 m^3

[a]Coal ash content assumed to be 6.2%, as fired. Data apply to "dry bottom" combustion.
[b]Natural gas emission factors apply for utility boilers ($>10^8$ J h^{-1} heat input).
[c]Heating value data from the U.S. Energy Information Administration (EIA, 1997).
[d]Assumes no emission control devices. Emission factor data from Buonicore and Davis (1992).
[e]Nitrogen oxide emission factors specified as if all NO_x were emitted as NO_2, even though the large majority is emitted as NO.
[f]Emissions for normal operating conditions. Values can be one or two orders of magnitude higher with improper operation or poor maintenance.

EXAMPLE 7.B.1 ***Nitrogen Oxide Emissions from Utility Boilers***

Using the data in Table 7.B.3, determine the nitrogen oxide (NO_x as NO_2) emissions from a utility boiler that burns (*a*) bituminous coal, (*b*) residual fuel oil, or (*c*) natural gas to produce electricity. Determine the emissions on the basis of grams emitted per kilowatt-hour of electricity produced. Assume that 10.9 MJ of thermal energy in fuel must be used to produce 1 kWh of electricity (implying a conversion efficiency of 33 percent).

SOLUTION First determine the mass or volume of fuel that must be consumed to produce 1 kWh in each case. For example, burning 450 g of bituminous coal produces 1 kWh:

$$\text{coal burned per kWh} = 10.9 \text{ MJ kWh}^{-1} \div 24.2 \text{ kJ g}^{-1} \times 1000 \text{ kJ MJ}^{-1} = 450 \text{ g/kWh}$$

Similarly, 0.261 L of fuel oil or 0.285 m^3 of natural gas must be burned to produce 1 kWh.

Nitrogen oxide emissions are determined by multiplying the fuel use by the emission factor. For coal, the result is 4.7 g NO_x per kWh:

$$NO_x \text{ emissions} = (10.5 \times 10^3 \text{ g per } 10^6 \text{ g fuel}) \times 450 \text{ g fuel/kWh} = 4.7 \text{ g } NO_x\text{/kWh}$$

Likewise, residual oil would produce 2.1 g NO_x per kWh, and natural gas would produce 2.5 g NO_x per kWh.

Emission Inventories

Emission inventories contain information on all significant sources of air pollutants in an air basin or a larger region. Accurate emission inventories are essential for developing air pollution control strategies. A comprehensive inventory contains information on the spatial and temporal distributions of pollutant emissions among individual source classes. Table 7.B.1 presents a summary of information contained in the U.S. nationwide emission inventory for some of the criteria air pollutants.

Emission sources may be broadly grouped into mobile (e.g., vehicular) and stationary sources. Stationary sources are commonly subdivided into large point sources (e.g., an industrial plant), area sources (e.g., drying paint), and biogenic sources (e.g., reactive organics emitted from vegetation). These source groupings are generally subdivided into other categories to permit better analysis and planning.

Much work to develop, maintain, and apply emission inventories occurs at the level of local air quality management districts. Information on pollutant emission factors is combined with a detailed accounting of the intensity of activities in the air basin to generate overall emissions estimates. Table 7.B.4 shows emissions data for one source category, basic petroleum refining, generated by the San Francisco Bay Area Air Quality Management District. The full inventory for this area contains more than a hundred such data sets.

7.B.2 Pollutant Generation by Combustion

As shown in Figure 7.B.2, combustion, whether for transportation or in a stationary source such as a factory or power plant, is a dominant source of carbon monoxide, nitrogen oxides, and sulfur dioxide. Combustion is also an important source of volatile organic gases that are precursors to photochemical smog and of airborne particulate matter, especially $PM_{2.5}$. Combustion can be an important source of many species regulated as hazardous air pollutants, such as formaldehyde, benzene, polycyclic organic

Table 7.B.4 Average Daily Pollutant Emissions for "Basic Petroleum Refining Processes" in the San Francisco Bay Area Air Quality Management District[a]

Species	Average emission rate (10^6 g d^{-1})
Particulate matter	1.59
Total organics[b]	0.14
Reactive organics[b]	0.10
Nitrogen oxides	8.8
Sulfur dioxide	28.1
Carbon monoxide	—

County factor		Year factor		Month factor		Day factor	
Alameda	0.00	1988	0.97	Jan	1.00	Sun	1.00
Contra Costa	0.60	1990	1.00	Feb	1.00	Mon	1.00
Marin	0.00	1992	1.02	Mar	1.00	Tue	1.00
Napa	0.00	1994	1.04	Apr	1.00	Wed	1.00
San Francisco	0.00	1996	1.07	May	1.00	Thu	1.00
San Mateo	0.00	1998	1.09	Jun	1.00	Fri	1.00
Santa Clara	0.00	2000	1.10	Jul	1.00	Sat	1.00
Solano	0.40	2002	1.10	Aug	1.00		
Sonoma	0.00	2004	1.10	Sep	1.00		
		2006	1.10	Oct	1.00		
		2008	1.10	Nov	1.00		
		2010	1.10	Dec	1.00		

[a]To obtain emissions for a specific county, year, month, or day, multiply the average emission rate from the upper table by the appropriate factor from the lower table.
[b]Reactive organics may contribute to urban photochemical smog; total organics include species considered nonreactive, such as methane.
Source: BAAQMD, 1993.

matter, polychlorinated biphenyls, cadmium, and mercury. Some HAPs are products of incomplete combustion. Others are emitted because they are trace elements in the fuel. In this section, we explore the factors that control the production of air pollutants in combustion.

Energy and Fuel Use in the United States

The types and quantities of fuel used in the United States are summarized in Figure 7.B.6. The bar on the left shows that most of the energy consumption (88 percent) is derived from burning three types of fossil fuels: coal, natural gas, and petroleum. The bar on the right shows that energy uses are approximately equally divided among three sectors in society: transportation, industrial use, and residential/commercial settings.

The largest use of coal is to produce electricity. Petroleum products are mainly used for transportation. Natural gas is used mostly for domestic and commercial purposes (e.g., space heating) or for industrial processes. These categories account for two-thirds of total energy conversion in the United States.

Pulverized coal combustion is a common method for burning coal in power plants. The coal is crushed to a powder with a characteristic particle diameter in the range 40–80 μm. The coal powder is dispersed in air and then blown into a furnace, where it is burned at a temperature of ~2000 K. The thermal energy liberated by burn-

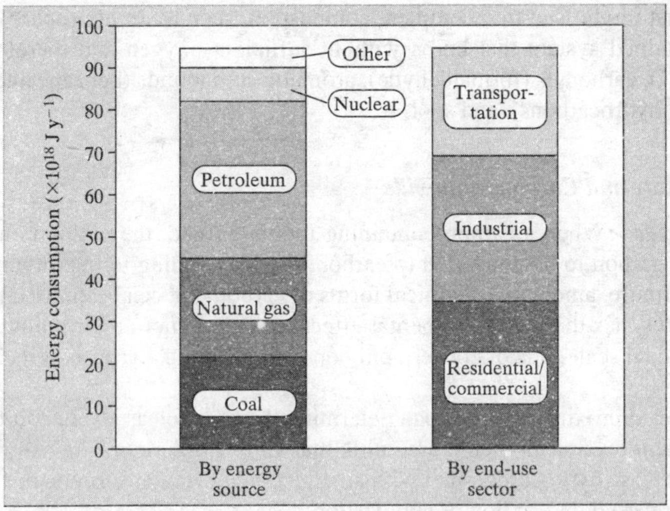

Figure 7.B.6 Overall energy and fuel use in the United States for 1996 (EIA, 1997).

ing the coal is used to convert water to steam, which drives a turbine to generate electricity. Important air pollution issues associated with coal combustion derive from impurities in the coal: Sulfur leads to the release of sulfur dioxide; trace metals and soil minerals lead to the release of particulate matter.

Petroleum for transportation is mainly burned as gasoline in spark ignition engines and as diesel fuel in compression ignition engines. Relative to coal, gasoline and diesel are clean fuels. The amounts of sulfur and incombustible minerals and metals in these fuels are relatively low, so emissions of sulfur dioxide and ash are generally not a problem. However, significant air pollution problems are associated with fuel combustion in motor vehicles (see Table 7.B.5). Emissions of CO, hydrocarbons, particulate soot, and nitrogen oxides are all serious concerns.

In comparison with pollutant generation by combustion of coal and petroleum, emissions from natural gas combustion are usually relatively low. Natural gas fuel contains low impurity levels. Significant concerns include nitrogen oxide emissions, for which this fuel offers negligible advantages relative to others. Also of concern are

Table 7.B.5 Some Characteristics of Internal Combustion Engines That Contribute to Combustion-Generated Air Pollutants

Characteristic	Consequence
Engines operate intermittently and under variable load	High CO and hydrocarbon emissions when cool and when under heavy load
Combustion occurs at peak pressures of 15–40 atm	Increased flame temperature caused by compression heating creates more NO_x
Combustion is followed by rapid expansion	Rapid cooling can freeze CO levels at high values
Small combustion chambers	Thermal quenching at walls causes incomplete combustion, increasing CO and hydrocarbon emissions
Cylinders are oil lined	Increased hydrocarbon emissions
Combustion occurs in ~10^8 engines (in United States)	Problems can arise from control device failures, poor tuning, or tampering in small fraction of fleet

conditions that might lead to incomplete combustion, such as an improperly tuned or poorly maintained system that burns without sufficient oxygen and therefore emits substantial CO, carbonyls (formaldehyde), aromatic compounds (benzene and polycyclic aromatic hydrocarbons), and soot.

Carbon Dioxide and Carbon Monoxide

Carbon Dioxide When a carbon-containing fuel is burned, the optimal outcome is for all of the carbon to be converted to carbon dioxide, leading to maximum heat release. Furthermore, among the chemical forms of carbon that can be emitted to the environment, CO_2 has the least detrimental effect, since it is not an air pollutant on an urban or regional scale. It is of concern only on a global scale, because of its influence on climate.

To a good approximation, one can determine the CO_2 release from carbonaceous fuel combustion by stoichiometry, assuming that all of the carbon is released as CO_2. Consider a generic hydrocarbon fuel with an effective molecular formula of C_nH_m and the following partial description of combustion:

$$C_nH_m + \text{air} \rightarrow n\, CO_2 + \cdots \tag{7.B.2}$$

Since the molecular weights of H, C, and CO_2 are 1, 12, and 44 g/mol, respectively, we can see that $44n/(12n + m)$ g of CO_2 is emitted per g of fuel burned.

The main option for reducing carbon dioxide emissions is to reduce the rate of fossil fuel consumption. In the short term, this goal is best achieved by reducing overall energy demand—for example, by improving energy efficiency—and by increasing the use of renewable energy sources, such as solar and wind power. Some benefit can also be realized by converting from high-carbon fossil fuels such as coal to lower-carbon fossil fuels such as natural gas (see Example 7.B.2). In the longer term, other approaches, such as the use of a hydrogen-based fuel system, may become practical.

For large-scale combustion processes, the possibility of capturing CO_2 from the exhaust gas and disposing of it (e.g., by injection into the deep ocean) is also being studied. Compared with particulate matter and SO_2, which can be successfully removed from flue gases, the mass emission rate of CO_2 is enormous. For example, consider coal combustion, where the mass fraction of C is 80 percent. Because of the addition of oxygen to C during combustion, the mass of CO_2 emitted would be three times the mass of coal burned! The capture, handling, and disposal of such a large mass of sequestered CO_2 pose enormous challenges.

EXAMPLE 7.B.2 ***Natural Gas and Global Warming***

The American Gas Association has advertised that burning natural gas substantially reduces emissions of CO_2 relative to burning other types of fuel. Determine whether this is true. Compute CO_2 emission factors (g CO_2 emitted per MJ of thermal energy released) for natural gas (NG) and bituminous coal (BC). Data on fuel composition is provided below.

NG composition	CH_4, 99.2%; N_2, 0.6%; CO_2, 0.2%
NG heating value	36.3×10^6 J m^{-3} (volume determined at 298 K, 1 atm)
BC composition (mass %)	C, 80.7%; H, 4.5%; N, 1.1%; O, 2.4%; S, 1.8%; moisture, 3.3%; ash, 6.2%
BC heating value	33.3×10^6 J kg^{-1}

EXAMPLE 7.B.2 ***Natural Gas and Global Warming (continued)***

SOLUTION Consider natural gas first. At 298 K and 1 atm, an ideal gas contains 40.9 mol m^{-3}. Therefore, the heating value of natural gas is 36.3 MJ m^{-3} $\div$ 40.9 mol m^{-3} = 0.89 MJ mol^{-1}. Combustion of 1 mol of NG liberates 0.994 mol of CO_2. Therefore, the emission factor for CO_2 from natural gas combustion is

$$E_{ng} = 0.994 \text{ mol (mol NG)}^{-1} \times 44 \text{ g/mol} \div 0.89 \text{ MJ (mol NG)}^{-1} = 49 \text{ g } CO_2 \text{ MJ}^{-1}$$

Each kg of coal contains 807 g of C and therefore its combustion liberates $807 \div 12 \times 44 = 2960$ g CO_2. The thermal energy release from burning 1 kg of coal is 33.3 MJ. Therefore, the emission factor is

$$E_{coal} = 2960 \div 33.3 = 89 \text{ g } CO_2 \text{ MJ}^{-1}$$

The advertisement is correct. Per unit of thermal energy, BC combustion emits about 80 percent more CO_2 than NG combustion. However, pay attention to one important caution. On a molar basis, methane is much more potent than CO_2 in influencing climate (see Table 7.A.6). A complete assessment of the climate impact of coal versus natural gas combustion should also consider the relative emissions of methane, for example, from leaks in the distribution and handling of the fuel.

Carbon Monoxide When a carbonaceous fuel is burned, the carbon is first oxidized to carbon monoxide before it becomes carbon dioxide. Two phenomena can interfere with the process, causing C to be emitted as CO rather than as CO_2. The more important case results from an improper mixture of fuel and air. If the combustion system burns with insufficient oxygen for complete oxidation (i.e., fuel rich), products of incomplete combustion are emitted. The second process, *quenching*, occurs when the flame is rapidly cooled. At high temperature, significant CO exists at equilibrium with CO_2, even with excess oxygen. As combustion products cool, the equilibrium condition shifts, strongly favoring CO_2. However, time is needed for CO to be oxidized to CO_2. If the combustion by-products are cooled too rapidly, the kinetics of oxidation rapidly slow and significant CO may be emitted.

We can use stoichiometry to make an upper-bound estimate on the emission rate of carbon monoxide from fuel-rich combustion. Assume that carbon is the only incompletely oxidized element in the exhaust and that no oxygen remains in the exhaust. For the combustion of a pure hydrocarbon, C_nH_m, in air, this assumption leads to the following stoichiometric balance (see also §3.D.3):

$$C_nH_m + \frac{1}{\phi}\left(n + \frac{m}{4}\right)(O_2 + 3.78\ N_2) \rightarrow b\ CO + (n - b)\ CO_2 + \frac{m}{2} H_2O + \frac{3.78}{\phi}\left(n + \frac{m}{4}\right) N_2 \tag{7.B.3}$$

Here ϕ is the equivalence ratio (>1 for fuel-rich conditions). The coefficient b represents an estimate of the moles of CO produced per mole of fuel burned and is determined by balancing O on both sides of the reaction:

$$b = 2\left(\frac{\phi - 1}{\phi}\right)\left(n + \frac{m}{4}\right) \tag{7.B.4}$$

So, for example, if methane (CH_4) is burned at an equivalence ratio of $\phi = 1.05$, the estimated CO production from this equation is 0.19 mol per mole of fuel, and the exhaust gas could contain as much as 0.19/10.2 = 1.9 percent CO. (Reaction 7.B.3 shows that 10.2 mol of exhaust are produced per mole of methane burned at $\phi = 1.05$.)

Stationary-source combustion processes are typically operated slightly fuel lean ($\phi < 1$) to ensure complete combustion of fuel. Modern automobiles are designed to run in stoichiometric balance ($\phi = 1$) to achieve proper operation of the three-way catalytic converter on the exhaust.

Sulfur Dioxide

Sulfur is a minor constituent of fossil fuels as they are extracted from the ground. Raw natural gas contains gaseous hydrogen sulfide; liquid and solid fuels may contain sulfur in organic compounds. Coal also contains inorganic sulfur, mainly as pyrite, FeS_2. When a sulfur-containing fuel is burned, almost all S is converted to SO_2.

There are three broad options for controlling SO_2 emissions: (*a*) remove S from the fuel before it is burned, (*b*) capture S during the combustion process, or (*c*) remove SO_2 from the flue gas. In this section, we will explore option *a*. Option *c* is discussed in §7.C. An example of option *b* is fluidized-bed combustion, in which gravel-sized pieces of coal are burned in a hot fluidized bed of limestone particles. The limestone surface, which is oxidized to CaO, reacts rapidly with SO_2 emitted from coal combustion to produce gypsum ($CaSO_4$), which would be disposed of in a landfill.

Hydrogen sulfide is routinely removed from natural gas prior to distribution by means of a gas absorber (de Nevers, 1995). The captured H_2S is then converted to elemental sulfur, which is used to manufacture sulfuric acid. For liquid fuels, the sulfur in organic compounds is converted to hydrogen sulfide through a catalytic reaction. The resulting gas stream can be treated as in the case of natural gas. With these technologies, the sulfur level in delivered fuels can be maintained below the regulatory limit of 4 ppm (by volume) for natural gas and 500 ppm (by mass) for diesel fuel (de Nevers, 1995).

Fuel cleaning can be applied to partially remove sulfur from coal. Inorganic S can be removed by sedimentation, based on the density difference between coal (1.1–1.3 g cm^{-3}) and pyrite (5 g cm^{-3}) (see §6.C.1 and Problem 6.20). Removing organic S from coal is technically feasible using solvent-refining techniques; however, this approach is more expensive than removing SO_2 from the flue gas. Because of these features of the fuels, coal combustion is the dominant contributor to ambient SO_2 emissions. Stoichiometry can be used to estimate uncontrolled SO_2 emissions, as shown in Example 7.B.3.

EXAMPLE 7.B.3 ***Sulfur Dioxide Emissions from Coal Combustion***

In 1996 the total coal consumption in the United States was 890 Tg (1 Tg = 10^{12} g = 1 million metric tons) (EIA, 1997). Estimate the total SO_2 emissions from coal combustion assuming that the average S content of coal as burned was 1.3 percent (de Nevers, 1995) and that the emissions were uncontrolled. Compare the result with total SO_2 emissions from stationary fuel combustion (Table 7.B.1).

SOLUTION The mass of S in the combusted coal is $0.013 \times 890 = 11.6$ Tg. If all S in coal were emitted as SO_2, the total mass emission rate of SO_2 from coal combustion would be $11.6 \times (64/32) = 23.2$ Tg y^{-1}, where the factor 64/32 represents the molecular weight ratio of SO_2 to S. Table 7.B.1 shows that SO_2 emissions from U.S. stationary

EXAMPLE 7.B.3 ***Sulfur Dioxide Emissions from Coal Combustion (continued)***

combustion sources are 14.2 Tg y^{-1}. This is of the same magnitude as, although moderately smaller than, our estimate, reflecting that some SO_2 emissions are controlled. Combustion of other types of fuel also contributes to SO_2 emissions.

Reducing SO_2 emissions to control acid deposition is an important feature of the Clean Air Act amendments enacted by the U.S. Congress in 1990. That legislation mandated a reduction of total emissions to 8.9 Tg y^{-1} by the year 2000. This target was to be achieved by a combination of applying stricter emission limits for existing large-scale coal-burning combustion systems and establishing a market for trading SO_2 emission rights.

Nitrogen Oxides

Globally, fossil fuel combustion is responsible for a significant portion of total nitrogen oxide (NO_x) emissions to the atmosphere. In the United States, emissions from large-scale industrial fuel consumption and from internal combustion engines used for transportation are approximately equal and account for more than 90 percent of anthropogenic NO_x emissions.

Most NO_x emissions, typically ~90 percent, occur in the form of nitric oxide (NO); the remainder are NO_2. Emission rates and emission factors are usually reported as though nitrogen oxides were all emitted as NO_2 because NO is oxidized to NO_2 fairly rapidly in the atmosphere and because NO_2 is the regulated pollutant.

Combustion-generated NO_x is divided into three types according to the origin of the nitrogen and the mechanism of formation. Figure 7.B.7 indicates the relative contributions of the three mechanisms for coal combustion.

Fuel NO_x refers to emissions caused by the presence of nitrogen in the fuel. Like sulfur, nitrogen is a significant minor constituent of coal and some fuel oils, typically present at a level ranging from 0.1 percent to a few percent. When a fuel that contains N is burned, depending on combustion conditions, the nitrogen may be converted into N_2 or may be oxidized and emitted as NO.

The nitrogen for *thermal* NO_x originates in the air mixed with the fuel. At high temperature, atomic oxygen radicals (important actors in the overall combustion process) can split the N_2 molecule, forming NO. Thermal NO_x formation is very sensitive to flame temperature but otherwise does not depend on the nature of the fuel. It is the dominant form of transportation NO_x emissions and an important form for stationary-source combustion, whether fueled by coal, oil, or natural gas.

Prompt NO_x is formed by a reaction between N_2 and the CH• radical, which occurs in fuel-rich regions of hydrocarbon-air flames. The resulting hydrogen cyanide (HCN) and nitrogen atoms are readily converted to NO if excess oxygen is encountered while the combustion gases remain hot. This mode is the least important of the three types of NO_x emissions.

The standard engineering approach to determine the rate of NO_x emission from combustion is to multiply an emission factor from AP-42 (USEPA, 1985) by the fuel consumption rate. The emission factors are usually based on limited experimental data and are reported as a single, best-estimate value for a given type of device and fuel. Combustion conditions can have a large effect on NO_x emissions, so the use of standard emission factors should be practiced with caution.

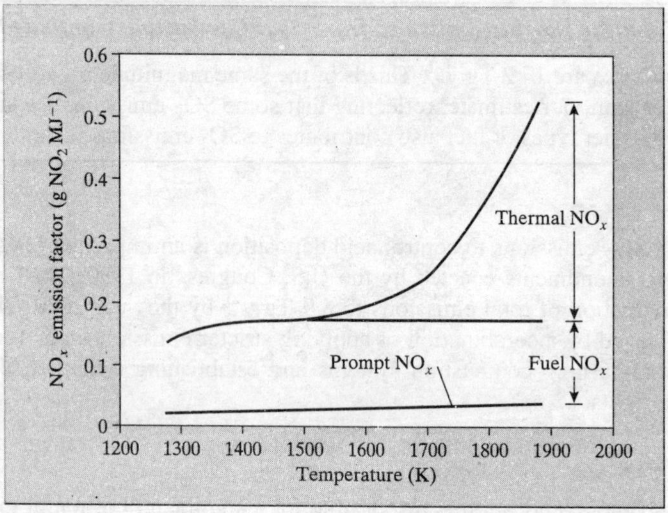

Figure 7.B.7 Schematic representation of NO_x emissions from coal combustion as a function of combustion temperature. Note that the data for NO_x emissions from bituminous coal listed in Table 7.B.3 correspond to an emission factor of 0.43 g NO_2 per MJ of thermal energy. (M. Hupa, R. Backman, and S. Boström, Nitrogen oxide emissions of boilers in Finland, *JAPCA: Journal of the Air Pollution Control Association*, **39**, 1496–1501, 1989. Reproduced with permission of the Air & Waste Management Association.)

Fuel NO_x Formation and Control Nitrogen, present as a trace element in some liquid and solid fuels, tends to be found in organic ring structures, either as pyridine (C_5NH_5) or pyrrole (C_4NH_5). Removing the N from fuel before burning is comparably difficult to removing the organic S from coal. High-molecular-weight liquid fuels and solid fuels tend to have the greatest amounts of nitrogen, with typical ranges of <0.4 percent for lightweight fuel oil, 0.3–2.2 percent for heavy petroleum distillates and residual oil, 1.1–1.7 percent for bituminous coal, and 0.25–2.5 percent for shale oil (Bowman, 1991). Organic nitrogen is not a significant constituent of natural gas or of gasoline.

Fuel N may be emitted either as NO or as N_2, depending on combustion conditions. Clearly, from an air pollution perspective, emission as N_2 is favorable and emission as NO is not. Manipulating combustion conditions to favor the conversion of fuel N to N_2 rather than to NO is a useful control approach.

The conversion of fuel N to NO is almost independent of temperature. In a flame, radicals attack the fuel nitrogen to form hydrogen cyanide (HCN). Additional fast reactions involving radicals maintain a pool of *fixed* nitrogen species (compounds that contain only one N atom) in an approximate equilibrium state during combustion. In addition to HCN, these species include N, NH, NH_2, NH_3, and NO. Reactions can then occur between two fixed-N species to form N_2. However. these reactions are relatively infrequent because they involve two fixed-N species that are not highly reactive and are present at low concentrations. The conversion of fixed-N species to N_2 can be encouraged by maintaining fuel-rich combustion conditions in part of the combustion zone so that the absolute concentrations of fixed-N species is relatively high. A system that uses this principle is *staged combustion* (see Figure 7.B.8). Here the

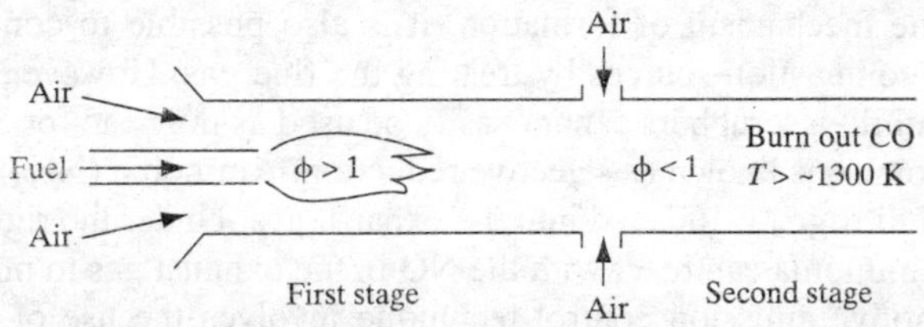

Figure 7.B.8 Schematic of a staged-combustion burner for controlling NO_x formation.

fuel is first burned under fuel-rich conditions, and then air is added to convert to fuel-lean conditions. In the fuel-rich region, fuel N can be converted to N_2. The fuel-lean zone completes the oxidation of C and H.

For industrial boilers that burn low-grade fuel with high fuel N content, fuel NO_x may dominate thermal NO_x. Combustion modifications such as staged combustion may yield 20–60 percent reductions in fuel NO_x relative to uncontrolled conditions. If this level of control is insufficient, fuel substitution or flue gas treatment may also be required.

Thermal NO_x Formation and Control Figure 7.B.7 shows that thermal NO_x formation depends strongly on temperature. Thermal NO_x is insignificant when peak combustion temperatures are below 1500 K but can become high for $T > 1800$ K. This phenomenon can be explained by the kinetics of thermal NO_x formation. The formation pathway, known as the Zeldovich mechanism, involves three reversible elementary reactions:

$$N_2 + O\bullet \rightarrow NO + N\bullet \qquad k_1 = 1.8 \times 10^8 \exp(-38{,}370/T)\ \mathrm{m^3\ mol^{-1}\ s^{-1}} \qquad (7.B.5)$$

$$N\bullet + O_2 \rightarrow NO + O\bullet \qquad k_2 = 1.8 \times 10^4 \exp(-4680/T)\ \mathrm{m^3\ mol^{-1}\ s^{-1}} \qquad (7.B.6)$$

$$N\bullet + OH \rightarrow NO + H\bullet \qquad k_3 = 7.1 \times 10^7 \exp(-450/T)\ \mathrm{m^3\ mol^{-1}\ s^{-1}} \qquad (7.B.7)$$

where T is the temperature in K. This pathway is limited by the rate of reaction 7.B.5 (Flagan and Seinfeld, 1988). The highly reactive oxygen radical fractures the strong N_2 bond to form one NO molecule. The second nitrogen atom then reacts rapidly by either reaction 7.B.6 or 7.B.7 to generate a second NO molecule. Since reaction 7.B.5 is rate limiting, the overall rate of NO production is just twice the rate of this reaction.

Because of the strong bond between the nitrogen atoms in N_2, the rate constant for reaction 7.B.5 is highly temperature dependent. For example, at a flame temperature of 2200 K the rate constant k_1 is 10 times higher than it is at 1950 K. In addition, the mole fraction of O• radicals in the flame increases significantly with increasing temperature. Thus, a moderate increase in flame temperature can lead to a much higher rate of thermal NO_x formation.

Because of its strong temperature dependence, a key approach to controlling thermal NO_x is to limit the peak flame temperature. A common method for achieving a lower flame temperature is to dilute the air-fuel mixture with an inert compound. The three main choices applied in practice are (1) air (i.e., reduce the equivalence ratio, ϕ), (2) recycled (and cooled) exhaust gas from the flue, and (3) steam or water.

In addition to these techniques, thermal NO_x control can also be achieved with staged combustion, as is used for fuel NO_x. Burning fuel under either fuel-rich or fuel-lean conditions leads to lower flame temperatures.

Independent of the mechanism of formation, it is also possible to control NO_x emissions from large combustion sources by treating the flue gas. However, because NO has low water solubility, scrubbers cannot easily be used as they can for SO_2. One method that shows promise is known as selective reduction. Ammonia (NH_3) or some other form of reduced nitrogen is injected into the exhaust gas. Under the right conditions, the nitrogen in ammonia can react with the NO in the exhaust gas to make N_2.

The major automotive emission control technique involves the use of catalytic converters to reduce NO to N_2.

Particulate Matter

Combustion-generated particles may be broadly classified according to their chemical composition and means of formation, as summarized in Table 7.B.6. Here we will focus on the generation of ash particles by pulverized coal combustion (for power generation and industrial processes) and on the generation of soot (e.g., from diesel engines).

Ash Particles from Pulverized Coal Combustion Ash is the dominant constituent of particulate mass generated by pulverized coal combustion. Most of this material is emitted as coarse particles that can be effectively removed by particle prefilters such as scrubbers or cyclones. A small portion of the mass is present as fine particles that may be collected only by high-efficiency control devices such as electrostatic precipitators or baghouse filters.

The composition of ash particles depends directly on the composition of mineral inclusions and incombustible elements in the fuel. Figure 7.B.9 shows data on the mean composition of ~100 coal samples. Nine relatively nontoxic mineral elements were present at levels of 0.05–2.5 percent. An additional 23 elements were found at levels of 0.2–300 ppm, including toxic metals such as arsenic, cadmium, chromium, lead, and nickel. Except for lead, coal combustion is a major source of each of these metals in the northeastern United States.

Detailed examination of the combustion process explains why both coarse and fine particles appear in the size distribution of coal ash. In pulverized coal, the ash is present as dispersed mineral inclusions with a characteristic size of ~1 μm. During combustion, the temperature rises to the point where the minerals melt. The molten mineral beads may coalesce on the surface of the char (burning coal particle), forming larger droplets. As the char continues to burn, the size of the pores grows and the original coal grain fractures into several pieces. The char then burns completely, leaving the coarse fly ash particles.

Table 7.B.6 Classification of Particles Generated in Combustion

Type	Description
Ash	Incombustible materials in the fuel
Soot	Carbonaceous particles formed by pyrolysis, a product of incomplete combustion of carbon-containing fuels
Char	Unburnt, carbonaceous, nonvolatile material in pulverized coal that burns more slowly than the volatiles and may escape combustion
Coke	Large (1–50 μm) porous carbonaceous shells that are formed from the spray droplets of fuel oil
Acid droplets	Originating from S in fuel and produced by a gas-to-particle conversion process involving sulfuric acid and water in the exhaust

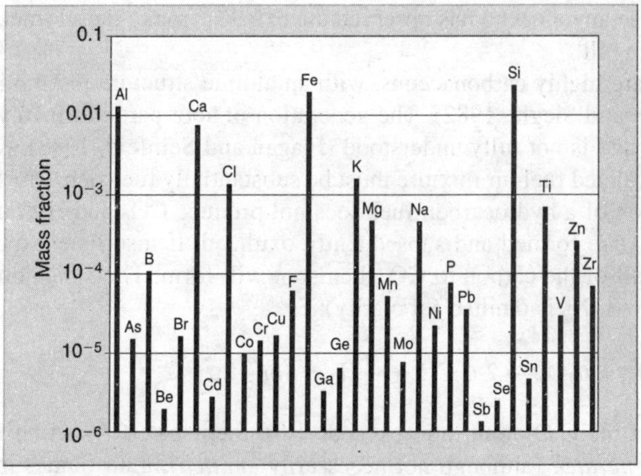

Figure 7.B.9 Average elemental composition of the incombustible portion of 101 samples of coal. The average total ash content was 11.4 percent. The remaining elements (and their average mass fractions) were C (0.703), O (0.087), H (0.050), S (0.033), and N (0.013). Data are from Flagan and Friedlander (1978).

The formation of fine particles probably involves a gas-to-particle conversion process. Volatile components of the ash, such as sodium, potassium, and arsenic, evaporate during combustion and condense as the exhaust gas cools to form new particles. Because of the manner in which the particles are formed, ash particle composition is not uniform with particle size. Many trace elements, having greater health consequences, are concentrated in fine particles, which are more difficult to remove from stack gases.

For pulverized coal, combustion conditions do not strongly influence the production rate of ash particles. A large fraction of the ash in the combusted coal is converted to fly ash that can be emitted as airborne particles. There are two main control techniques for reducing ash emissions from coal combustion. One method is coal cleaning: gravity separation of the crushed coal from minerals. This approach typically reduces the ash content of coal by about 50 percent at a cost of about 5 percent of the fuel value (Kilgroe, 1984). The second main option is to use control devices, such as cyclones, baghouse filters, or electrostatic precipitators, to remove particles from the stack gases (see §7.C.1).

More advanced coal combustion technologies are under development. Coal gasification is a technique that may produce significant energy efficiency and air quality benefits. In this process, coal is exposed to steam and oxygen at moderately high temperature and pressure. The carbon and hydrogen in coal are converted mainly to carbon monoxide and hydrogen gas (H_2). The sulfur is converted to H_2S. The gaseous fuel can then be cleaned before being burned in a combined-cycle gas turbine.

Soot Generation in Combustion In contrast to ash, whose formation is an inevitable result of mineral impurities in fuel, the formation of soot is very sensitive to combustion conditions. Soot is mostly emitted from small-scale combustion activities rather than in large-scale industrial processes. For example, diesel engine exhaust is the dominant source of soot generation in the Los Angeles air basin (Gray et al., 1986).

This will not surprise anyone who has observed the exhaust from a diesel truck or bus as it accelerates up a hill.

Soot particles are highly carbonaceous, with an atomic structure resembling that of graphite (Amann and Siegla, 1982). The generation of soot particles involves several complex steps and is not fully understood (Flagan and Seinfeld, 1988). One key factor is that the localized fuel-air mixture must be substantially fuel-rich. Mechanistically, the combustion of a hydrocarbon fuel does not produce CO_2 and H_2O directly. Instead, CO and H_2 are formed and subsequently oxidized. If insufficient oxygen is present to convert all of the carbon to CO, then soot will form. The stoichiometry is represented as follows (N_2 is omitted for clarity):

$$C_mH_n + \alpha\, O_2 \rightarrow 2\,\alpha\, CO + \frac{n}{2} H_2 + (m - 2\alpha)\ C_s \tag{7.B.8}$$

If $m > 2\,\alpha$ (i.e., if the C:O atom ratio exceeds 1.0), then soot (C_s) can be formed. Soot tends to be *generated* (although not necessarily *emitted*) in all flames in which the fuel and air are not premixed, because, inevitably, there are localized regions with high C:O ratios. However, the soot itself may burn as it moves from fuel-rich to fuel-lean regions.

In diesel engines the combustion of soot particles is quenched during the engine's expansion and exhaust strokes. Consequently, soot particles are emitted with the exhaust gas. Controlling diesel engine soot emissions is a challenging task. One approach is to operate with higher engine temperatures to promote more rapid combustion of the soot. Unfortunately, this would lead to an increase in thermal NO_x emissions, which is probably unacceptable. An alternative control approach involves the use of particle traps on the exhaust (DeKiep and Patterson, 1984).

7.B.3 Motor Vehicle Emissions

Motor vehicles are the dominant emission source of carbon monoxide, and an important source of nitrogen oxides and volatile organic compounds. In this section we will explore how these pollutants are emitted from motor vehicles and what can be done to control these emissions.*

Gasoline-fueled, spark ignition, four-stroke engines most commonly power passenger vehicles. Typically, on-board computers control spark timing, exhaust gas recirculation, and fuel injection to maintain stoichiometric combustion conditions. Figure 7.B.10 depicts a single cylinder. The typical engine speed is 3000 revolutions of the crankshaft per minute with two piston strokes generated per revolution. A complete combustion cycle is divided into four intervals or "strokes" during which the piston moves alternately in or out of the cylinder: (1) intake, (2) compression, (3) power, and (4) exhaust. During the intake stroke, the piston is withdrawn from the cylinder with the exhaust valve closed and the intake valve open. Air is drawn into the cylinder through the intake valve, and a dispersion of fine droplets of gasoline is injected into the cylinder to evaporate. The valves are then closed for the compression stroke. The piston is depressed into the cylinder, causing approximately adiabatic compression of the fuel-air mixture. The compression ratio, that is, the initial air volume in the cylinder divided by the minimum compressed-air volume, is typically about 10. Ignition occurs at the end of the compression stroke, induced by a surge of electrical current

*Note that in this section, we use gallons and miles rather than liters and kilometers for volume and length, respectively. This is done to be consistent with U.S. regulatory practice and current U.S. convention.

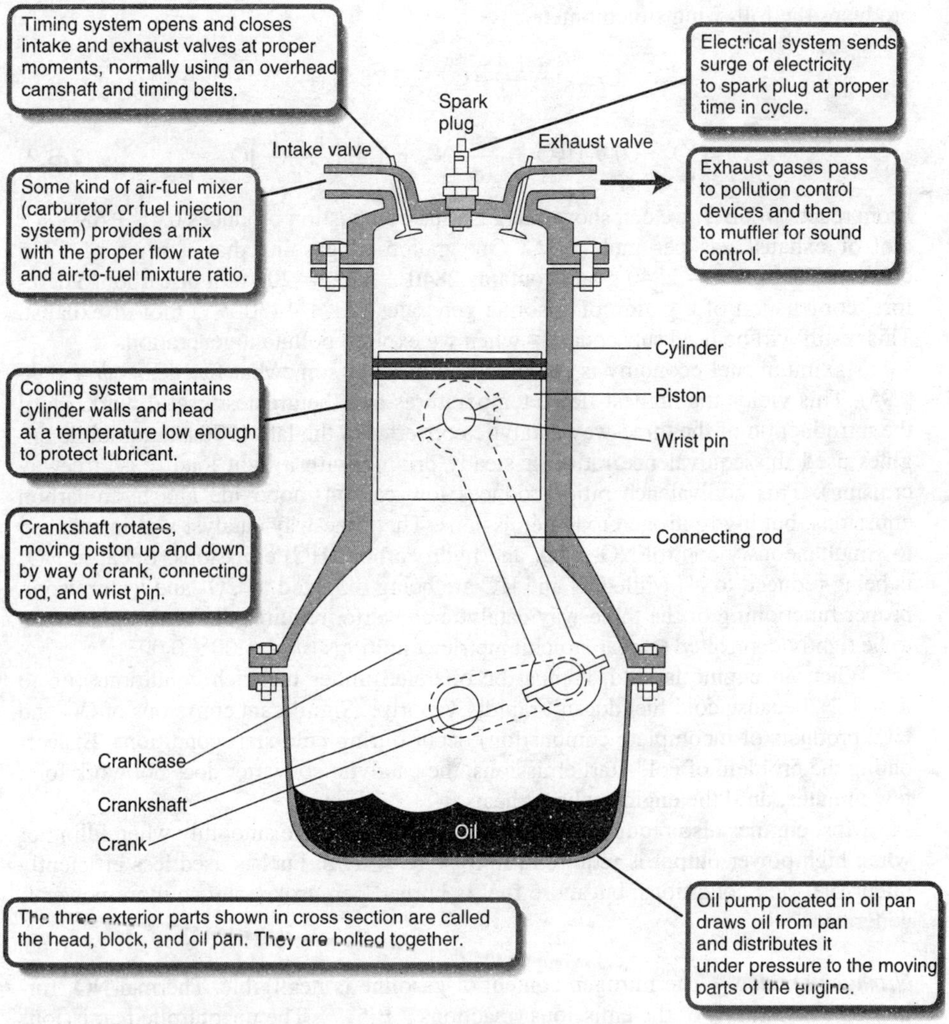

Figure 7.B.10 Simplified schematic of a single cylinder in a motor vehicle engine. (N. de Nevers, *Air Pollution Control Engineering*, McGraw-Hill, New York, 1995; reproduced with permission of The McGraw-Hill Companies.)

sent through the spark plug. As the flame propagates through the cylinder, the gas temperature rises to 2400–2800 K and the air pressure increases to a peak of 15–40 atm, driving the piston outward in the power stroke. The burnt fuel-air mixture cools because of adiabatic expansion. During the fourth stroke, the exhaust valve is opened, permitting further expansion and rapid cooling of the exhaust gases. The piston is pushed into the cylinder, causing most of the remaining exhaust gases to be discharged.

Gasoline is primarily a blend of hydrocarbon compounds, with an effective overall composition of $CH_{1.8}$ and a density of 0.75 g cm^{-3} (Singer and Harley, 1996).* Complete combustion in air with an equivalence ratio, ϕ, less than or equal to 1.0

*Reformulated gasoline includes a small percentage of oxygen (by mass).

produces the following stoichiometry:

$$CH_{1.8} + \frac{1.45}{\phi}(O_2 + 3.78\ N_2) \rightarrow$$

$$CO_2 + 0.9\ H_2O + \frac{5.48}{\phi}N_2 + 1.45\left(\frac{1-\phi}{\phi}\right)O_2 \qquad (7.B.9)$$

From reaction 7.B.9, we can show that gasoline combustion produces $0.45 + 6.93\ \phi^{-1}$ mol of exhaust gas per mole of C. One gallon of gasoline has a mass of 3785 $cm^3 \times 0.75\ g\ cm^{-3} = 2840$ g and contains $2840 \div 13.8 = 206$ mol of carbon. Therefore, combustion of a gallon of gasoline generates $(93 + 1430\ \phi^{-1})$ mol of exhaust. This result will be used subsequently when we explore pollutant generation.

Maximum fuel economy is promoted by burning somewhat fuel lean (ϕ = 0.9–0.95). This yields the highest flame temperatures and the greatest useful work. Until the introduction of the three-way catalytic converter in the late 1970s, automobile engines used this equivalence ratio for steady driving with a light load (e.g., freeway cruising). This equivalence ratio produces low carbon monoxide and hydrocarbon emissions, but high nitrogen oxide emissions. The three-way catalyst was introduced to simultaneously control NO_x, CO, and hydrocarbon (HC) emissions. Because NO_x is being reduced to N_2 while CO and HC are being oxidized to CO_2 and water vapor, proper functioning of the three-way catalytic converter requires the equivalence ratio to be tightly controlled to near stoichiometric conditions ($\phi = 1.00 \pm 0.003$).

When an engine is cold, it must be operated under fuel-rich conditions (up to $\phi = 1.2$), because cold fuel doesn't rapidly vaporize. Significant emissions of CO and HC (products of incomplete combustion) occur during cold-start conditions. Exacerbating the problem of cold-start emissions, the catalytic converter does not work for a few minutes, until the engine exhaust heats it.

Most engines also require a fuel-rich mixture to operate smoothly when idling or when high power output is required ($1.05 < \phi < 1.2$). Fuel is used less efficiently under fuel-rich conditions, but more fuel is burned per stroke, and so more power is generated.

Nitrogen Oxides The nitrogen content of gasoline is negligible. Thermal NO_x formation causes most of the emissions (reactions 7.B.5–7). The uncontrolled emissions of NO_x from automobiles in the late 1960s and early 1970s were about 5 g (as NO_2) per mile traveled (Calvert et al., 1993). The fleet-averaged fuel economy for passenger cars in 1970 was 13.5 miles per gallon (EIA, 1997). Assuming that combustion occurred at an equivalence ratio of 0.95, a car from that period would emit about 118 total moles of exhaust (reaction 7.B.9) and 0.11 mol of NO_x (5 g ÷ 46 g/mol) per mile traveled. This result implies that the uncontrolled exhaust contained an average NO_x mole fraction of about 900 ppm.

The current new-car emission standard for NO_x is 0.4 g (as NO_2) per mile, to be met for 5 years or 50,000 miles, and 0.6 g (as NO_2) per mile, to be met for 10 years or 100,000 miles. Assuming a fuel efficiency of 25 miles traveled per gallon of gasoline and stoichiometric combustion ($\phi = 1.0$), then 61 mol of exhaust are generated per mile traveled. The 0.4 g/mi limit means that only 0.0087 mol of NO_x can be emitted per mile traveled, so the average mole fraction in the exhaust must be limited to ~140 ppm (i.e., only about 15 percent of the 1970 uncontrolled level). Two main approaches are used to control NO_x emissions from motor vehicles: (1) limiting flame temperature through exhaust gas recirculation (EGR) and management of spark timing, and (2) exhaust gas treatment using a catalytic converter.

Carbon Monoxide Quench and fuel-rich combustion conditions contribute to CO emissions from motor vehicles. Under normal operating conditions with $\phi = 1.0$, there is substantial CO formed at the high equilibrium flame temperatures. The peak equilibrium mole fraction of CO in the cylinders can exceed 1 percent. As the cylinder expands during the power stroke, the exhaust gases cool and the equilibrium CO content diminishes. However, the rates of reaction slow as the gases cool. When the exhaust valve is opened, the exhaust gas expands and rapidly cools, freezing the CO content typically at several thousand ppm.

When the engine operates under fuel-rich conditions, CO levels in the exhaust can be very high. As described in §7.B.2, we can use stoichiometry to estimate the CO content of the exhaust under fuel-rich conditions if we assume that C is the only incompletely oxidized element and that no O_2 is present in the exhaust. Then the stoichiometry becomes

$$CH_{1.8} + \frac{1.45}{\phi}(O_2 + 3.78\ N_2) \rightarrow (1-x)\ CO_2 + x\ CO + 0.9\ H_2O + \frac{5.48}{\phi} N_2 \qquad \phi > 1 \tag{7.B.10}$$

The parameter x is determined by applying a mole balance on oxygen:

$$x = 2.9\left(\frac{\phi - 1}{\phi}\right) \tag{7.B.11}$$

The mole fraction of CO in the exhaust that results from fuel-rich, incomplete combustion is estimated to be

$$Y_{CO} = \frac{2.9(\phi - 1)}{1.9\phi + 5.48} \tag{7.B.12}$$

According to this expression, the carbon monoxide mole fraction in the exhaust is 3900 ppm for $\phi = 1.01$, 1.9 percent for $\phi = 1.05$, and 7.5 percent for $\phi = 1.2$. These values are much larger than the mole fraction of CO expected from quenching under stoichiometric or fuel-lean combustion conditions. Clearly, then, the frequency and extent of fuel-rich combustion conditions is an important factor influencing automotive CO emissions.

Uncontrolled CO emissions from automobiles in the late 1960s were approximately 84 g per mile traveled (see Table 7.B.7). Applying a fleet-average fuel economy

Table 7.B.7 U.S. Federal Emission Standards (1994 Model Year and Later) and Uncontrolled Pollutant Emissions for Light-Duty Motor Vehicles

	Emission factor	
Pollutant	Federal standard	Pre-control (~1965)
Hydrocarbons	0.25 g/mi	10.6 g/mi
Evaporative hydrocarbons	2 g/test	47 g/test
CO	3.4 g/mi	84 g/mi
NO_x (as NO_2)	0.4 g/mi	4.1 g/mi
Particles	0.2 g/mi	
Formaldehyde	0.015 g/mi	

Source: Calvert et al., 1993.

of 13.5 miles per gallon, a car from that period would emit about 118 total moles of exhaust and 3 moles of CO (84 g ÷ 28 g/mol) per mile, implying an average exhaust mole fraction of CO of about 2.5 percent.

The current emission standard for CO for new cars is 3.4 g (= 0.12 mol) per mile. Assuming a current fuel efficiency of 25 miles per gallon and combustion under stoichiometric conditions, the average mole fraction of CO in the exhaust must be controlled to 0.12 mol CO per mile ÷ 61 mol exhaust per mile = 0.2 percent (2000 ppm).

Carbon monoxide emission controls are achieved through a combination of three methods: (1) refined engine design and combustion management (via on-board computers) to ensure complete combustion; (2) fuel reformulation, through the addition of oxygenated organic compounds to promote more complete combustion; and (3) exhaust gas treatment with a catalytic converter.

Carbon monoxide emissions vary substantially within the vehicle fleet. Figure 7.B.11 shows CO emission data measured by remote sensing (see Figure 7.B.5) for on-road passenger vehicles. Expressed on a gram per mile basis, the CO emission factor for 15-year-old cars is more than 10 times the corresponding value for late-model vehicles: 45 g/mi in 1975 versus 3.7 g/mi in 1990.* Older cars emit more CO because emissions control technology has improved and also because, as vehicles age, the effectiveness of emission controls diminishes.

In addition to varying with vehicle age, emissions are highly variable among different vehicles of the same age. Remote sensing measurements have shown that the 10 percent of the vehicle fleet with the highest emissions are responsible for about 50 percent of the total motor vehicle emissions (Zhang et al., 1995). Although, on average, emissions are higher for older vehicles, the lowest-emitting 15-year-old cars emit less than the highest-emitting new cars. A comprehensive strategy for controlling air pollutants from motor vehicles must consider how to find and fix the high emitters.

Hydrocarbons Many mechanisms contribute to the emission of unburned hydrocarbons and partially oxidized organic compounds from automobiles. For example,

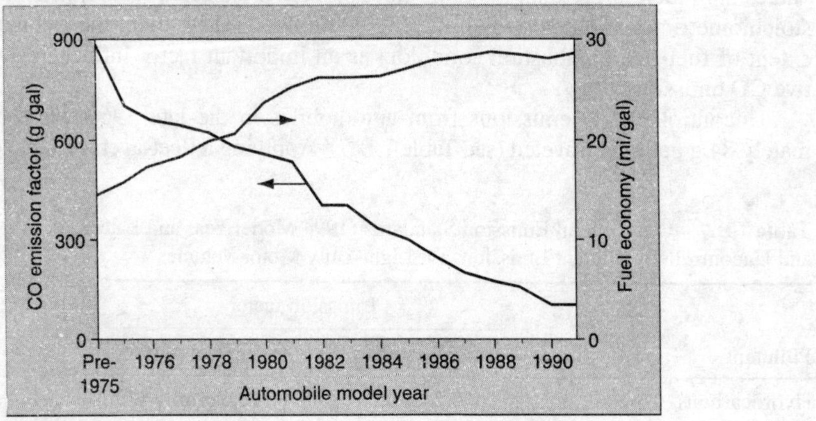

Figure 7.B.11 Mean carbon monoxide emission factors and fuel economy as measured by remote sensing of >70,000 automobiles in on-road driving conditions in Southern California (Singer and Harley, 1996).

*Divide the CO emission factor in g/gal by the fuel economy in mi/gal.

flame quenching at the cylinder walls and in the crevice between the piston and the cylinder can reduce the rate of combustion reactions and lead to the emission of products of incomplete combustion.

Similarly, leaks from the cylinder can lead to the emission of unburned gasoline vapors. "Blowby," which refers to leaks that occur between the piston ring and the cylinder walls, causes hydrocarbon vapors to enter the crankcase. Because of mechanical wear, blowby increases as the engine ages. It is effectively controlled by positive crankcase ventilation, which recycles crankcase vapors into the intake manifold. Leaky exhaust valves are another source of emissions of unburned fuel vapors. Such leaks can develop because of valve warping at high temperature, or because of particle deposits on valves or valve seats.

Similar to CO, significant hydrocarbon emissions can occur when the engine is started cold and the fuel-air mixture is rich. Cold-start emissions are especially significant because the catalytic converter does not work on the exhaust when cold.

Gasoline emissions can also occur during fuel handling and storage. For example, ambient temperature fluctuation causes the fuel tank to "breathe." As air warms, the vapors in the gas tank expand and constitute an emission source if allowed to escape untreated. The fuel tank is vented through a charcoal canister to control this problem. Gasoline vapors are also displaced when fuel tanks are filled. Vapor recovery systems are now common on dispensers at gasoline filling stations so that displaced vapors are actively drawn back into the storage tank as gasoline is pumped into the car's fuel tank.

Overall, several steps have been taken to achieve reductions in hydrocarbon emissions from motor vehicles. Engine design and combustion management systems (via computer control) have been improved to ensure more complete combustion. The replacement of the carburetor with a fuel injection system reduces evaporative emissions and allows better control of combustion stoichiometry. In some areas fuels have been reformulated to reduce emissions, for example, by reducing the vapor pressure. The use of on-board activated-carbon canisters helps reduce evaporative emissions from fuel tanks. Vapor capture devices at refueling stations reduce displacement losses. In addition, cars are equipped with catalytic converters for exhaust gas treatment. The net effect of these technologies is marked improvement in the emissions of hydrocarbons from motor vehicles.

Summary of Automotive Emission Control Strategies Automobile emissions are much less now, on a per-mile basis, than they were 25–30 years ago, before emissions were controlled. Standards for new gasoline-powered internal combustion engines permit only about 10 percent of the NO_x, 4 percent of the CO, and about 2 percent of the tailpipe hydrocarbon emissions that were prevalent in the 1960s (see Table 7.B.7). However, air quality standards are not being met in many areas of the United States. Consequently, automotive emissions are being subjected to further scrutiny, especially to reduce urban CO and ozone concentrations. Current efforts toward additional automotive emission reductions emphasize (a) gasoline reformulation, (b) the introduction of low-emission and zero-emission vehicles into the automotive fleet, and (c) inspection and maintenance of used vehicles.

For example, beginning in 1996, California has required the use of reformulated gasoline to reduce automotive emissions. Reformulated fuel has these characteristics:

- Reduced sulfur content (from ~ 150 to an upper limit of 40 ppm)
- Reduced vapor pressure (from ~ 8.5 psi at 100 °F, to an upper limit of 7.0 psi)

- Lower benzene content (because benzene is a hazardous air pollutant)
- Lower aromatic and olefin content (to reduce atmospheric reactivity of emitted hydrocarbons)
- Increased oxygen content of the fuel (to promote more complete combustion)

Currently, the most practical "zero-emission vehicles" (ZEVs) are electric cars, based on lead-acid battery technology. There are a host of environmental questions about widespread us of electric vehicles that have not yet been resolved. Among them are the environmental impacts of electricity generation to charge the batteries and of increased production of lead-acid batteries and their ultimate disposal. Low-emission vehicles include gasoline-electric hybrids that take advantage of energy-efficient combinations of technologies. Fuel cells are another technology that shows promise for sizable future reductions in automotive emissions. Rather than burning a fuel in an engine, fuel cells generate electrical current by chemically combining fuel and air at low temperature.

Inspection and maintenance ("smog check") programs have been a feature of automotive emission control for decades. The persistent existence of "superemitters" suggests that current practice is ineffective. Governments are exploring ways to implement more effective inspection and maintenance programs, including programs that encourage citizens to anonymously report observed high emitters.

Given the need to reduce emissions from the automotive fleet, a central policy question is whether we should focus on further reductions in new car emissions or instead emphasize maintaining low emission levels over the entire life cycle of vehicles. Since a small fraction of the vehicle fleet is responsible for a large proportion of total motor vehicle emissions, it seems logical that we should focus more on the latter. If automobiles emitted contaminants at or below the federal standards over their lifetimes, their contributions to air pollution would be much smaller. Motor vehicles would change from a dominant contributor to only a moderate contributor to total air pollutant emissions.

7.C TREATMENT TECHNOLOGIES

In §7.A many classes of air pollution problems were explored. In §7.B it was argued that solutions to these problems usually must focus on reducing emissions from sources. Broadly, options for reducing emissions can be divided into two categories: *pollution prevention* and *treatment technologies*.

In air pollution control, pollution prevention refers to any effort to prevent or minimize the generation of gaseous waste streams. At industrial facilities, for example, pollution prevention activities include changes in manufacturing processes, improved maintenance, and altered methods for storing volatile solvents. Pollution prevention should always be considered a more desirable control option than treatment (Freeman et al., 1992). However, even when all reasonable pollution prevention steps have been adopted, emission control measures often must include the application of treatment technologies. For various reasons, an air stream may contain undesirable levels of gaseous or particulate pollutants. Treatment technologies can be used to remove the pollutant from the gas stream or to change its chemical or physical form to make it less harmful.

As in the case of water and wastewater treatment, air pollution treatment technologies can be applied both to treat ambient air before use and to treat air after use, before it is released back to the environment (see §6.B). When viewed in this way, the dominant difference between air treatment and water/wastewater treatment is that air is present everywhere in the terrestrial environment. It is not necessary to collect, store, and convey air from environmental sources in the same way that we must collect, store, and convey water.

Treatment technologies are applied to improve the quality of an air stream before use. For example, high-efficiency filtration is applied to remove particles from air for clean-room manufacturing of electronic circuits (Cooper, 1986). Particle and gas filters are employed to minimize air pollution hazards for works of art displayed in museums (Baer and Banks, 1985). Automobiles are an example of a system in which air is treated both before and after use. To protect the engine from damage, the air used for combustion first passes through a filter. The objective of this filter is to remove coarse particles, such as soil dust, that could cause mechanical damage to the engine. After the air passes through the engine, the exhaust gases are subject to treatment by a catalytic converter to reduce the amounts of CO, nitrogen oxides, and unburned hydrocarbons discharged to the atmosphere.

This section briefly explores properties of air treatment technologies. Tables 7.C.1 and 7.C.2 summarize technologies that are used for controlling particulate and gaseous pollutants, respectively. A common feature of all these techniques is that they are applied only when the air is confined in some manner, most commonly while flowing through a flue or duct. Selected control techniques are described more substantially in §7.C.1 (cyclone, electrostatic precipitator, and fabric/fibrous filters) and §7.C.2 (absorption).

7.C.1 Particle Control Devices

Particle removal from air streams is often needed. Large coal-fired boilers require particle control equipment to remove fly ash from the exhaust gas. Particles must be filtered from air streams associated with certain industrial processes such as spraying and grinding. Municipal, medical, and hazardous waste incinerators generate exhaust gas with significant particle levels. In electronics and pharmaceuticals manufacturing, ventilation supply air is commonly passed through high-efficiency particle filters to establish particle-free environments.

Table 7.C.1 Control Devices for Capturing Particulate Air Pollutants

Device	Particle size	Collection mechanism and application
Settling chamber	>~20 μm	Separates particles from a gas stream by gravity; used to treat very dirty air streams that contain very coarse particles
Cyclone	>~1 μm	Separates particles by inertia in a vortex flow; common pretreatment process ahead of electrostatic precipitator or fabric filter
Scrubber	>~1 μm	Wet collector; induces collisions between particles and water droplets to remove particles from gas stream by inertia; may be used for combined collection of particles and water-soluble gases
Electrostatic precipitator	All	Creates electrostatic charge on particles so they can be removed by an electric field; high-efficiency device that is used to treat stack gases in industrial processes
Filter	All	Air flow is forced through matrix of fibers, capturing particles by a combination of Brownian motion, physical straining, interception, and impaction; high efficiency possible; applied for treating waste gases and for removing particles from air before use

Table 7.C.2 Treatment Technologies for Gaseous Air Pollutants

Technology	Pollutants[a]	Description and comment
Absorption	H_2S, SO_2, HCl, VOCs	A spray scrubber or packed column maintains a high gas-liquid contact area; especially effective for water-soluble species that can be converted to nonhazardous form in water
Adsorption	VOCs	Contact is promoted between gas and granular sorbent material, such as activated carbon, so that pollutant molecules adhere to surfaces; often the method of choice for controlling nonpolar organics; can be effective when low trace levels of contamination (ppb–ppm) must be achieved; effective in processing large air volumes with dilute contaminants
Incineration	VOCs	Waste gases are burned to convert H to H_2O, C to CO_2; commonly applied for low to medium levels of contamination with pure hydrocarbons or oxygenated organics
Catalytic redox	NO, CO, VOCs	Solid catalyst is used to increase rate of reaction and convert elements to less hazardous forms; common application is the three-way catalyst used in motor vehicles
Condensation	VOCs	Phase change from gas to liquid is caused either by cooling or by increasing pressure; requires high gas-phase concentration of species with significant recovery value and high boiling point; cannot achieve very low gas-phase concentrations, so sometimes used as pretreatment technique
Membrane recovery	VOCs	Organic vapors are separated from air by flowing gas past membranes that are more permeable to organics than to air; advanced, newly emerging technology

[a]Illustrative examples rather than an exhaustive list.

Almost all particle control devices operate on the same basic principle: Use body forces or physical obstacles to cause particles to collide with and adhere to solid or liquid surfaces, removing them from a gas stream. The physical mechanisms that cause collisions are mainly inertial drift, electrostatic drift, Brownian motion, and interception.

Table 7.C.1 summarizes five types of particle removal devices that may be used to clean gas streams. The following general statements apply. Settling chambers are inexpensive to build and operate, but they are effective only on very large particles. In most practical circumstances, airflow rates are large. Consequently, only short residence times (seconds) are possible, greatly limiting the effectiveness of a gravitational settler for airborne particle removal. Scrubbers and cyclones are routinely designed to work well against all coarse-mode particles (diameter $\geq 2\ \mu m$) and may be effective for collecting particles as small as 0.5 μm. They are ineffective, however, against smaller particles. They are commonly used as prefilters, upstream of electrostatic precipitators or fabric/fiber filters. Electrostatic precipitators (ESPs) and fabric/fiber filters can achieve very high efficiencies over the full range of particle sizes. ESPs cause particles to become electrically charged and then collect them on metal plates using strong electric fields. Fabric/fiber filters use Brownian diffusion, physical straining, interception, and impaction to remove particles from gas streams. Both classes of devices exhibit minimum collection efficiency in the accumulation mode, typically in the vicinity of 0.3 μm diameter. Particles that are smaller and larger than this size are collected more efficiently.

Characterizing Particle Control Efficiency

The efficiency of a control device treating a fluid stream is commonly defined as the fraction of pollutants removed from the fluid that flows through the device. If the rate of pollutant entry into a device is $Q \times C_{\text{in}}$ and the rate of pollutant flow out of a device is $Q \times C_{\text{out}}$, then by mass conservation, the rate of pollutant removal by the device is $Q \times (C_{\text{in}} - C_{\text{out}})$. Therefore, the efficiency, η, is related to the inlet and outlet concentrations by the expression

$$\eta = \frac{Q \times (C_{\text{in}} - C_{\text{out}})}{Q \times C_{\text{in}}} = 1 - \frac{C_{\text{out}}}{C_{\text{in}}} \tag{7.C.1}$$

Conversely, knowing the pollutant removal efficiency and the inlet pollutant concentration, we can predict the outlet concentration by rearranging equation 7.C.1:

$$C_{\text{out}} = (1 - \eta)C_{\text{in}} \tag{7.C.2}$$

Generally, for particle control devices, the removal efficiency, η, depends strongly on particle size. Therefore, to predict performance, it is important to know the removal efficiency as a function of particle diameter, $\eta(d_p)$. Given this and information on the size of particles in the untreated exhaust gas, the overall particle mass removal efficiency can be computed.

Caution must be exercised to properly present and interpret the removal efficiency of particle control devices. In general, coarse-particle removal is far easier to achieve than fine-particle removal. Because of this, the total mass removal efficiency may overstate the effectiveness of the control device in reducing the health risk associated with particle emissions. For example, as discussed in §7.B.2, toxic metals emitted from coal combustion tend to be concentrated in fine particles while a large fraction of the emitted particle mass is associated with coarse particles. Consequently, particle control devices applied to coal combustion could have a very high total mass removal efficiency and only a low to moderate removal efficiency for toxic metals.

Cyclones

Cyclones are widely used to remove coarse particles from gas streams. They exploit the inertia of particles to separate them from air. Relative to gravity settlers, cyclones achieve much better collection efficiencies for small coarse particles (down to about 2 μm) with smaller device volumes. Cyclones are often used upstream of fine-particle control devices, such as electrostatic precipitators or fabric filters.

The strengths of a cyclone include simple design and maintenance, a small floor area requirement, low to moderate pressure drop (typically 500–2500 Pa), and the ability to handle high particle loading rates. The key limitations are their ineffectiveness against small particles and the sensitivity of performance to airflow rate.

Figure 7.C.1 is a schematic diagram of a cyclone. In this design, air introduced at the top, with velocity V_i, swirls in an outer vortex around the perimeter and downward into the cone, then in an inner vortex up and out the top port. Inertia causes particles to drift toward the outer walls of the cyclone. This inertial drift is opposed by the drag exerted by air on the particles. Particles that reach the near vicinity of the outer wall settle under the influence of gravity and are collected at the bottom.

For industrial air cleaning, typical body diameters, D, are in the range 10 cm to 1 m; all other dimensions scale in proportion to the body diameter. The volumetric flow rate through a single unit can be up to a few cubic meters per second. Variations on the conventional design have also been developed. High-efficiency units achieve better

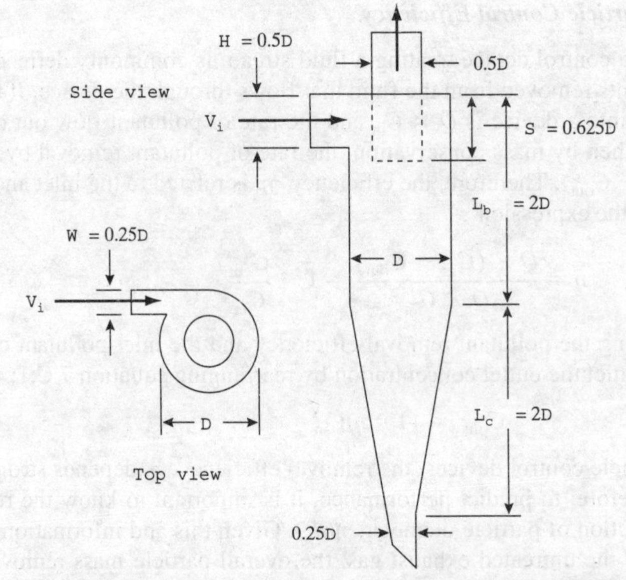

Figure 7.C.1 Schematic diagram of a cyclone of conventional design (Lapple, 1951). All dimensions are scaled to the body diameter, *D*.

performance at the expense of greater pressure drop. Conversely, high-throughput devices sacrifice efficiency but require a lower pressure difference to operate.

Particle removal in a cyclone is conceptually similar to that in a settling chamber (see §6.C.1). However, in place of gravity, centrifugal acceleration acts on the particles, causing them to drift outward. The magnitude of the centrifugal acceleration is v_θ^2/R, where v_θ is the tangential particle velocity and R is the radial distance of the particle from the axis of the cyclone. Commonly, the centrifugal acceleration is much greater than gravitational acceleration (i.e., $v_\theta^2/R \gg g$), which explains why cyclones can be much more efficient than gravity settlers.

The "cut-point diameter," d_{cut}, is defined as the particle size at which the collection efficiency is 50 percent. For larger particles, the efficiency will be greater; and for smaller particles, the efficiency will be smaller. Let's estimate the dependence of d_{cut} on the geometry and operating parameters of the cyclone. For the situation depicted in Figure 7.C.1, the residence time of gas in the outer vortex can be estimated as the ratio of the path length divided by the air speed:

$$\Theta \sim \frac{\pi D N_e}{V_i} \tag{7.C.3}$$

where N_e is the number of turns that the air makes while in the outer vortex and V_i is the air speed. The radial drift velocity of a particle in the outward direction is obtained by balancing drag force with the centrifugal force:

$$F_D \sim \frac{m V_i^2}{R} \tag{7.C.4}$$

where F_D ($= 3\pi\mu d_p v_r$ according to Stokes's law; see §4.B.1) is the drag force on the particle, v_r is the radial drift velocity of the particle, and m is the particle mass. We have approximated the tangential velocity in the vortex by the inlet velocity, V_i. We approximate the radial distance of the particle from the cyclone axis as $R = D/2$. Substituting for particle mass, m, in terms of particle diameter and rearranging yields this estimate for the drift velocity:

$$v_r \sim \frac{V_i^2 \rho_p d_p^2}{9 D \mu} \tag{7.C.5}$$

where ρ_p is the particle density. At the cut-point particle size, 50 percent of the particles will be collected. To achieve this removal effectiveness, particles would have to travel a radial distance of $\sim W/2$ (where W is the width of the outer vortex) during their residence time (Θ) in the outer vortex. This condition provides a basis for estimating the cut-point diameter:

$$\frac{V_i^2 \rho_p {d_{\text{cut}}}^2}{9 D \mu} \sim \frac{W}{2\Theta} \tag{7.C.6}$$

Substituting for Θ from 7.C.3 and solving for d_{cut} yields

$$d_{\text{cut}} \sim \left(\frac{9}{2\pi} \frac{\mu W}{\rho_p N_e V_i} \right)^{1/2} \tag{7.C.7}$$

The number of turns in the outer vortex is estimated as

$$N_e \sim \frac{1}{H}\left(L_b + \frac{L_c}{2} \right) \tag{7.C.8}$$

where the geometric parameters H, L_b, and L_c are, respectively, the height of the inlet, the length of the body, and the length of the cone region, as depicted in Figure 7.C.1. Although the procedure used to derive equation 7.C.7 was approximate, the equation has proven to be reasonably accurate (Lapple, 1951). Using the conventional cyclone dimensions of $H = 0.5D$, $L_b = L_c = 2D$, we find $N_e \sim 6$. An industrial cyclone might typically have $D = 0.5$ m, so $W = 0.125$ m $= 12.5$ cm, and $V_i = 20$ m s^{-1} = 2000 cm s^{-1}. Substituting into equation 7.C.7, with $\mu = 1.8 \times 10^{-4}$ g cm^{-1} s^{-1} and $\rho_p = 2.5$ g cm^{-3}, we obtain an estimated cut-point diameter of $d_{\text{cut}} \sim 3.3$ μm. Since the cut-point diameter varies weakly with cyclone diameter ($d_{\text{cut}} \sim D^{1/2}$) and with inlet air velocity ($d_{\text{cut}} \sim V_i^{-1/2}$), reducing d_{cut} to a value less than a micron would require large changes in design or operating parameters.

Given d_{cut} for a particular particle size and cyclone design, the efficiency for other particle sizes is well described by the following empirical relationship (see Figure 7.C.2):

$$\eta = \frac{(d_p/d_{\text{cut}})^2}{1 + (d_p/d_{\text{cut}})^2} \tag{7.C.9}$$

Electrostatic Precipitators

Electrostatic precipitators (ESPs) compete with fabric filters in large-scale industrial air-cleaning processes. Both technologies can achieve very high removal efficiencies, even for submicron particles. Relative to fabric filters, ESPs have small pressure drops (~100 Pa), which reduces operating costs, and ESPs can be used to treat high-temperature gases. Disadvantages include high initial cost, performance degradation when used on particles with low electrical conductivity, and potential safety hazards from high voltages.

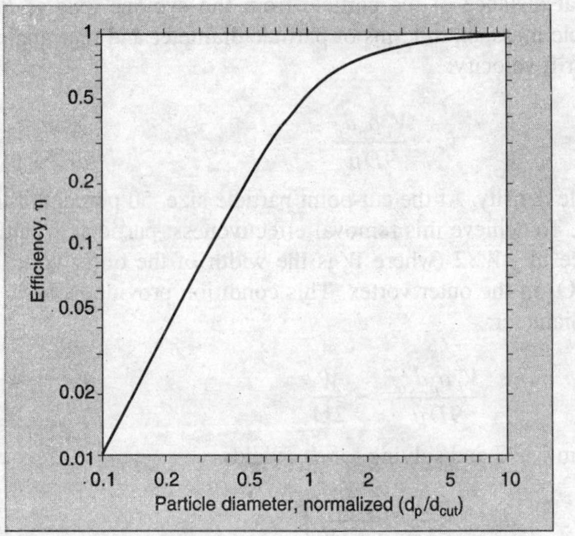

Figure 7.C.2 Collection efficiency as a function of particle size in a cyclone. The cut-point diameter, d_{cut}, is estimated by equation 7.C.7.

ESPs remove particles from gas streams in a two-step process. First, the particles are electrically charged. Then, an electric field is applied to the gas in the direction normal to the airflow. Charged particles are forced by this electric field to drift across the air stream, toward collection plates, to which they adhere. The collected particles are occasionally removed from the ESP by mechanically rapping the plates, causing the accumulated particles to fall into a hopper for disposal.

Figure 7.C.3 shows a schematic representation of an electrostatic precipitator design, in which the particle charging and removal steps occur within a single stage. In this design, the high-voltage wires generate the ions to charge the particles and establish the electric field (relative to the grounded plates) that induces electrostatic drift toward the plates. In some designs, these steps are separated.

Particle charging is achieved by means of corona discharge. A large DC voltage, typically tens of thousands of volts, is applied to a wire. Naturally occurring ions (e.g., from cosmic radiation) are accelerated in the strong electric field near the wire to a high velocity. With sufficient energy, the original ions create new ion pairs when they collide with gas molecules. If the electric field is strong enough, an avalanche of ions is produced within a thin zone surrounding the wire. Assume that a negative voltage is applied to the wire. Then the positive ions are attracted to the wire, but the negative ions are forced outward, toward the grounded plates. This process forms a large region that contains excess negative charges.

Particles that pass through this region acquire a negative charge by either of two mechanisms. *Diffusion charging* accounts for collisions between particles and ions

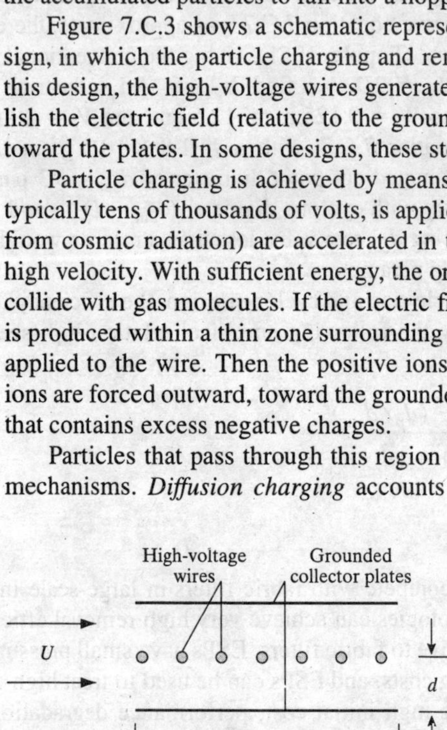

Figure 7.C.3 Schematic diagram of a single-stage electrostatic precipitator.

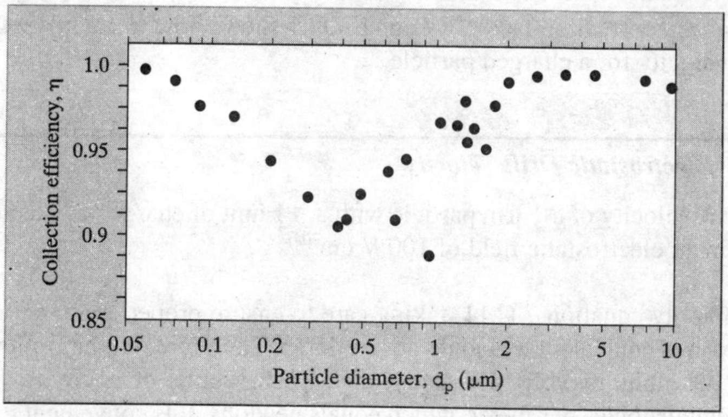

Figure 7.C.4 Measured collection efficiency as a function of particle size for an electrostatic precipitator installed on a pulverized coal boiler. (Reprinted with permission of the Air & Waste Management Association from J.D. McCain et al. [1975].)

due to their random motion. *Field charging* occurs when ions traveling along an electric field line collide with a particle through which the field line passes. Diffusion charging dominates the charging process for submicron particles, and field charging dominates for larger particles.

ESPs typically possess high particle removal efficiencies over a broad range of particle sizes, as illustrated by Figure 7.C.4. Large particles are removed efficiently because they become highly charged. Smaller particles do not acquire a large charge, but their electrostatic drift velocity remains high because the drag on them is small.

The efficiency of an electrostatic precipitator can be estimated using an analysis that is similar to that applied for a sedimentation basin (§6.C.1). However, here we assume that the flow is well mixed by turbulence in the direction normal to the flow. The situation is illustrated in Figure 7.C.5. A steady-state material balance applied to particles in the slice between x and $x + \Delta x$ produces this equation:

$$(UWd) \times C(x) = (UWd) \times C(x + \Delta x) + (V_e W \Delta x) \times C\left(x + \frac{\Delta x}{2}\right) \quad (7.C.10)$$

where W is the width of the collector plate, normal to the flow, and C is the particle concentration. The term on the left accounts for advective flow of particles into the slice. The two terms on the right account for, respectively, advective flow and electrostatic drift out of the slice. The electrostatic drift velocity, V_e, can be estimated by balancing the electrostatic force on a charged particle ($F_E = qE$, where q is the charge on the particle and E is the electric field) with the drag force ($F_D = 3\pi\mu d_p C_c^{-1} V_e$ [see equation 4.B.8]) to give

$$V_e = \frac{qEC_c}{3\pi\mu d_p} \quad (7.C.11)$$

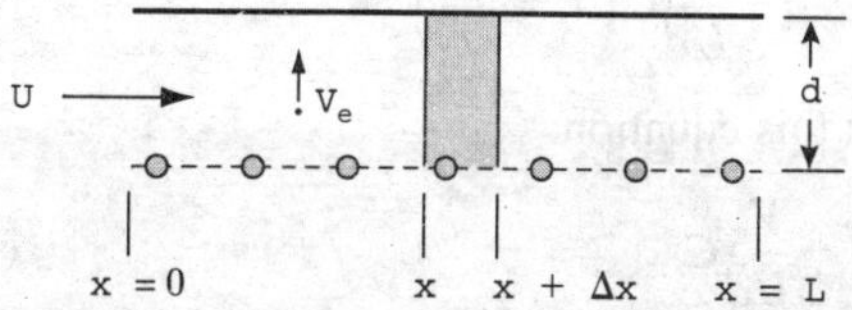

Figure 7.C.5 Schematic for modeling particle removal efficiency in an electrostatic precipitator.

where C_c is the slip correction factor. Example 7.C.1 shows how to calculate the electrostatic drift velocity for a charged particle.

EXAMPLE 7.C.1 ***Electrostatic Drift Velocity***

What is the drift velocity of a 1 μm particle with a +1 unit of charge (i.e., deficient by one electron) in an electrostatic field of 100 V cm^{-1}?

SOLUTION We can apply equation 7.C.11, taking care to ensure proper treatment of units. Note that the volt is equivalent to a joule per coulomb. Therefore, if q has units of coulombs and E has units of volts per meter, then qE has units of coulomb-volts per meter, which equals joules per meter, which equals newtons. It is convenient to use the mks system of units for evaluating electrostatic drift, later converting the velocity to cm s^{-1}, if desired.

For this problem we have

$$q = e = 1.6 \times 10^{-19}\ \mathrm{C}$$

$$E = 100\ \mathrm{V/cm} = 10^{4}\ \mathrm{V\ m^{-1}}$$

$$\mu = 1.81 \times 10^{-2}\ \mathrm{g\ m^{-1}\ s^{-1}} = 1.81 \times 10^{-5}\ \mathrm{kg\ m^{-1}\ s^{-1}} \quad \text{(Table 2.B.2)}$$

$$C_c = 1 + \frac{\lambda_g}{d_p}\left[2.51 + 0.80\exp\left(-\frac{0.55 d_p}{\lambda_g}\right)\right] \quad \text{(equation 4.B.9)}$$

$$\lambda_g = 0.066\ \mu\mathrm{m} \quad (\S 4.B.1)$$

$$d_p = 1\ \mu\mathrm{m} = 10^{-6}\ \mathrm{m}$$

So, substituting, we find

$$qE = 1.6 \times 10^{-15}\ \mathrm{N} = 1.6 \times 10^{-15}\ \mathrm{kg\ m\ s^{-2}}$$

$$C_c = 1.17$$

$$V_e = \frac{qEC_c}{3\pi\mu d_p} = 1.09 \times 10^{-5}\ \mathrm{m\ s^{-1}} = 1.1 \times 10^{-3}\ \mathrm{cm\ s^{-1}}$$

This value is about one-third of the gravitational settling velocity. In electrostatic precipitators, particles are strongly charged and the electric field can be much higher than assumed here. Consequently, it is possible for electrostatic drift velocities to be very much greater than the gravitational settling velocity of a particle.

A caution: If qE is large, the particle Reynolds number may exceed 0.3. In this case, Stokes's law underestimates drag, and the empirical relationships presented in §4.B.1 should be used to evaluate the drift velocity.

If we assume for simplicity that V_e is the same for all particles, then equation 7.C.10 can be rearranged to this form:

$$\frac{C(x + \Delta x) - C(x)}{\Delta x} = -\frac{V_e}{Ud} C\left(x + \frac{\Delta x}{2}\right) \quad (7.C.12)$$

Taking the limit as $\Delta x \to 0$, we obtain this equation:

$$\frac{dC}{dx} = -\frac{V_e}{Ud} C \quad (7.C.13)$$

This equation is solved subject to the initial condition $C = C(0)$ at $x = 0$ to yield

$$C(x) = C(0)\exp\left(-\frac{V_e x}{Ud}\right) \tag{7.C.14}$$

Using the definition of efficiency, η (equation 7.C.1), we derive a result known as the "Deutsch" equation or "Deutsch-Anderson" equation:

$$\eta = 1 - \frac{C(L)}{C(0)} = 1 - \exp(-\text{SCA} \times V_e) \tag{7.C.15}$$

where SCA $= L/(Ud)$ is known as the *specific collector area*. From Figure 7.C.3, we see that the area of each collector plate is $L \times W$, where W is the dimension normal to the page. The volumetric flow rate through a section bounded by the high-voltage wires and a single collector plate is $U \times d \times W$. Therefore, the specific collector area is the ratio of collector plate area to volumetric flow rate, in units of inverse velocity.

As shown in Figure 7.C.6, to achieve good performance, it is required that the specific collector area, $L/(Ud)$, be significantly higher than V_e^{-1}. Consider a 0.3 μm particle (roughly, the hardest size to remove, as shown in Figure 7.C.4) with a charge of 10 (giving $q = 10 \times 1.6 \times 10^{-19}$ C) and a typical ESP electric field strength of 2000 V cm^{-1}. Then the electrostatic drift velocity would be $V_e = 0.0099$ m s^{-1} = 1 cm s^{-1}. Note that this electrostatic drift velocity is about three orders of magnitude higher than the gravitational settling velocity for 0.3 μm particles (see Figure 4.B.4)! In practice, the electrostatic drift velocity, V_e, is determined empirically as the effective migration velocity for all suspended particles in a gas stream. The particle-size dependence of collection efficiency is often ignored, and only the overall mass collection efficiency is reported.

Given a drift velocity of 1 cm/s, good removal efficiency for 0.3 μm particles would require a specific collector area greater than about 1 s cm^{-1}. In practice, design values are usually in the range 0.2–2 s cm^{-1}. Common designs have channel widths of 15–40 cm and gas velocities, U, of 1.2–2.5 m s^{-1}. Operating expenses for ESPs are generally dominated by the electrical power requirements needed to operate the corona discharge and to maintain the strong electric field.

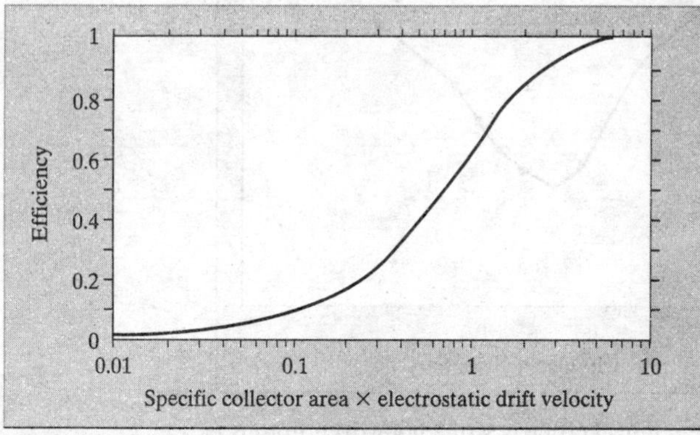

Figure 7.C.6 Efficiency of an electrostatic precipitator, as predicted by equation 7.C.15.

Fabric and Fiber Filters

Fabric and fiber filters operate by causing air to flow through a network of small obstacles, typically fine fibers. Particles that strike the fibers tend to adhere and be removed from the air stream. At a fundamental level, this process is similar to the use of a granular filter for removing particles from water (see §6.C.2). Any of four mechanisms can cause particle-fiber collisions (see Figure 6.C.6). *Straining* refers to the case where a particle is too large to pass between adjacent fibers and therefore is captured. *Interception* occurs when the fluid streamline that a particle is on passes within one particle radius of a fiber, causing a collision and particle capture. *Impaction* occurs when the inertia of a particle causes drift across bending fluid streamlines. *Diffusion* refers to the case where random motion of a particle causes it to migrate across fluid streamlines and collide with a fiber.

The mechanisms of straining, interception, and impaction become more effective as particle size increases. On the other hand, diffusion becomes increasingly effective with decreasing particle size. When all mechanisms are considered together, the overall collection efficiency is typically least effective for particles with $d_p \sim 0.1$–0.3 μm (see Figure 7.C.7).

An important application of fabric filtration is the use of "baghouse" filters in coal-fired power plants. The first application of this sort was in 1973, motivated by the Clean Air Act. At present, baghouse filters compete with electrostatic precipitators as the primary fine-particle control technology for large-scale utility and industrial applications. As of December 1989, units representing 4 percent of the U.S. coal-fired electric generating capacity were equipped with baghouse filters.

A baghouse filter unit typically consists of a large number of filter bags mounted in an array within an enclosure. A common material used is Teflon-coated glass fiber woven into a fabric. Filter bags have typical dimensions of 20–30 cm diameter by 7–11 m length.

Baghouse filters are equipped with automated mechanisms for routinely dislodging the accumulated cake of particles. The traditional techniques employed are mechanical shakers and flow reversal. In either case, an entire baghouse enclosure is taken out of service during cleaning. A relatively new alternative is "pulse-jet" clean-

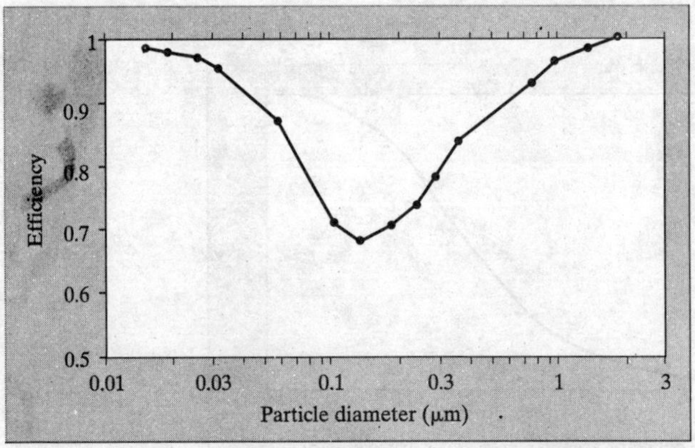

Figure 7.C.7 Particle removal efficiency of a nonwoven fibrous filter, measured under clean conditions with an approach velocity of 1.3 m s^{-1} (Hanley et al., 1994).

ing, in which a sudden blast of air in the reverse direction is applied to a single bag while the remainder of the baghouse enclosure continues to operate.

A key design parameter of a baghouse filter unit is the "air-to-cloth" ratio, defined as the volumetric air flow rate divided by the total surface area of the filter bags. A typical value is 1 cm s^{-1}. Units operate with pressure drops in the range 1000–2000 Pa, much larger than the pressure drop in an ESP (Donovan, 1985; Cushing et al., 1990).

There is a special category of fibrous filters known as HEPA (high-efficiency particulate air) filters. These were originally developed for military use and for use in nuclear reactor containment vessels. They are now widely available for a number of conventional commercial applications, including air treatment for semiconductor manufacturing clean rooms and respiratory protective equipment. HEPA filters have a minimum efficiency of 99.97 percent for 0.3 μm particles. These filters are made of very fine glass fibers (diameters less than a micron) held together within a matrix of larger fibers by an organic binder.

7.C.2 Absorption for Gaseous Pollutant Control

An absorber, or *scrubber*, is a device that removes a pollutant species from the gas phase by dissolving it into a liquid phase. The reverse of a scrubber, known as an *air stripper*, is a device for transferring a volatile pollutant from water to air. The same basic device configuration can be used either as an air stripper or as a scrubber. Analysis of these devices is included in §8.C.1. Here we will focus on a brief qualitative description of air pollution control applications.

Scrubbers are effectively used on pollutants such as HCl, which are highly soluble in water and are less hazardous in solution than in the gas phase. Chlorine is a major constituent of plastics and of many common solvents. When wastes containing chlorine are burned, such as in the incineration of medical wastes, most of the Cl is converted to HCl. If released to the atmosphere, hydrochloric acid could cause material and health damage. However, because it is highly soluble in water, it is easily removed by passing the waste gases through a scrubber. Once in the aqueous phase, HCl can be neutralized with a strong base, such as NaOH.

The key function of a scrubber is to cause contact between the contaminated air and the scrubbing solution, promoting efficient mass transfer from one phase to the other. A typical configuration, known as a *packed-tower countercurrent-flow scrubber*, is shown in Figure 7.C.8. Contaminated air enters the bottom of the tower and is cleaned as it flows upward through the column. The scrubbing solution is sprayed over the top of the column packing, and, as it flows downward, gradually becomes contaminated. In this system, the column is packed with inert rings or other objects that maintain a large air-water interfacial area and increase the contact time by slowing the downward flow rate of the scrubbing solution.

Flue Gas Desulfurization (FGD) by Wet Lime or Limestone Scrubbing

Sulfur dioxide generated by the combustion of coal can be scrubbed from flue gas in a process known as *flue gas desulfurization*. A lime/limestone slurry that is ~90 percent water and 10 percent finely ground slaked lime ($Ca(OH)_2$) or limestone ($CaCO_3$) is typically used as the scrubbing solution.

The basic chemistry of SO_2 scrubbing using lime is as follows. First, SO_2 dissolves into the water, forming sulfonic acid (H_2SO_3). Then the sulfonic acid gives up a

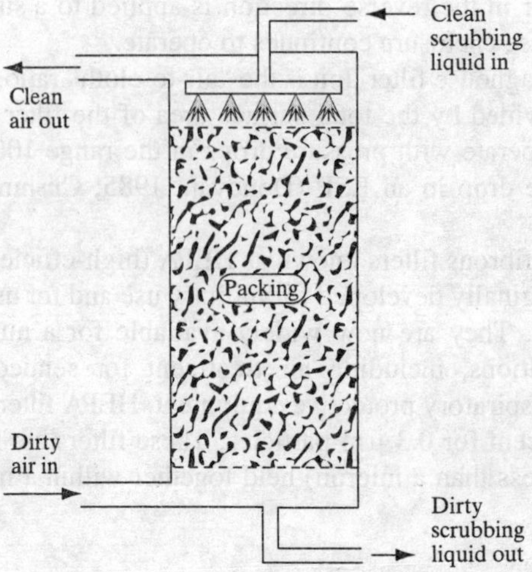

Figure 7.C.8 Schematic diagram of countercurrent scrubber for air treatment.

proton, forming bisulfite (HSO_3^-). The bisulfite combines with dissolved calcium (Ca^{2+}) from lime to form calcium sulfite ($CaSO_3$). Ultimately, the calcium sulfite in the scrubbing solution is oxidized to calcium sulfate ($CaSO_4$), which is a chemically inert, nontoxic, slightly water-soluble mineral that is widely found in natural environments. The liquid waste stream is then dewatered and mixed with dry fly ash to improve handling before landfill disposal.

Hence, the lime serves as a pH buffer, increasing the capacity of the water for SO_2. If pure water were used as the scrubbing solution rather than the lime slurry, the SO_2 absorption capacity would decrease because the pH drop caused by the dissolution of SO_2 would cause the H_2SO_3 to remain protonated rather than forming bisulfite.

7.D AIR QUALITY MODELS

In air quality engineering, mathematical models are widely used for understanding the causes and consequences of air pollution, predicting the impact of new or modified sources on air pollutant levels, and exploring the effects of proposed control strategies before they are implemented.

Many air quality modeling efforts are motivated by regulatory concerns. Consider the proposed constructi of a new facility, such as a manufacturing plant, that could emit significant quanti of air pollutants into the urban atmosphere. An environmental impact assessment is commonly required before such a facility can be built. Evaluation of the air pollution impact of this facility would involve predicting concentrations of and potential human exposures to any toxic air pollutants emitted from the facility.

Another common application of air quality models is in the formulation of strategies for meeting air quality standards. For example, as discussed in §7.A.1, ozone concentrations exceed health-based air quality standards in many urban areas. By federal law, it is necessary for air quality management agencies to develop and implement strategies to reduce excessive concentrations to meet the standards. Control of ozone in urban air can be practically achieved only by reducing the emissions of precursor species—nitrogen oxides (NO_x) and volatile organic compounds (VOCs)—which are emitted from many sources. Decisions as to which sources should be controlled, and to what extent, can be made with the assistance of mathematical models that predict the concentrations of photochemical pollutants based on emissions data and meteorology.

7.D.1 Summary of Modeling Approaches

Two broad families of air pollution models are currently in use: statistical models and deterministic models. Within each family are several model types and many specific implementations (see, for example, Zannetti, 1990). Many codes have been developed with government support and are in the public domain. The National Technical Information Service acts as a distributing repository (http://www.ntis.gov/fcpc/ntcalle.htm).

Statistical Models

Statistical models use tools from probability and statistics to analyze measured pollutant concentrations. Often the goal is to make near-term predictions. For example, statistical models might be used to answer these questions: (1) On how many days during the next year will the federal ozone standard be exceeded? (2) Given predicted weather conditions tomorrow, what are the expected air pollutant levels? Statistical models tend to have large data requirements. Although they are incapable of predicting the consequences of significantly changed conditions, they have the advantage of requiring no more than a superficial knowledge of the underlying chemistry and physics of air pollution.

Deterministic Models

Deterministic models are based on the principle of material balance. These models aim to describe the chain of cause-and-effect processes that link emissions from sources to concentrations at receptors. Deterministic air pollution models generally require information on meteorological conditions and pollutant emissions as input. They then predict concentrations, taking account of the most important atmospheric transport and transformation processes. Deterministic air pollution models can be further divided into two main groups: Gaussian plume models and transport and transformation models.

Gaussian Plume Models Gaussian plume models, which are explored further in §7.D.2, are used to predict the relationship between emissions from a single source (or small group of sources) and downwind concentrations. The greatest strength of these models is their ease of use. Their weaknesses include (a) some "leaps of faith" in deriving the models from first principles, (b) limited applicability in complex terrains, and (c) the inability to accommodate any chemistry beyond first-order decay. These models are useful mainly for source-oriented assessments of pollutants emitted from localized sources.

Transport and Transformation Models Transport and transformation models generally predict species concentrations while taking into consideration all important emission sources. Whereas the Gaussian plume models use approximate analytical solutions, transport and transformation models employ numerical methods to solve the governing equations.

Transport and transformation models have been widely applied toward understanding and controlling the problem of urban and regional photochemical smog, especially ozone. Generally, in these applications the models produce time-dependent predictions for periods of a few days. The level of spatial detail varies considerably. *Trajectory models* predict concentrations at a single receptor site. *Airshed models* predict the spatial distribution of concentrations throughout an air basin or over an entire region, with resolution in the horizontal direction on the order of 10 km.

A trajectory model simulates pollutant concentrations in single air parcels as these parcels are transported with the wind through an air basin or over a region. Good wind data are required to predict the movement of the parcel. Pollutant emissions into the parcel are included from all sources, necessitating an accurate, spatially resolved emissions inventory. The parcel is typically subdivided vertically into layers, and mixing between layers by turbulent transport is included in the analysis. However, to maintain parcel integrity, it is assumed that the wind speed is uniform with height. Also, horizontal turbulent transport of pollutants into or out of the parcel is not considered. Separate material-balance equations are written for each species of interest in each layer of the parcel. These differential equations are then solved numerically. A key advantage of these models, relative to Gaussian plume equations, is that reactive chemistry can be included in great detail. Trajectory models have intermediate computational requirements between Gaussian plume and airshed models.

In airshed models, an air basin (or larger region) is subdivided into a set of contiguous subdomains that are fixed in space. Material-balance equations are written and solved numerically for each pollutant species in each of the subdomains. The model predicts pollutant concentrations as a function of position and time throughout the modeling region. Airshed models of photochemical air pollution combine many complicated elements:

a. Spatially and temporally resolved emissions of key pollutants for the entire air basin

b. Chemical mechanisms that may include tens of species and hundreds of reactions

c. A solar simulator to predict photolysis rates as a function of time

d. A wind field generator that predicts or infers wind speed and direction at all locations within the air basin

e. A transport simulator that accounts for the effects of wind, turbulent dispersion in the horizontal direction, atmospheric stability and mixing in the vertical direction, and surface topography

f. A pollutant deposition simulator to account for the loss of pollutants by uptake at the ground surface

g. A computational algorithm that combines these elements into a set of mathematical equations that can be solved numerically to predict pollutant species concentrations

This modeling approach is computationally intensive and has enormous input data requirements. In a typical application, the physical domain may be divided into ~10,000 subdomains. The chemical mechanism may need to track 20–50 chemical species. Separate material-balance equations are needed for each species in each subdomain, so the number of equations that must be solved is on the order of 10^5 to 10^6.

Throughout the past three decades, substantial effort has been devoted to the development of airshed models for predicting photochemical smog (Seinfeld, 1988; National Research Council, 1991). Despite the enormous effort required to apply these models, they have become important tools in efforts to develop and evaluate ozone control strategies. With continued advances in computational power and information management, the importance of these modeling tools is likely to grow.

7.D.2 Gaussian Plume Dispersion Modeling

In §5.B equations were derived to predict air pollutant concentrations downwind of sources. These equations were obtained by solving the general material-balance equation for species concentrations. In these derivations, we assumed that the mean wind speed was constant as a function of height and that the turbulent diffusivities, ε_x, ε_y, and ε_z, were independent of position. However, these assumptions do not conform well to reality. In addition, turbulent diffusivities are difficult (if not impossible!) to measure directly.

In practice, these problems are resolved by writing the Gaussian plume equations in terms of dispersion parameters σ_x, σ_y, and σ_z instead of turbulent diffusivities. As described below, these dispersion parameters are then evaluated as functions of (a) distance downwind from the emission source and (b) atmospheric stability. Dispersion parameters can then be estimated from empirical observations.

The relationship between the dispersion parameters, σ, and the turbulent diffusivities, ε, are defined by this equation:

$$\sigma_j = \sqrt{2\varepsilon_j \frac{x}{U}} \qquad (7.D.1)$$

where j represents a coordinate direction (x, y, or z).

Consider, for example, equation 5.B.35, which predicts the concentration downwind of a point source that emits a pollutant at a constant rate m (mass per time). The coordinate system is configured so that the wind (mean speed = U) is aligned with the x-axis (Figure 7.D.1). Equation 7.D.1 is used to replace ε_y and ε_z with σ_y and σ_z. To account for the elevated emissions, the height coordinate z in equation 5.B.35 is replaced by $z - H$. The result is

$$C = \left(\frac{m}{2\pi U \sigma_y \sigma_z}\right) \times \exp\left(-\frac{y^2}{2\sigma_y^2}\right) \times \exp\left(-\frac{(z-H)^2}{2\sigma_z^2}\right) \qquad (7.D.2)$$

The species concentration predicted by equation 7.D.2 depends on atmospheric dispersion parameters, σ_y and σ_z, which depend on atmospheric stability. With appropriate modification, the Gaussian plume approach can be extended to take account of the effects of the ground on the spreading of the plume. It can also be modified to model a

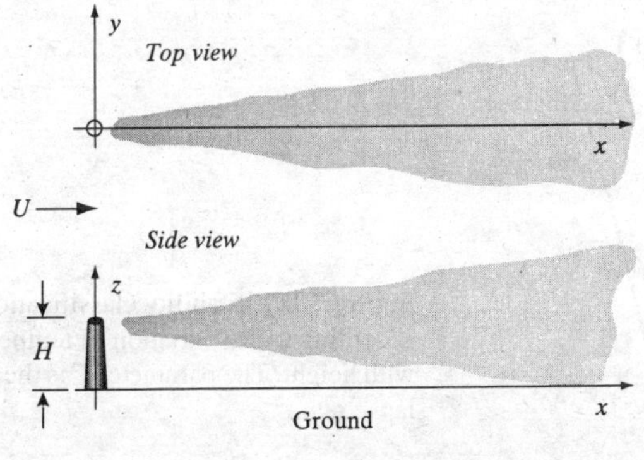

Figure 7.D.1 Coordinate system for applying the Gaussian plume model from a continuous point source. The emissions occur from point (0, 0, *H*).

more complex source geometry than a single point. These issues are addressed in the following subsections.

Atmospheric Stability

Mixing processes in the atmosphere are related to vertical stability, which, in turn, depends on how the air density varies with height. Unstable conditions lead to more rapid mixing, especially in the vertical direction. Stable conditions correspond to slower rates of mixing.

Air temperature generally decreases with height because air pressure decreases with height. As air rises, it expands in response to the lower pressure and, because it does work to expand against the surrounding air pressure, it cools. Atmospheric stability classifications can be related to the rate of temperature decrease with height ($-dT/dz$), which is known as the *lapse rate*. For neutral stability, the lapse rate is given by

$$-dT/dz = \Gamma \tag{7.D.3}$$

where Γ, known as the *adiabatic lapse rate*, is approximately 1 K per 100 m (Seinfeld and Pandis, 1998). When the atmospheric lapse rate matches the adiabatic lapse rate, air that begins in thermal equilibrium with adjacent air remains in thermal equilibrium, regardless of whether it is displaced vertically.

If the air temperature decreases more rapidly with height than the adiabatic lapse rate ($-dT/dz > \Gamma$), then the atmosphere is unstable (see Figure 7.D.2). Consider air that begins at the same temperature as the adjacent air but is suddenly lifted upward. As it encounters the lower pressure aloft, the air expands and cools, but not as much as the atmosphere. Being warmer than the adjacent air, the displaced air is buoyant and continues to rise. Therefore, under unstable conditions, small displacements are amplified by buoyancy, and mixing is enhanced. Conversely, when the temperature decreases less rapidly with height than the adiabatic condition ($-dT/dz < \Gamma$), the atmosphere is stable. That is, air that is suddenly displaced upward is cooler than the adjacent air and settles back to its original elevation. Mixing is impeded under stable conditions.

Pasquill Stability Class As we have seen, atmospheric stability can be related to the atmospheric lapse rate. The dominant features that control the atmospheric lapse rate

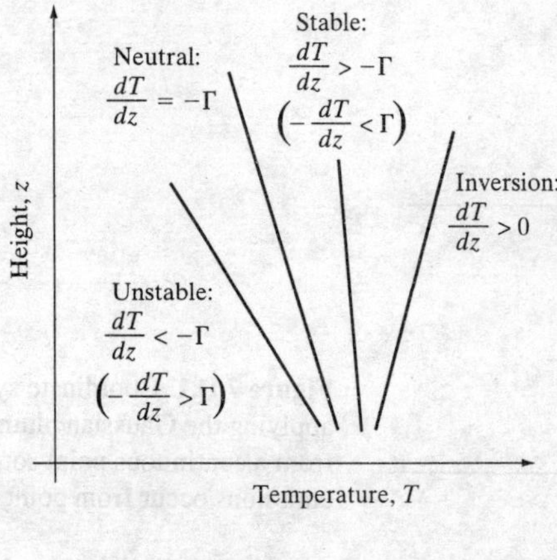

Figure 7.D.2 Stability classification of air according to the variation of temperature with height. The parameter Γ is the adiabatic lapse rate.

are incoming radiation from the sun, outgoing radiation from the earth's surface to the sky, and turbulence generated by winds.

Sunshine heats the earth's surface much more effectively than it heats air. During a clear summer day, the ground warms. Air adjacent to the ground is then heated by conduction. This air expands and becomes buoyant. Warm plumes rise from the ground, promoting vertical mixing. In the absence of other effects, intense sunshine causes unstable conditions. The shimmering that one sees above distant asphalt on a hot summer day is an indication of fluctuating temperatures caused by irregular vertical air movement.

At night, without cloud cover, the earth's surface radiates energy to the sky, causing the ground to cool. Air adjacent to the surface is cooled by conduction, causing it to become denser and remain near the ground, promoting a stable lapse rate. If the ground is sloped, the cool air will tend to flow downhill, collecting in channels and ravines much as water does in a storm. The cooling effect of surface radiation can cause frost to form on cool mornings, even when the air temperature remains above 0 °C. Cloud cover greatly diminishes the cooling effect by reducing the net energy loss rate from the surface.

Turbulence generated by strong winds flowing over rough surfaces tends to mix the air vertically. In the absence of other effects, this mixing pushes the temperature profile toward the adiabatic condition.

Pasquill (1961) proposed a relatively simple scheme for classifying atmospheric stability based on readily observed parameters. His framework is still widely used in modeling air pollutant transport. Pasquill's system defines six stability classes based on three observable characteristics: (a) the intensity of incoming solar radiation (daytime), (b) the extent of cloud cover (nighttime), and (c) the near-surface wind speed. The six stability classes are known by the letter designations A–F, where A–C are unstable, D is neutral, and E–F are stable. Table 7.D.1 shows how observable parameters are used to assign the stability class. The table shows that high wind speeds tend to push the stability class toward neutral (class D) while strong solar radiation increases instability. The absence of nighttime cloud cover tends to make conditions more stable.

Inversions Although the air temperature usually decreases with height in the troposphere, sometimes it increases with height. This condition, known as an *inversion*, is very stable with respect to vertical air movement (see Exhibit 7.D.1). Atmospheric inversions are important contributors to poor air quality.

Table 7.D.1 Estimation of Stability Class from Observable Parameters[a]

	Daytime: solar radiation			Nighttime: cloud cover	
Wind speed (m/s)[b]	Strong	Moderate	Slight	≥4/8 of sky	≤3/8 of sky
<2	A	A–B	B	—	—
2–3	A–B	B	C	E	F
3–5	B	B–C	C	D	E
5–6	C	C–D	D	D	D
>6	C	D	D	D	D

[a]For fully overcast skies, use stability class D for day or night.
[b]Measured at 10 m above ground.
Source: Pasquill, 1961.

EXHIBIT 7.D.1 *Inversions*

Inversions are classified according to the means by which they are generated and according to the height of their base. A *ground-based inversion* occurs when temperature increases with height beginning at the ground surface. In an *elevated inversion*, the temperature decreases with height from the ground to some altitude, above which the temperature increases with height.

Several mechanisms can generate inversions. The two classes of inversions that cause the greatest difficulty in terms of air pollution are radiation inversions and subsidence inversions.

A *radiation inversion* forms overnight when the winds are weak and the sky is clear (see Figure 7.D.3). The emission of electromagnetic (infrared) radiation from the ground to the night sky causes air adjacent to the earth's surface to cool. Beginning near sunset, a ground-based inversion is formed and gradually becomes deeper through the night. In the morning, when the sun rises, the ground is heated and the base of the inversion lifts. As the air below the inversion base continues to be heated, the inversion layer becomes progressively thinner until it is finally eliminated. Radiation inversions are strongest in the winter, when incoming solar radiation is weak. They can also be strong if their generation is accompanied by fog formation. Then much of the incoming radiation following sunrise is scattered rather than absorbed, and the air below the inversion base heats very slowly.

Stagnant high-pressure regions in the atmosphere cause *subsidence inversions*. An anticyclone is established with winds that flow in an almost closed circle aloft. As the surface is approached, ground friction causes a progressively increasing velocity component to point outward from the circle. In the upper part of this high-pressure cell there is a net downward flow of air, as required to replace the air that leaks out of the cell near the ground. As this air subsides, it is compressed by the higher pressure, which causes adiabatic heating and can generate an inversion.

The air quality problems in Los Angeles are strongly linked to inversions that keep pollutant emissions in the air basin trapped below the tops of adjacent mountains. In the summertime, the semipermanent north Pacific high-pressure region lies just offshore from Los Angeles. The high-elevation air is heated by subsidence. The ocean keeps the air near the ground cool. The consequences are little summer cloud cover or precipitation, generally weak summertime winds, and frequent elevated inversions. These factors contribute substantially to Los Angeles's air pollution problems.

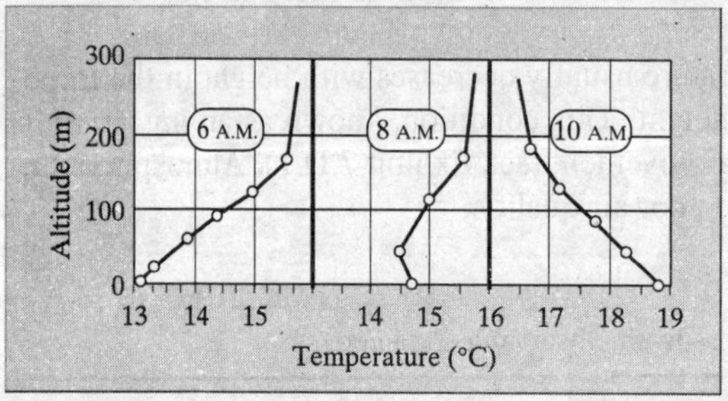

Figure 7.D.3 Averaged temperature profiles with height for September and October 1950, measured in Oak Ridge, Tennessee. The 6 A.M. profile shows a ground-based radiation inversion that extends strongly up to about 150 m above ground. At 8 A.M., the base of the inversion has lifted to about 50 m above the ground. Continued ground heating by the sun leads to complete destruction of the inversion by 10 A.M. (*Source:* U.S. Weather Bureau, cited in Wark and Warner, 1981.)

Atmospheric Dispersion Parameters

Atmospheric dispersion parameters vary with distance downwind of the source, x, and with the Pasquill stability class. They can be estimated from empirically based graphs or analytical expressions. For the graphical method, information on the Pasquill stability class is used to choose the correct curve in Figures 7.D.4 and 7.D.5 for estimation of σ_y and σ_z, respectively.

An analytical method for estimating dispersion parameters employs Tables 7.D.2 and 7.D.3. These tables provide the parameter values *a*–*f* to be used in the following expressions:

$$\sigma_y = a x^{b + c \ln x} \tag{7.D.4}$$

$$\sigma_z = d x^{e + f \ln x} \tag{7.D.5}$$

In these expressions, x has units of km, and σ_y and σ_z have units of m.

Physically, the dispersion parameters represent magnitude estimates of plume extent in the horizontal (σ_y) and vertical (σ_z) crosswind directions. So, for example, we see directly from Figure 7.D.4 that at 1 km downwind of a point source, a plume spreads 30 m to 200 m horizontally away from the centerline, depending on stability class. Figure 7.D.5 shows that the corresponding vertical spread at 1 km downwind is in the range 10–500 m. Note that the horizontal dispersion parameter, σ_y, is a much weaker function of stability class than the vertical dispersion parameter, σ_z. This finding is consistent with expectations, given the enormous importance of stability on vertical mixing.

Equation 7.D.1 indicates that σ_y and σ_z should scale with downwind distance as $x^{1/2}$, provided that the turbulent diffusivities, ε_y and ε_z, are constant with position. In most cases, the empirical dispersion parameters σ_y and σ_z increase more rapidly with x than would be predicted for constant values of ε. The use of empirical data for the

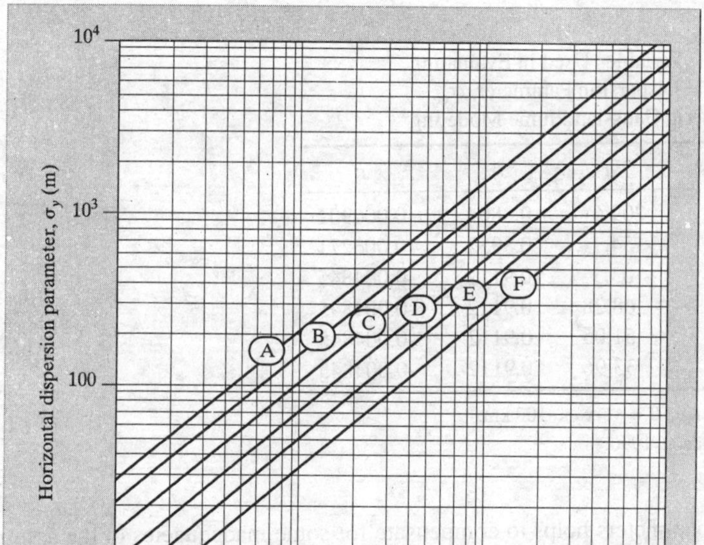

Figure 7.D.4 Horizontal dispersion parameters, σ_y, for atmospheric Gaussian plume modeling.

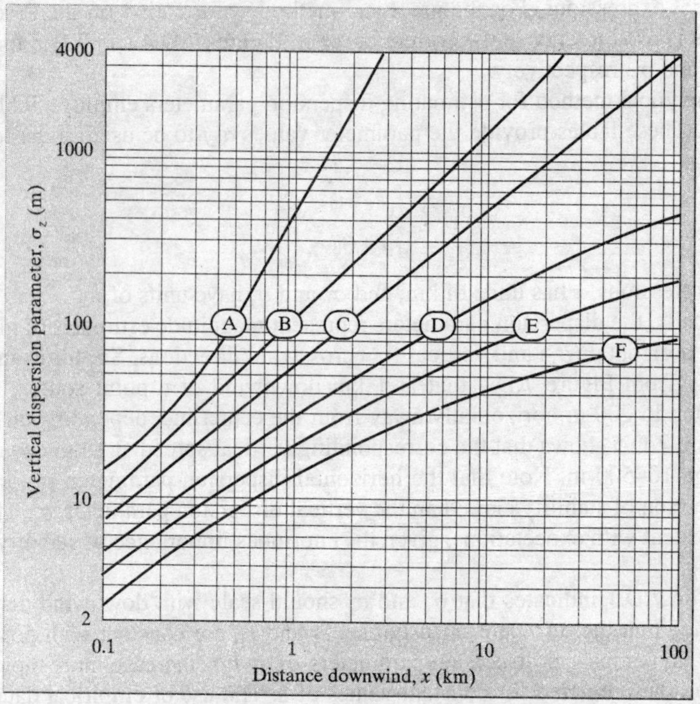

Figure 7.D.5 Vertical dispersion parameters, σ_z, for atmospheric Gaussian plume modeling.

Table 7.D.2 Variables Used in Evaluating the Horizontal Dispersion Parameter, σ_y, for Atmospheric Gaussian Plume Modeling[a]

Stability class	*a*	*b*	*c*
A	209.6	0.8804	−0.006902
B	154.7	0.8932	−0.006271
C	103.3	0.9112	−0.004845
D	68.28	0.9112	−0.004845
E	51.05	0.9112	−0.004845
F	33.96	0.9112	−0.004845

[a]Applicable range: $0.1 < x < 100$ km.

Source: Davidson, 1990.

dispersion parameters helps to compensate for some inadequacies of the assumptions made in deriving the model.

Effect of the Ground in Gaussian Plume Models

So far we have considered only unbounded spaces. However, air pollution sources and receptors are frequently close to the ground, which may have a significant influence on pollutant transport and dispersion.

Table 7.D.3 Variables Used in Evaluating the Vertical Dispersion Parameter, σ_z, for Atmospheric Gaussian Plume Modeling

Stability Class	d	e	f	Applicable range
A	310.4	1.773	0.1879	$0.1 < x < 0.45$ km
A	453.9	2.117	0.0	$0.45 < x < 3.1$ km
B	109.8	1.064	0.01163	$0.1 < x < 32$ km
C	61.14	0.9147	0.0	$0.1 < x < 100$ km
D	30.38	0.7309	−0.03200	$0.1 < x < 100$ km
E	21.14	0.6802	−0.04522	$0.1 < x < 100$ km
F	13.72	0.6584	−0.05367	$0.1 < x < 100$ km

Source: Davidson, 1990.

In this subsection we briefly explore the treatment of boundaries in Gaussian plume models. The tools we use are superposition and the method of images (see Exhibits 7.D.2 and 7.D.3).

We limit our attention to cases in which the ground is flat. We also restrict our attention to one limiting case for pollutant-surface interaction: *Perfect reflection* occurs when no transformation of a species occurs at a surface. Instead, the surface acts as a physical barrier that prevents further spread.

As in previous cases, the ground level is defined by $z = 0$. Mathematically, the perfect reflection boundary condition is described by setting the z-component of contaminant flux to zero at this surface. Since the z-component of flux is given by $J_z = -\varepsilon_z \, \partial C/\partial z$, the boundary condition becomes $\partial C/\partial z = 0$ at $z = 0$.

EXHIBIT 7.D.2 *The Superposition Principle*

The general idea in superposition is that, subject to some restrictions, we can predict species concentrations that are caused by multiple sources by analyzing the contributions from separate sources individually and then summing those contributions.

Superposition can be applied only for linear, homogenous systems. For application to species concentration (C) in a fluid, this means that each term in the governing equation must vary in proportion to C. Transport terms that vary with the first or second derivative of concentration (e.g., those accounting for advection and turbulent diffusion processes) satisfy this requirement because, for example, doubling C everywhere will double $\partial C/\partial x$ everywhere. Linearity requires that any transformation processes be first order. In addition, the boundary conditions must be homogeneous. This means that at all boundaries, some equation of this form must be satisfied: $a\,\partial C/\partial x_n + bC = 0$, where a, b are constants and $\partial C/\partial x_n$ is the partial derivative of the concentration with respect to the coordinate direction normal to the boundary.

EXHIBIT 7.D.3 *Method of Images*

The goal in the method of images is to map a complex physical system to a model description for which a solution can be more easily obtained. The concept is illustrated in Figure 7.D.6 for the case of a perfectly reflecting boundary at the ground. On the left, the physical system is depicted. Since no contaminant mass can travel across the boundary, and since turbulent diffusion is assumed to be constant, the contaminant that would have spread below the boundary if it were absent is reflected back and added to the plume above the boundary.

EXHIBIT 7.D.3 *Method of Images (continued)*

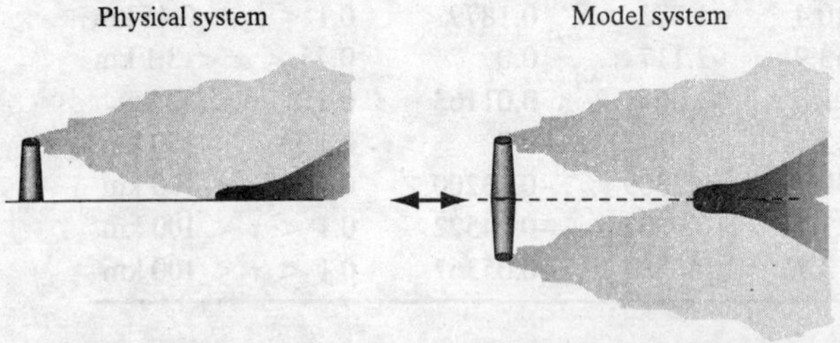

Figure 7.D.6 Application of the method of images in Gaussian plume modeling. The physical system has a continuous emission source at point $z = H$ and a perfectly reflecting ground surface. In the model system, an image source is added at $z = -H$, which emits at a rate equal to the true source. The species concentration is computed for the model system in unbounded space. The no-flux boundary condition at the ground surface is automatically satisfied by symmetry.

We can solve this problem by using an unbounded model system to represent the physical system. We do this by creating a mirror image of the contaminant source on the other side of the boundary. We then assume that there is no boundary between the sources and use the superposition principle to solve for the contaminant concentration in the model system that results from these two sources operating together. The contributions in the model system of the image source to concentrations above the boundary are identical to the contributions in the physical system caused by reflection from the boundary. The no-flux boundary condition is satisfied in the model since transport downward from the true source is equal to transport upward from the image source.

To illustrate the application of the superposition principle and the method of images in Gaussian plume modeling, we consider constant emissions at rate m from an elevated source located at position (0, 0, H) (see Figure 7.D.1). The wind has a constant velocity, U, and the ground is perfectly reflecting. We seek the steady-state mean concentration as a function of position.

The governing equation is as follows (see equation 5.B.32):

$$U\frac{\partial C}{\partial x} - \varepsilon_y\frac{\partial^2 C}{\partial y^2} - \varepsilon_z\frac{\partial^2 C}{\partial z^2} = 0 \qquad (7.D.6)$$

with boundary conditions

$$C(0, y, z) = \frac{m}{U}\,\delta(y)\,\delta(z - H) \qquad (7.D.7)$$

$$C(x, y, z) \to 0 \text{ as } y \to \pm\infty \qquad (7.D.8)$$

$$C(x, y, z) \to 0 \text{ as } z \to +\infty \qquad (7.D.9)$$

$$\frac{\partial C}{\partial z} = 0 \text{ at } z = 0 \qquad (7.D.10)$$

To apply the method of images, we consider two sources, each of which emits at rate m. The real source is located at (0, 0, H), and the image source is placed at (0, 0,

$-H$). The boundary condition 7.D.10 is replaced by the condition $C \rightarrow 0$ as $z \rightarrow -\infty$. The contribution to concentration from each source is first determined independently, and then they are summed as per the superposition principle. Here is the result:

$$C = \left\{\frac{m/(2\pi U)}{\sigma_y \sigma_z}\right\} \times \left\{ \exp\left[-\left[\left(\frac{y^2}{2\sigma_y^2}\right) + \left(\frac{(z-H)^2}{2\sigma_z^2}\right)\right]\right] + \exp\left[-\left[\left(\frac{y^2}{2\sigma_y^2}\right) + \left(\frac{(z+H)^2}{2\sigma_z^2}\right)\right]\right]\right\} \qquad (7.D.11)$$

Two terms appear in the second set of braces. The first one represents the contribution from the real source, and the second one represents the contribution from the image source. Example 7.D.1 explores the concentration profiles described by this equation.

For environmental impact assessments, Gaussian plume models may be used to predict the long-term average concentrations in the vicinity of a source. This goal can be accomplished by using an equation such as 7.D.11 together with appropriate meteorological data (wind speed and stability class) and emission estimates (m) to predict the concentration field on an hour-by-hour basis for, say, an entire year. Typically, one would compute the ground-level concentrations for each hour at an array of fixed locations and then directly compute the time-weighted average values at each of these locations.

EXAMPLE 7.D.1 ***Ground-Level Concentrations Downwind of a Point Source***

A nonreactive air pollutant is released from an industrial stack at a rate of $m = 10 \text{ g s}^{-1}$. The effective height of the release is $H = 50$ m. The mean wind speed is $U = 3 \text{ m s}^{-1}$. Predict and plot the ground-level concentrations downwind of the source for two stability classes: B (moderately unstable) and E (slightly stable).

SOLUTION First, the dispersion parameters are obtained from Tables 7.D.2 and 7.D.3 and equations 7.D.4 and 7.D.5:

Class B: $\sigma_y = 154.7x^{0.8932 - 0.006271 \ln(x)}$ $\qquad \sigma_z = 109.8x^{1.064 + 0.01163 \ln(x)}$

Class E: $\sigma_y = 51.05x^{0.9112 - 0.004845 \ln(x)}$ $\qquad \sigma_z = 21.14x^{0.6802 - 0.04522 \ln(x)}$

We then substitute into equation 7.D.11 to obtain an expression for the concentration as a function of position. We are interested only in ground-level concentrations ($z = 0$), so the expression simplifies to the following form:

$$C = \left(\frac{m}{\pi U \sigma_y \sigma_z}\right) \times \exp\left(-\frac{y^2}{2\sigma_y^2}\right) \times \exp\left(-\frac{H^2}{2\sigma_z^2}\right)$$

The parameters m, U, and H are given in the problem statement, and expressions for σ_y and σ_z as functions of x are presented above. Figure 7.D.7 plots the ground-level concentration along the plume centerline ($y = 0$) for the two stability classes, and Figure 7.D.8 shows contours of equal pollutant concentration at ground level. For class B, the peak concentration is ~200 $\mu\text{g m}^{-3}$ and occurs ~350 m downwind of the source. Because of the greater atmospheric stability, the peak ground-level concentration for class E does not occur until the plume has traveled further from the source, ~2 km. By

EXAMPLE 7.D.1 ***Ground-Level Concentrations Downwind of a Point Source (continued)***

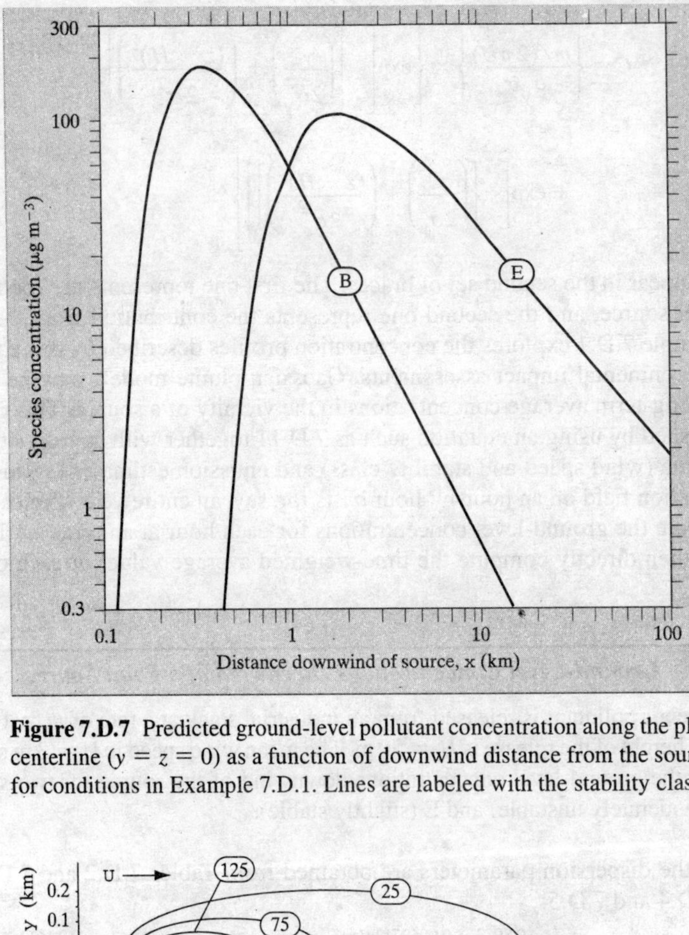

Figure 7.D.7 Predicted ground-level pollutant concentration along the plume centerline ($y = z = 0$) as a function of downwind distance from the source for conditions in Example 7.D.1. Lines are labeled with the stability class.

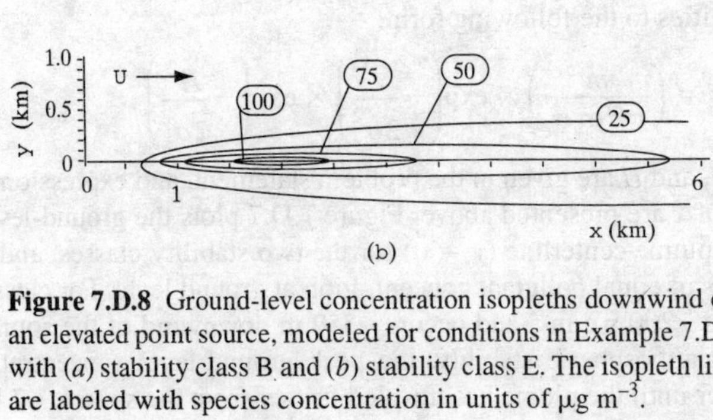

Figure 7.D.8 Ground-level concentration isopleths downwind of an elevated point source, modeled for conditions in Example 7.D.1, with (*a*) stability class B and (*b*) stability class E. The isopleth lines are labeled with species concentration in units of $\mu g\ m^{-3}$.

EXAMPLE 7.D.1 ***Ground-Level Concentrations Downwind of a Point Source (continued)***

this time the plume has had a greater opportunity to spread in the y-direction (note from Figures 7.D.4 and 7.D.5 that σ_y is less sensitive to stability class than σ_z), and so the peak concentration is only ~100 μg m^{-3}, about half that for class B.

Pollutant Concentrations Downwind of a Line Source

As shown by the superposition principle, pollutant concentrations caused by multiple sources can be computed by summing the contributions from the separate sources. This idea can be applied for distributed as well as discrete sources. For example, consider the case of a line source of pollution, such as the carbon monoxide emissions from a heavily traveled roadway. Figure 7.D.9 depicts the case where the wind direction is normal to the line source. Let m_{line} be the rate of pollutant emission per unit length of the line (g m^{-1} s^{-1}). Then we can arrange the coordinate axes so that a receptor located at some distance x downwind of the line is positioned at $y = 0$.

The line source can be broken into discrete intervals of length dy. The mass emission rate from each interval is then given by $m_{\text{line}}\ dy$. Substituting ($m_{\text{line}}\ dy$) for m in equation 7.D.11 gives the concentration at the receptor contributed by a single increment of the line source. Setting $H = 0$ (assuming that the line source is at ground level), we can express the concentration from one increment as

$$dC = \left(\frac{m_{\text{line}}\, dy}{\pi U \sigma_y \sigma_z}\right) \times \exp\left(-\frac{y^2}{2\sigma_y^2}\right) \times \exp\left(-\frac{z^2}{2\sigma_z^2}\right) \qquad (7.D.12)$$

The total pollutant concentration at the receptor is then obtained by summing the contributions over all line increments. If we assume that the emission rate is independent of position and that the line source extends infinitely in both directions (practically, this means that the length of the source must be much greater than the distance to the receptor), we can integrate as follows:

$$C = \int_{-\infty}^{\infty} dC = \left(\frac{m_{\text{line}}}{\pi U \sigma_y \sigma_z}\right) \times \exp\left(-\frac{z^2}{2\sigma_z^2}\right) \times \int_{-\infty}^{\infty} \exp\left(-\frac{y^2}{2\sigma_y^2}\right) dy \qquad (7.D.13)$$

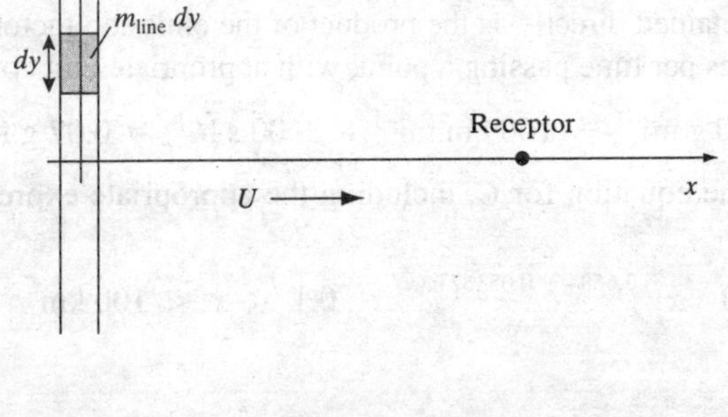

Figure 7.D.9 Coordinate orientation for application of the Gaussian plume modeling approach for a line source oriented normal to the wind.

To evaluate the integral, we substitute $s = \sqrt{1/2} \times (y/\sigma_y)$ and note that

$$\int_{-\infty}^{\infty} \exp(-s^2)\, ds = \sqrt{\pi} \tag{7.D.14}$$

The result is

$$C = \left(\frac{\sqrt{2} m_{\text{line}}}{\sqrt{\pi} U \sigma_z}\right) \times \exp\left(-\frac{z^2}{2\sigma_z^2}\right) \tag{7.D.15}$$

Note that this result does not depend on dispersion in the *y*-direction. Since the source is assumed to extend infinitely in the *y*-direction, there is no concentration gradient and therefore no net transport by turbulent diffusion in that direction. Example 7.D.2 illustrates the use of this result in a practical problem.

EXAMPLE 7.D.2 ***Ground-Level CO Concentrations Downwind of a Busy Freeway***

On a heavily traveled freeway, a total of 10,000 vehicles pass a given point per hour in each direction, traveling at an average speed of 50 miles per hour. Assume that the average CO emission factor for these vehicles is 20 g per mile. Assume that moderately stable conditions (class F) prevail, with a steady crosswind of 2 m s^{-1}. Determine the CO concentration as a function of distance downwind of the freeway. Compare the results with the appropriate air quality standard.

SOLUTION For ground-level emissions and ground-level concentrations, $H = z = 0$, and so equation 7.D.15 simplifies to

$$C = \left(\frac{\sqrt{2} m_{\text{line}}}{\sqrt{\pi} U \sigma_z}\right)$$

What is the CO emission rate per unit freeway length, m_{line}? Consider a small increment, 100 m, of the length of the freeway. The average vehicle speed is 22.3 m s^{-1} (= 50 miles per hour × 1609 m per mile ÷ 3600 s per hour). The time required for a single vehicle to travel through this increment is 4.5 s (100 m ÷ 22.3 m s^{-1}).

Assume that the vehicles are uniformly distributed along the freeway. Since 20,000 vehicles must pass through this slice per hour, and each vehicle spends an average of 4.5 s in the slice, there must be 25 vehicles in the slice at any time (= 20,000 ÷ 3600 × 4.5). These vehicles emit an average of 0.28 g s^{-1} of CO (20 g mi^{-1} × 50 mi h^{-1} ÷ 3600 s h^{-1}). Therefore, from this 100 m slice, CO is emitted at a rate of 7 g s^{-1} (= 25 × 0.28). Dividing by the length of the slice yields the result

$$m_{\text{line}} = 0.07 \text{ g m}^{-1} \text{ s}^{-1}$$

This result could also be obtained directly as the product of the emission factor (g mi^{-1}) times the number of vehicles per time passing a point, with appropriate unit conversion:

$$m_{\text{line}} = 2 \times 10^4 \text{ h}^{-1} \times 20 \text{ g mi}^{-1} \div (1609 \text{ m mi}^{-1} \times 3600 \text{ s h}^{-1}) = 0.07 \text{ g m}^{-1} \text{ s}^{-1}$$

Now, substitute into the equation for C, including the appropriate expression for σ_z from equation 7.D.5:

$$C = 2.04 \times 10^{-3} x^{-0.6584 + 0.05367 \ln(x)} \qquad 0.1 < x < 100 \text{ km}$$

EXAMPLE 7.D.2 *Ground-Level CO Concentrations Downwind of a Busy Freeway (continued)*

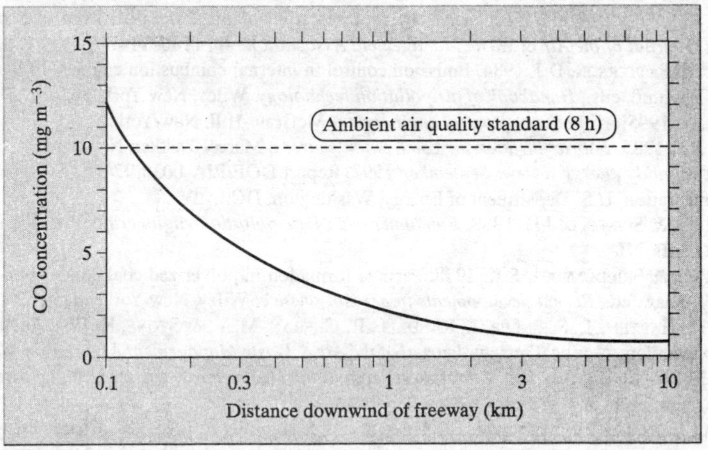

Figure 7.D.10 Predicted ground-level concentrations of carbon monoxide as a function of distance downwind of a heavily traveled freeway.

where x and C are in units of km and g m^{-3}, respectively. The results are plotted in Figure 7.D.10. The national ambient air quality standard (NAAQS) for carbon monoxide is 10 mg m^{-3} for an 8-hour average and 40 mg m^{-3} for a 1-hour average. According to the model predictions, the CO concentration exceeds the 8-hour standard at locations $\leq$125 m downwind of the freeway. At a distance of 1 km, the ground-level CO concentration is greatly reduced, to about 2 mg m^{-3}.

REFERENCES

ABBATT, J.P.D., & MOLINA, M.J. 1993. Status of stratospheric ozone depletion. *Annual Review of Energy and the Environment*, **18**, 1–29.

AMANN, C.A., & SIEGLA, D.C. 1982. Diesel particulates—What they are and why. *Aerosol Science and Technology*, **1**, 73–101.

ASHRAE. 1990. *Ventilation for acceptable indoor air quality*, ANSI/ASHRAE 62-1989. American Society of Heating, Refrigerating, and Air-Conditioning Engineers, Atlanta.

BAAQMD. 1993. *Emission inventory source category methodologies—base year: 1990*. Bay Area Air Quality Management District, October 1.

BAER, N.S., & BANKS, P.N. 1985. Indoor air pollution: Effects on cultural and historic materials. *International Journal of Museum Management and Curatorship*, **4**, 9–20.

BENEDICK, R.E. 1991. *Ozone diplomacy: New directions in safeguarding the planet*. Harvard University Press, Cambridge, MA.

BISHOP, G.A., & STEDMAN, D.H. 1990. On-road carbon monoxide emission measurement comparisons for the 1988–1989 Colorado oxy-fuels program. *Environmental Science & Technology*, **24**, 843–847.

BOWMAN, C.T. 1991. Chemistry of gaseous pollutant formation and destruction. In W. Bartok and A.F. Sarofim, eds., *Fossil fuel combustion: A source book*. Wiley, New York, pp. 215–260.

BUONICORE, A.J., & DAVIS, W.T. (eds.). 1992. *Air pollution engineering manual*. Van Nostrand Reinhold, New York.

CALVERT, J.G., HEYWOOD, J.B., SAWYER, R.F., & SEINFELD, J.H. 1993. Achieving acceptable air quality: Some reflections on controlling vehicle emissions. *Science*, **261**, 37–45.

COBB, N., & ETZEL, R.A. 1991. Unintentional carbon monoxide–related deaths in the United States, 1979 through 1988. *Journal of the American Medical Association*, **266**, 659–663.

COOPER, D.W. 1986. Particulate contamination and microelectronics manufacturing: An introduction. *Aerosol Science and Technology*, **5**, 287–299.

CUSHING, K.M., MERRITT, R.L., & CHANG, R.L. 1990. Operating history and current status of fabric filters in the utility industry, *Journal of the Air & Waste Management Association*, **40**, 1051–1058.

DAVIDSON, G.A. 1990. A modified power-law representation of the Pasquill-Gifford dispersion coefficients. *Journal of the Air & Waste Management Association*, **40**, 1146–1147.

DEKIEP, E., & PATTERSON, D.J. 1984. Emission control in internal combustion engines. In S. Calvert and H.M. Englund, eds., *Handbook of air pollution technology*. Wiley, New York, pp. 489–512.

DE NEVERS, N. 1995. *Air pollution control engineering*. McGraw-Hill, New York.

DONOVAN, R.P. 1985. *Fabric filtration for combustion sources*. Marcel-Dekker, New York.

EIA. 1997. *Monthly energy review: September 1997*. Report DOE/EIA-0035(97/09). Energy Information Administration, U.S. Department of Energy, Washington, DC.

FLAGAN, R.C., & SEINFELD, J.H. 1988. *Fundamentals of air pollution engineering*. Prentice-Hall, Englewood Cliffs, NJ.

FLAGAN, R.C., & FRIEDLANDER, S.K. 1978. Particle formation in pulverized coal combustion—A review. In D.T. Shaw, ed., *Recent developments in aerosol science*. Wiley, New York, chapter 2.

FREEMAN, H., HARTEN, T., SPRINGER, J., RANDALL, P., CURRAN, M.A., & STONE, K. 1992. Industrial pollution prevention: A critical review. *Journal of the Air & Waste Management Association*, **42**, 618–656.

GOODY, R.M., & YUNG, Y.L. 1989. *Atmospheric radiation: Theoretical basis*. 2nd ed., Oxford University Press, New York, p. 4.

GRAY, H.A., CASS, G.R., HUNTZICKER, J.J., HEYERDAHL, E.K., & RAU, J.A. 1986. Characteristics of atmospheric organic and elemental carbon particle concentrations in Los Angeles. *Environmental Science & Technology*, **20**, 580–589.

HANLEY, J.T., ENSOR, D.S., SMITH, D.D., & SPARKS, L.E. 1994. Fractional aerosol filtration efficiency of in-duct ventilation air cleaners. *Indoor Air*, **4**, 169–178.

HUPA, M., BACKMAN, R., & BOSTRÖM, S. 1989. Nitrogen oxide emissions of boilers in Finland. *JAPCA: Journal of the Air Pollution Control Association*, **39**, 1496–1501.

IPCC. 1996. *Climate change 1995: The science of climate change* (J.T. Houghton, L.G. Meira Filho, B.A. Callander, N. Harris, A. Kattenberg, and K. Maskell, eds.). Cambridge University Press, Cambridge, pp. 13–49.

IRVING, P.M. (ed.). 1991. *Acidic deposition: State of science and technology (summary report of the U.S. National Acid Precipitation Assessment Program)*. NAPAP Office of the Director, Washington, DC, September, p. 185.

JENKINS, P.L., PHILLIPS, T.J., MULBERG, E.J., & HUI, S.P. 1992. Activity patterns of Californians: Use of and proximity to indoor pollutant sources. *Atmospheric Environment*, **26A**, 2141–2148.

KILGROE, J.D. 1984. Coal cleaning. In S. Calvert and H.M. Englund, eds., *Handbook of air pollution technology*. Wiley, New York, pp. 435–488.

LAPPLE, C.E. 1951. Processes use many collector types. *Chemical Engineering*, **58** (5), 144–151.

LAYTON, D.W. 1993. Metabolically consistent breathing rates for use in dose assessments. *Health Physics*, **64**, 23–36.

LEE, B. 1991. Highlights of the Clean Air Act amendments of 1990. *Journal of the Air & Waste Management Association*, **41**, 16–19.

MCCAIN, J.D., GOOCH, J.P., & SMITH, W.B. 1975. Results of field measurements of industrial particulate sources and electrostatic precipitator performance. *Journal of the Air Pollution Control Association*, **25**, 117–121.

MOLINA, M.J., & ROWLAND, F.S. 1974. Stratospheric sink for chlorofluoromethanes: Chlorine-atom catalysed destruction of ozone. *Nature*, **249**, 810–812.

NATIONAL RESEARCH COUNCIL, Committee on Tropospheric Ozone Formation and Measurement. 1991. *Rethinking the ozone problem in urban and regional air pollution*. National Academy Press, Washington, DC.

PASQUILL, F. 1961. The estimation of the dispersion of windborne material. *Meteorological Magazine*, **90**, 33–49.

POPE, C.A., DOCKERY, D.W., & SCHWARTZ, J. 1995. Review of epidemiological evidence of health effects of particulate air pollution. *Inhalation Toxicology*, **7**, 1–18.

ROWLAND, F.S. 1990. Stratospheric ozone depletion by chlorofluorocarbons. *Ambio*, **19**, 281–292.

SEINFELD, J.H. 1988. Ozone air quality models: A critical review. *Journal of the Air Pollution Control Association*, **38**, 616–645.

SEINFELD, J.H. 1989. Urban air pollution: State of the science. *Science*, **243**, 745–752.

SEINFELD, J.H., & PANDIS, S.N. 1998. *Atmospheric chemistry and physics: From air pollution to climate change*. Wiley, New York.

SINGER, B.C., & HARLEY, R.A. 1996. A fuel-based motor vehicle emission inventory. *Journal of the Air & Waste Management Association*, **46**, 581–593.

SMITH, K.R., AGGARWAL, A.L., & DAVE, R.M. 1983. Air pollution and rural biomass fuels in developing countries: A pilot village study in India and implications for research and policy. *Atmospheric Environment*, **17**, 2343–2362.

SWAP, R., GARSTANG, M., GRECO, S., TALBOT, R., & KÅLLBERG, P. 1992. Saharan dust in the Amazon Basin. *Tellus*, **44B**, 133–149.

USEPA. 1981. *Guidelines for use of city-specific EKMA in preparing ozone SIP's*. Report EPA-450/4-80-027. U.S. Environmental Protection Agency, Research Triangle Park, NC.

USEPA. 1985. *Compilation of air pollutant emission factors. Vol. I: Stationary point and area sources*. 4th ed. AP-42. U.S. Environmental Protection Agency, Office of Air Quality Planning and Standards, Research Triangle Park, NC, September (electronic version maintained on-line at http://www.epa.gov/ttn/chief/ap42.html).

USEPA. 1993. *National air pollutant emission trends. 1900–1992*. Report EPA 454/R-93-032. U.S. Environmental Protection Agency, Office of Air Quality Planning and Standards, Research Triangle Park, NC, October.

USEPA. 1994. *National air quality and emission trends report. 1993*. Report EPA 454/R-94-026. U.S. Environmental Protection Agency, Office of Air Quality Planning and Standards, Research Triangle Park, NC, October.

USEPA. 1996. *National air pollutant emission trends. 1900–1995*. Report EPA-454/R-96-007. U.S. Environmental Protection Agency, Office of Air Quality Planning and Standards, Research Triangle Park, NC, October.

WALLACE, L.A. 1989. The exposure of the general population to benzene. *Cell Biology and Toxicology*, **5**, 297–314.

WALLACE, L.A., PELLIZZARI, E.D., HARTWELL, T.D., SPARACINO, C., WHITMORE, R., SHELDON, L., ZELON, H., & PERRITT, R. 1987. The TEAM study: Personal exposures to toxic substances in air, drinking water, and breath of 400 residents of New Jersey, North Carolina, and North Dakota. *Environmental Research*, **43**, 290–307.

WARK, K., & WARNER, C.F. 1981. *Air pollution: Its origin and control*. 2nd ed. Harper & Row, New York.

WESCHLER, C.J., SHIELDS, H.C., & NAIK, D.V. 1989. Indoor ozone exposures. *JAPCA*, **39**, 1562–1568.

WMO. 1990. *Scientific assessment of stratospheric ozone: 1989*. Vol. 1. Global Ozone Research and Monitoring Project, Report No. 20. World Meteorological Organization, Geneva.

ZANNETTI, P. 1990. *Air pollution modeling: Theories, computational methods, and available software*. Van Nostrand Reinhold, New York.

ZHANG, Y., STEDMAN, D.H., BISHOP, G.A., GUENTHER, P.L., & BEATON, S.P. 1995. Worldwide on-road vehicle exhaust emissions study by remote sensing. *Environmental Science & Technology*, **29**, 2286–2294.

PROBLEMS

7.1 Criteria air pollutants

(a) Which pollutant, when present in high concentration, causes the sky to appear brown-orange?

(b) No volatile organic compounds (VOCs) are "criteria pollutants," and yet VOC emissions are subject to aggressive controls. Name two criteria pollutants whose concentrations are significantly influenced by VOCs.

(c) For which criteria pollutant is the United States farthest from compliance with ambient air quality standards, when measured by the number of people who live in nonattainment areas?

(d) Which "criteria pollutant" is also listed as a "hazardous air pollutant"?

(e) Name the criteria pollutant(s) that contribute significantly to acid deposition.

7.2 A potpourri of air quality issues

(a) What specific human process or activity is the dominant cause of acid deposition?

(b) Why do emissions of SO_2 in the upper midwest (e.g., Ohio) lead to acid deposition in the eastern United States? Specifically, why doesn't the acid deposition occur in the near vicinity of the sources?

(c) The use of chlorofluorocarbons that cause stratospheric ozone depletion has been greatly reduced in the past decade. Yet even if releases were completely stopped now, the problem of stratospheric ozone depletion would be with us for many more decades. Why?

(d) Nitric oxide can be emitted from combustion processes. Name three adverse air quality effects that result from NO emissions.

(e) What class of air pollutants is primarily responsible for visibility impairment?

(f) What is the characteristic residence time of air in an urban air basin?

(g) Greenhouse gases include CO_2, CH_4, and N_2O. Name two important physical or chemical properties of these molecules that cause them to be of concern with respect to climate.

(h) From the perspective of how they are regulated, what is the major difference between *criteria pollutants* and *hazardous air pollutants?*

(i) What is the primary cause of human inhalation exposure to benzene?

(j) What pollutant emissions lead to the formation of ozone in urban air?

(k) Name three options available for reducing SO_2 emissions from coal combustion at electricity-generating stations.

(l) How are *emission factors* related to *emission inventories?*

(m) What device or devices would typically be used to control particle emissions from a coal-fired utility boiler?

(n) When fuels containing sulfur are burned, the sulfur is oxidized to SO_2. In the absence of controls, the SO_2 is emitted with the flue gas. Explain briefly the chemical and physical processes that account for how SO_2 emissions contribute to the following air pollution problems.

(i) Respiratory health concerns such as asthma

(ii) Acid deposition

(iii) Visibility impairment

7.3 Benzene: Sources, concentrations, and effects

A main purpose of this problem is to motivate you to become acquainted with the bibliographic resources available in your university library system. Sign up to attend a library-sponsored workshop if one is available.

(a) Identify at least five recent articles in the archival literature that have, as a major theme, benzene as an air pollutant. These may include articles on health effects, emissions, airborne concentrations, exposures, and control measures. Make a list of these articles in an acceptable citation form, such as this: K.J. Krost, E.D. Pellizzari, S.G. Walburn, and S.A. Hubbard, Collection and analysis of hazardous organic emissions, *Analytical Chemistry*, **54**, 810, 1982.

(b) Use these literature sources or other references to answer the following questions:

- What human health hazards are associated with inhalation exposure to benzene?
- What are typical airborne concentrations to which people are exposed?
- At what concentration level are standards or guidelines set to limit adverse effects?
- What are the sources of airborne benzene?

Cite the sources of your information.

7.4 Do you know more than a Harvard graduate?
The following was printed in the December 7, 1992, issue of *Newsweek:*

> *In a documentary film . . . , 23 gowned graduates at a Harvard University commencement are asked the question "Why is it hotter in summer than in winter?" Only two come up with the right answer.*

Explain why it is hotter in the summer than in the winter.

7.5 To air is human
(a) What volume is occupied by 1 mol of methane (CH_4) at a pressure of 1 atm and a temperature of 298 K?
(b) The methane is burned in dry air under stoichiometric conditions. The exhaust gas is cooled and captured. What volume is occupied by 1 mol of exhaust gas at a pressure of 1 atm and a temperature of 298 K?

7.6 Primary versus secondary air pollutants
(a) State the distinction between primary and secondary air pollutants.
(b) What is the significance of this distinction for air quality engineering?
(c) For each of the following pollutants, state whether they are primary, secondary, or both. Briefly justify your answers.

(i) Soot
(ii) HCHO
(iii) NO_2

7.7 On the politics of air pollution control
Automobile manufacturers have argued that further restrictions on automotive emissions of NO_x would be counterproductive to the nation's efforts to reduce ozone concentrations in urban areas as needed to meet the federal standard. In fact, they suggest that relaxing the emissions standard might help meet the goal.
(a) Explain the technical basis of this claim.
(b) Apart from ozone, list three important secondary pollutants whose concentrations would be affected by increased automotive emissions of NO_x. Indicate whether these species would increase, decrease, or remain the same if the automotive NO_x emissions were increased. Give one important health or welfare consequence associated with each species.

7.8 Smog and more smog
Urban-scale air pollution problems exist in many parts of the world. With respect to gaseous pollutants, two major classes of problems are found. One class is the photochemical smog system, as occurs in Los Angeles. The other class of problems is associated with extensive uncontrolled combustion of low-grade fuels, such as coal. The serious air pollution episode in London during 1952 was an example of the latter type of problem. Currently, many urban areas in China suffer air pollution due to uncontrolled coal combustion for activities ranging from industrial processes to home heating. Compare and contrast these two classes of air pollution problems. Consider the following issues in your discussion:

- What specific gaseous pollutants are of concern?
- What are the key features governing the generation of these pollutants?
- What roles do atmospheric transport and transformation play?

- What types of health effects might result from exposure in these environments?
- What are the major elements of a control strategy to manage these air pollution problems?

7.9 Why do cars emit so much CO?

Use the fundamentals of pollutant formation in hydrocarbon combustion systems to explain why internal combustion engines are responsible for 80 percent of anthropogenic CO emissions in the United States although they represent a much smaller fraction of the fuel consumed.

7.10 SO_2 emission factor

A sample of fuel oil (no. 2) was analyzed and found to have the characteristics presented below. Derive the SO_2 emission factor for combustion of this fuel. Your factor should have units of kg (SO_2 emitted) per m^3 (fuel burned).

Species	Percent by mass
C	86.4
H	12.7
N	0.1
O	0.1
S	0.7
Ash	Trace

Density: 0.865 g cm^{-3}

7.11 Determining pollutant emission rates

Certain building materials emit formaldehyde (HCHO), a toxic air pollutant. The following technique is proposed to determine the emissions rate.

1. Place a sample of the material to be tested into a CMFR and then close the chamber. The chamber volume is V (m^3).
2. Supply uncontaminated air into the chamber at a flow rate Q (m^3 s^{-1}). Allow air to flow out of the chamber at the same rate.
3. Monitor the formaldehyde concentration in the outlet air until a steady-state value, C_{ss} (μg m^{-3}), is reached.

Given V, Q, and C_{ss} from such an experiment, what is the formaldehyde emission rate from the sample in μg s^{-1}? Assume that the formaldehyde is nonreactive.

7.12 Gasoline vapor emissions from pumping gas

As a car's gas tank is filled, gasoline vapors in the tank are displaced and may be emitted into the atmosphere. For this problem, assume that the fuel pump is *not* equipped with a vapor recovery nozzle (as are pumps in many areas with poor air quality).

Gasoline is a complex mixture of hydrocarbons. The Reid vapor pressure (RVP) of gasoline represents the equilibrium vapor pressure of hydrocarbon vapors above fresh gasoline, measured at 100 °F (38 °C). Assume that the RVP of gasoline is 5.2×10^4 Pa (0.51 atm) and that gasoline vapors are well represented as pentane (C_5H_{12}). Also, assume that the gasoline vapors are in equilibrium with liquid gasoline in the fuel tank. Consider a day that is warm and sunny so that the gasoline vapor pressure is well represented by the RVP.

(a) Determine the emission rate of gasoline vapors associated with refueling, in units of grams emitted per gallon of gasoline pumped.

(b) Estimate the total nationwide emission rate of gasoline vapors for the United States by this mechanism in units of kg per day. Compare your result with the U.S. Environmental Protection Agency emission estimate of 62 million kg per day of volatile organic compounds emitted into the U.S. atmosphere. The gasoline consumption rate in U.S. automobiles is 70×10^9 gallons per year.

7.13 Impact of unvented kerosene space heaters on indoor sulfur dioxide levels
Unvented combustion appliances have become popular as space heaters because they are almost 100 percent energy efficient. Unfortunately, when these heaters are used indoors, not only is all of the energy in the fuel delivered to the room, but all of the combustion by-products are delivered, too. Several air quality problems may result from the use of an unvented combustion appliance. Your analysis is to be based on the following data:

Basic effective formula of kerosene: $CH_{1.8}$

Sulfur content of the fuel: 0.035% by mass

Energy delivery rating of the heater: 3 kW

Heating value of kerosene: 46.4 kJ g^{-1}

Size of the room to be heated: 40 m^3

(a) Assume that the heater is operated at its rated limit and that all sulfur in the fuel is liberated as SO_2. What is the emission rate of SO_2 into the room?

(b) Assume that the room can be modeled as a CMFR, that SO_2 is chemically inert, and that the outdoor level of SO_2 is negligible. In steady state, what flow rate of ventilation air is required to maintain the SO_2 concentration below the primary 24-hour standard? (*Hint:* See Appendix F, Table F.3, for the standard.)

(c) If enough ventilation is supplied to ensure that the concentration of SO_2 is maintained below the standard, how much of an increase in air temperature between outdoors and indoors is possible if there is no loss of energy from the room other than the outward flow of heated air?

7.14 Measuring CO emissions from a kerosene heater
An experiment is conducted in an experimental room to determine the emissions of carbon monoxide from a kerosene heater. The ventilation rate and temperature of the room are maintained at constant values. The CO level in the ventilation supply air is negligible. For $t < 0$, the indoor concentration of CO is negligible. At $t = 0$, the heater is ignited and operated at a constant rate. The concentration of CO in the room is monitored and the results plotted versus time. Use the figure at the top of page 476 along with the following data to determine the CO emission factor in μg kJ^{-1}. (*Hint:* You may assume that the emission rate is constant.)

$V = 40\ m^3$	Volume of chamber
$Q = 30\ m^3\ h^{-1}$	Ventilation rate
$F = 6000\ kJ\ h^{-1}$	Fuel consumption rate

7.15 Material balance, stoichiometry, and air pollution
A bituminous coal has the following effective chemical formula:

$$CH_{0.823}N_{0.016}O_{0.068}S_{0.0036}$$

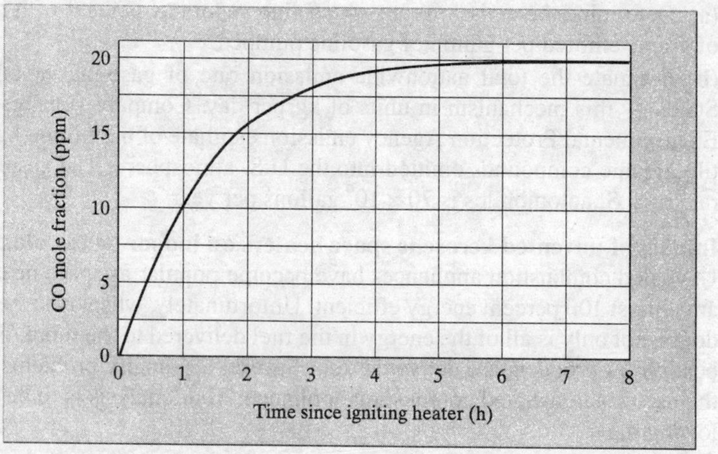

When operating at full capacity, an electric power–generating boiler consumes this fuel at a rate of 44 kg s^{-1}. Over an annual cycle, the boiler operates at 75 percent of its rated capacity. The sulfur in the fuel is converted to SO_2 during combustion. It is removed from the flue gas by means of a scrubber with 80 percent efficiency. What is the SO_2 emission rate from this power plant in metric tons (10^3 kg) per year?

7.16 Combustion of landfill gas

In landfills, microbes act on biodegradable components of the solid waste to produce gases that are depleted in oxygen. Under uncontrolled conditions, these gases escape slowly to the atmosphere, where they are gradually oxidized. A landfill operator has hired you to help investigate the economic feasibility of extracting these gases and burning them to produce usable energy. The composition of the landfill gas (mol %) is as follows:

Methane	CH_4	85%
Ammonia	NH_3	7%
Hydrogen sulfide	H_2S	5%
Water vapor	H_2O	3%

(a) Assume that an equivalence ratio of 0.9 is to be used. Write an overall reaction for complete combustion of this gas in dry air and determine the exhaust gas composition (i.e., the product species and their respective mole fractions in the exhaust). Assume that 30 percent of the fuel nitrogen is converted to NO.

(b) Determine the fuel-air mass ratio for conditions in part (*a*).

(c) For complete combustion, which component(s) of the exhaust gas would be considered urban air pollutants? Describe briefly the effects of the pollutant(s).

7.17 Air pollution impacts of electricity production

(a) Consider electricity generation by conventional fossil fuel combustion. Briefly summarize the relative merits of pulverized coal combustion versus natural gas combustion, with respect to each of the following categories of air pollution problems.

(i) Climate

(ii) Acid deposition

(iii) Urban air quality

(b) In light of your response to *(a)*, why is coal still widely used for electricity generation?

(c) What pollution control devices or techniques would be applied in a well-controlled pulverized coal combustor for electricity generation? Justify your answer.

7.18 Fuel selection and air pollution

Presented in Table 7.B.3 are emission factors and heating value data for selected fuels burned in utility boilers. Consider a thermal electric power plant that generates an average of 400 MW of electricity with an overall efficiency of 35 percent.

(a) Determine for each fuel the annual uncontrolled mass of each of the six listed pollutants that would be emitted from such a power plant. Give your results in the form of a table, using units of metric tons per year.

(b) Discuss briefly the significance of your results.

(c) The coal-fired power plant is equipped with an electrostatic precipitator that removes 98 percent of the particles from the flue gas. How much ash mass must be disposed of annually?

7.19 Control requirements for automobile engines

The 1994 federal emissions standard for light-duty passenger cars is 3.4 g per mile of CO and 0.4 g per mile of NO. In the absence of controls, the mole fractions of CO and NO in the exhaust gas are 0.01 and 0.0025, respectively. Consider an automobile that has a fuel economy rating of 25 miles per gallon. Assume that gasoline may be represented by octane, C_8H_{18} (density 0.70 g cm^{-3}). Also assume that combustion occurs under stoichiometric conditions.

(a) Determine the emission rate on a gram-per-mile basis for CO and NO in the absence of controls. What fractional reductions in emissions, if any, are required in order to meet the standard?

(b) Assuming that the mole fractions of CO and NO in untreated exhaust air are constant, discuss the relationship between fuel economy and the level of treatment of exhaust gas that is required to meet the emissions standard. Why would it not be a good idea to specify the emission standard on the basis of grams of contaminant emitted per gram of fuel burned?

7.20 Not a tornado

Consider a cyclone designed for treating particle-laden gas from an industrial process. The uncontrolled source emits particles that are distributed broadly over the size range 1–30 μm. Overall, the cyclone exhibits efficiency for this source of 90 percent on a mass basis. The suggestion is made that the treatment efficiency can be improved by adding a second cyclone, identical to the first, in series. Approximately what overall particle mass removal efficiency would you expect from the two units in series? Explain your answer.

7.21 Effectiveness of air cleaning for indoor odor control

Malodorous molecules are emitted from some source continuously into a room at rate E_o (g h^{-1}). We seek to achieve odor control by passing air through a wall-mounted filtration unit at the rate Q_f (m^3 h^{-1}). Malodorous molecules are removed from this air stream with a single-pass efficiency η_f (-). The conditions are shown schematically in the figure at the top of page 478. The room is modeled as a well-mixed reactor of volume *V*. Ventilation air flows through the room at a volumetric rate *Q*. Odorous molecules "*o*" are absent from the ventilation air. The mass concentration of odor molecules is C_o.

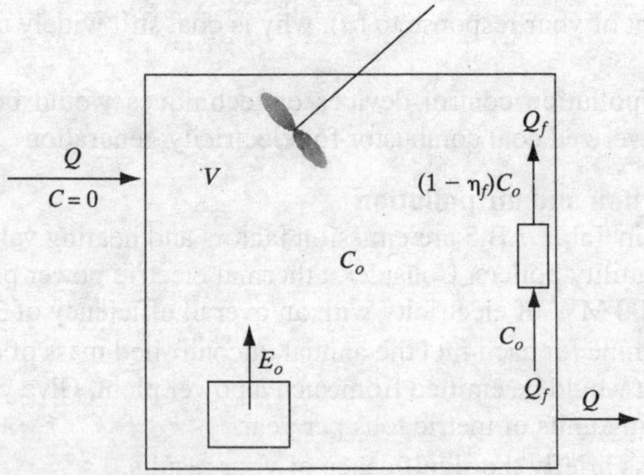

We define the effectiveness of odor control as follows. Let C_o^* be the steady-state concentration of odor molecules under uncontrolled (base-case) conditions, and let C_o be the steady-state concentration with the control device (recirculating air filter) operating. The odor control effectiveness of the control device, ε, is then

$$\varepsilon = 1 - \frac{C_o}{C_o^*}$$

(a) What are the numerical limits on the effectiveness, ε?

(b) Which is more desirable, a high value of ε or a low value?

(c) Derive an expression for the odor control effectiveness, ε, in terms of the parameters E_o, Q, Q_f, V, and η_f.

7.22 The case of the nitrogen oxides

Consider nitrogen oxides as a class of air pollutants. For the purposes of this problem, consider nitrogen oxides to be all gaseous species that contain both nitrogen and oxygen. Answer the following questions.

(a) In what primary chemical form are nitrogen oxides emitted?

(b) What is (are) the main mechanism(s) of formation of nitrogen oxides in combustion systems?

(c) What is (are) the main control method(s), and how does it (do they) work?

(d) Two key nitrogen oxide species are formed mainly in the atmosphere. Name them and describe the key formation pathways.

(e) What are the main adverse effects associated with each of the species named in *(d)*?

7.23 Zamboni!

At indoor ice-skating rinks, the resurfacing machine, known as a Zamboni, is typically powered by means of an internal combustion engine that emits air pollutants into the rink building. Let's model the rink as a CMFR, with volume V = 20,000 m^3 and ventilation rate Q = 250 m^3 min^{-1}. Assume that the CO concentration in outdoor air is negligible. Also assume that the indoor concentration is zero before the Zamboni begins operation.

Consider the carbon monoxide concentration that results from one operation cycle of the Zamboni. To resurface the ice, the Zamboni is operated for 15 minutes and emits 100 g/min of CO during operation.

(a) What is the peak CO concentration in the air in the rink during and after the single Zamboni operation cycle?

(b) Sketch the CO concentration in the rink air versus time for a 2-hour period beginning with the start of the resurfacing operation.

(c) A skater exercises on the rink for 1 hour beginning immediately after the resurfacing operation is complete. What is the CO concentration at the end of that hour?

(d) What is the 1-hour average concentration encountered by the skater? Compare your answer with the 1-hour ambient standard of 40 mg/m^3 (= 35 ppm).

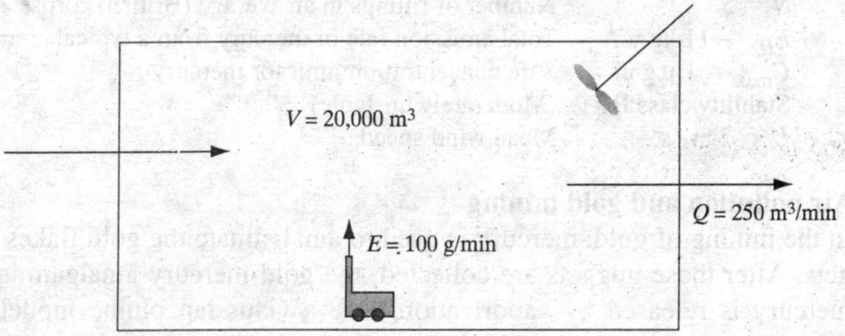

7.24 Gaussian plumage?

Sulfur dioxide is continuously emitted from the exhaust stack of a power plant. Assume that the steady-state downwind concentration is described by the Gaussian plume equation. Assume that SO_2 does not react either in the air or on the ground.

(a) Plot the SO_2 concentration at ground level downwind of the stack versus distance along the plume centerline (i.e., directly downwind). Use a linear scale for the concentration and a logarithmic scale for distance. Be sure to include a large enough range of distances to clearly show the rise and fall near the peak concentration.

(b) At what distance downwind does the maximum concentration occur?

(c) At the distance downwind determined in part (*b*), plot the ground-level concentration versus distance from the centerline along a crosswind line. Use linear scales for both concentration and distance.

(d) The U.S. national ambient air quality standard for SO_2 is 365 μg m^{-3}, averaged over a 24-hour period. Determine the area over which the ground-level SO_2 concentration exceeds the standard.

Data

Emission rate	m = 150 g s^{-1}
Wind speed	U = 2 m s^{-1}
Stability class	B
Stack height	H = 30 m

7.25 Mercury poisoning from crematoria?

An article in the journal *Nature* suggested that crematoria might be sources of excessive mercury emissions. Mercury is used in silver-colored tooth fillings. When a corpse is burned, the mercury is vaporized and may be released with the flue gas. Use the data below to answer the following questions:

(a) What is the rate at which corpses are burned in the crematorium?

(b) At the given average emission rate, and for the meteorological and dispersion parameters specified below, what must the effective stack height be to avoid concentrations in excess of the safe limit? Use the appropriate Gaussian plume equation, assuming that the ground is perfectly reflecting and that there is no elevated inversion. (Assume that the dispersion parameter information in Tables 7.D.2 and 7.D.3 is valid for $x < 0.1$ km.)

(c) If the minimum stack height determined in (*b*) were used, and if the atmospheric conditions were more stable (i.e., class C or E), would the maximum downwind concentration be increased or decreased? Explain your reasoning.

Data

$M_{Hg} = 0.6$ g	Mass of mercury in an average filling
$N = 5$	Number of fillings in an average (British) corpse
$E_{Hg} = 11$ kg y^{-1}	Total emission rate of mercury from a typical crematorium
$C_{max} = 1\ \mu$g m^{-3}	Safe concentration limit for mercury
Stability class B	(Moderately unstable)
$U = 2$ m/s	Mean wind speed

7.26 Air pollution and gold mining

In the mining of gold, mercury is used to amalgamate the gold flakes into larger nuggets. After these nuggets are collected, the gold-mercury amalgam is heated and the mercury is released by vaporization. Use a Gaussian plume model to analyze the downwind concentrations of mercury that result from this process. (Assume that the dispersion parameter information in Tables 7.D.2 and 7.D.3 is valid for $x < 0.1$ km.)

(a) What is the distance downwind of the processing point at which the predicted ground-level concentration falls to 1 μg/m^3?

(b) A second, identical processing point is established 100 m from the first. What is the predicted ground-level concentration at the point determined in part (*a*) if the second point is located in the crosswind direction?

Data

Wind speed	2.0 m/s
Emission height	Ground level
Hg uptake at ground	None
Incoming solar radiation	Moderate
Mercury vaporization rate	1 g min^{-1}

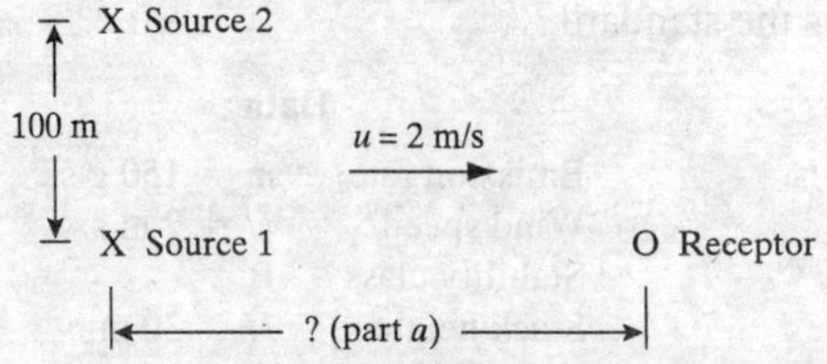

7.27 Control strategy for ambient particulate matter

Recent research has shown an association between mortality and airborne particle levels. Assume that a new stringent air quality standard is adopted for fine particulate matter ($PM_{2.5}$). The local area is out of compliance, so the Air Quality Management District retains your consulting firm to draft a plan that would reduce concentrations at all monitoring stations to below the new standard. Discuss the central features of your response. Address the following points in your discussion as well as any others that you consider to be of central importance.

- Overall control strategy
- Quantification of emission sources
- Available control technologies

7.28 Saharan dust in the Central Amazon Basin

Recently, researchers have argued that there may be a link between drought in western Africa and soil fertility in South America (Swap et al., 1992). The occasional windstorms that sweep through the Sahara Desert cause the suspension of large amounts of soil and dust into the troposphere. This material can then be transported by the prevailing winds westward, across the Atlantic Ocean. Frequent rainstorms facilitate its deposition in the Central Amazon Basin (CAB). There the dust contributes essential nutrients to help support plant growth. A key question arises: Is it possible for wind-blown soil dust to travel oceanic distances? In this problem, you are to explore this question using simple tools to model particle transport in the system. In your analyses, use the following symbols:

N = particle number concentration

x = distance along the direction of airflow, measured from the Sahara

U = wind speed, assumed to be uniform with height

H = height of the surface layer through which particles are well mixed

v_d = particle settling velocity

L = distance between the Sahara and the Central Amazon Basin

You may make the following assumptions:

- The only particle removal mechanism between the Sahara and the CAB is gravitational settling onto the earth's surface.
- Particles are continuously mixed in the vertical direction so that the concentration remains constant as a function of height throughout the mixed zone, H.
- Mixing in the horizontal plane may be neglected.
- The only particle source in the system is wind-blown dust in the Sahara.
- Wind blows on a straight line from the Sahara to the CAB.
- The entire system is in steady state.

(a) Using the symbols defined above, determine characteristic time scales τ_a for advection of particles from the Sahara to the CAB and τ_s for particle settling to the earth's surface. Explain how the relative magnitude of these time scales determines the likelihood of particles of a given size being transported across the Atlantic.
(b) Now set up a simple PFR model. Write a differential equation that gives dN/dx in terms of the defined symbols. Solve the differential equation to obtain $N(L)/N_0$ as a function of these symbols. N_0 is the particle concentration in the Sahara (at $x = 0$).
(c) Given the results of your analysis in (*a*) and (*b*) plus the data below, evaluate $N(L)/N_0$ separately for clay, silt, and sand particles. Comment on the significance.

Clay	$d_p = 1\ \mu m$	$v_d = 0.009\ cm\ s^{-1}$
Silt	$d_p = 10\ \mu m$	$v_d = 0.8\ cm\ s^{-1}$
Sand	$d_p = 100\ \mu m$	$v_d = 60\ cm\ s^{-1}$
Distance from Sahara to CAB		$L = 6000$ km
Mixed depth of air		$H = 4$ km
Average wind speed		$U = 30\ km\ h^{-1}$

7.29 Indoor smog

Indoor lighting levels are so much lower than outdoor daylight conditions that photolytic reactions can practically be ignored in indoor air. Considering only the primary photolytic cycle, evaluate the effect of the change in lighting levels on the indoor ozone concentration. Assume that the following conditions prevail, corresponding to an average morning in Los Angeles: Outdoor levels of NO and NO_2 are 10 and 50 ppb, respectively, and the outdoor photolytic reaction rate for NO_2 dissociation is 0.25 min^{-1}. The temperature is 298 K.

(a) Assuming that the photostationary state relation applies, determine the outdoor ozone concentration.

(b) Assume that an air parcel is brought into a building and does not mix with the indoor air. What are the steady-state concentrations of ozone, NO, and NO_2 in the parcel?

(c) Estimate the characteristic time required for the concentrations of ozone, NO, and NO_2 to relax to their steady-state values.

7.30 Environmental tobacco smoke

Consider a small building that can be modeled as a single CMFR. At 9:00 A.M., a single cigarette is smoked. It takes 6 minutes to smoke the cigarette, and the total mass of smoke particles emitted into the air during that period is 10 mg. Let $E(t)$ represent the emission source in units of particle mass per time. This is the only particle source you are to consider. Here are other relevant data for the building.

$V = 120\ m^3$	Volume
$\lambda_v = 0.5\ h^{-1}$	Air exchange rate
$k_r = 0.2\ h^{-1}$	Particle loss rate by deposition onto indoor surfaces

(a) Write an appropriate governing equation that describes the rate of change of tobacco smoke particle mass concentration with time in the building. Express your relationship in terms of the variables defined above.

(b) Solve the governing equation and plot the indoor tobacco smoke particle mass concentration from 8:00 A.M. to 8:00 P.M.

(c) What is the mass of tobacco smoke that deposits onto indoor surfaces?

(d) Assume that a nonsmoker is present in the building during this 12-hour period. The nonsmoker inhales air at a rate of 10 L min^{-1}. Assume that the overall lung deposition efficiency for environmental tobacco smoke is 11 percent. Evaluate, for the nonsmoker, the deposited mass of tobacco smoke particles. Compare your result with the dose of mainstream smoke particles deposited in the lungs of the smoker, which is likely to be in the range of 5–10 mg per cigarette.

7.31 Carbon monoxide emissions from motor vehicles

As illustrated in Figure 7.B.5, remote sensing can be used to characterize the emissions of some pollutants from motor vehicles under real driving conditions. This approach has been used extensively to characterize CO emissions. The measurement system can detect the mole fractions of CO and CO_2 in the exhaust.

The emission standard for CO from motor vehicles is expressed in terms of grams of CO emitted per mile driven. Determine the CO emission factor based on the measurement results from the infrared detection system and the other basic data about gasoline and fuel economy given below. (*Hint:* Assume that all of the carbon in the burned fuel is emitted as either CO or CO_2.)

$Y_{CO} = 150$ ppm	Average mole fraction of CO measured in the plume
$Y_{CO_2} = 10{,}000$ ppm	Average mole fraction of CO_2 measured in the plume

$CH_{2.2}$	Effective molecular composition of gasoline
M = 25 mi/gal	Fuel economy (miles driven per gallon of gasoline consumed)
ρ = 2840 g/gal	Fuel density

7.32 Measuring automotive CO emissions in a roadway tunnel

One method for quantifying motor vehicle emissions is to conduct experiments in a tunnel. Given the following data, calculate an average CO emission factor in grams emitted per liter of fuel burned.

Hints:

1. It may not be necessary to use all of the data specified.
2. Assume that all C in gasoline is converted to either CO or CO_2 when burned.

Data

Species mole fractions in tunnel air	CO_{in} = 32 ppm
	CO_{2in} = 1000 ppm
Species mole fractions in outside air	CO_{out} = 1 ppm
	CO_{2out} = 365 ppm
Gasoline composition	$CH_{2.2}$
Gasoline density	750 g/L
Length of tunnel	L = 1100 m
Average vehicle fuel economy	E = 5 km/L
Vehicles traveling through tunnel	F = 4300 vehicles/h
Average vehicle speed	V = 65 km/h

7.33 Atmospheric stability

(a) For air quality, is stable stratification generally good or bad? Explain.

(b) Describe the underlying physical reasons that ground-based inversions in the atmosphere occur on calm, clear nights.

(c) Sketch the air temperature profile with height in the atmosphere for the case of a ground-based inversion extending up to 100 m, with neutral stability conditions above.

(d) A ground-based temperature inversion exists up to a height of 200 m. The rate of temperature increase with height within the inversion is 1 K per 100 m. Above the inversion, the lapse rate is adiabatic. If the temperature at the ground surface is 15 °C, what is the temperature 1 km above the ground?

8

Hazardous Waste Management

8.A OVERVIEW

Hazardous wastes are produced by diverse activities, ranging in nature from highly industrial (such as materials manufacturing and chemical processing) to domestic (such as backyard gardening and household painting). A major challenge facing environmental engineers is to develop methods for properly managing hazardous waste, including minimizing generation, treating hazardous wastes that have been generated, and disposing of residuals.

8.A.1 History and Case Studies

The earliest hazardous wastes were natural materials, such as the toxic metals mercury and lead, that were mined, refined, and processed for various human uses. With the arrival of the industrial revolution came the large-scale extraction and utilization of potentially hazardous energy-related materials such as coal and petroleum. The need for materials with special properties during World War II, coupled with advances in chemistry, led to the development and production of synthetic organics such as artificial rubber, structural plastics, herbicides, pesticides, solvents, lubricants, and insulators. Following the war, a lucrative commercial industry emerged to satisfy a growing demand for chemical products. Improvements in sanitation and advances in medical sciences increased human life span and productivity. In turn, this generated an increase in personal consumption, as more discretionary income became available.

Before the 1960s, little attention was given to the production or disposal of hazardous wastes. In fact, when the concept of hazardous waste was introduced, people thought that it was an industrial problem that primarily affected workers. High-profile events that occurred in the 1960s and 1970s heightened public awareness of the widespread potential dangers of hazardous wastes. Several of these events are summarized in the following paragraphs.

Love Canal, New York

To support his vision of a large industrial complex in Niagara Falls, NY, William T. Love began an effort in 1892 to harness hydropower to generate direct current (DC) electricity at the junction of Lake Erie and Lake Ontario. Because DC energy cannot be transported long distances, it was necessary to generate the electricity near where it would be used. Love began construction of a 12 km canal to divert water from the upper Niagara River, around Niagara Falls, to the lower Niagara River in order to drive turbines to produce the electricity. In the late 1890s the development of alternating current (AC) electricity, which can be transported long distances, together with an economic slump derailed Love's plan. He abandoned the canal project after digging a 900 m × 30 m × 3–5 m (deep) hole approximately 450 m from the upper Niagara River.

During the early 1900s the hole filled with water and was used for swimming, ice skating, and other types of local recreation. Meanwhile, industries grew in the vicinity of Niagara Falls, including Hooker Chemical, a producer of pesticides, caustic, fertilizers, industrial chemicals, and plastics. Hooker became one of the biggest employers in Niagara Falls. In 1942 Hooker obtained permission from Niagara Power to dump chemical waste at the defunct canal site and subsequently bought the site in 1946. From 1947–1952, Hooker Chemical and others, including the city of Niagara Falls, dumped 22,000 tons of chemical and municipal waste, including lindane, dioxin, pesticides, chlorinated benzene, sulfides, and other compounds at the canal site.

In the early 1950s, the population of Niagara Falls expanded and the land surrounding the canal site was developed with tracts of middle-class homes. In 1952, desperate for land and strapped for cash, the local school board approached Hooker Chemical to acquire the canal site for the purpose of building a school. Hooker Chemical capped the canal hole, backfilled trenches with layers of clay and dirt, and sold the land to the local school board for $1 with a contract clause disclaiming future liability with regard to the chemicals buried on-site. In 1954, after breaking ground for the new school, the building location had to be shifted and the basement eliminated due to the dangers posed by the buried chemicals. The school operated from 1955 to the 1970s. During this time, there were anecdotal references to children being burned by exposure to chemicals on the playground, small explosions caused by reactant "pebbles" found on playing fields, and puddles with colorful and smelly surfaces. In 1968, the relocation of a street near the old canal caused a rupture in the canal cap, and Hooker Chemical agreed to haul away 40 truckloads of contaminated waste and dirt.

In the middle of the 1970s, a series of events elevated local concerns about the buried wastes. A backyard pool spontaneously lifted 0.8 m out of the ground and filled with colorful oozing fluid. Solvent smells and black fluid flowed into basements and city drains, corroding sump pumps. Following a series of heavy rains, chemical drums surfaced in the canal area. In the late 1970s, the USEPA and New York State began sampling air, soil, and groundwater, and a series of epidemiological studies was conducted on the local residents. In 1978 the school was closed and a partial evacuation was begun of houses surrounding the canal site. After federal studies reported chromosomal damage in the local population, the entire housing area surrounding Love Canal was evacuated and remediation of the site was initiated. Local houses were demolished and buried along with contaminated soils from the surrounding area and nearby creeks. The site was covered with a 40-acre clay and plastic cap. Drains were built adjacent to the site to collect leachate that flowed from the contaminated area, and an on-site leachate treatment facility was built. It is expected that this leachate treatment facility will be needed in perpetuity. In 1994, following years of litigation, Occidental Petroleum, the company that had bought Hooker Chemical, agreed to pay a portion of the remediation costs ($98 million) to the state of New York and to take responsibility for the perpetual leachate treatment system (estimated cost $10–25 million). Evacuated citizens were awarded $25 million for health and property damages.

The final chapter in this story may not yet have been written. In 1988 the USEPA pronounced major portions of land surrounding the canal site as "habitable." The Love Canal Area Revitalization Association (LCARA) began to fix up and sell houses in the community at 20 percent below market value, allowing a new generation of families to populate the Love Canal site (Figure 8.A.1).

Times Beach, Missouri

In the early 1970s the Bliss Waste Oil Co. accepted contracts from chemical industries to haul away waste oils. These oils were contaminated with chemicals such as 2,3,7,8-tetrachlorodibenzo-*p*-dioxin (TCDD), commonly known as dioxin, and polychlorinated biphenyls (PCBs). At the same time, the company entered into contracts to spray oil on roads for dust control in eastern Missouri. Dirt roads located in the small town of Times Beach, 20 miles southwest of St. Louis, were sprayed yearly by Bliss. The first signs of trouble arose when animals began dying of mysterious causes: sparrows, mice, cats, dogs, chickens, and finally horses at a horse-breeding farm that had been sprayed by the Bliss Co. Evidence of human health effects soon followed. Initially, the dioxin was probably mostly contained within the sprayed roadways and fields. However, in the early 1980s the town of Times Beach was flooded by the nearby Meramec River, an event that caused broad dispersal of the dioxin-laden sediments and forced

(*a*)

(*b*)

Figure 8.A.1 Photographs of the Love Canal area (*a*) in 1978, just prior to the initial evacuations, and (*b*) in 1988, just prior to the declaration that the area surrounding the canal site is once again habitable, following the major remediation effort. (Reprinted with permission from M. Kadlecek, Love Canal—10 years later, *The Conservationist*, Nov–Dec, 1988. Photographs are from The New York State Department of Environmental Conservation.)

evacuation of most of its residents. In 1983 the U.S. Centers for Disease Control and the EPA called for permanent relocation of all 2200 residents of Times Beach due to dioxin contamination. After evacuation, the town was closed and fenced with round-the-clock patrol guards. Private properties were purchased by the Federal Emergency Management Agency (FEMA) and the state of Missouri. Remediation of the contaminated soil and structures commenced. Contaminated soils and sediments were excavated both from the town and from the Meramec River. These materials were treated in a thermal unit before on-site burial. The town is presently closed to habitation but still contains a controversial hazardous waste incinerator used for remediating both local contaminated soil and dioxin-contaminated soils from 27 other sites in eastern Missouri.

Valley of Drums, Kentucky

The owner of a drum-cleaning business located on a 13-acre property in Kentucky saw the potential for making extra money by accepting hazardous wastes for disposal on his land. Between 1967 and 1977, Arthur L. Taylor received thousands of drums filled with hazardous materials, including paint manufacturing wastes, xylenes, chlorophenols, and polynuclear aromatic hydrocarbons. Taylor emptied the material from these drums into pits excavated on his property, onto surface soils, or into a nearby creek and then piled up the drums for recycle. The pits were eventually covered with soil. Following the owner's death in 1977, over 4000 drums were found on the site along with extensive soil, groundwater, and surface-water contamination that included heavy metals, volatile organics such as ketones, plastics such as phthalates, and polychlorinated biphenyls. Heavy contamination of the local creek, a tributary of the Ohio River, prompted state authorities to take remedial action. Approximately 30 percent of the remaining wastes at the site were removed in 1980 by industries identified as hazardous waste generators who had contracted with Mr. Taylor for disposal. The USEPA took responsibility for removing the remaining wastes and for overall site remediation, including excavation of contaminated soils and sediments and installation of interceptor trenches and a clay cap.

Minamata Bay, Japan

A chemical manufacturing facility in southern Japan used mercury oxide as a catalyst to produce acetaldehyde. During the 1950s and 1960s, large quantities of inorganic and methylated mercury were discharged from the facility into Minamata Bay. Although limited exposure to inorganic mercury is not significantly toxic (indeed, pure liquid mercury was once prescribed to cure constipation!), the organic forms of mercury can cause paralysis and sensory loss. Furthermore, organic mercury bioaccumulates in the fatty tissues of animals, resulting in long ecosystem residence times and significant accumulation in the upper portions of the food chain. Although some organic mercury was directly discharged into the bay, an important source of this material in Minamata Bay was the anaerobic conversion of settled inorganic mercury to methylated mercury in bottom sediments. The organic mercury in the bay accumulated in the local shellfish, a major food staple of the people living in the region. This caused epidemic mercury poisoning, with large numbers of neural disorders, sensory loss, paralysis, and the deaths of over 100 people. Minamata Bay underwent a huge dredging operation to remove the majority of sediment-bound mercury between 1977 and 1990. It is estimated that natural processes will finally complete the remediation by the year 2011.

High-profile events such as these made the public aware of the potential dangers associated with hazardous wastes. It became apparent that legislation was needed for several reasons:

1. To track the generation of hazardous wastes
2. To regulate the means by which hazardous wastes are treated, stored, transported, and disposed of
3. To develop mechanisms to detect and remediate past and present hazardous waste contamination of the environment

The result of this emerging public concern in the United States was the passage of regulations aimed specifically at the management of hazardous wastes and the releases of hazardous materials into the environment. The two major laws that address these issues are discussed in the next section.

8.A.2 Hazardous Waste Regulatory Framework

When Congress passes a law, it is written into a document called the U.S. code. The general terms of the law are then translated into specific regulations by an appropriate federal agency (e.g., USEPA). These regulations are added to the *Code of Federal Regulations* (CFR), which is a compilation of all current federal regulations of the United States. The federal agency interprets the general terms of the congressional law to produce specific enforceable regulations.

The Resource Conservation and Recovery Act (RCRA)

The Resource Conservation and Recovery Act (RCRA) was the first comprehensive federal effort to mandate the handling, treatment, and disposal of hazardous wastes. The primary goals of RCRA are to protect human health and the environment, to reduce waste, and to conserve energy and natural resources. RCRA establishes a legal definition of hazardous waste and requires cradle-to-grave management. Its regulations affect generators, transporters, and facilities for storage, treatment, and disposal. RCRA gives the USEPA primary regulatory authority over solid waste, hazardous waste, medical waste, and underground storage tanks.

After Congress passed RCRA in 1976, the USEPA took 4 years to write and begin enforcement of the associated regulations. RCRA is one of the most comprehensive regulatory programs ever instituted, occupying 500 pages in the *Federal Register* (FR 5/19/1980) and imposing an estimated paperwork burden for industry of 1.5 million hours per year. In 1984 Congress passed the Hazardous and Solid Waste Amendments (HSWA) to RCRA. The HSWA was written in a more detailed manner than RCRA, because Congress was unhappy with the way the EPA had interpreted RCRA. Specifically, Congress felt that the RCRA regulations were too lax and that the EPA took too long to write and implement them. The HSWA broadened the RCRA mandates to include new requirements for land disposal facilities, small-quantity waste generators such as auto mechanics and dry cleaners, and underground storage tanks. The HSWA also includes a land ban for liquid hazardous waste (see the discussion below).

RCRA addresses five major elements for the management of hazardous waste:

1. Definition and classification of hazardous and nonhazardous waste
2. Regulation by a cradle-to-grave manifest system, record keeping, and reporting requirements

3. Establishment of requirements to be followed by generators, transporters, and owners of treatment, storage, and disposal (TSD) facilities
4. Enforcement of standards through a permitting program and civil penalty policies
5. Authorization of state programs to operate in lieu of the federal program

In the late 1970s and early 1980s, new laws (such as the Clean Water Act) prohibiting the disposal of hazardous wastes to surface waters and increasing costs for land disposal of wastes led to illegal "midnight" dumping of wastes and other shady operations. For example, a money-making scheme that became popular with unscrupulous operators at the time was to steal a large truck, accept a load of hazardous waste—promising proper disposal for a hefty fee—and then abandon the waste-laden truck on the side of the road. Implementation of the manifest system halted such practices by requiring hazardous waste generators to fill out a form upon generation of a hazardous waste. Items to be specified on the form include a description of the waste; the generator's EPA number; the name and address of the generator; transporter(s); and treatment, storage, and disposal facilities to which the waste is sent. Following passage of the HSWA, manifests also include a waste minimization certification. The generator retains one copy of the form, and a second copy is sent to the USEPA. Other copies of the form accompany the waste until its ultimate disposal, with additional portions filled out by any transporters or treatment, storage, and disposal facilities that may handle the waste. Each entity that handles the waste retains a copy of the manifest, and the ultimate treatment, storage, and disposal facility sends a copy of the completed manifest back to the generator and the USEPA. In this manner, both the generator and the USEPA can track a waste from generation to ultimate disposal, and the generator can be held liable for any mishaps along the way. In fact, under current RCRA rules, the generator is also responsible for any long-term problems associated with the chosen disposal method for the waste. Stiff fines are imposed by the USEPA for omissions or mistakes on manifests (e.g., \$100 for simply omitting an area code, zip code, or signature).

RCRA requires hazardous waste generators to obtain an EPA identification number and to maintain proper records of their wastes. RCRA also specifies methods for proper handling of hazardous waste while it remains on-site and for packaging and labeling waste properly for transport.

Requirements for transporters of hazardous wastes include obtaining an EPA identification number and maintaining a proper manifest of wastes. Additional requirements specify how to deal with hazardous waste spills.

Treatment, storage, and disposal (TSD) facilities must obtain operating permits from the USEPA by fulfilling both design and operating requirements. Proper procedures must be established for spill prevention, for responding to chemical release emergencies, for ongoing groundwater monitoring, and for closure and postclosure maintenance of the facility. As of 1995, there were 1983 RCRA-regulated TSD facilities in the United States (USEPA, 1997a).

Corrective Action RCRA includes a set of regulations aimed at cleaning up spills or accidents that occur at USEPA-permitted facilities. These regulations are referred to as corrective action requirements. All USEPA facilities that are permitted to manage hazardous wastes must have a plan to address hazardous waste releases. They must also prove that they have the financial resources to cover the cost of implementing corrective measures. RCRA provides for enforcement of the corrective action regulations by means of administrative actions (fines and permit restrictions), civil actions (formal lawsuits), and criminal actions (possibly including large fines and jail time for responsible individuals).

Land Ban Introduced by Congress as part of the HSWA of 1984, the land ban requires a national phase-out of land disposal of untreated liquid hazardous wastes. The land ban was established to protect groundwater and surface water from contamination by runoff or drainage following land disposal of wastes. Restricted wastes under the land ban include solvents and dioxins, cyanide wastes, dissolved metal wastes, chlorinated organics, and all listed RCRA wastes. For all restricted hazardous wastes that were to be disposed of on land, the USEPA was required to establish treatment standards based on concentration limits and best demonstrated available technologies that were protective of human health and the environment. The intent of the land ban was to make land disposal a waste management alternative of last resort and to encourage waste minimization and effective waste treatment to reduce the quantity and toxicity of hazardous wastes.

The Comprehensive Environmental Response, Compensation, and Liability Act (CERCLA)

The Comprehensive Environmental Response, Compensation, and Liability Act (CERCLA), passed by Congress in 1980, is commonly referred to as Superfund. CERCLA was passed in response to the growing number of cases like Love Canal, where contamination from hazardous waste posed a direct threat to human health and the environment. The primary goals of CERCLA are (1) to locate, assess, and clean up sites contaminated due to past hazardous waste activities; (2) to provide a mechanism for reporting new releases of hazardous chemicals to the environment; and (3) to finance these activities. By targeting sites previously contaminated with hazardous wastes, CERCLA complements RCRA, which targets new hazardous wastes and current waste management practices. The major approaches of CERCLA are as follows:

1. Establish a fund for investigations and remedial actions at sites where responsible parties cannot be found or will not voluntarily contribute.
2. Create and maintain the National Priority List (NPL) of abandoned or inactive hazardous waste disposal sites for remediation.
3. Implement the National Contingency Plan (NCP), a mechanism to determine the appropriate action to take for accidental hazardous releases and for treatment of abandoned or inactive hazardous waste disposal sites.
4. Identify and determine the liability of potentially responsible parties (PRPs) for site remediation. Possible PRPs include owners or operators of hazardous waste generating facilities that contributed to the contamination, owners or operators of the contaminated site at the time of waste release or disposal, and current owners or operators of the contaminated site. Other PRPs include persons who arranged for disposal, treatment, or transport of wastes and persons who accepted the material for transport to disposal or treatment facilities.

CERCLA calls for "strict liability" of the PRPs. This means that parties can be held responsible to pay for site cleanup regardless of their claims of good-faith practices, lack of negligence, or ignorance regarding the wastes. The regulations also call for a "pay now, argue later" policy, where PRPs are asked to pay for site remediation even if they are disputing their liability or negotiating proportional liability with other PRPs. Finally, CERCLA mandates "joint and several liability." This provision allows EPA to force any PRP to assume liability for the total cost of site remediation, regardless of their level of involvement. PRPs are then allowed to use civil action to recover costs from other involved parties.

CERCLA established the Hazardous Substance Response Trust Fund (popularly known as the Superfund), money that is available to cover immediate response actions as well as long-term remediation actions when PRPs are not available to do so. This fund is financed by taxes on domestic crude oil, imported petroleum products, and feedstock chemicals, and by a broad-based tax on business income.

In 1986 CERCLA was amended by the Superfund Amendments and Reauthorization Act (SARA). SARA represented a complete rewrite of CERCLA and was nearly four times as long. The provisions for strict, joint, and several liability were clarified and reaffirmed. The amount of money in the Superfund was increased from $1.6 billion to $8.5 billion. The remedial standards for site cleanup were made more stringent, with emphasis on permanent cleanup. Community-right-to-know provisions were added requiring industries to plan for emergencies and to inform the public of hazardous substances that are in use.

The National Contingency Plan (NCP) is the set of regulations that implement CERCLA. The NCP specifies a series of steps for identifying and remediating hazardous waste releases. These steps are summarized below (see also Figure 8.A.2).

1. Site Identification Any spill of a hazardous substance that exceeds a specified minimum must be reported to the National Response Center (1-800-424-8802). Reportable quantities are listed for more than 700 hazardous materials and wastes and for 1500 radionuclides. For spills of hazardous substances that don't have a listed reportable quantity, a default quantity of one pound applies. Furthermore, all present or former owners and operators of hazardous substance treatment, storage, and disposal facilities or transporters to such facilities are required to report their existence to the USEPA. State and local governments and the public are also encouraged to identify known facilities or sites.

In response to site identification, a short-term removal action is authorized if the USEPA determines that an imminent and substantial danger exists to human health or to the environment. Short-term actions typically involve removal or containment and immediate cleanup of the site. Discharges of oil into navigable waters are specifically addressed under this provision. Removal actions are always accompanied by public notification and may include evacuation of the local population, provision of alternative water supplies, and limiting site access.

Sites that are deemed significant chronic threats that would require long-term remedial action are ranked according to the Hazard Ranking System (HRS), a method of quantifying risk to human health based on the hazardous nature of the contamination, contaminant quantity, potential routes of exposure, and the exposed population. Sites with the highest HRS score are placed on the National Priority List (NPL) to undergo long-term remedial action. Once a site is listed on the NPL, attempts are made to identify PRPs willing to undertake proper site remediation.

2. Remedial Investigation Sites listed on the NPL undergo a detailed investigation to characterize the contamination and to identify applicable local and state cleanup requirements that may be more stringent than the federal requirements.

3. Feasibility Study A feasibility study is performed on the site. The feasibility study includes a risk assessment for the site contamination, development of a list of

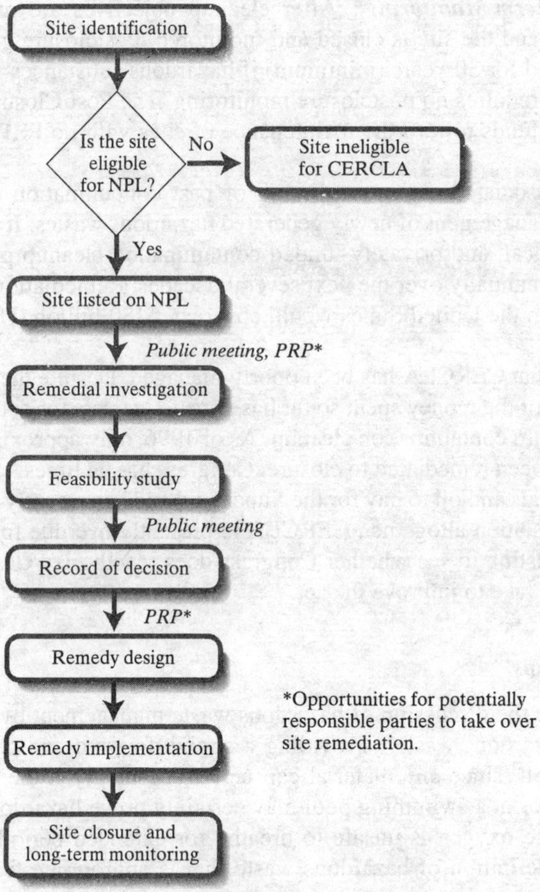

Figure 8.A.2 Outline of steps designated in the NCP for identification and remediation of hazardous waste sites. (Reprinted with permission of The McGraw-Hill Companies from M.D. LaGrega, P.L. Buckingham, and J.C. Evans, *Hazardous Waste Management*, McGraw-Hill, New York, 1994.)

alternative remediation options, performance of treatability investigations to test those options, and recommendations for the most promising and cost-effective site remedy.

4. Record of Decision Based on the remedial investigation and feasibility study, the USEPA chooses the most promising and cost-effective remediation alternative. Their selection is described and justified in the Record of Decision (ROD). The ROD is presented for public comment and for inspection by PRPs. This step represents the final opportunity for the PRPs to assume responsibility for remedial design and implementation. Otherwise the USEPA, the U.S. Army Corps of Engineers, or the state can implement the remedial measures with funding from either the PRPs or from the Superfund. There are advantages for the PRP to undertake the remediation itself, including participation in remedy selection and cleanup level determination. In addition, the PRP may save considerable money by designing a more frugal remediation plan. Any remediation plan is subject to EPA approval.

5. Remedy Design and Implementation The chosen remediation alternative is designed, compliance with local regulations is rechecked, bids are sought, and construction commences. The remediation operation continues until site restoration objectives are met. This step can require decades to complete.

6. Site Closure and Long-Term Monitoring After cleanup objectives are met, remediation is discontinued, and the site is closed and monitored. Postclosure groundwater monitoring is required for 30 years (minimum) if hazardous substances remain at the site. "Clean closure" requires no postclosure monitoring. The Post Closure Liability Trust Fund allocates funds toward the maintenance of sites with no PRPs after remediation is complete.

Currently, the costs associated with remediation of past contamination dwarfs those associated with the management of newly generated hazardous wastes. It is estimated that federal, state, local, and privately funded contamination cleanup projects will cost billions of dollars annually over the next several decades. Remediation of all known contaminated sites in the United States would cost over $180 billion (USEPA, 1997c)!

It is common opinion that CERCLA has been poorly managed. Despite the rhetoric of the original act, most of the money spent so far has gone to lawyers for determining liability rather than toward contamination cleanup. As of 1996, only approximately 300 of 2000 NPL sites had been remediated to closure. Congress has, at times, stopped levying the taxes on chemicals and oil to pay for the Superfund and has recently threatened to shut down this legislation altogether. CERCLA is presently overdue for reauthorization. It will be interesting to see whether Congress does reauthorize CERCLA and, if so, what steps it will take to improve the act.

Hazardous Waste Definitions

It is appropriate to continue the discussion of hazardous waste management by defining the overused term "hazardous waste." Providing a straightforward and comprehensive definition is difficult, since any material can be hazardous given the proper situation. For example, water in a swimming pool may certainly prove hazardous to a nonswimmer, and even pure oxygen is unsafe to breathe for extended periods. The challenge is to provide a definition of hazardous waste that is appropriate from the perspective of waste handling and treatment. For the purposes of this text, hazardous wastes will be defined as described by RCRA (40 CFR §261):

> *The term "hazardous waste" means a solid waste, or combination of solid wastes, which because of its quantity, concentration, or physical, chemical, or infectious characteristics may:*
>
> *(a) cause, or significantly contribute to an increase in mortality or an increase in serious irreversible, or incapacitating reversible, illness*
>
> *(b) pose a substantial present or potential hazard to human health or the environment when improperly treated, stored, transported, or disposed of, or otherwise managed.*

This definition refers to solid waste, which is actually a misnomer since RCRA's definition of solid waste applies to wastes of many forms other than "solid":

> *Any garbage, refuse, or sludge from a waste treatment plant, water supply treatment plant or air pollution control facility and other discarded material including solid, liquid, semi-solid or contained gaseous material resulting from industrial, commercial, mining and agricultural operations and from community activities or material that is discarded, abandoned, or "inherently waste-like."*

These definitions would be cumbersome to use in day-to-day determinations of the hazardous nature of wastes. So the USEPA has developed a straightforward series of rules for determining whether a given waste would be considered, under law, a hazardous waste. These rules are based on a combination of lists and characteristics. A waste is considered hazardous by the USEPA if it is specifically noted on one of the lists or if it exhibits any of the characteristics described below.

Listed Wastes The USEPA compiled three lists of waste sources that are automatically considered hazardous based on the toxicity, reactivity, corrosivity, or ignitability of given materials. Discarded materials that fall within any of the following three categories are considered hazardous wastes regardless of contaminant species concentration (40 CFR §261.31–§261.33):

1. Nonspecific sources (F code) are wastes commonly produced by manufacturing and industrial processes. Examples include spent halogenated solvents used in degreasing and spent cyanide plating bath solutions from electroplating.
2. Specific sources (K code) are wastes from specifically identified industries such as wood preserving and petroleum refining. Examples include spent pickle liquor from steel finishing and distillation bottoms from 1,1,1-TCA production.
3. Commercial chemical products (U code and P code) are specific commercial chemical products or manufacturing chemical intermediates. This list is segmented into toxic (U code) and acutely toxic (P code) categories. Examples include chloroform, creosote, and kepone.

Characteristic Wastes Any discarded material that is not specifically described on one of the EPA's above lists may still be considered a hazardous waste if it exhibits the characteristics of ignitability, corrosivity, reactivity, or toxicity as defined in Table 8.A.1.

Sixty percent of the hazardous waste generated in the United States is designated by one or more of the four defined characteristics, while 28 percent is designated by a list, and the remaining 12 percent is designated by both (Figure 8.A.3).

Many materials are explicitly excluded from the RCRA definition of hazardous wastes. These exclusions fall into two general categories: (1) large-volume wastes that are thought to pose no significant threat to human health or the environment, or (2) wastes that are regulated elsewhere. Examples of excluded wastes are listed in Table 8.A.2.

A generator may petition the EPA to delist a designated hazardous waste if it can show that none of four RCRA characteristics apply and that the waste is sufficiently stable and inert that no hazardous constituents can leach out following disposal. Delisting requires extensive analytical testing of the wastes as well as formal public notices and hearings.

8.A.3 Magnitude of the Hazardous Waste Problem

Listed in Table 8.A.3 are EPA estimates of the amount of hazardous waste generated in the United States over approximately the past three decades. The data were compiled from EPA biennial reports of hazardous waste generation. These reports are produced by EPA-permitted hazardous waste generators and are considered the most comprehensive survey of hazardous wastes generated in the United States. In 1990 the hazardous waste definition was broadened, causing additional wastes to be classified as hazardous and resulting in higher estimates of generation rates. As can be seen from the trends in Table 8.A.3, hazardous waste generation increased during the 1970s and early 1980s, but began a steady decline in the late 1980s and 1990s. The decline is

Table 8.A.1 System Used by RCRA to Define Hazardous Wastes According to Characteristics

Characteristic	Definition
Ignitability	Could readily catch fire and sustain combustion during transport, storage, or disposal: • Liquid (except <24% alcohol) with flash point <60 °C (<140 °F) • Nonliquid that, at standard temperature and pressure, can ignite by friction, absorption of moisture, or spontaneous chemical changes (e.g., alkali metals) • Flammable compressed gas • *Oxidizer* as defined by the Department of Transportation (DOT) *Examples:* H_2O_2, O_3, Cl_2, F_2, waste oil, spent solvents
Corrosivity	Acidic or alkaline wastes that can readily corrode or dissolve flesh, metal, or other materials: • Aqueous material with pH < 2 or pH > 12.5 • Nonaqueous material with pH < 2 or pH > 12 when mixed with an equal weight of water • Material that corrodes steel faster than 6.35 mm/y at 55 °C *Examples:* acidic bath waste, spent pickle liquor, lye rinsewater
Reactivity	Normally unstable, readily explodes, or reacts violently without external detonation: • Capable of detonation at standard temperature and pressure • Reacts violently with water • Forms explosive mixtures with water • Generates toxic gases or vapors when mixed with water (e.g., FeS) • Contains cyanide or sulfide and generates toxic gases when at $2 < \text{pH} < 12.5$ • Detonates or explodes when ignited under heat or confinement • Listed by DOT as Class A or B explosive *Examples:* water from TNT manufacturing, cyanide solvents
Toxicity	Likely to leach dangerous concentrations of certain known toxic chemicals into groundwater: • Waste or waste extract analyzed by the Toxicity Characteristic Leaching Procedure (TCLP), a test designed to simulate landfill conditions. If the TCLP residue contains contaminants in excessive concentrations, the waste is considered toxic.

Source: 40 CFR §261 Subpart C.

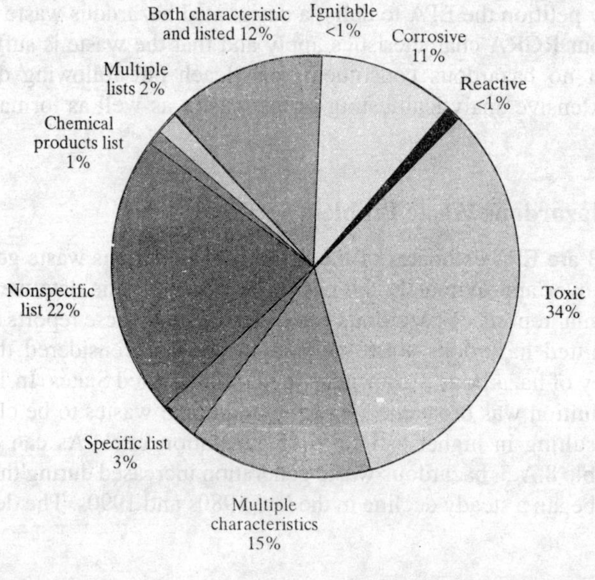

Figure 8.A.3 Relative amounts of U.S. generated hazardous waste defined according to RCRA-designated characteristics or lists (USEPA, 1997a).

Table 8.A.2 Examples of Wastes That Are Specifically Excluded from the RCRA Definition of Hazardous Waste

Household waste
Domestic sewage
Reclaimed or reused secondary materials
Wastes discharged to publicly owned sewage treatment works
Discharges of industrial wastewater subject to regulation under the Clean Water Act
Irrigation return flows
Solid wastes returned to soils as fertilizers (composting, manure)
Radioactive wastes
Fly ash waste generated from the combustion of coal or other fossil fuels
Mining overburden returned to the mine site
Drilling fluids and other fluids associated with crude oil, natural gas, geothermal exploration, development, and production
Ash, slag or flue gas emissions from coal or fossil fuel burning

Source: 40 CFR §261.4, 1997.

Table 8.A.3 EPA Estimates of the Quantity of Hazardous Waste Generated in the United States

	Hazardous waste generated (million tons per year)	
Year	Pre-1990 definition	Post-1990 definition
1973	10	
1975	48	
1981	306	
1985	430	
1987	238	
1989	198	
1991	144	306
1993	123	258
1995	109	214

Source: USEPA biennial reports (http://www.epa.gov/epaoswer/hazwaste/data/), 1995 data.

largely due to heightened awareness, strengthened regulations, and waste minimization efforts. A shortcoming of the reports is the exclusion of wastes produced by small-quantity generators (less than 1000 kg per month).

To put the numbers in Table 8.A.3 in perspective, it is estimated that the United States generated 208 million tons of municipal solid waste in 1996 (USEPA, 1997b). The solid waste number does not include wastewater. In 1995, 96 percent of the generated hazardous waste was in the form of wastewater (USEPA, 1997a). It is estimated that up to 10 percent of the industrial wastes generated in the United States are RCRA hazardous wastes (Allen and Rosselot, 1997).

Although hazardous wastes are generated throughout the United States, the top five hazardous waste–generating states (Texas, Tennessee, Louisiana, Michigan, and Illinois) accounted for over 70 percent of the total mass of hazardous waste generated in the United States in 1995, with Texas far outpacing the rest of the country (see Table 8.A.4).

Table 8.A.4 Quantity of Hazardous Waste Generated and Number of Large-Quantity Generators (LQGs) in the 10 Top-Ranking States

State	HW generated[a] (×1000 tons/y)	Rank	Percent of U.S. total	Number of LQGs
Texas	68,500	1	32.0	1,329
Tennessee	38,700	2	18.1	467
Louisiana	17,500	3	8.2	359
Michigan	13,500	4	6.3	718
Illinois	12,800	5	6.0	1,156
California	11,100	6	5.2	1,640
New Jersey	10,300	7	4.8	1,178
West Virginia	8,490	8	4.0	117
Pennsylvania	6,450	9	3.0	1,134
Washington	3,090	10	1.4	748
U.S. total	214,000		100.0	20,867

[a]Reported amounts rounded to three significant digits.

Source: USEPA biennial reports (http://www.epa.gov/epaoswer/hazwaste/data/), 1995 data.

8.A.4 Sources of Hazardous Wastes

Major sources of hazardous waste can be loosely categorized into three groups: industrial, research/medical, and household. Industrial sources can be further divided into large- and small-quantity generators. Large generators include chemical producers, electronics and electrical equipment manufacturers, petroleum refineries, and metal industries. Examples of small generators are film processors, dry cleaners, and automotive repair firms. Research/medical sources of hazardous wastes include universities, national and local laboratories, and hospitals. Typical household hazardous wastes include batteries, used automobile oil, pesticides and herbicides, paints and solvent thinners, chemical drain cleaners, septic tank cleaners, household cleaners, and nail polish remover.

It is estimated that there are 63,000 different chemicals in general use, and that roughly 220 million tons of chemicals are used per year in the United States (Englande, 1994). Table 8.A.5 lists the industries with the largest environmental chemical releases and off-site transfers in the United States.

Although the Toxics Release Inventory (TRI) data presented in Table 8.A.5 gives some indication of the industries that are most associated with hazardous waste generation, the data are actually a compilation of dry-weight chemical releases to the environment and transfers for off-site recycling, treatment, storage, and disposal. As such, TRI data are not reflective of industries that treat most of their hazardous waste on-site or that generate large quantities of hazardous waste in the form of wastewater, such as the petroleum industry. As can be seen in Table 8.A.6, which is a listing of the top 15 individual hazardous waste–generating companies in the United States, chemical and petroleum companies tend to be the largest hazardous waste generators. Remarkably, these top 15 companies accounted for 63 percent of the total hazardous waste generated in the United States. Actually, a fact that is not apparent from the presented data is that metal industries account for the highest *number* of hazardous waste generators while chemical industries account for the greatest *volume* of hazardous waste generation. This is due to the high degree of consolidation in the chemical industry, which has resulted in a small number of large-scale producers.

Table 8.A.5 General Industrial Categories for Industries with the Largest Chemical Releases and Off-Site Transfers

Industry	Percent of TRI releases and transfers
Chemicals and allied products	30
Primary metal industries	24
Fabricated metal products	6.9
Electrical and electronic components	6.9
Transportation equipment	5.3
Paper and allied products	5.1
Petroleum refineries and related industries	3.3
Rubber and associated products	2.7
Industrial and commercial machinery and computer equipment	1.5
Furniture and fixtures	1.0
Food and kindred products	1.0

Source: Reported on a dry-weight basis by the Toxics Release Inventory (TRI) in 1994; adapted from Allen and Rosselot, 1997.

Table 8.A.6 Top 15 Hazardous Waste Generators in the United States, 1995

Rank	Hazardous waste generator	Waste generated[a] (× 1000 tons/y)
1	Tenn Eastman Div. of Eastman Chemical (Kingsport, TN)	38,200
2	Amoco Oil Company (Texas City, TX)	17,800
3	E.I. Dupont de Nemours & Co. (Deepwater, NJ)	9,780
4	Dow Chemical Co.—Midland Plant Site (Midland, MI)	9,640
5	Shell Wood River Refining Co. (Roxana, IL)	8,630
6	Phillips 66 Co. (Old Ocean, TX)	8,620
7	Shell Oil Co. (Martinez, CA)	8,510
8	Rhone-Poulenc Institute Plant (Institute, WV)	7,470
9	BP Oil Co. Marcus Hook Refinery (Marcus Hook, PA)	4,910
10	E.I. Du Pont de Nemours & Co. (Victoria, TX)	4,500
11	Monsanto Co. (Alvin, TX)	4,070
12	Union Carbide Corp. Taft Plant (Taft, LA)	3,560
13	Crown Central Petroleum Corp. (Pasadena, TX)	3,140
14	BP Oil Co. Alliance Refinery (Belle Chasse, LA)	3,140
15	Coastal Refining & Marketing, Inc. (Corpus Christi, TX)	2,820
Total		135,000

[a]Reported weights and total have been rounded to three significant digits. The total applies only to the 15 generators listed.
Source: USEPA, 1997a.

8.B HAZARDOUS WASTE MINIMIZATION

All manufacturing and industrial processes generate wastes in the form of liquids, solids, or gases (Figure 8.B.1). As discussed in the previous section, some wastes are considered hazardous. Traditional waste control strategies have relied upon "end-of-the-pipe" solutions, such as treatment and disposal. However, new and increasingly

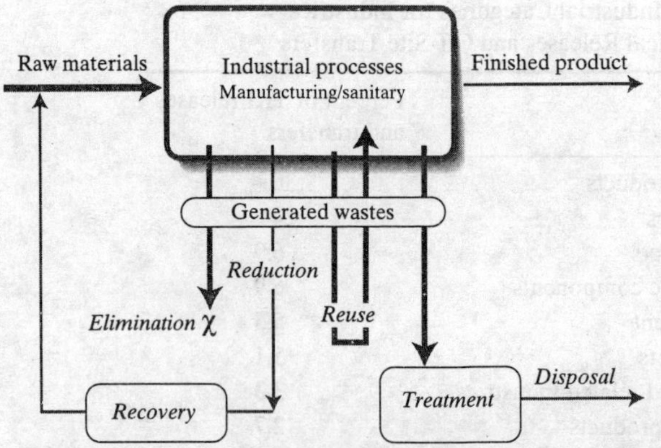

Figure 8.B.1 Schematic of generated wastes and potential waste management strategies for a generic industrial process.

stringent regulations have made waste disposal expensive, highly regulated, and liability prone, rendering the alternative of waste minimization more attractive.

8.B.1 What Is Waste Minimization?

The USEPA has outlined the following hierarchy of hazardous waste management priorities. These are ranked according to environmental preference, from most desirable to least desirable.

1. Eliminate waste generation
2. Reduce waste generation
3. Reuse, recover, or recycle waste materials
4. Treat waste to diminish quantity and to detoxify
5. Dispose of waste residuals

Originally, the USEPA classified any activity as *waste minimization* if it reduced the quantity or toxicity of wastes to be handled by TSD facilities. This broad definition was subsequently changed to exclude end-of-the-pipe processes that specifically treated waste streams. The USEPA's current definition of waste minimization includes the first three items in the above list, specifically excluding methods of treatment and disposal.

8.B.2 Why Should We Do It?

The human and environmental risks that are associated with the generation of large amounts of hazardous waste have become an issue of great concern in recent years. It has therefore become increasingly important to develop and promote methods of reducing the amount of hazardous waste that must be treated and disposed of. Congress and the USEPA have promoted waste minimization, initially by means of words alone (they entitled the first major federal legislation dealing explicitly with hazardous wastes the Resource Conservation and Recovery Act) and more recently by means of stricter regulations.

The Hazardous and Solid Waste Amendments (HSWA) to RCRA, passed in 1984, state that "Congress declares it to be the national policy of the United States that, wherever feasible, the generation of hazardous waste is to be reduced or eliminated as expeditiously as possible." The regulations associated with this act require biannual reporting on waste minimization by generators. Hazardous waste manifests must include the following waste minimization and detoxification declaration:

> *I certify that I have a program in place to reduce the volume and toxicity of waste generated to the degree I have determined to be economically practicable and I have selected the method of treatment, storage, or disposal currently available to me which minimizes the present and future threat to human health and the environment.*

However, despite early efforts, in the 1980s there were greater incentives for industry to treat hazardous waste than to minimize its generation. The costs associated with hazardous waste treatment and disposal were generally not significant enough to justify the costs associated with implementing hazardous waste minimization strategies. The late 1980s and early 1990s brought fresh motivation to industries. Stricter regulations were applied to hazardous waste disposal (e.g., the land ban) and waste disposal costs increased (e.g., land disposal costs in California increased 445 percent from 1983 to 1986, as reported by LaGrega et al. [1994]). During this time, companies became aware of the potential for long-term liability associated with future hazardous waste releases from land-based disposal facilities. In 1990 the U.S. Congress passed the Pollution Prevention Act (PPA), outlining the waste management hierarchy given above and reemphasizing the importance of source reduction over treatment and disposal strategies.

As a further regulatory incentive to promote waste minimization, the EPA instituted a voluntary "33/50" program in 1991, asking hazardous waste generators to implement a 33 percent reduction for 17 high-priority toxic chemicals by 1992, and a 50 percent reduction by 1995. Participation in the program was to be rewarded with public relations benefits for the industries. The program had high participation from the nation's largest waste generators, and the participating companies surpassed the waste minimization targets.

More recently, Carol Browner, EPA administrator under President Clinton, instituted an industry-by-industry approach to waste minimization and pollution prevention. This approach entails comprehensive, integrated audits of individual industries, simultaneously addressing air, water, and soil pollution.

Industries that have adopted waste minimization strategies have often been rewarded with initially unanticipated benefits, including lower spill and accident potential, decreased future waste-related liability, improved employee working conditions, and enhanced public relations. A well-known example of waste minimization strategies that have benefited a large company is the 3P program (Pollution Prevention Pays) of the 3M Corporation. This program reduced the company's generation of environmental pollutants by 50 percent from 1975 to 1992 and saved the company over $600 million during that time in its U.S. operations alone (McCoy and Associates, 1995).

8.B.3 Techniques for Waste Minimization

The most effective techniques for waste minimization utilize a combination of common sense and good engineering. Implementation of a comprehensive waste minimization plan requires a companywide effort, including commitment from top-level decision makers, appropriate financial and technical resources, and involvement at each level of the production process: procurement and inventory management, design, operations, and waste management. The greatest success can be achieved when the entire organizational structure is made aware of the costs associated with waste generation, both to the company and to the environment, and is motivated to develop and implement waste minimization goals and strategies.

Waste minimization can be implemented in stages that provide the most return for the least effort initially, with increased returns requiring increased investment and resources over time. The following list provides a logical, staged approach to waste minimization:

1. Commonsense waste reduction (e.g., avoid mingling hazardous waste with nonhazardous waste)
2. Information-driven reduction (e.g., use alternative raw materials)
3. Formal audit requiring some capital investment and potential process change (e.g., incorporate waste recovery)
4. Reductions that depend on research and development, requiring extensive study before implementation (e.g., adopt a new manufacturing technique)

Waste Minimization Audit Waste minimization audits can be implemented informally, using a checklist and direct observation, or formally, for example, by hiring outside consultants who specialize in preparing audits. The need for and efficacy of informal or formal audits depend on the degree to which waste minimization is already practiced in the company. A comprehensive waste minimization audit would typically be implemented using the following approach:

1. Inspect process flows using material and energy balances along with process flow schematics.
2. Inspect the physical layout of processes, looking for potential process modifications to increase efficiency and for material substitution or waste segregation opportunities.
3. Inspect the operating condition of equipment, looking for potential maintenance problems or upgrade potential.
4. Evaluate actions of operators, maintenance personnel, and managers for potential improvements in operating procedures.
5. Evaluate management strategies such as inventory control and operator training.
6. Determine real and hidden costs associated with each waste stream.

As part of the audit, all waste streams should be well characterized. Factors to consider include sources, generation rates, physical and chemical characteristics, and treatment and disposal costs. On the basis of the audit, minimization options should be identified evaluated, and compared so that the best strategy can be selected and implemented.

Waste minimization techniques can be categorized into four overlapping types of approaches: (1) inventory management, (2) production process modification, (3) volume reduction, and (4) reuse/recovery/recycle. Each of these approaches is briefly summarized next.

Inventory Management

Inventory management involves strict control over the storage and consumption of raw materials, intermediate products, final products, and associated waste streams. A key goal is to reduce quantities of out-of-date, off-specification, contaminated, or unnecessary materials and products. Although efficient inventory management may seem an obvious competitive corporate practice, one of the most common sources of hazardous waste production by industry is the disposal of off-specification products and excess raw materials. Remarkably, when excess or out-of-date chemicals are disposed of as hazardous wastes, the costs may be higher than the original purchase cost.

Inventory management may involve simple changes in ordering procedures. Sometimes labor may be invested to reduce waste generation, for example, by using enhanced inventory control programs and material tracking systems to decrease the accumulation of excess, out-of-date, and no-longer-used materials. Enhanced inventory control programs may involve purchasing the exact quantities of raw materials needed for each production run or rotating stored stock according to the expiration date.

An example of a highly controlled inventory management system is the "just-in-time" (JIT) manufacturing technique. When JIT manufacturing is implemented, raw materials are used promptly after delivery and finished products are shipped immediately upon completion. In this manner, inventories of raw materials and finished products are reduced. The success of JIT manufacturing depends heavily on reliable sources of raw materials and on advance planning to anticipate demand for the finished product. JIT works well in some, but not all, industries. When used effectively, it can result in significant waste reduction and cost savings.

Production Process Modification

Production processes can be modified to make better use of raw materials and to minimize waste generation. The efficiency of production processes can sometimes be improved through simple operational and maintenance procedures. In other cases, more involved material changes and process equipment modifications are needed.

Operational and Maintenance Procedures Improvements in operational and maintenance procedures are generally easy and inexpensive to implement. Overall knowledge of the production process is needed, as is employee training to ensure proper implementation. Simple operational changes such as increasing the drainage time for parts suspended over drip trays or processing light dyes prior to dark dyes to reduce the need for tank cleaning between runs may result in significant waste reduction. Incorporating the costs of waste treatment and disposal into the operational design of a process may also lead to significant waste reduction. In conventional past practice, the operational parameters for a process, such as temperature, pressure, and incubation time, were often designed to maximize the efficiency of product throughput and process cost, without regard to waste generation.

Improved maintenance procedures such as dry sweeping prior to mopping, segregating wastes, and improved materials handling techniques to minimize spills, leaks, and contamination are also effective waste minimization methods. The implementation of corrective and preventive maintenance programs that maintain process efficiency and diminish equipment failure, coupled with sufficient employee training, may decrease waste generation significantly.

Material Changes Hazardous materials used in a production process can sometimes be replaced with less-hazardous materials without a significant decrease in process efficiency. For example, heavy-metal-based pigments in inks, dyes, and paints can be replaced with organic pigments, and water-based paints and finishes can be used in place of solvent-based materials. However, the use of water-based materials may lead

Table 8.B.1 Examples of Material Changes Implemented to Minimize Waste Generation

Industry	Material change
Household appliance manufacturer	Replace chlorinated solvent with alkaline degreaser
Printing	Substitute water-based ink for solvent-based ink
Office furniture producer	Replace solvent-based paints with water-based paints
Air conditioner manufacturer	Replace solvent-based adhesives with water-based adhesives
Aerospace	Replace cyanide cadmium plating with non-cyanide bath
Plumbing fixture manufacturer	Substitute hexavalent chrome-plating bath with low-concentration trivalent chrome-plating bath
Electronic component manufacturer	Replace organic biocides with ozone in cooling towers
Pharmaceuticals	Replace solvent-based tablet-coating process with water-based process

Source: Adapted from Hunt and Schecter, 1989.

to higher-volume aqueous wastes, illustrating the need to evaluate the effects that material changes may have on all steps in a process train. Examples of successful material changes instituted by companies to achieve waste minimization are presented in Table 8.B.1.

Process Equipment Modification Installation of new, state-of-the-art process equipment, or modification of existing equipment to operate more efficiently, may result in reduced waste generation. For example, higher product throughput may be achieved with the same amount of raw materials and the number of rejected or off-specification products may be reduced by the use of high-efficiency process equipment. It is wise to factor in waste disposal costs when choosing new process equipment, since capital investments can be partly offset by the operational savings associated with decreased waste generation. A good example of this is a power tool manufacturer that replaced a spray solvent system with a water-based, electrostatic immersion unit for painting. Use of the water-based unit increased productivity and reduced costs for both raw material and waste disposal, allowing the company to recoup its capital investment in just over one year and providing operational savings in each additional year (Huisingh, 1985).

Simple equipment modification may also be effective in reducing waste generation. Examples include the installation of drip trays, dragout racks, and air knives (which blow cleaning solution off pieces) to recover excess cleaning solution following bath cleaning processes. Other examples are capture and reuse of rinsewater, installation of more effective seals to decrease spillage, and installation of lids on volatile solvent tanks to reduce evaporative losses.

Volume Reduction

The technique of volume reduction for waste minimization involves segregating hazardous and recoverable wastes from total waste streams and concentrating wastes, often by removing water. Segregating hazardous and nonhazardous wastes is worthwhile since the storage, transport, and disposal costs are much greater for hazardous than for nonhazardous wastes. Waste segregation may also make it easier to reuse and recover materials from individual waste streams. An example of a successful waste segregation operation is a printing firm that collects and segregates waste toluene from press and roller cleanup by ink color so that the waste toluene can be reused as ink thinner (Huisingh, 1985). In the metals fabrication industry, it has become common practice to segregate wastes by metal type, to facilitate metal recovery from the waste sludge.

Waste concentration is a viable waste minimization technique only if the resultant material is more amenable to reuse or recovery. Otherwise, decreasing waste volume by means of concentration is considered a treatment process. There are many methods for waste concentration, including gravity separation, dissolved air flotation, evaporation, distillation, solvent extraction, reverse osmosis, filter press, centrifugation, and compaction. Some of these technologies are discussed in §6.C and §8.C. An example is the use of filter presses to dewater copper-bearing sludge from a circuit board manufacturer. This process generates material from which copper can be economically recovered.

Reuse, Recovery, and Recycle

Once methods of waste elimination and reduction have been implemented, the next tier of waste management options involves reuse, recovery, and recycling. The meaning of these three terms is often confused, and indeed, they are not used consistently. For the purposes of this text, we will adopt the following definitions:

Reuse — The use of waste material generated from one process directly in a second process without significant treatment or processing (e.g., high-purity solvents used originally for the production of microelectronics can be reused as metal degreasers)

Recovery — Employment of a treatment process to extract material from the waste of one process to use again in that process or in a second process (e.g., copper and tin from the waste streams of a printed circuit board manufacturer can be recovered using electrolysis for application in plating operations)

Recycle — Regeneration of a product from waste material following significant reprocessing (e.g., glass from waste bottles can be ground and reformed into new bottles)

Reuse, recovery, and recycle operations can decrease waste disposal costs, reduce raw materials costs, and possibly provide income from the sale of waste materials. The following two examples illustrate successful waste minimization through reuse, recovery, and recycle. Metal-plating industries produce wastewater from plating solutions that contain high levels of copper, nickel, chromium, and phosphoric acid. Some metal platers have utilized ion exchange processing to recover the metals and return them to the plating process while the effluent water is used as boiler feed (Nemerow and Dasgupta, 1991). In a second example, a color portrait processor installed two units: an ion exchange system to regenerate color developer solution by contaminant removal, and an electrolysis system to recover silver from the fixer solution. The recovered silver was sold to generate revenue.

Waste reuse, recovery, and recycle processes can occur on-site or at centralized off-site locations. Advantages to on-site operations include reduced waste handling, transport, and reporting requirements. Off-site operations also offer several advantages: potential economies of scale associated with accepting wastes from many generators, economically viable material transfers between industries, and flexibility in quantity and timing. If an off-site operation accepts wastes from many generators, it is less important that flows from any specific waste generator be consistent. Recently, regional waste exchanges have begun to operate in many areas. This practice facilitates reuse, recovery, and recycling by providing a database of hazardous waste generated in an area. Some waste exchanges simply maintain a database of waste information, while others actually accept the waste and process it to a customer's purity specifications. Waste exchanges may be publicly or privately run. Local chambers of commerce, industrial consortia, or state agencies may set them up. An example of a successful waste exchange operation is an acetylene gas manufacturer that generates calcium hydroxide (lime) as a waste (from the calcium carbide process) that is pure enough to be used in water treatment applications. Another example is a pickle factory that generates waste brine solution that can be used as a source of acetic acid for a textile factory.

8.B.4 Management Tools

Management tools are useful for evaluating and comparing the overall effectiveness of waste minimization strategies. Three commonly applied tools are discussed here: present-worth analysis, cost-effectiveness analysis, and life-cycle analysis.

Present-Worth Analysis

Present-worth analysis permits a comparison of present and future costs by adjusting them to an equivalent present sum. When applied to industrial process costs, this technique requires estimates of future interest rates and equipment life.

Consider a fixed payment, F, which must be made at a time n years in the future. The "present worth" of that payment, PW, is calculated by

$$\text{PW} = \frac{F}{(1+i)^n} \tag{8.B.1}$$

where i is the interest rate in decimal form (e.g., 5 percent = 0.05). Note that the future worth of money (F) is always greater than the present worth (PW). An investment of PW for n years will yield F if the interest, i, is compounded annually. Present worth can be calculated for both liabilities and expenses, and for assets or receipts.

If an annual payment of amount A must be made at the end of each of the next n years, the present worth of the total expense is

$$\text{PW} = A \times \frac{(1+i)^n - 1}{i(1+i)^n} \tag{8.B.2}$$

Annual payment calculations are useful for assessing the present worth of operating costs that recur. These operating expenses may increase over time because of inflation and other factors. The rate of increase, s, may be different from the interest rate, i. In this case, the total present worth of the expense is given by the following equation:

$$\text{PW} = A \times \frac{(1+s)^n(1+i)^{-n} - 1}{s-i} \tag{8.B.3}$$

In this equation, it is assumed that the expense is incurred at the end of the year and that the increase due to inflation occurs after the first year. Hence, the payment in year n is $A(1+s)^{n-1}$. The long-term cost associated with the installation of new equipment includes both capital and operating expenses. If a capital investment C is due immediately and annual expenses of A must be made for n years, the total present worth is

$$\text{PW} = A \times \frac{(1+s)^n(1+i)^{-n} - 1}{s-i} + C \tag{8.B.4}$$

An application of present-worth analysis for comparison of waste minimization options is given in Example 8.B.1.

Cost-Effectiveness Analysis

A cost-effectiveness analysis allows process modifications to be compared based on cost (incorporating present-worth analysis) as well as additional factors that are more difficult to quantify. The additional factors are sometimes termed "intangible criteria" because it is difficult to ascribe specific values to them. Intangible criteria can include potential benefits and drawbacks associated with process changes, such as changes in productivity, liability, public relations, employee exposure risk, paperwork, and regulatory fines. Evaluation of the effects of each of these changes is based on a rating scale with priority weighting. That is, the relative importance of each criterion is evaluated and an appropriate weighting factor assigned. The effectiveness of each process modification option is then judged with respect to each criterion and compared with the option of making no change. Effectiveness values are assigned for each criterion and then summed. Effectiveness values can also be incorporated into operation research analyses, such as objective function maximization and linear programming. The advantages of cost-effectiveness analyses are that both quantitative and qualitative factors can be incorporated for comparison using engineering tools and an accounting perspective without the need to assign specific dollar values to intangible criteria. The major disadvantage to this approach is that much subjective judgment is necessary to assign weighting factors and effectiveness coefficients for each case. Subjective values are always subject to debate and are the weakest element in this quantitative approach. Example 8.B.1 illustrates cost-effectiveness analysis for comparison of waste minimization options.

EXAMPLE 8.B.1 ***Using Present-Worth and Cost-Effectiveness Analyses to Evaluate Waste Minimization Options***

A diesel engine manufacturing facility is looking for ways to minimize hazardous waste generation within its metal-parts cleaning line. A waste minimization audit was conducted to compare the current practice (option A) with a modification that applies a modest capital investment toward new equipment (option B). The new equipment, which has a 10-year lifetime, eliminates hazardous waste generation, resulting in lower disposal costs. The following data apply:

Option A: No change (operating and maintenance [O&M] cost = \$15,000/y)

Option B: Install new equipment that replaces solvent-based cleaning solution with water-based cleaners (capital cost = \$50,000, O&M cost = \$10,000/y)

Use present-worth calculations and a cost-effectiveness analysis to evaluate whether implementing the waste minimization option is a good strategy for the company.

SOLUTION To calculate the costs associated with both options on a consistent basis, a present-worth analysis is performed on the operating and capital costs over the estimated 10-year lifetime of the new equipment. Estimating that an interest rate of 8 percent applies ($i = 0.08$) and that there will be no significant inflation in operating costs

EXAMPLE 8.B.1 ***Using Present-Worth and Cost-Effectiveness Analyses to Evaluate Waste Minimization Options (continued)***

Table 8.B.2 Cost-Effectiveness Analysis for Example 8.B.1

		Effectiveness	
Criterion	Weight	Option A	Option B
Productivity	5	0	1
Future liability	2	0	5
Weighted total effectiveness		0	15
Cost ($ thousands)		101	117

($s = 0$), we use equation 8.B.4 to calculate the results:

Option A: $C = 0, A = \$15{,}000, n = 10$

$$\text{PW} = \$15{,}000 \times \frac{(1+0.08)^{-10} - 1}{-0.08} = \$101{,}000$$

Option B: $C = \$50{,}000, A = \$10{,}000, n = 10$

$$\text{PW} = \$10{,}000 \times \frac{(1+0.08)^{-10} - 1}{-0.08} + \$50{,}000 = \$117{,}000$$

Option B has a large capital cost. However, the lower annual operation expenses offset most of the present worth of this cost. Next we choose the criteria for the cost-effectiveness analysis (Table 8.B.2). Since this diesel engine manufacturing facility works with small profit margins, productivity is the most important criterion, and it will be assigned the maximum weight of 5. Future liability is also important but less tangible and definite, so it gets a weight of 2. Effectiveness scores for option B are chosen relative to option A, the no-change option. Since the new equipment is more efficiently designed than the old equipment, productivity is increased slightly, rating an effectiveness factor of 1. In addition, since the water-based cleaners are not hazardous wastes and can be safely discharged to the sanitary sewer, future liability is virtually eliminated, rating a maximum effectiveness score of 5. To complete the analysis, scores for each option are multiplied by the respective weighting factor and totaled. In the case of the diesel engine manufacturing facility, although the long-term cost of option B is somewhat higher than that of the no-change option, the effectiveness is significantly higher. This suggests that when less tangible factors are considered, option B may be the more desirable alternative.

Life-Cycle Assessment

Life-cycle assessment (LCA) is a method for evaluating the total environmental impact associated with a product or process by tracking the flows of energy, material, and wastes throughout its entire existence, from raw material to ultimate disposal. There are three components to a complete LCA:

1. Inventory analysis: Determination of materials, energy, and wastes associated with a product (Figure 8.B.2)

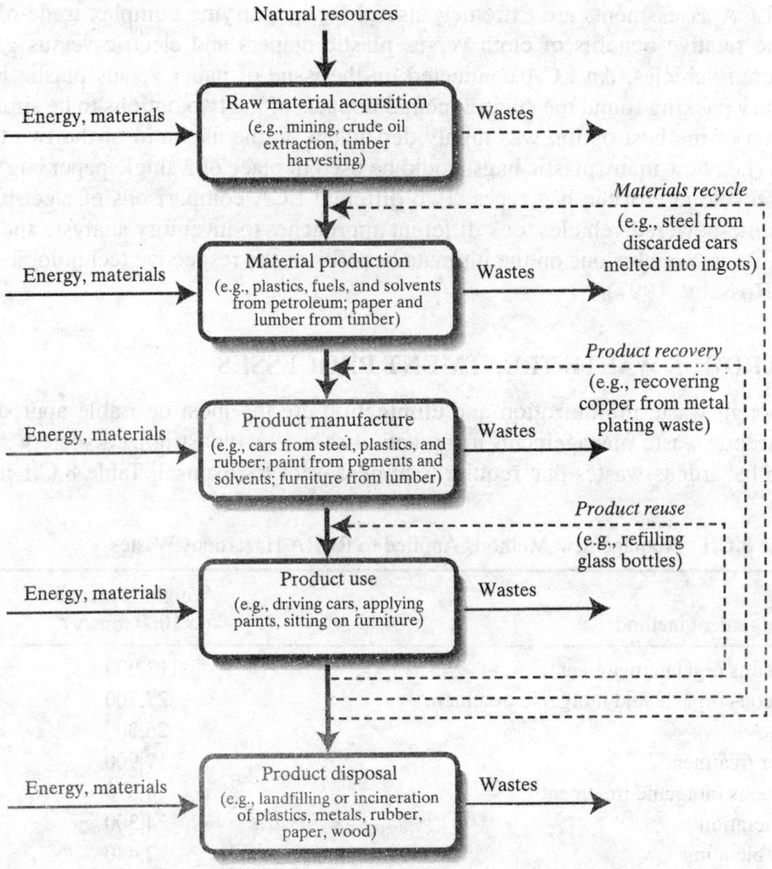

Figure 8.B.2 Framework for the inventory analysis component of life-cycle assessment. (Reprinted by permission of John Wiley & Sons, Inc., from D.T. Allen and K.S. Rosselot, *Pollution Prevention for Chemical Processes*. Copyright © 1997 Wiley, New York.)

2. Impact assessment: Evaluation of the impacts that these materials, energy, and wastes have upon human health and the environment
3. Improvements assessment: Exploration of mechanisms for reducing identified adverse environmental impacts

Although LCA can be usefully applied at any point within a product lifetime, employing LCA at the design stage of a product is an especially powerful waste minimization tool. LCA-facilitated improvements that have been adopted include the following examples (LaGrega et al., 1994).

- Car design has been changed to make cars easier to disassemble and to simplify recycling of the metals, plastics, and rubber.
- Numbers are stamped on plastic containers to simplify segregation for recycling.
- Compact disks are distributed in small cardboard cases rather than larger plastic ones.
- Fabric softeners are sold as a concentrate in smaller containers, instead of incorporating recycled material into larger containers.

LCA assessments are extremely useful for quantifying complex trade-offs such as the relative benefits of cloth versus plastic diapers and electric versus gasoline-powered vehicles. An LCA conducted on the issue of paper versus plastic bags for grocery packing found the environmental impacts of the two options to be similar. Selection of the best option was highly dependent on the use ratio of the two types of bags (i.e., how many plastic bags would be used in place of a single paper bag) and on the recycle rate of the bag types. Two different LCA comparisons of electric versus gasoline-powered vehicles took different approaches to inventory analysis and arrived at different conclusions on the ultimate benefits of the respective technologies (Allen and Rosselot, 1997).

8.C HAZARDOUS WASTE TREATMENT PROCESSES

Although waste minimization and elimination are the most desirable approaches to hazardous waste management, many activities in our industrialized society still produce hazardous wastes that require treatment prior to disposal. Table 8.C.1 indicates

Table 8.C.1 Management Methods Applied to RCRA Hazardous Wastes

Management method	Weight managed ($\times$1000 tons/y)[a]	Percent
Aqueous organic treatment[b]	117,000	56.0
Aqueous organic and inorganic treatment[c]	27,700	13.3
Disposal[d]	26,300	12.6
Other treatment[e]	17,900	8.6
Aqueous inorganic treatment[f]	8,370	4.0
Incineration	4,300	2.1
Fuel blending	2,440	1.2
Energy recovery (reuse as fuel)	1,910	0.9
Stabilization	1,020	0.5
Metals recovery for reuse	610	0.3
Sludge treatment[g]	481	0.2
Other recovery[h]	422	0.2
Solvents recovery[i]	356	0.2
Unknown (invalid code)	0.02	0.0
Total	208,000	100

[a]Rounded to three significant digits.
[b]Specific methods: biological treatment, carbon sorption, air/steam stripping, wet air oxidation.
[c]Specific methods: chemical precipitation/biological treatment, chemical precipitation/carbon sorption, wet air oxidation.
[d]Specific methods: deep well/underground injection, landfill, surface impoundments, land treatment/farming, other disposal.
[e]Specific methods: neutralization, evaporation, settling/clarification, phase separation.
[f]Specific methods: chrome reduction/precipitation, cyanide destruction/precipitation, cyanide destruction only, chemical oxidation/precipitation, chemical oxidation only, wet air oxidation, chemical precipitation, ion exchange, reverse osmosis.
[g]Specific methods: sludge dewatering, lime addition, absorption/adsorption, solvent extraction.
[h]Specific methods: acid regeneration, waste oil recovery, nonsolvent organics recovery, etc.
[i]Specific methods: fractionation/distillation, thin film evaporation, solvent extraction.
Source: USEPA, 1997a.

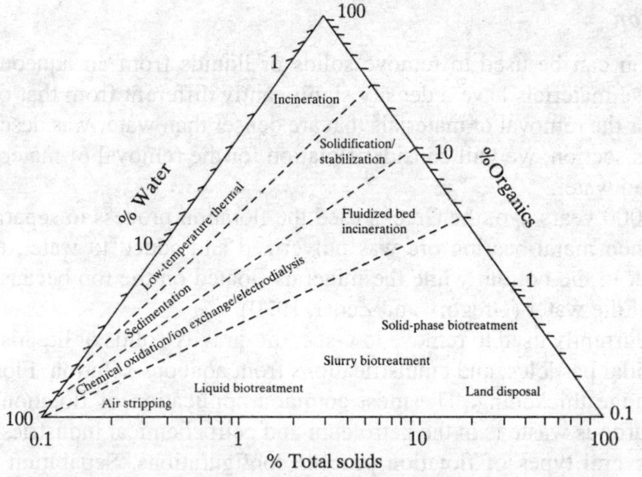

Figure 8.C.1 Applicability of hazardous waste management methods based on waste composition. (Reproduced with permission of The McGraw-Hill Companies from M.D. LaGrega, P.L. Buckingham, and J.C. Evans, *Hazardous Waste Management*, McGraw-Hill, New York, 1994.)

that the majority of hazardous wastes generated in the United States are treated using physical, chemical, and biological methods. Only a small portion (12.6 percent) undergoes eventual disposal and even less (1.5 percent) is recovered for reuse in other processes. However, it must be noted that the amount of potentially hazardous waste that is recovered or reused is seriously underestimated by these data. In conforming to RCRA's definition, the USEPA does not designate a significant amount of material with recovery or reuse potential as hazardous waste.

The major goals of hazardous waste treatment are to decrease the volume of the waste and to reduce or eliminate its hazardous characteristics. These objectives can be accomplished by physical separations and by chemical, biological, or thermal transformation processes (Figure 8.C.1). Examples of these processes are described in this section.

Since most hazardous waste is generated in aqueous form, many processes described in Chapter 6 for water and wastewater treatment can be applied to the treatment of hazardous waste. Examples include sedimentation, filtration, membrane processes, activated-carbon sorption and biological activated sludge. Extrapolation of the design of these processes to hazardous waste applications is based on the relevant waste characteristics.

Although the preliminary design of hazardous waste treatment processes can be based on theoretical considerations, final design for application to a specific waste stream usually requires bench- or pilot-scale testing.

8.C.1 Physical Separation

Physical separations are useful for concentrating or changing the phase of the hazardous components of waste, making it more amenable to further treatment or disposal.

Gravity Separation

Gravity separation can be used to remove solids or liquids from an aqueous waste stream when those materials have a density significantly different from that of water. Sedimentation for the removal of materials that are denser than water was described in Chapter 6. In this section, we will consider flotation for the removal of materials that are less dense than water.

More than 2000 years ago, the Greeks used the flotation process to separate metals from ore. When metal-bearing ore was pulverized and added to water, the rock grains would sink to the bottom while the minerals floated on the top because of the surface tension of the water (Gregory and Zabel, 1990).

Flotation is currently used to remove low-specific-gravity solids or liquids such as oil, grease, colloidal particles, and emulsifications from aqueous solution. Flotation is also used for sludge thickening. The most common application of flotation for the treatment of hazardous waste is in the petroleum and petrochemical industries.

There are several types of flotation process configurations. Separation may be achieved simply by quiescent gravity flow, or it may be assisted by the introduction of air bubbles.

Gravity Flotation Flotation by means of quiescent gravity flow is similar to sedimentation. In the case of gravity flotation, primary sludge removal occurs at the surface of the tank rather than at the bottom. Flotation tanks are often referred to as API tanks, after the American Petroleum Institute, which conducted a series of studies on their design. API tanks are typically rectangular in shape, with aqueous detention times of 0.5–1 h and typical horizontal fluid velocities of 20 m h^{-1}. The primary application of gravity flotation for hazardous waste treatment is for petroleum wastewaters contaminated with oil or grease. In 1991, 76 petroleum refineries in the United States generated 210,000 wet tons of API separator sludge (Allen and Rosselot, 1997).

Flotation efficiency is a function of particle size, specific gravity, and fluid viscosity. Tank design is based on droplet rise velocity, which is calculated using equation 4.B.15:

$$v_r = \left[\frac{4}{3}\frac{g d_p}{C_d}\left(\frac{\rho_f - \rho}{\rho_f}\right)\right]^{1/2} \qquad \rho < \rho_f \tag{4.B.15}$$

where v_r is the droplet rise velocity (m s^{-1}), ρ_f is the fluid (water) density (kg m^{-3}), ρ is the droplet density (kg m^{-3}), g is gravitational acceleration (9.8 m s^{-2}), d_p is the droplet diameter (m), and C_d is the drag coefficient (–). Under the low Reynolds-number conditions typical of a flotation unit, the rise velocity expression can be simplified to this form:

$$v_r = \frac{g(\rho_f - \rho)d_p^2}{18\mu} \tag{8.C.1}$$

where μ is the viscosity of water (kg m^{-1} s^{-1} or N s m^{-2}). This expression gives reasonable magnitude estimates of rise velocities. However, buoyant liquids commonly coalesce upon rising, and are not easily characterized in terms of size or density. It is therefore important in the design of gravity flotation units to use bench-scale studies to experimentally measure the rise velocities associated with specific wastes.

Dissolved Air Flotation Dissolved air flotation (DAF) is a physical separation process for removing suspended liquids or solids that do not sink or float at sufficiently

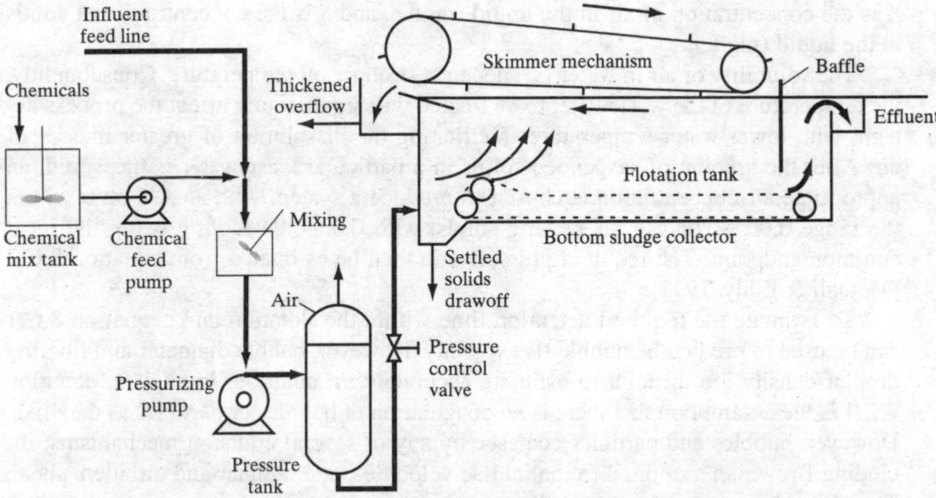

Figure 8.C.2 Schematic of a dissolved air flotation unit. (Reproduced with permission of The McGraw-Hill Companies from Metcalf & Eddy, *Wastewater Engineering: Treatment, Disposal, Reuse*, 3rd edition, McGraw-Hill, New York, 1991.)

high rates. DAF is based upon specific-gravity differences between the aqueous liquid and the suspended material. It can be applied for the removal of oil and grease, suspended solids, and fibrous materials from water and for sludge thickening.

Removal is achieved by producing extremely fine gas bubbles within the waste liquid. As the bubbles rise, they attach to the suspended material, causing it to float to the surface, where it can be removed by skimming (Figure 8.C.2). The gas bubbles are produced by pressurizing the influent flow in the presence of air so that the solution becomes supersaturated with dissolved gas, taking advantage of the fact that the solubility of air in water is proportional to pressure. The air-saturated liquid is then exposed to atmospheric pressure, causing the dissolved gas to evolve as minute bubbles with diameters in the range of 10–120 μm. The bubble-laden waste then undergoes quiescent flotation, allowing the bubbles and associated contaminants to float to the surface of the water. The suspended material that is skimmed from the top of a DAF unit is referred to as "float," and for petroleum wastewater applications is considered a hazardous waste. In 1991, 44 petroleum refineries in the United States generated 405,000 wet tons of DAF flotation sludge (Allen and Rosselot, 1997).

DAF units are designed around two major parameters: the ratio of air to mass of suspended material and the bubble rise velocity. Air can be introduced into the waste flow stream by pressurizing the entire influent stream, a portion of the influent stream, or a recycle stream. The advantage to pressurizing the entire influent stream is that the largest ratio of air to mass of suspended material can be achieved and that the piping system is simplified over a recycle system. The ratio of air to mass of suspended material can be estimated using the following equation for systems that pressurize the entire influent flow:

$$\frac{A}{S} = \frac{s_a}{S}\left(f_{\text{sat}}\frac{P_{\text{abs}}}{P_{\text{atm}}} - 1\right) \qquad (8.C.2)$$

where s_a is the solubility of air in water (25 mg/L at 20 °C and 1 atm), f_{sat} is the efficiency of the air saturation reactor (typically 0.5–0.9), P_{abs} is the absolute air pressure within the saturation reactor (typically 2–4 atm), P_{atm} is atmospheric pressure (1 atm),

A is the concentration of air in the liquid (mg/L), and S is the concentration of solids in the liquid (mg/L).

The solubility of air in water (s_a) depends strongly on temperature. Consequently, the temperature of the wastewater to be treated will significantly affect the process design, with lower water temperatures facilitating the dissolution of greater masses of air. After the amount of suspended solids in a particular wastewater is measured, an appropriate air concentration is chosen to produce a system with an A/S ratio within the range 0.007–0.08 mg air per mg solids, with 0.03–0.04 mg/mg being the most common endpoint. The required pressure can then be estimated from equation 8.C.2 (Metcalf & Eddy, 1991).

To estimate the required detention time within the flotation tank, equation 8.C.1 can be used to predict the bubble rise velocity. However, bubble diameter and floating droplet density are difficult to estimate accurately. In addition, implicit in equation 8.C.1 is the assumption that there is no coagulation of bubbles or particles as they rise. However, bubbles and particles coalesce by any of several collision mechanisms, including Brownian motion, differential rise velocities, and laminar and turbulent shear. Therefore, laboratory flotation cell tests are generally necessary for confirmation of waste-specific parameters prior to final design.

DAF tanks can be rectangular or circular in design, and generally have depths of approximately 1.5 m and overflow rates (see equation 6.C.3) of 8–12 m h^{-1} (Gregory and Zabel, 1990). The pressurization tanks in DAF systems typically have a detention time of 3–20 minutes, with flotation tank residence times of 20–40 minutes. Typical suspended solids and BOD removal efficiencies for DAF processes are given in Table 8.C.2. Suspended solids removal efficiency is often significantly higher than the BOD removal efficiency, since much of the BOD may be dissolved, which is a form that is not effectively removed by density separation. However, some dissolved BOD may be removed in these units by biological activity promoted by the availability of dissolved oxygen (Nemerow and Dasgupta, 1991).

Some applications of DAF require the addition of chemical amendments such as surfactants to remove surface-active compounds. This is common in the metals industry for treatment of aqueous wastes containing cadmium, chromium, copper, iron, lead, manganese, nickel, or zinc (Berkowitz et al., 1978). In addition, coagulants such as lime, alum, or organic polymers are sometimes added to sludge-thickening or petroleum DAF processes to increase floc size and stabilize float composition.

Table 8.C.2 Suspended Solids (SS) and Biochemical Oxygen Demand (BOD) Removal Efficiencies for DAF Units Treating a Variety of Waste Types

Waste source	Influent SS (mg/L)	SS removal (%)	Influent BOD (mg/L)	BOD removal (%)
Petroleum processing	440	95		
Meat packing	1400	86	1230	67
Paper manufacturing	1180	98	210	63
Vegetable oil processing	890	95	3050	92
Fruit and vegetable canning	1350	80	790	60
Soap manufacture	390	92	310	92
Septic tank sludge	6450	96	3400	87
Primary sewage sludge	250	69	330	49
Glue production	540	94	1820	92

Source: Adapted from Nemerow and Dasgupta, 1991.

Air Stripping

Air stripping is a physical process commonly applied to hazardous and industrial wastewaters to remove dilute concentrations (<100 mg/L) of volatile contaminants (Henry's law constant, $H_g > 0.1$ atm M^{-1}; see §3.B.2) such as organic solvents or hydrogen sulfide. Air stripping transfers the contaminants from aqueous phase to gas phase, which can then be treated further by processes such as activated-carbon sorption.

Air stripping can be achieved using various unit configurations, including aeration tanks, spray towers, and crosscurrent towers. The most common configuration for hazardous waste applications is the countercurrent packed tower (Figure 8.C.3). In this configuration, contaminated water trickles down through packing that generally consists of plastic spheres, rings, or saddles (Table 8.C.3) while clean air is blown upward through the packing to exit at the top of the tower. The packing facilitates contact between the water and the air, allowing the volatile contaminants to be transferred to the gas phase for removal with the air. This process is the complement to scrubbing, which is commonly used for transferring sulfur dioxide and hydrogen chloride from gas streams into water (see §7.C.2).

To derive the equations needed to describe contaminant transfer from liquid to gas phases within a countercurrent tower, we can apply a material balance on the contaminant in the liquid phase within a differential control volume at some height z along the air stripping tower (Figure 8.C.4).

Assuming that the transport of contaminants across the air-water interface can be described by the two-film model (equation 4.C.11) introduced in §4.C.2, the material

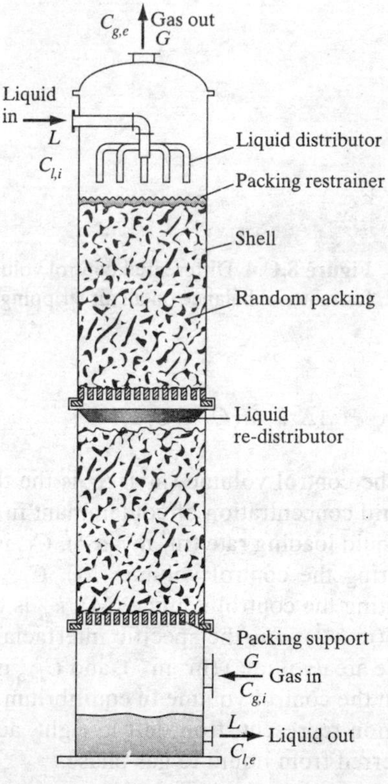

Figure 8.C.3 Typical configuration of a countercurrent packed tower used for air stripping. (Reproduced with permission of The McGraw-Hill Companies from R.D. Treybal, *Mass-Transfer Operations,* McGraw-Hill, New York, 1980.)

Table 8.C.3 Characteristics of Packing Materials for Countercurrent Air Strippers

		Sherwood-Holloway constants		Onda constants	
Packing type	Packing diameter, d (m)	Empirical constant, n (–)	Empirical constant, m (–)[a]	Packing area, a_t (m^2 m^{-3})	Empirical constant, c (–)
Raschig rings	0.013	0.28	1600	364	2.00
	0.025	0.22	1100	190	5.23
	0.038	0.22	970	125	5.23
	0.05	0.22	860	92	5.23
	0.075			62	5.23
Berl saddles	0.013	0.28	1600	466	2.00
	0.025	0.28	1800	249	5.23
	0.038	0.28	1700	144	5.23
	0.05			105	5.23
Tri-packs	0.05	0.14	440		
Pall rings	0.016			341	5.23
	0.025			206	5.23
	0.038			128	5.23
	0.05			102	5.23

[a]Converted to SI units using a factor of 10.76.

Source: LaGrega et al., 1994; Cornwell, 1990; Roberts et al., 1985; Treybal, 1980.

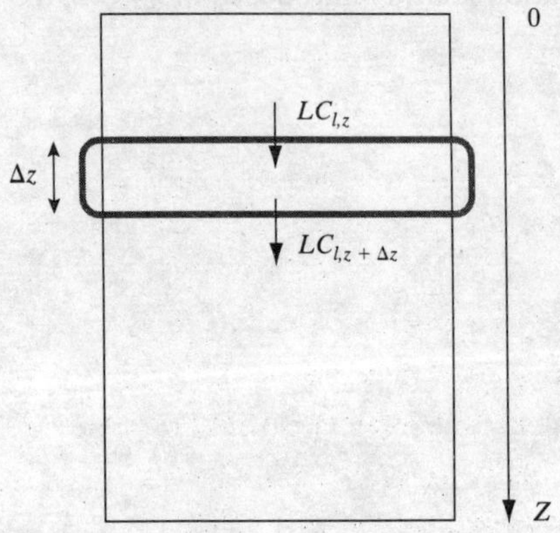

Figure 8.C.4 Differential control volume used for material balance on an air stripping tower.

balance has this form:

$$A\Delta z\frac{dC_l}{dt} = LAC_{l,z} - LAC_{l,z+\Delta z} - A\Delta z k_{gl} a(C_l - C_{l,\text{eq}}) \tag{8.C.3}$$

where A is the cross-sectional area of the control volume (m^2), Δz is the thickness of the control volume (m), C_l is the liquid concentration of contaminant in the control volume (M), t is time (h), L is the liquid loading rate (m^3 m^{-2} h^{-1}), $C_{l,z}$ is the liquid concentration of contaminant entering the control volume (M), $C_{l,z+\Delta z}$ is the liquid concentration of contaminant exiting the control volume (M), k_{gl} is the overall gas-liquid mass transfer coefficient (m h^{-1}), a is the specific interfacial surface area between liquid and gas (i.e., surface area/volume) (m^2 m^{-3}), and $C_{l,\text{eq}}$ is the liquid concentration of contaminant within the control volume in equilibrium with the gas phase (M). The terms in this equation represent, from left to right, accumulation, flow in, flow out, and mass transferred from liquid to gas phase.

If we assume that the column is operating at steady state, then $dC_l/dt = 0$; that is, there is no contaminant accumulation within the control volume. Dividing by $A\Delta z$ and taking the limit as Δz approaches zero yields

$$-L\frac{dC_l}{dz} = (k_{gl}a)(C_l - C_{l,\text{eq}}) \tag{8.C.4}$$

Rearranging, we have

$$dz = -\frac{L}{(k_{gl}a)}\frac{dC_l}{(C_l - C_{l,\text{eq}})} \tag{8.C.5}$$

We can use Henry's law to make the following substitution, replacing the equilibrium aqueous concentration with a term that depends on the gas-phase concentration:

$$C_{l,\text{eq}} = \frac{C_g RT}{H_g} \tag{8.C.6}$$

where C_g is the concentration of the contaminant in the gas phase (mol L^{-1}), R is the gas constant (0.0821 atm L mol^{-1} K^{-1}), T is the temperature (K), and H_g is Henry's constant (atm M^{-1}) of the contaminant. Applying an overall material balance from the control volume to the bottom of the column for the contaminant within the liquid and gas phases yields

$$A(LC_l + GC_{g,i}) = A(LC_{l,e} + GC_g) \tag{8.C.7}$$

where G is the gas loading rate for the column (m^3 m^{-2} h^{-1}), and $C_{l,e}$ is the concentration of contaminant in the liquid effluent of the column. We'll assume that the air entering the column contains no contaminant ($C_{g,i} = 0$), so equation 8.C.7 can be simplified and rearranged as

$$C_g = \frac{L}{G}(C_l - C_{l,e}) \tag{8.C.8}$$

Finally, we introduce the stripping factor, S, which is defined as follows:

$$S = \frac{H_g G}{RTL} \tag{8.C.9}$$

By substituting equations 8.C.6, 8.C.8, and 8.C.9 into 8.C.5 and integrating from the top to the bottom of the column ($z = 0$ to Z; $C_l = C_{l,i}$ to $C_{l,e}$), we obtain

$$Z = \frac{L}{k_{gl}a} \times \text{NTU} = \text{HTU} \times \text{NTU} \tag{8.C.10}$$

where

$$\text{NTU} = \frac{C_{l,i}}{C_{l,e}} - 1 \qquad S = 1 \tag{8.C.11}$$

$$\text{NTU} = \frac{S}{S-1}\ln\left(\frac{\dfrac{C_{l,i}}{C_{l,e}}(S-1) + 1}{S}\right) \qquad S \neq 1 \tag{8.C.12}$$

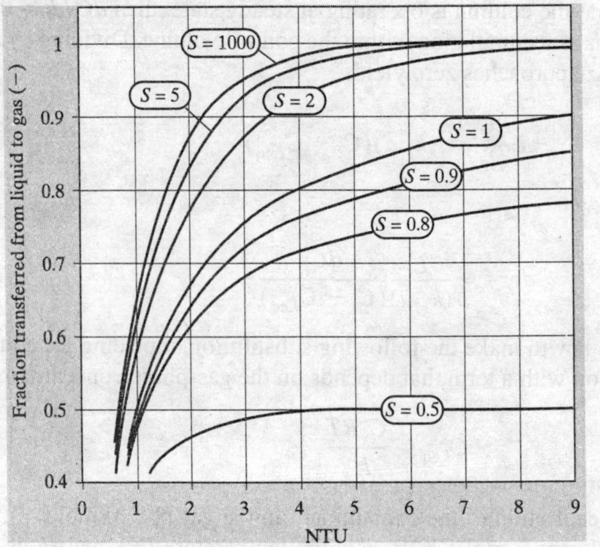

Figure 8.C.5 Effect of stripping factor (S) on contaminant removal within an air stripper. The fraction transferred from liquid to gas is equal to $1 - (C_{l,e}/C_{l,i})$.

In these equations, the variables have the following definitions. The height of the packing in the column is Z (m). The variable HTU ($= L/[k_{gl}a]$) is known as the *height of a transfer unit* (m). The variable NTU is known as the *number of transfer units* (–). The influent and effluent liquid concentrations of contaminant are $C_{l,i}$ and $C_{l,e}$, respectively (M). S is the stripping factor (–). G is the volumetric gas flow rate through the column, normalized by the column cross-sectional area ($m^3\ m^{-2}\ h^{-1}$). L is the volumetric liquid flow rate through the column, normalized by the column cross-sectional area ($m^3\ m^{-2}\ h^{-1}$).

In equation 8.C.10, the column height is expressed as the product HTU × NTU to differentiate between the kinetics of the process (*how fast* mass transfer occurs), which is represented by HTU, and the equilibrium endpoint (*how far* mass transfer will ultimately proceed), represented by NTU. The kinetics are controlled by the rate of mass transfer between the liquid and gas phases; hence, HTU is a function of ($k_{gl}a$), the overall mass-transfer coefficient. The equilibrium endpoint is a function of the volatility of the contaminant, and NTU therefore depends on S, which in turn is a function of Henry's constant, H_g. For an air stripper to function effectively, S should be much greater than 1. For stripping factors significantly less than 1, the driving force for mass transfer between the liquid and gas phases would not be sufficient to promote significant amounts of stripping (Figure 8.C.5).

A typical countercurrent air stripper for the treatment of hazardous wastewater has a height, Z, of 2–15 m and a diameter of 0.5–4 m. It is operated with a gas-to-liquid flow rate ratio (G/L) of 5–200 (volume per volume) and a stripping factor, S, of 2–20 (LaGrega et al., 1994; Boegel, 1989; Roberts and Levy, 1985). Example 8.C.1 illustrates the application of these relationships.

EXAMPLE 8.C.1 ***Designing a Countercurrent Air Stripper for TCE Removal from Water***

Groundwater contaminated with TCE is to be treated in a countercurrent packed air-stripping column at 20 °C. The column is operated with a liquid loading rate, L, of 36

EXAMPLE 8.C.1 ***Designing a Countercurrent Air Stripper for TCE Removal from Water (continued)***

$m^3\ m^{-2}\ h^{-1}$ and an air/water flow rate ratio (G/L) of 25 (m^3/m^3). Henry's constant for TCE is $H_g = 9.1$ atm M^{-1}. The gas-liquid mass-transfer coefficient is $k_{gl}a = 36\ h^{-1}$.

(a) Compute the stripping factor and estimate the percentage removal of TCE from groundwater in a 2 m column.

(b) How tall would the column need to be to achieve 99 percent removal?

SOLUTION

(a) Begin by calculating the stripping factor using equation 8.C.9:

$$S = \frac{(9.1 \text{ atm M}^{-1})}{(8.21 \times 10^{-2} \text{ atm K}^{-1} \text{ M}^{-1})(293 \text{ K})}(25) = 9.5$$

Then calculate the height of a transfer unit (HTU) from equation 8.C.10:

$$\text{HTU} = \frac{L}{k_{gl}a} = \frac{36 \text{ m}^3 \text{ m}^{-2} \text{ h}^{-1}}{36 \text{ h}^{-1}} = 1.0 \text{ m}$$

Since $Z = \text{HTU} \times \text{NTU} = 2$ m, the number of transfer units is NTU = 2. Use equation 8.C.12 to calculate the fraction of contaminant remaining in the effluent liquid ($C_{l,e}/C_{l,i}$):

$$2 = \frac{9.5}{(9.5 - 1)} \ln\left(\frac{\frac{C_{l,i}}{C_{l,e}}(9.5 - 1) + 1}{9.5}\right)$$

Solving this equation, we find

$$\frac{C_{l,e}}{C_{l,i}} = 0.15$$

The percentage of TCE transferred from water to air is $(1 - 0.15) \times 100 = 85$ percent.

(b) For 99 percent removal, we would need $C_{l,e}/C_{l,i} = 0.01$. With the given stripping factor, $S = 9.5$, we can apply equation 8.C.12 to find that NTU = 5 would be required. (Note that HTU remains unchanged from part (*a*), since the kinetics of mass transfer is unchanged.) The new column height can then be directly calculated:

$$Z = \text{HTU} \times \text{NTU} = 1 \text{ m} \times 5 = 5 \text{ m}$$

Now that we have been introduced to a model that can predict the efficiency of a countercurrent air stripper, it is necessary to understand how to estimate the mass-transfer coefficient that is required in order to apply this model.

The best method for estimating $k_{gl}a$ is to use a pilot plant experiment with the actual contaminated water. However, in the absence of experimental data, there are two widely used empirically based methods that will give adequate estimates. The first method, developed by Sherwood and Holloway (1940), involves a single equation that is based solely on liquid-phase mass-transfer resistance:

$$k_{gl}a = D_l m\left(\frac{0.3048L_m}{\mu_l}\right)^{1-n}\left(\frac{\mu_l}{\rho_l D_l}\right)^{0.5} \tag{8.C.13}$$

where $k_{gl}a$ is the overall mass-transfer rate coefficient (s^{-1}), D_l is the diffusivity of the contaminant in liquid ($m^2\ s^{-1}$), L_m is the liquid mass loading rate ($kg\ m^{-2}\ s^{-1}$), m and n are empirical constants that depend on the type of packing (see Table 8.C.3); μ_l is the liquid viscosity ($kg\ m^{-1}\ s^{-1}$), and ρ_l is the liquid density ($kg\ m^{-3}$). Note that the Sherwood-Holloway correlation is not dimensionally consistent. The values of the empirical constants in Table 8.C.3 are valid in equation 8.C.13 only for the specific dimensions indicated for the variables.

A second set of empirical equations for estimating $k_{gl}a$ are the Onda correlations (Onda et al., 1968). This set of equations is more complex than the Sherwood-Holloway model but gives more accurate predictions since the gas- and liquid-phase resistances are both considered (Roberts et al., 1985). The key equations are as follows:

$$a = a_t\left\{1 - \exp\left[-1.45\left(\frac{\sigma_c}{\sigma}\right)^{0.75}\left(\frac{L_m}{a_t\mu_l}\right)^{0.1}\left(\frac{L_m^2 a_t}{\rho_l^2 g}\right)^{-0.05}\left(\frac{L_m^2}{\rho_l \sigma a_t}\right)^{0.2}\right]\right\} \tag{8.C.14}$$

$$k_l = 0.0051\left(\frac{L_m}{a\mu_l}\right)^{2/3}\left(\frac{\mu_l}{\rho_l D_l}\right)^{-1/2}(a_t d)^{2/5}\left(\frac{\rho_l}{\mu_l g}\right)^{-1/3} \tag{8.C.15}$$

$$k_g = (a_t D_g c)\left(\frac{G_m}{a_t\mu_g}\right)^{0.7}\left(\frac{\mu_g}{\rho_g D_g}\right)^{1/3}(a_t d)^{-2} \tag{8.C.16}$$

where a is the wetted specific surface area of packing ($m^2\ m^{-3}$), a_t is the total specific surface area of packing ($m^2\ m^{-3}$) (see Table 8.C.3), k_l is the liquid-phase mass-transfer coefficient ($m\ s^{-1}$), k_g is the gas-phase mass-transfer coefficient ($m\ s^{-1}$), D_g is the diffusivity of contaminant in gas ($m^2\ s^{-1}$), μ_g is the gas viscosity ($kg\ m^{-1}\ s^{-1}$), ρ_g is the gas density ($kg\ m^{-3}$), L_m is the liquid mass loading rate ($kg\ m^{-2}\ s^{-1}$), G_m is the gas mass loading rate ($kg\ m^{-2}\ s^{-1}$), d is the nominal diameter of the packing material (m), σ is the liquid surface tension ($kg\ s^{-2} = N\ m^{-1}$), σ_c is the critical surface tension for the packing material ($kg\ s^{-2}$) (see Table 8.C.4), and c is an empirical constant (–) (see Table 8.C.3). The other parameters are as previously defined.

To use the Onda correlations, the wetted specific surface area, a, is first calculated from equation 8.C.14, and then the liquid- and gas-side mass-transfer coefficients, k_l

Table 8.C.4 Critical Surface Tension of Packing Materials Used in Countercurrent Air Strippers

Material	σ_c ($kg\ s^{-2}$)
Ceramic	0.061
Glass	0.073
Polyethylene	0.033
PVC	0.04

Source: Cornwell, 1990.

and k_g, are computed from equations 8.C.15 and 8.C.16. Finally, k_{gl} is computed using a variant of equation 4.C.19 (§4.C.2):

$$\frac{1}{k_{gl}} = \frac{1}{k_l} + \frac{RT}{H_g k_g} \tag{8.C.17}$$

Example 8.C.2 demonstrates the use of the Onda correlations.

Another important design consideration for countercurrent air-stripping towers is the air pressure drop. Typical pressure drops through the tower should be kept at 50–400 Pa per meter of packing depth. Pressure drops higher than this can cause an accumulation of water at the top of the tower and possible flooding or water blowout. Pressure drops below this range would probably not result in cost-effective treatment because of the low volumetric air flow rate (Treybal, 1980).

EXAMPLE 8.C.2 ***Estimating Mass-Transfer Coefficients for the Removal of TCE from Water in an Air-Stripping Column***

Estimate mass-transfer coefficients for the countercurrent packed air-stripping column described in Example 8.C.1. Assume that the column is packed with 50 mm (d = 0.05 m) polyethylene Raschig rings. The diffusivity of TCE in water and air, and the viscosity and density of water and air at 20 °C are provided here: $D_l = 9.3 \times 10^{-10}$ m^2 s^{-1}, $D_g = 1.0 \times 10^{-5}$ m^2 s^{-1}, $\mu_l = 1 \times 10^{-3}$ kg m^{-1} s^{-1}, $\mu_g = 1.8 \times 10^{-5}$ kg m^{-1} s^{-1}, ρ_l = 998 kg m^{-3}, and ρ_g = 1.2 kg m^{-3}. The surface tension of water is σ = 0.073 kg s^{-2}. The liquid loading rate is L_m = 10 kg m^{-2} s^{-1}.

(a) Estimate the values of k_l, k_g, and $k_{gl}a$ using the Onda correlations.

(b) How would your estimate of the overall mass-transfer coefficient differ using the Sherwood-Holloway correlation?

SOLUTION

(a) We begin by calculating the wetted surface area of the packing using equation 8.C.14 and values from Tables 8.C.3 and 8.C.4.

$$a = 92 \text{ m}^2 \text{ m}^{-3}[1 - \exp(-1.45\, A^{0.75} B^{0.1} C^{-0.05} D^{0.2})] = 50 \text{ m}^2 \text{ m}^{-3}$$

where the capital letters signify dimensionless groups of variables:

$$A = \frac{\sigma_c}{\sigma} = \frac{0.033 \text{ kg s}^{-2}}{0.073 \text{ kg s}^{-2}} = 0.45$$

$$B = \frac{L_m}{a_t \mu_l} = \frac{10 \text{ kg m}^{-2} \text{ s}^{-1}}{(92 \text{ m}^2 \text{ m}^{-3})(1 \times 10^{-3} \text{ kg m}^{-1} \text{ s}^{-1})} = 109$$

$$C = \frac{L_m^2 a_t}{\rho_l^2 g} = \frac{(10 \text{ kg m}^{-2} \text{ s}^{-1})^2 (92 \text{ m}^2 \text{ m}^{-3})}{(998 \text{ kg m}^{-3})^2 (9.8 \text{ m s}^{-2})} = 0.00094$$

$$D = \frac{L_m^2}{\rho_l \sigma a_t} = \frac{(10 \text{ kg m}^{-2} \text{ s}^{-1})^2}{(998 \text{ kg m}^{-3})(0.073 \text{ kg s}^{-2})(92 \text{ m}^2 \text{ m}^{-3})} = 0.015$$

EXAMPLE 8.C.2 ***Estimating Mass-Transfer Coefficients for the Removal of TCE from Water in an Air-Stripping Column (continued)***

To proceed, we next calculate the gas loading rate from the given air-to-water flow ratio, $G/L = 25$:

$$G_m = L_m \frac{G}{L}\left(\frac{\rho_g}{\rho_l}\right) = 10 \text{ kg m}^{-2} \text{ s}^{-1} (25) \frac{1.2 \text{ kg m}^{-3}}{998 \text{ kg m}^{-3}} = 0.3 \text{ kg m}^{-2} \text{ s}^{-1}$$

We can now use equations 8.C.15 and 8.C.16 to calculate k_l and k_g:

$$k_l = 0.0051 E^{2/3} F^{-1/2} G^{2/5} H^{-1/3} = 2.1 \times 10^{-4} \text{ m s}^{-1}$$

where the capital letters represent the following groups:

$$E = \frac{L_m}{a\mu_l} = \frac{10 \text{ kg m}^{-2} \text{ s}^{-1}}{(50 \text{ m}^2 \text{ m}^{-3})(1 \times 10^{-3} \text{ kg m}^{-1} \text{ s}^{-1})} = 200$$

$$F = \frac{\mu_l}{\rho_l D_l} = \frac{1 \times 10^{-3} \text{ kg m}^{-1} \text{ s}^{-1}}{(998 \text{ kg m}^{-3})(9.3 \times 10^{-10} \text{ m}^2 \text{ s}^{-1})} = 1077$$

$$G = a_t d = (92 \text{ m}^2 \text{ m}^{-3})(0.05 \text{ m}) = 4.6$$

$$H = \frac{\rho_l}{\mu_l g} = \frac{998 \text{ kg m}^{-3}}{(1 \times 10^{-3} \text{ kg m}^{-1} \text{ s}^{-1})(9.8 \text{ m s}^{-2})} = 1.0 \times 10^5 \text{ s}^3 \text{ m}^{-3}$$

And

$$k_g = IJ^{0.7} K^{1/3} G^{-2} = 9.9 \times 10^{-3} \text{ m s}^{-1}$$

where the capital letters are defined as follows (G is defined and evaluated above):

$$I = a_t D_g c = (92 \text{ m}^2 \text{ m}^{-3})(1.0 \times 10^{-5} \text{ m}^2 \text{ s}^{-1})(5.23) = 0.0048 \text{ m s}^{-1}$$

$$J = \frac{G_m}{a_t \mu_g} = \frac{0.3 \text{ kg m}^{-2} \text{ s}^{-1}}{(92 \text{ m}^2 \text{ m}^{-3})(1.8 \times 10^{-5} \text{ kg m}^{-1} \text{ s}^{-1})} = 181$$

$$K = \frac{\mu_g}{\rho_g D_g} = \frac{(1.8 \times 10^{-5} \text{ kg m}^{-1} \text{ s}^{-1})}{(1.2 \text{ kg m}^{-3})(1.0 \times 10^{-5} \text{ m}^2 \text{ s}^{-1})} = 1.5$$

Finally, $k_{gl}a$ is calculated using equation 8.C.17:

$$k_{gl} = \left(\frac{1}{k_l} + \frac{RT}{H_g k_g}\right)^{-1}$$

$$= \left(\frac{1}{2.1 \times 10^{-4} \text{ m s}^{-1}} + \frac{(8.21 \times 10^{-2} \text{ atm K}^{-1} \text{ M}^{-1})(293 \text{ K})}{(9.1 \text{ atm M}^{-1})(9.9 \times 10^{-3} \text{ m s}^{-1})}\right)^{-1}$$

$$= 2.0 \times 10^{-4} \text{ m s}^{-1}$$

EXAMPLE 8.C.2 ***Estimating Mass-Transfer Coefficients for the Removal of TCE from Water in an Air-Stripping Column (continued)***

So

$$k_{gl}a = (2.0 \times 10^{-4}\ \text{m s}^{-1})(50\ \text{m}^2\ \text{m}^{-3}) = 1.0 \times 10^{-2}\ \text{s}^{-1} = 36\ \text{h}^{-1}$$

(b) To calculate $k_{gl}a$ using the Sherwood-Holloway correlation, apply equation 8.C.13 with values from Table 8.C.3 for 50 mm polyethylene Raschig rings:

$$k_{gl}a = (9.3 \times 10^{-10}\ \text{m}^2\ \text{s}^{-1})(860)\left(\frac{0.3048(10\ \text{kg m}^{-2}\ \text{s}^{-1})}{1 \times 10^{-3}\ \text{kg m}^{-1}\ \text{s}^{-1}}\right)^{1-0.22} F^{0.5}$$

$$= 1.4 \times 10^{-2}\ \text{s}^{-1}$$

where F is defined and evaluated above. In this case, the two correlations produce values of the overall mass-transfer coefficient $k_{gl}a$ that are of the same magnitude but differ by about 30 percent (difference divided by the mean). Considering the uncertainty involved in the parameters used to calculate these values, this degree of agreement is about what one should expect.

Steam Stripping

Steam stripping is a mass-transfer process that is similar to the air-stripping process discussed above. However, application of steam stripping is more appropriate than air stripping for treating wastes with high concentrations of organics (from low parts per million to several percent by weight) or with less volatile contaminants.

The typical configuration of a countercurrent steam-stripping unit is shown in Figure 8.C.6. The influent wastewater is partially heated by energy from the effluent

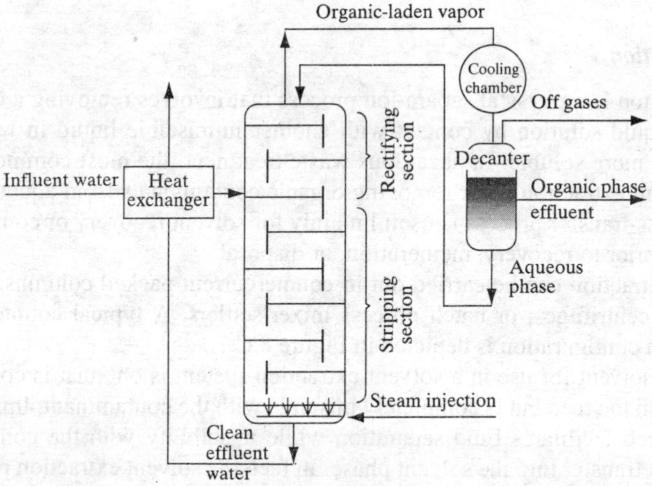

Figure 8.C.6 Typical configuration of a countercurrent steam-stripping column for the recovery of hydrocarbons from water.

water in a heat exchanger. The influent then enters the tower, where the waste flows downward through a series of trays. An upward flow of steam heats the waste, causing contaminants to volatilize into the gas phase. The contaminants rise through the tower with the steam and are carried out of the column with the steam vapors. The column is divided into two sections. The stripping section functions in a similar manner to an air-stripping unit. The rectifying section, located at the top of the tower, causes the steam vapors to become more highly enriched with contaminant so that it is possible to produce a separate organic phase in the condenser. The contaminant can then be recovered by removing the organic phase from the condenser. The trays inside the steam stripper are generally constructed of stainless steel, with sieves or two-way valves to allow steam to rise through the downward-flowing liquid phase. Columns are also sometimes constructed with stainless steel or ceramic random packing elements rather than trays. Steam towers operate at temperatures of 101–103 °C and at pressures slightly above atmospheric (Rogers and Brant, 1989; LaGrega et al., 1994).

Due to the high temperatures associated with steam stripping, this process has higher capital and operating costs than air stripping. However, steam stripping is especially attractive for solvent recovery from hazardous waste mixtures since solvents are recovered in a separate liquid phase. Demonstrated applications for steam stripping include wastewaters containing petroleum aromatics and aliphatics, ketones, esters, PCBs, pesticides, organosulfur compounds, pentachlorophenol, and other chlorinated solvents. Steam stripping is not appropriate for recovery of nonvolatile contaminants or for noncondensable gases (e.g., propane or vinyl chloride) when they are the only contaminants present in wastewater since they will not form a separate recoverable phase in the condenser. However, these types of contaminants can be removed when they are present in wastewater mixtures containing other low-solubility volatile organics (e.g., toluene or hexane) (LaGrega et al., 1994; Boegel, 1989).

A typical countercurrent steam stripper uses a steam application rate of approximately 0.2 kg steam per kg influent waste stream. The volume of water leaving the column is generally slightly higher (~5 percent) than the volume of the entering waste stream since some of the applied steam condenses. Suspended solids or potential precipitates (e.g., iron or sulfides) should be removed from the waste stream before steam stripping to prevent fouling of the tower (Boegel, 1989; Rogers and Brant, 1989).

Solvent Extraction

Solvent extraction is a physical separation process that involves removing a contaminant from a liquid solution by contact with another immiscible liquid in which the contaminant is more soluble. In hazardous waste treatment, the most common application of solvent extraction is for removing organic contaminants from aqueous solutions. This mass-transfer process is useful mainly for solvent recovery or contaminant concentration prior to recovery, incineration, or disposal.

Solvent extraction can be carried out in countercurrent packed columns, concurrent columns, centrifuges, or batch process mixer-settlers. A typical countercurrent packed column configuration is depicted in Figure 8.C.7.

The ideal solvent for use in a solvent extraction system is one that is completely immiscible with the feed but is completely miscible with the contaminant. Immiscibility with the feed facilitates fluid separation while miscibility with the contaminant promotes mass transfer into the solvent phase. In fact, the solvent extraction process is analogous to air stripping in that equilibrium and kinetics of the transfer between fluid phases drive the system design. However, instead of using Henry's constant and the

Figure 8.C.7 Countercurrent configuration for a solvent extraction process, showing the terminology commonly used for the influent and effluent waste streams (feed and raffinate), and solvent streams (solvent, extract).

two-film theory for gas-liquid mass transfer, multicomponent phase diagrams are used to predict equilibrium concentrations.

Beneficial solvent properties include the following (Berkowitz, 1989; Treybal, 1980):

- *High solvent selectivity.* Solvents that have a much greater affinity for the contaminant than for the feed promote more efficient contaminant transfer and produce higher-quality extract and raffinate.
- *High saturation solubility.* The higher the solubility of the contaminant in the solvent, the greater the mass of contaminant that can be removed per mass of solvent.
- *Low solubility of solvent in feed.* The lower the solubility of the solvent in the feed, the easier the separation of the raffinate from the extract.
- *Density difference between solvent and feed.* Large differences in liquid densities facilitate separation of the raffinate from the extract.

Suitable solvents for application to hazardous waste treatment include crude oil, light oil, octanol, pentane, hexane, isopropyl ether, and other compounds listed in Table 8.C.5. Some reported applications of solvent extraction for hazardous waste

Table 8.C.5 Candidate Extraction Solvents for Removing Six EPA Designated Priority Pollutants from Wastewater

	Contaminant (EPA-designated priority pollutants)					
Extraction solvents	Acrolein	Acrylonitrile	2-Chlorophenol	Isophorone	Bis(2-chloroethyl) ether	Bis(2-chloroethoxy) ether
n-Butyl acetate	✓					
Methyl isobutyl ketone	✓		✓	✓	✓	
Toluene	✓			✓		
Tributyl phosphate		✓	✓	✓	✓	✓
Hydrocarbons			✓	✓	✓	✓
Isobutyl heptyl ketone			✓		✓	
Di-isobutyl ketone				✓		✓

Source: Adapted from Joshi et al., 1984.

treatment include the following (Berkowitz, 1989):

- Removal of 90–98 percent of phenols from petroleum refinery wastes with influents of 0.01–1 percent organic content using di-isopropyl ether or methyl isobutyl ketone
- Recovery of ~16 percent acetic acid from wastewater using ethyl acetate
- Removal of 99 percent of pentachlorophenol present at 100 mg/L in wood-preserving wastewater using fuel oil and amyl alcohol still bottoms
- Removal of 90–98 percent of PCBs from mineral oil using diethylene glycol monomethyl ether

There are two major disadvantages associated with the use of solvent extraction for the treatment of hazardous wastes. First, this process is simply a mass-transfer process that results in two waste streams (the raffinate and the extract) with potentially significant concentrations of contaminant and solvent in each. Total separation of fluids is rarely achieved after extraction is complete, requiring postextraction treatment on one or both fluids. Second, the extraction solvents are expensive and sometimes exhibit hazardous characteristics themselves. However, with proper attention to the management of residuals, solvent extraction can be a valuable hazardous waste treatment technology.

Sorption on Activated Carbon

Sorption on activated carbon is a physical separation process that is commonly applied for hazardous waste treatment, especially for treating water contaminated with hydrocarbons or chlorinated solvents. Activated carbon is commonly used to treat the effluent gas from air-stripping processes. It is also used in combination with biological treatment for removing organics that are difficult to biodegrade. The theoretical and practical issues associated with the application of activated carbon for the treatment of hazardous wastes are similar to those for the treatment of water and wastewater (see §6.D.3).

8.C.2 Chemical Treatment

Unlike physical separation processes that merely concentrate or change the phase of hazardous wastes, chemical treatment processes can reduce the toxicity, ignitability, reactivity, or corrosivity of wastes. Chemical treatment is commonly applied on-site to avoid transporting large quantities of potential hazardous waste, but it is also practiced at off-site commercial facilities. Applications include metal-bearing wastes, inorganic contaminants, and dissolved organics. Chemical processes are often used on concentrated solutions containing mixtures of contaminants, such as spent acidic plating solutions or pickling baths, alkali from removal of sulfur from petroleum products, and hazardous landfill leachates.

Neutralization

Acid/base neutralization is a common hazardous waste treatment method and is sometimes the only treatment required to declassify a waste as hazardous. Virgin chemicals (acids or bases) can be added to the waste to achieve neutralization, or acidic waste can be mixed with basic waste.

Neutralization processes can be carried out in batch or continuous flow systems. Dissolved gases are typically delivered through diffusers or mixers, while solids are granularized before addition, and liquids can be added directly to the flow stream. Chemical addition rates and waste flows are generally regulated with pH sensors coupled to feed meters. Potential problems inherent to neutralization processes include excess heat from exothermic reactions, corrosivity of the reactants, clogging or scaling caused by precipitation of by-products, and evolution of volatile contaminants (Dawson and Mercer, 1986; Haas and Vamos, 1995).

Acidic Wastes Although the acidic nature of a waste is detected by measuring its pH, the actual amount of acid present in a wastewater that will require neutralization is measured by the acidity of the waste. The acidity of a waste is its capacity to neutralize a strong base. It is analogous to the concept of alkalinity, which measures the capacity of a solution to neutralize a strong acid (see §3.C.4). Acidity is measured by titrating the waste with a strong base to a desired endpoint, such as pH 7. Compounds that contribute to the acidity of wastes include strong acids such as nitric (HNO_3), sulfuric (H_2SO_4), and hydrochloric (HCl), and weak acids such as acetic (CH_3COOH), hydrocyanic (HCN), hypochlorous (HOCl), and carbonic (H_2CO_3) (Haas and Vamos, 1995). The stoichiometry of acid neutralization is illustrated in Examples 8.C.3 and 8.C.4.

Commonly used neutralization agents for acidic wastes include limestone ($CaCO_3$), quicklime (CaO), hydrated lime (or slaked lime, $Ca(OH)_2$), soda ash (Na_2CO_3), and sodium hydroxide (or caustic soda, NaOH). Quicklime is cheaper than hydrated lime; however, the equipment and processing necessary to convert quicklime to the hydrated form prior to use often make it more economical to use the more expensive hydrated lime. Limestone is generally very inexpensive, but it dissolves more slowly than lime and produces a larger quantity of residual sludge for treatment. One method of limestone neutralization involves passing the waste through a bed of crushed limestone. Use of a weak base such as soda ash instead of a strong base such as sodium hydroxide can be advantageous, since it avoids the potential production of a strongly basic waste due to overdosing. In addition, weak bases are safer to handle. However, weak bases may be more expensive on a base-equivalent scale than strong bases, and the addition of weak bases results in larger increases in waste volume (Dawson and Mercer, 1986). Advantages to using sodium hydroxide for waste neutralization include rapid reaction rates, low volume increases, and minimal sludge production.

Basic Wastes A measure of the amount of base that requires neutralization in a basic waste is the alkalinity (or basicity) of the waste. Potential sources of alkalinity in wastes include strong bases such as sodium hydroxide (NaOH) and weak bases such as ammonia (NH_3), carbonates, and heavy-metal salts.

Commonly used neutralization agents for basic wastes include strong acids such as hydrochloric and sulfuric, weak acids such as acetic, and carbon dioxide gas. Sulfuric acid is less expensive than hydrochloric acid but requires a slightly longer reaction time (15–30 minutes versus 5–20 minutes) and has the potential for producing significant amounts of precipitate sludge when used to treat calcium-bearing wastes. Hydrochloric acid, on the other hand, can potentially form a corrosive acid mist during the neutralization reaction (Haas and Vamos, 1995). Carbon dioxide dissolves in water to form carbonic acid, which in turn dissociates to bicarbonate and carbonate, liberating protons that can neutralize alkaline wastes (see §3.C.4). Carbon dioxide is safe to handle and dosing is easy to control due to the inherent buffering capacity of the carbonates. In addition, on-site flue gas from hydrocarbon fuel combustion can often be used as an economical source of carbon dioxide, reducing emissions of a gaseous waste. However, some disadvantages associated with the use of carbon dioxide as a neutralizing agent are its relatively high cost (if it must be purchased) and its potential for forming sludge precipitates with calcium-bearing wastes (Haas and Vamos, 1995).

EXAMPLE 8.C.3 ***Acid Neutralization with a Strong Base***

An aqueous waste that is considered hazardous because of its steel corrosion capabilities has a measured pH of 3 caused by the presence of one or more strong acids. It is desired to treat the waste with a strong base, NaOH, to raise the pH to 8. Assuming that there are no weak acids (or buffers) in this waste, calculate the mass of NaOH that must be added per liter of waste to carry out this neutralization at 25 °C.

SOLUTION The strong acid in the untreated waste can be represented generically as HA, which undergoes the following dissociation reaction:

$$HA \Leftrightarrow H^+ + A^-$$

Since HA is a strong acid, we know that the above reaction will proceed far to the right, and there will be negligible undissociated HA present in either the untreated or the treated waste.

The electroneutrality relationship in the untreated waste can be represented by

$$[H^+] = [OH^-] + [A^-]$$

There may be other cations and anions present in the solution. If so, they will balance and will not participate in the acid-base reactions since it was given that no buffers are present.

Since the untreated waste has a pH of 3 and

$$[H^+][OH^-] = K_w = 10^{-14}$$

then $[H^+] = 10^{-3}$ M, $[OH^-] = 10^{-11}$ M, and consequently $[A^-] = 10^{-3}$ M.

After the addition of NaOH, the electroneutrality relationship becomes

$$[Na^+] + [H^+] = [OH^-] + [A^-]$$

The addition of the strong base does not influence $[A^-]$, and since the pH of the final solution is 8, we determine that

$$[Na^+] = [OH^-] + [A^-] - [H^+] = 10^{-6} + 10^{-3} - 10^{-8} = 10^{-3}\text{ M}$$

In words, we must add 10^{-3} mol of NaOH per liter of waste to achieve the desired neutralization. The molecular weight of NaOH is 40 g mol^{-1}; therefore, the required dose of NaOH is $10^{-3} \times 40 = 0.04$ g/L or 40 mg/L.

EXAMPLE 8.C.4 ***Acid Neutralization with a Weak Base***

What dose of weak base, soda ash (Na_2CO_3), is required to raise the pH of the waste in Example 8.C.3 from 3 to 8? Assume that the temperature is 25 °C, the total air pressure is 1 atm, and the reactor is in equilibrium with the atmosphere.

EXAMPLE 8.C.4 ***Acid Neutralization with a Weak Base (continued)***

SOLUTION From Example 8.C.3, we have as the key composition of the untreated waste $[H^+] = 10^{-3}$ M, $[OH^-] = 10^{-11}$ M, and $[A^-] = 10^{-3}$ M.

When soda ash is added to water, it will dissolve completely, liberating sodium (Na^+) and carbonate (CO_3^{2-}) ions. The carbonate can accept a proton to become bicarbonate (HCO_3^-) which, in turn, can accept a proton to become carbonic acid ($H_2CO_3^*$) (see §3.C.4). The following electroneutrality relationship applies to the treated waste:

$$[Na^+] + [H^+] = [OH^-] + [A^-] + [HCO_3^-] + 2[CO_3^{2-}]$$

Recall the following equilibrium relationships that apply to the reactions of carbonate species in water (§3.C.4):

$$\text{Carbonic acid dissociation: } [HCO_3^-] = \frac{K_1[H_2CO_3^*]}{[H^+]}$$

$$\text{Bicarbonate dissociation: } [CO_3^{2-}] = \frac{K_2[HCO_3^-]}{[H^+]}$$

$$\text{Henry's law: } [H_2CO_3^*] = (1 + K_m)K_{HC}P_{CO_2}$$

The partial pressure of CO_2 in air is $P_{CO_2} = 350 \times 10^{-6}$ atm. The equilibrium constants at 25 °C are $K_m = 1.58 \times 10^{-3}$, $K_1 = 4.47 \times 10^{-7}$ M, $K_2 = 4.68 \times 10^{-11}$ M, and $K_{HC} = 0.034$ M atm^{-1}.

Substituting from these equilibrium relationships for bicarbonate and carbonate, the electroneutrality relationship becomes

$$[Na^+] + [H^+] = [OH^-] + [A^-] + \frac{K_1(1 + K_m)K_{HC}P_{CO_2}}{[H^+]}\left(1 + 2\frac{K_2}{[H^+]}\right)$$

The addition of the weak base does not influence $[A^-]$, so it remains at 10^{-3} M. Since the pH of the final solution is 8, we can substitute $[H^+] = 10^{-8}$ M and $[OH^-] = 10^{-6}$ M. Rearranging and substituting for the partial pressure of CO_2 plus the equilibrium constants, we obtain the following result:

$$[Na^+] = 10^{-6} + 10^{-3} + 5.32 \times 10^{-4}(1 + 9.4 \times 10^{-3}) - 10^{-8} = 1.5 \times 10^{-3}\text{ M}$$

The molecular weight of Na_2CO_3 is 106 g mol^{-1}, which corresponds to 53 g per mol Na^+; therefore, we must add $1.5 \times 10^{-3} \times 53 = 0.08$ g/L or 80 mg/L of Na_2CO_3 to neutralize the waste. Note that the amount of weak base required to neutralize this waste is two times the required amount of strong base in Example 8.C.3. Part of the difference occurs because of the lower strength of soda ash as a base: More equivalents of the weak base (1.5 meq/L) must be added to achieve the same level of neutralization that is achieved with a smaller amount of the strong base (1.0 meq/L). The other factor contributing to the difference is the greater molecular weight of soda ash (53 g per equivalent) compared with caustic soda (40 g per equivalent).

Chemical Precipitation

Precipitation is a chemical treatment process that involves removing soluble contaminants from solution by causing a reaction that forms an insoluble product. Once insoluble precipitates are formed, they can be removed from solution by settling or filtration.

Precipitation is the most commonly applied process for the removal of inorganic contaminants from aqueous hazardous wastes. Typical applications include wastewaters that contain arsenic, barium, cadmium, chromium, copper, lead, mercury, nickel, selenium, or zinc. The wastewater sources include metal plating, steel production, paint and dye processing, mining, electronics manufacturing, and landfill leachate (Chung, 1989). In these cases, precipitation is a concentration technique rather than a destruction method, since the residuals that are created contain the hazardous contaminants and are considered to be hazardous waste. Precipitation is also used in wastewater treatment to remove hardness, phosphorus, nitrogen, and BOD (Metcalf & Eddy, 1991). Precipitation as a water treatment technology was discussed in §6.D.4. Here we emphasize issues specific to hazardous waste treatment.

Recall that the precipitation of dissolved cations and anions to form insoluble salts can be described by the following stoichiometric relationship (see §3.B.2):

$$xA + yB \Leftrightarrow A_xB_y \qquad (3.B.8)$$

Here A and B represent cationic and anionic species dissolved in water, and A_xB_y represents an ionic solid. The solubility product is given by

$$[A]^x[B]^y = K_{sp} \qquad (3.B.9)$$

where [ ] denotes the activity of species in solution. (As elsewhere in this text, we assume that the activity equals the molar concentration, which is a good approximation for dilute solutions.)

In hazardous waste treatment, the toxic waste ion is usually a cation and the treatment technique involves adding an anion to cause chemical precipitation. Processes are sorted according to the added anion. The most common processes involve hydroxide, sulfide, or carbonate reactions. Each of these three types of reactions has associated advantages and disadvantages with respect to specific wastes, making the choice of precipitant anion highly dependent on waste composition.

Hydroxide Precipitation Hydroxide precipitation using lime is by far the most commonly applied precipitation reaction. It is effective for heavy-metal removal as well as for removing hardness and phosphorus. Hydroxide precipitation is performed by adding a strong or weak base to the wastewater of interest to raise the pH and cause formation of insoluble hydroxide species. Bases typically used for this purpose are lime ($Ca(OH)_2$) and sodium hydroxide (NaOH). A precipitation reaction involving a generic dissolved metal (M^{2+}) and lime is given as follows:

$$M^{2+} + Ca(OH)_2 \Leftrightarrow M(OH)_2(s) + Ca^{2+} \qquad (8.C.18)$$

If the amount of metal hydroxide formed is greater than the aqueous solubility of that species, then a precipitate will form that can be removed from solution by settling or filtration. Reaction 8.C.18 suggests that the removal efficiency of the undesired metal ion, M^{2+}, can always be improved by adding more lime. However, due to the potential for most metals to form various soluble hydroxide complexes, the solubility of metal hydroxides in solution tends to decrease with pH only up to a certain point, above which the solubilities increase with pH, as illustrated in Figure 8.C.8.

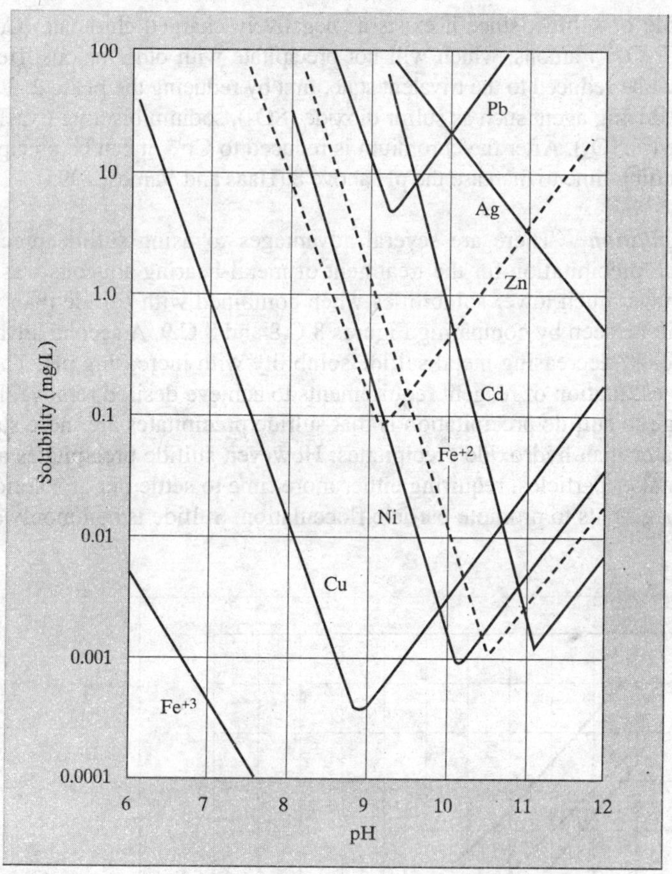

Figure 8.C.8 Indicative information on the solubilities of selected metal hydroxides as a function of pH (USEPA, 1983).

In a waste that contains a mixture of metal ions, this type of precipitation reaction poses a challenge. The optimal pH to maximize removal can vary among metals. In addition, potential impurities within the waste, such as EDTA (ethylenediaminetetra-acetic acid), can form chelates with the metals, inhibiting precipitation reactions.

Hydroxide precipitation is generally carried out at pH values of 9–11 and can typically achieve metal concentrations of 1 mg/L in effluent streams, depending on the particular metal solubility. Effluent metal levels as low as 0.1 mg/L are possible when precipitation is coupled with solids removal by filtration (Chung, 1989). Precipitates formed by lime addition are easier to settle and dewater than those formed by sodium hydroxide addition. However, for wastes containing sulfates, lime addition can cause scaling problems by producing insoluble gypsum ($CaSO_4$). Bench-scale jar tests (see §6.D.2) are required for precise estimation of precipitant doses, especially for mixed wastes. In addition, wastes containing more than one metal may require several process stages that allow precipitation to occur at separately optimized pH values (Chung, 1989; Haas and Vamos, 1995).

Chromium removal is a common application for hydroxide precipitation but must be considered as a special case because chromium can be precipitated only in the trivalent form (Cr(III)). The highly toxic hexavalent chromium (Cr(VI)) form cannot be treated using hydroxide or sulfide, since it exists as negatively charged chromate (CrO_4^{2-}) or dichromate ($Cr_2O_7^{2-}$) anions, which will not precipitate with other anions. Hexavalent chromium must be reduced to the trivalent state, first by reducing the pH to 2–3 and then by adding a reducing agent such as sulfur dioxide (SO_2), sodium bisulfite ($NaHSO_3$), or ferrous sulfate ($FeSO_4$). After the chromium is reduced to Cr^{3+}, it can be precipitated as $Cr(OH)_3$ by adding lime to increase the pH above 8 (Haas and Vamos, 1995).

Sulfide Precipitation There are several advantages to using sulfide precipitation over hydroxide precipitation for the treatment of metal-bearing aqueous waste. First, most metals have much lower solubilities when combined with sulfide than with hydroxide, as can be seen by comparing Figures 8.C.8 and 8.C.9. A second advantage is the monotonically decreasing metal sulfide solubility with increasing pH. This facilitates accurate estimation of reagent requirements to achieve desired removal levels. A third advantage to sulfide precipitation is that sulfide precipitates are more stable and easier to dewater than hydroxide precipitates. However, sulfide precipitates also tend to occur as smaller particles, requiring either more time to settle out of solution or the addition of coagulants to promote particle flocculation. Sulfide is commonly added in

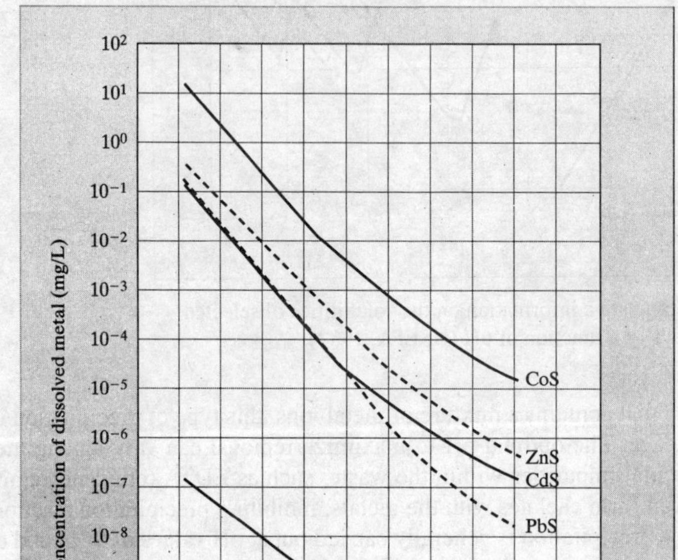

Figure 8.C.9 Curves based on experimental data that indicate the solubilities of selected metal sulfides as a function of pH (USEPA, 1983).

the form of ferrous sulfide (FeS), sodium sulfide (Na_2S), or sodium hydrosulfide (NaHS). A precipitation reaction involving a generic metal ion (M^{2+}) and ferrous sulfide is

$$M^{2+} + FeS \Leftrightarrow MS(s) + Fe^{2+} \tag{8.C.19}$$

Sulfide is a weak base and can exist as S^{2-}, HS^-, or H_2S in aqueous solution. Hydrogen sulfide (H_2S) is highly volatile and can be released as a toxic gas when it is preferentially formed under pH conditions of 7 and below. Therefore, to take advantage of the low solubilities of metal sulfides and to minimize generation of hydrogen sulfide gas, sulfide precipitation is carried out at high pH values.

Sulfide precipitation can reduce hexavalent chromium to the trivalent form and precipitate it in a single step. This is illustrated in the following reaction, in which dichromate ions react with ferrous sulfide to produce insoluble chromium hydroxide:

$$\begin{aligned} &Cr_2O_7^{2-} + 2\ FeS + 7\ H_2O \\ &\quad \Leftrightarrow 2\ Fe(OH)_3(s) + 2\ Cr(OH)_3(s) + 2\ S(s) + 2\ OH^- \end{aligned} \tag{8.C.20}$$

From a redox perspective, each chromium atom is being reduced from +6 to +3 in this reaction by the oxidation of iron from +2 to +3 and sulfur from −2 to 0.

When operated effectively, a sulfide precipitation process can produce effluent with metal contaminant levels of 0.1 mg/L or less (Chung, 1989). As with hydroxide precipitation, bench-scale jar tests are required to precisely estimate precipitant doses.

Carbonate Precipitation Carbonate precipitation is sometimes used instead of hydroxide precipitation to remove certain metals (e.g., Cd, Ni, Pb) because it operates at lower pH values and generates a denser precipitate that is easier to dewater. For example, cadmium and lead removal by hydroxide precipitation requires a pH of 10 but can be accomplished by carbonate precipitation at pH 8 (Chung, 1989).

Carbonate is typically added in the form of soda ash (Na_2CO_3). A generic carbonate precipitation reaction is

$$M^{2+} + Na_2CO_3 \Leftrightarrow MCO_3(s) + 2\ Na^+ \tag{8.C.21}$$

Since carbonate is a weak base that can be present as carbonic acid (H_2CO_3), bicarbonate (HCO_3^-), or carbonate (CO_3^{2-}), its form in water is dictated by pH, with higher pH values promoting greater formation of the carbonate ion. Since high pH levels also result in hydroxide formation, carbonate precipitation is often carried out as a coordinated process with hydroxide precipitation.

Solidification and Stabilization

Solidification and stabilization were originally developed as processes to treat industrial, hazardous, and certain radioactive wastes. The goal of solidification and stabilization is not to transform or destroy wastes, but rather to contain and minimize the release of wastes into the environment. Solidification and stabilization process objectives include detoxifying waste and limiting its mobility while improving its material properties such as increasing its structural strength and decreasing its compressibility and permeability (LaGrega et al., 1994). These processes are applied to radioactive wastes, heavy-metal wastes, ash residues, highly toxic sludges, oily sludges, contaminated soils, and complex mixtures. Solidification and stabilization are most effective when applied to inorganic contaminants; however, they are sometimes used for treating organics or inorganic-organic mixtures (Wiles, 1989).

Solidification and stabilization may reduce the volume and mass of the waste—for example, by dewatering—to minimize the size and disposal cost of the residual. Solidification and stabilization residuals can sometimes be buried on-site or in landfills, or employed as construction grade materials, landfill covers, backfill for landfills, or structural fill.

Although the definitions of the terms *solidification* and *stabilization* overlap significantly and are sometimes used interchangeably, it is useful to distinguish them (LaGrega et al., 1994). *Stabilization* is a fixation process that converts the contaminant to an insoluble form by means of binding additives. Stabilization processes can include chemical conversion such as precipitation, redox reactions, complexation, or ion exchange. *Solidification* is a process in which the waste is encapsulated or otherwise encased in a solid of high structural integrity. Materials are added to the waste to produce a solid in the form of individual fine waste particles (microencapsulation) or large blocks (macroencapsulation). Solidification processes either decrease the surface area available for leaching or completely isolate the waste by mechanical binding rather than by chemical conversion.

Solidification and stabilization are often used in combination. For example, a chemical reaction used to stabilize a waste may result in a solidified residual, as occurs when Portland cement is added to an aqueous metal-bearing waste.

Solidification and stabilization can be categorized as inorganic-based or organic-based processes. Inorganic-based solidification and stabilization processes have the general characteristics of low cost, good long-term physical and chemical stability, and inertness to both UV radiation and microbial degradation. Some examples are cement-based, lime-based, sorption-based, and glassification processes.

Cement-based solidification/stabilization generally involves the addition of cement to produce concrete. This application is extremely versatile and can be used to treat a wide range of inorganic contaminants. It is not suitable for treating organic wastes unless a binder such as clay or vermiculite is added, since organics interfere with the concrete setting process. Residuals have extremely high structural stability and, except for arsenic, low leaching potential.

Lime-based fixation is also referred to as *Pozzolanic fixation*. *Pozzolanic* is a term used to describe material that does not exhibit cementing ability by itself, but does when combined with other materials. For example, fine noncrystalline silica from fly ash can be mixed with lime to produce cement. Blast-furnace slag and cement-kiln dust will also produce cement when mixed with lime. Lime-based fixation significantly increases the volume of the waste and produces residuals that are not as stable as those from the cement-based process, but it is cheaper to apply. Lime-based and cement-based processes are sometimes applied in combination by adding both lime and cement to wastes.

Sorption-based solidification or microencapsulation involves the use of activated carbon or modified clays to remove contaminants from solution. Modified clays are made more hydrophobic by substituting quaternary ammonium compounds for inorganic cations between the silica and alumina layers of the clay, increasing the layer spacing and providing a nonpolar phase for the stabilization of organic compounds. Microencapsulated residuals are either disposed of in landfills or treated in a second solidification/stabilization process, such as those described here for macroencapsulation. Applications for sorption-based processes include hydrophobic organics such as petroleum hydrocarbons, pesticides, transformer liquid, and chlorinated solvents.

Glassification is an expensive solidification/stabilization process in which the waste material is mixed with sand, borax, and nitric acid before being subjected to high temperatures (>1600 °C) to fuse the mixture into glass. This process is energy intensive and requires specialized equipment. Although it would be an effective method for treating a wide range of hazardous wastes, because of the high costs, it is used mainly to treat radioactive waste.

Organic-based solidification/stabilization processes are not as commonly applied as inorganic processes. They are characterized by relatively high cost, questionable stability in the presence of microorganisms or UV light, potential for contaminant volatilization from treated residual material, and limited commercial experience. Examples of organic solidification/stabilization processes include thermoplastic encapsulation and organic polymer fixation.

Thermoplastic encapsulation physically entraps the waste in inert material, which is then fused into a hard block by heating. Thermoplastic materials include paraffin, bitumin, asphalt, polypropylene, and polyethylene. Potential applications include treatment of radioactive wastes, petroleum hydrocarbons, organic solvents, and oxidizing salts (such as nitrates, chlorates, and perchlorates). The volume increase of the final residual is small, but compatibility of the waste with the encapsulation material can be a major problem. For example, greases soften asphalt, and xylene and toluene diffuse through asphalt, making any of these combinations inappropriate. Furthermore, waste mixtures containing reducing or oxidizing species that are reactive at high temperatures cannot be treated with this process.

Organic polymer fixation produces a spongy, immobilized residual by blending the waste with reagents at ambient temperature and allowing the polymer to form within the waste. There is no specific reaction between the waste and the polymer, and the polymer will not encapsulate all liquids. Typical polymers used in this process include urea-formaldehyde, polybutadiene, polyester-epoxy, acrylamide gel, and polyolefin. Organic polymer fixation is applicable for radioactive wastes, metal-bearing wastes, hazardous sludges, and nonvolatile organic wastes.

Oxidation

Oxidation is a chemical reaction that causes net removal of electrons from an ion, atom, or molecule (see §3.D). The goal of an oxidation treatment process is to reduce the hazardous nature of waste by transforming the contaminants to less harmful products. Common applications for chemical oxidation treatment include wastewaters containing iron, manganese, cyanide, sulfides, hydrocarbons, solvents, phenols, chlorophenols, amines, and mercaptans (Fochtman 1989; LaGrega et al., 1994). Typical oxidizing agents used to treat hazardous waste include dissolved oxygen (O_2), ozone (O_3), chlorine (HOCl, supplied as $Cl_2(g)$, NaOCl(l), or $Ca(OCl)_2(s)$), chlorine dioxide (ClO_2), hydrogen peroxide (H_2O_2), and potassium permanganate ($KMnO_4$). The choice of an appropriate oxidizing agent depends on the electrode potential of the contaminant to be oxidized (see Appendix E.3). That is, an effective oxidizing agent must have a higher electrode potential than the target contaminant. Consequently, oxidizing agents with high electrode potentials, such as ozone and hydrogen peroxide, are capable of oxidizing a broader range of contaminants than lower-electrode-potential oxidants such as oxygen and hypochlorite. (See Table 8.C.6.) A common hazardous waste application for oxidation is the treatment of cyanide-bearing wastewater. Cyanide (CN^-) is a highly toxic material. It is also a complexing agent that inhibits precipitation of metals. Example 8.C.5 illustrates the calculations

Table 8.C.6 Electrode Potential ($E°$) of Chemical Oxidizing Agents That Are Commonly Applied for Hazardous Waste Treatment

Oxidizing agent	Half-reaction	$E°$ (V)
Chlorine	$Cl_2(g) + 2e^- \longrightarrow 2Cl^-$	1.36
Hypochlorous acid	$HOCl + H^+ + 2e^- \longrightarrow Cl^- + H_2O$	1.49
Hypochlorite	$OCl^- + H_2O + 2e^- \longrightarrow Cl^- + 2OH^-$	0.90
Chlorine dioxide	$ClO_2 + e^- \longrightarrow ClO_2^-$	1.15
Oxygen	$O_2 + 4H^+ + 4e^- \longrightarrow 2H_2O$	1.23
Ozone	$O_3 + 2H^+ + 2e^- \longrightarrow O_2 + H_2O$	2.07
Hydrogen peroxide	$H_2O_2 + 2H^+ + 2e^- \longrightarrow 2H_2O$	1.78
Permanganate	$MnO_4^- + 4H^+ + 3e^- \longrightarrow MnO_2 + 2H_2O$	1.68

Source: Adapted from Glaze, 1990; LaGrega et al., 1994.

that can be done to determine whether a particular redox reaction is thermodynamically favorable.

Oxidation reactions in wastewater are nonspecific; any reduced compound in the waste may be oxidized by the added oxidant. Therefore, the presence of nontarget compounds in the waste, such as nontoxic organics, may lower the effectiveness of oxidation for treating the target compound. The choice of oxidant should be based on cost and material handling concerns. Bench-scale testing is generally required before implementation.

EXAMPLE 8.C.5 *Oxidation of Sulfide-Contaminated Wastewater*

A manufacturing process produces two waste streams, one containing hydrogen sulfide (H_2S) at pH 3 and a second containing permanganate (MnO_4^-). You would like to treat these wastes by mixing them to oxidize the sulfide to sulfate. There are no other competing reactions.

(a) Write the balanced redox reaction from the two half-reactions for the oxidation of hydrogen sulfide (H_2S) to sulfuric acid (H_2SO_4) by permanganate (MnO_4^-).

(b) Calculate the standard electrode potential ($\Delta E°$) for the above reaction given that the standard electrode potential for the H_2SO_4-H_2S half-reaction is 0.34 V.

SOLUTION

(a) First write the oxidation half-reaction:

$$H_2S + 4\,H_2O \longrightarrow H_2SO_4 + 8\,H^+ + 8\,e^-$$

Then write the reduction half-reaction:

$$MnO_4^- + 4\,H^+ + 3\,e^- \longrightarrow MnO_2 + 2\,H_2O$$

Combine the two in the correct proportion to cancel electrons, yielding the balanced redox reaction:

$$H_2S + 8/3\,MnO_4^- + 8/3\,H^+ \longrightarrow H_2SO_4 + 8/3\,MnO_2 + 4/3\,H_2O$$

(b) The standard electrode potential for the reaction is calculated as described in Appendix E (§E.3). $E°$ for the MnO_4^--MnO_2 pair is 1.68 V as given in Table 8.C.6. The electrode potential for the H_2S-H_2SO_4 pair is –0.34 V (see Table E.2 in Appendix E),

EXAMPLE 8.C.5 ***Oxidation of Sulfide-Contaminated Wastewater (continued)***

with the negative sign added from the given electrode potential to reverse the direction of the reaction from H_2SO_4-H_2S. Then

$$\Delta E^{\circ} = 1.68 \text{ V} + (-0.34 \text{ V}) = 1.34 \text{ V}$$

Oxidation reactions are commonly kinetically limited, even if thermodynamically favored. The rates of oxidation reactions are a function of reactant concentrations, temperature, interfering impurities, and pH. Oxidation reactions are typically carried out in completely mixed flow reactors or batch reactors. In both cases, high mixing efficiency is essential to ensure sufficient contact between oxidant and contaminant (LaGrega et al., 1994).

The mechanism and kinetics of chlorination, a commonly applied oxidation process for the treatment of biologically contaminated wastes, were discussed in detail in Chapter 6 (§6.D.1). Two additional oxidants commonly used to treat hazardous wastes, ozone and hydrogen peroxide combined with UV light, are discussed below.

Ozonation Ozone, an unstable, slightly water-soluble gas (the Henry's law constant is 0.01 M atm^{-1} at 25 °C), is a strong general oxidizing agent that is commonly used to treat hazardous wastes that contain organics, cyanide, sulfides, nitrite, iron, or manganese (Fochtman, 1989; Novak, 1989). The instability of ozone necessitates its generation on site. Ozone can be generated by passing a high-voltage electrical discharge (5–30 kV) through dry air or oxygen gas. The discharge splits the oxygen molecule into two oxygen radicals (O•) that then combine with diatomic oxygen to form ozone (O_3) (Fochtman, 1989). The oxygen radicals needed to form ozone can also be produced by ultraviolet photodissociation of O_2.

Ozone can react directly with a contaminant or indirectly by means of generating hydroxyl radicals, as shown in Figure 8.C.10. Ozone reacts with hydroxide ions in water

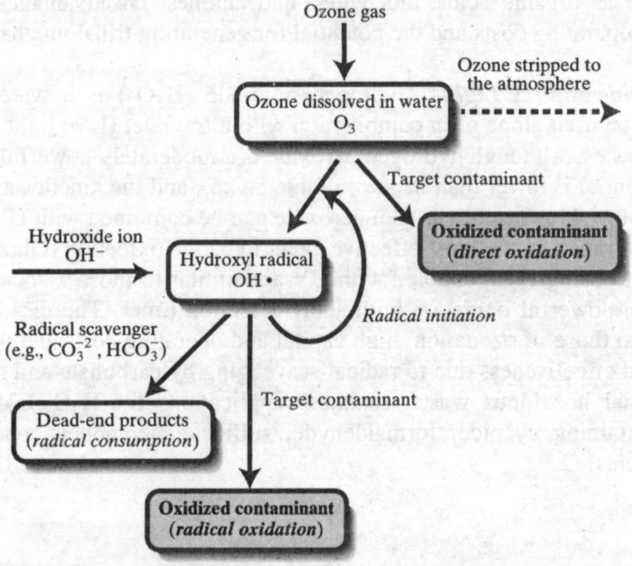

Figure 8.C.10 Oxidation reactions of ozone in water (adapted from Hoigne and Bader, 1983; Aieta et al., 1988).

Table 8.C.7 Second-Order Rate Constants for Aqueous-Phase Oxidation Reactions Involving Ozone Directly or Hydroxyl Radicals at Temperatures of 20–25 °C

Compound	$k_{O_3}(M^{-1}s^{-1})$	$k_{OH\bullet}(M^{-1}s^{-1})$
Benzene	2 ± 0.4	670×10^7
Ethanol	0.37 ± 0.04	185×10^7
Acetone	0.032 ± 0.006	7×10^7
Tetrachloroethylene	<0.1	230×10^7
Trichloroethylene	17 ± 4	400×10^7

Source: Adapted from Haas and Vamos (1995) as compiled from Hoigne and Bader (1983, 1976), and Farhataziz and Ross (1977).

to form hydroxyl radicals, which are very powerful nonselective oxidizing agents. Hydroxyl radicals have oxidation rate constants that are orders of magnitude higher than those associated with ozone alone (Table 8.C.7). Since high pH conditions generate high concentrations of hydroxide ions, high pH promotes increased formation of hydroxyl radicals. However, the presence of bicarbonate (HCO_3^-) or carbonate (CO_3^{2-}) ions in solution decreases the effectiveness of ozone by scavenging the hydroxyl radical.

Due to the multitude of possible reactions and pathways involved in ozonation, kinetic predictions of the destruction of target compounds are difficult. Pilot plant or bench studies are required to collect data needed for process design. Ozonation processes are generally carried out in closed vessels, where the ozone gas is injected into the water under pressure to promote dissolution and to decrease losses to the atmosphere. Typical contact times within ozone reaction vessels are 1–60 minutes with ozone dosages of 10–100 mg/L (Haas and Vamos, 1995).

The use of ozone as a chemical oxidant offers the benefits of a high oxidation potential and favorable kinetics. A side benefit of ozonation is that it causes some organics in water to be transformed to more water-soluble, less volatile, more degradable compounds such as organic acids, aldehydes, and ketones. Disadvantages include high capital and operating costs and the potential for generating trihalomethanes.

Hydrogen Peroxide and UV Light Hydrogen peroxide (H_2O_2) is a water-soluble oxidant that can be used alone or in combination with ultraviolet (UV) light for treating hazardous wastes. Although hydrogen peroxide is a moderately powerful oxidant, its electrode potential is lower than ozone's (Table 8.C.6), and the kinetics associated with its use are slow. However, hydrogen peroxide can be combined with UV light to produce hydroxyl radicals, the most effective agent in ozone oxidation (Glaze, 1990). The advantages to using H_2O_2 coupled with UV are similar to those associated with ozonation: rapid, powerful oxidation with short detention times. The disadvantages are also similar to those of ozonation: high capital and operating costs and the potential for decreased effectiveness due to radical scavenging by carbonate and bicarbonate ions. Potential hazardous waste treatment applications for H_2O_2-UV include wastewaters containing cyanide, formaldehyde, sulfides, mercaptans, phenols, or chlorinated solvents.

Reduction

Chemical reduction is a treatment process that is applied to decrease the hazardous nature of contaminants by adding electrons. Electron addition, or reduction, de-

Table 8.C.8 Reduction Reactions Used to Treat Hexavalent Chromium Wastewaters

Reducing agent	Reduction reaction
Sulfur dioxide (SO_2)	$2\ CrO_3 + 3\ SO_2 \longrightarrow Cr_2(SO_4)_3$
Bisulfite ($NaHSO_3$)	$4\ CrO_3 + 6\ NaHSO_3 + 3\ H_2SO_4 \longrightarrow 2\ Cr_2(SO_4)_3 + 3\ Na_2SO_4 + 6\ H_2O$
Ferrous sulfate ($FeSO_4$)	$2\ CrO_3 + 6\ FeSO_4 + 6\ H_2SO_4 \longrightarrow Cr_2(SO_4)_3 + 3\ Fe_2(SO_4)_3 + 6\ H_2O$

Source: Adapted from Fochtman, 1989.

creases the valence charge of the species. The most common application of chemical reduction for hazardous waste treatment is the reduction of highly toxic hexavalent chromium (Cr (VI)) to the less toxic trivalent chromium (Cr (III)) form using one of the reductants listed in Table 8.C.8. Sulfur dioxide applied under acidic conditions is the most common process. Ferrous sulfate is the least common reagent because of the large amount of iron sludge produced by this process. Other, less prevalent applications include the treatment of tetraethyl lead and chelated metals such as mercury, silver, and cadmium using sodium borohydride ($NaBH_4$) (Dawson and Mercer, 1986). Reduction reactions are generally carried out in completely mixed batch or flow reactors with typical contact times of less than one hour (Haas and Vamos, 1995).

8.C.3 Thermal Treatment

Thermal treatment applies high temperature to convert hazardous wastes to forms that are generally significantly less toxic, have lower volume, and are more easily disposed of. Two classes of thermal technologies are used. *Incineration* involves the combustion of wastes in the presence of oxygen. *Pyrolysis*, or cracking, is thermal decomposition of molecules in the absence of oxygen.

Waste gases, liquids, sludges, soils, and other solids are all potential candidates for thermal destruction. Among treatment technologies, thermal processes are capable of achieving the highest overall degree of destruction for the broadest range of hazardous waste streams. However, because of the large expense involved in properly applying these processes and because of the lack of public acceptance, thermal technologies have historically not been widely applied for hazardous waste treatment. Recent changes in laws that increase the long-term liabilities and costs of land disposal, together with advances in thermal technology design, have led to renewed interest in this treatment option.

Major advantages associated with thermal treatment technologies include the potential for nearly complete contaminant destruction (99.99–99.9999 percent removal), applicability to complex mixtures of contaminants in various forms (solids, liquids, or gases), and the potential for heat recovery from wastes with high energy content. Disadvantages include the potential for emitting odors, particulate matter, and hazardous substances in the gaseous effluent. Other disadvantages are the potential for producing hazardous ash and control device residuals, the need for highly trained operators to manage the shifting chemical and physical composition of the waste stream, high capital and maintenance costs, and adverse public reaction.

Adverse public reaction stems in part from a lack of adequate regulation and enforcement of hazardous waste thermal treatment technologies in the past. There is a perception that thermal waste residues, such as stack gas emissions, are impossible to monitor and control effectively, since they are more transient and transportable than potential environmental releases from alternative disposal options, such as those from leaking hazardous waste landfills.

Incineration

Hazardous waste streams containing significant quantities of organics and minimal quantities of inorganics are the most suitable candidates for incineration. It has been estimated that 25 to 60 percent of generated hazardous waste can be practically and cost-effectively incinerated (Haas and Vamos, 1995; Schaefer and Albert, 1989; USEPA, 1987). Wastes containing high concentrations of halogenated compounds or volatile metals such as mercury and lead are unsuitable for incineration, because of the potential for emitting toxic air contaminants and for producing toxic ash residuals.

Incineration is an oxidation process in which wastes are heated to high temperatures in the presence of oxygen within a heat-resistant reactor. Organics within the waste are converted to CO_2 and water, nitrogen and sulfur are converted to inorganic gases, halogens are converted to acid gases or salts, and metals are oxidized and precipitated into ash or volatilized. Ash is an inherent by-product of incineration that must be collected and ultimately disposed of. Ash may consist of carbon, salts, metals, and refractory minerals. The applicability of incineration for treating waste is a function of the heating value, volatile matter, ash, halogen, moisture, and heavy-metal content of the waste.

Contaminant destruction within an incineration unit depends on temperature, time, turbulence, and oxygen (TTTO). Sufficient temperature is required to convert the waste to gaseous form and to overcome the activation energy of the oxidation reactions. Sufficient residence times within the reactor are required to allow the oxidation reactions to proceed to completion. High turbulence is required to enable adequate heat transfer to the reactants and to ensure contact between the contaminants and oxygen within the combustion zone. Finally, oxygen is the major reactant required for the combustion reactions. If oxygen is present in insufficient quantities, the production of harmful intermediate combustion by-products can result.

For complete combustion of a hydrocarbon in air, the reaction stoichiometry can be described as follows (see §3.D.3):

$$C_nH_m + \left(n + \frac{m}{4}\right)(O_2 + 3.78\ N_2) \longrightarrow n\ CO_2 + \frac{m}{2}H_2O + 3.78\left(n + \frac{m}{4}\right)N_2 \quad (3.D.7)$$

To ensure sufficient oxygen for complete combustion, a 5–100 percent stoichiometric excess of oxygen may be provided, with the value depending on the mixing efficiency of the incinerator. It is necessary to balance the need for sufficient oxygen with the objective to limit air addition, because excess air increases the volume of gaseous emissions, decreases the residence time of waste in the incinerator, increases the need for supplementary fuel, and decreases combustion temperatures.

For stoichiometric combustion calculations involving elements other than H and C, the assumption of complete combustion usually implies the following. Sulfur is converted to sulfur dioxide (SO_2). Nitrogen either is emitted as NO or is converted to N_2. Halogens in the fuel (Cl, F, Br, or I) are converted to acid form, such as HCl. Phosphorus is converted to P_2O_5. Metals are converted to their elemental forms or to metal oxides.

The potential for producing undesirable combustion by-products such as carbon monoxide (CO), halogen gases (Cl_2 or F_2), nitrogen oxides (NO, NO_2), and vaporized metals (e.g., arsenic, antimony, cadmium, lead, mercury, and selenium) can be diminished by restricting the types of wastes burned in the incinerator. The temperature and oxygen content of the system can also be manipulated to promote complete combustion (LaGrega, 1994; Dawson and Mercer, 1986).

Overall, combustion reactions are highly exothermic. That is, the chemical oxidation reactions of combustion produce substantial net heat. However, combustion reactions must be initiated by supplying enough heat to initially overcome the activation energy associated with breaking chemical bonds in the waste or fuel molecules. The heat generated by the subsequent oxidation reactions suffices to overcome the activation energy of later reactions.

Energy generated by combustion reactions heats the incoming air and waste to the combustion temperature, converts the waste to gaseous form, and heats the refractory materials surrounding the combustion chamber. Excess heat is lost in the ash residue and flows out in the effluent gases. Wastes with low heating values may need to be blended with auxiliary fuels to achieve necessary combustion temperatures. Many organic wastes have enough heating value to support sustained combustion and need auxiliary fuel only for ignition. Typical auxiliary fuels used in hazardous waste incineration include natural gas, propane, waste fuel oils, and spent solvents.

Incinerator performance depends on heat transfer to the wastes and to the incinerator walls. Three mechanisms contribute to heat transfer. *Conduction* is analogous to molecular diffusion. Heat moves down a temperature gradient by molecular motion. Conduction occurs only between materials that are in physical contact. *Convection* is an advective transport process that transfers heat by the movement or mixing of fluids. *Radiation* has no analogue in contaminant transport. It involves the transfer of heat through space by electromagnetic waves and requires no intervening solid, liquid, or gaseous medium. Although all three of these processes occur simultaneously, the relative importance of each depends on the incinerator configuration and operating conditions, with lower temperatures favoring conduction and convection and higher temperatures favoring radiation.

Prior to incineration, it is necessary to characterize the waste stream to evaluate the feasibility of thermal destruction, the need for auxiliary fuel, and potential ash and effluent residual production. Waste characterizations rely on two types of tests, *proximate analysis* and *ultimate analysis*. Proximate analysis measures the moisture, ash, volatile solids, and fixed carbon content of the waste. Ultimate analysis includes all of these plus evaluations of elemental content (carbon, nitrogen, hydrogen, halogens, metals, etc.) and specific principal organic hazardous constituent (POHC) content (e.g., PCBs, dioxin).

Incinerator Configurations Several different incinerator configurations are used for the treatment of hazardous waste (Table 8.C.9). The choice of an appropriate configuration depends on the characteristics and volume of the waste, incinerator availability, and cost. Physical characteristics of the waste, such as the solids content and viscosity, are more important for choosing a reactor configuration than the chemical characteristics, such as elemental composition or heating value. Many configurations include heat recovery, and almost all include air pollution control devices (APCDs) because of strict RCRA effluent emission regulations.

All incinerator configurations consist of some type of refractory lining within the combustion unit. Refractory materials, such as high-alumina brick and ceramic, resist changes caused by extreme heat. The walls of a combustion chamber usually consist of a 20–30 cm thickness of refractory material backed by an 8–15 cm layer of insulating material such as firebrick.

Hazardous waste incineration has been in use for more than 30 years. Early hazardous waste incineration units were grate-type municipal incinerators that performed inefficiently, with significant effluent gas release into the environment. These early units are part of the reason that public opposition to hazardous waste incineration has become so strong over the years. There are several additional reasons for public opposition to this technology:

- Adverse environmental and health impacts associated with incineration residuals
- Poor site selection
- Hazardous material spills at privately owned and operated incineration facilities
- Distrust of incinerator owners and operators
- Inability of government agencies to adequately enforce regulatory compliance

There are currently 166 USEPA-permitted hazardous waste incinerators operating in the United States, the majority of which are privately owned and operated commercial units (USEPA, 1997a).

Liquid Injection Liquid injection incinerators are simple units composed of atomization devices in refractory-lined combustion chambers (Figure 8.C.11). The purpose of the atomizers is to disperse the liquid waste or fuel into fine droplets that have a large surface area to facilitate heat and mass transfer. High-efficiency atomizers create large numbers of droplets per unit of liquid flow, which results in extremely high liquid surface area and low required residence times (Table 8.C.9). Atomizers may also mix air into the waste stream, producing an oxygen-rich zone around the liquid droplets.

Liquid injection is the most common hazardous waste incinerator configuration in use in the United States (Santoleri, 1989). It is a proven technology associated with

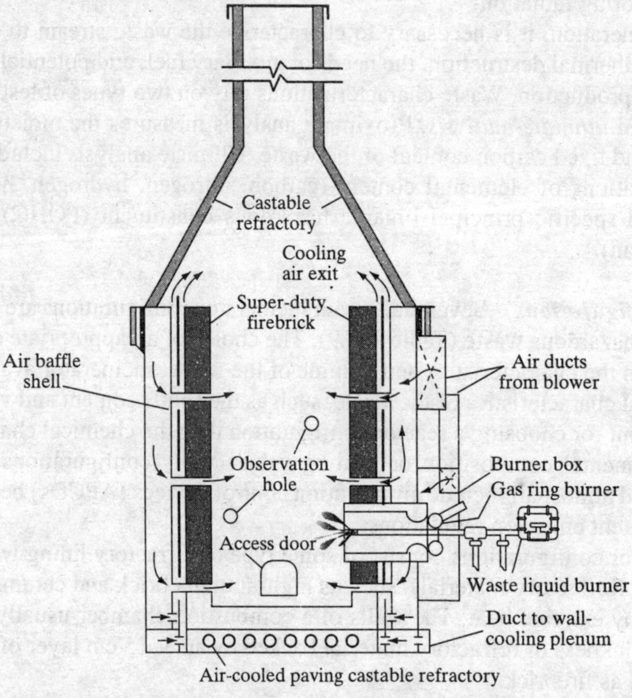

Figure 8.C.11 Typical configuration of a liquid injection incinerator. (Reprinted with permission from C.R. Brunner, *Incineration Systems Handbook*, Van Nostrand Reinhold, 1996.)

Table 8.C.9 Typical Incinerator Operating Conditions

Incinerator type	Temperature range (°C)	Solid residence time (min)	Liquid or gas residence time (s)
Liquid injection	1000–1700	NA[a]	0.1–2
Rotary kiln	820–1600	20–60	1–3
Fluidized bed	760–980	600	1–12
Multiple hearth	720–980	30–90	0.25–3
Pyrolysis	480–820	12–15	1–3

[a]Form of waste not applicable to this incinerator type.
Source: Crumpler and Martin, 1987; Santoleri, 1989.

substantial operational experience and is applicable to the treatment of any atomizable waste, including gases and nonviscous liquids. Atomization of the waste is achieved by pressure induced by hydraulic or mechanical means, or by steam or compressed air. The degree of atomization achievable for a given waste depends on the viscosity and amount of solid impurities present; however, 40–100 μm diameter droplets are typical. Heat is transferred to droplets within the combustion chamber primarily by radiation and convection.

Advantages of the liquid injection configuration include very short residence times, fast temperature response to changes in waste-fuel composition and flow rate, minimal ash removal, no moving parts, low maintenance costs, and a practice-proven technology. The major disadvantage is its applicability only to atomizable wastes.

Rotary Kiln A rotary kiln incinerator is a long, cylindrical, refractory-lined tube poised at a slight downward angle. It rotates slowly along its horizontal axis. Waste and auxiliary fuels are fed into the top of the tube. The gaseous effluent flow and solid residuals (ash) are collected at the bottom of the tube (Figure 8.C.12). Air is injected either at the top of the tube, near the waste feed, or along the tube using several injector nozzles. The tube rotation rate and the incline angle are controlled to mix the waste with air and to move the waste toward the ash and residual collection system at a controlled rate. An internal dam may be added to the end of the kiln tube to increase the solids retention time. Chains are sometimes suspended from the kiln walls to inhibit agglomeration of the waste materials. An afterburner similar to a liquid injection unit is commonly located in the effluent gas stream following the kiln combustion unit. Heat transfer within a rotary kiln occurs primarily by radiation and conduction from the kiln walls. Because of this, rotary kilns need to be preheated to operating temperature by auxiliary fuel prior to injecting the waste stream. In addition, it is common to cofire solid waste with fuel or liquid waste to ensure adequate combustion temperatures. Rotary kilns are the second most common thermal destruction technology for hazardous waste and are applicable for treating diverse forms of waste, including solids, semisolids, viscous and nonviscous liquids, sludges, complex mixtures, and containerized wastes.

Rotary kilns were originally developed for use in the cement, lime, phosphate, iron ore, and coal industries, and they are still widely applied for those purposes. Some cement kilns are now being used for the thermal destruction of hazardous wastes under specific RCRA regulations, significantly increasing the incineration capacity in the United States (Chadbourne, 1989).

Typical rotary kiln furnaces are 1.5–5 m diameter and 3–15 m long, with a length:diameter ratio between 2:1 and 10:1. Rotary kilns are operated at temperatures

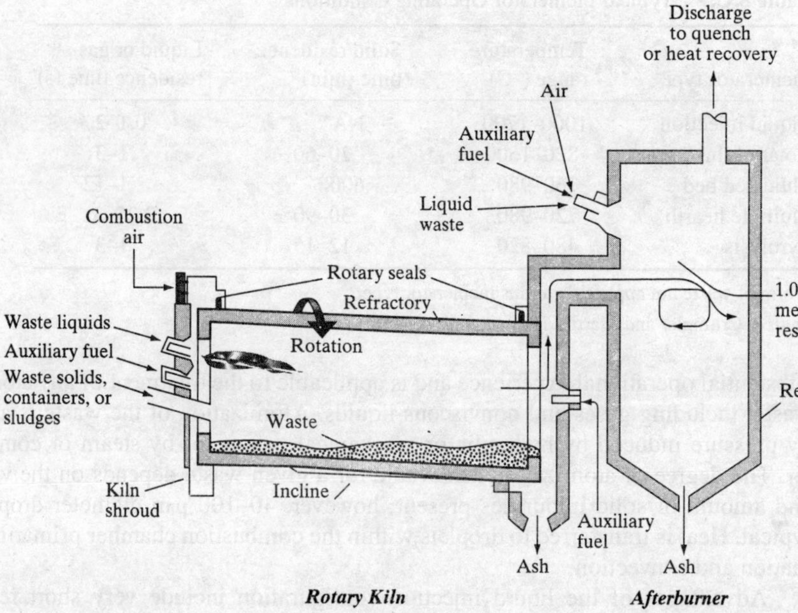

Figure 8.C.12 Typical configuration of a rotary kiln incinerator. (Reprinted with permission of the Air & Waste Management Association from E.T. Oppelt, Incineration of hazardous waste: A critical review, *JAPCA*, **37**, 558–586, 1987.)

of 820–1600 °C and have solids retention times of 20–60 minutes without an internal dam, or greater than one hour with a dam (Dawson and Mercer, 1986; Schaefer and Albert, 1989).

Advantages associated with the use of rotary kilns for incinerating hazardous waste include applicability to a wide variety of solid and liquid wastes, continuous ash removal, few moving parts, minimal waste preparation requirements, and a practice-proven technology. Disadvantages include high capital costs, large potential for particulate emissions, necessity of an afterburner to treat gaseous emissions, and low thermal efficiency.

Fluidized Bed Fluidized-bed incinerators are refractory-lined chambers containing an inert solid granular material such as sand, which is fluidized and expanded in the reactor by means of air blown into the chamber bottom (Figure 8.C.13). The inert media facilitates turbulent mixing within the reactor and provides a large surface area for radiative and conductive heat transfer. Wastes and air that are injected into the bed are mixed with and heated by the fluidized material to promote effective combustion. Large waste particles remain suspended in the fluidized material until combustion is complete. Ash and gaseous effluents exit together at the top of the unit. Auxiliary fuel is required to preheat the inert fluidized bed material and the refractory prior to introduction of the waste stream. The air distribution system at the bottom of the reactor produces a bed that expands from ~1 m at rest to ~2 m during combustion. Wastes are added from above or directly into the fluidized bed.

Fluidized bed incineration was introduced relatively recently for hazardous waste treatment, and presently represents only a small fraction of the applied thermal technologies. This technology is applicable for treating gaseous, liquid, and sludge wastes

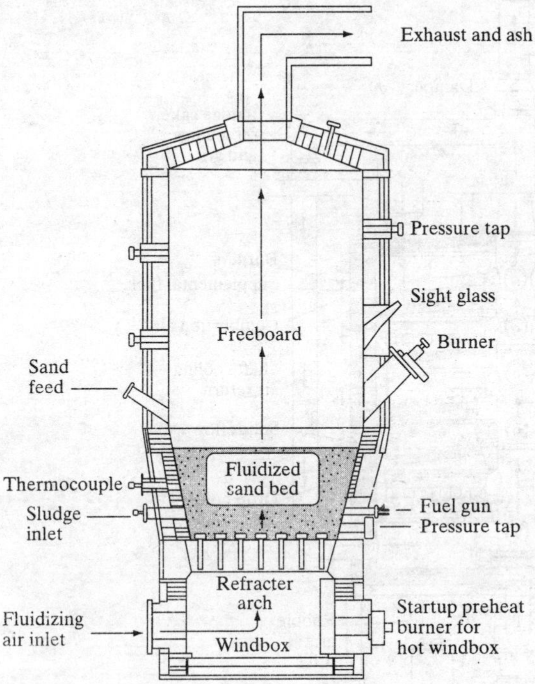

Figure 8.C.13 Typical configuration of a fluidized-bed incinerator. (From Water Pollution Control Federation, *Incineration Manual of Practice OM-11*, 1988. Copyright © Water Environment Federation. Reprinted with permission.)

that include grain diameters up to ~1.3 mm (Rasmussen et al., 1989). Typical operating temperatures of fluidized-bed incinerators are 750–850 °C since bed fusion can occur when silica sand is used at temperatures >900 °C. Crushed refractory can be substituted for sand if higher operating temperatures are required. Other important reactor design parameters include gas velocity (typically 1.5–2.4 m/s), bed diameter (typically 2.5–7.5 m), bed depth (typically 0.4–1 m), and waste residence times (typically 5–8 s for liquids and 1–10 h for solids) (Crumpler and Martin, 1987).

There are several advantages to using fluidized-bed configurations for hazardous waste incineration. A variety of waste forms (solid, liquid, and gas) can be treated and variable waste composition can be easily handled. Intrinsically high heat retention in the fluidized bed dampens the potential for high or low temperature shocks. Superior turbulence characteristics ensure good heat transfer and reactant mixing. Effective waste destruction can be achieved at lower temperatures than with other processes. The process is simple and has low maintenance requirements, with capital costs that are generally less than those for rotary kilns but more than those for liquid injection. Disadvantages of fluidized beds include difficult removal of solid residuals retained in the bed, limited operational temperature range, and little operational experience (O'Brien and Gere Engineers, Inc., 1995).

Multiple Hearth (Hereshoff Furnace) Multiple-hearth incinerators were originally designed and used for the treatment of dewatered sludge from sewage treatment plants. However, they have recently begun to be used for treating hazardous wastes. A multiple-hearth incinerator consists of a series of flat hearths radiating from a central rotating shaft, surrounded by a refractory-lined vessel (Figure 8.C.14). Waste is introduced to the uppermost hearth and is pushed by rotating rabble arms to successive

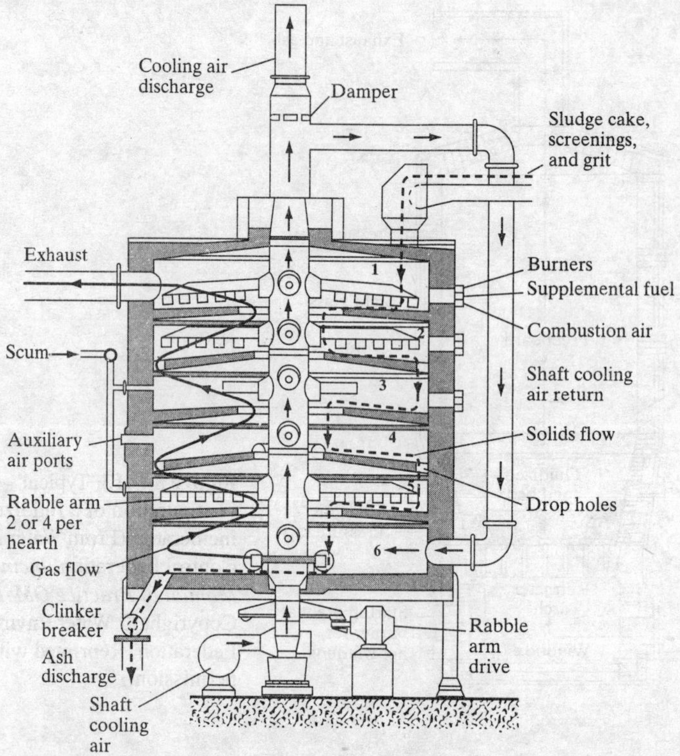

Figure 8.C.14 Typical configuration of a multiple-hearth incinerator. (From *Operation of Municipal Wastewater Treatment Plants*, Fifth Edition, Manual of Practice 11, Vol. 3 (1996). Copyright © Water Environment Federation. Reprinted with permission.)

hearth levels below. Air is introduced at the bottom of the hearth and is removed with the gaseous effluent at the top of the unit while ash residuals are removed at the bottom of the unit. Heat transfer occurs primarily by radiation and conduction from each hearth surface, as well as by convection induced by the flow of air. Liquid or gaseous wastes and auxiliary fuels can also be injected into the center zone of the incinerator using auxiliary burner nozzles.

Current hazardous waste applications of multiple-hearth incinerators include chemical sludges and tars, and activated-carbon regeneration (Crumpler and Martin, 1987). Multiple-hearth incinerators work best with wastes that do not have large water content (i.e., <85 percent water) because of their limited ability to handle the evaporation of water to steam. Typical multiple-hearth configurations have heights of 2–8 m, diameters of 4–20 m, and 5–12 hearths. Operating temperatures are in the range 430–650 °C for the drying zone near the top of the unit, 720–980 °C for the combustion zone in the center of the incinerator, and 150–230 °C in the cooling zone near the bottom of the unit (WPCF, 1988). Typical retention times within muitiple-hearth incinerators are 30–90 min for solids and 0.25–3 s for liquids (Table 8.C.9).

Multiple-hearth incinerators are applicable to a wide variety of physical and chemical waste types, including low-volatility, high-viscosity wastes. Disadvantages include slow temperature response, making control of auxiliary fuel firing difficult; complex machinery with lots of moving parts, which requires high maintenance and specially trained operators; and the requirement of a secondary combustion chamber for hazardous waste applications.

Regulatory Requirements Regulatory requirements governing the incineration of most hazardous wastes are defined by RCRA. The Toxic Substances Control Act defines additional standards for the incineration of PCBs. A trial burn must be performed at the proposed facility using a range of wastes typical of those that will be treated. The facility must meet standards that include waste analyses prior to incineration, specific operating requirements, and adherence to federal performance standards.

As part of the federal performance standards, RCRA requires the identification of principal organic hazardous constituents (POHCs) present in any hazardous waste that is to undergo incineration. POHCs are specific organics that are known to be difficult to incinerate. The federal performance standards are based on destruction and removal efficiencies (DRE) defined by RCRA for incineration as

$$\text{DRE} = \frac{Q_{\text{in}} - Q_{\text{out}}}{Q_{\text{in}}} \times 100 \tag{8.C.22}$$

where DRE is the destruction and removal efficiency (percent), Q_{in} is the mass flow rate of POHC into the incinerator (kg/h), and Q_{out} is the mass emission rate of POHC out of the incinerator (kg/h).

Specific federal performance standards are as follows:

- DRE ≥ 99.99 percent for most organics and specifically for POHCs
- DRE ≥ 99.9999 percent for PCBs, dioxins, and dibenzofurans
- Particulate emissions < 180 mg per m^3 dry effluent gas, corrected to 7 percent oxygen by the following expression:

$$E_c = \frac{0.14E_m}{(0.21 - Y_{O_2})} \tag{8.C.23}$$

  where E_c is the corrected emission concentration (mg/m^3), E_m is the measured emission concentration (mg/m^3), and Y_{O_2} is the mole fraction of molecular oxygen in the dry effluent gas

- HCl emissions < 1.8 kg/h or removal of 99 percent from the effluent gas
- Carbon monoxide ≤ 100 ppm by volume in dry effluent gas corrected to 7 percent oxygen by equation 8.C.23
- Risk-based guidelines for 10 priority metals: arsenic, beryllium, cadmium, chromium, antimony, barium, lead, mercury, silver, and thallium

Trial burns as defined by RCRA must be conducted for each incinerator to acquire an EPA permit to treat hazardous waste. Trial burns include detailed waste analyses of the full range of potential wastes to be treated by the specific incinerator and actual data from trial burns of the waste with typical feed rates conducted over a range of incinerator operating conditions. Data must show compliance with all federal performance standards for all types of hazardous waste to be incinerated.

Incinerator Emission Controls To meet regulatory requirements for flue gas emissions, most hazardous waste incinerators are operated with air pollution control devices (APCDs). APCDs are used to remove a range of contaminants from the flue gas effluent (LaGrega et al., 1994; Brunner, 1991):

- Fugitive volatile organics—compounds that pass through the combustion chamber without being oxidized
- Products of incomplete combustion (PICs)—organics that are formed within the incinerator
- Particulate matter—ash and solids from the waste or that are formed within the incinerator
- Metals—in gaseous or particulate form
- Acid gases—such as SO_2 formed from sulfur-containing waste, and HCl and HF formed from halogenated organics

If the level of fugitive organics or PICs from the primary combustion chamber of an incinerator is significant, then the first APCD in the effluent treatment train would be a secondary combustion chamber or afterburner, where auxiliary fuel and oxygen are added to promote complete oxidation of the organics. Secondary combustion chambers are often employed following rotary kilns, fluidized beds, and multiple-hearth incinerators, and are capable of removing volatiles and PICs with extremely high efficiency.

If APCDs other than afterburners are to be used, incinerator effluent gases often must first be cooled to temperatures <250 °C. Effluent cooling can be achieved using either a wet-gas or a dry-gas quench system. A typical wet-gas quench system is the adiabatic saturation quench chamber, which simultaneously saturates effluent gas with water and cools it from 1000 °C to 80 °C by spraying water into a concurrent effluent gas flow chamber with a 1 s residence time. Wet-gas quench systems can be used for acid removal as well as temperature reduction. They produce a low-pH (3–5.5) liquid waste residual that must be disposed of. Typical quench systems that do not saturate the effluent gases (dry-gas) include heat exchangers such as waste heat boilers, and tempering chambers, which are downflow refractory-lined vessels that use a small amount of spray water to decrease gas temperatures from 1000 °C to 400 °C. In this case, the goal is for all of the injected water to evaporate into the gas flow, leaving no liquid residual (Gill and Quiel, 1993).

The other specific APCDs applied to hazardous waste incineration systems are the same as those used for conventional air pollution control applications (see §7.C).

Pyrolysis

Pyrolysis is a thermal decomposition process that occurs in the absence of oxygen. This reductive process is also called *cracking* because it converts large saturated hydrocarbons to smaller saturated and unsaturated molecules. Pyrolysis occurs mainly due to radical attack. That is, molecules with unpaired electrons chemically react to destabilize contaminants, promoting their degradation to smaller, chemically reduced products: carbon (in the form of char), CO, CH_4, H_2, and additional radicals. Pyrolysis products may then be combusted (oxidized) in an afterburner, where heat recovery can be achieved. The cracking portion of the pyrolysis reaction is endothermic and therefore requires energy input. The combustion portion is highly exothermic, more than making up for the energy required in the first stage, so that the overall process usually generates thermal energy. The main function of the reductive step is for gasification of the waste so that it can be easily and completely incinerated in the later step.

Examples of the chemical reactions that occur during the reductive pyrolysis of octane and PCBs are given below.

$$C_8H_{18}\text{ (octane)} \longrightarrow C_4H_{10}\text{ (butane)} + C_4H_8\text{ (butene)} \tag{8.C.24}$$

$$C_{12}H_7Cl_3\text{ (PCBs)} \longrightarrow 12\text{ C (carbon char)} + 3\text{ HCl} + 2\text{ H}_2 \tag{8.C.25}$$

Pyrolysis processes are applicable for the treatment of wastes that are not amenable to conventional incineration (Shah et al., 1989):

- Viscous liquids or sludges that are too abrasive or heterogeneous for atomization in an incinerator
- Wastes such as plastics that undergo phase changes during thermal treatment
- High-residue wastes that would generally require substantial stack-gas treatment
- Wastes containing salts or metals that melt and volatilize at normal incineration temperatures and cause refractory buildup, fouling of heat exchanger surfaces, or submicron particle emissions
- Wastes in containers or drums that cannot easily be drained

Pyrolysis units are available in many configurations, including those that employ indirect heating for the endothermic step and those that employ direct heating. Indirect heating can be achieved by means of radiant tubes, electrical heaters, or induction-heated metallic walls. Direct heating is achieved by partial oxidation of the waste or an auxiliary fuel in a mixture that has less than the stoichiometric amount of oxygen needed for complete combustion. Direct heating is also referred to as "starved-air gasification" since the major products are gaseous and burnable. Specific pyrolysis configurations include rotary or roller hearths (high capacity, low maintenance, semicontinuous flow), rotary kilns (lower capacity, continuous flow), fluidized beds (continuous flow, for noncontainerized wastes only), and plasma arc units (very high operating temperature [~5000 °C], direct heating by electrical discharge). Typical pyrolytic operating temperatures are 400–750 °C. Residence times are a function of the waste form and range from minutes to hours. Generally, particulate emissions from pyrolysis units are low enough to make emission control equipment unnecessary; however, acid scrubbers are often required.

The major advantages associated with pyrolytic processes for hazardous waste treatment include more efficient energy recovery compared with incineration due to the separately controlled endothermic and exothermic stages, applicability to a broad range of wastes, low particulate emissions, lower operating temperatures relative to incineration, and consistently high destruction efficiency. Major limitations of this process include the requirement of auxiliary heating for the endothermic stage, long residence times compared with incineration units, potentially hazardous products in the gaseous emissions, and potentially hazardous leachable residues resulting from treatment of metals and salt-bearing wastes. Consequently, pyrolysis is currently not applied for the treatment of hazardous waste nearly as often as incineration.

8.C.4 Biological Treatment

Biological treatment processes are used to decrease the hazardous nature of wastes by transforming the contaminants to less harmful forms. Biological treatment relies on useful microbial reactions including degradation and detoxification of hazardous organics, inorganic nutrient and electron acceptor consumption, metal transformations both to increase solubility for potential recovery and to decrease solubility for precipitation, and pH adjustments. Biological treatment processes are applied to gaseous, aqueous, and solid wastes containing biodegradable organics; inorganic ions such as sulfate, nitrate, phosphate, and ammonia; and, to a lesser extent, metals such as iron, chromium, manganese, lead, selenium, and mercury. General advantages associated with the use of biological processes for the treatment of hazardous wastes include relatively low cost, simple and generally well understood technology, and the potential for complete contaminant destruction. General disadvantages associated with these processes include their ineffectiveness for treating recalcitrant (difficult to degrade) compounds and the potential for process disruption due to inconsistent loadings or toxic shocks.

Microorganisms involved in biological hazardous waste treatment processes include bacteria, fungi, protozoa, and, to a limited extent, algae. The most active and diverse group are the bacteria. For microorganisms to degrade hazardous contaminants, conditions that promote their growth and reproduction must be maintained. Required for this are sources of energy, carbon, nitrogen, and assorted trace compounds, along with proper environmental conditions such as temperature, pH, and moisture content. In addition, the success of biological treatment relies on the biodegradability of the contaminants of interest. Several major factors affect the biodegradability of a specific contaminant:

- Presence of an appropriate microbial population
- Chemical structure of the contaminant (e.g., reduction potential, presence of halogens, presence of single or double bonds in the structure)
- Physical characteristics of the contaminant (e.g., concentration; solid, liquid, or gas phase)

Each of these factors must be considered in the choice of the biological reactor configuration to promote the desired degradation reactions.

To start a biological process, seed microorganisms or a microbial inoculum is required. Suitable microbial inocula can be derived from sources that sustain rich, diverse microbial populations, such as activated sludge units, anaerobic digester fluid, fluid from the rumen (digestive system) of a cow, soil extracts, and proprietary commercially available cultures. Acclimation times required to grow a sufficient microbial community following inoculation range from days to months, depending on the inoculum and on the process configuration, with anaerobic processes and immobilized cell processes requiring longer acclimation times than aerobic and suspended cell processes.

As discussed in §3.D, microorganisms derive energy from redox reactions, so energy sources for microbial populations are appropriate electron donors and electron acceptors. For wastes containing organic contaminants, electron donors are often the contaminants themselves. Appropriate electron acceptors include oxygen for aerobic degradation reactions, or nitrate, ferric iron, sulfate, carbon dioxide, or other oxidized compounds for anaerobic degradation reactions. Aerobic degradation processes are generally faster, more stable, and less sensitive to toxic shocks than anaerobic processes. However, anaerobic processes require less energy, produce less microbial cell mass, are applicable to higher contaminant concentrations, avoid stripping of volatiles, and can be used to treat some compounds that cannot be degraded aerobically. Nitrogen sources required for biological growth can include nitrate or ammonia that is present as part of the waste stream or is added specifically for the process, or, for very specialized applications, nitrogen gas can be fixed into usable forms by particular microbial populations. Required sources of trace nutrients can also be derived from a suitably formulated waste stream or can be added as needed.

Contaminant Suitability

Biological treatment processes are commonly applied to contaminants that can be used by microorganisms as carbon or energy sources, such as petroleum hydrocarbons, alcohols, ketones, carboxylic acids, and other reduced organics (Table 8.C.10).

Table 8.C.10 Some Contaminants That Can Be Treated by Biological Degradation Processes

Chemical class	Mechanism and notes	Limitations
Gasoline aliphatics, fuel oils	Degradable aerobically by most microbial populations; carbon and energy source	May form nonaqueous-phase liquids
Aromatics	Degradable aerobically and anaerobically by most microbial populations; carbon and energy source	May be toxic at high concentrations; volatile
Polycyclic aromatic hydrocarbons	Degradable aerobically by specific microbial populations; carbon and energy source	Sorbs very strongly to biomass; may form solid phase; low solubility
Creosote	Degradable aerobically by specific microbial populations; carbon and energy source	Sorbs very strongly to biomass; may form NAPL
Alcohols, ketones, esters	Degradable aerobically by most microbial populations; carbon and energy source	May be toxic at high concentrations
Ethers	Degradable aerobically by specific microbial populations; some used as carbon and energy sources	Volatile
Phenols	Degradable aerobically by specific microbial populations; carbon and energy source	May be toxic at high concentrations
Highly chlorinated aliphatics	Degradable anaerobically as electron acceptor or aerobically as cometabolic substrate by specific microbial populations	May form NAPLs; sorbs to biomass; volatile
Less chlorinated aliphatics	Degradable aerobically and anaerobically by specific microbial populations	May form NAPLs; sorbs to biomass; volatile
Highly chlorinated aromatics	Degradable aerobically or anaerobically as cometabolic substrate by specific microbial populations	May form NAPLs or solids; sorbs strongly to biomass
Less chlorinated aromatics	Degradable aerobically and anaerobically by specific microbial populations	May form NAPLs or solids
Highly chlorinated biphenyls	Degradable anaerobically as cometabolic substrate by specific microbial populations	Sorbs strongly to biomass
Less chlorinated biphenyls	Degradable aerobically by specific microbial populations	Sorbs strongly to biomass
Ammonia	Degradable aerobically by specific microbial populations; used as energy source	Volatile
Nitroaromatics	Some aerobically and anaerobically degradable by specific microbial populations	Some are recalcitrant
Metals	Some aerobically transformed; others anaerobically transformed	May produce highly toxic products; controlled by redox reactions

Source: Committee on In Situ Bioremediation, 1993; Torpy, 1989.

However, it is also possible to use biological processes to transform oxidized contaminants that serve as alternate electron acceptors (e.g., nitrate, chlorinated solvents), contaminants that are transformed as a cellular detoxification mechanism (e.g., metals), or contaminants that are transformed without providing any benefit to the microbial population (*cometabolic* transformations). Cometabolic transformations are catalyzed by nonspecific enzymes, which have been produced to carry out beneficial biological reactions. The active site on these nonspecific enzymes accepts the cometabolic substrate and catalyzes the reaction even though the reaction yields no benefit to the cell. To promote cometabolic reactions, some other substrate must be provided to the cell to produce the carbon and energy necessary to sustain microbial activity. Fortunately for microorganisms (but unfortunately for environmental engineers), most enzymes are very specific, so cometabolic reactions are rather uncommon. Nevertheless, cometabolic reactions are very important for the degradation of man-made organic contaminants such as chlorinated solvents and pesticides, since, from an evolutionary perspective, microbial populations have had little time to develop enzymes to degrade these recently developed molecules.

Although metals are typically not specifically targeted for removal in biological treatment processes, some metals removal may occur due to settling of particulate matter, uptake by microbial biomass, and stripping of volatiles. Metals removal efficiencies of 40–98 percent have been reported in activated sludge units for cadmium, silver, zinc, lead, and copper, with lower removal efficiencies for nickel, arsenic, and selenium.

Chlorinated Solvents Chlorinated solvents are anthropogenic compounds used as solvents, degreasers, and chemical feedstocks. They are common hazardous waste constituents and groundwater contaminants. Chlorinated solvents represent a good class of hazardous waste contaminants for illustrating three biological degradation mechanisms: use as a primary substrate, use as an alternate electron acceptor, and cometabolic transformation. Chlorinated aliphatics with few chlorine atoms per molecule (e.g., vinyl chloride [C_2H_3Cl], chloroethane [C_2H_5Cl], and methylene chloride [CH_2Cl_2]) can be aerobically degraded as primary carbon and energy substrates. However, highly chlorinated aliphatics are generally degraded as electron acceptors under anaerobic conditions or by cometabolic degradation reactions under a variety of conditions. Under anaerobic conditions, chlorinated aliphatics can be degraded by sequential reductive dechlorination (the replacement of a chlorine atom with a hydrogen atom), as depicted in Figure 8.C.15.

Microorganisms that perform reductive dechlorination use the chlorinated aliphatic as an electron acceptor in a redox reaction. That is, the cells essentially use the chlorinated solvents to *breathe* in a manner analogous to the way humans use oxygen. For this process to occur, an appropriate electron donor must be available and energetically favorable conditions must be present. That is, there must be no usable electron acceptor present with a higher electrode potential than that of the chlorinated aliphatic being transformed. As can be seen from Table 8.C.11, the electrode potentials for perchloroethylene (PCE), trichloroethylene (TCE), vinyl chloride (VC), and dichloroethylene (DCE) are each well below those of oxygen and nitrate and above those of sulfate and carbon dioxide. This indicates that these chlorinated compounds may be used as potential electron acceptors only in the absence of oxygen and nitrate. Reductive dechlorination is a significant biological contaminant transformation mechanism in both engineered and natural systems.

Chlorinated solvents can also be degraded by cometabolic reactions under aerobic conditions. Aerobic cometabolic degradation of chlorinated solvents involves

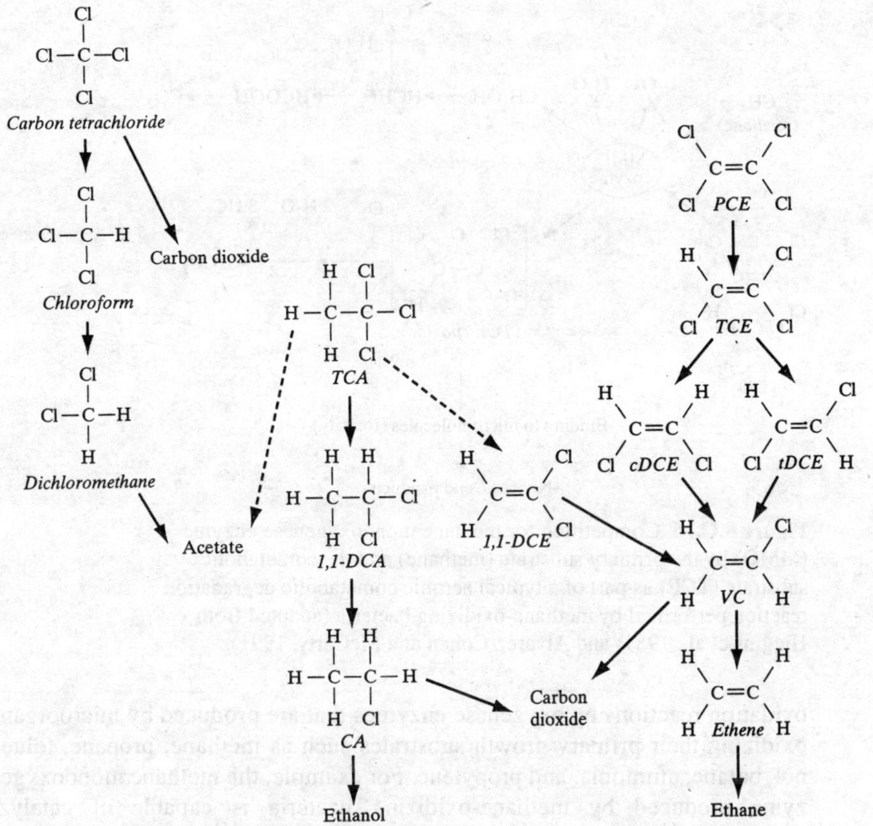

Figure 8.C.15 Degradation patterns for chlorinated aliphatics under anaerobic conditions. Solid arrows represent reactions that are biologically catalyzed under anaerobic conditions, and dashed arrows represent chemical reactions that occur under most redox conditions. *Abbreviations:* TCA, trichloroethane; DCA, dichloroethane; CA, chloroethane; PCE, tetrachloroethylene; TCE, trichloroethylene; DCE, dichloroethylene; VC, vinyl chloride. (Adapted from Vogel et al., 1987.)

Table 8.C.11 Standard Electrode Potentials at pH 7 ($E^{o\prime}$) of Chlorinated Solvents and Compounds Typically Used as Electron Acceptors in Microbial Redox Reactions

Half-reaction[a]	$E^{o\prime}$ (V)
$O_2 + 4\,H^+ + 4\,e^- \longrightarrow 2\,H_2O$	+ 0.81
$NO_3^- + 6\,H^+ + 5\,e^- \longrightarrow 0.5\,N_2 + 3\,H_2O$	+ 0.71
$PCE + 2\,H^+ + 2\,e^- \longrightarrow TCE + HCl$	+ 0.55
$TCE + 2\,H^+ + 2\,e^- \longrightarrow cDCE + HCl$	+ 0.54
$VC + 2\,H^+ + 2\,e^- \longrightarrow ethene + HCl$	+ 0.51
$cDCE + 2\,H^+ + 2\,e^- \longrightarrow VC + HCl$	+ 0.44
$Fe(OH)_3 + 3\,H^+ + e^- \longrightarrow Fe^{+2} + 3\,H_2O$	+ 0.16
$SO_4^{2-} + 9\,H^+ + 8\,e^- \longrightarrow HS^- + 4\,H_2O$	− 0.22
$CO_2 + 8\,H^+ + 8\,e^- \longrightarrow CH_4 + 2\,H_2O$	− 0.26

[a]Electron acceptors are listed as the first compound in each reaction.

Source: Wiedemeier et al., 1996.

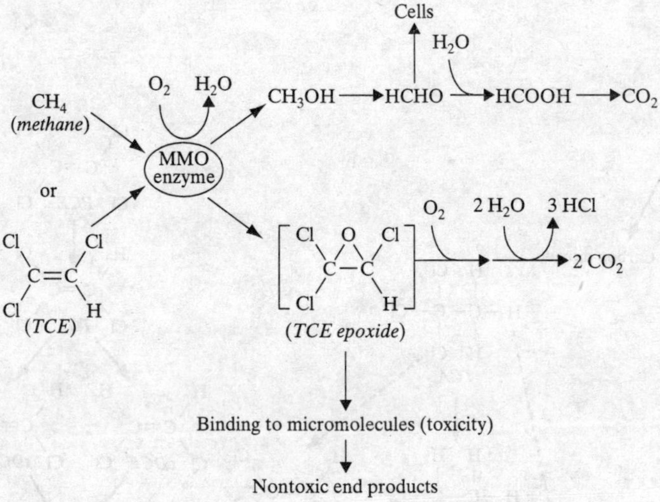

Figure 8.C.16 Competition for methane monooxygenase enzyme (MMO) by the primary substrate (methane) and the cometabolic substrate (TCE) as part of a typical aerobic cometabolic degradation reaction performed by methane-oxidizing bacteria (adapted from Higgins et al., 1981; and Alvarez-Cohen and McCarty, 1991).

oxidation reactions by oxygenase enzymes that are produced by microorganisms for oxidizing their primary growth substrates such as methane, propane, toluene, phenol, butane, ammonia, and propylene. For example, the methane monooxygenase enzyme produced by methane-oxidizing bacteria is capable of catalyzing the cometabolic degradation of chlorinated ethenes, ethanes, methanes, and aromatics (Figure 8.C.16). These cometabolic reactions require oxygen as the electron acceptor and generate products that are further degraded to carbon dioxide and salts in aqueous solution. The major limitations to the application of these cometabolic reactions for the destruction of hazardous wastes are as follows:

- An appropriate primary substrate must be provided to sustain the cometabolic reaction.
- Since both primary substrate oxidation and the cometabolic reaction are catalyzed by a single enzyme, competition between substrates for the enzyme can be a problem.
- Chlorinated solvent oxidation can exert product toxicity on the responsible microbial population.

However, each of these limitations can be overcome using appropriate reactor configurations and operational conditions. Although this process is still largely in the research stage, aerobic cometabolic degradation reactions may soon represent a rapid and inexpensive method for the treatment of hazardous wastes containing chlorinated aliphatics.

Inhibition and Toxicity

One important consideration when applying biological processes to the treatment of hazardous waste is the potential effect of toxicity caused by the influent contaminants on the resident microbial population. Toxicity of contaminants can take several different forms, including substrate toxicity, inhibition effects, and product toxicity.

Substrate toxicity involves direct damage to the microbial cells caused by a contaminant itself. Examples of contaminants that exert substrate toxicity on microbial populations are toluene, benzene, and methanol.

Inhibition effects tend to be reversible processes that are aimed at a specific enzyme responsible for catalyzing a reaction. Inhibition effects can be caused by high concentrations of the substrate itself, in which case it is similar to substrate toxicity. Alternatively, inhibition effects can involve a separate inhibitory compound that affects the degradation of the substrate by reacting with the responsible microbial enzyme.

Product toxicity involves adverse effects to microbial cells caused by products of the degradation reaction. Reactions that result in product toxicity include the aerobic degradation of chlorinated solvents, which produce unstable intermediates that attack and damage the substrate degradation enzyme and general cellular components.

Biological Treatment Configurations

Biological processes for the treatment of hazardous wastes are similar to those used for the treatment of wastewater (§6.E) and can be categorized into two general configuration types: suspended cell and immobilized cell (Figure 8.C.17). The goal in either

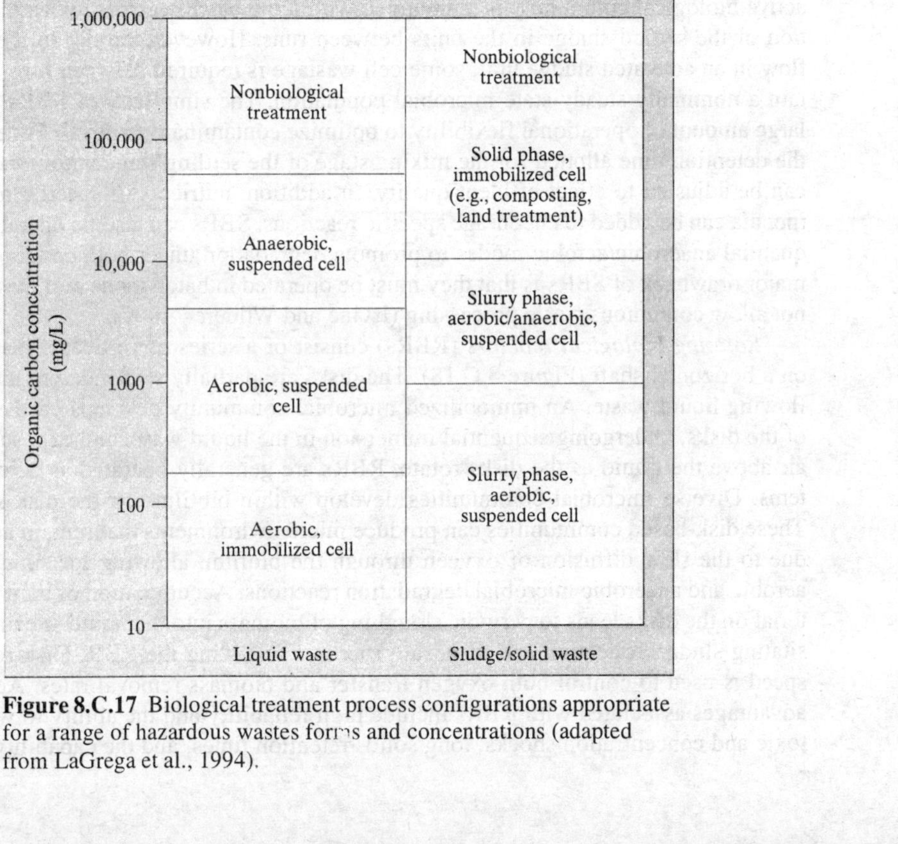

Figure 8.C.17 Biological treatment process configurations appropriate for a range of hazardous wastes forms and concentrations (adapted from LaGrega et al., 1994).

configuration type is to bring the contaminants into contact with a microbial population under conditions that foster degradation. To that end, each configuration must also provide a source of energy, carbon, and nutrients to the microbial population.

Immobilized cell configurations have the advantages of highly efficient contaminant removal rates and the presence of microenvironments that promote diverse microbial communities that can withstand toxic shocks and degrade wider ranges of contaminants. One disadvantage of immobilized cell processes is the accumulation of inert cell material that must be removed and disposed of as a residual. Other disadvantages include mass-transfer limitations for contaminant transport to active microorganisms and inherent difficulties in assessing and controlling the diverse biological population. Suspended cell configurations have the advantages of more uniform microbial populations, easily controlled cell retention times, and rapid mass transfer between contaminant and microorganism. The major disadvantage of suspended cell configurations is a requirement for an independent cell separation step in the process train.

Biological process configurations for the treatment of wastewaters and sludges, including activated sludge units, trickling filters, and anaerobic sludge digesters, were discussed in detail in §6.E and will not be repeated here. Additional reactor configurations used for the treatment of hazardous wastes include sequencing batch reactors, rotating biological reactors, fluidized-bed reactors, and biofilters.

Sequencing batch reactors (SBRs) are single-tank units that are sequentially filled with liquid waste, mixed to promote biological contaminant degradation, settled under quiescent conditions, and finally emptied of treated liquid effluent. Oxygen or air may be added during the mixing phase if aerobic treatment is desired. Otherwise, anaerobic treatment can be promoted by adding alternate electron acceptors as needed. An active biological community is maintained within the batch reactors by leaving a portion of the settled sludge in the units between runs. However, similar to the recycle flow in an activated sludge unit, some cell wastage is required between runs to maintain a nominally steady-state microbial population. The simplicity of SBRs allows a large amount of operational flexibility to optimize contaminant removal. For example, the detention time allotted for the mixing stage or the settling stage within the reactor can be adjusted to affect effluent quality. In addition, nutrients or specific microbial inocula can be added to encourage specific reactions. SBRs can also be operated in sequential anaerobic/aerobic modes to promote degradation under both conditions. The major drawback of SBRs is that they must be operated in batch mode and therefore do not allow continuous waste processing (Irvine and Wilderer, 1989).

Rotating biological reactors (RBRs) consist of a series of vertical disks rotating on a horizontal shaft (Figure 8.C.18). The disks are partially suspended in the slowly flowing liquid waste. An immobilized microbial community develops on the surface of the disks, undergoing sequential immersion in the liquid waste and exposure to the air above the liquid as the disks rotate. RBRs are generally operated as aerobic systems. Diverse microbial communities develop within biofilms on the disk surfaces. These disk-based communities can produce microenvironments that remain anaerobic due to the slow diffusion of oxygen through the biofilm, allowing for simultaneous aerobic and anaerobic microbial degradation reactions. Accumulation of inert cell material on the disks leads to periodic sloughing of biomass into the liquid stream, necessitating sludge collection and disposal processes following the RBR. Disk rotational speed is used to control both oxygen transfer and biomass removal rates. Additional advantages associated with RBRs include high reliability and the ability to withstand toxic and concentration shocks, long solids retention times, and the capability of pro-

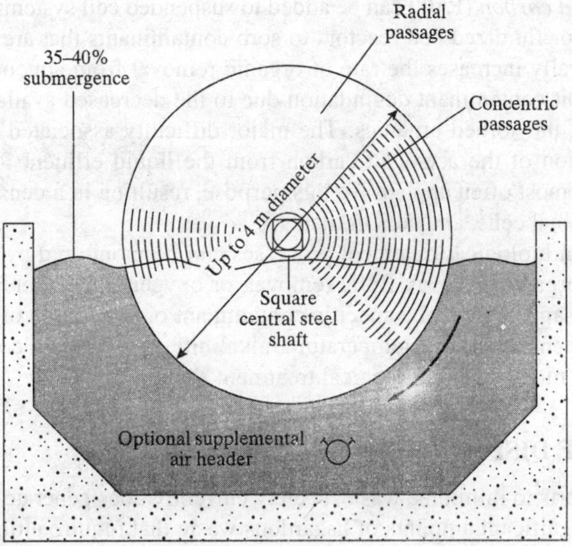

Figure 8.C.18 The configuration of a rotating biological reactor for the treatment of hazardous waste liquids. (Reprinted with permission of The McGraw-Hill Companies from Metcalf & Eddy, *Wastewater Engineering: Treatment, Disposal, Reuse*, 3rd edition, McGraw-Hill, New York, 1991.)

ducing low-concentration effluents. Disadvantages include long startup times and potentially persistent operational problems, such as shaft failures and odor problems (Metcalf & Eddy, 1991).

A biological fluidized-bed reactor (BFBR) consists of solid support material such as sand, coal, or activated carbon suspended in an upward fluid flow within a closed vessel. A microbial community that grows on the support material degrades contaminants within the liquid waste stream supplied to the reactor. BFBRs can be operated as aerobic or anaerobic processes, depending on the nature of the hazardous waste to be treated. Aerobic operation requires injection of oxygen. This can be achieved by supplying bubbles, which add to turbulent mixing within the reactor but can also strip volatile contaminants from the water before they are biodegraded. The use of activated carbon as the support medium for BFBRs is a recent innovation that has shown great promise since it promotes the removal of recalcitrant organics by sorption onto the carbon as well as providing a highly beneficial attachment site for the microbial population. Advantages associated with BFBRs include the potential for highly efficient contaminant removal rates, the capability of treating high organic loadings and wastes containing particulate matter, the development of a diverse biological community, and the presence of microenvironments allowing contaminant degradation to occur under multiple concurrent redox conditions. The major disadvantage associated with this configuration is the requirement for a constant high liquid velocity to maintain the fluidized state of the bed, which generally necessitates flow recycle.

Biofilters are used for the treatment of gaseous wastes and consist of a microbial population grown on support material within either an upward- or downward-flow packed-bed gas reactor. Support media used in biofilters can include inert materials such as plastics or diatomaceous earth (diatom skeletons!); nutrient-providing material such as peat, soil, or compost; or actively sorbing material such as granular activated carbon. A small amount of liquid is typically trickled over the support media to provide moisture and nutrients to the microbial population. Efficient contaminant removal is often achieved within biofilters with gas residence times of 15–60 s.

Powdered activated carbon (PAC) can be added to suspended cell systems such as activated sludge units or fluidized bed reactors to sorb contaminants that are difficult to degrade. PAC generally increases the rate of organic removal from aqueous waste streams, but may inhibit contaminant degradation due to the decreased availability to the microorganisms of the sorbed organics. The major difficulty associated with the use of PAC is separation of the activated carbon from the liquid effluent following treatment. Filtration is most often applied for this purpose, resulting in a dense sludge containing PAC, microbial cells, and contaminants.

Cell activity within biological treatment processes can be monitored by measuring cell density, total organic removal, BOD removal, or oxygen uptake, or by direct measurement of the disappearance of a specific contaminant or generation of specific products. Oxygen concentration, pH, temperature, alkalinity, and nutrient concentrations are also typically monitored in biological treatment systems.

8.D HAZARDOUS WASTE DISPOSAL

Despite aggressive efforts to minimize waste, society currently does generate and will continue to generate significant amounts of hazardous waste that require ultimate disposal. In fact, even the treatment of hazardous waste produces residual wastes that require disposal. For example, chemical precipitation and biological treatment both produce waste sludges while incineration produces ash and scrubber residuals. These treatment residuals generally must be disposed of as hazardous wastes.

Historically, hazardous waste disposal has taken many forms, including releases to water, air, and land. As concerns in the United States over environmental contamination increased in the 1960s and 1970s, regulations that strictly limited air disposal (Clean Air Act) and water disposal (Clean Water Act) were adopted. Disposal to land, as the last venue to be brought under strict regulation (Resource Conservation and Recovery Act), continues to be the most common method of ultimate hazardous waste disposal. Two methods of hazardous waste land disposal are commonly practiced in the United States: burial within strictly controlled landfills and injection in deep underground wells.

8.D.1 Landfills

Hazardous waste landfills are in-ground units constructed to contain hazardous wastes for indefinite periods while minimizing the release of contaminants to the environment. Although the use of landfills for containment of hazardous wastes has not always been successful (for example, see the Love Canal case study in §8.A.1), increased regulation and technological advances during the past 15 years have resulted in vastly improved landfill design and operation.

Under RCRA legislation, the USEPA controls the permitting process for hazardous waste landfills. Wastes are typically buried in 55-gallon drums or as bulk soils and sludges. Currently, untreated liquid wastes and noncontainerized liquid wastes are prohibited from landfills, according to RCRA's land-ban statute (§8.A.2). Batteries and certain small containers are exempted and may be buried in landfills.

Location

Site selection is perhaps the most difficult aspect of hazardous waste landfill design. Environmental, economic, and political issues must all be considered in the site selection process. The brief discussion here will focus on the technical environmental issues involved since the relevant economic and political issues are location specific and can be emotionally charged and highly controversial.

Two of the most important environmental factors to consider when siting a hazardous waste landfill are the geology and hydrogeology of the site. Ideal characteristics for landfill placement include existing natural or excavated depressions or pits surrounded by flat terrain. From a geological perspective, formations consisting of siliceous sandstone, siltstone, and consolidated alluvial bedrock are more desirable than limestone, dolomite, or heavily fractured crystalline rock formations. Geologic faults and fractures as well as flood plains should be avoided for obvious reasons. Locations with underlying low-permeability clay formations are ideal.

With respect to hydrology and hydrogeology, locations with low rainfall and high evaporation rates are desirable. Locations that are upgradient of high-quality surface water or groundwater should be avoided. Ideally, hazardous waste landfills should be located on sites where the underlying groundwater is nonpotable due to high concentrations of natural constituents. In addition, a proposed location should be at least 1.5 m higher than the historical high groundwater level, and the direction of groundwater recharge should be away from any potential drinking water sources.

In summary, the location of a hazardous waste landfill should be chosen to provide as many levels of redundancy as possible for containment to maximize the long-term protection of human health and the environment.

Control Cells

Wastes are segregated into control cells of compatible types (Figure 8.D.1). Documentation is maintained recording the amount and composition of waste placed into each control cell as well as the generator responsible for the waste. These types of records are useful for several reasons. First, they help the landfill operators to place compatible wastes together and avoid potentially hazardous incompatibility problems. Second, they make it possible to pinpoint specific waste burial locations to aid in future excavation for resource recovery or alternative treatment. Finally, in the event of contaminant release due to a defect in the landfill structure, it may be

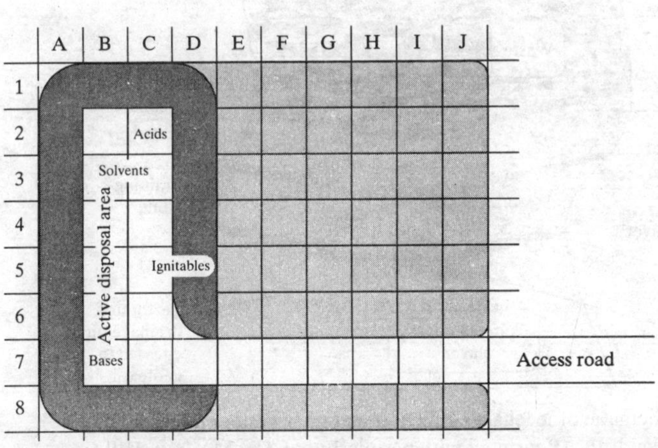

Figure 8.D.1 Typical grid system for a hazardous waste landfill (Wright et al., 1989).

possible to locate and fix the defect by identifying the specific location of the contamination source.

Control cells are commonly 5 to 6 meters in height. Waste buried in drums is typically covered and surrounded by bulk wastes and contaminated soils. A 0.3 m layer of clean compacted soil is placed over the waste layer every night to minimize odor and potential airborne contamination.

Leachate Control

Leachate is generated by direct precipitation during the active life of the landfill (before the permanent cover is installed), by the generation or release of liquids within the landfill, and by infiltration through the landfill cover following closure. Leachate must be collected, treated, and properly disposed of. Leachate collection systems consist of double flexible membrane liners placed on the bottom and side slopes of the landfill in alternating layers with perforated plastic collection pipes (Figure 8.D.2). A low-permeability compacted clay layer is generally constructed (unless it occurs there naturally) beneath the entire two-layer leachate control system. Landfill bottoms are sloped to control leachate flow, and a sufficient number of leachate collection pipes are provided to avoid leachate depths greater than 0.3 m. Flexible membrane liners are commonly made of plastic or rubber and must be chemically inert, impermeable, and durable.

Landfill Cap

A low-permeability cap is constructed over completed portions of the landfill to reduce leachate production. A landfill cap must have an estimated permeability as low

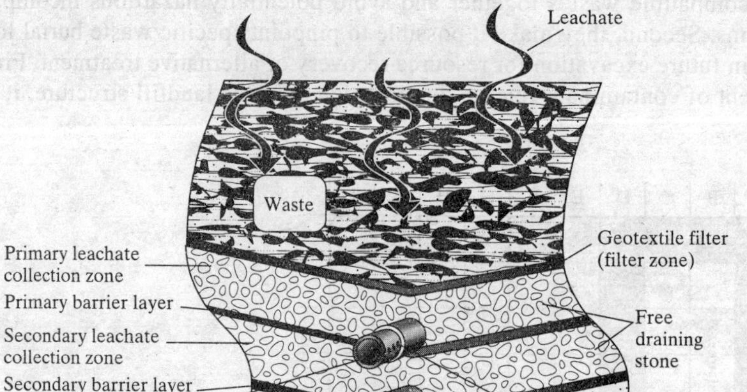

Figure 8.D.2 Schematic of leachate collection and liner system required in hazardous waste landfills. (Reprinted with permission of The McGraw-Hill Companies from M.D. LaGrega, P.L. Buckingham, and J.C. Evans, *Hazardous Waste Management*, McGraw-Hill, New York, 1994.)

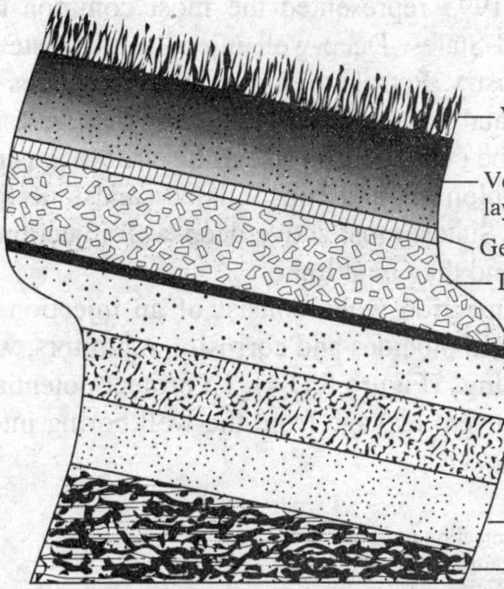

Figure 8.D.3 Schematic of typical landfill cap with gas collection system used for hazardous waste landfills. (Reprinted with permission of The McGraw-Hill Companies from M.D. LaGrega, P.L. Buckingham, and J.C. Evans, *Hazardous Waste Management*, McGraw-Hill, New York, 1994.)

as that of the bottom liners to prevent water accumulation within the landfill. Therefore, caps also generally include flexible membrane liners. Perforated pipes below the liner are used to collect landfill gas for subsequent treatment. A topsoil layer that is capable of supporting vegetation is placed over the cap, and shallow rooted vegetation such as grass and bushes is planted (Figure 8.D.3). Surface water from off site is diverted around the landfill by a series of ditches while runoff from the landfill is directed to localized holding sumps, where it is tested for contamination and treated if necessary. Long-term cap maintenance must be provided to control water runoff; prevent erosion; counteract any effects of subsidence, desiccation, or freezing; and avoid interference from burrowing animals or deep-rooted plants.

Monitoring Facilities

Hazardous waste landfill sites are monitored before, during, and long after the active operational life of the landfill. A monitoring system is designed to inspect the atmosphere, soil gas, groundwater, and surface water for contamination from landfill gas or leachate. Atmospheric gases are monitored at property boundaries as well as within on-site structures. Soil gas is monitored using a series of shallow wells positioned completely around the landfill. Groundwater is monitored using at least four wells installed 3 m below the groundwater depth. One well is positioned upgradient of the groundwater flow to monitor background groundwater characteristics, and three are located downgradient to detect potential releases. Leachate that is generated within the landfill is also collected and monitored for composition or flow changes that may indicate a leak or compatibility problem within the landfill.

8.D.2 Deep-Well Injection

The injection of liquid hazardous waste into a deep, vertically confined, porous, saline-water-bearing rock stratum is currently an EPA-permitted method of hazardous waste disposal, and as recently as 1995 represented the most common hazardous waste disposal method in the United States. Deep-well injection of wastewater was first practiced by the petroleum industry. Petroleum exploration generates extensive knowledge of the deep substratum, and it was found that wells that were originally drilled for oil production could later be converted for use as hazardous waste disposal wells. Locations within limestone, dolomite, and consolidated sandstone formations are the most desirable. In the United States, most active deep-well injection facilities are clustered around the Gulf Coast and the Great Lakes.

Underground hazardous waste injection wells consist of an injection pipe surrounded by an annulus filled with liquid biocides and corrosion inhibitors, which is in turn surrounded by cement-filled casings (Figure 8.D.4). To protect potential sources of drinking water, care is taken to prevent leakage from the well boring into shallow

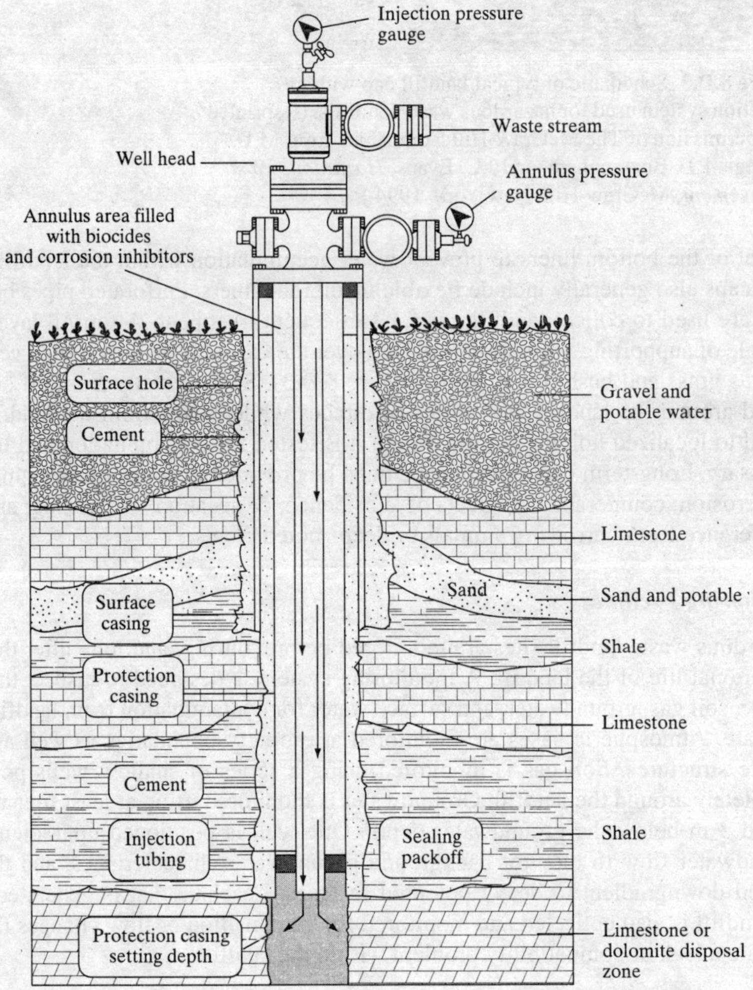

Figure 8.D.4 Schematic of a well used for deep-well injection of liquid hazardous waste. (Reprinted with permission of The McGraw-Hill Companies from C.A. Wentz, *Hazardous Waste Management*, 2nd edition, McGraw-Hill, New York, 1995.)

formations through which the well is constructed. Well depths (typically 350–3000 m) must be sufficient to reach a formation that is not hydrologically connected to potential drinking water sources. Minimum well depths, therefore, are site specific and are determined on a case-by-case basis.

The EPA recognizes a variety of injection well classes, as outlined in Table 8.D.1. Only EPA-permitted, Class I wells may legally be used for hazardous waste disposal. The use of Class IV wells is illegal. Specifically, the underground injection of hazardous waste is not permitted into or above a formation that is within a quarter mile of an underground drinking water source.

A wide variety of liquid wastes from diverse industries are amenable to deep-well disposal (Tables 8.D.2 and 8.D.3). Although it is not typically necessary to know the exact chemical composition of wastewater to dispose of it by deep-well injection, it is important to determine the reactivity and stability of the liquid to avoid potential adverse reactions within both the well and the receiving formation. In addition, potential

Table 8.D.1 Well Classes Designated by EPA for Regulation

Class I	Used for injection of liquid hazardous waste beneath any potential underground sources of drinking water
Class II	Used for enhanced recovery or storage of oil, or for disposal of fluids produced in conjunction with oil and gas production
Class III	Used in mining or recovery of minerals
Class IV	Used for injection of liquid hazardous or radioactive waste above or into a formation containing an underground source of drinking water
Class V	All other injection wells

Table 8.D.2 Distribution, by Mass, of Types of Liquid Hazardous Wastes Disposed of by Deep-Well Injection in the United States

Waste type	Percentage
Acids	41
Organic chemicals	36
Heavy metals	1.4
Inorganic chemicals	0.1
Others	21

Source: USEPA, 1985.

Table 8.D.3 Distribution, by Industry, of Users of Deep-Well Injection for Hazardous Waste Disposal

Industry	Percentage of wells	Waste volume (%)
Organic chemical manufacturers	44	50
Petroleum refiners	20	25
Other chemical manufacturers	18	13
Metals and minerals industry	8	6
Aerospace industry	1	1.5
Commercial off-site disposal	9	4.4

Source: USEPA, 1985.

problems with carbonate dissolution, clay interactions, swelling, precipitate formation, and polymerization can largely be avoided by testing the geochemical compatibility of the waste with the formation water and core material.

It is unknown what happens to wastes following deep-well injection. It is expected that the isolation of the receiving formation together with the dilution and chemical reactions that occur within the deep substrata are sufficient to ensure long-term containment of the hazardous constituents. Regardless, all liquid hazardous wastes must be treated according to the RCRA land-ban requirements (§8.A.2) prior to disposal by deep-well injection.

Monitoring Facilities

Monitoring wells are not frequently used for deep-well injection due to the high cost of installation and the potential for increasing waste migration routes. However, pressure changes in the injection fluid are monitored to detect formation plugging or hydrocracking, and annulus liquids are monitored to detect potential leaks in the well casings.

8.E ENVIRONMENTAL RELEASES AND REMEDIATION

Under EPA authorization, site owners or operators, hazardous waste transporters, state and local government officials, or the public may identify potential hazardous waste sites. In addition, releases of hazardous substances to the environment must be reported as outlined by the National Contingency Plan, described in §8.A.2.

Of the more than 400,000 potential hazardous waste sites that have been reported to state or federal authorities over the past 15 years, 217,000 sites require remediation under current state and federal regulations (Table 8.E.1). This number excludes sites

Table 8.E.1 Estimated Number of Hazardous Waste Sites That Have Not Yet Been Remediated in the United States

Category	Number of sites	Comments
National Priority List sites (nonfederal)	547	Includes proposed and final nonfederal National Priority List sites that still required further remedial action as of September 10, 1996
RCRA corrective action sites	2600–3700	This EPA estimated range represents approximately half of the nearly 6200 currently operating RCRA treatment, storage, and disposal facilities
RCRA underground storage tank sites	165,000	Sites officially designated as "cleanup has been initiated" are not included even though remediation action may not yet have begun
Department of Defense sites	8336	Includes 130 sites that are on the National Priority List
Department of Energy sites	10,500	Includes 30 sites on the National Priority List; full characterization of 46% of these sites has been completed, and cleanup of several hundred may have been completed
Other federal agency sites	>700	The number of facilities is given, and each facility may contain more than one site
State-regulated sites	29,000	The number of sites that "need attention" is given, although all may not need to be remediated
Total	217,000–218,000	Sites where cleanup work is ongoing or complete have not been included

Source: Derived from USEPA, 1997c.

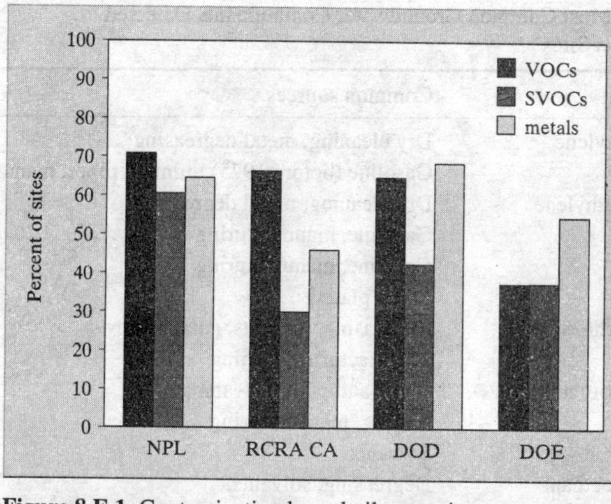

Figure 8.E.1 Contamination by volatile organic compounds (VOCs), semivolatile organic compounds (SVOCs), and metals at hazardous waste sites listed on the National Priority List (NPL), RCRA corrective action facilities (RCRA CA), Department of Defense facilities (DOD), and Department of Energy facilities (DOE) (derived using data from USEPA, 1997c).

where remediation is ongoing or had been completed as of 1996. Among those included are large sites contaminated by past practices (National Priority List sites) and sites where active hazardous waste treatment, storage, or disposal is practiced (RCRA corrective action sites). Also included in this list are sites in federal installations and smaller currently active sites such as local gas stations, dry cleaners, and metal plating shops (state sites). The cost for remediation of these sites under current regulations with today's technology is estimated at $187 billion (1996 dollars). The time required for remediation is estimated at 10–30 years for most sites, with some expected to take considerably longer (USEPA, 1997c). Not included in Table 8.E.1 or the cited cost estimate are "brownfield" sites, which are defined by the EPA as abandoned, idle, or underused industrial and commercial facilities where expansion or redevelopment is complicated by real or perceived environmental contamination. The U.S. General Accounting Office estimates the number of brownfield sites to be between 130,000 and 450,000.

The most common contaminants at hazardous waste sites are volatile organic compounds, including solvents and petroleum hydrocarbons (Figure 8.E.1). Commonly found semivolatile organic compounds (SVOCs) include polynuclear aromatic hydrocarbons (PAHs) and polychlorinated biphenyls (PCBs). Common metal contaminants found at NPL sites include lead, chromium, zinc, and arsenic (Table 8.E.2).

8.E.1 Site Characterization

After a hazardous waste release has been identified, it is necessary to characterize the subsurface conditions, identify the contaminant source, and define the nature and location of subsurface contamination to determine if and how remediation should be attempted. Existing site data are used in combination with both direct and indirect field measurements to evaluate the nature and extent of contamination and to identify the

Table 8.E.2 The 15 Most Common Groundwater Contaminants Detected at National Priority List Sites

Rank	Contaminant	Common sources
1	Trichloroethylene	Dry cleaning, metal degreasing
2	Lead	Gasoline (before 1975), mining, pipes, manufacturing
3	Tetrachloroethylene	Dry cleaning, metal degreasing
4	Benzene	Gasoline, manufacturing
5	Toluene	Gasoline, manufacturing
6	Chromium	Metal plating
7	Methylene chloride	Degreasing, solvents, paint removal
8	Zinc	Manufacturing, mining
9	1,1,1-Trichloroethane	Metal and plastic cleaning
10	Arsenic	Mining, manufacturing
11	Chloroform	Solvents
12	1,1-Dichloroethane	Degreasing, solvents
13	Trans-1,2-Dichloroethene	Degradation product of trichloroethylene
14	Cadmium	Mining, plating
15	Manganese	Manufacturing, mining

Source: National Research Council, 1994.

contaminant source. Existing data categories that are especially beneficial include chronologies of hazardous material use on the site and documentation of past hazardous waste practices. Information from facility records can be augmented with additional data from state and local agencies such as permitting records, hazardous waste manifests, and emergency response records. Historical photographs and verbal descriptions from site workers or neighboring citizens can also provide useful information about hydrologic characteristics and contaminant releases. All existing data should be subjected to the same level of quality-assurance/quality-control (QA/QC) procedures that are to be applied to the ongoing investigation.

Direct Measurement Methods

Direct field measurements, involving the use of boreholes and monitoring wells, are used to evaluate the aquifer and vadose zone characteristics and to provide samples of aquifer material, groundwater, and soil gas to define the horizontal and vertical distribution of contaminants in the subsurface. Essential data required to characterize hydrogeologic aquifer properties include hydraulic conductivities and groundwater gradients that delineate groundwater flow patterns. Additionally, specific recharge and discharge zones, subsurface heterogeneity, occurrence of confining layers and hydraulic connections between discrete aquifers need to be evaluated. In situ analyses conducted by measuring water levels in monitoring wells during pumping tests are typically used to estimate hydrologic parameters. Air photos may help to identify drainage patterns and potential recharge and discharge areas. Well logs from new or existing wells may contain data on water table depth, geologic formations, and subsurface heterogeneities.

Assessing the nature and three-dimensional distribution of contamination involves collecting data on the contaminant source, the magnitude and location of the contaminant plume, and the contaminant partitioning characteristics. In addition, the

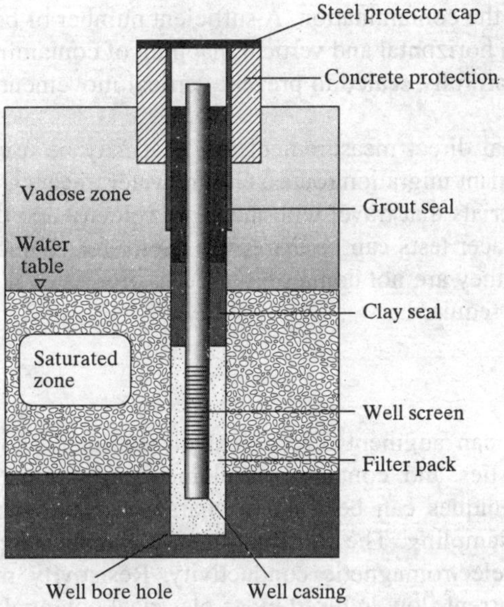

Figure 8.E.2 Schematic of monitoring well for characterizing subsurface contamination.

geochemical characteristics of groundwater from background areas and from the contaminant plume need to be measured. Boreholes and monitoring wells are drilled for direct sampling of aquifer material, groundwater, and soil gas (Figure 8.E.2). The most important issues related to the design of an appropriate monitoring system are the number and location of wells required, the borehole diameters, and the borehole and screen depths (Riggs and Hathaway, 1988). In determining well placement for site assessment, the cost associated with each boring must be weighed against the information lost in the spaces between borings. Optimal well placement is highly dependent on the heterogeneity of the aquifer. It is necessary to place wells upgradient of the contamination source to provide information on background water quality, as well as downgradient to gain information on plume magnitude and migration route. In the case of volatile contaminants, it may be possible to decrease the number of soil and groundwater samples that are required to establish the plume location and migration route by using relatively inexpensive soil-gas surveys. Gas samples are extracted from the unsaturated zone to detect volatile contaminants as they migrate upward from the water table.

Borehole diameters depend on the size of the devices that will be used to collect the subsurface samples and on whether the well will be used for extraction or injection of water. Common subsurface sampling devices include down-hole pumps, bailers, and split-barrel samplers. Although large boreholes are more expensive to drill and require greater volumes of purge water prior to sampling, they may be advantageous for accommodating several well casings to sample at multiple depths, or for subsequent use as production or injection wells for hydrodynamic control of contamination (see §8.E.3).

The predicted migration behavior of the contaminants of interest will help to determine placement depths and lengths of monitoring well screens. A well screen is the portion of the well that is open to the subsurface formation and through which water flows to the well. A screen over the entire water-producing zone of an aquifer will sample water averaged over the entire formation, whereas more discrete screening intervals within a saturated zone will produce more specific information on the stratification and distribution of the contamination. A sufficient number of boreholes must be drilled to provide both a horizontal and vertical mapping of contaminant distribution. Also, wells must be carefully sealed to prevent vertical movement of contaminant within the borehole.

Tracer tests are an additional direct measurement tool that may be used to increase understanding of contaminant migration routes. Groundwater tracers are chemically inert, easily detected materials that travel with the same velocity and direction as the groundwater. Although tracer tests can be extremely useful for characterizing subsurface transport properties, they are not commonly used in professional practice because of the time and expense required.

Indirect Measurement Methods

Indirect measurement methods can augment direct field measurements to further characterize aquifer characteristics and contaminant plume distribution. Remote sensing using geophysical techniques can be used to characterize the subsurface without extensive drilling and sampling. The most commonly applied geophysical techniques are resistivity and electromagnetic conductivity. Resistivity measurements detect the resistance to current flow induced by an electrical potential applied to the subsurface matrix. The electromagnetic conductivity technique is similar in that an electromagnetic current is applied directly to the subsurface matrix and the resulting magnetic field is detected. These methods may be used to detect variations in the subsurface soil and rock, to identify impermeable zones and fractures, and, in some cases, to map the three-dimensional distribution of contaminants in unsaturated and saturated zones. Since both methods rely on variations in the specific conductance of pore fluids to map subsurface contamination, plume detection using these methods is limited to inorganic contaminants and, in some cases, nonaqueous-phase liquids (NAPLs). Both of these methods can be applied at the surface or within bore holes, and can be repeated over time for long-term plume observation. Surface measurements are less expensive and require less equipment than borehole studies. Conversely, borehole measurements can produce higher-resolution mapping and may avoid surface interference from overlying clay layers and low-resistivity zones.

Although direct measurements involving drilling and sampling are required to verify the results of indirect measurement techniques, considerable time and cost savings can be realized by using indirect measurements to plan well placement and to minimize the required number of direct measurements.

8.E.2 Quantitative Risk Analysis

To make rational decisions regarding remediation of environmental contamination, potential risks to human health and the environment posed by hazardous releases must be quantified. A properly conducted risk analysis can facilitate objective decision making and help to prioritize the use of limited resources by determining levels of cleanup that will provide the most protection for the money. Currently, most risk analyses conducted at contaminated sites are based on potential hazards to human health. Although other environmental hazards could also be incorporated into a comprehensive risk analysis, uncertainty with respect to methods for quantifying such effects and lack of regulatory pressure have largely precluded them from use. Potential hazards to human health are defined as injury, disease, or death. A risk assessment seeks to quantify the likelihood of such effects occurring due to human exposure to environmental contaminants.

The USEPA has promoted a quantitative approach to risk analysis at hazardous waste sites that is divided into four tasks (USEPA, 1989):

1. Hazard identification
2. Exposure assessment
3. Toxicity assessment
4. Risk characterization

Hazard Identification

The general objective of the hazard identification step is to identify the potentially harmful contaminants present at the site and to determine their concentrations and spatial distributions. Contaminants present in each medium, including soil, groundwater, surface water, and soil gas, should be evaluated for the potential to pose a hazard to human health. Hazard identification is therefore linked to the site characterization process discussed above. Site characterization provides information on contaminant identification, concentration, and location, and on site characteristics that may affect the fate and transport of the contaminants. Contaminants present at the site can then be rapidly evaluated so that only those with the potential for causing adverse human health effects are considered further.

Exposure Assessment

The overall goal of the exposure assessment is to estimate the potential contaminant doses affecting defined populations by all possible exposure pathways. The exposure assessment as defined by the USEPA has three components. The first component involves characterizing the exposure setting and identifying potentially exposed populations. The second component entails identifying potential exposure pathways by considering contaminant source, means of release, transport routes, and exposure points. The third component involves estimating exposure concentrations at the endpoint of each pathway and calculating potential contaminant intakes for current and future land uses.

Characterizing the exposure setting requires determining present and future land use plus associated demographic information. Populations that may be exposed to contaminants from the site can then be identified. Populations in the vicinity of the site may include residents, workers, recreational users, and visitors. Typical exposure pathways involve groundwater, surface water, and soil-based transport (Table 8.E.3). Three specific intake routes are generally considered: ingestion, inhalation, and dermal contact. When possible, estimates of exposure concentrations at the endpoint of each pathway should be based on actual monitoring data (e.g., current contaminant concentrations in a drinking water well or dust contaminant concentrations). However, when projections of future contaminant exposures are required, concentrations must be extrapolated by means of modeling contaminant behavior. Contaminant transport models such as those introduced in §5.B can be calibrated for use at the specific site of interest and then applied for exposure assessment.

Once exposure concentrations for each potential pathway have been estimated, they can be used to calculate contaminant intake rates for representative members of the potentially exposed population. Contaminant intake rates are commonly expressed

Table 8.E.3 Typical Pathways for Human Exposure to Environmental Contaminants at Hazardous Waste Sites

Medium	Residential land use	Industrial land use
Groundwater	Ingestion from drinking	Ingestion from drinking
	Inhalation of volatiles	Inhalation of volatiles
	Dermal absorption from bathing	Dermal absorption from direct contact
Surface water	Ingestion from drinking	Ingestion from drinking
	Inhalation of volatiles	Inhalation of volatiles
	Dermal absorption from bathing or swimming	Dermal absorption from direct contact
	Ingestion during swimming	
	Ingestion of contaminated fish	
Soil	Ingestion	Ingestion
	Inhalation of particles	Inhalation of particles
	Inhalation of volatiles	Inhalation of volatiles
	Exposure to indoor air from soil gas	Exposure to indoor air from soil gas
	Exposure to groundwater contaminated by soil	Exposure to groundwater contaminated by soil
	Ingestion via plant, meat, or dairy products	Inhalation of particles from trucks and heavy equipment
	Dermal absorption from gardening	Dermal absorption from direct contact

Source: Derived from USEPA, 1991.

in units of contaminant mass per body mass per time (mg kg^{-1} d^{-1}), as calculated using the following equation:

$$I = \frac{(\mathrm{CR})(\mathrm{EF})(\mathrm{ED})}{(\mathrm{BW})(\mathrm{AT})} C \qquad (8.E.1)$$

where I is the intake rate (mg contaminant [kg body weight]$^{-1}$ d^{-1}), CR is the rate of contact with the contaminated medium (L d^{-1}, mg d^{-1}, or m^3 d^{-1}, depending on the medium), EF is the frequency of exposure to contaminant (d y^{-1}), ED is the duration of exposure to contaminant (y), BW is the average body weight (kg), AT is the period over which exposure is averaged (d), and C is the average contaminant concentration during the exposure period (mg L^{-1} for a contaminant in water, mg mg^{-1} for a contaminant in soil, and mg m^{-3} for an airborne contaminant).

Intake rates for each contaminant are calculated for each potential exposure pathway. Exposures would vary among the exposed population because of variability in the parameters that appear in equation 8.E.1. When applied to hazardous waste site evaluation, the risk assessment process is designed to generate exposure estimates that are higher than average. The USEPA provides default parameter values for use in equation 8.E.1 (Table 8.E.4). The average contaminant concentration (C) used for calculating intake is, by convention, based on the 95 percent upper confidence limit of the arithmetic average measured or estimated concentration at the exposure endpoint. The units and values of C and CR in equation 8.E.1 are adjusted to correspond to the exposure route of interest. Intake rates for exposure routes that are not addressed in Table 8.E.4, such as dermal contact or fish intake, can be calculated by adjusting the units of C and CR accordingly. The exposure frequency (EF) is given as 350 d y^{-1} for a resident, to allow for vacations, while 250 d y^{-1} is used to estimate intake rates for a

Table 8.E.4 EPA Default Values for Use in Exposure Assessment Calculations for Residents and Workers

Parameter	Resident	Worker
CR	$2\ L\ d^{-1}$ drinking water $100\ mg\ d^{-1}$ soil and dust ingestion $30\ m^3\ d^{-1}$ air inhalation	$1\ L\ d^{-1}$ drinking water $50\ mg\ d^{-1}$ soil and dust ingestion $30\ m^3\ d^{-1}$ air inhalation
EF	$350\ d\ y^{-1}$	$250\ d\ y^{-1}$
ED	Actual event duration or 30 years for chronic effects	Actual event duration or 25 years for chronic effects
BW	70 kg	70 kg
AT	Actual event duration for noncarcinogenic effects or $365\ d\ y^{-1} \times 70\ y$ for cancer calculations	Actual event duration for noncarcinogenic effects or $365\ d\ y^{-1} \times 70\ y$ for cancer calculations

Source: USEPA, 1989; Arulanantham et al., 1995.

laborer (5 days per week × 50 weeks per year). The choice of 70 years for AT in cancer calculations is based on the assumption that cancer effects are cumulative over a lifetime and that high doses applied over a short time are equivalent to low doses spread over a longer period. Although the validity of this and other assumptions may be debatable, they are currently incorporated in calculations conducted under EPA risk assessment guidelines. For noncarcinogenic effects, AT values are chosen to match ED values. A sample exposure assessment calculation is presented in Example 8.E.1.

EXAMPLE 8.E.1 ***Exposure Assessment for Disinfection By-products***

When chlorine is used for disinfection of drinking water, chloroform can be produced by the reaction of chlorine with residual organics in the water. Estimate the ingestion intake rate for noncarcinogenic and carcinogenic effects on an adult resident of a home receiving tap water with an average chloroform concentration of 65 μg/L.

SOLUTION To calculate the ingestion intake for noncarcinogenic effects, apply equation 8.E.1 with ED and AT values of 30 years. CR, EF, and BW values are taken from Table 8.E.4, and the value for C is given in the problem statement.

$$I_{\text{noncarc}} = (0.065\ \text{mg L}^{-1})\frac{(2\ \text{L d}^{-1})(350\ \text{d y}^{-1})(30\ \text{y})}{(70\ \text{kg})(365\ \text{d y}^{-1} \times 30\ \text{y})} = 1.8 \times 10^{-3}\ \text{mg kg}^{-1}\ \text{d}^{-1}$$

For carcinogenic effects, an AT value of 70 years is chosen to reflect lifetime effects associated with 30 years of exposure:

$$I_{\text{carc}} = (0.065\ \text{mg L}^{-1})\frac{(2\ \text{L d}^{-1})(350\ \text{d y}^{-1})(30\ \text{y})}{(70\ \text{kg})(365\ \text{d y}^{-1} \times 70\ \text{y})} = 7.6 \times 10^{-4}\ \text{mg kg}^{-1}\ \text{d}^{-1}$$

Toxicity Assessment

Toxicity assessments quantify the relationship between specific contaminant exposures or doses and the resulting adverse human health effects. This quantification is

Table 8.E.5 Estimated Reference Dose Factors (RfD) and Slope Factors (SF) for Oral Ingestion of a Variety of Chemicals

Species	Oral RfD (mg kg^{-1} d^{-1})	Oral SF (mg kg^{-1} d^{-1})$^{-1}$
Arsenic (inorganic)	3.0×10^{-4}	1.5
Benzene	No data	2.9×10^{-2}
Benzo(*a*)pyrene	No data	7.3
Cadmium	5.0×10^{-4}	No data
Chlordane	5.0×10^{-4}	3.5×10^{-1}
Chloroform	1.0×10^{-2}	6.1×10^{-3}
Chromium(VI)	3.0×10^{-3}	No data
1,1-Dichloroethylene	9.0×10^{-3}	6.0×10^{-1}
Methyl mercury	1×10^{-4}	No data
Naphthalene	2×10^{-2}	No data
Polychlorinated biphenyls (PCBs)	No data	1.0
2,3,7,8-TCDD (dioxin)	No data	1.5×10^{5}
Tetrachloroethylene	1.0×10^{-2}	5.2×10^{-2}
Trichloroethylene	6×10^{-3}	1.1×10^{-2}
Toluene	2.0×10^{-1}	No data
Vinyl chloride	No data	1.9

Source: Compiled from the EPA's IRIS database: www.epa.gov/iris/.

typically based on statistical analyses of contaminant exposure studies conducted with laboratory animals (rats or mice). The adverse health effects considered are divided into carcinogenic and noncarcinogenic effects. The results of the animal studies are used to produce slope factors (SF) that quantify carcinogenic effects and reference dose factors (RfD) for noncarcinogenic effects. A compilation of estimated slope factors and reference dose factors is published and updated frequently by the USEPA (Table 8.E.5). The EPA's Integrated Risk Information System (IRIS) database can be accessed electronically at www.epa.gov/iris/.

Risk Characterization

The final task in the risk analysis process is risk characterization. The objectives are to estimate the potential for adverse human health effects in populations of interest due to exposure to specific contaminants, and to evaluate the uncertainties, assumptions, and qualitative issues involved in the overall risk assessment. In practice, however, the risk characterization task often simply involves estimating the potential carcinogenic and noncarcinogenic effects due to exposure to the suite of chemicals chosen in the hazard identification step.

Noncarcinogenic health effects are estimated by calculating the hazard quotient (HQ) from the noncarcinogenic intake rate and the reference dose factor (Table 8.E.5) as follows:

$$HQ = \frac{I_{noncarc}}{RfD} \quad (8.E.2)$$

HQ is unitless since both the intake rates and reference dose factors are given in units of mg kg^{-1} d^{-1}. A cumulative hazard index for a contaminated site is calculated by adding the HQs for all chemicals of concern over all potential exposure pathways. A hazard quotient ≥ 1 is considered an unacceptable level of risk for noncarcinogenic health effects (see Example 8.E.2).

Potential carcinogenic health effects are estimated by calculating the individual excess lifetime cancer risk (IELCR) from the carcinogenic intake rate and the cancer slope factor (Table 8.E.5) according to the following equation:

$$\text{IELCR} = I_{\text{carc}} \times \text{SF} \tag{8.E.3}$$

IELCRs are also unitless since the cancer slope factor units are the inverse of those for the intake rate. The total lifetime cancer risk associated with a particular site is estimated by summing the IELCRs for all chemicals over all potential exposure pathways.

The USEPA considers IELCRs of $\leq 10^{-6}$–10^{-4} to be acceptable for regulatory purposes (see Example 8.E.2). These numbers represent *excess* cancer risks, that is, the incremental risk of developing cancer due to contaminant exposure that is above the general background. To put these numbers into proper perspective, it is helpful to consider the following. Approximately 2 million deaths occur each year in the United States. If the entire U.S. population received some exposure that caused a lifetime cancer risk of 10^{-6}, there would be an average of two more deaths per year (if the cancers were fatal). Furthermore, approximately one-quarter of all deaths in the United States are caused by cancer. So contaminant exposure to the entire population that resulted in an average IELCR of 10^{-4} could cause the proportion of deaths caused by cancer to increase from 25 percent to 25.01 percent. Because the risk assessment process incorporates elements of health-protective conservatism, the true increases in cancer death rates would probably be lower than indicated by these estimates.

EXAMPLE 8.E.2 ***Risk Characterization for Disinfection By-products***

Consider again the household that receives drinking water containing chloroform, as in Example 8.E.1. Calculate the potential noncarcinogenic and carcinogenic health risks associated with ingesting the tap water for an adult resident of the house.

SOLUTION For noncarcinogenic effects, calculate HQ for the ingestion pathway. Use the noncarcinogenic intake calculated in Example 8.E.1 and RfD from Table 8.E.5.

$$\text{HQ}_{\text{ing}} = \frac{I_{\text{ing,noncarc}}}{\text{RfD}_{\text{oral}}} = \frac{1.8 \times 10^{-3}}{1.0 \times 10^{-2}} = 0.18$$

The noncarcinogenic effect from ingestion is < 1, suggesting that the EPA would not consider this concentration of chloroform as posing a significant toxic risk to the residents of the household.

For carcinogenic effects, calculate the IELCR for the ingestion pathway using the carcinogenic intake calculated in Example 8.E.1 and the ingestion SF for chloroform from Table 8.E.5.

$$\text{IELCR}_{\text{ing}} = (7.6 \times 10^{-4})(6.1 \times 10^{-3}) = 4.6 \times 10^{-6}$$

The carcinogenic effect from ingestion is near the bottom of the range of potentially acceptable risk defined by the EPA as 10^{-6}–10^{-4}. Therefore, the EPA would probably consider the potential carcinogenic effects from this concentration of chloroform in the tap water to be insignificant and not subject to regulatory action.

EXAMPLE 8.E.2 ***Risk Characterization for Disinfection By-products (continued)***

It is important to note that in this example, we have considered only one exposure route, ingestion. For a volatile contaminant in tap water, other important exposure routes, such as inhalation and skin absorption associated with showering and bathing, could contribute to the overall risk (Moya et al., 1999). A comprehensive risk characterization for chloroform in tap water would include summations of the HQs and IELCRs associated with all potential exposure routes.

Risk characterization should include a description of the major assumptions, judgments, and uncertainties involved in calculating the cumulative HQ and IELCR, including those involved in the toxicity and exposure assessment steps. Due to the great uncertainties associated with many of the estimates included in a risk assessment, and since scientific and engineering judgment must be applied in the quantification of intakes, pathways, and the related health effects, specific explanations must be included to properly interpret the results. The EPA has provided documents to assist in risk analysis decision-making (USEPA, 1989, 1991). However, there are a number of site-specific assumptions (e.g., projected transport rates, relevant exposure pathways, substitution of oral factors for inhalation factors) that require proper description and documentation.

8.E.3 Site Remediation

When it has been determined that contamination at a site could potentially pose a threat to human health or the environment, site remediation is undertaken. The first steps in a comprehensive site remediation process involve mitigating the contaminant source and removing easily accessible accumulations of contaminant, such as pooled free product. If source mitigation and short-term removal techniques are not sufficient to eliminate the potential risk, additional steps may be needed, such as containment of the contaminant plume and application of remediation techniques for destruction or removal of the remaining contamination.

Source Mitigation

Corrective measures applied to a contaminant source may involve simple and obvious actions such as the excavation of leaking underground storage tanks, the repair of damaged pipes, or the suspension of treatment, storage, and disposal practices that cause contaminant releases. However, source mitigation may also require more involved or long-term actions such as the repair of leaking landfill liners or the removal from a site of nonaqueous-phase liquids (NAPLs). It may be possible to remove some NAPLs from an aquifer by direct pumping from extraction wells. Alternatively, a combination of injection and extraction wells can be used to cause directional migration of the floating or sinking product. However, most pumping strategies are capable of removing only a portion of NAPLs from the subsurface. Much of the organic material remains trapped in pores as residual free product, dissolved in the surrounding groundwater as a contaminant plume, sorbed onto the solid subsurface material, or volatilized into the gas-filled pores of the unsaturated zone (see §6.A.3), constituting a long-term source of subsurface contamination. In these cases, techniques may be applied to address source mitigation, containment, and remediation simultaneously. In some cases, source mitigation may not be practical with present technology, so that contaminant releases must be managed by means of long-term (sometimes indefinite) containment and remediation methods. An example is the Love Canal site, where an impermeable containment cap and a leachate collection system have been installed to manage the buried chemical wastes (see §8.A.1). A second example is Iron Mountain, a mine site in California, where acid mine drainage must be continually diverted from natural waterways for treatment.

Containment Techniques

Remediation efforts may require years to decades to reduce residual contamination to acceptable levels. In addition, for some types of contamination, there may be no known effective methods of remediation. Therefore, containment techniques are required to restrict the spread of subsurface contamination.

Physical and hydrodynamic controls can be used to contain a subsurface contaminant plume. Physical controls include surface-water diversions, such as covers, caps, and drainage ditches, as well as low-permeability vertical subsurface walls to physically block the transport of a contaminant plume or to inhibit the flow of clean groundwater into a contaminated zone. The most commonly applied subsurface containment barrier is the slurry trench wall. Slurry walls are built by excavating a trench while using liquid slurry of bentonite clay and water to stabilize the trench walls. The trench is then backfilled with a bentonite clay–soil mixture to form a low-permeability barrier (Figure 8.E.3). Grout curtains and sheet pile walls are also used on a limited basis for physical containment in rock formations or when the structural stability of the ground must be maintained (Table 8.E.6). A grout curtain is made by injecting grout into hollow forms forced into the subsurface or directly into subsurface fractures. Sheet pile walls are built by forcing steel sheets into the subsurface. Ideally, a containment wall is "keyed in" or embedded within an impermeable layer below the zone of contamination to ensure that the contaminant doesn't migrate below the containment. A containment wall that is keyed in to a confining impermeable layer can

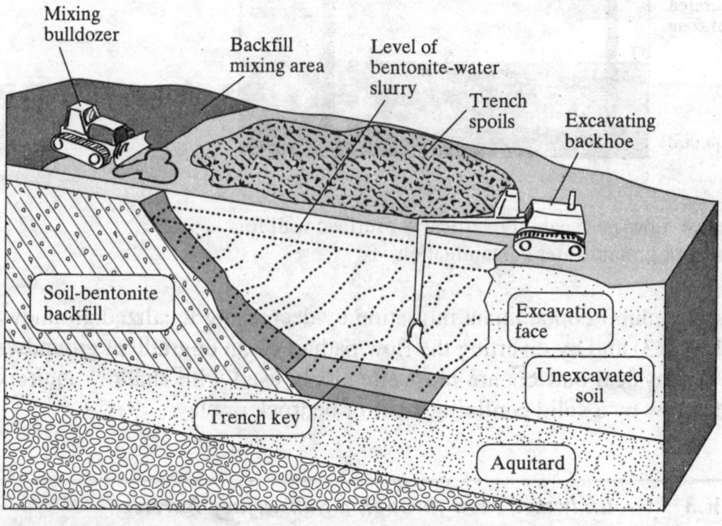

Figure 8.E.3 Construction of a slurry trench wall for subsurface plume containment. (Reprinted with permission of The McGraw-Hill Companies, from M.D. LaGrega, P.L. Buckingham, and J.C. Evans, *Hazardous Waste Management*, McGraw-Hill, New York, 1994.)

Table 8.E.6 Summary of Common Barrier Walls Used for Physical Containment of Groundwater Contamination

Technology	Advantages	Disadvantages
Slurry wall	Most versatile; low operation and maintenance costs; commonly applied technology	Requires excavation and space to mix backfill; difficult to verify wall integrity; decreases ground structural integrity; not applicable in rock formations
Grout curtain	Minimal site disturbance; no excavation required; applicable in rock formations; low operation and maintenance costs	Grout chemicals may pose leaching hazard; difficult to verify wall integrity; relatively expensive to install
Sheet pile	No excavation required; retains ground structural integrity; low operation and maintenance costs	Relatively expensive to install

Source: Adapted from Wentz, 1995.

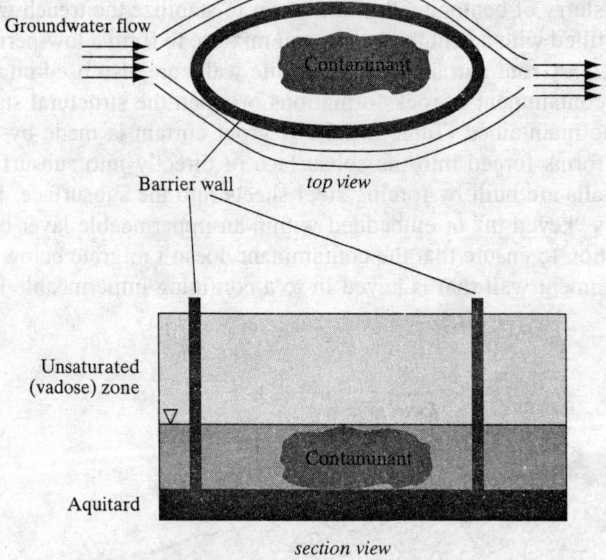

Figure 8.E.4 Low-permeability vertical subsurface wall for containment of groundwater contamination.

significantly inhibit contaminant migration by decreasing localized groundwater flow (Example 8.E.3) and by diverting the flow path of clean upgradient groundwater (Figure 8.E.4). Physical barriers are most effectively used with shallow aquifers that are bounded below by a solid confining layer of bedrock or clay.

EXAMPLE 8.E.3 ***Groundwater Flow through a Subsurface Barrier***

Calculate the reduction in groundwater flow rate achieved by placing a slurry wall with a hydraulic conductivity of 1×10^{-7} cm s^{-1} around a contaminant source in an aquifer with a hydraulic conductivity of 1×10^{-4} cm s^{-1}.

EXAMPLE 8.E.3 ***Groundwater Flow through a Subsurface Barrier (continued)***

SOLUTION From Darcy's law (§4.D.1), we recall that the flow rate of fluid through porous media is directly proportional to the hydraulic conductivity of the media:

$$U = -K\frac{dh}{dl}$$

where U is the groundwater flow velocity (m s^{-1}), K is the hydraulic conductivity, and dh/dl is the rate of change of pressure head with distance. If the pressure head remains constant across the barrier, then the groundwater flow velocity and associated groundwater flow rate (m^3 s^{-1}) through the wall will be decreased in proportion to the hydraulic conductivity, giving a flow rate that was $10^{-7}/10^{-4} = 1/1000$ that of the original flow rate.

Hydrodynamic containment is often used in conjunction with physical containment and is particularly conducive to use as part of a remediation strategy. Hydrodynamic controls involve the active manipulation of the groundwater flow to prevent undesirable plume movement. Hydrodynamic control is typically achieved by using a combination of injection and extraction wells or infiltration ponds. For example, extraction wells may be situated downgradient from the zone of contamination to capture the contaminant plume for aboveground treatment (Figure 8.E.5). Wells are ideally situated so that their radii of influence overlap, ensuring that the minimum amount of contaminant is able to migrate beyond the extraction wells. Plume direction, shape, and migration speed can be manipulated hydrodynamically, preventing further spread of contaminated water. The major drawback to hydrodynamic containment is that it is an active system that may require operation over indefinitely long periods. For a wide variety of subsurface contaminants, such as petroleum hydrocarbons, solvents, and pesticides, mass-transfer limitations may cause extremely long-term (on the order of hundreds of years) contaminant release from the subsurface matrix, necessitating indefinite containment. Therefore, hydrodynamic containment is typically used in conjunction with active remediation techniques designed to attack the concentrated contaminant source.

Remediation Techniques

Although source mitigation and containment is sometimes sufficient to eliminate significant risk from contaminant releases, it is frequently necessary to apply additional site remediation methods to decrease threats to human health and the environment.

A range of innovative and conventional remediation technologies can be applied to contaminated sites. These technologies include in situ methods that are applied directly

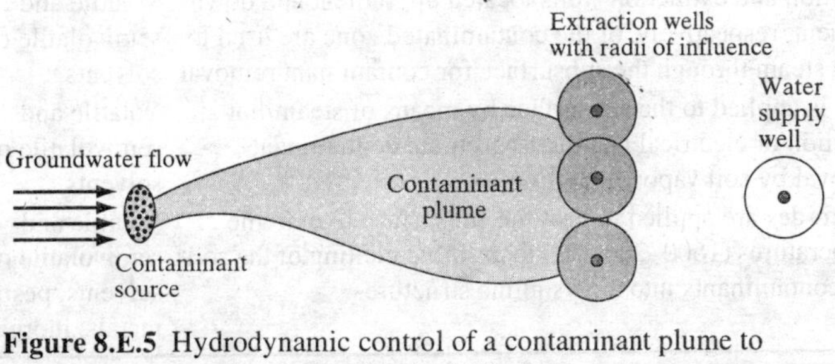

Figure 8.E.5 Hydrodynamic control of a contaminant plume to protect a downgradient water supply well.

to the subsurface environment (Table 8.E.7) and ex situ methods that involve aboveground treatment of excavated soil, groundwater, or soil gas. When in situ technologies are not appropriate or are inadequate for achieving the required level of remediation, or in the remediation of a small area containing extremely high contaminant concentrations (as may be found directly below the source of contamination), soil excavation followed by ex situ treatment may be necessary. Many physical, chemical, thermal, and biological processes can be used to treat contaminated soils (Table 8.E.8), depending

Table 8.E.7 In Situ Remediation Methods for Subsurface Contamination

Technology	Description	Application
	Physical/chemical processes	
Soil flushing	Injection and extraction wells located upgradient and downgradient, respectively, of the contaminated zone are used to flush water through the subsurface for contaminant removal; surfactants may be added to the flushing water	Volatile and semivolatile organics, solvents, pesticides, metals, inorganics
Soil vapor extraction	Extraction wells are used to apply a vacuum to the vadose zone, withdrawing soil gases containing volatile contaminants; resultant gases must be treated prior to release	Volatile organics, solvents
Air sparging	Air is injected through wells into the saturated aquifer, stripping contaminants by volatilization; operated concurrently with soil vapor extraction	Volatile organics, solvents
Passive treatment walls	Permeable reactive wall is installed across the groundwater flow path of a contaminant plume; the contaminant is chemically degraded by the wall components	Volatile and semivolatile solvents, metals, inorganics
	Biological processes	
Bioremediation	Naturally occurring microorganisms within the saturated and unsaturated subsurface degrade the contaminants; oxygen and nutrients may be supplied to enhance the biological activity	Volatile and semivolatile organics, solvents, pesticides
Bioventing	Wells are used to deliver oxygen to the unsaturated zone to promote biological degradation of contaminants	Volatile and semivolatile organics, solvents
Biosparging	Injection wells are used to deliver oxygen into the saturated aquifer to promote biological degradation of contaminants	Volatile and semivolatile organics, solvents
	Thermal processes	
Steam stripping	Injection and extraction wells located upgradient and downgradient, respectively, of the contaminated zone are used to flush steam through the subsurface for contaminant removal	Volatile and semivolatile organics, solvents
Thermally enhanced soil vapor extraction	Heat is applied to the subsurface by means of steam/hot air injection or electrical heating to promote contaminant removal by soil vapor extraction	Volatile and semivolatile organics, solvents
Vitrification	Electrodes are applied to heat the subsurface to extreme temperatures (1600–2000 °C) to promote melting of the soil and contaminants into a crystalline structure	Volatile and semivolatile organics, solvents, pesticides, metals, inorganics

Source: Adapted from USEPA, 1993.

Table 8.E.8 Ex Situ Remediation Methods for Contaminated Soils

Technology	Description	Application
	Physical/chemical processes	
Soil washing	Excavated soils are washed with water that may be amended with surfactants, chelating agents, or leaching agents to remove sorbed contaminants; resultant liquids must be treated prior to disposal	Semivolatile organics, solvents, pesticides, metals, inorganics
Soil vapor extraction	Soils are piled onto perforated pipes that are used to apply a vacuum, withdrawing soil gases containing volatile contaminants; resultant gases must be treated prior to emission	Volatile organics, solvents
Solidification/ stabilization	Contaminants are physically or chemically bound or enclosed within a stabilized material to reduce their mobility	Pesticides, metals, inorganics
Soil oxidation/ reduction	Oxidizing agents such as ozone, hydrogen peroxide, and chlorine or reducing agents such as alkaline polyethylene glycolate are mixed with soils in reaction vessels that may be heated; results in contaminant destruction	Semivolatile organics, solvents, pesticides, metals, inorganics
Solvent extraction	Excavated soils are mixed with solvents, allowing the contaminants to partition into the solvent phase; treated soils are then separated from the solvent; resulting solvent must be treated prior to recovery or disposal	Semivolatile organics, solvents, pesticides
	Biological processes	
Slurry-phase biological treatment	Excavated soils are mixed with water, nutrients, and microbial inocula in reactors or lined lagoons to promote contaminant biodegradation	Volatile and semivolatile organics, solvents, pesticides
Composting	Soils are mixed with amendments and piled on perforated pipes that are used to apply a vacuum, withdrawing soil gases containing volatile contaminants and drawing air into the soil pores to promote biodegradation; amendments may include bulking materials, nutrients, and microbial inocula	Volatile and semivolatile organics, solvents, pesticides
	Thermal processes	
Thermal desorption	Soils are placed in reactors and heated to temperatures of 95–300 °C (low *T* desorption) or 300–500 °C (high *T* desorption) to volatilize water and contaminants and to promote limited oxidation; gases are treated prior to release	Volatile and semivolatile organics, solvents
Incineration	Soils are heated to temperatures of 850–1200 °C in reactors in the presence of oxygen to volatilize water and contaminants and to promote oxidation; gases are treated prior to release	Volatile and semivolatile organics, solvents, pesticides
Pyrolysis	Soils are heated to temperatures of 400–750 °C in reactors in the absence of oxygen to volatilize water and contaminants and to promote reductive decomposition; gases are treated prior to release	Volatile and semivolatile organics, solvents, pesticides
Vitrification	Excavated soils are heated to extreme temperatures (1600–2000 °C) to promote melting of the solids and contaminants into a crystalline structure	Volatile and semivolatile organics, solvents, pesticides, metals, inorganics

Source: Adapted from USEPA, 1993.

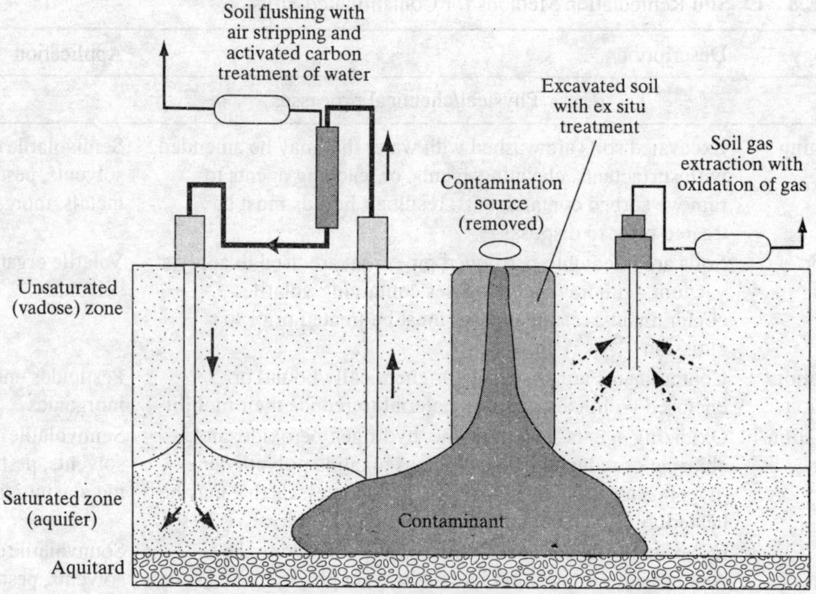

Figure 8.E.6 Commonly applied remediation techniques for subsurface contamination.

on the soil and contaminant characteristics. However, soil excavation is expensive and results in increased potential exposure to humans and the environment. Furthermore, regulatory control may prevent replacement of the treated soil, resulting in the additional expense and liability associated with transport and disposal in an appropriate landfill.

Several of the methods listed in Table 8.E.7 are *pump-and-treat* technologies, where the contaminant is removed from the subsurface by an extraction method, such as soil vapor extraction, soil flushing, or steam stripping (Figure 8.E.6). In these methods, the contaminant is brought to the surface in gaseous, dissolved aqueous, or NAPL form. With the exception of contaminants in NAPL form that can potentially be recycled, the resultant contaminant stream must be treated using an ex situ technology. Common ex situ treatment methods for liquid and gas streams include the physical, chemical, biological, and thermal processes discussed in §8.C.

An important consideration with pump-and-treat remediation methods is that contaminant removal efficiency generally declines significantly over time due to mass-transfer limitations within the subsurface. That is, rates of contaminant transport into pore fluids by means of dissolution from NAPLs, desorption from the solid phase, diffusion from micropores, or volatilization from liquid phases tend to be slower than the advective rate of fluid flushing through the media. As a result, contaminant levels in extraction fluids tend to become more dilute with time. However, when pumping is halted, transport of contaminants into pore fluids continues, resulting in a reestablishment of the contaminant plume at elevated concentrations (Figure 8.E.7). Mass-transfer-induced contaminant rebounds will recur until the mass of sequestered contaminant is substantially removed, which may require tens, hundreds, or thousands of years to achieve by flushing methods alone. For this reason, pump-and-treat methods should be applied in conjunction with containment or in situ contaminant destruction techniques such as bioremediation.

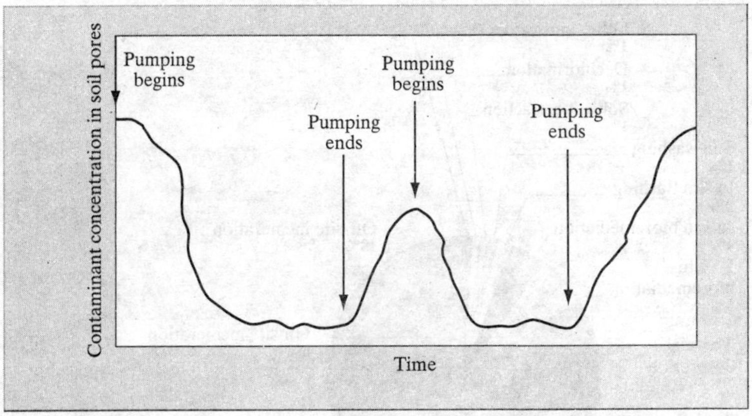

Figure 8.E.7 Mass-transfer-induced rebound of contaminant in soil pores when pumping of flushing fluids (gases or liquids) is discontinued.

With the wide variety of remediation alternatives, choosing the appropriate combination of technologies for a specific site requires careful consideration. The EPA has designated nine performance criteria that must be used to evaluate and select environmental remediation technologies for use at Superfund sites (USEPA, 1988):

1. Overall protection of human health and the environment
2. Compliance with all other applicable or relevant and appropriate requirements
3. Long-term effectiveness and permanence of the remediation
4. Reduction of toxicity, mobility, or volume through treatment
5. Short-term effectiveness
6. Implementability
7. Cost
8. State acceptance
9. Community acceptance

Items 1 and 2 are threshold criteria; they must be completely met for a technology to be considered further. Items 3–7 are balancing criteria, which do not have to be fully satisfied to implement the technology. They instead serve as indicators of technology performance and are used to help decide among alternative technologies. Items 8 and 9 are modifying criteria used to ensure that state and community concerns are given adequate attention. In the past, adherence to these criteria has tended to favor application of conventional and expensive processes: soil excavation followed by incineration or solidification/stabilization (Figure 8.E.8). However, recent changes in the regulatory environment have allowed increased application of innovative in situ and ex situ technologies. Two of the fastest-growing innovative technologies are soil vapor extraction and in situ bioremediation.

Soil Vapor Extraction Soil vapor extraction (SVE), also referred to as soil venting or vacuum extraction, removes volatile contaminants from soil by flushing air through soil pores. SVE can be applied ex situ, to excavated soils, by using a series of perforated pipes installed within soil piles, or in situ, by using gas extraction wells installed in the unsaturated zone. For in situ application, flushing air may be drawn directly

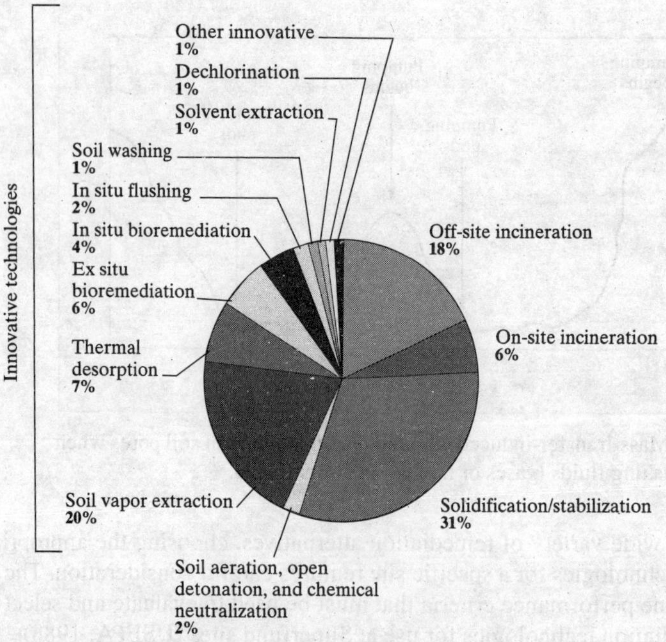

Figure 8.E.8 Established and innovative remediation technologies chosen for application at Superfund sites from 1982 to 1995 (USEPA, 1997c).

from the ground surface or from injection wells installed upgradient of the contamination zone. Extracted gases must be treated to remove contaminants prior to release. Typical off-gas treatment systems include activated-carbon sorption, catalytic or thermal oxidation, and biofilters. Major advantages associated with SVE are that it can be implemented with minimal surface disturbance, is easily installed and operated, and is not unduly expensive.

The fundamental principle of SVE is that volatile contaminants will diffuse out of the soil matrix into the flushing air as it moves through the larger vadose zone pores. Hence, the performance of an SVE system depends on the magnitude of the induced airflow, the length of the diffusion path for the contaminants, and other contaminant transport properties. Soil properties that significantly affect SVE performance include permeability, porosity, heterogeneity, moisture content, and organic carbon content. Soils with higher permeabilities and porosities will accommodate larger flow rates of flushing air and are more amenable to SVE. Heterogeneities decrease SVE efficiency by providing low-permeability zones within which contaminants remain sequestered. Increased soil moisture can diminish SVE performance by decreasing effective permeabilities and contaminant diffusion, while simultaneously enhancing performance by reducing contaminant sorption affinities for the soil. Organic carbon associated with soils negatively affects SVE performance by providing additional sites for contaminant partitioning out of the gas phase.

Contaminant properties that significantly influence SVE performance include solubility, volatility, vapor pressure, sorption properties, and diffusivity. Contaminants in the subsurface will potentially partition between a NAPL phase, solid or sorbed phase, dissolved aqueous phase, and gas phase (Figure 8.E.9). The first four of the properties listed above define the affinity of the contaminant to partition among each of these

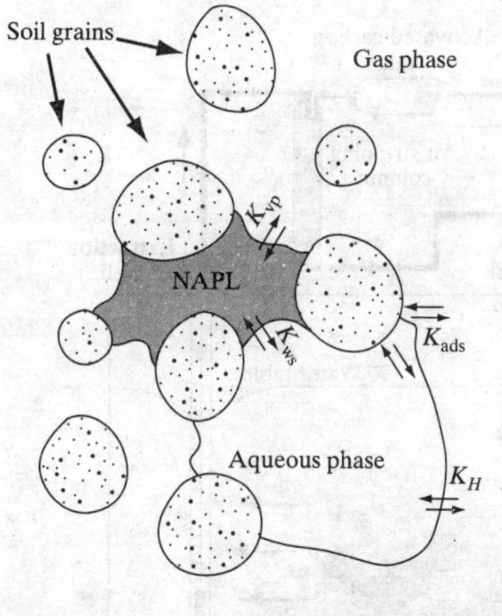

Figure 8.E.9 Partitioning of volatile organic contaminants between NAPL, sorbed, aqueous, and gas phases in the subsurface is governed by vapor pressure (K_{vp}), aqueous solubility (K_{ws}), Henry's constant (K_H), and sorption (K_{ads}) coefficients.

phases. The diffusivity of a contaminant dictates how quickly the contaminant can be transported through small pores within the soils to the flushing gas for removal. Diffusion coefficients tend to decrease with increasing molecular size of the contaminant and are very much smaller in water than in soil gas (§4.A.3).

Although models based on fundamental transport and partitioning relationships can be developed to predict the performance of SVE systems, the complexity and inherent heterogeneities of natural soils limit the applicability of fundamental models to actual systems. Instead, empirical models based on estimates of the potential radius of influence of each extraction well and calibrated with field tests are typically used to design SVE layouts.

In Situ Bioremediation In situ bioremediation involves the stimulation of microorganisms within a subsurface aquifer or vadose zone to promote contaminant degradation. The major advantage associated with in situ bioremediation is that the contaminants are destroyed in place, with minimal transport to the surface. A wide variety of contaminants are amenable to bioremediation, including gasoline hydrocarbons, jet fuels, oils, aromatics, phenols, creosote, chlorinated phenols, solvents, nitrotoluenes, and PCBs.

When contaminants are organic compounds that can be used by microorganisms as growth substrates, it may be possible for the indigenous microbial population to biodegrade the contaminants at significant rates in the absence of engineered intervention. This is recognized as a legitimate form of remediation and is referred to as "intrinsic bioremediation," "natural attenuation," or "bioattenuation." Intrinsic bioremediation is considered a suitable remediation technology only when natural contaminant biodegradation occurs faster than migration, resulting in a stable or shrinking contaminant plume. Site requirements for intrinsic bioremediation include an adequate supply of electron acceptors and trace nutrients to promote microbial growth, as well as adequate pH buffering to counter the acidification that results from microbial activity.

If intrinsic bioremediation occurs too slowly or is inhibited due to lack of substrate, nutrients, or other requirement, engineered bioremediation may be applied by using subsurface water or gas to transport compounds to the bioactive zone (Figure 8.E.10). Since engineered bioremediation involves injection or extraction of subsurface fluids, it has higher associated costs than intrinsic bioremediation. However, it

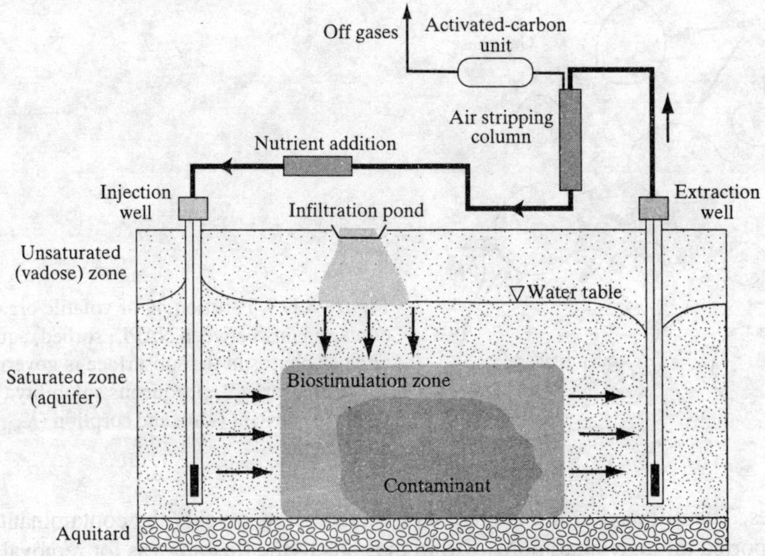

Figure 8.E.10 In situ bioremediation process employing an injection well and infiltration pond for nutrient delivery and air stripping with activated carbon for treatment of extracted water.

involves less reliance on naturally occurring subsurface conditions and growth factors and is therefore applicable over a broader range of site and contaminant characteristics. In addition, engineered bioremediation may be the more appropriate option when time or liability constraints are of concern, or when it is necessary to show good faith to the surrounding community or regulators by "engineering" a solution.

The feasibility of applying in situ bioremediation at a specific site depends on several factors, including aquifer hydrologic and geochemical characteristics, the indigenous microbial population, and the nature and distribution of the contaminants. The fundamental requirement for bioremediation to occur is that the contaminant, the microorganisms, and any other required reactant (e.g., oxygen) be brought into contact so that the biodegradation reactions can proceed. As a rule of thumb, aquifers with a hydraulic conductivity of 10^{-4} cm s^{-1} or greater are the most amenable to in situ bioremediation (Thomas and Ward, 1989). However, microbial growth in aquifer material can cause permeability to decrease by a factor of a thousand (Taylor et al., 1990). In addition, aquifer heterogeneity may significantly hinder transport of contaminants and reactants to the microorganisms, severely reducing remediation rates.

Since a wide variety of subsurface contaminants are reduced organics, the most common amendment required to stimulate bioremediation is oxygen. Because of the low solubility of oxygen in water (44 mg/L at 20 °C and 1 atm O_2), maintaining a sufficient oxygen concentration in the bioactive zone by means of injection wells alone is often a challenge. Alternative methods for delivering oxygen to the bioactive zone include bioventing (withdrawing soil gas using extraction wells to pull air into the subsurface), biosparging (using injection wells to deliver oxygen directly into the saturated zone), hydrogen peroxide injection, and the use of down-well solid-phase oxidizers.

Microbial metabolism is also significantly affected by temperature: Microbial reaction rates tend to decrease with decreasing subsurface temperatures. Although temperatures within the top 10 meters of the subsurface may fluctuate seasonally, subsurface temperatures down to 100 meters typically remain within 1–2 °C of the mean annual surface temperature (Freeze and Cherry, 1979). This fact suggests that subsurface bioremediation should occur more quickly in warm-weather areas than in cold climates. Additional factors that may limit microbial activity include pH values outside the neutral range (pH < 6 or pH > 8), desiccating moisture conditions, and extreme redox potentials.

Indigenous microbial populations from many aquifers have been shown to degrade a wide range of organic contaminants, obviating the need to introduce an exogenous culture in many bioremediation applications, including those involving simple hydrocarbon contamination. However, in the absence of appropriate indigenous strains, the introduction of laboratory-enriched microbial populations or even genetically engineered microorganisms may be possible.

Many organic contaminants can be used as growth substrates for microbial metabolism, such as gasoline and oil as well as other petroleum derivatives. Bioremediation of a growth substrate can be highly efficient since it involves a built-in termination mechanism: When the contaminant is consumed, growth of the microbial population ceases. On the other hand, bioremediation of contaminants that do not provide growth or energy to microorganisms (cometabolism) requires an alternative growth substrate to stimulate microbial activity (§8.C.4). Cometabolic bioremediation represents a challenge since additional substrates must be provided to the microorganisms, and it is difficult to ensure that the contaminant, growth substrate, and appropriate electron donors/acceptors are all brought into contact with the microbial cells simultaneously. Methane, toluene, and propane are among the compounds that have been used to stimulate the cometabolic degradation of chlorinated solvents (e.g., trichloroethylene, dichloroethylene, vinyl chloride, trichloroethane, and chloroform) for in situ bioremediation applications.

An important factor that can limit the feasibility of in situ bioremediation is the availability of the contaminant for microbial attack. That is, contaminants that have extremely low solubilities (e.g., polynuclear aromatic hydrocarbons), sorb strongly to solids (e.g., PCBs), or are otherwise physically inaccessible, are protected from microbial degradation and will tend to persist in the environment. Methods to enhance the bioavailability of contaminants by employing surfactants or elevated temperature to increase solubilities are being studied.

Finally, proving that contaminant destruction by microbial degradation is occurring can be challenging due to the inaccessibility of the subsurface, aquifer heterogeneities, the complexity of differentiating biological and nonbiological processes, and the wide range of potential contaminant fates. Overlapping lines of evidence from a range of field monitoring techniques are required to measure the success of in situ bioremediation. Examples of monitoring observations that provide evidence for in situ bioremediation include contaminant disappearance in the bioactive zone, increased biological activity, generation of degradation intermediates, depletion of electron acceptors, and changes in stable isotope ratios of degradation products.

Remediation Goals and Closure Criteria

One of the most important decisions that must be made with respect to site remediation is the determination of how clean the site must be before remediation is considered complete. The general rule that the USEPA uses to guide the choice of appropriate closure criteria is stated in the National Contingency Plan as follows:

> *EPA expects to return usable ground waters to their beneficial uses wherever practicable, within a time frame that is reasonable given the particular circumstances of the site. When restoration of ground water to beneficial uses is not practicable, EPA expects to prevent further migration of the plume, prevent exposure to the contaminated groundwater, and evaluate further risk reduction (Federal Register, 1990, §300.430 (a)(1)(iii)(F)).*

In practice, regulators adopt various approaches to setting remediation goals, and these approaches are commonly negotiated on a site-by-site basis. Potential approaches include cleanup to

- Background or nondetectable levels
- Drinking water standards (MCLs)
- Technology-based levels (best available technology)
- Cost-based levels
- Risk-based levels

While the decision to require cleanup to background levels may seem attractive to regulators due to its simplicity and high likelihood of public acceptance, it is often impossible or impractical to achieve. The use of drinking water standards as de facto cleanup goals also serves to simplify regulatory implementation, but may be overprotective for sites that contain no potential drinking water sources. Technology-based levels are also easy to implement but are insensitive to site-by-site variation. Cost-based levels are complicated for regulators and have low public acceptance. The use of risk-based remediation targets has been growing significantly over the past several years. In fact, the USEPA recommends the use of risk-based analyses as practical, technically based approaches to site remediation and closure. These methods, which are based on the techniques outlined in §8.E.2, can be consistently applied to diverse hazardous waste releases and tend to have high public acceptance, but are more complicated to implement than most other types of cleanup goals. The use of risk analysis represents an important improvement over practices in the past, when decisions about hazardous waste cleanup methods and levels of contamination allowed to remain following site closure were made on a site-by-site basis, without specific or consistent guidelines.

REFERENCES

AIETA, E.M., REAGAN, K.M., LANG, J.S., MCREYNOLDS, L., KANG, J., & GLAZE, W.H. 1988. Advanced oxidation processes for treating groundwater contaminated with TCE and PCE: Pilot scale evaluations. *Journal of the American Water Works Association*, **80**, 64–72.

ALLEN, D.T., & ROSSELOT, K.S. 1997. *Pollution prevention for chemical processes*. John Wiley & Sons, New York.

ALVAREZ-COHEN, L., & MCCARTY, P.L. 1991. Product toxicity and cometabolic competitive inhibition modeling of chloroform and trichloroethylene transformation by methanotrophic resting cells. *Applied and Environmental Microbiology*, **57**, 1031–1037.

ARULANANTHAM, R., SALHOTRA, A.M., & STANLEY, C. 1995. *Risk and decision making at petroleum contaminated sites*. University of California Extension, Berkeley.

BERKOWITZ, J.B., FUNKHOUSER, J.T., & STEVENS, J.I. 1978. *Unit operations for the treatment of hazardous industrial wastes*. Noyes Data Corp., Park Ridge, NJ.

BERKOWITZ, J.B. 1989. Solvent extraction. In H.M. Freeman, ed., *Standard handbook of hazardous waste treatment and disposal*. McGraw-Hill, New York, §6.6.

BOEGEL, J.V. 1989. Air stripping and steam stripping. In H.M. Freeman, ed., *Standard handbook of hazardous waste treatment and disposal*. McGraw-Hill, New York, §6.8.

BRUNNER, C.R. 1991. *Handbook of incineration systems*. McGraw-Hill, New York.

CHADBOURNE, J.F. 1989. Cement kilns. In H.M. Freeman, ed., *Standard handbook of hazardous waste treatment and disposal*. McGraw-Hill, New York, §8.5.

CHUNG, N.K. 1989. Chemical precipitation. In H.M. Freeman, ed., *Standard handbook of hazardous waste treatment and disposal*. McGraw-Hill, New York, §7.2.

COMMITTEE ON IN SITU BIOREMEDIATION. 1993. *In situ bioremediation: When does it work?* National Academy Press, Washington, DC.

CORNWELL, D.A. 1990. Air stripping and aeration. In F.W. Pontius, ed., *Water quality and treatment: A handbook of community water supplies*. 4th ed. McGraw-Hill, New York, Chapter 5.

CRUMPLER, E., & MARTIN, E.J. 1987. Incineration of hazardous waste. In E.J. Martin and J.H. Johnson, eds., *Hazardous waste management engineering,* Van Nostrand Reinhold, New York, Chapter 4.

DAWSON, G.W., & MERCER, B.W. 1986. *Hazardous Waste Management*. John Wiley & Sons, New York, Chapters 9–10.

ENGLANDE, A.J. 1994. Status and direction of waste minimization in the chemical and petrochemical industries. *Water Science and Technology*, **29**, 25–36.

FARHATAZIZ, P.C., & ROSS, A.B. 1977. Selective specific rates of reactions of transients in water and aqueous solutions. *National Standard Reference Data Series 59*, U.S. National Bureau of Standards, P-III.

FOCHTMAN, E.G. 1989. Chemical oxidation and reduction. In H.M. Freeman, ed., *Standard handbook of hazardous waste treatment and disposal*. McGraw-Hill, New York, §7.4.

FREEZE, R.A., & CHERRY, J.A. 1979. *Groundwater*. Prentice-Hall, Englewood Cliffs, NJ.

GILL, J.H., & QUIEL, J.M. 1993. *Incineration of hazardous, toxic, and mixed wastes*. North American Mfg. Co., Cleveland.

GLAZE, W.H. 1990. Chemical oxidation. In F.W. Pontius, ed., *Water quality and treatment: A handbook of community water supplies*. 4th ed. McGraw-Hill, New York, Chapter 12.

GREGORY, R., & ZABEL, T.F. 1990. Sedimentation and flotation. In F.W. Pontius, ed., *Water quality and treatment: A handbook of community water supplies*. 4th ed. McGraw-Hill, New York, Chapter 7.

HAAS, C.N., & VAMOS, R.J. 1995. *Hazardous and industrial waste treatment*. Prentice Hall, Englewood Cliffs, NJ, Chapters 5–7.

HIGGINS, I.J., BEST, D.J., HAMMOND, R.C., & SCOTT, D. 1981. Methane oxidizing microorganisms. *Microbiological Reviews*, **45**, 556–590.

HOIGNE, J., & BADER, H. 1976. The role of hydroxyl radical reactions in ozonation processes in aqueous solutions. *Water Research*, **10**, 377–384.

HOIGNE, J., & BADER, H. 1983. Rate constants of reactions of ozone with organic and inorganic compounds in water. *Water Research*, **17**, 173–194.

HUISINGH, D. 1985. *Profits of pollution prevention: A compendium of North Carolina case studies*. North Carolina Board of Science and Technology, Raleigh.

HUNT, G.E., & SCHECTER, R.N. 1989. Minimization of hazardous-waste generation. In H.M. Freeman, ed., *Standard handbook of hazardous waste treatment and disposal*, McGraw-Hill, New York, §5.1.

IRVINE, R.L., & WILDERER, P.A. 1989. Aerobic processes. In H.M. Freeman, ed., *Standard handbook of hazardous waste treatment and disposal*. McGraw-Hill, New York, §9.1.

JOSHI, D.K., SENETAR, J.J., & KING, C.J. 1984. Solvent-extraction for removal of polar-organic pollutants from water. *Industrial & Engineering Chemistry Process Design and Development*, **23**, 748–754.

KADLECEK, M. 1988. Love Canal—10 years later. *Conservationist*, Nov.–Dec.

LAGREGA, M.D., BUCKINGHAM, P.L., & EVANS, J.C. 1994. *Hazardous waste management*. McGraw-Hill, New York, Chapters 1–2, 6–13.

MCCOY & ASSOCIATES. 1995. Waste minimization: What, why and how. *Hazardous Waste Consultant*, **13**, D1–D35.

METCALF & EDDY. 1991. *Wastewater engineering: Treatment, disposal, reuse*. 3rd ed. McGraw-Hill, New York.

MOYA, J., HOWARD-REED, C., & CORSI, R.L. 1999. Volatilization of chemicals from tap water to indoor air from contaminated water used for showering. *Environmental Science & Technology*, **33**, 2321–2327.

NATIONAL RESEARCH COUNCIL. 1994. *Alternatives for groundwater cleanup*. National Academy Press, Washington, DC.

NEMEROW, N.L., & DASGUPTA, A. 1991. *Industrial and hazardous waste treatment*. Van Nostrand Reinhold, New York, Chapters 7, 12.

NOVAK, F.C. 1989. Ozonation. In H.M. Freeman, ed., *Standard handbook of hazardous waste treatment and disposal*. McGraw-Hill, New York, §7.6.

O'BRIEN & GERE ENGINEERS, INC. 1995. *Hazardous waste site remediation: The engineer's perspective*. Van Nostrand Reinhold, New York.

ONDA, K., TAKEUCHI, H., & OKUMOTO, Y. 1968. Mass transfer coefficients between gas and liquid phases in packed columns. *Journal of Chemical Engineering of Japan*, **1**, 56–62.

OPPELT, E.T. 1987. Incineration of hazardous waste: A critical review. *JAPCA*, **37**, 558–586.

RASMUSSEN, G.F., BENEDICT, R.W., & YOUNG, C.M. 1989. Fluidized-bed thermal oxidation. In H.M. Freeman, ed., *Standard handbook of hazardous waste treatment and disposal*. McGraw-Hill, New York, §8.3.

RIGGS, C.O., & HATHAWAY, A.W. 1988. Ground-water monitoring field practice—An overview. In A.G. Collins and A.I. Johnson, eds., *Ground-water contamination: Field methods*. ASTM Special Technical Publication 963. American Society for Testing and Materials, Philadelphia.

ROBERTS, P.V., HOPKINS, G.D., MUNZ, C., & RIOJAS, A.H. 1985. Evaluating two-resistance models for air stripping of volatile organic contaminants in a countercurrent packed column. *Environmental Science & Technology*, **19**, 164–173.

ROBERTS, P.V., & LEVY, J.A. 1985. Energy requirements for air stripping trihalomethanes. *Journal of the American Water Works Association*, **77**, 138–146.

ROGERS, T.N., & BRANT, G. 1989. Distillation. In H.M. Freeman, ed., *Standard handbook of hazardous waste treatment and disposal*. McGraw-Hill, New York, §6.2.

SANTOLERI, J.J. 1989. Liquid-injection incinerators. In H.M. Freeman, ed., *Standard handbook of hazardous waste treatment and disposal*. McGraw-Hill, New York, §8.1.

SCHAEFER, C.F., & ALBERT, A.A. 1989. Rotary kilns. In H.M. Freeman, ed., *Standard handbook of hazardous waste treatment and disposal*, McGraw-Hill, New York, §8.2.

SHAH, J.K., SCHULTZ, T.J., & DAIGA, V.R. 1989. Pyrolysis processes. In H.M. Freeman, ed., *Standard handbook of hazardous waste treatment and disposal*. McGraw-Hill, New York, §8.7.

SHERWOOD, T.K., & HOLLOWAY, F.A.L. 1940 Performance of packed towers—Liquid film data for several packings. *Transactions of the American Institute of Chemical Engineers*, **36**, 39–70.

TAYLOR, S.W., MILLY, P.C.D., & JAFFE, P.R. 1990. Biofilm growth and the related changes in the physical properties of a porous medium: Permeability. *Water Resources Research*, **26**, 2161–2169.

THOMAS, J.M., & WARD, C.H. 1989. *In situ* biorestoration of organic contaminants in the subsurface. *Environmental Science & Technology*, **23**, 760–766.

TORPY, M.F. 1989. Anaerobic digestion. In H.M. Freeman, ed., *Standard handbook of hazardous waste treatment and disposal*. McGraw-Hill, New York, §9.2.

TREYBAL, R.D. 1980. *Mass-transfer operations*. McGraw-Hill, New York, Chapters 6, 10.

USEPA. 1983. *Development document for effluent limitations guidelines and standards for the metal finishing point source category*. EPA440/1-83/91. National Service Center for Environmental Publications, Cincinnati.

USEPA. 1985. *Report to Congress on injection of hazardous waste*. EPA570/9-85-003. Office of Drinking Water, United States Environmental Protection Agency, Washington, DC.

USEPA. 1987. *A compendium of technologies used in the treatment of hazardous wastes.* EPA/625/8-87/014. Center for Environmental Research Information, United States Environmental Protection Agency, Cincinnati.

USEPA. 1988. *Guidance for conducting remedial investigations and feasibility studies under CERCLA: Interim Final.* EPA/540/G-89/004. Office of Emergency and Remedial Response, United States Environmental Protection Agency, Washington, DC.

USEPA. 1989. *Risk assessment guidance for Superfund. Volume 1. Human health evaluation manual. (Part A)*. EPA/540/1-89/002. Office of Solid Waste and Emergency Response, United States Environmental Protection Agency, Washington, DC.

USEPA. 1991. *Risk assessment guidance for Superfund. Volume 1. Human health evaluation manual (Part B)*. OSWER-9285.7-01BFS. Office of Emergency and Remedial Response, United States Environmental Protection Agency, Washington, DC.

USEPA. 1993. *Remediation technologies screening matrix and reference guide.* Version I. EPA/542-B-93-005. Solid Waste and Emergency Response, United States Environmental Protection Agency, Washington, DC.

USEPA. 1997a. *1995 national biennial RCRA hazardous waste report*. EPA530-S-97-022. Office of Solid Waste, United States Environmental Protection Agency, Washington, DC.

USEPA. 1997b. *Characterization of municipal solid waste in the United States: 1996 update*. EPA530-R-97-015. Office of Solid Waste, United States Environmental Protection Agency, Washington, DC.

USEPA. 1997c. *Cleaning up the nation's waste sites: Markets and technology trends, 1996 edition*. EPA/542/R-96/005A, Office of Solid Waste and Emergency Response, United States Environmental Protection Agency, Washington, DC.

VOGEL, T.M., CRIDDLE, C.S., & MCCARTY, P.L. 1987. Transformation of halogenated aliphatic compounds. *Environmental Science & Technology*, **21**, 722–736.

WENTZ, C.A. 1995. *Hazardous waste management*. 2nd ed. McGraw-Hill, New York, Chapters 3, 5, 8, 12.

WIEDEMEIER, T.H., SWANSON, M.A., MONTOUX, D.E., GORDON, E.K., WILSON, J.T., WILSON, B.H., & KAMPBELL, D.H. 1996. *Technical protocol for evaluating natural attenuation of chlorinated solvents in groundwater*. Air Force Center for Environmental Excellence, Brooks Air Force Base, San Antonio, TX.

WILES, C.C. 1989. Solidification and stabilization technology. In H.M. Freeman, ed., *Standard handbook of hazardous waste treatment and disposal*. McGraw-Hill, New York, §7.8.

WPCF. 1988. *Incineration manual of practice OM-11*. Water Pollution Control Federation, Alexandria, VA.

WRIGHT, T.D., ROSS, D.E., & TAGAWA, L. 1989. Hazardous-waste landfill construction: The state of the art. In H.M. Freeman, ed., *Standard handbook of hazardous waste treatment and disposal*. McGraw-Hill, New York, §10.1.

ADDITIONAL SOURCES OF INFORMATION

AHRENS, R.A. 1989. The Times Beach trial. *Pollution Engineering*, **21**, 66–69.

ATCHESON, J. 1995. Voluntary pollution prevention programs. In H.M. Freeman, ed., *Industrial pollution prevention handbook*. McGraw-Hill, New York.

BALL, W.P., JONES, M.D., & KAVANAUGH, M.C. 1984. Mass transfer of volatile organic compounds in packed tower aeration. *Journal of the Water Pollution Control Federation*, **56**, 127–136.

BARCELONA, M., WEHRMANN, W., KEELY, J.F., & PETTYJOHN, W.A. 1990. *Contamination of groundwater: prevention, assessment, restoration*. Noyes Data Corporation, Park Ridge, NJ.

BENSON, R.C., TURNER, M., TURNER, P., & VOGELSONG, W. 1988. *In situ*, time-series measurements for long-term groundwater monitoring. In A.G. Collins and A.I. Johnson, eds., *Ground-water contamination: Field methods*. ASTM Special Technical Publication 963. American Society for Testing and Materials, Philadelphia, pp. 58–72.

BLACKMAN, W.C. 1996. *Basic hazardous waste management*. CRC Press, Boca Raton, FL, Chapter 7.

BOUWER, E.J. 1992. Bioremediation of organic contaminants in the subsurface. In R. Mitchell, ed., *Environmental microbiology*. Wiley-Liss, New York, Chapter 11.

BRITTON, G. 1994. *Wastewater microbiology*. Wiley-Liss, New York.

BROWN, M.H. 1981. *Laying waste*. Washington Square Press, New York.

CHAPELLE, F.H. 1993. *Ground-water microbiology and geochemistry*. John Wiley & Sons, New York.

COHEN, Y., & GIRALT, F. 1996. Strategies in pollution prevention: Waste minimization and source reduction. *Afinidad LIII*, **462**, 80–92.

DEEGAN, J. 1987. Looking back at Love Canal. *Environmental Science & Technology*, **21**, 421–426.

EDITOR. 1982. Cleaning up the Valley of the Drums. *Chemical Engineering*, **89**, 25–26.

GRASSO, D. 1993. *Hazardous waste site remediation: Source control*. CRC/Lewis Publishers, Boca Raton, FL.

GUJIKI, M., & TAJIMA, S. 1992. The pollution of Minamata Bay by mercury. *Water Science and Technology*, **25**, 133–140.

HALLAS, L.E., & HEITKAMP, M.E. 1995. Microbiological treatment of chemical process wastewater. In L.Y. Young and C.E. Cerniglia, eds., *Microbial transformation and degradation of toxic organic chemicals*, Wiley-Liss, New York, Chapter 10.

HIGGINS, T. 1989. *Hazardous waste minimization handbook*. Lewis Publishers, Chelsea, MI.

HIRSCHORN, J.S., & OLDENBURG, K.U. 1989. Waste reduction audit: Matching types to stages. In R.A. Conway et al., eds., *Hazardous and industrial solid waste minimization practices*. ASTM Special Technical Publication 1043. American Society for Testing and Materials, Philadelphia.

HUNT, G.E. 1995. Overview of waste reduction techniques leading to pollution prevention. In H.M. Freeman, ed., *Industrial pollution prevention handbook*. McGraw-Hill, New York.

KIRSCHNER, E. 1994. Love Canal settlement. *Chemical and Engineering News*, June 27.

KUDO, A., & MIYAHARA, S. 1991. Case history: Minamata mercury pollution in Japan. *Water Science and Technology*, **23**, 283–290.

MCKONE, T.E. 1986. Dioxin risk management at Times Beach, Missouri: An evaluation. *Environmental Professional*, **8**, 13–24.

MCVEIGH, T.P. 1989. Sampling and monitoring of remedial-action sites. In H.M. Freeman, ed., *Standard handbook of hazardous waste treatment and disposal*. McGraw-Hill, New York, §12.2.

NATIONAL RESEARCH COUNCIL. 1997. *Innovations in groundwater and soil cleanup*. National Academy Press, Washington, DC.

REINHARDT, J.R. 1989. Summary of Resource Conservation and Recovery Act legislation and regulation. In H.M. Freeman, ed., *Standard handbook of hazardous waste treatment and disposal*. McGraw-Hill, New York, §1.2.

RITTMANN, B.E. 1992. Innovations in biological processes for pollution control. In R. Mitchell, ed., *Environmental microbiology*. Wiley-Liss, New York, Chapter 10.

SPEECE, R.E. 1996. *Anaerobic biotechnology*. Archae Press, Nashville, TN.

STANFORD R., & YANG, E.C. 1989. Summary of CERCLA legislation and regulations and the EPA Superfund program. In H.M. Freeman, ed., *Standard handbook of hazardous waste treatment and disposal*. McGraw-Hill, New York, §1.3.

SUTHERSAN, S.S. 1997. *Remediation engineering: Design concepts*. Lewis Publishers, New York, Chapter 5.

VIGON, B. 1995. Life-cycle assessment. In H.M. Freeman, ed., *Industrial pollution prevention handbook*. McGraw-Hill, New York.

WARNER, D.L. 1989. Subsurface injection of liquid hazardous wastes. In H.M. Freeman, ed., *Standard handbook of hazardous waste treatment and disposal*. McGraw-Hill, New York, §10.3.

PROBLEMS

8.1 To be or not to be: Hazardous waste definitions

Determine whether each of the following materials would be defined as a hazardous waste under RCRA and explain why or why not.

(a) Ash from a coal-burning electrical generation plant

(b) Wastewater generated by Aunt Tilly flushing spot remover (carbon tetrachloride) down her toilet

(c) Spent methanol used for removing dirt and dust from manufactured parts

(d) Residual pesticide being disposed of by a farmer on his or her own property

(e) Discarded cartons of saccharin from a Weight Watchers food preparation plant

(f) Radioactive tracers that have been used in a university laboratory

(g) Selenium- and nitrate-laden drainage water from irrigated land

(h) Domestic sewage containing discarded waterproofing solution (1,1,1-trichloroethane)

(i) Gasoline stored in a tank underneath a local gas station

(j) Mothballs (made from naphthalene) discarded from a clothes manufacturing operation

(k) Vodka-contaminated wastewater from a liquor producer, which is discharged to the municipal sewage treatment plant

(l) Residues resulting from the incineration or thermal treatment of soil contaminated with EPA hazardous waste number F023

(m) Groundwater extracted from a Superfund site contaminated with trichloroethylene at a concentration of 750 μg/L

8.2 A potpourri of hazardous waste issues

(a) What is the purpose of classifying hazardous wastes by lists as well as by characteristics?

(b) Explain the difference between a RCRA solid waste and hazardous waste.

(c) Describe the "land ban" and explain why it was established.

(d) What percentage of hazardous waste is produced by the top five hazardous waste–generating states in the United States? Are the top five hazardous waste–generating states different from the ones you would have expected? Explain.

(e) What is the manifest system, where is it defined, and who are the players that must deal with it?

(f) What is the hierarchy of hazardous waste management priorities designated by the U.S. federal government? How are they promoted through legislation?
(g) Describe how a contaminated site becomes eligible for the Superfund.

8.3 TCLP and acid-base chemistry
The toxicity characteristic leaching procedure (TCLP) test for solids involves the incubation of solids with 20 times their mass of pH 5 acetic acid solution. After 18–24 hours of agitated incubation, the solids are separated from the solution, and the solution is adjusted to a pH of 2 and sampled for nonvolatiles. You would like to determine whether a soil sample containing 0.1 percent by mass lead hydroxide $Pb(OH)_2(s)$ particles would be considered a hazardous waste under the TCLP toxicity test. If the lead content of the TCLP solution exceeds 5 mg/L, the waste would be designated as hazardous under the toxicity characteristic. The following reactions and equilibrium constants apply:

Acetic acid	$CH_3COOH \Leftrightarrow CH_3COO^- + H^+$	$pK_A = 4.7$
Lead hydroxide	$Pb(OH)_2 \Leftrightarrow Pb^{2+} + 2\,OH^-$	$pK_{sp} = 14.9$

(a) What molar concentration of acetic acid must be added to the initially pure water to produce a pH 5 solution (prior to incubation with solids)?
(b) What is the pH of the acetic acid solution after incubation with the solids (before it is adjusted to pH 2)?
(c) What is the molar lead concentration of the solution in *(b)*?
(d) Will this liquid pass the TCLP test either before or after adjustment to pH 2? Explain.

8.4 Waste minimization decisions based on present worth analysis
Placebo Chemicals has asked you to implement waste minimization modifications into the redesign of a manufacturing process that presently costs $450,000 per year to operate. You identify two minimization options as alternatives to the present process. Option A can be implemented immediately with a capital cost of $1.4 million and an annual operating cost of $250,000. Option B would not be implemented for three years and would cost $1.5 million in capital costs (in the third year) plus $200,000 per year in operating costs.
(a) Compare the overall costs for the three options (no action, minimization option A, and minimization option B) over a 10-year period with an interest rate of 5 percent.
(b) Which option would be the most economically attractive to Placebo Chemicals if the calculation time period were substantially increased?

8.5 Saving some money in Silicon Valley
Grintel, an extremely happy computer chip production factory, wants to consider methods for minimizing its hazardous waste generation to reduce disposal costs. They hire a waste minimization consulting firm to conduct an audit, which results in the following three options:

Option A	No change: operating and maintenance (O&M) cost $350,000/y, disposal cost $100,000/y
Option B	Modify operations and maintenance to decrease hazardous waste generation over time by increasing labor costs: O&M cost $400,000/y, disposal cost $80,000/y for 3 years and $30,000/y after that
Option C	Purchase new equipment that consumes lesser quantities of chemicals and generates less hazardous waste: capital cost $1,000,000, O&M cost $300,000/y, disposal cost $10,000/y

(a) Using an operation time horizon of 15 years and an interest rate of 6 percent, evaluate the present worth of the three options and determine which is the most economically attractive.

(b) How high would the interest rate have to be to change the present worth ranking of the three options?

8.6 Dissolved air flotation used by the petroleum industry

Wastewater from the ChevMobiCo petroleum plant flows at 11,000 m^3/d and 15 °C, and contains 3000 mg/L suspended solids. Design a dissolved air flotation unit to treat this water.

(a) Compute the pressure required to achieve an optimum *A/S* ratio of 0.015 mg/mg given 75 percent saturator efficiency.

(b) You realize that the suspended solids are so fragile that the maximum pressure they can be exposed to is 3.0 atm. What pressure would be required to retain the same *A/S* ratio using 40 percent recycle? Is this ratio practical?

(c) Calculate the maximum bubble size possible to maintain conditions of viscosity-dominated drag within this unit ($Re < 1$). Is this a reasonable bubble size to use?

(d) Calculate the surface area of the required flotation tank to achieve 98 percent bubble removal with 80 μm diameter bubbles.

8.7 Performance of a dissolved air flotation unit

An ideal dissolved air flotation unit, operating at a flow of 10,000 $m^3\ d^{-1}$, is capable of producing 50 μm diameter air bubbles. If the system has a removal efficiency of 85 percent, and the unit is 10 m long, 10 m wide, and 3 m deep, determine the average density of the contaminant-laden bubbles.

8.8 Air stripping for water cleaning

Purity Springs, a town that depends on groundwater for its water supply, finds carbon tetrachloride (CT) in its groundwater wells. The town decides to use a countercurrent packed tower to remove CT from an average flow rate of 3000 $m^3\ d^{-1}$ of contaminated groundwater. The Onda correlation is applied to determine individual liquid- and gas-phase mass-transfer coefficients to be $2 \times 10^{-4}\ m\ s^{-1}$ and $1.5 \times 10^{-3}\ m\ s^{-1}$, respectively, for 50 mm Raschig rings with a wetted specific surface area of 120 m^2/m^3.

(a) Estimate the removal of CT within a 10 m countercurrent air stripping tower with a 2 m^2 cross-sectional area operating at a 10:1 gas-to-liquid volume ratio at 25 °C.

(b) Given that Purity Springs is actually in a cool enough climate that the extracted groundwater averages 15 °C during treatment, estimate the new tower height required for the same removal calculated in (*a*).

8.9 Fundamentals of steam stripping

Steam stripping is a mass-transfer process that can be used to treat certain hazardous waste streams.

(a) Describe the fundamental operation of a steam stripper (include a sketch). Specifically address the functions of different sections within the stripper, as well as additional required operational units.

(b) Would steam stripping be practical for removal of 10 percent ethanol from water? Why or why not?

(c) Give an additional example of a waste stream that would be appropriate for steam stripping and one that would be inappropriate, and explain why.

8.10 Wastes to riches

You are hired by Jewels-R-Us, a schlock jewelry company that cheaply plates all of its products in silver. To save money, they would like to use precipitation to remove dissolved silver from their waste stream. You need to decide whether a hydroxide, sulfide, or carbonate precipitation strategy would be most effective. The waste stream contains 350 mg/L of dissolved silver. The following chemical information applies.

Ag_2CO_3	$K_{sp1} = 8.1 \times 10^{-12}\ M^3$
Ag_2S	$K_{sp2} = 2 \times 10^{-49}\ M^3$
$AgOH$	$K_{sp3} = 2 \times 10^{-8}\ M^2$
$Ca(OH)_2$	$K_{sp4} = 5.5 \times 10^{-6}\ M^3$

(a) Calculate the minimum pH required to achieve the desired effluent silver concentration of 0.05 mg/L in the wastewater by hydroxide precipitation.

(b) Is it possible to achieve this pH by adding only lime if the industrial waste stream is initially at pH 7 and contains no appreciable acidity or alkalinity?

(c) Calculate the dissolved carbonate concentration required to achieve the target silver concentration of 0.05 mg/L in the wastewater by carbonate precipitation.

(d) How much soda ash (Na_2CO_3) is required to achieve the target silver concentration? (*Hint:* Assume all carbonate in solution remains present in the CO_3^{2-} state.) Does this seem practical?

(e) How much sodium sulfide (Na_2S) must be added to achieve the target silver concentration?

(f) Why would you probably want to consider a combination of lime and sodium sulfide to achieve the most efficient and safe removal?

8.11 Precipitation of cadmium waste

A battery manufacturing facility, Bunnies-Run-Forever, produces an aqueous waste stream that is high in cadmium. Cadmium removal from the waste stream can be carried out by hydroxide or sulfide precipitation.

(a) Given the following solubility constants, compute the theoretical effluent cadmium concentration for addition of 2 μM ferrous sulfide (FeS) or lime ($Ca(OH)_2$) to a reactor maintained at pH 9.

$Cd(OH)_2$	$K_{sp1} = 5.9 \times 10^{-15}\ M^3$
CdS	$K_{sp2} = 7.8 \times 10^{-27}\ M^2$

(b) Is there a pH value ($0 < pH < 14$) at which cadmium removal by hydroxide precipitation would be more effective than sulfide precipitation? Justify your answer.

8.12 Carbonate precipitation of chromium waste

Chromium is a heavy metal that can sometimes be found in water due to contamination from abandoned mining operations, metal plating facilities, or fossil fuel combustion. The maximum contaminant level for chromium is 0.1 mg/L. The simplest method for removing chromium from water is to precipitate it with hydroxide as described by the reverse of this reaction:

$$Cr(OH)_3(s) \Leftrightarrow Cr^{3+} + 3\,OH^- \qquad K_{sp} = 6.7 \times 10^{-31}\ M^4$$

Water that contains excess chromium is in contact with the atmosphere, containing 350 ppm CO_2, and with solid limestone ($CaCO_3$). The chromium contaminant was originally in the form of $CrCl_3$ at a concentration of 0.1 mM, which dissolved totally within the water.

(a) Calculate the pH needed to remove chromium to below the maximum contaminant level, and indicate whether it is the maximum or minimum pH required.

(b) Sketch the equilibrium chemical system with the solid/liquid/gas phases complete with each dissolved and nondissolved species and the governing equilibrium constants.

(c) Set up an electroneutrality equation for this water.

(d) Given that you observe some precipitation of $Cr(OH)_3(s)$, rewrite the electroneutrality equation in terms of known constants and $[H^+]$ only.

8.13 Methods of solidification/stabilization

Describe one advantage, one disadvantage, and one suitable waste for each of the following solidification/stabilization methods.

(a) Pozzolan

(b) Thermoplastic encapsulation

(c) Glassification

(d) Self-cementing

(e) Organic polymer

(f) Cement-based

8.14 There's gold in those hills!

Mercury was widely used for separating gold from ore in the California foothills. Waste mercury was dumped for many years into Fools-Rush pond in the gold country, near Grass Valley. The anaerobic sediments at the bottom of this lake contain high concentrations of solid mercury sulfide (HgS) in equilibrium with the water of the pond.

$$HgS \Leftrightarrow Hg^{2+} + S^{2-} \qquad K_{sp} = 4 \times 10^{-23}\ M^2$$

The local fauna absorb dissolved mercury, so that the aqueous concentration is maintained at a steady-state value of 10^{-5} M. The sulfide (S^{2-}) reacts very rapidly with protons in the water at the bottom of the pond to produce hydrogen sulfide in the simplified reaction shown below:

$$S^{2-} + 2\,H^+ \Leftrightarrow H_2S \qquad K_1 = 10^{20}\ M^{-2}$$

The H_2S produced by the pond sediments diffuses through the water to the surface of the pond, where it is given off as H_2S gas. The movement of H_2S from the bottom of the pond to the top is driven only by diffusion with a diffusion coefficient of $1.2 \times 10^{-9}\ m^2\ s^{-1}$. There is a steady breeze blowing above the pond surface, so that the concentration of H_2S gas in the air above the pond is approximately zero. The surface area of Fools-Rush pond is 10 m^2. This pond can be modeled vertically as a one-dimensional steady-state system with the sediments at $z = 0$ m, the pond surface at $z = 2$ m, and no water flow.

(a) Set up a material balance on H_2S within the water column of the lake (use C for concentration of H_2S), and derive a differential equation to describe the change in concentration of H_2S with depth (z). (You may leave this answer in terms of constants and variables.)

(b) Give the boundary conditions for H_2S within the water column (i.e., the concentrations of H_2S at $z = 0$ and at $z = 2$ m) in terms of the proton concentration $[H^+]$ within the water.

(c) Derive an expression to describe the concentration of H_2S as a function of depth ($C(z)$) in terms of $[H^+]$.

(d) Given that the rate of H_2S emission into the air is 20 mg d^{-1}, calculate the pH of the water and the concentration of H_2S at 0.25 m below the water surface.

8.15 The thermodynamics of chromium reduction

The aqueous waste from Deep-Hole Mining Co. contains 432 mg/L chromium in the form of the toxic +6 oxidation state: $Cr_2O_7^{2-}$. You are considering using sulfite, SO_3^{2-}, as a reductant to convert the Deep-Hole Mining chromium to the less harmful +3 state.

(a) Construct a balanced overall reaction from the two half-reactions for this oxidation-reduction process.

(b) Calculate the standard free energy for this redox reaction and determine whether the reaction would occur spontaneously under standard conditions. (*Hint:* See Appendix E.)

(c) Your treatment goal is to reduce 99 percent of the chromium to the +3 state within a batch process. Evaluate whether it is thermodynamically possible to achieve this reduction by adding double the stoichiometric SO_3^{2-} required to carry out the reaction in a solution buffered at pH 4.

(d) Briefly describe how would you deal with the treated effluent containing chromium +3.

8.16 Oxidation of cyanide wastewater from an electroplating company

You are asked to design an industrial waste process for Highly Shiny Electroplating, Inc., a company that produces a wastewater stream containing moderately high cyanide concentrations (500 mg/L). You decide that the best approach for treatment of this waste would be to oxidize the cyanide.

(a) You choose sodium hypochlorite (NaOCl) for the cyanide oxidation. Given that the full reaction for sodium cyanide (NaCN) oxidation by sodium hypochlorite is as follows, set up the two half-reactions that would lead to this full reaction.

$$2\,NaCN + 5\,NaOCl + H_2O \longrightarrow N_2 + 2\,NaHCO_3 + 5\,NaCl$$

(b) Calculate the dose of sodium hypochlorite (mg/L) that is necessary to stoichiometrically oxidize the cyanide to the desired treatment level of 10 mg/L.

(c) The standard electrode potentials (pH 0, 25 °C, etc.) for the cyanide and hypochlorite redox pairs are as follows: $HCO_3^-/CN^- = -0.31$ V, and $OCl^-/Cl^- = 1.49$ V. Calculate the free energy change (ΔG) for the cyanide oxidation reaction and determine whether this reaction would proceed spontaneously to the desired endpoint with the use of 30 percent excess hypochlorite in a solution buffered at pH 10. (*Hint:* See Appendix E.)

8.17 Incineration of a paint-thinning solvent waste

The Burn-Baby-Burn Company has been hired to design a hazardous waste incinerator to treat paint-thinning waste containing a mixture of 54 percent pentane (C_5H_{12}) and 46 percent toluene (C_7H_8) by weight. This waste is contaminated with significant amounts of dissolved metals from paint pigments.

(a) Calculate the maximum flow of this waste (g/min) that can be stoichiometrically burned given that you can deliver 420 mol/min of air into the incinerator.

(b) The particulate matter concentration produced from burning this waste with 1.5 times the stoichiometric oxygen is measured at 240 mg/m³ in the dry gas at 20 °C. The regulations require the oxygen-corrected emissions in dry flue gas at 20 °C to be

$\leq$180 mg m^{-3} before gases are released to the atmosphere. Calculate whether the above operation would exceed the particulate emission level, assuming no particle control units are used.

(c) Draw and label the treatment train that you would recommend for controlling effluent emissions to the atmosphere from this incinerator. Include a brief description of the purpose for each unit.

8.18 Comparing incineration and pyrolysis

(a) Explain the difference between pyrolysis and incineration treatment processes.

(b) Write a balanced equation for a reaction that would occur during the (i) incineration and (ii) pyrolysis of wastewater that is 25 percent hexane in water (by weight).

8.19 Treatment of PCE-contaminated soils by incineration

You are the operator of a hazardous waste incinerator and have just taken delivery of soil that is contaminated with perchloroethylene (PCE, C_2Cl_4). After running an analysis on the contaminated soil, you determine that it contains 20 g PCE per kg soil. Since the energy available from burning the PCE alone is not sufficient to maintain exothermic conditions within the incinerator, you must add a supplementary fuel. Fortuitously, the Lucky-Six corporation has just dropped off a shipment of hexanol ($C_6H_{13}OH$) that was used to clean machined metal pieces and therefore contains several trace metals. A back-of-the-envelope analysis suggests that 50 g hexanol per kg soil would be sufficient to ensure exothermic conditions.

(a) What configuration of incinerator would you choose to treat this contaminated soil? Explain.

(b) Calculate the exact amount of air required for burning this waste (give the answer in m^3 of air per kg soil at 25 °C, 1 atm).

(c) The default flow delivered by your air blowing system is 0.73 m^3 air per kg soil. At this flow, the particulate matter produced from waste incineration is measured at 230 mg/m^3 in the dry gas. The regulations require the oxygen-corrected emissions in dry flue gas to be $\leq$180 mg/m^3 before gases are released to the atmosphere. Would the default flow of air be sufficient to incinerate the given wastes without exceeding the particulate emission standard?

(d) Draw and label the treatment train that you would recommend for treating the gaseous effluent from this incinerator. Include a brief description of the function and mechanism of operation for each unit.

8.20 Biological treatment of wood preservative waste

TreesToPoles Industries is a company that produces telephone poles and generates significant amounts of hazardous waste containing pentachlorophenol (PCP, an agent used on the poles as a wood preservative). This company sends the majority of the spent pure-phase PCP to be incinerated in Utah but is left with wastewater containing a dilute level (15 mg/L) of the toxic material. As the resident environmental engineer, you have decided to treat the PCP-contaminated wastewater by aerobic biodegradation in a two-stage reactor: a CMFR followed by a PFR. The first stage and second stage have detention times of 6 h and 2 h, respectively. The first stage is aerated to promote oxygen transfer into solution for degradation of the PCP. An unfortunate side effect of the aeration is stripping of the PCP from the CMFR, which occurs at a rate of r_{st}, where

$$r_{st} = -k_{st}(C - C_g^*)$$

Here C and C_g^* represent the PCP concentrations in the liquid and at equilibrium with the gas phase, respectively. The minus sign indicates that the net transfer is from water to air (a loss mechanism for water) when $C > C_g^*$. The biochemical degradation rate of PCP in both reactors can be expressed by

$$r_s = -k_m \frac{XC}{K_s}$$

Since the microbial population can use PCP as a primary substrate, the microbial growth rate can be expressed as

$$r_x = -r_s Y - k_d X$$

Assume that the air movement above the CMFR is fast enough to set $C_g^* = 0$. Also assume that, although there are no microbial cells in the influent flow, cell growth has reached steady state within the reactors (think about why this is possible). The following parameters apply:

$$k_{st} = 10\ \text{d}^{-1} \qquad k_m = 2\ \text{d}^{-1} \qquad K_s = 0.25\ \text{mg/L}$$

$$Y = 0.8\ \text{g/g} \qquad k_d = 0.1\ \text{d}^{-1}$$

(a) Calculate the PCP concentration in the CMFR effluent liquid.

(b) Calculate the cell concentration within the CMFR.

(c) Assume that the cell concentration remains constant in the PFR (i.e., PCP degradation occurs with negligible cell growth or decay). Calculate the PFR effluent PCP concentration (remember, there is no stripping in the second stage).

(d) What are two major advantages of placing the CMFR before the PFR?

(e) Would the waste sludge from this process (settled microbial cells) be considered a hazardous waste?

(f) What treatment/disposal options would you consider for the sludge?

8.21 Microbiology and chemistry: A winning combination

Brazenly Byzantine Industries (BBI) produces an aqueous waste stream from a particularly noxious industrial process. The waste stream contains organics and metals. BBI seeks to pretreat this waste on-site, enabling them to route the effluent to the local, municipal wastewater treatment facility. Careful inspection by the local authorities combined with back-of-the-envelope risk calculations lead them to decide that since the influent benzene (C_6H_6) and Fe(II) concentrations are 9.4 ppm and 25 ppm, respectively, BBI is required to treat its effluent to achieve levels of ≤ 20 ppb benzene and ≤ 2 ppm Fe(II) prior to disposal to the sewer. BBI wishes to carry out the entire treatment process in one reactor-and-clarifier combination. Since benzene is rapidly biodegradable under aerobic conditions, and Fe(II) can be precipitated by oxidation to $Fe(OH)_3$ with molecular oxygen, it was proposed that an aerated CMFR may be adequate for the desired treatment. The precipitation occurs as follows:

$$4\ Fe^{2+} + O_2 + 10\ H_2O \longrightarrow 4\ Fe(OH)_3(s) + 8\ H^+$$

At neutral pH, the rate of iron precipitation is a function of oxygen and Fe^{+2} concentrations (C_{O_2}, C_{Fe}), and can be estimated by

$$r_{Fe} = -k_{Fe} C_{O_2} C_{Fe}$$

where $k_{Fe} = 2.9\ \text{L mg}^{-1}\ \text{d}^{-1}$. The rate of oxygen transfer into the reactor solution is estimated by

$$r_{O_2} = k_{O_2}(C_{sat} - C_{O_2})$$

where $k_{O_2} = 27\ d^{-1}$ and $C_{sat} = 8$ mg/L. The rate of benzene degradation is given by

$$r_b = -\frac{k_m C_b X}{K_s + C_b}$$

where $k_m = 6.3\ d^{-1}$, $K_s = 0.3$ mg/L, and X represents the cell concentration (mg/L). The rate of cell growth on benzene is expressed by

$$r_X = \left(Y\frac{k_m C_b}{K_s + C_b} - k_d\right)X$$

where $Y = 0.8$ g/g and $k_d = 0.1\ d^{-1}$.

(a) Write a stoichiometrically balanced reaction for the oxidation of benzene to CO_2, and compute the stoichiometric ratio of oxygen to benzene consumption in molar and mass ratios.

(b) Compute the necessary detention time within the steady-state reactor to achieve the benzene removal goal and the corresponding concentration of cells within the reactor.

(c) Compute the minimum necessary concentration of oxygen within the reactor that would result in achieving the iron removal goals.

(d) Compute the steady-state oxygen concentration in the reactor given that the waste has an initial oxygen concentration of 3 mg/L and the reactor is operating as described above with both treatment goals met.

(e) Does this reactor design seem feasible? What specific design changes could you make to improve the reactor performance?

8.22 Quantifying the risk associated with kidnapping

In many of the great old movies from the 1940s and 1950s, the bad guys would kidnap some unsuspecting character by applying chloroform to a handkerchief and holding it over the victim's face (this still occurs quite commonly in soap operas, or so we are told . . .). As the unsuspecting character breathes in the chloroform, he or she is rendered unconscious and can then be easily carted away. Taking into consideration the vapor pressure of chloroform, this one-time inhalation exposure would cause a carcinogenic risk of 8×10^{-7}. The chloroform cancer slope factor is 8.1×10^{-2} (mg kg^{-1} d^{-1})$^{-1}$ and the reference dose factor is 0.01 mg kg^{-1} d^{-1}.

(a) Compute the relevant chronic daily intake and hazard quotient for determining the level of chloroform toxicity.

(b) The unsuspecting character is kidnapped in this fashion on three different occasions (over the course of a very exciting 2 years). Determine the applicable kidnapping cancer risk and hazard quotient, and explain whether the EPA would be concerned about these exposures.

(c) Are there other routes of exposure that one should consider for the chloroform during these kidnapping events? Explain.

8.23 Site remediation for mixed groundwater contamination

PaintsForPeople, Inc. in Berkeley, California, has been using toluene (C_7H_8) as a paint thinner and general solvent for many years. Unfortunately, due to careless storage and disposal methods, the groundwater below PaintsForPeople is contaminated with high concentrations of lead (10 mg/L), which was once used in pigments, and toluene (150 mg/L). Unimaginative Engineers, Inc. has been called in to remediate the site and decides to pump 500 m^3/d of groundwater from below the site and treat the extracted water with a two-stage unit, precipitation followed by biological treatment. To dispose of this water to the municipal wastewater treatment facility, the local water board has mandated that the proposed water be treated to 5 μg/L lead and 2 mg/L toluene.

Precipitation

Since the required effluent level for lead is quite low, sulfide precipitation is a reasonable option. Sulfide can be added in the form of sodium sulfide (Na_2S), which dissolves totally in water. Since sulfide is a weak base, it will combine with protons as follows, resulting in the undesirable production of hydrogen sulfide:

$$S^{2-} + H^+ \Leftrightarrow HS^- \quad K_1 = 10^{12.9}\ M^{-1}$$
$$HS^- + H^+ \Leftrightarrow H_2S \quad K_2 = 10^{7.1}\ M^{-1}$$

(a) Given that the solubility constant for lead sulfide (PbS) is $K_{sp} = 8 \times 10^{-28}\ M^2$, calculate the stoichiometric amount (mol/L) of Na_2S that would be needed to carry out the required removal at pH 7 and the amount (mol/L) of H_2S gas that would be produced.

(b) Evaluate whether the pH of the groundwater would be significantly affected by this reaction.

Biological Treatment

A separate biotreatment process is to be used for toluene removal: a 50 m^3 CMFR with oxygen addition.

(c) Calculate the stoichiometric amount of oxygen (moles O_2 per moles toluene) that you would need to oxidize the toluene completely to CO_2.

(d) Toluene is readily used as a primary substrate for microbial growth. Consequently, the rate of mass transfer of oxygen into the bioreactor limits the rate of toluene degradation. Oxygen transfer can be represented as follows:

$$r_{O_2} = -k_{O_2}(C_{O_2,\text{sat}})$$

where r_{O_2} = oxygen uptake rate within reactor (mg L^{-1} d^{-1}), k_{O_2} = oxygen mass-transfer coefficient within the reactor [d^{-1}], and $C_{O_2,\text{sat}}$ = saturation concentration of oxygen in water = 8 mg/L. Furthermore, since toluene is volatile, it will be stripped out of the reactor at a rate proportional to the air mass-transfer rate, given by

$$r_{\text{st}} = -k_{\text{air}}\alpha C_t$$

where α is a dimensionless constant equal to 2, k_{air} is the mass transfer constant of air into the reactor, r_{st} = toluene stripping rate (mg L^{-1} d^{-1}), and C_t = toluene concentration within the reactor (mg/L). Derive an expression for the fraction of the influent toluene that will be stripped out of the reactor when it is operating at the specified removal efficiencies.

(e) For the actual design, in which order would you place these two units? Explain.

8.24 In situ bioremediation of hydrocarbons and trichloroethylene

A groundwater contamination plume contains petroleum hydrocarbons mixed with TCE. The growth rate of microbial populations can be described by

$$\frac{dX}{dt} = -Y\frac{dS}{dt} - Xk_d$$

The degradation rate of petroleum hydrocarbons can be described as

$$-\frac{dS}{dt} = \frac{k_m XS}{K_s + S}$$

(a) Calculate the maximum possible growth rate of a subsurface microbial population that is maintained on petroleum hydrocarbons, if the population's measured decay rate (k_d) is 0.1 d^{-1} and the Monod parameters are k_m = 0.8 mg mg^{-1} d^{-1}, K_s = 1.5 mg/L, and Y = 0.6 g/g.

(b) Calculate the minimum amount of petroleum hydrocarbons that this microbial population requires to sustain itself at steady state.

(c) This population can also cometabolically degrade TCE with measured parameters of k_m = 0.25 mg $mg^{-1}d^{-1}$ and K_s = 2.6 mg/L. Calculate the rate of TCE degradation per cell mass (mg TCE [mg cells]$^{-1}$ d^{-1}) and cell growth (mg cells [mg TCE]$^{-1}$ d^{-1}) when the cells are exposed to 10 mg/L TCE in the absence of the petroleum hydrocarbons.

8.25 Superfund closure criteria

List three different types of closure criteria and describe how final closure levels are ultimately determined for a Superfund site.

A

Basic Data for Environmental Engineering Science

Table A.1 Symbols, Atomic Numbers, and Atomic Masses of the Elements

Element	Symbol	Atomic number	Atomic mass (g/mol)[a]
Actinium	Ac	89	227.03
Aluminum	Al	13	26.98
Americium	Am	95	(243)
Antimony	Sb	51	121.76
Argon	Ar	18	39.95
Arsenic	As	33	74.92
Astatine	At	85	(210)
Barium	Ba	56	137.33
Berkelium	Bk	97	(247)
Beryllium	Be	4	9.01
Bismuth	Bi	83	208.98
Bohrium	Bh	107	(264)
Boron	B	5	10.81
Bromine	Br	35	79.90
Cadmium	Cd	48	112.41
Calcium	Ca	20	40.08
Californium	Cf	98	(251)
Carbon	C	6	12.01
Cerium	Ce	58	140.12
Cesium	Cs	55	132.91
Chlorine	Cl	17	35.45
Chromium	Cr	24	52.00
Cobalt	Co	27	58.93
Copper	Cu	29	63.55
Curium	Cm	96	(247)
Dubnium	Db	105	(262)
Dysprosium	Dy	66	162.50
Einsteinium	Es	99	(252)
Erbium	Er	68	167.26
Europium	Eu	63	151.96
Fermium	Fm	100	(257)
Fluorine	F	9	19.00
Francium	Fr	87	(223)

Table A.1 Symbols, Atomic Numbers, and Atomic Masses of the Elements (continued)

Element	Symbol	Atomic number	Atomic mass (g/mol)[a]
Gadolinium	Gd	64	157.25
Gallium	Ga	31	69.72
Germanium	Ge	32	72.61
Gold	Au	79	196.97
Hafnium	Hf	72	178.49
Hassium	Hs	108	(265)
Helium	He	2	4.00
Holmium	Ho	67	164.93
Hydrogen	H	1	1.008
Indium	In	49	114.82
Iodine	I	53	126.90
Iridium	Ir	77	192.22
Iron	Fe	26	55.85
Krypton	Kr	36	83.80
Lanthanum	La	57	138.91
Lawrencium	Lr	103	(262)
Lead	Pb	82	207.2
Lithium	Li	3	6.94
Lutetium	Lu	71	174.97
Magnesium	Mg	12	24.31
Manganese	Mn	25	54.93
Meitnerium	Mt	109	(268)
Mendelevium	Md	101	(258)
Mercury	Hg	80	200.59
Molybdenum	Mo	42	95.94
Neodymium	Nd	60	144.24
Neon	Ne	10	20.18
Neptunium	Np	93	237.05
Nickel	Ni	28	58.69
Niobium	Nb	41	92.91
Nitrogen	N	7	14.01
Nobelium	No	102	(259)
Osmium	Os	76	190.23
Oxygen	O	8	16.00
Palladium	Pd	46	106.42
Phosphorus	P	15	30.97
Platinum	Pt	78	195.08
Plutonium	Pu	94	(244)
Polonium	Po	84	(209)
Potassium	K	19	39.10
Praseodymium	Pr	59	140.91
Promethium	Pm	61	(145)
Protactinium	Pa	91	231.04
Radium	Ra	88	(226)
Radon	Rn	86	(222)
Rhenium	Re	75	186.21
Rhodium	Rh	45	102.91

Table A.1 Symbols, Atomic Numbers, and Atomic Masses of the Elements (continued)

Element	Symbol	Atomic number	Atomic mass (g/mol)[a]
Rubidium	Rb	37	85.47
Ruthenium	Ru	44	101.07
Rutherfordium	Rf	104	(261)
Samarium	Sm	62	150.36
Scandium	Sc	21	44.96
Seaborgium	Sg	106	(263)
Selenium	Se	34	78.96
Silicon	Si	14	28.09
Silver	Ag	47	107.87
Sodium	Na	11	22.99
Strontium	Sr	38	87.62
Sulfur	S	16	32.07
Tantalum	Ta	73	180.95
Technetium	Tc	43	(99)
Tellurium	Te	52	127.60
Terbium	Tb	65	158.93
Thallium	Tl	81	204.38
Thorium	Th	90	232.04
Thulium	Tm	69	168.93
Tin	Sn	50	118.71
Titanium	Ti	22	47.87
Tungsten	W	74	183.84
Ununbium	Uub	112	(277)
Ununhexium	Uuh	116	(289)
Ununnilium	Uun	110	(269)
Ununoctium	Uuo	118	(293)
Ununquadium	Uuq	114	(289)
Unununium	Uuu	111	(272)
Uranium	U	92	238.03
Vanadium	V	23	50.94
Xenon	Xe	54	131.29
Ytterbium	Yb	70	173.04
Yttrium	Y	39	88.91
Zinc	Zn	30	65.39
Zirconium	Zr	40	91.22

[a]Rounded to two decimal places. For radioactive elements, the mass number of an important isotope is given in parentheses.

Sources: Lide, 1991; http://www.webelements.com/.

Table A.2.a Common Unit Conversions and Derived SI Units[a]

Length (L)
1 Å = 10^{-10} m
1 in = 2.54 cm
1 ft = 0.3048 m
1 mi = 1.609 km

Area (L^2)
1 acre = 0.405 hectare
1 mi^2 = 640 acres
1 hectare = 10^4 m^2 = 10^{-2} km^2

Volume (L^3)
1 L = 10^{-3} m^3 = 1000 cm^3
1 acre-foot = 1233.5 m^3
1 ft^3 = 0.0283 m^3 = 28.3 L
1 gal = 0.003785 m^3 = 3.785 L

Concentration (L^{-3})
1 M = 1 mol L^{-1}
1 N = 1 eq L^{-1}

Mole fraction (or volume fraction) (for gases) (—); mass fraction (for liquids and solids) (—)
1% = 10^{-2}
1‰ = 10^{-3}
1 ppm = 10^{-6}
1 ppb = 10^{-9}
1 ppt = 10^{-12}

Pressure ($M L^{-1} T^{-2}$)
1 atm = 101.3 kPa
1 bar = 100 kPa
1 psi (lb ft^{-2}) = 6.89 kPa
1 mmHg = 133.3 Pa
1 torr = 1 mmHg
1 in H_2O = 249 Pa
1 Pa = 1 N m^{-2} = 1 kg m^{-1} s^{-2}

Radioactive decay (T^{-1})
1 Bq = 1 decay per second
1 Ci = 36.7 × 10^9 Bq

Temperature
To convert K to °C, subtract 273.15
To convert °F to °C, subtract 32 and multiply by 5/9

Time (T)
1 min = 60 s
1 h = 3600 s
1 d = 8.64 × 10^4 s
1 y = 3.15 × 10^7 s

Velocity (L/T)
1 cm s^{-1} = 0.6 m min^{-1}
1 fps (ft s^{-1}) = 18.3 m min^{-1}
1 fpm (ft min^{-1}) = 18.3 m h^{-1}
1 mph (mi h^{-1}) = 0.447 m s^{-1}

Volumetric flow rate (L^3/T)
1 cfm (ft^3 min^{-1}) = 472 cm^3 s^{-1}
1 cfm (ft^3 min^{-1}) = 1.70 m^3 h^{-1}
1 MGD = 10^6 gal d^{-1} = 0.0438 m^3 s^{-1}

Permeability (L^2)
1 darcy = 0.987 × 10^{-12} m^2

Mass (M)
1 lb_m = 454 g
1 ton = 907 kg (= 1 short ton)
1 tonne = 1000 kg (= 1 metric ton)

Force ($M L T^{-2}$)
1 lb_f = 4.45 N
1 N = 1 kg m s^{-2}
1 dyne = 10^{-5} N = 1 g cm s^{-2}

Energy ($M L^2 T^{-2}$)
1 cal = 4.186 J
1 BTU = 1.055 kJ
1 kWh = 3.6 MJ
1 J = 1 N m = 1 kg m^2 s^{-2}
1 erg = 10^{-7} J = 1 g cm^2 s^{-2}
1 MeV = 1.6 × 10^{-13} J

Power ($M L^2 T^{-3}$)
1 hp = 746 W
1 W = 1 J s^{-1} = 1 N m s^{-1} = 1 kg m^2 s^{-3}

Electrical conversion
1 V = 1 J C^{-1}
1 A = 1 C s^{-1}

[a]Notation for dimensions: L = length, M = mass, T = time.

Table A.2.b Unit Abbreviations[a]

Abbreviation	Full name	Measures
A	ampere	Current ($X\ T^{-1}$)
Å	angstrom	Length (L)
atm	atmosphere	Pressure ($M\ L^{-1}\ T^{-2}$)
Bq	becquerel	Radioactive decay rate (T^{-1})
BTU	British thermal unit	Energy ($M\ L^2\ T^{-2}$)
cal	calorie	Energy ($M\ L^2\ T^{-2}$)
C	coulomb	Electrical charge (X)
Ci	curie	Radioactive decay rate (T^{-1})
d	day	Time (T)
erg	erg	Energy ($M\ L^2\ T^{-2}$)
eq	equivalent	Moles of charge on ions
eV	electron volt	Energy ($M\ L^2\ T^{-2}$)
ft	foot	Length (L)
g	gram	Mass (M)
gal	gallon (U.S.)	Volume (L^3)
h	hour	Time (T)
hp	horsepower	Power ($M\ L^2\ T^{-3}$)
in	inch	Length (L)
J	joule	Energy ($M\ L^2\ T^{-2}$)
K	kelvin	Temperature (K)
kWh	kilowatt-hour	Energy ($M\ L^2\ T^{-2}$)
L	liter	Volume (L^3)
lb_f	pound force	Force ($M\ L\ T^{-2}$)
lb_m	pound mass	Mass (M)
M	moles per liter	Concentration (L^{-3})
m	meter	Length (L)
MGD	million gallons per day	Volumetric flow rate ($L^3\ T^{-1}$)
mi	mile	Length (L)
min	minute	Time (T)
mmHg	millimeters of mercury	Pressure ($M\ L^{-1}\ T^{-2}$)
mol	mole	Count (—)
N	newton	Force ($M\ L\ T^{-2}$)
NTU	nephelometric turbidity unit	Turbidity of water
Pa	pascal	Pressure ($M\ L^{-1}\ T^{-2}$)
psi	pounds (force) per square inch	Pressure ($M\ L^{-1}\ T^{-2}$)
s	second	Time (T)
TCU	true color unit	Color of water
V	volt	Electrical potential ($M\ L^2\ T^{-2}\ X^{-1}$)
W	watt	Power ($M\ L^2\ T^{-3}$)
°C	degrees centigrade	Temperature (K)
°F	degrees Fahrenheit	Temperature (K)

[a]*Notes:*

1. See Table A.2.c. for unit prefixes.
2. By convention, SI units named after individuals are capitalized when abbreviated but written in lower case when spelled out.
3. Derived units are related to primary dimensions with the notation L = length, M = mass, T = time, K = degrees of temperature, X = charge.

Table A.2.c Abbreviations for Decimal Prefixes

Symbol	Prefix	Multiplicative factor
E	exa	10^{18}
P	peta	10^{15}
T	tera	10^{12}
G	giga	10^{9}
M	mega	10^{6}
k	kilo	10^{3}
c	centi[a]	10^{-2}
m	milli	10^{-3}
μ	micro	10^{-6}
n	nano	10^{-9}
p	pico	10^{-12}
f	femto	10^{-15}
a	atto	10^{-18}

[a]SI guidelines discourage the use of prefixes that do not correspond to even multiples of 3 in the exponent. However, in environmental systems, when dealing with units that include area or volume, the use of cm^2 or cm^3 often proves far more intuitive than the alternatives permitted by strict SI convention.

Table A.3 Physical, Chemical, and Mathematical Constants

$c = 3.0 \times 10^{8}$ m s^{-1}	Speed of light (in vacuum)
$e = 2.7183$	Base of natural logarithm
$F = 96{,}490$ C mol^{-1}	Faraday constant
$k = 1.38 \times 10^{-23}$ J K^{-1}	Boltzmann's constant
$N_{av} = 6.02 \times 10^{23}$ mol^{-1}	Avogadro's number
$R = 8.314$ J K^{-1} mol^{-1}	Universal gas constant
$= 82.05 \times 10^{-6}$ mol^{-1} atm K^{-1} m^3	
$= 1.987$ cal mol^{-1} K^{-1}	
$= kN_{av}$	
$X_e = 1.603 \times 10^{-19}$ C	Elementary charge (charge on an electron)
$\pi = 3.1416$	Pi (circumference: diameter ratio for circle)

Table A.4 Properties of the Earth

$A_e = 5.1 \times 10^{14}$ m^2	Surface area
$g = 9.8$ m s^{-2}	Acceleration of gravity
$M_e = 6.0 \times 10^{24}$ kg	Mass of the earth
$M_{hydro} = 1.4 \times 10^{21}$ kg	Mass of the hydrosphere
$M_{atm} = 5.3 \times 10^{18}$ kg	Mass of the atmosphere
$R_e = 6370$ km	Radius of the earth

REFERENCE

LIDE, D.R. (editor-in-chief). 1991. *CRC handbook of chemistry and physics.* 72nd ed. CRC Press, Boca Raton, FL, pp. 1–11.

B

Primer on Ionizing Radiation

Radioisotopes and Types of Radioactive Emissions

When *ionizing radiation* passes through matter, it breaks electron bonds in atoms. Ionizing radiation is emitted when a radionuclide undergoes radioactive decay; in this decay process, the nucleus of the atom is transformed. Radionuclides are described by their chemical element name (which specifies the number of protons in the nucleus) and by their atomic mass (which specifies the sum of the number of protons and neutrons in the nucleus). A given chemical element may have many separate radioactive *isotopes*, that is, atoms with different numbers of neutrons associated with the given number of protons. For example, radon has three naturally occurring isotopes, with atomic masses 219, 220, and 222. A radioisotope is typically specified by providing both the chemical symbol (name) and the atomic mass number, such as ^{222}Rn or, equivalently, radon-222.

There are several distinct types of ionizing radiation, mainly characterized by the mass and electrical charge of the emission. *Alpha* particles (α) consist of a cluster of two protons plus two neutrons (i.e., the same composition as the nucleus of a helium atom). An α decay reduces the radionuclide's atomic number by 2 and its atomic mass by 4. *Beta* particles (β) are electrons that are ejected from the nucleus as a neutron is converted to a proton. When a radionuclide undergoes β decay, the atomic number *increases* by 1 and the atomic mass is unchanged. Both α and β emissions have electrical charge: β has the charge associated with a single electron ($q_e = -1.6 \times 10^{-19}$ C) and α has a positive charge equal to twice that magnitude. Among the various forms of ionizing radiation, alpha particles are by far the most massive, at 6.6×10^{-24} g. Beta particles have the small mass associated with an electron, 9×10^{-28} g. The third major type of ionizing radiation is *gamma* radiation (γ), which is purely electromagnetic, with no charge or mass. Gamma radiation originates from a transformation of the state of energy within the nucleus, much as light is emitted from a molecule when an electron changes its energy state. Because no mass is ejected, gamma radiation does not result in any change in the chemical element or the atomic mass of the radioisotope. Gamma radiation commonly accompanies α and β decay.

Radioactivity, Decay Constants and Half-Lives

Radioactivity refers to the decay of radioisotopes and the emission of ionizing radiation. The rate of this process for a given species is referred to as the *activity* of the species. In kinetic terms, we can think of radioactivity as a first-order reaction (however, it is a nuclear reaction, not a chemical reaction). In contrast with chemical reactions, the rate constants for nuclear reactions are independent of environmental conditions such as temperature. Nuclear reaction rates also do not depend on the chemical or physical state of the element. The amount of a radionuclide in a system is commonly expressed in terms of its activity. The following relationships hold for any specific radionuclide in a given system:

$$I = \lambda N = \left(-\frac{dN}{dt}\right)_{\text{radioactive decay}} \tag{B.1}$$

where I is the activity of the radionuclide, λ is the radioactive decay constant for that species (units of inverse time), and N is the number of radioactive atoms of that species in the sample.

Fundamentally, activity has units of atoms decaying per time, or disintegrations per time. Since atomic decay can be expressed as an integer count, this part of the unit can be dropped, so activity may simply be considered to have units of inverse time. Two sets of activity units are in common use. The traditional activity unit is the curie (Ci). One curie of any substance has the same activity as 1 g of pure radium-226, which corresponds to 37 billion disintegrations per second. From an environmental health perspective, this is an enormously high level of radioactivity; ordinary levels are expressed in traditional units in terms of millicuries (mCi = 10^{-3} Ci), microcuries (μCi = 10^{-6} Ci), or picocuries (pCi = 10^{-12} Ci). The SI unit of activity is the becquerel, abbreviated Bq. One Bq of a radioactive substance implies an average decay rate of one atom per second and is equal to 27.2 pCi. Example B.1 illustrates the relationship between the activity and mass of a radioactive substance.

EXAMPLE B.1 ***Units of Radioactivity***

Given that one gram of radium-226 has an activity of one curie, determine its radioactive decay constant.

SOLUTION The number of atoms of radium-226 in one gram is

$$N_{226} = \frac{1\text{ g}}{226\text{ g/mol}} \times 6.02 \times 10^{23}\text{ g mol}^{-1} = 2.7 \times 10^{21}$$

The rate of radioactive decay is equal to the decay constant times the number of atoms:

$$1\text{ Ci} = 3.7 \times 10^{10}\text{ s}^{-1} = \lambda_{226} \times (2.7 \times 10^{21}) \quad \Rightarrow \quad \lambda_{226} = 1.4 \times 10^{-11}\text{ s}^{-1}$$

It is helpful to keep in mind that the radioactive decay of atoms is a random, discrete process. The same is true, of course, for chemical reactions. But, typically, we are interested in chemical reactions only when very large numbers of molecules are involved so that the difference between being discrete and continuous becomes negligible. On the other hand, radioactivity is such a potent process in terms of its potential to cause damage that even small rates of radioactive decay may be important. Consider what it means, for example, for a material to have an activity of 1 Bq. This implies that the expected rate of radioactive decay is one atom per second. But in any given second, the true number of decays may be zero, one, two, and so on. When the decay constant is small relative to 1, it can be interpreted as the probability of a single atom of that isotope decaying during the time interval associated with the decay constant. For example, the decay constant of radium-226 (1.4×10^{-11} s^{-1} = 4.4×10^{-4} y^{-1}) implies that the probability of any given atom of radium-226 decaying in one year is 0.044 percent. The probability of a specific radioactive atom decaying does not depend on its history, so the probability of a radium-226 atom decaying is 0.044 percent per year, independent of the number of years since its formation.

It can be difficult to gain an intuitive understanding of rate constants in units of inverse time. So instead of reporting the radioactive decay constant for an isotope, the rate of decay is often indicated by means of the *half-life* of the element. The half-life is defined to be the period required for half of the original amount of the radioisotope to undergo radioactive decay. This measure is directly related to the decay constant, as will now be shown.

Consider a sealed container that contains some initial quantity, N_0, of a radioisotope that decays with a rate constant λ. The rate of change of the quantity of the isotope in the container is given by

$$\frac{dN}{dt} = -I = -\lambda N \tag{B.2}$$

with initial condition $N(0) = N_0$. The solution to this equation is

$$N(t) = N_0 e^{-\lambda t} \tag{B.3}$$

The time required for the amount to decay to half the initial value (i.e., the half-life) is denoted $t_{1/2}$ and is obtained by solving the following expression:

$$N(t_{1/2}) = N_0/2 = N_0 e^{-\lambda t_{1/2}} \tag{B.4}$$

So the half-life and the radioactive decay constant are related by this expression:

$$t_{1/2} = \frac{\ln(2)}{\lambda} = \frac{0.693}{\lambda} \tag{B.5}$$

Radioactive Decay Chains

Radioactive decays follow an almost completely deterministic sequence (some species have two or three possible decay paths, but these cases are unusual). Most radioisotopes possess a single mode of decay, have a fixed half-life (or decay constant), and produce a specific isotope as a product. The product isotope may itself be radioactive, or it may be stable. A sequence of radioisotopes that are produced one after another in succession is referred to as a radioactive decay chain.

Two symbolic representations of a single step in a decay chain are presented in Figure B.1. Style (*a*) shows the mode of decay and the product species. Style (*b*) is more complete, as it gives the half-life and the energy associated with the radioactive emission. The traditional units for expressing the energy released by radioactive decay are based on the electron volt, abbreviated eV. The electron volt is the kinetic energy that results from accelerating an electron through a potential difference of 1 V. The SI

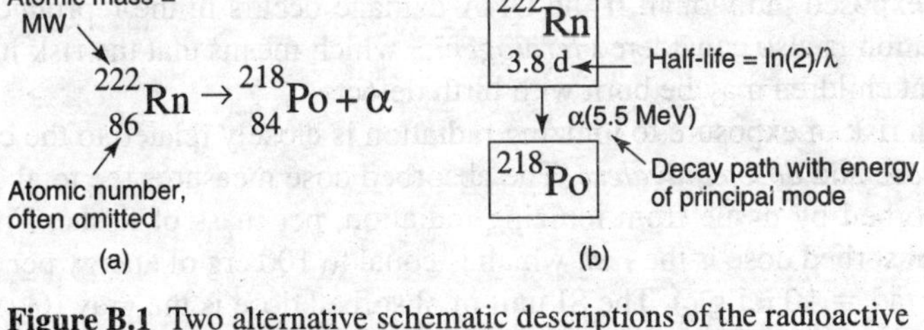

Figure B.1 Two alternative schematic descriptions of the radioactive decay of ^{222}Rn to ^{218}Po. Note that the emission of an alpha particle reduces the atomic number by 2, causing radon to be converted to polonium, and reduces the atomic mass by 4.

unit of energy, the joule, is equal to 6.24×10^{18} eV. Practical radioactive energy units are keV (10^3 eV) and MeV (10^6 eV). Frame (*b*) in Figure B.1 shows that radon-222 decay emits an α particle with a kinetic energy of 5.5 MeV. Since the mass of an alpha particle is known, the initial speed of these alpha particles is determined to be 1.6×10^7 m s^{-1}, using the relationship that the kinetic energy is equal to one-half the mass of the object times the square of its speed.

Interaction of Radiation with Matter

Ionizing radiation causes damage by transferring energy to matter. Much of the energy transfer initially generates ion pairs. Typically, about 10 eV is required to ionize a single molecule. Consequently, a 5.5 MeV α particle would generate about a half-million ion pairs before coming to rest.

Alpha and beta particles, because of their electrical charge, interact relatively strongly with matter. They steadily lose energy as they move through matter, and they only travel relatively short distances. Gamma radiation, being uncharged, does not interact steadily with matter. Instead, its interactions tend to occur with atomic nuclei and are destructive. Since nuclei occupy a very small fraction of the space in an atom, gamma radiation travels much farther, on average, through matter than either α or β particles.

The distance that radiation travels through a given material is called the *range*. The range varies strongly with the type of radiation, with the radiation energy, and also with the density and composition of the material. Alpha particles exhibit only a short range. In air, a 5.5 MeV α particle will travel approximately 4 cm. In human tissue, which is about 1000 times denser than air, the range of an alpha particle is only 40 μm, less than the thickness of dead skin cells. Because of this, alpha-emitting radionuclides are harmful only if they are present inside the body when they decay. Furthermore, the damage caused by α emitters is highly localized to the vicinity of the decay site. By contrast, the range of β particles in condensed matter is on the order of millimeters, and γ emissions can travel on the order of 10 cm. Both β and γ emissions can cause harm to tissue even if the radionuclides are outside the body. These emissions also cause more diffuse damage over a larger volume of tissue than α particles.

Radiation and Health

The adverse health effects of radiation exposure can be divided broadly into two classes. At high doses, a significant fraction of the cells in tissue can be directly killed, leading to a short-term risk of radiation sickness or even death in severe cases. At low to moderate doses, the primary concern is damage to DNA in chromosomes. Ionizing radiation is considered *carcinogenic* because this damage may increase the risk of cancer in the exposed individual. If the DNA damage occurs in the reproductive organs, the radiation is also considered *teratogenic,* which means that the risk increases that subsequent children may be born with birth defects.

The health risk of exposure to ionizing radiation is closely related to the concepts of *absorbed dose* and *dose equivalent*. The absorbed dose measures the total quantity of energy absorbed by tissue from ionizing radiation, per mass of tissue. The traditional unit of absorbed dose is the *rad,* which is equal to 100 erg of energy per gram of tissue (also, 1 rad = 10 μJ g^{-1}). The SI unit of absorbed dose is the gray (Gy), where 1 Gy = 1 J kg^{-1} = 100 rad.

The dose equivalent is obtained by multiplying the absorbed dose by a *quality factor* (QF), which accounts for the fact that the type of ionizing radiation also influences risk. Typically, a QF of 20 is applied for α particles, and the QF is taken as 1 for

Table B.1 Significance of the Dose Equivalent Associated with Exposure to Ionizing Radiation

Dose equivalent	Significance
600 rem	LD_{50} (fatal to one-half of population if delivered rapidly)
50 rem	Maximum dose with no visible short-term effects
5 rem/y	Maximum allowed dose rate for atomic workers
500 mrem/y	Maximum allowed dose to any individual in public (industrial sources)
300 mrem/y	Effective dose equivalent from average indoor radon
170 mrem/y	Maximum allowed dose averaged over population (industrial sources)
100–150 mrem/y	Background natural radiation, excluding radon
10 mrem	Dose equivalent caused by a typical chest X-ray

β and γ radiation. The traditional unit of dose equivalent is the rem, which is numerically equal to the absorbed dose in rad times the quality factor. The SI unit of dose equivalent is the seivert (Sv), which is numerically equal to the absorbed dose in gray times the quality factor. Thus, 1 Sv = 100 rem.

Studies of health risk suggest that the likelihood of a fatal malignancy during one's lifetime is directly related to the dose equivalent; for small fractional risks, the risk is considered to be directly proportional to the dose equivalent, with no safe threshold. A recent estimate of the risk factor is 1.65×10^{-4} rem^{-1}, which means that a dose equivalent of 1 rem confers a 0.0165 percent chance of developing fatal cancer. Table B.1 provides a summary of some dose equivalent benchmarks. Table B.2 summarizes some properties of radioisotopes that are of environmental concern.

Table B.2 Some Radionuclides of Environmental Interest

Isotope	Radioactive half-life	Principal mode of decay[a]	Dominant sources	Exposure pathway	Notes
^{3}H	12.3 y	β (18 keV)	Weapons testing, fuel reprocessing	Hydrologic cycle	b
^{14}C	5730 y	β (155 keV)	Natural production, weapons testing	Various	c
^{40}K	1.3×10^{9} y	β (1.31 MeV)	Natural, primordial	Food chain	d
^{85}Kr	10.7 y	β (672 keV)	Nuclear power, (fuel reprocessing)	External gas exposure	e
^{90}Sr	28.8 y	β (540 keV)	Weapons testing	Food chain	f
^{131}I	8.04 d	β (610 keV) γ (360 keV)	Weapons testing, reactor accidents	Inhalation, food chain	g
^{137}Cs	30.2 y	β (510 keV) γ (660 keV)	Weapons testing	Food chain	h
^{222}Rn	3.85 d	α (5.49 MeV)	Natural: earthen materials	Inhalation	i
^{226}Ra	1600 y	α (4.78 MeV)	Natural	Water and food chain	j
^{235}U	7×10^{8} y	α (4.6 MeV) γ (186 keV)	Natural, enriched for reactors and weapons	Industrial and military activities	k
^{239}Pu	24,100 y	α (5.15 MeV)	Used in weapons and as fuel for reactors	Industrial and military activities	l
^{241}Am	432 y	α (5.49 MeV)	Smoke detectors	External radiation	m

(continued on next page)

Table B.2 Some Radionuclides of Environmental Interest *(continued)*

[a]Energy of primary emission is shown in parentheses.

[b]Hydrogen-3 (tritium) substitutes for ^{1}H in water and becomes distributed through the environment by means of the hydrologic cycle. The testing of weapons in the atmosphere was the largest source of environmental tritium. The human doses are relatively small.

[c]Carbon-14 is produced continuously in the atmosphere by the bombardment of nitrogen by cosmic radiation. (An energetic electron is absorbed by a nitrogen nucleus, converting a proton to a neutron, thereby reducing the atomic number from 7 to 6.) Organisms accumulate ^{14}C and the dominant, stable isotope ^{12}C in approximately constant proportions throughout their lives. Because of the continuous radioactive decay of ^{14}C, the ratio of ^{14}C to ^{12}C serves as a reliable means of establishing the date of death of organisms that lived thousands of years ago. Nuclear weapons testing noticeably increased the atmospheric abundance of ^{14}C beginning in the 1950s.

[d]Potassium-40 occurs naturally as a trace constituent of all potassium, at a mole fraction of about 10^{-4}. Potassium is a substantial minor element in muscle tissue. A typical adult male has a ^{40}K activity of 0.1 μCi (3700 Bq). The internal presence of ^{40}K delivers a dose of 15–20 mrem y^{-1} to adults.

[e]Krypton is a noble gas and does not participate in metabolic processes. It is produced in nuclear reactor fuel rods during reactor operation. It is released to the atmosphere during normal operation, from accidents, and especially during fuel reprocessing. It accumulates and becomes distributed throughout the entire atmosphere until removed by radioactive decay.

[f]Strontium is the most hazardous radionuclide in fallout from weapons testing. Strontium is chemically similar to calcium. It enters the food chain via uptake from soil or foliar deposition, is concentrated in cow's milk and other dietary sources of calcium, and accumulates in bones following ingestion. Exposure to strontium via ingestion increases the risk of bone cancer.

[g]Because of its short half-life, the radiation exposure hazard from ^{131}I is primarily associated with short-term, intensive releases. Radioiodine released from weapons tests or reactor accidents can be directly inhaled. It can also deposit on foliage and subsequently be ingested (with locally grown food or locally produced fresh milk). Iodine is an essential human nutrient that accumulates in the thyroid gland; the radiation dose from ingesting radioiodine contributes to the risk of thyroid cancer. Uptake of radioiodine can be blocked by timely administration of a large dose of stable iodine.

[h]Cesium-137 is chemically analogous to potassium. Atmospheric testing of atomic weapons was the dominant source. Cesium is tightly bound to clay soils. It enters the food chain mainly by foliar deposition from the atmosphere. Ingestion of cow's milk is the dominant human exposure route. Unusually large body burdens of ^{137}Cs have been detected in Laplanders and other inhabitants of the far north. Their exposure was caused by the ingestion of meat from reindeer that had grazed on lichens, which had concentrated the cesium from the atmosphere.

[i]Radon is a chemically inert gas that is produced naturally in soil and other earthen materials throughout the earth's crust. Summed over the entire human population, inhalation of the radioactive decay products of radon is the largest source of radiation exposure.

[j]Radium-226 occurs naturally throughout the earth's crust. Chemically similar to calcium, it is somewhat soluble in water and, if ingested, accumulates in the bones.

[k]Uranium-235 is a key isotope in the nuclear fuel and weapons cycles. It is naturally present in the earth's crust. To use uranium as a fuel, the abundance of ^{235}U relative to ^{238}U must be increased through an enrichment process. The mining, processing, and waste disposal of uranium poses many opportunities for human exposure, but the long half-life of uranium isotopes means that the radiation hazard per unit mass of uranium ingested or inhaled is small.

[l]Plutonium-239 is produced by a fusion reaction involving ^{238}U in a class of nuclear reactors known as "breeders." Although breeder reactors hold the promise of improved fuel utilization relative to conventional nuclear reactors, their development is controversial because it is relatively easy to produce nuclear weapons from plutonium. Plutonium is highly toxic chemically but binds tightly to soils and so does not move readily through the environment.

[m]About 0.5–1 μCi (~20–40 kBq) of ^{241}Am is present in the smoke detectors that are widely used in residences and other indoor environments. Exposure to the external radiation from these devices is believed to pose a negligible risk in comparison with the benefit of reducing the rate of injury or death from fires.

Source: Eisenbud, 1987.

REFERENCE

EISENBUD, M. 1987. *Environmental radioactivity: From natural, industrial, and military sources*. 3rd ed. Academic Press, Orlando, FL.

C

Primer on Environmental Organic Chemicals

Organic chemicals are built on a structure of carbon. Hydrogen and oxygen are common constituents. Other elements frequently encountered in environmentally significant organic chemicals are nitrogen, sulfur, chlorine, and phosphorus. Fluorine, bromine, iodine, and metals are also present in some organic species.

An important fundamental characteristic of an organic chemical is its structure. The structure of a chemical is defined by the presence of covalent bonds between atoms. These bonds result from the sharing of electrons in the valence, or outer shell, of the respective atoms. The number of bonds that each atom can have in a molecule is generally the smaller of either the number of electrons in the valence band or the number of electrons needed to fill the valence band (see Table C.1). Note that the innermost shell of atomic orbitals can contain 2 electrons, the next two shells each can contain 8 electrons, and the fourth and fifth shells can each hold 18 electrons.

In diagrams showing the structure of organic chemicals, a covalent bond is represented as a single line connecting two atoms. The atoms themselves are represented by their elemental symbols.

Not all bonds between atoms are based on a single shared pair of electrons. Sometimes two or even three pairs of electrons are shared. In these cases, two or three parallel lines are used to represent the bonds, and each line (bond) is counted toward the total required for both atoms.

Organic chemicals are grouped into families that are defined by some aspects of their structure plus the presence of minor elements. The members of these families tend to share properties that govern their environmental behavior and, to some extent,

Table C.1 Number of Bonds Associated with Elements in Organic Molecules

Element	Atomic number	Typical number of bonds
H	1	1
C	6	4
N	7	3
O	8	2
F	9	1
P	15	3
S	16	2
Cl	17	1
Br	35	1
I	53	1

adverse effects. For the remainder of this primer, characteristics and example molecules are presented for elementary and important families of environmental organic chemicals.

C.1 PURE HYDROCARBONS

Alkanes

In addition to containing only hydrogen and carbon atoms, alkanes, also known as *paraffins*, are characterized by having open chains of carbon atoms (either branched or straight) in which pairs of carbon atoms share only single bonds. Alkanes have a relatively high vapor pressure, are fairly stable in the environment, and are not very soluble in water. Alkanes are convenient fuels and are not especially hazardous.

Alkanes are named according to the number of carbon atoms in the longest unbranched chain. The general chemical formula for an alkane is C_nH_{2n+2}, where n represents the number of carbon atoms in the molecule. The first five straight-chain alkanes are illustrated in Figure C.1. Note that the "-ane" ending is characteristic of all alkanes. Respectively, "meth-," "eth-," "prop-," "but-," and "pent-" indicate the presence of one through five carbon atoms in the group. The prefix *n*- indicates that the carbon atoms make a straight chain. Larger straight-chain alkanes can be formed by successively replacing a hydrogen atom at one end with a methyl group, $—CH_3$. Branched alkanes can be constructed by replacing one or more of the hydrogen atoms attached to an interior carbon with a methyl group.

Alkenes

A second class of pure hydrocarbons is known as *alkenes* or *olefins*. Alkenes differ from alkanes in that at least one pair of carbon atoms shares a double bond. Like alkanes, low-molecular-weight alkenes have a high vapor pressure and low solubility in water. Alkenes are somewhat more reactive than alkanes because it is easier to sever one branch of a double bond than a single bond.

Alkenes are named in a manner that is similar to the convention for alkanes. The number of carbon atoms in the longest straight chain determines the prefix of the main part of the name. The suffix that characterizes alkenes is "-ene" or "-ylene." Figure C.2 shows the structures of three alkenes. Ethene (or ethylene) has two carbon atoms ("eth-") that share a double bond; all other bonds are between carbon and hydrogen. To form propene (propylene), one of the hydrogen atoms in ethene is replaced by a methyl ($—CH_3$) group. Butadiene has four carbon atoms forming a straight chain

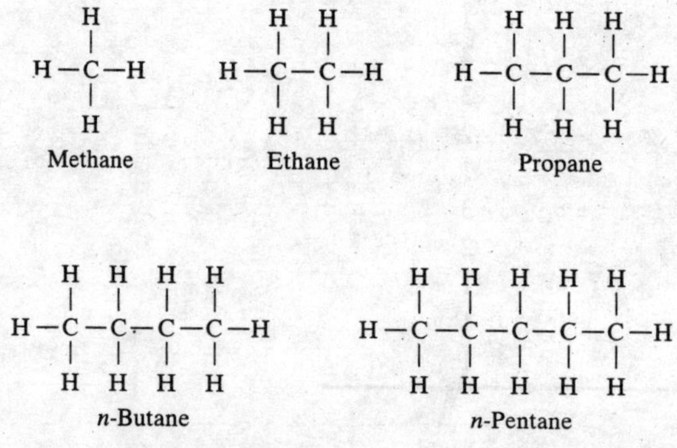

Figure C.1 Structures and names of straight-chain alkanes with carbon numbers 1–5.

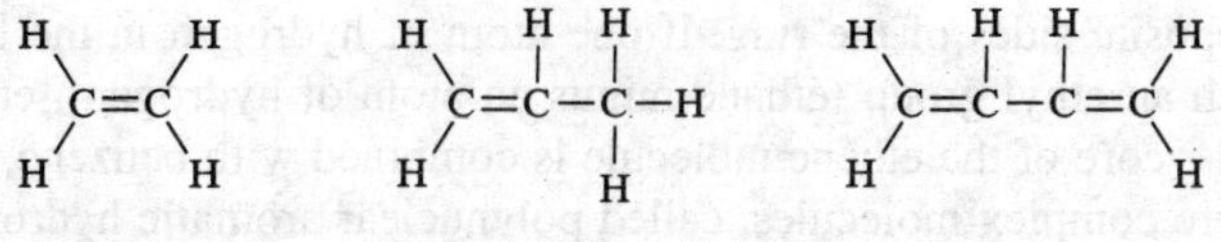

Figure C.2 Structures of some alkenes.

("but-"), and there are two pairs of double bonds between carbon atoms ("-diene," where "di" indicates 2). Alkenes are widely used in the chemical manufacturing of plastics (polyethylene and polypropylene are polymers made from chains of ethene and propene molecules), synthetic rubber, and latex paints.

Aromatics

The primary building block of an aromatic compound is the benzene ring. As depicted in the upper left of Figure C.3, benzene consists of six carbon atoms in a stable ring configuration with alternating single and double bonds. Thus, in a single ring structure, each carbon atom has three bonds with other carbon atoms and one bond free to combine with another atom. The simplest aromatic is benzene itself (C_6H_6), in which each of the additional bonds beyond those in the core ring are formed with hydrogen. More complex aromatic hydrocarbons with a single ring can be formed by replacing one or more of the atoms of hydrogen with fragments of alkanes or alkenes. For example, replacing one atom of hydrogen in benzene with the methyl group creates toluene. Replacing two atoms of hydrogen with methyl groups forms xylene. There are three forms of xylene, distinguished by the relative position of the methyl groups on the benzene ring. In *o*-xylene the methyl groups occupy adjacent positions; in *m*-xylene, they are separated by a single atom of hydrogen; and in *p*-xylene, depicted in Figure

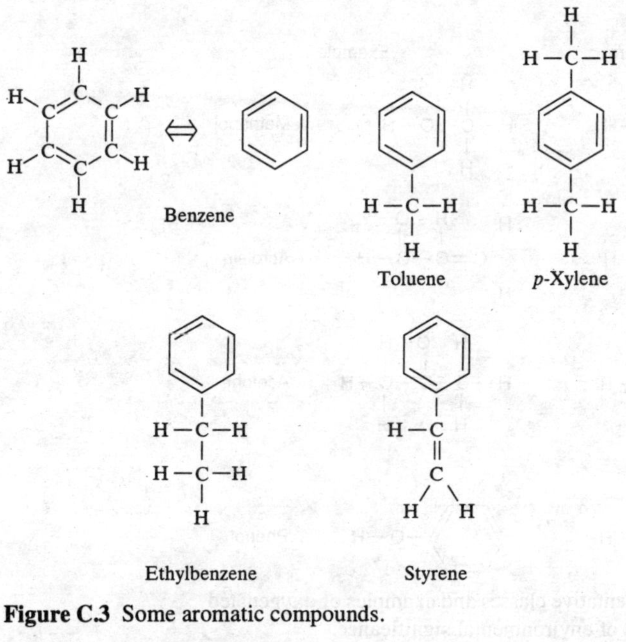

Figure C.3 Some aromatic compounds.

C.3, they are on opposite sides of the ring. If one atom of hydrogen in the benzene ring is replaced with an ethyl group (ethane minus an atom of hydrogen), ethylbenzene is formed. If the core of the ethene molecule is combined with benzene, styrene is created. Even more complex molecules, called polynuclear aromatic hydrocarbons (PAHs or sometimes PNAs), can be formed by combining two or more rings.

Aromatic compounds are of great interest as environmental contaminants. Many single-ring compounds are extensively used in fuels or in industrial processes. For example, benzene, toluene, ethylbenzene, and xylene (known in combination as BTEX) are blended to form an important component of gasoline. Aromatic compounds make effective solvents and are important feedstock chemicals in the manufacture of plastics and synthetic rubber. All five of the aromatic compounds depicted in Figure C.3 are included on federal lists, both as hazardous air pollutants and as regulated drinking water contaminants.

C.2 OXYGENATED ORGANIC COMPOUNDS

Some classes of organic compounds that include oxygen are depicted in Figure C.4. For each class, the general structure is shown and a specific example is given. In the general structure, R and R′ represent hydrocarbon fragments or "moieties," typically alkanes or alkenes that are missing one hydrogen atom. In the examples shown, the hydrocarbon moieties are the methyl group (—CH_3) and the ethene group (—C_2H_3). The symbol ϕ represents a benzene molecule that is missing a hydrogen atom.

Aliphatic *alcohols* are characterized by an —OH group attached to an alkane moiety. The main portion of the name indicates the number of carbon atoms in the alkane, with an "-ol" suffix to designate an alcohol. Thus, the next higher-order alcohols above methanol are ethanol (C_2H_5OH) and propanol (C_3H_7OH). The —OH group in alcohols gives them a high solubility in water. They are useful as solvents for cases in which it is necessary to keep other organic molecules blended in a water solution. Methanol and ethanol are also used as motor vehicle fuels.

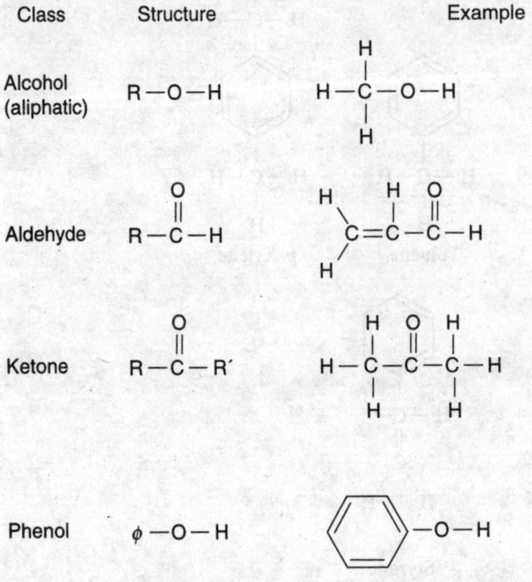

Figure C.4 Representative classes and examples of oxygenated organic compounds of environmental significance.

Aldehydes have a characteristic —CHO group, with a double bond between the carbon and oxygen atoms. The simplest aldehyde is formaldehyde (HCHO). Higher-order aldehydes in which the hydrocarbon moiety is an alkane are acetaldehyde (CH_3CHO) and propionaldehyde (C_2H_5CHO). Acrolein, the aldehyde depicted in Figure C.4, has the aldehyde group attached to an ethene group. Each of these four aldehydes is classified as a hazardous air pollutant. Aldehydes are highly irritating to mucous membranes. They are widely used as industrial chemicals. They are also created in the atmosphere by photochemical oxidation processes, and they are formed when hydrocarbon fuels are burned under poor combustion conditions. *Ketones* have a similar structure to aldehydes, but with carbon atoms on both sides of the oxygen-bonded carbon. Together, aldehydes and ketones are grouped under the broader heading of "carbonyls," organic compounds that have a carbon atom double-bonded to oxygen. In acetone, the ketone depicted in Figure C.4, both hydrocarbon moieties are methyl groups. The next higher-order ketone is methyl ethyl ketone (MEK), a widely used industrial chemical in which one of the methyl groups in acetone is replaced by an ethyl group (—C_2H_5).

Phenolic compounds have the —OH group that is characteristic of alcohols coupled to a benzene ring. *Phenol* is the name of this class of compounds and of its simplest member. Another common member is cresol, in which a methyl group replaces a second atom of hydrogen on the benzene ring. Phenol and cresol are both classified as hazardous air pollutants.

C.3 ORGANOHALIDES

Halogens are the atoms in the column of the periodic table headed by fluorine. For environmental engineering, the most important atom in this class is chlorine. In addition to chlorine and fluorine, bromine and iodine also have some environmental significance. Halogens are missing a single valence-band electron, and so they form only one covalent bond, as does hydrogen. Therefore, halogens are common substitutes for hydrogen in organic molecules.

Several classes of *organohalides* are important in the environment. Chlorinated solvents constitute one group. The four most widely used members of this group are shown in Figure C.5. The worldwide production of these four chemicals combined exceeds 2 million metric tons (2×10^{12} g) per year. Improper past disposal of used solvents from this group plus leaks from underground storage tanks and transmission lines have caused a significant part of the hazardous waste problem in the United States.

The structure of simple organohalides can usually be determined directly from the name of the compound plus knowledge of the basic structure of simple pure hydrocarbons. In the examples shown in Figure C.5, the last part of each name describes the underlying structure of the alkane or alkene. The first part indicates how many substitutions of chlorine for hydrogen atoms are made. In the case of ethane, since there are

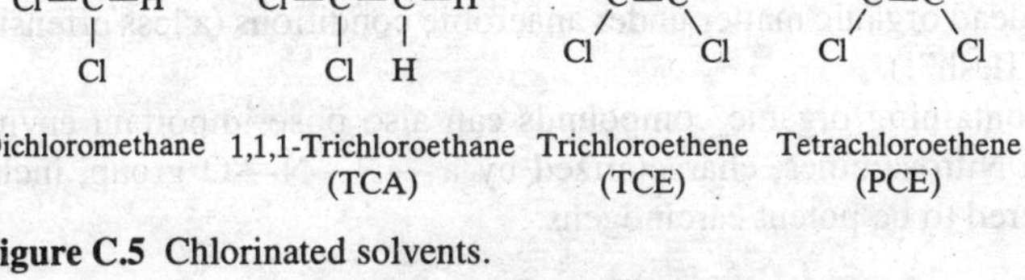

Figure C.5 Chlorinated solvents.

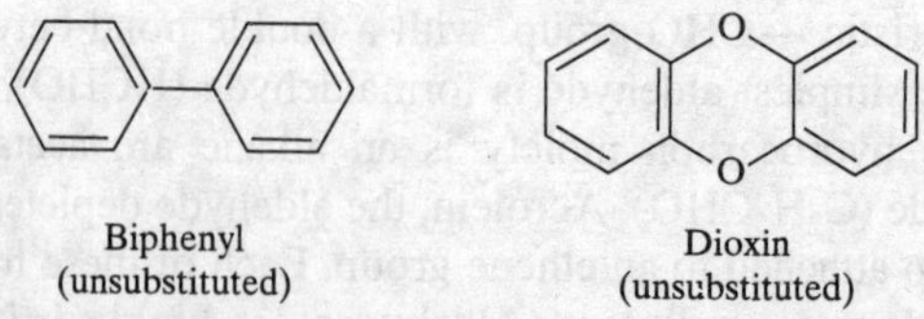

Figure C.6 Underlying organic structures of polychlorinated biphenyls (PCBs) and dioxins. In each case, the pollutant molecule is formed by substitution of chlorine for hydrogen atoms on the basic structure.

six hydrogen atoms in the molecule, there are many different places to put the three chlorine atoms to make trichloroethane. The designation "1,1,1-" indicates that each of these three substitutions occurs on the number 1 carbon atom. (Of course, orientation doesn't matter, so 2,2,2-trichloroethane would be the same molecule.) Trichloroethane, trichloroethene, and tetrachloroethene (also known as perchloroethene) are so common that they are frequently referred to by the abbreviations TCA, TCE, and PCE.

The treatment of drinking water with chlorine can produce members of a class of compounds known as *trihalomethanes* (THMs). As the name suggests, these compounds have a structure like methane, but with three halogen atoms replacing hydrogens. The most important compound in this group is trichloromethane, commonly known as chloroform ($CHCl_3$). Three other compounds are potentially important, too: dichlorobromomethane ($CHCl_2Br$), dibromochloromethane ($CHBr_2Cl$), and bromoform ($CHBr_3$).

Two other groups of organohalides have received considerable attention as environmental pollutants: polychlorinated biphenyls (PCBs) and chlorinated dioxins (dioxin). The basic structures of these compounds are shown in Figure C.6.

Another class of organohalides is the chlorofluorocarbons, including trichlorofluoromethane (CCl_3F) and dichlorodifluoromethane (CCl_2F_2). These compounds are gradually being phased out of production and use worldwide because of the hazard they pose to the stratospheric ozone layer (see §7.A.6).

Organohalides are also an important class of compounds used as pesticides, herbicides, fungicides, and soil fumigants. These are very important as environmental pollutants. Their chemical structure is often fairly complex and cannot be easily discerned from their common or trade names.

C.4 ORGANIC COMPOUNDS CONTAINING SULFUR OR NITROGEN

Organic compounds containing sulfur or nitrogen are often of interest because of their strong odors. For example, *mercaptans*, which are characterized by a —SH group, are so strong and sour smelling that they are added to natural gas in trace quantities so that leaks may be easily detected. *Amines*, compounds with a —NH_2 group, give rotten fish its odor. Both methyl mercaptan and methyl amine can be produced by the microbial degradation of dead organic matter under anaerobic conditions (a less offensive way of saying "rotting flesh"!).

Nitrogen-containing organic compounds can also pose important environmental health hazards. Nitrosamines, characterized by a —N—N=O group, include some species considered to be potent carcinogens.

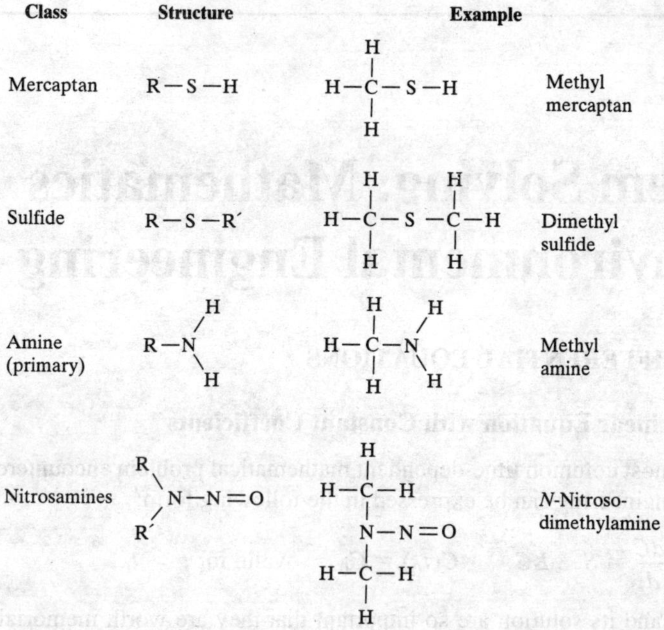

Figure C.7 General structures and specific examples of some organic compounds that contain either sulfur or nitrogen.

Figure C.7 illustrates the structures of a few of the many sulfur- and nitrogen-containing organic compounds.

D

Problem Solving: Mathematics for Environmental Engineering

D.1 ORDINARY DIFFERENTIAL EQUATIONS

D.1.a First-Order Linear Equation with Constant Coefficients

Probably the most common time-dependent mathematical problem encountered in environmental engineering can be expressed in the following form:

$$\frac{dC}{dt} = S - LC \qquad C(t_0) = C_0 \qquad \text{valid for } t > t_0 \tag{D.1}$$

This problem and its solution are so important that they are worth memorizing. It is also important to develop an awareness of situations that can be described by this equation.

Problems of this form are found in simple chemical kinetic systems. Such equations also describe the time-dependent behavior of species concentration in well-mixed reactors. In many circumstances, governing equations will be of this general form, but the S and L terms may be complex. Recognizing that this equation often appears and that it has a simple solution can save a lot of work. Because this equation and its solution have many applications in environmental engineering, it will be explored in detail here.

In words, the governing differential equation says that the rate of change of concentration (C) is given by the sum of two terms. The first term, S, represents all generation or "source" processes. This term accounts for mechanisms that tend to increase the species concentration. The second term, LC, represents all removal or "loss" processes. This term accounts for all mechanisms that tend to reduce the species concentration. In this equation, the loss processes are represented as first order, meaning that the rate of loss is proportional to the species concentration, C. The parameter L is sometimes called the loss-rate coefficient.

This equation is a first-order, ordinary differential equation. To have a unique solution, a first-order differential equation must have one initial or boundary condition. The second equation in D.1 provides this condition, stating that the species concentration at some time, t_0, is given by C_0.

All terms in an equation must have the same units. If concentration is expressed in mass per volume, then dC/dt has units of mass per volume per time. The source term, S, must also have units of mass per volume per time. The loss-rate coefficient, L, will always have units of inverse time for a first-order process.

In some circumstances, the parameters S and L vary with time or with the concentration of other species. In these cases, it may be difficult or impossible to obtain an analytical solution for the differential equation, and numerical approaches are useful (see §D.1.c). However, in many cases, S and L are constant, or at least approximately constant. If (and only if) S and L are constant, then the following solution to the differential equation applies:

$$C(t) = C_0 \exp[-L(t - t_0)] + \frac{S}{L}\{1 - \exp[-L(t - t_0)]\} \tag{D.2}$$

Substituting $t = t_0$ into this equation, we see that the initial condition, $C(t_0) = C_0$, is satisfied. Substituting $t = \infty$, we see that the steady-state solution is $C(\infty) = S/L$. By direct substitution, it is not difficult to show that this equation also satisfies the governing differential equation.

Because the solution of a linear differential equation is unique, the fact that the governing equation and initial condition are satisfied proves that this is the solution we seek. The problem can also be solved by direct methods, as shown in subsequent paragraphs.

Let's consider the physical significance of this solution. The equation is presented as the sum of two terms. The first term depends on the initial condition, C_0, but not on the source parameter, S. The second term depends on the source, but not on the initial condition. We may rewrite the solution as follows:

$$C(t) = C_1(t) + C_2(t) \tag{D.3}$$

where

$$C_1(t) = C_0 \exp[-L(t - t_0)] \tag{D.4a}$$

and

$$C_2(t) = \frac{S}{L}\{1 - \exp[-L(t - t_0)]\} \tag{D.4b}$$

The first term, C_1, is the solution to this problem:

$$\frac{dC_1}{dt} = -LC_1 \qquad C_1(t_0) = C_0 \tag{D.5}$$

And the second term, C_2, is the solution to this problem:

$$\frac{dC_2}{dt} = S - LC_2 \qquad C_2(t_0) = 0 \tag{D.6}$$

By adding these two governing equations and initial conditions, the original problem is recovered. Physically, $C_1(t)$ represents the concentration decay from the initial state to zero without resupply. $C_2(t)$ represents the growth of concentration from an initial state of zero toward the final steady-state condition, S/L. The component parts of the solution plus the aggregate solution are illustrated in Figure D.1.

The characteristic time, τ, is the same for both components of the solution and for the aggregate solution:

$$\tau \sim \frac{1}{L} \tag{D.7}$$

This is an interesting, important, and possibly counterintuitive result. Note that the characteristic time depends only on the rate of removal, L, not on the rate of supply, S. Changing the source term, S, affects the steady-state concentration, but not the length of time that is required to achieve it. Changing the loss rate coefficient, L, changes both the steady-state concentration and the characteristic time for the system to respond.

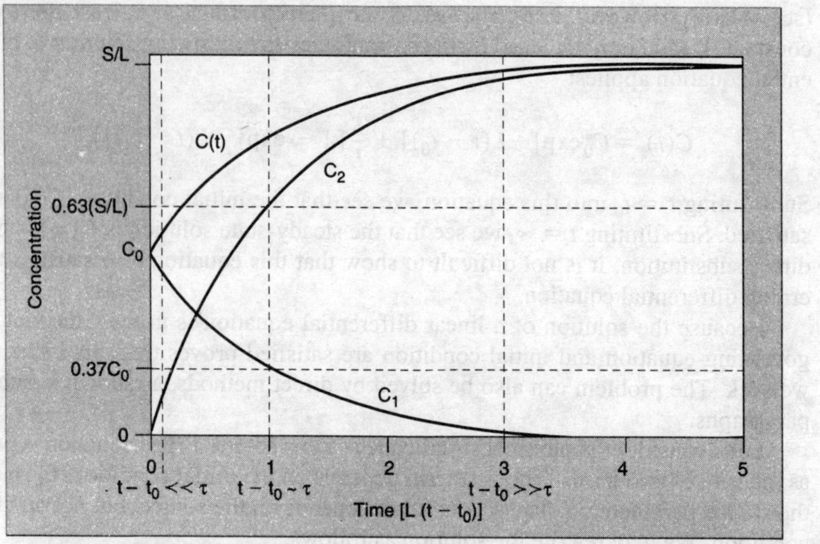

Figure D.1 Species concentration versus time in a first-order kinetic system with constant coefficients. The total response can be considered as the sum of C_1, representing the decay of the initial concentration, and C_2, representing growth from zero toward the final steady-state result, S/L. Both components exhibit a characteristic time $\tau \sim 1/L$.

The characteristic time is a magnitude estimate that has the following attributes in this case (valid only for $t > t_0$):

$t - t_0 \ll \tau \Rightarrow$ system is approximately unchanged from its initial state

$t - t_0 \sim \tau \Rightarrow$ system is near midway between the initial and final state

$t - t_0 \gg \tau \Rightarrow$ system has evolved to approximately steady-state conditions

To be specific about what is meant by $\gg$ and $\ll$, note that when $t - t_0 = 0.1\tau$, then $C = 0.90C_0 + 0.10(S/L)$, so that the concentration is heavily weighted by its initial condition. When $t - t_0 = \tau$, then $C = 0.37C_0 + 0.63(S/L)$; that is, the concentration has completed a substantial portion (63 percent) of its ultimate change from the initial condition C_0 to its final state S/L. When $t - t_0 = 3\tau$, then $C = 0.05C_0 + 0.95(S/L)$, so that the concentration is heavily weighted by its steady-state condition.

Figure D.1 shows these points for conditions in which S/L and C_0 are of the same magnitude. When little time has elapsed from the initial state, as measured relative to τ, then $C \sim C_0$. In this case, the species concentration does not depend on generation (or on loss, except to determine how long $t - t_0$ may be). When several multiples of the period τ has passed, then $C \sim S/L$. The concentration no longer depends on the initial condition; those molecules have all been removed from the system by the loss mechanisms. The dynamic balance between the rate of supply and the rate of removal determines the species concentration.

The notion of stocks, flows, and residence times, introduced in §1.C.8, can also be applied to this system. As shown in Figure D.2, C represents the stock, S represents the flow into the stock, and LC represents the flow out. In §1.C.8, we argued that such a system has a characteristic time of

$$\tau \sim \frac{\text{stock}}{\text{flow out}} \tag{D.8}$$

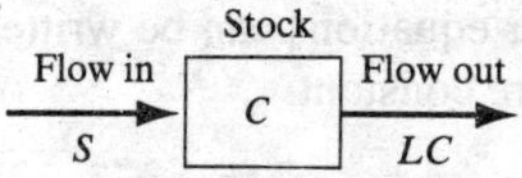

Figure D.2 Representation of first-order dynamics using stocks and flows.

Considering the system from this perspective yields the same estimate of characteristic time previously obtained:

$$\tau \sim \frac{C}{LC} \sim \frac{1}{L} \tag{D.9}$$

The formal solution of the governing equation can be developed from several approaches. The direct integration approach is presented here. Rearranging the differential equation and integrating from t_0 to t yields

$$\int_{t_0}^{t} dt = \int_{C_0}^{C(t)} \frac{dC}{S - LC} \tag{D.10}$$

The lower limits of the integrals describe the initial conditions, and the upper_limits correspond to some subsequent time, t. Evaluation of the integral on the left is trivial; it is equal to $t - t_0$. The integral on the right is evaluated by a change of parameters. Let

$$Y = S - LC \tag{D.11}$$

Then

$$dY = -L\, dC \tag{D.12}$$

And the right-hand integral becomes

$$\int_{C_0}^{C(t)} \frac{dC}{S - LC} = -\left(\frac{1}{L}\right)\int_{S - LC_0}^{S - LC(t)} \frac{dY}{Y} \tag{D.13}$$

Noting that $\int (dY/Y) = \ln(Y)$ and that $\ln(a) - \ln(b) = \ln(a/b)$, we have

$$(t - t_0) = -\frac{1}{L}\ln\left(\frac{S - LC(t)}{S - LC_0}\right) \tag{D.14}$$

Solving for $C(t)$ leads to the result presented in equation D.2.

D.1.b Coupled First-Order Ordinary Differential Equations

Equation D.1 arises when a single species is considered in a single, well-mixed reactor (batch or flow). Often, problems are encountered that involve either multiple species or multiple reactors with exchange through transport or transformation between them. These problems are often well described by a system of coupled ordinary differential equations, with one equation for each species in each reactor. If only two such equations need to be solved, it may be practical to use the general solution presented below. If more than two equations are involved, it is usually preferable to tackle the problem numerically rather than analytically.

The general problem for two coupled, first-order equations can be written as follows. We assume that the coefficients S_i, A_i, and L_i are constant.

$$\frac{dC_1}{dt} = S_1 + A_2C_2 - L_1C_1 \qquad C_1(0) = B_1 \tag{D.15}$$

$$\frac{dC_2}{dt} = S_2 + A_1C_1 - L_2C_2 \qquad C_2(0) = B_2 \tag{D.16}$$

The equations can be solved by means of converting the two first-order ODEs to a single second-order ODE. For example, rearrange D.16 to solve for C_1 and then substitute into D.15. The resulting equation is

$$\frac{d^2C_2}{dt^2} + (L_1 + L_2)\frac{dC_2}{dt} + (L_1L_2 - A_1A_2)C_2 = A_1S_1 + L_1S_2 \tag{D.17}$$

Since this is a second-order differential equation, two initial conditions are required. The first is given in the right-hand portion of D.16. The second is obtained by substituting the initial conditions on C_1 and C_2 into the left-hand portion of D.16, yielding

$$\frac{dC_2(0)}{dt} = S_2 + A_1B_1 - L_2B_2 \tag{D.18}$$

The solution to D.17 can be divided into a homogeneous and a particular part:

$$C_2(t) = C_{2,h} + C_{2,p} \tag{D.19}$$

where the homogeneous part is the general solution to

$$\frac{d^2C_{2,h}}{dt^2} + (L_1 + L_2)\frac{dC_{2,h}}{dt} + (L_1L_2 - A_1A_2)C_{2,h} = 0 \tag{D.20}$$

The particular part is any solution that satisfies D.17. A suitable particular solution is the following constant ratio, which is also the steady-state concentration:

$$C_{2,p} = \frac{A_1S_1 + L_1S_2}{L_1L_2 - A_1A_2} = C_{2,\mathrm{ss}} \tag{D.21}$$

The general solution to D.20 has this form:

$$C_{2,h} = F\exp(r_1t) + G\exp(r_2t) \tag{D.22}$$

where r_1 and r_2 are the roots of the characteristic equation obtained by substituting $C_{2,h} = e^{rt}$ into equation D.20, carrying out the differentiation, and then dividing by e^{rt}:

$$r^2 + (L_1 + L_2)r + (L_1L_2 - A_1A_2) = 0 \tag{D.23}$$

The parameters F and G are constants whose values are selected to satisfy the initial conditions. After carrying out the necessary algebra, these results are obtained:

$$r_1 = -\frac{L_1 + L_2}{2} + \frac{1}{2}\sqrt{(L_1 - L_2)^2 + 4A_1A_2} \tag{D.24}$$

$$r_2 = -\frac{L_1 + L_2}{2} - \frac{1}{2}\sqrt{(L_1 - L_2)^2 + 4A_1A_2} \tag{D.25}$$

$$F = \frac{(S_2 + A_1B_1) - B_2(r_2 + L_2) + r_2C_{2,ss}}{[(L_1 - L_2)^2 + 4A_1A_2]^{1/2}} \tag{D.26}$$

$$G = \frac{B_2(r_1 + L_2) - (S_2 + A_1B_1) - r_1C_{2,ss}}{[(L_1 - L_2)^2 + 4A_1A_2]^{1/2}} \tag{D.27}$$

The final solution for $C_2(t)$ is given by equation D.19, substituting from D.21–22 and D.24–27. A similar approach is used to solve for C_1. The result is

$$C_1(t) = H\exp(r_1t) + J\exp(r_2t) + C_{1,ss} \tag{D.28}$$

where

$$C_{1,ss} = \frac{A_2S_2 + L_2S_1}{L_1L_2 - A_1A_2} \tag{D.29}$$

The constants H and J are given by

$$H = \frac{(S_1 + A_2B_2) - B_1(r_2 + L_1) + r_2C_{1,ss}}{[(L_1 - L_2)^2 + 4A_1A_2]^{1/2}} \tag{D.30}$$

$$J = \frac{B_1(r_1 + L_1) - (S_1 + A_2B_2) - r_1C_{1,ss}}{[(L_1 - L_2)^2 + 4A_1A_2]^{1/2}} \tag{D.31}$$

The characteristic equation for C_1 is the same as for C_2, and so the roots, r_1 and r_2, are given by equations D.24–25.

These equations are less complicated than they look. Once numbers are substituted for the parameters (A_i, B_i, S_i, and L_i, $i = 1, 2$), the expressions for C_1 and C_2 reduce to the sum of two exponential transient terms (with characteristic times r_1^{-1} and r_2^{-1}), plus a steady-state term given by the particular solution ($C_{1,ss}$ and $C_{2,ss}$, respectively).

D.1.c Introduction to the Numerical Solution of Ordinary Differential Equations

The change in species concentrations over time in reactor models often yields ordinary differential equations. In §D.1.a and §D.1.b, we considered two common cases that can be solved analytically. However, in many circumstances, analytical solutions to the governing equations are not available. For example, the governing chemical kinetics may be complex, as in the case of the saturation reaction characteristic of microbial kinetics (§3.D.5). The generation and loss rates may be time dependent. Through flow or other exchange processes, several environmental compartments may be coupled such that species concentrations must be determined in all simultaneously. Or species may interact through transformation processes so that many species concentrations must be solved for simultaneously. In these situations, one can use numerical methods to solve the governing equations.

The basic idea in using numerical methods to solve the ordinary differential equations that describe species concentrations in environmental engineering is simple. Usually, we have one or more equations of this form:

$$\frac{dC_i}{dt} = F(C_i, C_j, t) \tag{D.32}$$

where C_i is the concentration of the species in the compartment of interest, and F is some function. In general, the function, F, depends on C_i plus C_j, the concentrations of other species in the same compartment and concentrations of the same species in other compartments. F may also depend on time, t. (In a plug-flow reactor, the governing differential equations are written either in terms of time since entry into the reactor or distance downstream of the entrance. The ideas explored here still apply.) One equation of form D.32 is required for each unknown concentration in the system. Usually, we know the concentrations of all species at some $t = t_0$ and wish to determine how these concentrations evolve.

To make the discussion more explicit, assume we have a single unknown species concentration, $C(t)$, that is governed by this equation:

$$\frac{dC}{dt} = S(t) - L(t)C \qquad C(t_0) = C_0 \tag{D.33}$$

where $S(t)$ is the rate of supply of the species to the system and $L(t)$ is the loss-rate coefficient. Both S and L may vary in an arbitrary manner with time, so the analytical solution developed in §D.1.a does not apply.

The essence of the numerical approach is to replace the differential term in equation D.33 with an algebraic approximation:

$$\frac{dC}{dt} \approx \frac{C(t + \Delta t) - C(t)}{\Delta t} \tag{D.34}$$

Substituting and solving for the concentration at time $t + \Delta t$ yields

$$C(t + \Delta t) \approx C(t) + \Delta t[S(t) - L(t)C(t)] \tag{D.35}$$

This equation is solved iteratively. To begin, a time step Δt is selected. The source term and loss-rate coefficient are evaluated at $t = t_0$, where, recall, $C = C_0$. The right-hand side of equation D.35 is computed to evaluate $C(t_0 + \Delta t)$. Next, the source term and the loss-rate coefficient are evaluated at the new time, $t_0 + \Delta t$. These values, along with $C(t_0 + \Delta t)$ from the first iteration, are substituted into the right-hand side of D.35 to evaluate $C(t_0 + 2\Delta t)$. The process is successively repeated until C has been determined for the entire period of interest.

This approach works only if C, S, and L change by small fractional amounts during the interval Δt. Nevertheless, the method is easy to understand and useful for some problems. More sophisticated solution methods feature two important improvements. First, equation D.35 is replaced by a more accurate expression that accounts for the fact that S, L, and C change with time and so should not be evaluated only at the beginning of a time step. Second, the size of the time step, Δt, is automatically adjusted to assure accuracy yet permit maximum computational speed (Press et al., 1986).

In §3.D.5, a simple numerical approach was used to solve a problem of microbial kinetics in a batch reactor. The problem is described by equations 3.D.22–23 and 3.D.25, reproduced below along with the algebraic approximations:

$$\frac{dX}{dt} = r_g X - k_d X \quad \Rightarrow \quad X(t + \Delta t) = X(t) + [(r_g(t) - k_d)X(t)]\Delta t \tag{D.36}$$

$$\frac{dS}{dt} = -\frac{r_g}{Y}X \quad \Rightarrow \quad S(t + \Delta t) = S(t) - \frac{r_g(t)}{Y}X(t)\Delta t \tag{D.37}$$

$$r_g(t) = \frac{k_m Y S(t)}{K_s + S(t)} \tag{D.38}$$

The problem was solved for these parameter values (see Figure 3.D.3): k_m = 3 ([g substrate]/[g cell]) d^{-1}, Y = 0.5 ([g cell]/[g substrate]), K_s = 50 (g substrate) m^{-3}, k_d = 0.2 d^{-1}, $X(0)$ = 1 (g cell) m^{-3}, and S = 100 (g substrate) m^{-3}. It was an easy matter to generate Figure 3.D.3 with the help of a spreadsheet program, using the algebraic equations D.36–38 with a time step of Δt = 0.1 d.

D.2 PARTIAL DIFFERENTIAL EQUATIONS THAT ARISE IN TRANSPORT PROBLEMS

In models based on the general material-balance equation (§5.B), species concentrations may depend on one to three spatial coordinates and on time. If the time-dependent concentration is sought, then the governing equation will contain a first-order term involving a time derivative of concentration, to account for accumulation. Advection contributes one or more terms that contain the first derivative of species concentration with respect to the direction(s) of fluid flow. Diffusion and dispersion add terms that contain the second derivative of concentration with respect to coordinate direction. Commonly, then, the governing model equations are partial differential equations involving some combination of a first derivative in time and first and second derivatives in space.

In §5.B, some commonly encountered problems were introduced in which an analytical solution is sought for the general material-balance equation. Here some of the missing mathematical details are provided.

Laplace Transform and Dimensional Analysis

In Case 2 (§5.B.2), we considered the problem of one-dimensional, transient diffusion from a point source. The governing equation and the boundary and initial conditions are as follows:

$$\frac{\partial C}{\partial t} - D\frac{\partial^2 C}{\partial x^2} = 0 \tag{D.39}$$

$$C(-\infty, t) = C(\infty, t) = 0 \tag{D.40}$$

$$C(x, 0) = M_a\, \delta(x) \tag{D.41}$$

The governing differential equation is Fick's second law of diffusion. The boundary conditions state that the species concentration must go to zero at infinite distance from the origin. The initial condition indicates that a quantity of contaminant (M_a, mass per area of y-z plane) is released at the origin (x = 0) at t = 0.

The most efficient method of solving this problem employs a Laplace transform to convert the partial differential equation into an ordinary differential equation, which is solved by direct integration. This approach, described in detail in Cussler (1984), is not illuminating unless one has studied Laplace transforms. An alternative approach, suggested by Fischer et al. (1979), is presented here. This method employs the Buckingham-Π theorem for the same purpose as the Laplace transform. The partial differential equation is converted to an ordinary differential equation, which is then solved. The analysis is fairly complicated. This should help convince you of the merits of solving transport problems, wherever possible, by transforming them into problems that have already been solved, rather than attempting to solve the governing differential equations by direct means.

The Buckingham-Π theorem states that the maximum number of independent dimensionless groups in any problem is equal to the difference between the number of parameters and the number of fundamental dimensions. For the present problem, there are five parameters (C, M_a, D, t, and x) and three dimensions (mass or moles, length, and time). This suggests that the problem posed by equations D.39–41 can be recast such that there are only two dimensionless groups: a dimensionless concentration that depends on a single dimensionless coordinate.

The Buckingham-Π theorem provides no guidance about how to select the dimensionless groups. The choices we make are guided by experience. The dimensionless concentration can be written by taking the ratio of the concentration (C) to the mass injected per area (M_a) and multiplying it by a length scale. The length scale is chosen to be $(Dt)^{1/2}$. The dimensionless coordinate is then taken as $\eta = x/(Dt)^{1/2}$. So, according to the Buckingham-Π theorem, we should be able to find some function $f(\eta)$ that satisfies the expression

$$\frac{C(x,t)(Dt)^{1/2}}{M_a} = f\left(\frac{x}{(Dt)^{1/2}}\right) \tag{D.42}$$

or

$$C(x,t) = \frac{M_a}{(Dt)^{1/2}} f(\eta) \tag{D.43}$$

To solve for the function f, we substitute D.43 into D.39. The partial derivative terms are evaluated as follows:

$$\frac{\partial C}{\partial t} = \left[\frac{\partial}{\partial t}\left(\frac{M_a}{(Dt)^{1/2}}\right)\right] f(\eta) + \frac{M_a}{(Dt)^{1/2}} f'(\eta) \frac{\partial \eta}{\partial t}$$

$$= \left(-\frac{1}{2t}\right) \frac{M_a}{(Dt)^{1/2}} [f(\eta) + \eta f'(\eta)] \tag{D.44}$$

$$\frac{\partial C}{\partial x} = \frac{M_a}{(Dt)^{1/2}} f'(\eta) \frac{\partial \eta}{\partial x} = \frac{M_a}{Dt} f'(\eta) \tag{D.45}$$

$$\frac{\partial^2 C}{\partial x^2} = \frac{M_a}{Dt} f''(\eta) \frac{\partial \eta}{\partial x} = \frac{M_a}{(Dt)^{3/2}} f''(\eta) \tag{D.46}$$

In these equations, the primes (′) represent derivatives of the function with respect to its argument (i.e., $f'(\eta) = df/d\eta$ and $f''(\eta) = d^2f/d\eta^2$). Substituting into D.39 and simplifying produces this ordinary differential equation for $f(\eta)$:

$$f''(\eta) + \frac{\eta}{2} f'(\eta) + \frac{1}{2} f(\eta) = 0 \tag{D.47}$$

This is still not an easy equation to solve. To make progress, we note that the equation can be rewritten as

$$\frac{d}{d\eta}\left[\frac{df}{d\eta} + \frac{\eta}{2} f(\eta)\right] = 0 \tag{D.48}$$

Since its derivative is equal to zero, the term in brackets must equal a constant:

$$\frac{df(\eta)}{d\eta} + \frac{\eta}{2} f(\eta) = A \tag{D.49}$$

From symmetry, we may argue that the constant, A, must be zero. The argument proceeds like this. Species must diffuse equally in both the $+x$ and $-x$ directions from the site of release at the origin. Therefore, the concentration must be a maximum at $x = 0$, which implies that $df/d\eta = 0$ at $\eta = 0$. However, by equation D.49, $df/d\eta = A$ at $\eta = 0$. Therefore, $A = 0$.

Now, we can rearrange equation D.49 and integrate to solve for f:

$$\int \frac{df(\eta)}{f(\eta)} = -\int \frac{\eta}{2} d\eta \tag{D.50}$$

So

$$f(\eta) = B \exp\left(-\frac{\eta^2}{4}\right) \tag{D.51}$$

We determine the value of the constant B by applying the condition of mass conservation: All of the mass originally released must remain in the system at all times t. Mathematically, this statement translates as follows:

$$M_a = \int_{-\infty}^{\infty} C(x, t)\, dx = \frac{M_a}{(Dt)^{1/2}} B \int_{-\infty}^{\infty} \exp\left(-\frac{x^2}{4Dt}\right) dx \tag{D.52}$$

The integral has a value $(4\pi Dt)^{1/2}$, so $B = (4\pi)^{-1/2}$, and the concentration is

$$C(x, t) = \frac{M_a}{(4\pi Dt)^{1/2}} \exp\left(-\frac{x^2}{4Dt}\right) \tag{D.53}$$

This is the solution to the problem (equation 5.B.22).

Separation of Variables

In Case 3 (§5.B.2), we considered the problem of three-dimensional transient diffusion from an instantaneous point source. Here are the governing equation and the boundary and initial conditions:

$$\frac{\partial C}{\partial t} - \varepsilon_x \frac{\partial^2 C}{\partial x^2} - \varepsilon_y \frac{\partial^2 C}{\partial y^2} - \varepsilon_z \frac{\partial^2 C}{\partial z^2} = 0 \tag{D.54}$$

$$C(x, y, z, t) \longrightarrow 0 \qquad \text{as } x \longrightarrow \pm\infty,\ y \longrightarrow \pm\infty, \text{ or } z \longrightarrow \pm\infty \tag{D.55}$$

$$C(x, y, z, 0) = M\delta(x)\delta(y)\delta(z) \tag{D.56}$$

The initial condition says that a fixed quantity of species, M, is released at the origin at $t = 0$.

We will apply the method of separation of variables to convert the partial differential equation (PDE) D.54 into three PDEs with the advantage that each equation will depend on one spatial coordinate (x, y, or z) and time. Furthermore, these equations will be of the same form as the problem specified in equations D.39–41, and so we will be able to write the solution directly.

We postulate that the concentration can be written as separable functions of x, y, and z, as follows:

$$C(x, y, z, t) = M \times X(x, t) \times Y(y, t) \times Z(z, t) \tag{D.57}$$

Substitute D.57 into D.54 and carry out the derivatives. Then divide the equation by C and rearrange to obtain this result:

$$\left\{\frac{1}{X}\left(\frac{\partial X}{\partial t} - \varepsilon_x \frac{\partial^2 X}{\partial x^2}\right)\right\} + \left\{\frac{1}{Y}\left(\frac{\partial Y}{\partial t} - \varepsilon_y \frac{\partial^2 Y}{\partial y^2}\right)\right\} + \left\{\frac{1}{Z}\left(\frac{\partial Z}{\partial t} - \varepsilon_z \frac{\partial^2 Z}{\partial z^2}\right)\right\} = 0 \quad \text{(D.58)}$$

The key to the separation of variables method is to recognize that the terms in each set of braces, { }, depend on only one spatial coordinate (x, y, and z, respectively) and time. Since the x, y, and z coordinates can be varied independently, and since the solution must satisfy the governing equation at all points in space, the solution must have the property that each of the terms in braces equals zero.

What about the boundary and initial conditions? The original problem has a solution of the form D.57 if the following conditions are met:

$$\begin{aligned} X(x, t) &\longrightarrow 0 \quad \text{as } x \longrightarrow \pm\infty \\ Y(y, t) &\longrightarrow 0 \quad \text{as } y \longrightarrow \pm\infty \\ Z(z, t) &\longrightarrow 0 \quad \text{as } z \longrightarrow \pm\infty \end{aligned} \quad \text{(D.59)}$$

Also,

$$\begin{aligned} X(x, 0) &= \delta(x) \\ Y(y, 0) &= \delta(y) \\ Z(z, 0) &= \delta(z) \end{aligned} \quad \text{(D.60)}$$

Equations D.58–60 specify problems for $X(x, t)$, $Y(y, t)$, and $Z(z, t)$ that are similar to the problem specified for $C/M_a(x, t)$ by equations D.39–41; we need only replace D by ε_x, ε_y, or ε_z, respectively. Using equation D.53, we can write

$$X(x, t) = \frac{1}{(4\pi\varepsilon_x t)^{1/2}} \exp\left(-\frac{x^2}{4\varepsilon_x t}\right) \quad \text{(D.61)}$$

$$Y(y, t) = \frac{1}{(4\pi\varepsilon_y t)^{1/2}} \exp\left(-\frac{y^2}{4\varepsilon_y t}\right) \quad \text{(D.62)}$$

$$Z(z, t) = \frac{1}{(4\pi\varepsilon_z t)^{1/2}} \exp\left(-\frac{z^2}{4\varepsilon_z t}\right) \quad \text{(D.63)}$$

So

$$C(x, y, z, t) = \frac{M}{(4\pi t)^{3/2} (\varepsilon_x \varepsilon_y \varepsilon_z)^{1/2}} \times \exp\left(-\frac{x^2/\varepsilon_x + y^2/\varepsilon_y + z^2/\varepsilon_z}{4t}\right) \quad \text{(D.64)}$$

This is the solution that was presented in §5.B, Case 3 (equation 5.B.26).

D.3 SOLVING EQUILIBRIUM PROBLEMS: ROOTS OF NONLINEAR ALGEBRAIC EQUATIONS

The mathematical problems that arise in analyzing chemical equilibrium are systems of algebraic equations. A problem is completely specified when there is one independent equation for each unknown. Typically, the unknowns are species concentrations, and the equations are based on some combination of the principle of material balance, equilibrium relationships, charge balance (electroneutrality), and stoichiometry. In equilibrium relationships, species concentrations are often combined as products or ratios. Consequently, the systems of equations that arise in equilibrium problems may be nonlinear. The methods of linear algebra, such as matrix elimination, do not apply.

Nevertheless, equilibrium problems can be solved using a fairly systematic procedure. Provided that the number of unknowns is not too large (fewer than 5–10), manual or spreadsheet-assisted solution methods are practical.

Relatively simple problems sometimes generate a quadratic equation:

$$aC^2 + bC + d = 0 \qquad \text{(D.65)}$$

where a, b, and d are constants and C is the unknown species concentration. Most students will recall that this equation has two solutions:

$$C = \frac{-b + \sqrt{b^2 - 4ad}}{2a} \quad \text{and} \quad C = \frac{-b - \sqrt{b^2 - 4ad}}{2a} \qquad \text{(D.66)}$$

For chemical equilibrium problems only one of the solutions, the correct one, will be physically meaningful. (For example, the other solution may predict a negative concentration.)

Closed-form solutions exist for cubic equations (Selby, 1974), but these require so much algebraic manipulation that the search procedure suggested below is more efficient.

Other equilibrium problems generate systems of equations that cannot be reduced to such a simple form as a single quadratic or cubic equation. The main purpose of this section is to describe a solution procedure that is fairly robust and practically useful. For illustration, the procedure is applied to the problems that arose in examples presented in §3.B and §3.C.

The general strategy in this procedure involves three steps. First, the system of algebraic equations is converted to a single equation in a single unknown of the form $f(C) = 0$. Second, lower and upper bounds on the plausible values of C are defined. Call these C_{min} and C_{max}, respectively. Third, concentration values bounded by C_{min} and C_{max} are systematically tested, in the search for $f(C) = 0$. With the assistance of a spreadsheet or programmable calculator, the computational part of this procedure can be carried out quickly.

The search procedure takes advantage of several properties of chemical equilibrium problems. First, only one physically meaningful concentration, C, is possible. If we have selected C_{min} and C_{max} properly, we know that only one value of C between these bounds will satisfy $f(C) = 0$. Call this concentration value C^*. The concentration C and the function $f(C)$ are continuous. Consequently, $f(C)$ must have a single algebraic sign for the entire domain $C_{min} < C < C^*$, and it must have the opposite algebraic sign for the domain $C^* < C < C_{max}$. It is this attribute that allows us to conduct an efficient search.

We begin by evaluating $f(C_{min})$ and $f(C_{max})$ to determine their sign. Say, for example, that $f(C_{min}) > 0$ and $f(C_{max}) < 0$. Then we select a trial value, C_1, between these limits and evaluate $f(C_1)$. If $f(C_1) > 0$, then $C_1 < C^* < C_{max}$. Alternatively, if $f(C_1) < 0$, then $C_{min} < C^* < C_1$. Based on the outcome of the first evaluation, we select a new trial value C_2 from the part of the domain that contains C^*. We repeat these steps until we converge on the value of C^* to whatever accuracy we seek. If the procedure is carried out on a programmable calculator, it is most efficient to conduct a binary search, in which each successive trial value is taken at the midpoint of the do-

Table D.1 Trial Results for Finding the Root of Equation D.67

Trial	C_i	$f(C_i)$	Minimum C^*	Maximum C^*
0	1.0×10^{-3}	+	0	1.0×10^{-3}
1	1.0×10^{-4}	−	1.0×10^{-4}	1.0×10^{-3}
2	3.2×10^{-4}	+	1.0×10^{-4}	3.2×10^{-4}
3	1.8×10^{-4}	−	1.8×10^{-4}	3.2×10^{-4}
4	2.4×10^{-4}	+	1.8×10^{-4}	2.4×10^{-4}
5	2.08×10^{-4}	+	1.8×10^{-4}	2.08×10^{-4}
6	1.93×10^{-4}	+	1.8×10^{-4}	1.93×10^{-4}
7	1.86×10^{-4}	+	1.8×10^{-4}	1.86×10^{-4}
8	1.83×10^{-4}	+	1.8×10^{-4}	1.83×10^{-4}
9	1.815×10^{-4}	+	1.8×10^{-4}	1.815×10^{-4}
10	1.807×10^{-4}	−	1.807×10^{-4}	1.815×10^{-4}
11	1.811×10^{-4}	+	1.807×10^{-4}	1.811×10^{-4}

main. With a spreadsheet, it is more efficient to conduct the search by evaluating $f(C)$ initially for widely spaced trial values over the whole domain, then with successively finer values in the subdomains that contain C^*. These points are illustrated in the following three cases, drawn from examples in the text.

Case 1: Using a Programmable Calculator

In §3.B.3, an example was presented in which we sought to determine the equilibrium partitioning of toluene between aqueous, gas, and sorbed phases in a closed system (Example 3.B.5). The mathematical problem reduced to finding the concentration of toluene in water (C_t in units of M) that satisfied the following expression:

$$f(C_t) = C_t + 0.0291\,C_t^{0.45} - 7.83 \times 10^{-4} = 0 \qquad \text{(D.67)}$$

Appropriate minimum and maximum values of C_t are $C_{\min} = 0$ and $C_{\max} = 10^{-3}$ M. The latter value is based on material balance, where it is assumed that all of the toluene in the system remains in the aqueous phase at equilibrium.

A calculator is programmed to return the value of $f(C_t)$ when C_t is entered. We find that $f(C_{\min}) < 0$ and $f(C_{\max}) > 0$. We proceed with a search in which the trial concentrations are successively reduced from $C_{\max}$ by a factor of 10 until the sign of $f(C_i)$ changes. Then, a binary search on a logarithmic scale is conducted to refine the estimate of C^*. The results are presented in Table D.1. For each trial, the table reports C_i (the trial value of C^*), whether the function $f(C_i)$ is positive or negative at that value C_i, and, based on the sign of f, the revised lower and upper bounds on the solution. The new guess for C_i is the geometric mean of the minimum C^* and maximum C^* values from the previous trial. The answer, correct to three significant figures, is $C_t = 0.181 \times 10^{-3}$ M. Eleven trials are required to achieve this level of precision.

Case 2: Using a Spreadsheet Program

In §3.C, acid-base equilibrium problems are considered. Example 3.C.4 generated the following cubic equation, in which C represents the equilibrium molar concentration of H^+ ions.

$$f(C) = C^3 - 5.34 \times 10^{-12} C - 4.99 \times 10^{-22} = 0 \qquad \text{(D.68)}$$

Table D.2 Spreadsheet Used to Solve Equation D.68

pH_i	C_i	$f(C_i)$
1	1.0E-01	1.0E-03
2	1.0E-02	1.0E-06
3	1.0E-03	1.0E-09
4	1.0E-04	1.0E-12
5	1.0E-05	9.5E-16
6	1.0E-06	– 4.3E-18
5.0	1.0E-05	9.5E-16
5.1	7.9E-06	4.6E-16
5.2	6.3E-06	2.2E-16
5.3	5.0E-06	9.9E-17
5.4	4.0E-06	4.2E-17
5.5	3.2E-06	1.5E-17
5.6	2.5E-06	2.5E-18
5.7	2.0E-06	–2.7E-18
5.60	2.51E-06	2.5E-18
5.61	2.45E-06	1.7E-18
5.62	2.40E-06	1.0E-18
5.63	2.34E-06	3.9E-19
5.64	2.29E-06	–1.9E-19
5.630	2.344E-06	3.9E-19
5.631	2.339E-06	3.3E-19
5.632	2.333E-06	2.7E-19
5.633	2.328E-06	2.1E-19
5.634	2.323E-06	1.5E-19
5.635	2.317E-06	9.3E-20
5.636	2.312E-06	3.6E-20
5.637	2.307E-06	–2.1E-20

Table D.2 shows a spreadsheet that was constructed to solve this equation. The search is conducted by selecting trial pH values, pH_i, evaluating C_i and $f(C_i)$ at each trial pH, and looking for the zero crossing in $f(C)$. Initially, the search is conducted with broad pH intervals, one unit per step. Succeeding groups of rows refine the search: The pH increment in the second group of rows is 0.1 unit and in the third group is 0.01 unit. The fourth group uses a 0.001 unit pH increment so that the final answer can be specified to ±0.01 precision. Each group continues until $f(C_i)$ changes sign from + to –, and the starting pH value for the next group of rows is the highest value from the current group that produced $f(C_i) > 0$. The solution is bounded by $5.636 < pH^* < 5.637$, so we write the answer as pH = 5.64.

Of course, a cubic equation has three roots. The other roots in this problem can be determined by dividing equation D.68 by $C - 2.31 \times 10^{-6}$ (the first root, determined from the spreadsheet) and then applying the quadratic formula, equations D.65–66. The other two roots are -9.34×10^{-11} and -2.31×10^{-6}, both negative and therefore not physically meaningful. In practice, this outcome is always obtained: If the physical problem has only one answer, then the mathematics will generate only one plausible solution.

Case 3: Spreadsheet Solution of a More Complex Problem

One of the advantages of using the spreadsheet method is that some of the algebraic manipulation that is required to convert the system of equations into a single equation of the form $f(C) = 0$ can be avoided. This can save time and tedium, plus reduce the likelihood of errors.

We will consider the acid-base problem presented as Example 3.C.5 to illustrate this point. In this problem, initially pure water was placed in contact with limestone ($CaCO_3$). Some of the limestone ions dissolved, and the carbonate participated in acid-base reactions. The goal of the problem is to determine the equilibrium pH.

Six species must be considered: H^+, OH^-, Ca^{2+}, H_2CO_3*, HCO_3^-, and CO_3^{2-}. The six equations that link them are based on the following principles:

- **a.** Charge neutrality
- **b.** Equilibrium dissociation of water
- **c.** Equilibrium dissociation of carbonic acid
- **d.** Equilibrium dissociation of bicarbonate
- **e.** Equilibrium precipitation/dissolution of limestone
- **f.** A stoichiometric requirement that the sum of the dissolved carbonate species in solution must equal the calcium ion concentration

Using (b), (c), and (e), the electroneutrality relationship is quickly rewritten into the following form, in which only H^+ and CO_3^{2-} concentrations are unknown:

$$2\frac{K_{sp}}{[CO_3^{2-}]} + [H^+] = \frac{K_w}{[H^+]} + \frac{[H^+][CO_3^{2-}]}{K_2} + 2[CO_3^{2-}] \qquad (D.69)$$

Condition (f) is combined with the equilibrium conditions (c), (d), and (e) such that, with modest algebraic effort, the following equation is generated that contains the same two unknowns.

$$[CO_3^{2-}] = K_{sp}^{1/2}\left(\frac{[H^+]^2}{K_1K_2} + \frac{[H^+]}{K_2} + 1\right)^{-1/2} \qquad (D.70)$$

With a spreadsheet, we can avoid the step of substituting D.70 into D.69 and rearranging to obtain a workable expression of the form $f([H^+]) = 0$. Table D.3 presents a few illustrative rows from the spreadsheet that was used to solve this problem. The columns headed by species names contain algebraic expressions that determine the concentration of the indicated ion, at the given pH. So, for example, the cells below $[OH^-]$ contain the expression $K_w/[H^+]$. Entries in the carbonate column contain equation D.70. The final column represents the electroneutrality relationship, equation D.69, rewritten so that all of the ions appear on one side of the expression. The problem is solved when EN = 0, which occurs at a pH in the range $9.95 < pH < 9.96$.

Table D.3 Sample Rows from Spreadsheet for Solving Equations D.69–70

pH	$[H^+]$	$[OH^-]$	$[HCO_3^-]$	$[CO_3^{2-}]$	$[Ca^{2+}]$	EN
9.94	1.15E-10	8.71E-05	9.33E-05	3.80E-05	1.31E-04	6.29E-06
9.95	1.12E-10	8.91E-05	9.20E-05	3.84E-05	1.30E-04	2.89E-06
9.96	1.10E-10	9.12E-05	9.06E-05	3.87E-05	1.29E-04	–5.54E-07

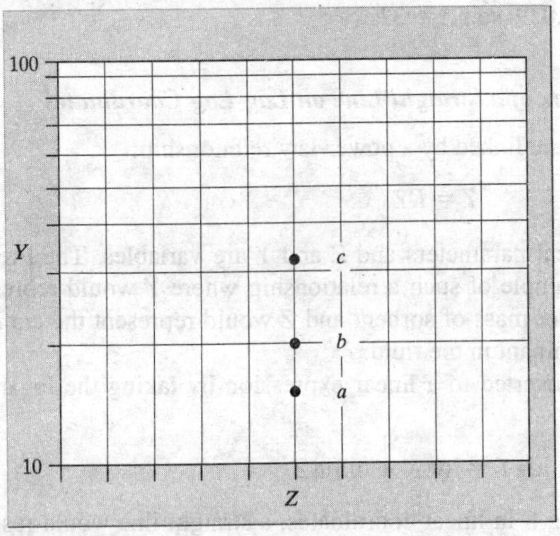

Figure D.3 Sample points plotted on a logarithmic axis, *Y*.

D.4 DATA MANIPULATION ON A LOGARITHMIC SCALE

Students sometimes make mistakes in reading data from figures plotted in logarithmic coordinates. The discussion in this section serves to point out a common pitfall and to demonstrate how to read data accurately.

Consider the *Y*-coordinates of the points plotted in Figure D.3.

Common Pitfall

The limits of the vertical axis are 10 (bottom) to 100 (top). Students often read point *a* as 10.5, rather than the correct value of 15. Likewise, point *b* is plotted at 20 along the vertical axis, not 11.

Accurately Reading Data

When data are presented with a full grid, then accurately reading the data is straightforward (provided one avoids pitfalls such as that listed above). However, frequently grid lines and even tick marks are missing. If so, then one can make an accurate determination using an ordinary ruler. (An "engineer's scale" or "architect's scale" is useful for this purpose.)

Consider point *c*. From the grid, it is clearly plotted between 30 and 40 along the vertical axis. To obtain a more precise determination of this point's *Y*-coordinate, y_c, follow these steps:

a. Identify the position of the next lower grid line, in this case 30. Label this value Y_{ref}.

b. Measure the vertical distance of point *c* above the height of Y_{ref}. Call this distance d_1.

c. Using the same linear scale, measure the vertical distance spanned by an order of magnitude (e.g., from Y_{ref} to $10 \times Y_{ref}$). Call this distance d_2.

d. Evaluate y_c from these relationships:

$$r = d_1/d_2 \qquad y_c = Y_{ref} \times 10^r \tag{D.71}$$

In this example, $r = 0.029$, so $y_c = 32$.

Determining the Parameters of a Straight Line on Log-Log Coordinates

Assume that two variables are linked by a power-law relationship:

$$Y = KZ^n \tag{D.72}$$

where K and n are empirical parameters and Z and Y are variables. The Freundlich sorption isotherm is an example of such a relationship where Y would represent the contaminant mass sorbed per mass of sorbent and Z would represent the equilibrium concentration of the contaminant in the fluid.

This relationship is converted to a linear expression by taking the logarithm of each side:

$$\log Y = \log K + n \log Z \tag{D.73}$$

If one plots log Y versus log Z in linear coordinates, a straight line would result with slope n and intercept log K. Alternatively, one can plot Y versus Z in log-log coordinates, again generating a straight line.

Frequently, one seeks to determine the parameters K and n from data presented as a straight-line plot in log-log coordinates. Here is an example of how to do so.

Consider the line in Figure D.4. First, mark two widely separated points, a and b. Then determine the Y- and Z-coordinates of a and b; call these (Z_a, Y_a) and (Z_b, Y_b). The parameters of equation D.72 are given by

$$n = \frac{\log(Y_b) - \log(Y_a)}{\log(Z_b) - \log(Z_a)} \tag{D.74}$$

$$K = \frac{Y_a}{Z_a^n} \tag{D.75}$$

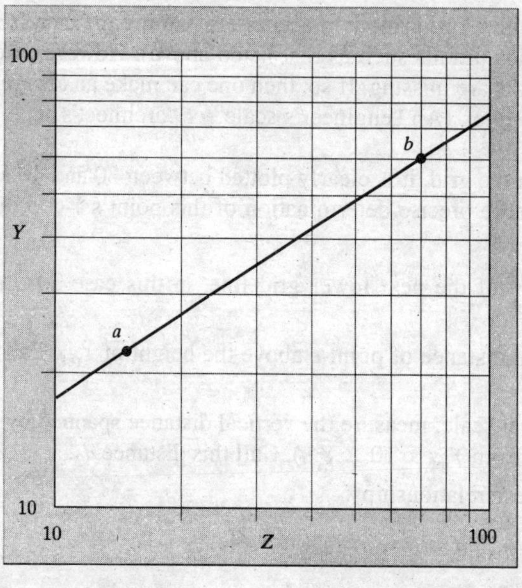

Figure D.4 Example of a straight line in log-log coordinates. The straight line represents the power-law relationship $Y = 4.0Z^{0.64}$.

For example, for the line drawn in this figure, the points have coordinates (Z_a, Y_a) = (15, 22.5) and (Z_b, Y_b) = (70, 60). Therefore, from equations D.74–75, $n = 0.64$ and $K = 4.0$, so $Y = 4.0Z^{0.64}$.

D.5 LINEAR REGRESSION

In many branches of engineering and applied science, one commonly encounters a situation in which an independent variable is expressed as a function of one or more dependent variables through a semiempirical equation. Often, such an equation includes parameters that must be determined experimentally.

In environmental engineering, this situation might occur in determining the rate of a kinetic process or in extracting the parameters of a sorption isotherm. The governing equations may be nonlinear, but they can be manipulated into a linear form. Usually, there are a small number of parameters that must be determined experimentally. For example, in the case of sorption isotherms, there may be two parameters that must be obtained by experiment. This means that a minimum of two separate experimental runs must be conducted. However, allowing for experimental uncertainties, we typically would run several more experimental tests than the strict minimum. We would then analyze the data using a statistical technique to obtain a best estimate of the parameters. This statistical technique is called "linear regression analysis" or "least-squares fitting."

Consider this situation. We have some number N of observations of paired data, Z and Y, each pair resulting from a separate experimental run. In the adsorption case, Z and Y would be measures related to the equilibrium concentration in the fluid, C_e, and the amount of contaminant sorbed per mass of sorbent, x/m. We seek to determine the parameters s (slope) and t (intercept) that yield a straight line that best fits the data, where the straight line is described by

$$Y = t + sZ \tag{D.76}$$

The linear regression method uses the individual measurements to determine the best values of t and s, where "best" means that we minimize the sum, over all measurements, of the squared distances between the measured values of Y and the predicted values of Y (from the linear equation).

Let z_i, y_i represent the paired results of the ith experimental measurement, where i is an integer that varies from 1 to N. Then, in the least-squares method, the parameters s and t are determined by

$$s = \frac{N\sum_{i=1}^{N}(z_i y_i) - \sum_{i=1}^{N} z_i \sum_{i=1}^{N} y_i}{N\sum_{i=1}^{N} z_i^2 - \left(\sum_{i=1}^{N} z_i\right)^2} \tag{D.77}$$

$$t = \frac{\left[\sum_{i=1}^{N} y_i - s\sum_{i=1}^{N} z_i\right]}{N} \tag{D.78}$$

For small data sets ($N \leq \sim 10$) it is not difficult to calculate these parameters by hand (see Example D.1). For the most part, though, engineers and scientists no longer do these calculations by hand, but rather rely on packaged computational tools such as spreadsheet programs or scientific calculators.

EXAMPLE D.1 ***Determining the Parameters for a Sorption Isotherm***

A sorption experiment was conducted by executing a series of five batch runs. In each run 10 mg of sorbent was added to 1 L of water. Different masses of the contaminant were added for each run. The equilibrium aqueous-phase concentration of the contaminant was measured. We seek to fit a Langmuir isotherm to the results and determine the parameters a and b:

$$\frac{x}{m} = a\frac{bC_e}{1 + bC_e}$$

In this expression, x/m represents the mass of sorbed contaminant per mass of sorbent, and C_e represents the equilibrium concentration of contaminant that remains dissolved in the water.

The original experimental data are as follows:

Run	Contaminant mass added (mg)	Equilibrium aqueous conc. (mg/L)
1	0.5	0.09
2	1.5	0.45
3	2.5	0.95
4	5	3.1
5	7	5.2

The isotherm is transformed into a linear relationship by (1) dividing both sides by (x/m) and (2) multiplying both sides by $(1 + bC_e)/ab$. Thus,

$$\frac{C_e}{x/m} = \frac{1}{ab} + \frac{C_e}{a}$$

A plot of $C_e/(x/m)$ versus C_e should yield a straight line with slope $1/a$ and intercept $1/ab$. The data are tabulated below, where x/m is obtained from a material balance as one-tenth of the difference between contaminant mass added and the equilibrium aqueous concentration:

		Z	Y
i	x/m (mg/mg)	C_e (mg/L)	$C_e/(x/m)$ (mg/L)
1	0.041	0.09	2.195
2	0.105	0.45	4.29
3	0.155	0.95	6.13
4	0.190	3.1	16.3
5	0.180	5.2	28.9

EXAMPLE D.1 ***Determining the Parameters for a Sorption Isotherm (continued)***

The data are plotted in Figure D.5, together with a line that represents the least-squares fit to the data. The data are well fitted by the straight line, which indicates that the Langmuir isotherm is a good description of equilibrium sorption for this contaminant/sorbent combination (in this concentration range). The slope is 5.14 and the intercept is 1.5. Therefore, the Langmuir parameters are

$$a = \frac{1}{5.14} = 0.195 \qquad b = \frac{1}{1.5a} = 3.4 \text{ L mg}^{-1}$$

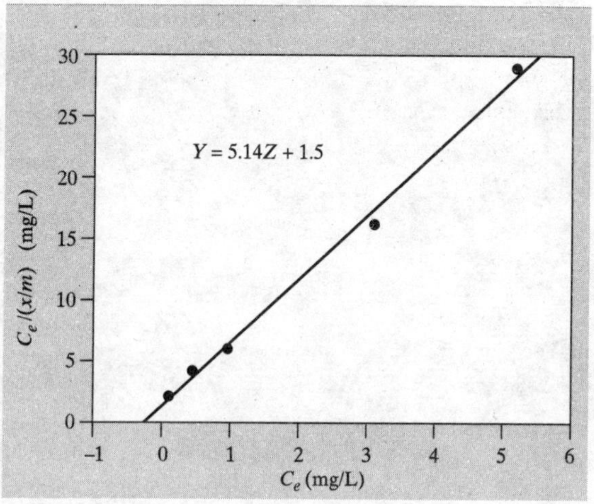

Figure D.5 Fitting a straight line to experimental data.

To assist in reproducing the regression calculation, here are the intermediate results:

$$\sum_{i=1}^{N} (z_i y_i) = 208.76$$

$$\sum_{i=1}^{N} z_i = 9.79$$

$$\sum_{i=1}^{N} y_i = 57.82$$

$$\sum_{i=1}^{N} z_i^2 = 37.76$$

REFERENCES

CUSSLER, E.L. 1984. *Diffusion: Mass transfer in fluid systems*. Cambridge University Press, Cambridge.

FISCHER, H.B., LIST, E.J., KOH, R.C.Y., IMBERGER, J., & BROOKS, N.H. 1979. *Mixing in inland and coastal waters*. Academic Press, New York.

PRESS, W.H., FLANNERY, B.P., TEUKOLSKY, S.A., & VETTERLING, W.T., 1986. *Numerical recipes: The art of scientific computing*. Cambridge University Press, Cambridge.

SELBY, S.M., 1974. *Standard mathematical tables*. 22nd ed. CRC Press, Cleveland, p. 103.

E

A Further Look at Transformation Processes

The material presented in Chapter 3 provides an introduction to the transformation processes that are important in environmental engineering. The goal of this appendix is to extend that material by summarizing some additional ideas from the theories of chemical equilibrium and kinetics. Activity coefficients, free energy, electrode potentials, activation energy, and the temperature dependence of kinetics are explored in this appendix.

E.1 ACTIVITY

For the equilibrium relationships developed and presented in Chapter 3, we assumed that, at a fixed temperature, a species' degree of participation in a reaction is a function only of its concentration (molar concentration in the aqueous phase or partial pressure in the gas phase). For many environmental engineering applications, species concentrations are sufficiently dilute that this assumption is reasonable and appropriate. However, the assumption fails under some conditions that will be described here.

Formally, chemical equilibrium relationships are expressed in terms of the *activity* of a species, where the activity is the product of an *activity coefficient* and the species concentration. Commonly, species activities are denoted by enclosing the chemical symbol in braces, {}, and the activity coefficient is represented by γ with a subscript to indicate the species name (e.g., γ_{H^+}). For impurities dissolved in water, the activity can be expressed as the product of the molar concentration and the activity coefficient. For example, the activity of protons in water is given as follows:

$$\{H^+\} = \gamma_{H^+}[H^+] \tag{E.1}$$

For ideal mixtures in water, activity coefficients are equal to 1, so equilibrium relationships in this case depend only on species concentrations. For gas-phase species, the activity is written as the product of the activity coefficient and the partial pressure of the species. Departures from ideal-gas behavior are rarely encountered in environmental engineering, so it is almost always accurate to represent the activity by the partial pressure alone. For pure liquid phases (e.g., liquid water) and solid phases the activity equals 1.

Nonidealities in aqueous chemistry may be caused by interactions between impurities. In an ideal solution, impurities are surrounded by water molecules and not significantly affected by the presence of other impurities. So, for example, under ideal conditions, the tendency of H^+ and OH^- to combine to form water would not be influenced by the levels of other ions, such as Na^+ and Cl^- in the solution. Given that the molarity of water is 55 M whereas typical freshwater impurity levels are in the mM range or less, this condition, called the *infinite dilution limit*, often holds. However, in some aqueous systems the impurities can reach a sufficiently high concentration to interact with one another. In that case, the equilibrium relationships must be altered to account for these interactions.

A common cause of nonideal behavior encountered in environmental engineering systems is ionic interactions in water. The effect is relatively weak for monovalent ions with impurities at levels typically present in drinking water, as demonstrated below. For multivalent ions in solutions with moderate ionic strengths, the effects can become substantial. For freshwater systems, the activity coefficient of an ion may be related to the ionic strength, I (see §2.C.4), using the Güntelberg equation (valid for $I \leq 0.1$ M):

$$\log_{10}(\gamma_i) = -\frac{0.5(Z_i)^2 I^{1/2}}{1 + I^{1/2}} \tag{E.2}$$

In this equation, I is expressed in molar units and Z_i represents the ionic charge on species i. In Example 2.C.2, the ionic strength of a sample of river water was found to be 1.7 mM. At this ionic strength, the Güntelberg equation predicts an activity coefficient of 0.96 for a singly charged ion such as OH^- and 0.66 for a triply charged ion such as Al^{3+}. Given that the accuracy and precision of environmental measurements are usually no better than 20 percent, activity coefficients for solutions with $I < 58$ mM can be ignored for monovalent ions, but become important for divalent ions at $I > 2.6$ mM and for trivalent ions at $I > 0.5$ mM. On the other hand, activity coefficients are extremely important for seawater ($I = 0.7$ M) or atmospheric haze droplets ($I \sim$ several M). Equations such as E.2 are not applicable for high-ionic-strength solutions like these. For seawater, separately determined thermochemical data are used.

For uncharged species in water, activity corrections are typically smaller than for charged species and are seldom needed in environmental engineering applications.

E.2 GIBBS FREE ENERGY AND EQUILIBRIUM CONSTANTS

Gibbs free energy can be used to determine the direction in which a reaction will tend to proceed (see Example E.1). If the change in free energy associated with a transformation process is negative (i.e., the products have less free energy than the reactants), then the reaction will tend to occur spontaneously in the forward direction. If the change in free energy is positive, the reaction will tend to occur spontaneously in the reverse direction. If there is no change in the free energy associated with the transformation process, the reactants and products are in equilibrium and no net transformation will spontaneously occur.

The change in free energy associated with a reaction is determined by subtracting the sum of the free energies ($G_{f,i}$) for all reactants from the sum for all products:

$$\Delta G = \sum_i G_{f,i}(\text{products}) - \sum_i G_{f,i}(\text{reactants}) \tag{E.3}$$

$G_{f,i}$ for each species i in a reaction is calculated from the free energy of formation (G°_{fi}) by the following equation:

$$G_{f,i} = n_i[G^\circ_{fi} + RT \ln \{i\}] \tag{E.4}$$

Table E.1 Free Energy of Formation (G°_{fi}) of Some Common Environmental Species[a]

Species	G°_{fi} (kJ mol^{-1})	Species	G°_{fi} (kJ mol^{-1})
Ca^{2+}(aq)	− 553.54	$Mg(OH)_2$ (brucite)	− 833.5
$CaCO_3$ (calcite)	− 1128.8	N_2(g)	0
$Ca(OH)_2$ (portlandite)	− 898.4	NH_3(aq)	− 26.57
C (graphite)	0	NH_3(g)	− 16.48
CH_4(g)	− 50.79	NH_4^+(aq)	− 79.37
CO_2(g)	− 394.37	NO(g)	86.57
CO_3^{2-}(aq)	− 527.9	NO_2(g)	51.3
H^+(aq)	0	NO_2^-(aq)	− 37.2
H_2(g)	0	NO_3^-(aq)	− 111.3
HCO_3^-(aq)	− 586.8	OH^-(aq)	− 157.29
H_2CO_3*(aq)	− 623.2	O_2(aq)	16.32
H_2O(l)	− 237.18	O_2(g)	0
H_2O(g)	− 228.57	O_3(g)	163.2
HS^-(aq)	12.05	S (rhombic)	0
H_2S(aq)	− 27.87	SO_2(g)	− 300.2
H_2S(g)	− 33.56	S^{2-}(aq)	85.8[b]
Mg^{2+}(aq)	− 454.8	SO_4^{2-}(aq)	− 744.6

[a]At reference conditions of T = 298 K, P = 1 atm.

[b]Stumm and Morgan (1996) caution that this value appears to be too low based on recently measured equilibrium constants for the acid dissociation of HS^-.

Source: Stumm and Morgan (1996, Table 3A). See this source for a much more extensive list for aquatic chemistry applications. Sawyer et al. (1994) provide a useful list of free energies of formation for compounds relevant to environmental engineering applications. Hill (1977) presents properties for an extensive list of industrially important chemicals, many of which are of interest as toxic and hazardous materials.

where n_i is the stoichiometric coefficient associated with the species in the reaction, R is the universal gas constant, T is temperature (K), and $\{i\}$ is the activity of the species. The free energies of formation for some environmentally relevant species are presented in Table E.1.

EXAMPLE E.1 ***Using Free Energy to Determine the Direction of a Reaction***

Water that is in contact with solid lime ($Ca(OH)_2$) at 25 °C contains concentrations of calcium and hydroxide ions of 10^{-5} M and 10^{-4} M, respectively. Use the free energy to determine whether the tendency is for more of the solid to form by precipitation of dissolved ions or for more of the solid to dissolve.

SOLUTION The lime dissolution reaction is written as

$$Ca(OH)_2 \Leftrightarrow Ca^{2+} + 2\ OH^-$$

If the change in free energy for this reaction is negative, the reaction will proceed in the forward direction, and more of the solid will dissolve. If the result is positive, then the tendency will be for the ions to form more solid.

EXAMPLE E.1 ***Using Free Energy to Determine the Direction of a Reaction (continued)***

The contribution to the change in free energy for each species is tabulated below. Note that the activity of the solid phase lime is 1.

$$G_{f,\mathrm{Ca(OH)_2}} = G^\circ_{fi} + RT\ln\{\mathrm{Ca(OH)_2}\} = -898.4\,\mathrm{kJ\,mol^{-1}}$$

$$G_{f,\mathrm{Ca^{2+}}} = G^\circ_{fi} + RT\ln\{\mathrm{Ca^{2+}}\} = -553.54 + 0.008314 \times 298 \times \ln(10^{-5})$$

$$= -582.1\,\mathrm{kJ\,mol^{-1}}$$

$$G_{f,\mathrm{OH^-}} = 2[G^\circ_{fi} + RT\ln\{\mathrm{OH^-}\}] = 2[-157.29 + 0.008314 \times 298 \times \ln(10^{-4})]$$

$$= -360.2\,\mathrm{kJ\,mol^{-1}}$$

$$\Delta G_{\mathrm{reaction}} = G_{f,\mathrm{Ca^{2+}}} + G_{f,\mathrm{OH^-}} - G_{f,\mathrm{Ca(OH)_2}} = -942.3 - (-898.4)$$

$$= -43.9\,\mathrm{kJ\,mol^{-1}}$$

Since the free-energy change is negative, the tendency is for the reaction to proceed in the forward direction, so additional solid will dissolve.

One of the more useful aspects of the concept of free energy is its link with the equilibrium constant of a transformation process. This is expressed in the following relationship:

$$K = \exp\left(-\frac{\Delta G^\circ}{RT}\right) \qquad \text{(E.5)}$$

The superscript ° in the Gibbs free-energy term designates standard activity conditions of temperature (298 K), pressure (1 atm), and all reactants and products at a concentration of 1 M (for aqueous-phase species) or at a partial pressure of 1 atm (for gases). This equation, in combination with data on the free energy of formation of species, can be used to compute equilibrium constants as illustrated in Example E.2.

Note that the form of E.5 suggests that K should be dimensionless. This is inconsistent with our experience with K (§3.A). This apparent inconsistency is discussed in Exhibit E.1.

To calculate the free-energy change (ΔG) for a reaction occurring under nonstandard conditions (i.e., molar concentrations $\neq 1$), the following equation is used:

$$\Delta G = \Delta G^\circ + RT\ln Q \qquad \text{(E.6)}$$

Here Q is the reaction quotient and is given in a form analogous to the equilibrium constant expression (equation 3.A.12). For example, for a generic reaction a A + b B ⇔ c C + d D, the reaction quotient, Q, would be

$$Q = \frac{\{C\}^c\{D\}^d}{\{A\}^a\{B\}^b} \qquad \text{(E.7)}$$

EXAMPLE E.2 ***Determining an Equilibrium Constant from Free-Energy Data***

Use the data in Table E.1 and equation E.5 to compute the equilibrium constant for the acid dissociation of bicarbonate at $T = 298$ K.

EXAMPLE E.2 ***Determining an Equilibrium Constant from Free-Energy Data (continued)***

SOLUTION The reaction can be written as

$$HCO_3^- \Leftrightarrow H^+ + CO_3^{2-}$$

The standard free-energy change associated with this reaction can be computed using equation E.3 written in terms of standard free energies:

$$\Delta G^\circ = \sum_i G^\circ_{fi} \text{ (products)} - \sum_i G^\circ_{fi} \text{ (reactants)}$$

$$= [(1)(0) + (1)(-527.9)] - [(1)(-586.8)] = 58.9 \text{ kJ mol}^{-1}$$

The equilibrium constant for the forward reaction is then given by equation E.5 as

$$K(298 \text{ K}) = \exp\left(-\frac{58.9 \text{ kJ mol}^{-1}}{8.314 \times 10^{-3} \text{ kJ mol}^{-1} \text{ K}^{-1} \times 298 \text{ K}}\right) = 4.74 \times 10^{-11}$$

This result agrees to within a few percentage points of the value given in equation 3.C.16. However, it appears that the units do not match (M versus dimensionless). This apparent discrepancy is reconciled in Exhibit E.1.

EXHIBIT E.1 ***Units on Equilibrium Constants***

Throughout the discussion of Gibbs free energy, we have worked with species activities expressed in units of molar concentration or partial pressure. However, formally, the equations should be applied using molar ratios or partial-pressure ratios for species. In each ratio, the numerator is the activity of the species and the denominator is the standard-state activity. Since the standard state is 1 M for species in water and 1 atm for gaseous species, the numerical value of the activity ratio is unchanged from that of the activity itself.

For example, in equation E.4, $\{i\}$ should properly be divided by the standard-state activity, 1 M. For notational simplicity, this detail is commonly omitted. The same point applies to the reaction quotient (Q) expressed in equation E.7: Each of the species activities should be, properly, the relative species activity normalized by the standard-state activity. Strictly, the equilibrium constant that is calculated in equation E.5 applies when Q is written in terms of molar ratios rather than molar concentrations. To convert back to molar concentration units, both sides of equation E.5 can be multiplied by the standard state activities of the reaction products divided by those of the reactants.

In Example E.2, the equilibrium constant would be formally written as

$$K = \frac{(\{H^+\}/[1 \text{ M}]) \times (\{CO_3^{2-}\}/[1 \text{ M}])}{\{HCO_3^-\}/[1 \text{ M}]}$$

Since activities have molar units, K is dimensionless. However, when we calculate the equilibrium constant in terms of product and reactant activities, as is commonly done, we get

$$\frac{\{H^+\}\{CO_3^{2-}\}}{\{HCO_3^-\}} = K \times 1 \text{ M}$$

By convention, we ascribe the units (M in this case) to the equilibrium constant, and therefore we obtain an answer similar to that given by equation 3.C.16.

Table E.2 Standard Electrode Potentials, $E°$, of Species in Water[a]

Half-reaction	$E°$ (V)
$H_2O_2 + 2\,H^+ + 2\,e^- \longrightarrow 2\,H_2O$	1.78
$ClO_2 + 2\,H_2O + 5\,e^- \longrightarrow Cl^- + 4\,OH^-$	1.71
$HOCl + H^+ + 2\,e^- \longrightarrow Cl^- + H_2O$	1.49
$Cl_2(g) + 2\,e^- \longrightarrow 2\,Cl^-$	1.36
$O_3 + H_2O + 2\,e^- \longrightarrow O_2 + 2\,OH^-$	1.24
$2\,NO_3^- + 12\,H^+ + 10\,e^- \longrightarrow N_2(g) + 6\,H_2O$	1.24
$O_2(g) + 4\,H^+ + 4\,e^- \longrightarrow 2\,H_2O$	1.23
$OCl^- + H_2O + 2\,e^- \longrightarrow Cl^- + 2\,OH^-$	0.90
$NO_3^- + 10\,H^+ + 8\,e^- \longrightarrow NH_4^+ + 3\,H_2O$	0.88
$NO_3^- + 2\,H^+ + 2\,e^- \longrightarrow NO_2^- + H_2O$	0.83
$NHCl_2 + 2\,H_2O + 4\,e^- \longrightarrow 2\,Cl^- + NH_3 + 2\,OH^-$	0.79
$Fe^{3+} + e^- \longrightarrow Fe^{2+}$	0.77
$NH_2Cl + H_2O + 2\,e^- \longrightarrow Cl^- + NH_3 + OH^-$	0.75
$Cu^{2+} + 2\,e^- \longrightarrow Cu$	0.34
$SO_4^{2-} + 10\,H^+ + 8\,e^- \longrightarrow H_2S(g) + 4\,H_2O$	0.34
$SO_4^{2-} + 9\,H^+ + 8\,e^- \longrightarrow HS^- + 4\,H_2O$	0.25
$CO_2(g) + 8\,H^+ + 8\,e^- \longrightarrow CH_4(g) + 2\,H_2O$	0.17
$S + 2\,H^+ + 2\,e^- \longrightarrow H_2S$	0.17[b]
$2\,H^+ + 2\,e^- \longrightarrow H_2(g)$	0.00
$Pb^{2+} + 2\,e^- \longrightarrow Pb$	−0.126
$Sn^{2+} + 2\,e^- \longrightarrow Sn$	−0.136
$Fe^{2+} + 2\,e^- \longrightarrow Fe$	−0.44
$Zn^{2+} + 2\,e^- \longrightarrow Zn$	−0.76
$Al^{3+} + 3\,e^- \longrightarrow Al$	−1.66
$Mg^{2+} + 2\,e^- \longrightarrow Mg$	−2.35

[a]Determined at $T = 298$ K, pH = 0.
[b]Sawyer et al. (1994) give 0.14.
Source: Sawyer et al. (1994, Table 3-4); Schwarzenbach et al. (1993, Table 12.16); Stumm and Morgan (1996, Table 8.3); LaGrega et al. (1994, Table 9.9).

E.3 ELECTRODE POTENTIALS AND FREE-ENERGY CHANGE IN REDOX REACTIONS

Equilibrium relationships for redox chemistry in aqueous systems are frequently described in terms of electrode potentials and electron activity. This section introduces these ideas and relates them to the Gibbs free energy.

The standard electrode potential is a measure of the tendency for a species to become reduced by taking up an electron. Species that are highly oxidized tend to take up electrons easily, and thus have high electrode potentials. On the other hand, relatively reduced species tend not to take up electrons easily, and have low electrode potentials.

Table E.2 presents data on the standard electrode potentials for many half-reactions of interest in environmental engineering. Each of these half-reactions is written as a reduction. The species that are the strongest chemical oxidizers, such as hydrogen peroxide, chlorine dioxide, and hypochlorous acid, are on the reactant side near the top of the list. The species that are the most easily oxidized, such as magnesium, aluminum, and zinc metal, are on the product side near the bottom of the list.

The standard electrode potentials ($E°$) presented in Table E.2 apply for standard conditions, that is, when the activity of each species in the reaction is 1 M (or 1), a very high value. For other activity levels, the nonstandard electrode potential, E, must be corrected from the standard value by means of the Nernst equation:

$$E = E° - \frac{RT}{n_e F} \ln Q \tag{E.8}$$

In this equation, n_e is the number of electrons transferred (i.e., the stoichiometric coefficient on electrons in the half-reaction); F is the Faraday constant, which is equal to the electrical charge possessed by one mole of electrons (96,490 C mol^{-1}); and Q is as defined in equation E.7. Note that the coulomb (C) is equivalent to one joule per volt.

The half-reactions listed in Table E.2 explicitly contain electrons as an apparent reactant, even though free electrons do not exist in aqueous systems. The term *half-reaction* reflects the fact that these reactions cannot occur alone. Instead, at least two half-reactions must occur simultaneously—a reduction in the forward direction, and an oxidation in the reverse direction—to produce an overall redox reaction.

The free-energy change associated with an overall redox reaction is related to the difference in electrode potentials of the oxidation and reduction half-reactions according to

$$\Delta G = -n_e F \, \Delta E \tag{E.9}$$

where $\Delta E = E_{red} - E_{ox}$, the electrode potential of the reduction half-reaction minus that of the oxidation half-reaction. Recall that a reaction is thermodynamically favored when the change in free energy is negative. Equation E.9 shows that this occurs when ΔE is positive.

When $\Delta G = \Delta E = 0$, the relative amounts of reactants and products correspond to their equilibrium levels, and there is no net reaction. From this result, it is readily shown that the equilibrium constant for the redox reaction is given by

$$K = \exp\left[\frac{\Delta E° n_e F}{RT}\right] \tag{E.10}$$

where $\Delta E°$ is the standard electrode potential for the reducing reaction minus that for the oxidizing reaction. The issue of units on K is analogous here to the case for the free-energy expression, equation E.5. (See Exhibit E.1.) The calculation of equilibrium constants for redox reactions from electrode potentials is demonstrated in Example E.3.

EXAMPLE E.3 ***Determining the Equilibrium Constant from Electrode Potentials***

Determine the equilibrium constant for the nitrification reaction given below (see 3.D.18) using standard electrode potentials. For water (pH 7) in equilibrium with the atmosphere (with a partial O_2 pressure of 0.21 atm), what is the equilibrium ratio of nitrate (NO_3^-) to ammonium (NH_4^+)?

SOLUTION The overall reaction can be written as the sum of two half-reactions:

	$NH_4^+ + 3\,H_2O \longrightarrow NO_3^- + 10\,H^+ + 8\,e^-$	Oxidation
	$2 \times [O_2(g) + 4\,H^+ + 4\,e^- \longrightarrow 2\,H_2O]$	Reduction
Overall:	$NH_4^+ + 2\,O_2(g) \Leftrightarrow NO_3^- + 2\,H^+ + H_2O$	

EXAMPLE E.3 ***Determining the Equilibrium Constant from Electrode Potentials (continued)***

The standard electrode potential for this reaction is

$$\Delta E^\circ = E_{red} - E_{ox} = 1.23 \text{ V} - 0.88 \text{ V} = 0.35 \text{ V}$$

Note that to determine the stoichiometry of the overall reaction, the reduction half-reaction was multiplied by a factor of 2 to balance the number of electrons needed for the oxidation reaction. The electrode potential was not multiplied, however, because E° is expressed in volts ($V = J\,C^{-1}$), which is a measure of energy *per unit charge.* Since $\Delta E^\circ > 0$, this reaction would proceed in the forward direction if all the reactants and products were at 1 M or 1 atm concentration.

The equilibrium constant for this reaction is obtained directly from equation E.10:

$$K = \exp\left(\frac{0.35 \text{ J C}^{-1} \times 8 \times 96{,}490 \text{ C mol}^{-1}}{8.314 \text{ J mol}^{-1} \text{ K}^{-1} \times 298 \text{ K}}\right) = 2 \times 10^{47}$$

The equilibrium relationship is written in terms of species concentrations and the partial pressure of O_2 as follows, assuming that the infinite dilution approximation applies:

$$K = \frac{[NO_3^-][H^+]^2}{[NH_4^+]\,P_{O_2}^2}$$

At a pH of 7 and with $P_{O_2} = 0.21$ atm, the equilibrium ratio of nitrate to ammonium is

$$\frac{[NO_3^-]}{[NH_4^+]} = 1 \times 10^{60}$$

What does this staggeringly large ratio mean? In the presence of dissolved oxygen, all ammonium in aqueous solutions would be oxidized to nitrate at equilibrium. However, ammonium is commonly found in oxygen-bearing waters, indicating that kinetics for oxidation processes are often slow.

E.4 ACTIVATION ENERGY AND THE TEMPERATURE DEPENDENCE OF REACTION RATES

The rate of almost every chemical reaction increases with temperature. A commonly used relationship describing this temperature dependence is called the Arrhenius equation:

$$k(T) = A(T)\exp\left(-\frac{E_a}{RT}\right) \tag{E.11}$$

where $A(T)$, sometimes called the Van't Hoff–Arrhenius coefficient, is typically a weak function of temperature and E_a is the activation energy (J mol^{-1}).

Figure E.1 is a sketch of the energy associated with a generic chemical transformation reaction. The vertical axis represents the sum of the free energy of all of the molecules that participate in the transformation, while the horizontal axis represents time. On the left, the species are present as reactants; on the right, they have become products. The reaction, as presented in the figure, is thermodynamically favored since the products have less free energy than the reactants. However, for the reaction to oc-

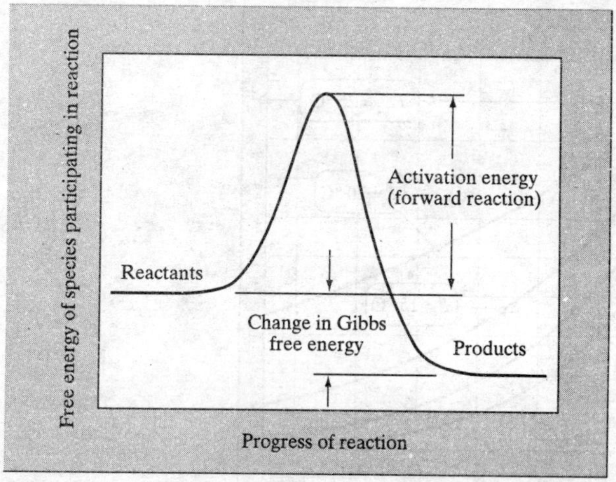

Figure E.1 Free energy of species participating in a chemical transformation reaction as the reactants are converted to products.

cur, an energy barrier, the activation energy, must be overcome. The activation energy represents the energy required to break the chemical bonds in the reactant molecules before new bonds can be formed and products created. The activation energy is not directly related to the free-energy change of the reaction once it is completed; however, the activation energy strongly influences the kinetics of the reaction. Reactions with low activation energy tend to proceed rapidly while reactions with high activation energy proceed more slowly. Chemical catalysts and biological enzymes function by reducing the activation energy of specific reactions. This speeds the approach to equilibrium by increasing reaction rates.

The influence of activation energy on rate constant is illustrated quantitatively in Figure E.2. The ratio of rate constants at two temperatures is plotted against the activation energy. The reaction rate is reduced by approximately a factor of 2 if the product of the activation energy and the temperature decrease (from reference conditions) equals 500 kJ K mol^{-1}. The reverse also holds: The rate of the reaction doubles if the product of activation energy and the temperature increase is 500 kJ K mol^{-1}.

In some environmental engineering applications, such as BOD kinetics (see §3.D.5), the temperature dependence of the rate constant is described relative to some reference temperature (T_{ref}) by using a ratio of the Arrhenius equations as follows:

$$\frac{k(T)}{k(T_{ref})} = \frac{A(T)}{A(T_{ref})} \exp\left(\frac{(T - T_{ref})E_a}{RTT_{ref}}\right) \tag{E.12}$$

Since the Van't Hoff–Arrhenius coefficient, A, is generally a weak function of temperature, we can approximate $A(T)/A(T_{ref})$ to be equal to 1. Furthermore, if the difference between the reference temperature and the temperature of interest is small ($|T - T_{ref}| \ll T_{ref}$), we can make the following approximation:

$$\frac{E_a}{RTT_{ref}} \cong \text{constant (independent of } T) \tag{E.13}$$

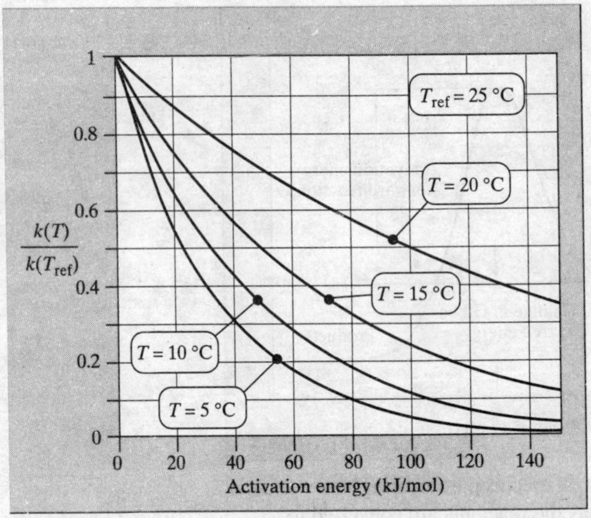

Figure E.2 Variation of reaction rate with activation energy for several temperatures. The Van't Hoff–Arrhenius coefficient, $A(T)$, is assumed to be independent of temperature. The horizontal axis spans a typical range of activation energies encountered in environmental transformations.

With these approximations, we can rewrite E.12 as

$$k(T) \cong k(T_{ref})\,\theta^{(T - T_{ref})} \tag{E.14}$$

where

$$\theta = \exp\left(\frac{E_a}{RTT_{ref}}\right) \tag{E.15}$$

Equation E.14 is used to correct the rate of BOD degradation from the standard measurement condition (T = 293 K) to the environmental condition of interest (e.g., temperature of the body of water into which wastewater is to be discharged). In this case, θ is treated as an empirical constant and is typically assigned a value of 1.047, which corresponds to an activation energy of ~33 kJ mol^{-1}. A temperature increase of 15 K approximately doubles the rate constant, and a decrease of 15 K halves the rate constant, relative to the reference temperature.

REFERENCES

HILL, C.G., Jr. 1977. *An introduction to chemical engineering kinetics & reactor design*. Wiley, New York.

LAGREGA. M.D., BUCKINGHAM. P.L., & EVANS, J.C. 1994. *Hazardous waste management*. McGraw-Hill, New York.

SAWYER, C.N., MCCARTY, P. L., & PARKIN, G.F. 1994. *Chemistry for environmental engineering*. 4th ed. McGraw-Hill, New York.

SCHWARZENBACH, R.P., GSCHWEND, P.M., & IMBODEN, D.M. 1993. *Environmental organic chemistry*. Wiley, New York.

STUMM, W., & MORGAN, J.J. 1996. *Aquatic chemistry: Chemical equilibria and rates in natural waters*. 3rd ed. Wiley, New York.

F

United States Federal Regulations for Water and Air Quality

Table F.1 National Primary Drinking Water Standards[a]

Contaminant	MCL, [MCLG] (mg/L)[b,c]	Potential health effects from ingestion in water	Sources of contaminant in drinking water
		Metals and inorganics	
Antimony	0.006 [0.006]	Blood cholesterol and glucose problems, cancer risk	Petroleum refineries, fire retardants, ceramics, electronics, solder
Arsenic	0.05 [none[d]]	Skin damage, circulatory system problems, cancer risk	Semiconductor manufacturing, petroleum refineries, wood preservatives, herbicides, natural deposits
Asbestos (fibers > 10 μm)	7 MFL[e] [7 MFL]	Intestinal polyps	Decay of asbestos cement in water mains, erosion of natural deposits
Barium	2 [2]	Increase in blood pressure	Drilling wastes, metal refineries, natural deposits
Beryllium	0.004 [0.004]	Intestinal lesions, bone and lung damage	Metal refineries, coal-burning factories, electrical, aerospace, and defense industries
Cadmium	0.005 [0.005]	Kidney damage	Corrosion of galvanized pipes, natural deposits, metal refineries, batteries and paints
Chromium (total)	0.1 [0.1]	Allergic dermatitis, liver or kidney disorders	Steel and pulp mills, natural deposits, electroplating
Copper	AL = 1.3[f] [1.3]	Gastrointestinal distress, liver and kidney damage	Corrosion of household plumbing, natural deposits, wood preservatives
Cyanide (free cyanide)	0.2 [0.2]	Nerve damage or thyroid problems	Steel/metal factories, plastic and fertilizer factories
Fluoride	4.0 [4.0]	Bone disease, mottled teeth	Water additive, natural deposits, fertilizer and aluminum factories
Lead	AL = 0.015[f] [0]	Developmental delays in children, kidney problems, high blood pressure	Corrosion of household plumbing systems, natural deposits

(continued)

Table F.1 National Primary Drinking Water Standards[a] (continued)

Contaminant	MCL, [MCLG] (mg / L)[b,c]	Potential health effects from ingestion in water	Sources of contaminant in drinking water
		Metals and inorganics	
Inorganic mercury	0.002 [0.002]	Kidney damage	Natural deposits, refineries and factories, runoff from landfills and cropland
Nitrate (measured as N)	10 [10]	"Blue baby syndrome" in infants under 6 months	Fertilizers; septic tanks, sewage; natural deposits
Nitrite (measured as N)	1 [1]	"Blue baby syndrome" in infants under 6 months	Fertilizers; septic tanks, sewage; natural deposits
Selenium	0.05 [0.05]	Hair or fingernail loss, circulatory problems	Petroleum refineries, natural deposits, mines
Thallium	0.002 [0.0005]	Hair loss; changes in blood; kidney, intestine, or liver problems	Ore-processing sites; electronics, glass, and pharmaceutical companies
		Organics	
Acrylamide	TT[g, h] [0]	Nervous system or blood problems, cancer risk	Flocculent used in water and wastewater treatment
Alachlor	0.002 [0]	Eye, liver, kidney, or spleen problems; anemia; cancer risk	Herbicide used on row crops
Atrazine	0.003 [0.003]	Cardiovascular system problems, reproductive difficulties, tumors	Herbicide used on row crops
Benzene	0.005 [0]	Anemia, decrease in blood platelets, cancer risk	Factories, gas storage tanks, landfills
Benzo(*a*)pyrene	0.0002 [0]	Reproductive difficulties, cancer risk	Linings of water storage tanks and distribution lines, coal tars, petroleum refineries, wood preserving
Carbofuran	0.04 [0.04]	Blood or nervous system problems, reproductive difficulties	Soil fumigant used on rice and alfalfa
Carbon tetrachloride	0.005 [0]	Liver problems, cancer risk	Chemical plants and other industrial activities
Chlordane	0.002 [0]	Liver or nervous system problems, cancer risk	Residue of banned termiticide
Chlorobenzene	0.1 [0.1]	Liver or kidney problems	Chemical and agricultural chemical factories, metal degreasing
2,4-D	0.07 [0.07]	Kidney, liver, or adrenal gland problems	Herbicide used on row crops
Dalapon	0.2 [0.2]	Minor kidney changes	Herbicide used on rights of way

(continued)

Table F.1 National Primary Drinking Water Standards[a] (continued)

Contaminant	MCL, [MCLG] (mg/L)[b,c]	Potential health effects from ingestion in water	Sources of contaminant in drinking water
		Organics	
1,2-Dibromo-3-chloropropane (DBCP)	0.0002 [0]	Reproductive difficulties, cancer risk	Soil fumigant used on soybeans, cotton, pineapples, and orchards
o-Dichlorobenzene	0.6 [0.6]	Liver, kidney, or circulatory system problems	Industrial chemical factories
p-Dichlorobenzene	0.075 [0.075]	Anemia; liver, kidney, or spleen damage	Industrial chemical factories
1,2-Dichloroethane	0.005 [0]	Cancer risk	Industrial chemical factories
1,1-Dichloroethylene	0.007 [0.007]	Liver problems	Industrial chemical factories
cis-1,2-Dichloroethylene	0.07 [0.07]	Liver and kidney problems	Industrial chemical factories
trans-1,2-Dichloroethylene	0.1 [0.1]	Liver and kidney problems	Industrial chemical factories
Dichloromethane	0.005 [0]	Liver problems, cancer risk	Pharmaceutical and chemical factories
1,2-Dichloropropane	0.005 [0]	Increased risk of cancer	Industrial chemical factories
Di(2-ethylhexyl) adipate	0.4 [0.4]	General toxic effects or reproductive difficulties	PVC plumbing systems, chemical factories
Di(2-ethylhexyl) phthalate	0.006 [0]	Reproductive difficulties, liver problems, cancer risk	Rubber and chemical factories
Dinoseb	0.007 [0.007]	Reproductive difficulties, thyroid problems	Herbicide used on soybeans and vegetables
Dioxin (2,3,7,8-TCDD)	3×10^{-8} [0]	Reproductive difficulties, cancer risk	Emissions from waste incineration and other combustion, chemical factories
Diquat	0.02 [0.02]	Cataracts, liver and kidney problems	Herbicide use
Endothall	0.1 [0.1]	Stomach and intestinal problems	Herbicide use
Endrin	0.002 [0.002]	Nervous system effects	Residue of banned insecticide
Epichlorohydrin	TT[g, h] [0]	Stomach problems, reproductive difficulties, cancer risk	Industrial chemical factories; added to water during treatment process
Ethylbenzene	0.7 [0.7]	Liver or kidney problems, nervous system effects	Petroleum refineries
Ethylene dibromide	0.00005 [0]	Stomach problems, reproductive difficulties, cancer risk	Petroleum refineries

(continued)

Table F.1 National Primary Drinking Water Standards[a] (continued)

Contaminant	MCL, [MCLG] (mg/L)[b,c]	Potential health effects from ingestion in water	Sources of contaminant in drinking water
		Organics	
Glyphosate	0.7 [0.7]	Kidney problems, reproductive difficulties	Herbicide use
Heptachlor	0.0004 [0]	Liver damage, cancer risk	Residue of banned termiticide
Heptachlor epoxide	0.0002 [0]	Liver damage, cancer risk	Breakdown of heptachlor
Hexachlorobenzene	0.001 [0]	Liver or kidney problems, reproductive difficulties, cancer risk	Metal refineries, agricultural chemical factories
Hexachloro-cyclopentadiene	0.05 [0.05]	Kidney or stomach problems	Chemical factories
Lindane	0.0002 [0.0002]	Liver or kidney problems, nervous and immune system effects	Insecticide used on cattle, lumber, gardens
Methoxychlor	0.04 [0.04]	Reproductive difficulties, liver and kidney effects	Insecticide used on fruits, vegetables, alfalfa, livestock
Oxamyl (Vydate)	0.2 [0.2]	Slight nervous system effects, kidney damage	Insecticide used on apples, potatoes, and tomatoes
Polychlorinated biphenyls (PCBs)	0.0005 [0]	Thymus gland problems; immune, reproductive, or nervous system difficulties; cancer risk	Landfills, waste chemicals
Pentachlorophenol	0.001 [0]	Liver or kidney problems, cancer risk	Wood preservative factories
Picloram	0.5 [0.5]	Liver and kidney problems	Herbicide runoff
Simazine	0.004 [0.004]	Blood problems, cancer risk	Herbicide runoff
Styrene	0.1 [0.1]	Liver, kidney, and circulatory problems	Rubber and plastic factories, leaching from landfills
Tetrachloroethylene	0.005 [0]	Liver problems, increased risk of cancer	Leaching from PVC pipes; factories and dry cleaners
Toluene	1 [1]	Nervous system, kidney, or liver problems	Petroleum factories, gasoline storage
Total trihalomethanes	0.1 [none[d]]	Liver, kidney, or central nervous system problems; cancer risk	By-product of drinking water disinfection
Toxaphene	0.003 [0]	Kidney, liver, or thyroid problems; cancer risk	Insecticide used on cotton and cattle
2,4,5-TP (Silvex)	0.05 [0.05]	Liver problems	Residue of banned herbicide
1,2,4-Trichloro-benzene	0.07 [0.07]	Changes in adrenal glands	Textile finishing factories

(continued)

Table F.1 National Primary Drinking Water Standards[a] (continued)

Contaminant	MCL, [MCLG] (mg/L)[b,c]	Potential health effects from ingestion in water	Sources of contaminant in drinking water
		Organics	
1,1,1-Trichloroethane	0.2 [0.2]	Liver, nervous system, or circulatory problems	Metal degreasing sites, factories
1,1,2-Trichloroethane	0.005 [0.003]	Liver, kidney, or immune system problems	Industrial chemical factories
Trichloroethylene	0.005 [0]	Liver problems, cancer risk	Petroleum refineries
Vinyl chloride	0.002 [0]	Cancer risk	Leaching from PVC pipes; plastic factories; degradation of solvents
Xylenes (total)	10 [10]	Nervous system damage	Petroleum factories; chemical factories; gasoline storage
		Radionuclides	
Beta particles and photon emitters	4 mrem/y [none[d]]	Cancer risk	Decay of natural and man-made deposits
Gross alpha particle activity	15 pCi/L [none[d]]	Cancer risk	Erosion of natural deposits
Radium-226 and radium-228 (combined)	5 pCi/L [none[d]]	Cancer risk	Erosion of natural deposits
		Microorganisms	
Giardia lamblia	TT[g, i] [0]	Giardiasis, gastroenteric disease	Human and animal fecal waste
Heterotrophic plate count	TT[g, i] [N/A]	Indicator organisms	Human and animal fecal waste, soil runoff
Legionella	TT[g, i] [0]	Legionnaire's disease (pneumonia)	Found naturally in water, multiplies in heating systems
Total coliforms	5.0%[j] [0]	Indicator organisms [k]	Human and animal fecal waste
Turbidity	TT[g, i] [N/A]	Interferes with disinfection	Soil runoff
Viruses (enteric)	TT[g, i] [0]	Gastroenteric disease	Human and animal fecal waste

[a]Primary standards are legally enforceable regulations for contaminants that can adversely affect public health and are known or anticipated to occur in public water systems.

[b]*Maximum contaminant level* (MCL): The maximum permissible level of a contaminant in water that is delivered to any user of a public water system. MCLs are enforceable standards. *Maximum contaminant level goal* (MCLG): The maximum level of a contaminant in drinking water at which no known or anticipated adverse effect on the health of persons would occur, and that allows for an adequate margin of safety. MCLGs are nonenforceable public health goals.

[c]Units are milligrams per liter (mg/L) unless otherwise noted.

[d]MCLGs were not established before the 1986 amendments to the Safe Drinking Water Act. Therefore, there is no MCLG for this contaminant.

[e]MFL = million fibers per liter.

(continued)

Table F.1 National Primary Drinking Water Standards[a] (continued)

[f]*Action level* (AL): The concentration of lead or copper in water that triggers the requirement for treatment techniques in public water systems. Systems are required to take tap water samples at sites with lead pipes, or copper pipes that have lead solder, or that are served by lead service lines. If the action levels are exceeded in more than 10% of tap water samples, water systems must undertake treatment steps.

[g]*Treatment technique* (TT): An enforceable procedure or level of technical performance, which public water systems must follow to ensure control of a contaminant.

[h]Each water system must certify, in writing, to the state (using third-party or manufacturer's certification) that when acrylamide and epichlorohydrin are used in drinking water systems, the combination (or product) of dose and monomer level does not exceed the levels specified:

Acrylamide: 0.05% dosed at 1 mg / L (or equivalent)
Epichlorohydrin: 0.01% dosed at 20 mg / L (or equivalent)

[i]The Surface Water Treatment Rule requires systems using surface water or ground water under the direct influence of surface water (1) to disinfect their water and (2) to filter their water unless they can meet certain water quality source requirements and site-specific criteria. The following treatment requirements must be met:

Giardia lamblia: 99.9% killed or inactivated
Viruses: 99.99% killed or inactivated
Legionella: No limit, but the EPA believes that if *Giardia* and viruses are inactivated, *Legionella* will also be controlled.
Turbidity: At no time can turbidity exceed 5 nephelometric turbidity units (NTU); systems that filter must ensure that the turbidity does not exceed 1 NTU (0.5 NTU for conventional or direct filtration) in at least 95% of the daily samples in any month.
Heterotrophic plate count: No more than 500 bacterial colonies per milliliter.

[j]No more than 5.0% of samples test positive for total coliforms in a month. (For water systems that collect fewer than 40 routine samples per month, no more than one sample can be total coliform positive.) Every sample that has total coliforms must be analyzed for fecal coliforms. There cannot be any fecal coliforms detected.

[k]Fecal coliforms and *E. coli* are bacteria whose presence indicates that the water may be contaminated with human and animal wastes. Microbes in these wastes can cause diarrhea, cramps, nausea, headaches, or other symptoms.

Source: Adapted from http://www.epa.gov/safewater/mcl.html and F.W. Pontius and S.W. Clark, Drinking water quality standards, regulations, and goals, in R.D. Letterman, ed., *Water quality and treatment: A handbook of community water supplies*, 5th ed., McGraw-Hill, New York, 1999.

Table F.2 National Secondary Drinking Water Standards[a]

Contaminant	Secondary standard	Aesthetic effects
Aluminum	0.05–0.2 mg / L	Discoloration of water
Chloride	250 mg / L	Taste, corrosion of pipes, toxic to plants
Color	15 (color units)	Aesthetic, staining
Copper	1.0 mg / L	Metallic taste, blue-green stain
Corrosivity	Noncorrosive	Aesthetic; can leach contaminants such as lead and copper into water
Fluoride	2.0 mg / L	Fluorosis or mottling, a brown discoloration of the teeth
Foaming agents	0.5 mg / L	Aesthetic
Iron	0.3 mg / L	Taste, staining, scaling, discoloration of water
Manganese	0.05 mg / L	Taste, staining, scaling, discoloration of water
Odor	3 threshold odor number	Aesthetic, if from H_2S; may cause stains and be corrosive
pH	6.5–8.5	Corrosivity of water, staining
Silver	0.10 mg / L	Argyria (discoloration of skin)
Sulfate	250 mg / L	Taste, laxative effects

(continued)

Table F.2 National Secondary Drinking Water Standards[a] (continued)

Contaminant	Secondary standard	Aesthetic effects
Total dissolved solids	500 mg/L	Taste, corrosivity; limits effectiveness of soap and detergents
Zinc	5 mg/L	Taste

[a]National Secondary Drinking Water Regulations (NSDWRs or secondary standards) are nonenforceable guidelines regulating contaminants that may cause cosmetic effects (such as skin or tooth discoloration) or aesthetic effects (such as taste, odor, or color) in drinking water. The EPA recommends secondary standards to water systems but does not require systems to comply. However, states may choose to adopt them as enforceable standards.

Source: Adapted from http://www.epa.gov/safewater/mcl.html

Table F.3 National Ambient Air Quality Standards for Criteria Pollutants

Pollutant	Standard[a]	Type[b]
Carbon monoxide (CO)		
8 h average[c]	9 ppm (10 mg m^{-3})	Primary
1 h average	35 ppm (40 mg m^{-3})	Primary
Nitrogen dioxide (NO_2)		
Annual arithmetic mean	53 ppb (100 μg m^{-3})	Primary and secondary
Ozone (O_3)		
8 h average[c]	80 ppb (157 μg m^{-3})	Primary and secondary
1 h average	120 ppb (235 μg m^{-3})	Primary and secondary
Particulate matter ≤ 10 μm (PM-10)		
Annual arithmetic mean	50 μg m^{-3}	Primary and secondary
24 h average	150 μg m^{-3}	Primary and secondary
Particulate matter ≤ 2.5 μm (PM-2.5)[c]		
Annual arithmetic mean	15 μg m^{-3}	Primary and secondary
24 h average	65 μg m^{-3}	Primary and secondary
Sulfur dioxide (SO_2)		
Annual arithmetic mean	30 ppb (80 μg m^{-3})	Primary
24 h average	140 ppb (365 μg m^{-3})	Primary
3 h average	500 ppb (1.3 mg m^{-3})	Secondary
Lead (Pb)		
Quarterly average	1.5 μg m^{-3}	Primary and secondary

[a]Parenthetical value is an approximately equivalent mass concentration.

[b]*Primary standards* set limits to protect public health, including the health of sensitive populations, such as asthmatics, children, and the elderly. *Secondary standards* set limits to protect public welfare, including protection against decreased visibility and damage to animals, crops, vegetation, and buildings.

[c]The USEPA proposed the 8-h standard for ozone and the PM-2.5 standards in 1997. A 1999 federal court ruling blocked implementation. As of the summer of 2000, the matter is unresolved.

Source: http://www.epa.gov/airs/criteria.html

Table F.4 Hazardous Air Pollutants Regulated by the 1990 Clean Air Act Amendments

Acetaldehyde	Acetamide	Acetonitrile
Acetophenome	2-Acetylaminofluorene	Acrolein
Acrylamide	Acrylic acid	Acrylonitrile
Allyl chloride	4-Aminobiphenyl	Aniline
o-Anisidine	Antimony compounds	Arsenic compounds (inorganic including arsine)
Asbestos	Benzene (including benzene from gasoline)	Benzidine
Benzotrichloride	Benzyl chloride	Beryllium compounds
Biphenyl	Bis(2-ethylhexyl)phthalate (DEHP)	Bis(chloromethyl) ether
Bromoform	1,3-Butadiene	Cadmium compounds
Calcium cyanamide	Caprolactam[a]	Captan
Carbaryl	Carbon disulfide	Carbon tetrachloride
Carbonyl sulfide	Catechol	Chloramben
Chlordane	Chlorine	Chloroacetic acid
2-Chloroacetophenone	Chlorobenzene	Chlorobenzilate
Chloroform	Chloromethyl methyl ether	Chloroprene
Chromium compounds	Cobalt compounds	Coke oven emissions
Cresols/cresylic acid (isomers and mixture)	*m*-Cresol	*o*-Cresol
p-Cresol	Cumene	Cyanide compounds
2,4-D, salts and esters	DDE	Diazomethane
Dibenzofurans	1,2-Dibromo-3-chloropropane	Dibutylphthalate
1,4-Dichlorobenzene(p)	3,3-Dichlorobenzidene	Dichloroethyl ether (bis(2-chloroethyl)ether)
1,3-Dichloropropene	Dichlorovos	Diethanolamine
N,N-Diethyl aniline (*N,N*-dimethylaniline)	Diethyl sulfate	3,3-Dimethoxybenzidine
Dimethyl aminoazobenzene	3,3′-Dimethyl benzidine	Dimethyl carbamoyl chloride
Dimethyl formamide	1,1-Dimethyl hydrazine	Dimethyl phthalate
Dimethyl sulfate	4,6-Dinitroso-*o*-cresol, and salts	2,4-Dinitrophenol
2,4-Dinitrotoluene	1,4-Dioxane (1,4-diethyleneoxide)	1,2-Diphenylhydrazine
Epichlorohydrin (1-chloro-2,3-epoxypropane)	1,2-Epoxybutane	Ethyl acrylate
Ethyl benzene	Ethyl carbamate (urethane)	Ethyl chloride (chloroethane)
Ethylene dibromide (dibromoethane)	Ethylene dichloride (1,2-dichloroethane)	Ethylene glycol
Ethylene imine (aziridine)	Ethylene oxide	Ethylene thiourea
Ethylidene dichloride (1,1-dichloroethane)	Formaldehyde	Glycol ethers
Heptachlor	Hexachlorobenzene	Hexachlorobutadiene

(continued)

Table F.4 Hazardous Air Pollutants Regulated by the 1990 Clean Air Act Amendments (continued)

Hexachlorocyclopentadiene	Hexachloroethane	Hexamethylene-1,6-diisocyanate
Hexamethylphosphoramide	Hexane	Hydrazine
Hydrochloric acid	Hydrogen fluoride (hydrofluoric acid)	Hydroquinone
Isophorone	Lead compounds	Lindane (all isomers)
Maleic anhydride	Manganese compounds	Mercury compounds
Methanol	Methoxychlor	Methyl bromide (bromomethane)
Methyl chloride (chloromethane)	Methyl chloroform (1,1,1-trichloroethane)	Methyl ethyl ketone[b] (2-butanone)
Methyl hydrazine	Methyl iodide (iodomethane)	Methyl isobutyl ketone (hexone)
Methyl isocyanate	Methyl methacrylate	Methyl *tert*-butyl ether
4-4′ Methylene bis(2-chloroaniline)	Methylene chloride (dichloromethane)	Methylene diphenyl diisocyanate (MDI)
4,4′-Methylenedianiline	Fine mineral fibers	Naphthalene
Nickel compounds	Nitrobenzene	4-Nitrobiphenyl
4-Nitrophenol	2-Nitropropane	*N*-Nitroso-*N*-Methylurea
N-Nitrosodimethylamine	*N*-Nitrosomorpholine	Parathion
Pentachloronitrobenzene (quintobenzene)	Pentachlorophenol	Phenol
p-Phenylenediamine	Phosgene	Phosphine
Phosphorus	Phthalic anhydride	Polychlorinated biphenyls (Aroclors)
Polycyclic organic matter	1,3-Propane sultone	*β*-Propiolactone
Propionaldehyde	Propoxur (Baygon)	Propylene dichloride (1,2-dichloropropane)
Propylene oxide	1,2-Propylenimine (2-methyl aziridine)	Quinoline
Quinone	Radionuclides (including radon)	Selenium compounds
Styrene	Styrene oxide	2,3,7,8-Tetrachlorodibenzo-*p*-dioxin
1,1,2,2-Tetrachloroethane	Tetrachloroethylene (perchloroethylene)	Titanium tetrachloride
Toluene	2,4-Toluene diamine	2,4-Toluene diisocyanate
o-Toluidine	Toxaphene (chlorinated camphene)	1,2,4-Trichlorobenzene
1,1,2-Trichloroethane	Trichloroethylene	2,4,5-Trichlorophenol
2,4,6-Trichlorophenol	Triethylamine	Trifluralin
2,2,4-Trimethylpentane	Vinyl acetate	Vinyl bromide
Vinyl chloride	Vinylidine chloride (1,1-dichloroethylene)	Xylenes (isomers and mixture)
m-Xylene	*o*-Xylene	*p*-Xylene

[a] Caprolactam was delisted by the USEPA in 1996.

[b] The USEPA is reviewing a petition to delist methyl ethyl ketone.

Source: http://www.epa.gov/reg5oair/toxics/haps-cas.txt; http://www.epa.gov/ttn/uatw/pollsour.html

Index

Some Important Equations

Description	Equation	For more information see:
Residence time	$\tau_r \sim \frac{S}{F_0}$	Equation 1.C.5
Ideal gas law	$\frac{n}{V} = \frac{P}{RT}$	Equation 2.B.1
Henry's law	$C_w = K_{H,g}P_g$ or $P_g = H_g C_w$	§3.B.2
Solubility product	$\mathrm{A}_x\mathrm{B}_y \Leftrightarrow x\,\mathrm{A} + y\,\mathrm{B} \Rightarrow [\mathrm{A}]^x[\mathrm{B}]^y = K_{\mathrm{sp}}(T)$	§3.B.2
Dissociation of acid	$K_\mathrm{A} = \frac{[\mathrm{H}^+][\mathrm{A}^-]}{[\mathrm{HA}]}$	Equation 3.C.8
Definition of pH	$\mathrm{pH} = -\log_{10}[\mathrm{H}^+]$	Equation 2.C.1
Definition of pK	$\mathrm{pK} = -\log_{10}(K)$	Equation 3.C.9
Microbial kinetics	$\frac{dX}{dt} = \mu X = Y\frac{k_m S}{K_s + S}X - k_d X$	§3.D.5
Advective flux (1-D)	$J_a = CU$	§4.A.2
Fick's law (1-D)	$J_d = -D\frac{dC}{dx}$	§4.A.3
Two-film model	$J_{gl} = k_{gl}(C_s - C)$	§4.C.2
Darcy's law (1-D)	$U = -K\frac{dh}{dl}$	§4.D.1
Gravitational settling		
$v_t = \left[\frac{4}{3}\frac{g d_p}{C_d}\left(\frac{\rho - \rho_f}{\rho_f}\right)\right]^{1/2}$	$C_d = \frac{24}{\mathrm{Re}_p}(1 + 0.14\mathrm{Re}_p^{0.7})$	$\mathrm{Re}_p = \frac{d_p v_t \rho_f}{\mu}$
Stokes's law regime $\Rightarrow$	$v_t = \left[\frac{C_c g d_p^2}{18}\left(\frac{\rho - \rho_f}{\mu}\right)\right]$ valid for $\mathrm{Re}_p < 0.3$	§4.B.2
Batch reactor material balance	$\frac{d(CV)}{dt} = rV$	§5.A.1
CMFR material balance	$\frac{d(CV)}{dt} = Q_{\mathrm{in}}C_{\mathrm{in}} - Q_{\mathrm{out}}C + rV$	§5.A.2
PFR material balance	$\frac{\partial C}{\partial t} = -U\frac{\partial C}{\partial x} + r$	§5.A.3
General material-balance equation (3-D)	$\frac{\partial C}{\partial t} + \left(\frac{\partial J_x}{\partial x} + \frac{\partial J_y}{\partial y} + \frac{\partial J_z}{\partial z}\right) - r = 0$	§5.B.1

Some Important Parameters and Data (see also Appendix A)

Avogadro's number	$N_{av} = 6.02 \times 10^{23}$ molecules mol^{-1}
Gas constant	$R = 82.05 \times 10^{-6}\ mol^{-1}\ m^3\ atm\ K^{-1} = 8.314\ J\ mol^{-1}\ K^{-1}$
Boltzmann's constant	$k = 1.38 \times 10^{-16}\ erg\ K^{-1} = 1.38 \times 10^{-23}\ J\ K^{-1}$
Volume of gas per mole	$V/n = 24.04$ L at 1 atm and 293 °C
Mass of the earth	$M_e = 6.0 \times 10^{24}$ kg
Mass of the hydrosphere	$M_{hydro} = 1.4 \times 10^{21}$ kg
Mass of the atmosphere	$M_{atm} = 5.3 \times 10^{18}$ kg
Radius of the earth	$R_e = 6370$ km
Acceleration of gravity	$g = 9.8\ m\ s^{-2}$
Charge on an electron	$X_e = 1.603 \times 10^{-19}$ C
Faraday constant	$F = 96{,}490\ C\ mol^{-1}$

Some Important Unit Conversions (see also Appendix A)

Length (L)	1 m = 100 cm = 3.28 ft = 39.4 in	1 km = 1000 m = 0.62 mi
Area (L^2)	1 hectare = $10^4 m^2$ = 2.47 acres	1 mi^2 = 640 acres = 259 hectares
Volume (L^3)	1 m^3 = 1000 L = $10^6\ cm^3$ = 264 gal = 35.3 ft^3	1 acre-foot = 1233.5 m^3
Temperature	20 °C = 293 K = 68 °F °C = 273 + K	°C = (°F − 32)/1.8
Time (T)	1 h = 60 min = 3600 s	1 y = 365 d = 3.15 × 10^7 s
Speed (L/T)	1 m s^{-1} = 2.24 mph (mi h^{-1}) = 3.28 ft s^{-1} = 3.6 km h^{-1}	1 cm s^{-1} = 1.97 ft min^{-1}
Vol. flow (L^3/T)	1 $m^3\ h^{-1}$ = 0.59 cfm ($ft^3\ min^{-1}$) = 6.3 × 10^{-3} MGD	1 MGD = 0.0438 $m^3\ s^{-1}$
Mass (M)	1 kg = 1000 g = 2.2 lb_m	1 tonne = 1000 kg = 1.1 ton
Force ($M\ L\ T^{-2}$)	1 N = 1 kg m s^{-2} = 10^5 dyne = 10^5 g cm s^{-2}	1 N = 0.225 lb_f
Pressure ($M\ L^{-1}\ T^{-2}$)	1 atm = 101.3 kPa = 1.013 bar = 14.7 psi = 760 mmHg = 407 in H_2O	
Energy ($M\ L^2\ T^{-2}$)	1 J = 1 kg $m^2\ s^{-2}$ = 10^7 erg = 10^7 g $cm^2\ s^{-2}$ = 0.239 cal = 9.48 × 10^{-4} BTU	
Power ($M\ L^2\ T^{-3}$)	1 W = 1 J s^{-1} = 1 N m s^{-1} = 1 kg $m^2\ s^{-3}$ = 3.41 BTU h^{-1} = 1.34 × 10^{-3} hp	